Biotechnology

Second Edition

Volume 5a

Recombinant Proteins, Monoclonal Antibodies and Therapeutic Genes

WILEY-VCH

Biotechnology

Second Edition

Fundamentals

Volume 1
Biological Fundamentals

Volume 2
Genetic Fundamentals and
Genetic Engineering

Volume 3
Bioprocessing

Volume 4
Measuring, Modelling and Control

Products

Volume 5a
Recombinant Proteins, Monoclonal
Antibodies and Therapeutic Genes

Volume 5b
Genomics

Volume 6
Products of Primary Metabolism

Volume 7
Products of Secondary Metabolism

Volumes 8a and b
Biotransformations I and II

Special Topics

Volume 9
Enzymes, Biomass, Food and Feed

Volume 10
Special Processes

Volumes 11a and b
Environmental Processes

Volume 12
Legal, Economic and
Ethical Dimensions

A Multi-Volume Comprehensive Treatise

Biotechnology

Second, Completely Revised Edition

Edited by
H.-J. Rehm and G. Reed
in cooperation with
A. Pühler and P. Stadler

Volume 5a

Recombinant Proteins, Monoclonal Antibodies and Therapeutic Genes

Edited by
A. Mountain, U. Ney and D. Schomburg

 **WILEY-VCH**

Weinheim · New York · Chichester · Brisbane · Singapore · Toronto

Series Editors:
Prof. Dr. H.-J. Rehm
Institut für Mikrobiologie
Universität Münster
Corrensstraße 3
D-48149 Münster
FRG

Prof. Dr. A. Pühler
Biologie VI (Genetik)
Universität Bielefeld
P.O. Box 1001 31
D-33501 Bielefeld
FRG

Dr. G. Reed
1029 N. Jackson St. #501-A
Milwaukee, WI 53202-3226
USA

Prof. Dr. P. J. W. Stadler
Artemis Pharmaceuticals
Geschäftsführung
Pharmazentrum Köln
Neurather Ring
D-51063 Köln
FRG

Volume Editors:
Dr. A. Mountain
Cobra Therapeutics Ltd.
The Science Park
University of Keele
Keele, Staffordshire ST5 5SP
UK

Dr. U. Ney
Celltech Therapeutics Ltd.
216 Bath Road
Slough Berkshire SL1 4EN
UK

Prof. Dr. D. Schomburg
Institut für Biochemie
Universität zu Köln
Zülpicher Str. 47
D-50674 Köln
FRG

Library of Congress Card No.: applied for

British Library Cataloguing-in-Publication Data:
A catalogue record for this book is available from the British Library

Die Deutsche Bibliothek – CIP-Einheitsaufnahme
Biotechnology : a multi volume comprehensive
treatise / ed. by H.-J. Rehm and G. Reed. In
cooperation with A. Pühler and P. Stadler. –
2., completely rev. ed. – VCH.
  ISBN 3-527-28310-2 (Weinheim ...)

NE: Rehm, Hans-J. [Hrsg.]

Vol. 5a: Recombinant Proteins, Monoclonal Antibodies and Therapeutic Genes / ed. by A. Mountain, U. Ney
and D. Schomburg – 1999
  ISBN 3-527-28315-3

© WILEY-VCH Verlag GmbH, D-69469 Weinheim (Federal Republic of Germany), 1999

Printed on acid-free and chlorine-free paper.

Composition and Printing: Zechnersche Buchdruckerei, D-67330 Speyer.
Bookbinding: J. Schäffer, D-67269 Grünstadt.
Printed in the Federal Republic of Germany

Preface

In recognition of the enormous advances in biotechnology in recent years, we are pleased to present this Second Edition of "Biotechnology" relatively soon after the introduction of the First Edition of this multi-volume comprehensive treatise. Since this series was extremely well accepted by the scientific community, we have maintained the overall goal of creating a number of volumes, each devoted to a certain topic, which provide scientists in academia, industry, and public institutions with a well-balanced and comprehensive overview of this growing field. We have fully revised the Second Edition and expanded it from ten to twelve volumes in order to take all recent developments into account.

These twelve volumes are organized into three sections. The first four volumes consider the fundamentals of biotechnology from biological, biochemical, molecular biological, and chemical engineering perspectives. The next four volumes are devoted to products of industrial relevance. Special attention is given here to products derived from genetically engineered microorganisms and mammalian cells. The last four volumes are dedicated to the description of special topics.

The new "Biotechnology" is a reference work, a comprehensive description of the state-of-the-art, and a guide to the original literature. It is specifically directed to microbiologists, biochemists, molecular biologists, bioengineers, chemical engineers, and food and pharmaceutical chemists working in industry, at universities or at public institutions.

A carefully selected and distinguished Scientific Advisory Board stands behind the series. Its members come from key institutions representing scientific input from about twenty countries.

The volume editors and the authors of the individual chapters have been chosen for their recognized expertise and their contributions to the various fields of biotechnology. Their willingness to impart this knowledge to their colleagues forms the basis of "Biotechnology" and is gratefully acknowledged. Moreover, this work could not have been brought to fruition without the foresight and the constant and diligent support of the publisher. We are grateful to VCH for publishing "Biotechnology" with their customary excellence. Special thanks are due to Dr. Hans-Joachim Kraus and Karin Dembowsky, without whose constant efforts the series could not be published. Finally, the editors wish to thank the members of the Scientific Advisory Board for their encouragement, their helpful suggestions, and their constructive criticism.

H.-J. Rehm
G. Reed
A. Pühler
P. Stadler

Scientific Advisory Board

Contents

Introduction XV
U. Ney, D. Schomburg

Structure and Function of Proteins

1 Types and Function of Proteins 1
   S. Wolf, H. G. Gassen
2 Sequence and Structure of Proteins 43
   F. Eisenhaber, P. Bork
3 Protein Interactions 87
   M. Meyer, D. Schomburg

Recombinant Proteins

4 In vitro Folding
   of Inclusion Body Proteins
   on an Industrial Scale 111
   R. Rudolph, H. Lilie, E. Schwarz
5 Medical Applications
   of Recombinant Proteins
   in Humans and Animals 125
   G. D. Wetzel
6 Enzymes for Industrial Applications 189
   W. Aehle, O. Misset

Monoclonal Antibodies

7 Antibody Engineering
   and Expression 219
   J. R. Adair
8 Manufacture of Therapeutic
   Antibodies 245
   A. J. Racher, J. M. Tong, J. Bonnerjea
9 Use of Antibodies for Immuno-
   purification 275
   D. J. King
10 Preclinical Testing of Antibodies:
   Pharmacology, Kinetics,
   and Immunogenicity 289
   R. Foulkes, S. Stephens
11 Preclinical Testing of Antibodies:
   Safety Aspects 303
   R. W. James
12 Therapeutic Applications
   of Monoclonal Antibodies:
   A Clinical Overview 311
   M. Sopwith

Case Studies

13 Antibodies for Sepsis:
   Some Lessons Learnt 329
   S. M. Opal
14 An Engineered Human Antibody
   for Chronic Therapy: CDP571 343
   M. Sopwith, S. Stephens
15 Antibody Targeted Chemotherapy 355
   M. S. Berger, P. R. Hamann, M. Sopwith

16 ReoPro Clinical Development:
A Case Study   365
H. F. Weisman

Gene Therapy

17 Overview of Gene Therapy   383
A. Mountain
18 Viral Vectors for Gene Therapy   395
B. J. Carter
19 Non-Viral Vectors for Gene Therapy   427
N. Weir
20 Issues for Large-Scale Plasmid DNA
Manufacturing   443
M. Schleef
21 Gene Therapy for HIV Infection   471
*M. Poznansky, Myra McClure,
Gregor B. Adams*

Regulatory and Economic Aspects

22 Regulation of Antibodies
and Recombinant Proteins   495
J. Foulkes, G. Traynor
23 Regulation of Human Gene Therapy   517
J. Parker
24 Economic Considerations   531
I. J. Nicholson

Index   543

Contributors

Dr. John R. Adair
Axis Genetics plc
Babraham
Cambridge, CB2 4A2
UK
Chapter 7

Dr. Gregor B. Adams
Department of Medicine
and Cummunicable Diseases
St Mary's Hospital
Praed Street, Paddington
London, W2 1PG
UK
Chapter 21

Dr. Wolfgang Aehle
Genencor International
P.O. Box 642
NL-2600 Delft
The Netherlands
Chapter 6

Dr. Mark Berger
Wyeth-Ayerst Research
145 King of Prussia Road
Radnor, PA 19087
USA
Chapter 15

Dr. Julian Bonnerjea
Development Division
Lonza Biologics
228 Bath Road
Slough, Berkshire SL1 4EN
UK
Chapter 8

Dr. Peer Bork
EMBL
Meyerhofstr. 1
D-69102 Heidelberg
Germany
and
Max-Delbrück-Centrum
für Molekulare Medizin
Robert-Rössle-Straße 10
13125 Berlin
Germany
Chapter 2

Dr. Barrie J. Carter
Targeted Genetics Corp
1100 Olive Way Suite 100
Seattle, WA 98101
USA
Chapter 18

Dr. Frank Eisenhaber
EMBL
Meyerhofstr. 1
D-69102 Heidelberg
Germany
and
Max-Delbrück-Centrum
für Molekulare Medizin
Robert-Rössle-Straße 10
13125 Berlin
Germany
Chapter 2

Julie Foulkes
Regulatory Affairs Consultant
Willow House, Dropmore Road
Burnham, Buckinghamshire SL1 8AY
UK
Chapter 22

Dr. Roly Foulkes
Biology Department, Research Division
Celltech Therapeutics Ltd.
216 Bath Road
Slough, Berkshire SL1 4EN
UK
Chapter 10

Prof. Dr. Hans Gassen
Institut für Biochemie
Technische Universität Darmstadt
Petersenstr. 22
D-64287 Darmstadt
Germany
Chapter 1

Dr. Philip Hamann
Wyeth-Ayerst Research
401 North Middletown Road
Pearl River, NY 10965
USA
Chapter 15

Dr. Ronald W. James
Soteros Consultants Ltd.
St. Johns House, Spitfire Close
Huntingdon, Cambridgeshire PE18 6XY
UK
Chapter 11

Dr. David J. King
Biology Department, Research Division
Celltech Therapeutics Ltd.
216 Bath Road
Slough, Berkshire SL1 4EN
UK
Chapter 9

Dr. Hauke Lilie
Institut für Biotechnologie
Martin-Luther-Universität Halle-Wittenberg
Kurt-Mothes-Straße 3
D-06120 Halle
Germany
Chapter 4

Dr. Michael Meyer
Institut für Molekulare Biotechnologie
Beutenbergstraße 11
D-07745 Jena
Germany
Chapter 3

Dr. Myra McClure
Department of Medicine and Cummunicable
Diseases
St Mary's Hospital
Praed Street, Paddington
London, W2 1PG
UK
Chapter 21

Dr. Onno Misset
Gist-Brocades
P.O. Box 1
NL-2600 Delft
The Netherlands
Chapter 6

Dr. Andrew Mountain
Cobra Therapeutics Ltd.
The Science Park
University of Keele
Keele, Staffordshire ST5 5SP
UK
Chapter 17

Ian J. Nicholson
Oxford Asymmetry International
151 Milton Park
Abingdon, Oxon OX14 4SD
UK
Chapter 24

Dr. Steven M. Opal
Infectious Diseases Division
Memorial Hospital of Rhode Island
111 Brewster Street
Pawtucket, RI 02860
USA
Chapter 13

Dr. James Parker
Strategic Bioscience Corporation
93 Birch Hill Road
Stow, MA 01775
USA
Chapter 23

Dr. Mark Poznansky
Division of Infectious Disease Medicine
Beth Israel Hospital
Harvard Medical School
Brookline Avenue
Boston, MA 02115
USA
Chapter 21

Dr. Andrew J. Racher
Development Division
Lonza Biologics
228 Bath Road
Slough, Berkshire, SL1 4EN
UK
Chapter 8

Prof. Dr. Rainer Rudolph
Institut für Biotechnologie
Martin-Luther-Universität Halle-Wittenberg
Kurt-Mothes-Straße 3
D-06120 Halle
Germany
Chapter 4

Dr. Martin Schleef
Qiagen GmbH
Max-Volmer-Straße 4
D-40724 Hilden
Germany
Chapter 20

Prof. Dr. Dietmar Schomburg
Institut für Biochemie
Universität zu Köln
Zülpicher Str. 47
D-50674 Köln
Germany
Chapter 3

Dr. Elisabeth Schwarz
Institut für Biotechnologie
Martin-Luther-Universität Halle-Wittenberg
Kurt-Mothes-Straße 3
D-06120 Halle
Germany
Chapter 4

Dr. Mark Sopwith
Director of Medicine
Celltech Therapeutics Ltd.
216 Bath Road
Slough, Berkshire SL1 4EN
UK
Chapters 12, 14, 15

Dr. Sue Stephens
Biology Department, Research Division
Celltech Therapeutics Ltd.
216 Bath Road
Slough, Berkshire SL1 4EN
UK
Chapters 10, 14

Dr. Jeremy M. Tong
Development Division
Lonza Biologics
228 Bath Road
Slough, Berkshire SL1 4EN
UK
Chapter 8

Gillian Traynor
Senior Regulatory Affairs Manager
PPD Pharmaco
Lockton House, Clarendon Road
Cambridge, CB2 2Bh
UK
Chapter 22

Dr. Neil Weir
Biology Department, Research Division
Celltech Therapeutics Ltd.
216 Bath Road
Slough, Berkshire SL1 4EN
UK
Chapter 19

Dr. Harlan F. Weisman
Vice-President
Clinical Research & Biomedical Operations
Centocor Inc., Great Valley Parkway
Malvern, PA 19355-1307
USA
Chapter 16

Dr. Gayle D. Wetzel
Preclinical Biology
Bayer Corp.
800 Dwight Way
Berkeley, CA 94701
USA
Chapter 5

Dr. Sabine Wolf
Institut für Biochemie
Technische Universität Darmstadt
Petersenstr. 22
D-64287 Darmstadt
Germany
Chapter 1

Introduction

ANDREW MOUNTAIN

Keele, UK

URSULA M. NEY

Slough, UK

DIETMAR SCHOMBURG

Köln, Germany

This volume reflects the progress that has been made in all aspects of the development and use of proteins during the last few years.

Proteins form without doubt, the most interesting class of biological molecules. Their diverse activities allow them to play a dominating role in almost any biological process. This diversity is complemented by a specificity of function which can only rarely be mimicked by a small molecule, and this can make proteins the molecules of choice when highly specific medical treatments are required. In addition, the catalytic functions of proteins make them ideal candidates for many technical and industrial processes such as in bioreactors, biosensors or as detergents.

The increasing use of proteins including enzymes and antibodies, has been stimulated by the progress made in the understanding of protein function and the ability to produce recombinant proteins. This has allowed both the large-scale production of proteins and the possibility to analyze and modify protein structure and sequence. We now have the opportunity to design protein molecules with properties optimized for their intended use. This is demonstrated by the replacement, for therapeutic use, of murine antibodies by chimeric and humanized antibodies, which offer advantages in terms of safety and efficacy and for widening the range of potential therapeutic applications.

This volume begins with an overview of protein structure and function and is followed by several chapters on recombinant proteins, their production and application for both medical and industrial use.

The next section of the book deals specifically with monoclonal antibodies and issues for design and production of new recombinant antibody-based molecules and for their subsequent development as therapeutic agents.

A series of case studies is included which illustrates both the problems and the successes with using antibodies as therapeutic agents and highlights some of the new approaches to therapy using humanized antibodies and antibodies as targeting agents.

The advances in our understanding of the human genome has brought with it the prospect of new ways to approach the treatment of disease. Here, the emergence of the potential for gene therapy has prompted the inclusion in this volume of a number of chapters which review the issues and problems currently facing this approach.

In the last section of the book the regulation of these products of biotechnology by the responsible authorities in Europe and the USA is reviewed. As progress is made in the understanding, production and use of these recombinant products, the regulations and guidelines are constantly changing, and this section provides a snapshot of current thinking.

The final chapter looks at the economics of the biotechnology revolution and the increasing use of recombinant products, proteins, antibodies and DNA for therapy. This trend is expected to continue, with some analysts forecasting that recombinant proteins will make up 20–30% of the total new products over the next five years. The cost of development of the biotech products is high and their use as therapies in a cost conscious health market is a challenge facing many companies. The options for producing cheaper, more cost effective recombinant proteins have been touched on in this volume, but for therapeutic use the protein agents still have to prove their competitiveness with new chemical entities, particularly in chronic indications. The next challenge facing the scientists is how to administer these highly specific agents in ways other than by parenteral injection.

Keele, Slough, Köln,   A. Mountain
November 1998   U. M. Ney
                         D. Schomburg

Structure and Function of Proteins

1 Types and Function of Proteins

SABINE WOLF, HANS GÜNTER GASSEN

Darmstadt, Germany

1 Introduction   2
2 Protein Types   5
  2.1 Proteins of Different Cell Compartments   6
    2.1.1 Membrane Proteins   6
  2.2 Conjugated Proteins   9
    2.2.1 Glycoproteins   9
    2.2.2 Lipoproteins   14
    2.2.3 Lipid-Linked Proteins   16
    2.2.4 Phosphoproteins   17
    2.2.5 Metalloproteins   17
    2.2.6 Hemoproteins   19
    2.2.7 Flavoproteins   20
  2.3 Proteins of Therapeutical Interest   21
3 Protein Function   21
  3.1 Enzymes   21
    3.1.1 Oxidoreductases   24
    3.1.2 Transferases   24
    3.1.3 Hydrolases   24
    3.1.4 Lyases   25
    3.1.5 Isomerases   25
    3.1.6 Ligases (Synthetases)   25
  3.2 Transport Proteins   25
  3.3 Contractile Proteins   30
  3.4 Structural Proteins   30
  3.5 Nutrient and Storage Proteins   32
  3.6 Defense Proteins   33
  3.7 Regulatory Proteins   37
4 Conclusions   39
5 References   39

List of Abbreviations

ACTH	adrenocorticotropic hormone
EGF	epidermal growth factor
ES	enzyme–substrate complex
FAD	flavin adenine dinucleotide
FMN	flavin mononucleotide
FSF	fibrin stabilizing factor, factor XIII
FSH	follicle stimulating hormone
G protein	guanine nucleotide-binding regulatory protein
GalNAc	N-acetyl galactosamine
GGT	γ-glutamyl transpeptidase
G_i protein	inhibiting G protein
Gla	γ-carboxy glutamic acid
GPI	glycosylphosphatidyl inositol
G_s protein	stimulating G protein
HDL	high-density lipoprotein
IgG	immunoglobulin G
LCAT	lecithin–cholesterol acyl transferase
LDL	low-density lipoprotein
LH	luteinizing hormone
LHRH	LH releasing hormone
MAC	membrane attack complex
matrix Gla protein	matrix γ-carboxy glutamic acid-containing protein
MGP	matrix Gla protein
MOAT	multispecific organic anion transfer
mw	molecular weight
NAD^+	nicotinamide adenine dinucleotide
NGF	nerve growth factor
TRH	TSH releasing hormone
TSH	thyroid stimulating hormone
VLDL	very low-density lipoprotein

1 Introduction

Almost every process occurring in a living organism involves one or more proteins. The structure and function of every individual cell, as well as metabolic activity is definitely influenced by proteins. Hence, it is not surprising that these biopolymers are also involved in their own synthesis and degradation. In view of this, one may pose the question which biopolymers existed first – RNA, DNA or proteins. Experiments undertaken by STANLEY MILLER and HAROLD UREY in 1953 illustrated that simple organic compounds like nucleotides and amino acids – the monomers of RNA/DNA and proteins – were formed simultaneously in a prebiotic atmosphere (for review, see LAZCANO and MILLER, 1996).

Over a period of 100 million years the prebiotic reactions led to an increase of these compounds in the oceans and in numerous places, such as drying lagoons, and the "prebiotic soup" became more and more concentrated. In such "concentrated environments" the component organic molecules could have condensed to form polypeptides and polynucleotides. Possibly, the reactants were adsorbed to minerals which acted as catalysts. It is widely believed that within this "prebiotic soup" a number of nucleic acid molecules combined to form a simple self-replicating system, thus enabling the synthesis of molecules complementary to themselves. A plausible hypothesis for the evolution of self-replicating systems is that they initially consisted entirely of RNA. The idea of the "RNA world" scenario is based on the observation that certain RNA species exhibit enzyme-like catalytic properties (GESTELAND and ATKINS, 1993; MCCORCKLE and ALTMANN, 1987). From this point of view, RNA was the primary substance of life. The Darwinian fitness was later increased by DNA and protein participation, thus refining the system. Through compartmentalization, the generation of cells was possible. The first true cells probably contained only RNA, proteins, and smaller molecules to form cell walls and provide metabolic energy.

Compartmentalization contributed another crucial advantage: the enclosure and protection of any self-replicating system. Within this protected environment of the first primitive cells, biochemical pathways developed including those enzymes required for RNA to DNA conversion, as well as from DNA to DNA. Subsequently, DNA was generated thus providing a more stable form of long-term information storage. Since the membranes of these primeval cells were lipid-based, the permeability for gases and water was possible. Other metabolites and ions, required for energy-dependent enzyme-catalyzed reactions may have entered the cell through porine-like structures. Porins, transmembrane channel forming proteins, which make membranes freely permeable to most small molecules up to a molecular weight (mw) of 10,000 also exist in the outer membranes of "modern" gram-negative bacteria and in the outer membrane of eukaryotic mitochondria. Obviously, more selective transport proteins developed much later in evolution. Another milestone during this period was the development of the genetic code, thus allowing specific protein sequences that exhibit useful properties to be reproduced. As cells required an increasing array of catalysts and structural components proteins became the focus of natural selection. At this stage of evolution the criteria of life – replication, catalysis, and mutability – were achieved.

The increasing number of organisms dramatically diminished energy-containing compounds within the "prebiotic soup". As a consequence, the organisms developed enzymatic systems that could synthesize those substances from simpler, abundant precursors, thus initiating the evolution of energy producing metabolic pathways. In summary, enzymes represent the first proteins appearing together with simple transport proteins following the development of the self-replicating system, thus guaranteeing the function of a basic replicative and metabolic apparatus.

Approximately 700 million years ago, after the development of photosynthesis and the establishment of the O_2-atmosphere on earth, multicellular organisms appeared. By this time, proteins were already involved as catalysts in an extraordinary range of chemical reactions. As working molecules of the cell they regulated metabolite concentrations, gene function control and membrane permeability, provided structural rigidity and were involved in cell motion (e.g., by cilia and flagella-like structures). The proteins controlling growth and differentiation and molecules conducting signals from cell to cell within an individual multicellular organism became increasingly more important. With the development of the nervous system, the number of

signal pathways and receptor proteins increased rapidly. Finally, the proteins of animal tissues enlarged the incredible diversity of these multifunctional biomolecules. Last but not least all vertebrates produce a large variety of proteins for defense tasks. Antibodies or immunoglobulins are able to distinguish between different molecules of foreign origin.

Despite many differences, prokaryotic and eukaryotic organisms including mammals have many biochemical pathways in common and most aspects, e.g., the translation of mRNA into proteins are similar. In many cases, individual proteins have presumably only one or a few ancestors, e.g., cytochrome c or histones, where the essential amino acids are highly conserved. All proteins are the result of 3.5 billion years of evolution (SCHOPF, 1993).

The genetic information of each individual cell, ultimately expressed as proteins, reflects their significance among the biopolymers. There are thousands of different proteins in a typical cell, each encoded by a gene (a specific DNA segment) and each performing a specific function. The differences in genetic organization and, therefore, in protein multiplicity between different organisms becomes obvious when considering the amount of DNA per cell (Tab. 1).

The genome of the bacterium *Escherichia coli* consists of $4 \cdot 10^6$ bp. Three DNA bases encode the position of each amino acid in a protein with an average size of 400 amino acids. Approximately 1,200 bp of DNA are used to encode a protein species. The maximum encoding capacity of the *E. coli* genome is about 3,300 different proteins. Although not all of the bacterial DNA encodes protein – a part of it is used as gene regulating DNA sequences – an *E. coli* cell may actually contain as many as 2,000 different protein species.

In general, eukaryotic cells contain more DNA than prokaryotic cells. Yeast cells possess one of the smallest eukaryotic genomes $(1.35 \cdot 10^7$ bp), which is still three times larger than that of *E. coli*. The cells of higher plants and animals contain 40–1,000 times more DNA than *E. coli*. However, only a small part of the total DNA in these cells usually en-

Tab. 1. DNA and Protein Content of Various Cells (Darnell et al., 1990)

Organism	DNA Number of bp	Maximum Number of Proteins Encoded[a]
Prokaryotic *Escherichia coli*	$4 \cdot 10^6$	$3.3 \cdot 10^3$
Eukaryotic *Saccharomyces cerevisiae* (yeast)	$1.35 \cdot 10^7$	$1.125 \cdot 10^4$
Homo sapiens (human)	$2.9 \cdot 10^9$	$2.42 \cdot 10^6$
Zea mays (corn)	$5.0 \cdot 10^9$	$4.0 \cdot 10^6$

[a] Assuming 1,200 bp per protein.

codes proteins, far less than the theoretical capacities of their genomes. It is widely believed that 50,000–100,000 proteins with different functions are translated in humans. Of course, in this calculation antibodies are considered as one polypeptide species. However, it should be noted that several million types of these molecules may be synthesized in vertebrates.

In keeping with the multiplicity of their functions, proteins are extremely complex molecules. They differ in structure, size, and in the components with which they are conjugated. Molecular weights of proteins range from 6,000 to over a million. In addition, many proteins are composed of two or more subunits. With 20 different amino acids as monomer compounds for a polypeptide of only 60 amino acids and a molecular weight of 6,000 D – without post-translational modification such as glycosylation – 20^{60} ($1.15 \cdot 10^{78}$) different species are theoretically possible. Taking into account the modifications mentioned above, this immense number would increase many times. In view of these possible variations it is easy to imagine that such a multiplicity of protein functions could develop. In general, proteins are classified according to their functions. Within these different classes different types of proteins, such as glycoproteins, lipoproteins or membrane proteins exist (Tab. 2).

One of the most important protein classes are enzymes. Virtually all chemical reactions of organic biomolecules in cells are catalyzed by enzymes. Many thousands of them, each capable of catalyzing a different kind of chemical reaction, have been discovered in different organisms.

In plasma membranes and intracellular membranes transport proteins are present. They are responsible, e.g., for the transport of nutrients such as amino acids and glucose or other substances across the membrane. Other transport proteins bind and transport specific molecules within the blood plasma of vertebrates from one organ to the other. For example, hemoglobin binds oxygen as blood passes the lungs and carries it to the peripheral tissues, whereas albumin is responsible for carrying fatty acids and other lipophilic compounds in the plasma.

Dynein, together with tubulin, the protein from which microtubuli are built, are components of flagella and cilia. They belong to the class of contractile or mobile proteins that allow cells to contract and change shape, or to move. Actin and myosin, proteins of the skeletal muscle also belong to this class. Strength, rigidity, and protection in multicellular organisms is given by structural proteins. They serve supporting filaments, fibers, or sheets. Collagen, the major protein in animals is a component of skin, tendon, and cartilage among other components in so-called connective tissue. Hair, fingernails, and feathers consist largely of the insoluble protein keratin.

Nutrient and storage proteins are found among animals and plants. Ovalbumin, the major component of egg white, is the main resource for amino acids for embryo growth. In mammals casein, the major milk protein, guarantees the essential supply of the infant with protein. The seeds of many plants store nu-

Tab. 2. Variability of Protein and Function

Name	Type	Function
β-Casein (milk)	mammalian phosphoprotein	nutrient protein
Insulin receptor	eukaryotic membrane protein	regulatory protein
Transferrin	glycoprotein	transport of iron to the tissues
Collagen type II	animal protein	structural protein (of tendons)
Fibrinogen	glycoprotein vertebrate	defense protein (blood clotting)
Lac repressor	prokaryotic protein	regulatory protein (DNA-binding protein)
γ-Glutamyl transpeptidase (GGT)	membrane protein (but also soluble)	enzyme
Na$^+$-K$^+$ ATPase	membrane protein	transport protein
Hemoglobin	hemoprotein	transport protein
Ovalbumin	glycoprotein	nutrient protein
Human IgG	glycoprotein	defense protein
Ribonuclease A	protein (only amino acids)	enzyme
Ribonuclease B	glycoprotein	enzyme
Xanthine oxidase	metalloprotein flavoprotein	enzyme
Renal dipeptidase	lipid-linked protein	enzyme
Serum response transcription factor	glycoprotein	regulatory (DNA-binding) protein
Murein lipoprotein	lipid-linked protein	structural protein

trient proteins required for the growth of the germinating seedling.

A further class of proteins regulates cellular and physiological activity. In addition to hormones and growth factors, DNA binding proteins that regulate the biosynthesis of enzymes and RNA molecules involved in cell division belong to this group of regulatory proteins.

Defense proteins provide protection from injury and defense against invasion by other species. Once injured, the blood clotting proteins fibrinogen and thrombin prevent the vascular system from loss of blood. Within the blood plasma, antibodies recognize foreign proteins, bacteria and viruses, leading to precipitation or neutralization of the invaders. Defense proteins are also found in plants and bacteria, e.g., ricin or bacteriotoxins.

In addition to the protein functions mentioned above, there are numerous proteins with rather exotic functions which are not easily classified.

Protein types are introduced in Sect. 2, before the different main functions of proteins are dicussed in more detail.

2 Protein Types

Similar to the classification of protein function as shown in Tab. 2, proteins can be divided into different types. A very simple distinction results, if purification is of interest. From this point of view, proteins are soluble or insoluble in the most common buffers. Most of the cytosolic proteins, but also plasma proteins such as antibodies and albumins as well as signal molecules (e.g., peptide hormones and growth factors) are soluble. Many membrane proteins are insoluble and can be released from the bilipid layer by detergents or chaotrophic buffers. Cross-linked collagens, the major components of fiber forming connective tissue, are also insoluble. This simple classification does not do justice to the different protein types.

Considering the structures of every individual cell, proteins may be classified into lysosomal, mitochondrial, etc. Since membranes lead to this compartmentalization, membrane proteins (Sect. 2.1.1) represent a major group of proteins.

Protein types can be divided more systematically according to their structure and/or their chemical constituents. Conjugated proteins, e.g., glyco- and lipoproteins are, therefore, described in more detail in Sect. 2.2.

In view of the topics covered in this volume Sect. 2 will close with a short discussion of therapeutically interesting proteins.

2.1 Proteins of Different Cell Compartments

Prokaryotic and eukaryotic cells are surrounded by a plasma membrane. In general, all internal membranes in prokaryotic cells are connected to the plasma membrane. In some bacteria, the plasma membrane has infoldings called mesosomes. The extensive mesosomes of photosynthetic bacteria contain proteins that utilize light energy to generate ATP. Some internal photosynthetic membranes (thylakoid vesicles) may not be connected to the plasma membrane. In contrast, eukaryotic cells contain extensive internal membranes that enclose and separate specific regions from the rest of the cytoplasm. These membranes define a collection of subcellular structures called organelles. The organelles are

- the nucleus that contains the chromosomes,
- the mitochondria in which oxidation of small molecules generates most of the cellular ATP,
- the rough and the smooth endoplasmic reticulum, a network of membranes in which glycoproteins and lipids are synthesized,
- the Golgi vesicles which direct membrane constituents to appropriate places in the cell,

- the peroxisomes in all eukaryotic cells and the glyoxisomes in plant seeds which metabolize hydrogen peroxide,
- the lysosomes involved in the degradation of proteins, nucleic acids and lipids,
- the chloroplasts found in plant cells, the site of photosynthesis,
- plant cells and certain eukaryotic microorganisms contain one or more vacuoles that store nutrient and waste molecules and also participate in the degradation of cellular proteins and other molecules.

Within this cellular organization membrane proteins not only facilitate molecular transport between different organelles, but they also cover the inner and outer membranes of the organelles with membrane-bound enzymes, specific for each compartment. Furthermore, the plasma membrane is covered with many different receptors for signal transduction, cell–cell attachment, and recognition.

2.1.1 Membrane Proteins

Proteins and polar lipids account for almost all of the mass of biological membranes. The small amount of carbohydrate present is generally part of glycoproteins and glycolipids. The protein composition of membranes of different sources varies even more widely than the majority of lipids, reflecting the functional specialization. Different staining procedures, tissue preparation methods, and last but not least electron microscopic techniques reveal important details of membrane structure. These studies show that each membrane protein is specifically oriented in the bilayer. The asymmetric arrangement of membrane proteins results in functional asymmetry. For example, glycoproteins of the plasma membrane are invariably situated with their oligosaccharide residues on the outer surface of the cell. All molecules of a given ion pump have the same orientation and, consequently, they pump in the same direction.

In 1972, the fluid mosaic model of biomembrane structure was proposed by SINGER and NICHOLSON (1972). They defined two classes of membrane proteins associated, to varying degrees, with the phospholipid bilayer. These two major classes are classified according to the mode of association with the lipid bilayer of diverse membranes such as cell membranes or the endoplasmatic reticulum, and mitochondrial or nuclear membranes. Such membrane proteins may be peripheral or integral. In addition, certain components of the cytoskeleton and the extracellular matrix are also directly or indirectly associated with membranes by the action of receptor molecules or via binding proteins. In this section, the two major classes, peripheral and integral membrane proteins, will be discussed in more detail.

Peripheral (or extrinsic) membrane proteins are associated with the membrane surface through interactions either with other proteins or with the exposed regions of phospholipids. They are only partially buried in the lipid matrix and are exposed on only one of the two membrane surfaces. Thus, in plasma membranes some of these proteins face the extracellular space and others face the cytosol. In view of their isolation and purification, these extrinsic membrane proteins can normally be released by mild treatments that do not involve the solubilization of the membrane itself. These conditions include the treatment with metal chelators (1–10 mM EDTA or EGTA), mild alkaline buffers (pH 8–11) at low ionic strength, dilute non-ionic detergents, low concentrations of organic solvents or high ionic strength (e.g., 1M NaCl) or the combination of two or more of these methods (PENEFSKY and TZAGOLOFF, 1971).

Integral (or intrinsic) membrane proteins are integrated in the hydrophobic phase of the membrane bilayer. They are subdivided into different types according to the proportion of their structure in contact with the hydrophobic phase of the bilayer (LOW, 1987; FINDLAY, 1990; REITHMEIER, 1996).

In Fig. 1 various mechanisms of membrane protein anchoring are shown. In Fig. 1a the substantial proportion of the membrane protein is buried in the bilayer. These intrinsic membrane proteins are designated as multispanning or polytopic. The folding ranges from two membrane spanning segments (bacterial signal peptidase) up to 7 (rhodopsin) or

more (glucose transporter). The second group of proteins (Fig. 1b) spans the bilipid layer once and contains a single transmembrane segment. Two types can be distinguished: one type where the amino terminus faces the cell exterior (or the lumen if located in intracellular organelles), whereas the other type has the opposite orientation. The receptors for insulin, growth factors, and LDL are representatives of this class (ULRICH et al., 1985; SÜDHOF et al., 1985). They have substantial portions of their mass on both membrane surfaces. Their full biological activity depends not only on the two hydrophilic domains, but also on the transmembrane segment. However, the globular portions in the aqueous phase can be freed from the rest of the molecule by proteolytic action. Under these conditions it is still possible for the extracellular domain, e.g., to retain its ligand-binding activity. Full biological activity is lost, however, for there is no means of communicating with the intracellular domain.

In Fig. 1c only one hydrophilic domain is present and consequently the orientation within the bilayer is less certain (e.g., cytochrome b_5; MARKELLO et al., 1985).

Figs. 1d–f represent proteins that are covalently attached to fatty acids or lipid. In Fig. 1d a myristic acid is amide-linked to the α-amino group on the N-terminal glycine residue. In Fig. 1e an amide-linked fatty acid and a thioether-linked diacylglycerol are attached to the α-amino and thiol groups of the N-terminal cysteine, respectively. Phosphatidyl inositol-linked membrane proteins are shown in Fig. 1f. The bulk of protein in these groups is in the aqueous phase and in general their biological activity does not involve the lipid bilayer. Examples include a large number of enzymes such as peptidases, esterases, and phosphatases. In these instances, the bilayer simply acts as a structural support or means of localization and/or organization. Proteins of this type can often be liberated from the membrane by use of proteases or phospholipases without destroying their native activities. The role of the cleavage enzymes is to separate the protein from its covalently attached protein or lipid membrane anchor which remains within the bilayer. Members of these last mentioned groups (Figs. 1d–f) belong to

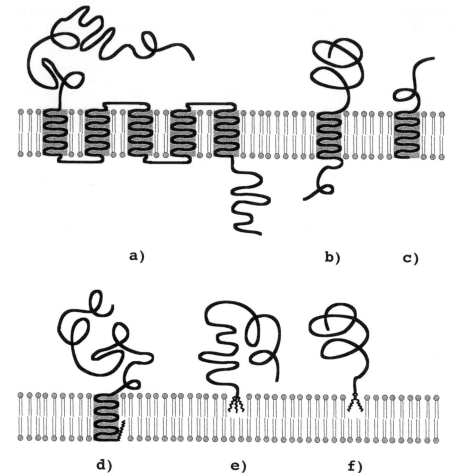

hydrophilic protein domains

membrane
bilayer

hydrophobic
transmembrane
domain

different
lipid
anchors

Fig. 1. Various mechanisms of membrane protein anchoring. Proteins with (a) two and more transmembrane segments, (b) one transmembrane segment and two hydrophilic domains at both membrane surfaces, (c) only one hydrophilic domain, which may be orientated extra- or intracellular (d)–(f) lipid-anchored membrane proteins.

the lipid-linked proteins and are discussed in more detail in Sect. 2.2.3.

2.2 Conjugated Proteins

Many proteins, such as the enzymes ribonuclease A and chymotrypsinogen, contain only amino acids. Proteins, which are composed of amino acids and other chemical components are described as conjugated proteins. On the basis of the nature of their non-amino acid part, usually called prosthetic group, they are classified as, e.g., glycoproteins or metalloproteins. In most cases, the prosthetic group plays an important role in the biological function of such proteins, and often a conjugated protein possesses more than one prosthetic group. With respect to the fundamental significance of conjugated proteins, they are discussed in Sects. 2.2.1–2.2.7 in more detail.

2.2.1 Glycoproteins

Glycoproteins, together with glycolipids, belong to the family of glycoconjugates. Sugar moieties, called glycans, are covalently linked to proteins. They occur in animals, plants, microorganisms, and viruses, mainly as membrane or serum proteins. For a long time, glycoprotein biochemistry lagged far behind that of nucleic acids and proteins in general, because of the complexities of carbohydrate chemistry. For example, two amino acids can form only two different dipeptides, whereas two monosaccharides can lead to more than 60 disaccharides. Furthermore, the investigation of glycoproteins requires knowledge of the complex primary structure of oligo- and polysaccharides on the one hand and of protein chemistry on the other.

With the increasing developments in biotechnology, the molecular biology of this important protein class is now firmly established. The biological roles of the glycan moieties in glycoproteins are (MONTREUIL, 1995)

- to interfere with protein folding, during biosynthesis, thus favoring secretion from the cell,

- to stabilize the conformation of proteins in biologically active forms by means of glycan–glycan and glycan–protein interactions,
- to act as "shields"; glycan moieties protect polypeptide side chains against proteolytic attack,
- to control the specific proteolysis of precursors leading to active proteins or peptides,
- to provide a protective mechanism by masking peptide epitopes which often cause weak antigenicity of glycoproteins,
- in the control of lifetime: desialylation induces capture of glycoproteins by hepatocytes and of erythrocytes by macrophages,
- to act as receptor sites for various proteins and additionally for microorganisms and viruses,
- similar to membrane lectins they are involved in cell–cell recognition, cell adhesion, cell differentiation and development, and cell contact inhibition,
- membrane glycoprotein glycans are epitopes of tissue antigens, of bacteria, fungi and parasite envelopes, and of tumor-associated antigens.

Almost all protein(and lipid)-linked sugars are localized on the exoplasmic face of eukaryotic plasma membranes and almost all secreted proteins of eukaryotic cells are glycosylated. Indeed, protein glycosylation is more abundant than all other post-translational modifications. The polypeptide chains of glycoproteins, like those of all proteins are synthesized under genetic control. Their carbohydrate chains are enzymatically generated within the endoplasmatic reticulum and the Golgi apparatus and covalently linked to the polypeptide without the rigid guidance of nucleic acid templates. Since processing enzymes are generally not available in sufficient quantities to ensure the synthesis of uniform products, glycoproteins often have a variable composition with respect to the carbohydrate domains, a phenomenon known as microheterogeneity. Purification and characterization difficulties are a consequence of this phenomenon.

In contrast, most cytosolic and nuclear proteins are not glycosylated. An exception is the nuclear pore complex (HANOVER, 1992), which is involved in the transport of mRNA from the place of origin, the nucleus, to the cytosol where mRNA directed protein synthesis takes place. Additionally, several transcription factors are also glycosylated. In the above mentioned cases, a single N-acetyl glucosamine residue is linked to a serine or threonine hydroxyl group (REASON et al., 1992).

In the majority of glycoproteins facing the exoplasmic membranes, sugar residues are commonly linked to two different classes of amino acid residues.

They are classified as O-linked, if they are joined to the hydroxyl oxygen of serine or threonine (or hydroxylysine in the case of collagen) via an α-O-glycosidic bond (Fig. 2). These O-glycosylated serine and threonine residues are not members of any common sequence. Apparently, the location of the glyco-

sylation sites is specified only by the secondary or tertiary structure of the polypeptide. The oligosaccharide side chains are generally short (e.g., the disaccharide side chain of collagen: glucose–galactose–hydroxylysine) but often rather variable. However, some O-linked oligosaccharides, such as those bearing the AB0 blood group antigens, can be very long (see Fig. 2c). Almost all humans can produce the 0-type oligosaccharide, but special enzymes are required for the addition of either galactose (type B) or N-acetyl galactosamine (type A). Some individuals possess one of these enzymes, some possess the other, and a few can produce both. The latter individuals have type AB, with both A and B oligosaccharides on the cell surfaces of erythrocytes.

Studies of the biosynthesis of mucin, an O-linked glycoprotein secreted by the submaxil-

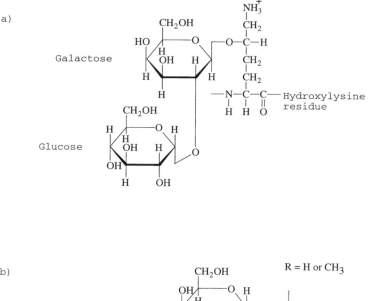

Fig. 2. O-glycosidic linked disaccharide in collagen to 5-hydoxylysyl residues (a). The most common O-glycosidic attachment in glycoproteins involves the dissaccharide core β-galactosyl-$(1\rightarrow3)$-α-N-acetyl galactosamine α linked to the hydroxyl group of either Ser or Thr (b). Blood group antigens as an example for more complex O-glycosylation of proteins (c). The O-oligosaccharide type forms the core structure of both, A and B type. The A and B antigens are formed by addition of GalNAc or Gal, respectively.

c)

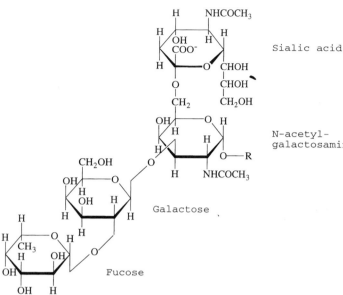

Sialic acid

N-acetyl-galactosamine

Galactose

Fucose

Type 0

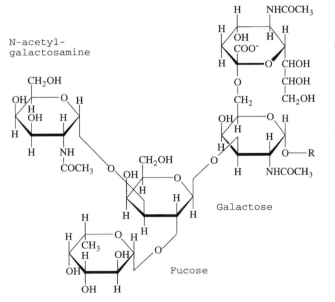

N-acetyl-galactosamine

Sialic acid

N-acetyl-galactosamine

Galactose

Fucose

Fig. 2. Continued.

Type A

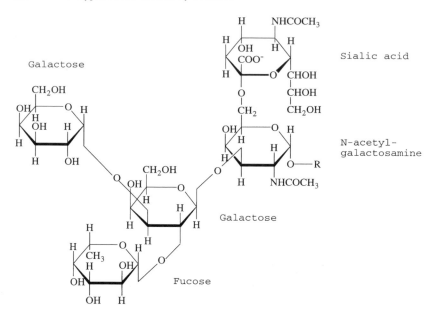

Type B

Fig. 2. Continued.

lary gland, have indicated that these types of glycoproteins are synthesized in the Golgi apparatus by serial addition of monosaccharide units to a complete polypeptide chain, beginning with the transfer of N-acetyl galactosamine (GalNAc) from UDP-GalNAc catalyzed by GalNAc transferase (ROTH et al., 1994).

N-linked glycosylation occurs through the amide nitrogen of asparagine side chains by β-N-glycosidic bonds to N-acetyl glucosamine. A common sequence surrounding the asparagine residue is Asn-Xaa-Thr or Asn-Xaa-Ser, where Xaa may be any amino acid residue, excluding Pro or Asn. All of the known N-linked oligosaccharides have a common structure: Manα 1→3 Manα 1→6 Manβ 1→4 GlcNAcβ1→4 GlcNAc Asn. N-linked oligosaccharide side chains often exhibit complex branched structures. These can be categorized in terms of three basic structures: high mannose, complex, and hybrid (Fig. 3). The structure found attached to human immunoglobulin G (IgG) belongs to the complex type. The carbohydrate domain is linked to every heavy chain of each immunoglobulin. These constant domains permit the recogni-

tion of different immunoglobulins for proper tissue distribution as well as for the interaction with phagocytic cells, which destroy the antigen–antibody complex. A part of this recognition is based on differences in the oligosaccharide side chains. The terminal sialin (N-acetyl-neuraminic acid) residue is also found in other glycoproteins circulating in the blood serving as a "timer" to signal old molecules for degradation (SCHAUER, 1985). The sialin residues at the termini of such oligosaccharides are cleaved during circulation. Receptors on the surfaces of liver cells recognize and bind these proteins, after which they are internalized and destroyed.

In contrast to O-linked glycoproteins, biosynthesis of N-linked oligosaccharides does not occur on the polypeptide chain, but on a lipid-linked intermediate (VERBERT et al., 1987; ELBEIN, 1987). A precursor is then transferred to the polypeptide chain, which is synthesized in a cotranslational action. Finally, the transferred oligosaccharide is subject to further processing steps as it passes from the rough and smooth endoplasmic reticulum through the Golgi apparatus.

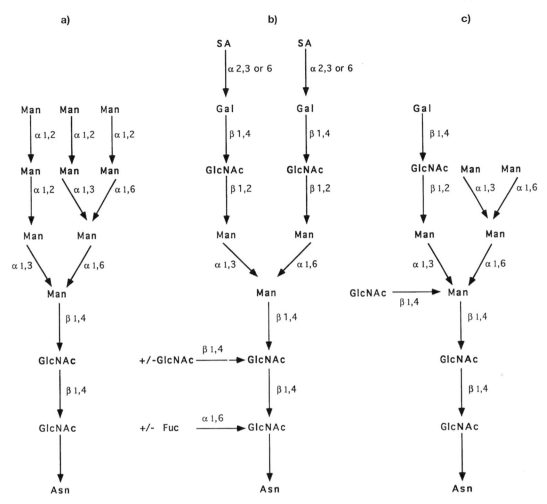

Fig. 3. Typical primary structures of N-linked oligosaccharides. (a) high-mannose, (b) complex, (c) hybrid. GlcNAc: N-acetyl glucosamine, Man: mannose, Gal: galactose, SA: sialic acid (after KORNFIELD and KORNFIELD, 1985).

A third type of glycoproteins is represented by glycosylphosphatidyl inositol (GPI) membrane-anchored proteins. Their polypetide chain is attached to the moiety via an amide bond between mannose-6-phosphoethanolamine and the C-terminal carboxyl group (Fig. 4). The core GPI structure is synthesized on the luminal face of the endoplasmic reticulum and often modified with a variety of additional sugar residues, depending on the species and the protein to which it is attached (CROSS, 1990; ENGLUND, 1993). The amino group of GPI phosphoethanolamine nucleo-philically attacks a specific amino acyl group of the protein near its C-terminus, resulting in a transamidation. Finally, a 20–30 residue hydrophobic C-terminal peptide is released. Since GPI-anchored proteins originate within the rough endoplasmic reticulum, they occur on the exoplasmic face of the plasma membrane. Since phospholipid is essential for the binding of the protein to the membrane, this type of glycoprotein also belongs to the class of lipoproteins and membrane proteins. Several cell surface proteins, including Thy-1, an antigen confined to thymocytes and T lym-

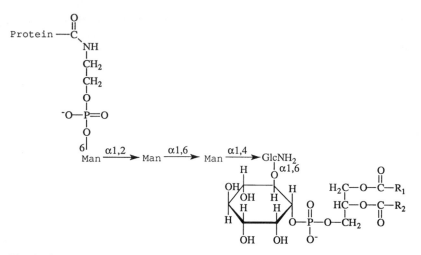

Fig. 4. The core structure of glycosylphosphatidyl inositol (GPI) anchors of proteins. R_1 and R_2 represent fatty acid residues, their identities varying with the protein. The sugar residues within the core tetrasaccharide also vary with the protein.

phocytes (ROBINSON et al., 1989), as well as many enzymes, e.g., alkaline phosphatase are anchored by a complex glycosylated phospholipid. In such cases, treatment with phospholipase C, an enzyme that cleaves the phosphate–glycerol bond, releases these proteins from the exterior surface.

Recently, the production of therapeutically interesting human glycoproteins (i.e., antibodies) by genetic engineering and recombinant techniques has been confronted by an enormous problem. The lack of endoplasmic reticulae and Golgi apparatus in prokaryotic cells is the reason for the inability of these organisms to synthesize glycoproteins. Eukaryotic, but non-human cells, produce different glycosylations than human cells. In the near future, eukaryotic cells have to be engineered to produce recombinant glycoproteins which are identical to the native ones. Variances in identity may impair therapeutic effectiveness and safety through

- different hydrophobicity,
- decrease or inhibition of secretion,
- decrease of stability towards proteases and consequently shortening of the *in vivo* life span of those molecules by increased clearance,

- changes in the affinity to specific receptors, and
- increase of antigenicity.

Taking this into account, as well as the pathogenicity or toxicity of microorganisms or cell line products, a general assessment of biosafety must be made (BRAUER et al., 1995).

2.2.2 Lipoproteins

In contrast to glycoproteins, lipoproteins are not covalently linked to lipids. Lipids and proteins form non-covalent aggregates which are responsible for the transport of lipids, triacylglycerols, cholesterol, and phospholipids in the blood plasma. Insoluble lipids are transported as lipoproteins from the place of their origin, either the liver where they were synthesized or the intestine where resorption takes place, to the tissues. This lipid transport system delivers lipids to cells which require them for anabolic or energy purposes. For this reason, lipoproteins represent the only protein type, where function is restricted to one task: transport of different lipids. For this reason, types and functions of different lipoproteins are discussed together in this section.

When plasma is subjected to high-speed centrifugation, lipoproteins float because lipid has a lower buoyant density than water. In contrast, nearly all plasma proteins sediment under these conditions. The protein moiety in lipoproteins is called apolipoprotein. At least 9 different protein components are known in human plasma. Together with different lipids they build up different classes of lipoprotein particles, spheric aggregates with hydrophobic lipids in the center and hydrophilic amino acid side chains at the surface. Each of these classes fulfills a specific function, depending upon

● type and quantity of apolipoproteins,
● type and amount of lipids,
● the origin of synthesis,
● the tissue which is supplied.

In Tab. 3 the major classes of human plasma lipoproteins are listed.

Chylomicrons are the largest lipoprotein particles (diameter 50–200 nm) with the lowest density, a consequence of the high amount of triacylglycerols. They are synthesized in the smooth endoplasmic reticula of epithelial cells which cover the small intestine. Exogenous nutrient lipids, triacylglycerols, and cholesterol are transferred as chylomicrons from the lymphatic system of the intestine (*chylus*) via the lymphatic system and are passed into circulation at the *A. subclavia*. Binding sites on the surface of the muscle capillary endothelial cells and fatty tissue are responsible for the attachment of these particles. Triacylglycerols are hydrolyzed in a process cata-

lyzed by extracellular lipoprotein lipase which is activated by apoliprotein ApoC-II. While the products of hydrolysis, monoacylglycerols and free fatty acids, are taken up by the tissue, the chylomicrons shrink to residual particles which are consequently rich in cholesterol. These particles leave the binding sites of the cells and are resorbed by the liver.

When the diet contains more fatty acids than needed, they are converted to triacylglycerols in the liver. Excess carbohydrates can also be converted into triacylglycerols within this tissue. Together with cholesterol, cholesteryl esters, and specific apolipoproteins (Tab. 4), these triacylglycerols are packed into very low-density lipoprotein (VLDL) particles in the liver. As in chylomicrons, apolipoprotein ApoC-II activates lipoprotein lipase and causes the release of fatty acids. The destination for VLDL is the adipose tissue. Uptake of fatty acids, resynthesis of triacylglycerols and storage in intracellular lipid droplets is then carried out by adipocytes.

During transport, the VLDL particles are converted into cholesterol and cholesteryl ester-rich low-density lipoprotein (LDL). The major protein component of these particles is apoB-100. Specific surface receptors of the peripheral tissue (dissimilar to those found in liver) recognize this major protein component of LDL, mediating the uptake of cholesterol and cholesteryl esters by LDL receptor presenting cells.

High-density lipoproteins (HDL), the fourth major type of lipoproteins, are protein-rich particles, which contain ApoC-I and ApoC-II. In addition to other lipoproteins they also contain the enzyme lecithin–cholesterol acyl transferase (LCAT). Synthesized in

Tab. 3. Major Classes of Human Plasma Lipoproteins (from KRITCHEVSKY, 1986)

Lipo-protein	Density [g mL^{-1}]	Composition (in % by Weight)				
		Protein	Phospho-lipids	Free Cholesterol	Cholesterol-esters	Triacyl-glycerols
Chylomicrons	<1.006	2	9	1	3	85
VLDL	0.95–1.006	10	18	7	12	50
LDL	1.006–1.063	23	20	8	37	10
HDL	1.063–1.210	55	24	2	15	4

Tab. 4. Function of Apolipoproteins in Human Plasma (VANCE and VANCE, 1985)

Apolipo-protein	Molecular Weight	Lipoprotein Particles	Function (as far as known)
ApoA-I	28331	HDL	activation of LCAT
ApoA-II	17380	HDL	
ApoA-IV	44000	chylomicrons, HDL	
ApoB-48	240000	chylomicrons	
ApoB-100	51300	VLDL, LDL	binding site of LDL receptor
ApoC-I	7000	VLDL, HDL	
ApoC-II	8837	chylomicrons, VLDL, HDL	activation of lipoprotein lipase
ApoC-III	8751	chylomicrons, VLDL, HDL	inhibition of lipoprotein lipase
ApoD	32500	HDL	
ApoE	34145	chylomicrons, VLDL, HDL	initiation of VLDL and chylomicron degradation

the liver, these particles are released into the bloodstream where nascent (newly synthesized) HDL collects cholesterol and phosphatidyl choline from other circulating lipoproteins. After the removal of triacylglycerols from the corresponding tissues, chylomicrons and VLDLs are rich in these components. The action of LCAT on the surface of nascent HDL induces cholesterol and phosphatidyl choline conversion to cholesteryl esters, which are able to enter the interior of nascent HDL particles. During uptake, the HDL particles change their shape from a flat disc to a sphere. Mature HDL, a cholesterol-rich lipoprotein, is then released in the liver. As summarized in Tab. 4, apolipoproteins have different functions. These range from activating (ApoC-II) or inhibiting (ApoC-III) lipoprotein lipase or activating LCAT (ApoA-I) to receptor-binding sites (ApoB-100, ApoE) for the uptake of LDL and chylomicrons by receptor-mediated endocytosis. Since this is a general mechanism whereby cells take up large molecules, it will be discussed later.

2.2.3 Lipid-Linked Proteins

A large number of proteins in eukaryotic cells contain covalently linked fatty acids or phospholipids that bind the protein to the plasma membrane. The latter, GPI-anchored proteins are discussed in Sect. 2.2.1.

Among the exclusively lipid-linked proteins, three different covalent attachments with three classes of lipids are known.

The most common is the attachment to isoprenoid groups of the farnesyl(C15)- and geranylgeranyl(C20)-type (Fig. 5a, b). These lipid-linked proteins occur mainly on the cytoplasmic surface of the plasma membrane. A major class of these proteins is formed by those that possess the C-terminal sequence CysAaaAaaXaa, in which the Aaa amino acid residue is generally aliphatic. If Xaa is represented by leucine, a cytosolic isoprenyl transferase transfers the C_{20} group of geranylgeranyl pyrophosphate to the cysteine sulfur atom via thioether linkage. The tripeptide AaaAaaXaa is then proteolytically excised and the newly exposed terminal group is esterified with a methyl group. Alternatively, the terminal residues Xaa such as alanine, methionine, serine or glutamine are farnesylated by a distinct cytosolic enzyme. The net effect of these modifications is the creation of a C-terminal S-isoprenyl cysteine α-methyl ester. Other consensus sequences for the C-terminal isoprenylation are -CysXaaCys and -CysCys precursor sequences (CLARKE, 1992).

Many of the small G proteins, the γ-subunits of the large G proteins, and the Ras proteins (transforming proteins that cause cells to become cancerous) contain these precursor sequence motifs. These protein species function in signal transduction processes across the plasma membrane and also in the control of cell division.

Another type of lipid-linked proteins is represented by fatty acylated proteins (Fig. 5c). Two fatty acids are known to be linked to eukaryotic proteins and the fatty acyl group is thought to function as a membrane anchor. The specific requirement for one of these residues suggests that these groups also participate in targeting their attached protein.

Like isoprenylated polypetides, palmitoylated proteins occur almost exclusively on the interior surface of the plasma membrane. Palmitic acid, a common saturated C_{16} fatty acid is bound via thioester linkage to a specific cysteine residue of the polypeptide near the C-terminus. This modification is reversible and occurs post-translational in the cytosol.

In contrast, the rare saturated myristic acid (C_{14} fatty acid) is appended to a protein in a stable amide linkage to the α-amino group of a N-terminal glycine residue in a cotranslational process (Fig. 5d). Myristoylated proteins are found in many subcellular compartments including cytosol, nucleus, and plasma membrane.

In Figs. 5e and 1e another type of lipid linkage is shown. The murein lipoprotein of *E. coli* is synthesized through a series of post-translational modifications and processing reactions. These lead to an aminoterminal acylated cysteine residue, whereby the thiol group is transferred to a glyceryl moiety (WU and TOKUNAGA, 1986).

2.2.4 Phosphoproteins

Protein phosphorylation is a principal regulatory mechanism in the control of almost all cellular processes. Reversible phosphorylation and dephoshorylation at specific Ser/Thr or Tyr residues results in conformational changes of target proteins and plays a crucial role in, e.g., signal transduction pathways.

Another reason for protein phosphorylation, mainly irreversible, is found in many proteins isolated from mineralized bone matrix. Synthesized by osteoblast cells, these acidic phosphoproteins are generally restricted to calcified tissues, suggesting an involvement in calcium binding and nucleation of calcium hydroxyapatite crystal formation (GORSKI, 1992). Investigations of posphorylation sites of matrix Gla protein (MGP) demonstrate that the protein is phosphorylated at three serine residues near the N-terminus within tandemly repeated Ser-Xaa-Glu sequences. This recognition motif (Ser-Xaa-Glu/Ser-P) was also found in secreted proteins (PRICE et al. 1994). Phosphoproteins secreted into milk (casein) or saliva are fully phosphorylated at each target serine residue and many are also capable of presenting calcium binding sites.

2.2.5 Metalloproteins

Many enzymes contain metal ions, usually held by coordinate-covalent bonds from the amino acid side chains, but sometimes bound to a prostheic group like heme. As shown in Tab. 5, these ions have various functions. Metal ions in enzymes

- act as metal catalysts, like the zinc ion in carboxypeptidase A;
- serve as a redox reagent; in the case of catalase, a Fe^{3+} ion is attached to the protoporphyrin IX, in the reaction, Fe^{3+} is alternately reduced and reoxidized;
- are found to be necessary for catalytic efficiency, even though they may not remain permanently attached to the protein; for a number of enzymes that couple ATP hydrolysis to other processes (e.g., most of the glycolytic enzymes) Mg^{2+} is required for efficient function, in some cases, the Mg^{2+}–ATP complex is a better substrate than ATP itself, and it is not clearly established whether these enzymes should be classified as metalloproteins.

Among the proteins involved in electron transfer, the hemoproteins (cytochromes) and the flavoproteins are discussed in Sects. 2.2.6–2.2.7. A third group is represented by iron–sulfur proteins or non-heme iron proteins. They are associated with inorganic sulfur and/or with sulfur atoms of distinct cysteine residues in the polypeptide chain. Fe—S centers range from simple structures where a single Fe atom is coordinated to 4 cysteines via the

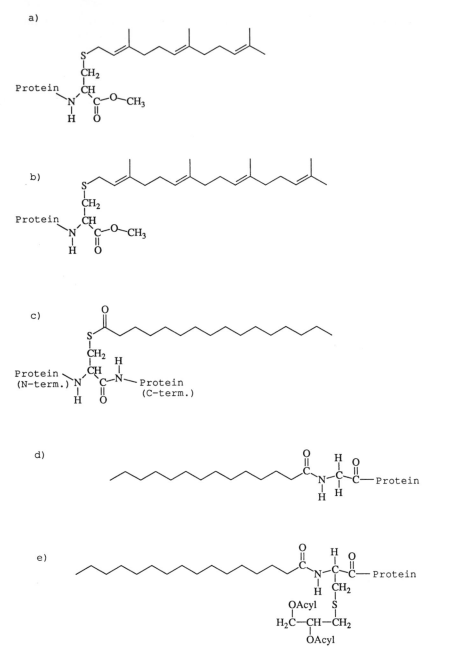

Fig. 5. Isoprenylated proteins (a, b). In both cases, the farnesyl (a) and geranylgeranyl (b) residue is linked to a C-terminal cysteine residue within a consensus sequence of CysAaaAaaXaa. During biosynthesis, the isoprenyl group is appended to the protein via a thioether link to the Cys residue, the tripeptide AaaAaaXaa is hydrolytically cleaved. The new carboxyl terminus is methyl esterified. Palmitic acid is linked to proteins via a thioester link to a cysteine residue near the C-terminus of the proteins (c). Myristoylation takes place at the N-terminal glycine residue via a stable amide bond (d). Structure of aminoterminal lipid link in bacterial murein lipoprotein (e).

Tab. 5. Metal Ions and Trace Elements of Importance as Enzymatic Cofactors

Metal	Enzyme	Role of Metal
Fe^{2+} or Fe^{3+}	cytochrome oxidase catalase peroxidase	oxidation–reduction
Cu^{2+}	lysyl oxidase	oxidation–reduction
Zn^{2+}	carboxypeptidase A	involved in hydrolysis
Mn^{2+}	histidine ammonialyase arginase ribonucleotide reductase	aids in catalysis by electron withdrawal
Co	glutamate mutase	part of cobalamin coenzyme
Ni^{2+}	urease	involved in catalysis
Mo	xanthine oxidase	oxidation–reduction
V	nitrate reductase	oxidation–reduction
Se	glutathione peroxidase	replaces S in one cysteine in active site

sulfur in the side chains, to more complex Fe—S centers with 2 or 4 Fe atoms. During electron transfer one of the iron atoms is oxidized or reduced (BEINERT, 1990). Another role of the ferric ions in proteins is to ensure the delivery of iron. In the serum, iron is carried by transferrin. Endocytosis of diferric transferrin, mediated by the transferrin receptor, guarantees the iron supply of the cells. Within the cells, iron is stored within ferritin (LASH and SALEEM, 1995). All ferritins have 24 subunits arranged to form a hollow shell with a cavity of 80 Å in diameter. Up to 4,500 Fe(III) atoms in the form of an inorganic complex (HARRISON and AROSIO, 1996) can be stored within this cavity. Ferritin-stored iron can be utilized in a number of iron-containing proteins. In contrast, iron and iron binding proteins can also act as regulators of immune function (DE SOUSA et al., 1988).

Furthermore, another important class of DNA binding proteins among the metalloproteins must be considered. The so-called zinc finger motifs contain highly conserved amino acids, histidine and/or cysteine residues. These ligands are tetrahedrally coordinated around the Zn^{2+} ions (SCHWABE and KLUG, 1994).

2.2.6 Hemoproteins

Within the class of hemoproteins, two different subtypes may be distinguished according to function: transport of oxygen or of electrons.

The first subtype is represented by the oxygen binding proteins myoglobin and hemoglobin. The prosthetic group of these proteins, the heme group – iron protoporphyrin IX – (Fig. 6) is identical to that of cytochrome c which belongs to the electron transporting hemoproteins. The central iron atom of the

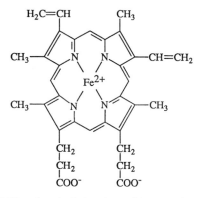

Fig. 6. The chemical structure of iron protoporphyrin IX, representing the prosthetic group in hemoglobin and myoglobin.

porphyrine system has 6 coordination bonds, 4 in the plane of the flat porphyrine molecule and 2 perpendicular to it. In myoglobin and hemoglobin, one of these is bound to a nitrogen of a histidine residue, the other serves as a binding site for the O_2 molecule. In both proteins the heme group is responsible for the deep brown color. Myoglobin, a single chain polypeptide (153 amino acid residues) with a single heme group, functions as an oxygen binding protein in muscle cells. Particulary in the muscles of diving mammals such as the whale, storage of oxygen by muscle myoglobin permits these animals to remain submerged for a long period of time. Hemoglobin was the first oligomeric protein to be subjected to X-ray analysis. This protein contains 4 polypeptide chains and 4 iron–porphyrine prosthetic groups. One heme is bound to each of the hemoglobin polypetide chains. Hemoglobin, the major protein of erythrocytes, is reponsible for oxygen binding in the lungs and oxygen release in the peripheral tissue.

The respiratory chain, located within the mitochondrial membrane, consists of a series of electron carrier proteins. Most of these are integral membrane proteins. Within the four types of electron transfer systems in biological systems, the direct transfer of electrons, as in the reduction of Fe^{3+} to Fe^{2+}, is carried out by cytochromes. These iron-containing electron transfer proteins of the mitochondrial inner membrane are also present in the thylakoid membranes of chloroplasts and in the plasma membrane of bacteria. They are intensely colored, an effect produced by their prosthetic heme group. According to differences in the light absorption spectra they are classified as a, b, and c. The heme groups of cytochromes a and b are tigthly but not covalently bound to their associated proteins; heme groups of type c cytochromes (heme c) are covalently attached via thioether bonds to two Cys residues of the polypeptide chain.

2.2.7 Flavoproteins

Many enzymes catalyzing oxidation–reduction reactions are flavoproteins. They use flavin mononucleotide (FMN) or flavin adenine dinucleotide (FAD) as a cofactor (Fig. 7). The fused ring structure of these nucleotides undergoes reversible reduction during

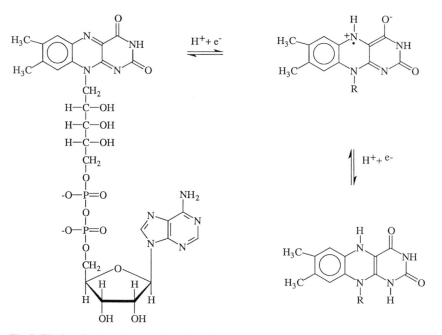

Fig 7. Flavin adenine dinucleotide (FAD) and reduced forms.

catalysis, accepting one or two electrons in the form of one or two hydrogen atoms. Participating in either one- or two-electron transfers, flavoproteins can serve as intermediates between reactions in which two electrons are donated (dehydrogenations) and those in which only one electron is accepted (GHISLA and MASSEY, 1989). The reduction of the flavin nucleotide is accompanied by a change in the major absorption band, a feature often used in analyzing reactions involving flavoproteins.

In most flavoproteins the prosthetic group is bound rather tightly and non-covalently to the enzyme. However, electron transport is not carried out by diffusion from one enzyme to the next. Covalent linkage is also present in some enzymes, e.g., succinate dehydrogenase. The variability of the standard reduction potential of the prosthetic group is an important feature of flavoproteins and depends on the associated protein. Reduction potentials of flavoproteins are, therefore, sometimes quite different from those of free flavin nucleotides.

In addition to the flavin nucleotides, the often very complex flavoproteins are tightly bound to inorganic ions such as iron or molybdenum, which are capable of participating in electron transfers.

2.3 Proteins of Therapeutical Interest

Therapeutically useful proteins may be divided into 4 groups: regulatory factors (including hormones, cytokines, lymphokines, and other regulatory factors of cellular growth and metabolism), blood products (including serum-derived blood factors and enzymatic fibrinogen activators), vaccines, and monoclonal antibodies. These 4 protein classes exhibit a wide range of biochemical structures. Cytokines and other regulatory factors include tripeptides such as thyroliberin, small well-characterized proteins such as insulin, growth hormones, and interferons, and larger, more complex molecules such as erythropoietin and gonadotropins, which may both be glycosylated and consist of multiple subunits.

Most proteins in the blood product category are complex, varying from enzymes such as urokinase, which may exist in a variable ratio of high-to-low molecular weight forms, to high molecular-weight assemblies, such as the clotting factor VIII complex.

Vaccines are generally high molecular-weight structures and may exist as a glycoprotein complexed with a lipid component, as in the case of hepatitis B vaccine.

Monoclonal antibodies are generally IgG, although IgM may also be considered for therapeutic use. It can be readily appreciated that with the advent of recombinant DNA technology, the range and quantity of proteins available to medicine have increased enormously. Furthermore, the situation now exists that more or less any protein considered likely to be of possible therapeutic benefit to man can be made available in sufficient quantity for clinical evaluation.

3 Protein Function

Another possibility for protein classification is given through the assignment to different functions. The main functions of proteins are catalysis, transport, defense, and regulation. Proteins also serve as nutrient and storage compounds as well as structural components. They build motility systems in pro- and eukaryotic cells as well as in the muscles of vertebrates. In Sects. 3.1–3.7 (on protein functions), a few examples, mainly of vertebrate proteins, will be discussed in more detail.

3.1 Enzymes

Most essential biological reactions proceeding at efficient rates under physiological conditions are catalyzed by proteins called enzymes. The following simple example illustrates the importance of these biochemical catalysts: for a bacterium that divides once every 20 min, a reaction taking many hours to approach completion is not metabolically useful. Under biologically relevant conditions, uncatalyzed reactions tend to be slow. Enzymes circumvent these problems through

- higher reaction rates; enzymatically catalyzed reactions are typically faster by factors of 10^6–10^{12} than those of the corresponding uncatalyzed reactions;
- milder reaction conditions; reactions catalyzed by enzymes occur below 100°C, at atmospheric pressure and at nearly neutral pH. Efficient chemical catalysis often requires elevated temperature, pressures, and extremes of pH;
- greater reaction specificity; enzymes have a greater degree of substrate (reactants) and product specificity than chemical catalysts and rarely form side products;
- regulation of activity by control of enzyme availability (synthesis vs. degradation) by structural or conformational alterations, by binding of small molecules (effectors), or by feedback inhibition, in which the product of a biosynthetic pathway controls the activity of an enzyme near the beginning of that pathway.

A catalyst is defined as a substance that increases the rate or velocity of a chemical reaction, itself remaining unchanged in the overall process. Substrates and other molecules bind to enzymes with identical forces that dictate conformations of proteins themselves: van der Waals, electrostatic, hydrogen-bonding forces, and hydrophobic interactions. In 1894, EMIL FISCHER discovered that glycolytic enzymes can distinguish between stereoisomeric sugars. He proposed the lock and key hypothesis: the enzyme accomodates a specific molecule as a lock does a specific key by binding the molecule (substrate) into a specific region of the enzyme (active site). Although this hypothesis explained specificity, it did not increase the understanding of catalysis itself. This was realized by an elaboration of FISCHER'S idea – the induced fit hypothesis. KOSHLAND postulated the distortion of the enzyme as well as the subtrate when forming the enzyme–substrate complex ES. The role of the enzyme is to keep the substrate under stress, thereby forcing it into a conformation approximating the transition state. A catalyst lowers the energy barrier for

a reaction, thereby increasing the reaction rate in either direction without influencing the position of the equilibrium. The difference in the barrier peaks between the two directions of a given chemical reaction (ground state substrate S vs. ground state product P) is exactly the same whether a catalyst is present or not (Fig. 8). ΔG_0 and K, the equilibrium constant of the process, remain unchanged. Since the alignment of reacting groups, the formation of transient unstable charges, bond rearrangements, and other transformations require energy, there is an energetic barrier between the ground states of S and P. Molecules that undergo reaction must overcome this barrier and, therefore, must be raised to a higher energy level, the transition state. Enzymes lower the energy level of the transition state and consequently enhance reaction rates.

The above mentioned ES complex is known as the Michaelis complex named by LENOR MICHAELIS, who together with MAUDE MENTEN, analyzed enzyme kinetics in the early 20th century. The analysis of the enzyme reaction was performed under the assumption that the overall reaction consists of two elementary reactions, whereby the substrate forms a complex with the enzyme, which subsequently decomposes to product and enzyme (Eq. 1).

$$E + S \rightleftharpoons [ES] \rightleftharpoons E + P \qquad (1)$$

This leads to the formulation of the Michaelis–Menten equation, a basic equation of enzyme kinetics (Eq. 2).

$$v_0 = \frac{V_{max}[S]}{K_m + [S]} \qquad (2)$$

The Michaelis–Menten equation, which has the functional form of a hyperbola, has two parameters: v_{max}, the maximum reaction rate, which occurs when the substrate concentration is saturated, and K_M, the Michaelis constant, which represents the substrate concentration $[S]$ at half-maximal reaction rate. Physically more realistic models of enzyme mechanisms assume that the enzymatic reaction is reversible (right side Eq. 1) and involves one or more intermediates. For more detailed in-

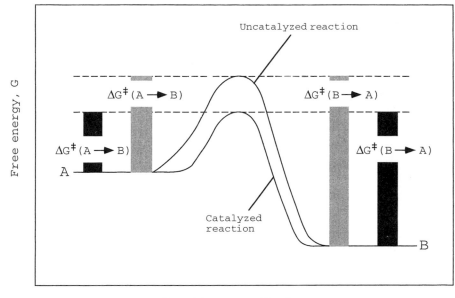

Fig. 8. Effect of a catalyst. The catalyst lowers ΔG^{+}, thus enabling a larger fraction of reactant molecules to possess sufficient energy to reach this lowered transition state.

formation in enzyme kinetics the reader is referred to the literature (SEGEL, 1975). It is important to note that measurements of the kinetic parameters of a given enzyme reaction are among the most powerful techniques for elucidating the catalytic mechanism of the enzyme.

Like other proteins, enzymes have molecular weights ranging from about 12,000 to over 1 million. Some of them require no chemical groups other than their amino acid residues for activity. Their functional groups can facilitate acid–base reactions and take part in charge–charge interactions. They are less suitable for many types of group transfer reactions or oxidation–reduction processes. For these purposes, enzymes need an additional chemical component known as a cofactor. In many cases, the enzyme may require the help of some ions (Mg^{2+}, Fe^{2+}, Mn^{2+} or Zn^{2+}) or other small organic molecules (coenzymes) to carry out a reaction. Coenzymes function as transient carriers of functional groups. A covalently linked coenzyme or ion is called a prosthetic group. The catalytically active enzyme together with the coenzyme and/or me-

tal ion is called holoenzyme, whereas the protein part of such an enzyme is called apoenzyme or apoprotein. Some important coenzymes are listed in Tab. 6.

Like enzymes, coenzymes remain unchanged during the overall process. Sometimes it is difficult to distinguish a coenzyme from a substrate – the substance an enzyme acts upon. For example, dehydrogenase enzymes all have a strong binding site for the oxidized form of NAD^{+}. After oxidation of the primary substrate NADH, the reduced form, leaves the enzyme and is reoxidized by other electron acceptor systems of the cell. NAD^{+} once reoxidized can now bind to another dehydrogenase molecule. In these cases NAD^{+} seems to function as a second substrate, rather than a true coenzyme.

Since metabolism is very complex, there are thousands of different kinds of enzymes. Many of these have been given common names, some of them are descriptive of the enzyme function, e.g., fructose-1,6-bisphosphatase, and others like pepsin are not. In 1961, the first Enzyme Commission devised a classification system for enzymes which also

Tab. 6. Some Important Coenzymes

Name	Reaction in Which the Coenzyme is Involved
Nicotinamide adenine dinucleotide (NAD$^+$)	oxidation–reduction
Flavin adenine dinucleotide (FAD)	oxidation–reduction
Coenzyme A	acyl transfer
Thiamine pyrophosphate	group transfer
Pyridoxal phosphate	group transfer; transamination
Tetrahydrofolate	group transfer
5′-Deoxyadenosyl-cobalamine (coenzyme B12)	group transfer
Biotin	carboxylation

serves as a basis for the allocation of code numbers. These code numbers, now widely in use, contain 4 elements separated by points, with the following significance:

(1) the first number shows to which of the 6 main divisions (classes) the enzyme is assigned,
(2) the second number indicates the sub-class,
(3) the third figure gives the sub-subclass,
(4) the fourth figure is the serial number of the enzyme in the sub-subclass.

The main divisions and subclasses are described in Sects. 3.1.1–3.1.6 (FASMAN, 1976).

3.1.1 Oxidoreductases

All enzymes catalyzing oxidation–reduction reactions belong to this class. The substrate which is oxidized is regarded as a hydrogen donor. The systematic name is based on donor–acceptor oxidoreductase. The recommended name will be dehydrogenase whenever this is possible. Alternatively, the term reductase may be used. Oxidase is only used in cases where O_2 is the acceptor.

The second figure indicates the group in the hydrogen donor molecule which undergoes oxidation (e.g., for an aldehyde or a keto group).

The third figure indicates the type of acceptor involved in the reaction. For example, 1 denotes NAD(P).

3.1.2 Transferases

This class of enzymes catalyzes the transfer of a group, e.g., the methyl group, from one compound (donor) to another (acceptor). The systematic names are formed according to the scheme donor–acceptor group transferase. The recommended names are normally formed according to acceptor group transferase or donor group transferase. The second figure in the code number of transferases indicates the group transferred: a one-carbon group (2.1.), an aldehyde, or a ketonic group (2.2), a glycosyl group (2.3.), and so on. The third figure gives further information on the group transferred, e.g., the formyl group. An exception is made in group 2.7, where the third figure indicates the nature of the acceptor group.

3.1.3 Hydrolases

These enzymes catalyze the hydrolytic cleavage of C—O, C—N, C—C, and some other bonds including phosphoric anhydride bonds. While the systematic name always includes hydrolase, the recommended name is, in many cases, formed by the name of the substrate with the suffix -ase.

The second figure in the code number of the hydrolases indicates the nature of the bond hydrolyzed: 3.1 are esterases, 3.2 are glycosidases. The third figure normally specifies the nature of the substrate: carboxyl ester hydrolases (3.1.1) or thiol ester hydrolases

(3.1.2). Exceptionally, the third figure is based on the catalytic mechanism elucidated by active site studies, exemplified by peptidyl peptide hydrolases.

3.1.4 Lyases

Enzymes that cleave $C-C$, $C-O$, $C-N$ and other bonds by elimination, leaving double bonds or conversely adding groups to double bonds, are called lyases. The systematic name is formed according to the pattern substrate group-lyase. To avoid confusion, the hyphen, as an important part of the name, should not be omitted (e.g., hydro-lyase not hydrolyase).

The second figure in the code numbers of lyases indicates which bond is cleaved (4.1. are carbon–carbon-lyases), whereas the third figure gives further information on the eliminated group.

3.1.5 Isomerases

These enzymes catalyze geometric and structural changes within a molecule (racemases, epimerases, etc.). The subclasses are formed according to the type of isomerism, the sub-subclasses indicate the type of substrate.

3.1.6 Ligases (Synthetases)

The combination of two molecules coupled with the hydrolysis of a pyrophosphate bond in ATP or a similar triphosphate is catalyzed by these enzymes. Systematic names are formed on the system X–Y ligase. In the recommended nomenclature the term synthetase may be used. The second figure of the code number indicates the bond formed (6.1 for $C-O$ bonds). Sub-sublasses are only specified for $C-N$ ligases. Tab. 7 lists one example from each major class of enzymes.

Among the different enzyme classes, 6 types of catalytic mechanisms have been defined:

(1) acid–base catalysis, a process in which partial proton transfer from a Brønsted acid (or by partial proton abstraction by a Brønsted base) lowers the energy of the reaction transition state;

(2) covalent catalysis involves rate acceleration through transient formation of a catalyst–substrate covalent bond;

(3) metal ion catalysis, where tightly or loosely bound metal ions participate in the catalytic process;

(4) electrostatic catalysis, in which the binding of the subtrate generally excludes water from the enzyme active site;

(5) catalysis through proximity and orientation effects, whereby molecules react most readily only if they have the proper relative orientation;

(6) catalysis by preferential transition state binding, where the enzyme has a greater affinity to the transition state than the corresponding substrates or products.

The great catalytic efficiency of enzymes arises from their simultaneous use of several of these mechanisms.

3.2 Transport Proteins

Biological membranes which surround cells and organelles protect the interior from certain exogenous toxic compounds. On the other hand, metabolites must be transported into the cell and in addition waste products must be removed. Only a few, mostly small molecules (e.g., O_2) or lipophilic compounds are able to penetrate biological membranes by molecular diffusion through the phospholipid bilayer. For many substances the slow transport by passive diffusion is insufficient. To fulfill the functional and metabolic needs, membranes of cells and organelles are provided with transport proteins. There are two different mechanisms of protein-mediated transport:

● Facilitated transport through protein carrier molecules or protein pores; the driving force is based upon the concentration gradient of the transported molecule. Transport ceases when an equal concentration of the transported substance on both sides of the membrane is reached.

Tab. 7. Examples for Each Major Class of Enzymes

Class	Example and Reaction Type	Reaction Catalyzed
1. Oxidore-ductases	Alcohol dehydrogenase (EC 1.1.1.1) oxidation with NAD^+	
2. Trans-ferases	Glucokinase (EC 2.7.1.2) phosphorylation	
3. Hydrolases	Trypsin (EC 3.4.21.4) peptide bond cleavage ($R_1 = $ Arg or Lys; $R_2 = $ non-specific)	
4. Lyases	Oxalate decarboxylase (EC 4.1.1.2) decarboxylation	

• Active transport takes place when substances are transported against even very unfavorable concentration gradients. According to thermodynamical considerations this kind of transport requires an input of free energy. Primary active transport uses energy from the hydrolysis of ATP but also from redox- or light-coupled processes (e.g., cytochrome oxidase or bacteriorhodopsin, respectively). It is estimated that cells spend 30–50% of their ATP (up to 70% in nerve cells) just on active transport. Active transport may be driven directly by ATP hydrolysis (ion pumps) or indirectly by the free energy that is stored in ion gradients (secondary active transport) built up through the action of an ion pump.

One of the best characterized processes of facilitated diffusion via protein carrier molecules is the transport of D-glucose across the erythrocyte membrane, which is essential for the survival of the cell. Passive transport of D-glucose by molecular diffusion through an artificial phospholipid bilayer is very slow, the permeability coefficient P is $4 \cdot 10^{-10}$ cm s^{-1}. The erythrocyte glucose transporter, a 55 kD glycoprotein with a bundle of 12 membrane spanning segments presumably forms a hydrophobic cylinder surrounding a hydrophilic

Tab. 7. Continued

Class	Example and Reaction Type	Reaction Catalyzed
5. Isomerases	Maleate isomerase (EC 5.2.1.1) *cis-trans* isomerization	(structure diagram) Maleate ⇌ Fumarat
6. Ligases	Glutathione synthetase (EC 6.3.2.3) glutathione synthesis	(structure diagram) γ-Glu-Cys + Gly → Glutathione

channel. Through this channel D-glucose, which is available from the blood plasma, is transported with a 50,000-fold increased rate $(P=2\cdot10^{-5}$ cm s^{-1}) in comparison to passive diffusion. In general, all known transport proteins are asymmetrically situated transmembrane proteins. They alternate between two conformational states whereby the ligand binding sites are exposed in turn to alternate sides of the membrane. D-glucose binds to the carrier protein at the exterior face of the membrane, followed by a conformational change that closes the outer site and exposes the inner membrane where dissociation takes place. Such a system is referred to as a gated pore (BALDWIN et al., 1982). Other types of gated pores play a major role in the propagation of nerve impulses.

A further well-studied example of a gated pore is the gap junction, a non-specific pore between adjacent cells in some animal tissues, allowing exchange of small molecules between these cells. In response to certain stimuli (e.g., increasing Ca^{2+} concentration as a consequence of cell damage), these gap junctions close.

Among the active, ATP hydrolyzing transport proteins, three cation transporting types have been identified (P-, F-, and V-type ATPases, see Tab. 8). The most common prototype for active transport systems is the Na$^+$–K$^+$ATPase found in plasma membranes. The protein belongs to P-type ATPases which are all reversibly phosphorylated as a part of the transport cycle. The protein is located in the plasma membrane and consists of two subunit types. A non-glycosylated α-subunit possessing catalytic activity and ion binding sites (110 kD) and a glycosylated β-subunit (55 kD) with unknown function. The function of the whole molecule of this so-called Na$^+$–K$^+$ pump, which is also an enzyme because

Tab. 8. The 3 Classes of Cation Transport ATPases

Transported Ion	Source	Membrane	Function
P type[a] ATPases			
Na^+K^+	higher eukaryotes	plasma	maintains the transmembrane electrical potential
H^+K^+	mammals	plasma	acidifies contents of stomach
H^+	fungi	plasma	creates a low pH in the compartment; activation of proteases and other hydrolytic enzymes
H^+	higher plants	plasma	creates a low pH in the compartment; activation of proteases and other hydrolytic enzymes
Ca^{2+}	higher eukaryotes	plasma	maintains low $[Ca^{2+}]$ in cytosol
Ca^{2+}	animal muscle cells	sarcoplasmatic reticulum	sequesters intracellular Ca^{2+}, keeping cytosolic $[Ca^{2+}]$ low
F type[b] ATPases			
H^+	eukaryotes	inner mitochondrial membrane	catalyzes formation of ATP from $ADP + P_i$
H^+	higher plants	thylakoid	catalyzes formation of ATP from $ADP + P_i$
H^+	prokaryotes	plasma	catalyzes formation of ATP from $ADP + P_i$
V type[c] ATPases			
H^+	animals	lysosomal, endosomal, secretory vesicles	creates a low pH in the compartment; activation of proteases and other hydrolytic enzymes
H^+	higher plants	vacuolar	creates a low pH in the compartment; activation of proteases and other hydrolytic enzymes
H^+	fungi	vacuolar	creates a low pH in the compartment; activation of proteases and other hydrolytic enzymes

The ATPases are named according to the following features:
[a] P types are reversibly phosphorylated during a transport cycle.
[b] F types were identified as energy coupling factors.
[c] V types occur in vacuolar membranes.

of the ATP hydrolyzing activity, is to set and to maintain the sodium and potassium gradients across the plasma membrane. The concentrations of both cations are quite different in the cytosol and in the extracellular fluid or blood plasma. The intracellular concentrations are approximately 12 mM (Na^+) and 140 mM (K^+). In contrast, extracellular concentrations are 145 mM (Na^+) and 4 mM (K^+), thus generating a transmembrane electric potential. This electric potential is central to electric signaling in neurons. In a variety of cell types, the Na^+ gradient drives the co-transport of various compounds against the concentration gradient.

Besides the ATP-dependent ion pumps other energy-dependent transporters have been identified. The multispecific organic anion transporter (MOAT) from rat liver plasma membranes is specific for the transport of non-bile acid organic anions into the bile (PiKULA et al., 1994a, b) and belongs to the family of mammalian plasma membrane P-glycoproteins. Another example of this type of transporter, also found in bacteria, is represented by the P-glycoprotein (GROS et al., 1986). This transporter accounts for the cellular phenotype of multidrug resistance, meaning the immunity against chemotherapeutic reagents. This export pump was first discovered in tumor cells and is responsible for lowering intracellular concentrations of carcinostatic agents, thereby reducing therapeutic effectiveness (PEARSON and CUNNINGHAM, 1993).

Other types of transport proteins are represented by blood plasma serum proteins. These function as transport vehicles for small lipophilic molecules, thereby increasing their solubility in plasma. For example, serum albumin is able to reversibly bind a wide range of diverse ligands with high affinity. Several binding regions of the molecule are discussed, which are specific for long-chain fatty acids or less specific for tryptophan, bilirubin, and hemin. Additional binding sites for high-affinity binding of drugs (KRAGH-HANSEN, 1990) are also discussed.

The ability to bind lipophilic intrinsic ligands as well as lipophilic drugs and certain steroids is a property of lipocalins, a family of small extracellular proteins. Members of this protein superfamily are retinol-binding protein, β-lactoglobulin, and human complement protein C8 gamma. Lipocalins bind and transport retinol (vitamin A), an essential nutrient factor with hydrophobic properties in blood plasma and milk (FLOWER, 1994; HAEFLIGER et al., 1991). Many proteins of the lipocalin family have been identified according to their common structure, but in many cases the specific ligand is not known.

In analogy to the lipophilic compounds, iron is less soluble in serum and, therefore, carried by a major serum glycoprotein transferrin. Diferric transferrin binds to a specific receptor (transferrin receptor), which is present on the target cell surface and is endocytosed (LASH and SALEEM, 1995). Receptor-mediated endocytosis is a general mechanism to import proteins from the surrounding medium into certain cells. These include low-density lipoprotein (LDL), peptide hormones, and circulating proteins that are destined to be degraded. Some toxins and viruses, such as diphtheria toxin, cholera toxin, and the influenza virus also enter cells in this way. The best characterized example is the transferrin system whereby a transferrin–iron complex is internalized and, after removal of the iron, the apotransferrin receptor complex is returned to the cell surface (DAUTRY-VARSAT et al., 1983). Tab. 9 summarizes some important proteins and particles, including their

Tab. 9. Receptor-Mediated Endocytosis of Proteins and Particles in Animal Cells (from DARNELL et al., 1990)

Ligand	Function of Ligand–Receptor Complex	Cell Type
Low-density lipoproteins	cholesterol supply	most
Transferrin	iron supply	most
Glucose- or mannose-terminal glycoproteins	remove injurious agents from circulation	macrophages
Galactose-terminal glycoproteins	remove injurious agents from circulation	hepatocyte
Immunoglobulins	transfer immunity to fetus	fetal yolk sac; intestinal epithelial cells of neonatal animals
Phosphovitellogenins	protein supply to embryo	developing oocyte
Fibrin	removal of injurious agents	epithelial
Insulin and other peptide hormones	alter cellular metabolism; ligand and often receptor are degraded after endocytosis	most

function, taken up by receptor-mediated endocytosis.

3.3 Contractile Proteins

The capacity for organized movement is one of the most striking characteristics of living organisms.

Vertebrates and many invertebrates have two classes of muscle, smooth and striated; a third class is represented by the cardiac muscle in vertebrates. Contraction and relaxation of smooth muscles which surround internal organs as well as large blood vessels is controlled subconsciously by the central nervous system.

The most familiar and one of the best understood motility systems is that of the striated muscle. At the light microscopic level skeletal muscles have a striated appearence that consists of long parallel bundels of muscle fibers with diameters ranging from 20–100 μm. The muscle fibers are composed of parallel bundles of around a 1,000 myofibrils of 1–2 μm in diameter. As shown in electron microscopic studies, the muscle fiber striations arise from a banded structure of myofibrils. Alternating bands of higher and lower electron density (A bands and I bands) build up repeating units of myofibrils, which are known as the sarcomere. Within the sarcomere, the light I bands are composed entirely of thin actin filaments, whereas the darker colored A bands are built up of thick myosin filaments. In the relaxed muscle, sarcomers are 2.5–3 μm long and they progressively shorten during muscle contraction. As shown in Fig. 9, the ends of the actin filaments are anchored at Z discs. During contraction, the I band shortens, while the width of the A band remains constant. Several experiments have established that these thick and thin filaments move relative to each other during contraction energized by the hydrolysis of ATP.

Myosin, actin, and other muscle proteins are also present in non-muscle cells, but in much lower ratios to actin than in muscle. In non-muscle cells these proteins possess both contractile and structural functions. For example, actin filaments in the intestinal brush border have primarily a structural role. Numerous actin-binding proteins, like villin and fibrin, help to form the plasma membrane into elongated microvilli. Actin filaments are also thought to participate in a number of motile events in non-muscle cells. In cytokinesis, the final separation of daugther cells during mitosis, a contracting ring of actin and myosin fibers runs just under the plasma membrane in the region of the cell constriction. Another example for the participation of actin filaments is the streaming of the cytoplasm of large algae.

A different type of biological motility system is typified by cilia and flagella. Both structures are found in many eukaryotic cells. They are about 0.25 μm in diameter and differ mainly in length (flagella are longer). For example, sperm cells use flagella for propulsion, and a large number of cilia cover the surfaces of the mammalian respiratory tract to sweep out particulate matter that collects in the mucous secretions of these tissues.

The principal structural elements of cilia, flagella, and the mitotic spindle are microtubules constructed of the protein tubulin. In view of the contractile properties of tubulin as well as actin, they are regarded as contractile proteins. Movement of these structures is not a result of shortening of a single molecule, but due to a movement of molecules relative to each other as exhibited during muscle contraction (Fig. 9). In the mitotic spindle apparatus movement of chromosomes during the anaphase is caused by the depolymerization of microtubules into free tubulin. Since actin (e.g., in microvilli) and microtubules and, therefore, tubulin also form more or less permanent structures like axonal filaments both these proteins are also designed to be structural proteins.

3.4 Structural Proteins

Collagen, the most abundant structural protein in vertebrates, performs a wide variety of functions. It is the major protein of connective tissue such as cartilage and tendons, it is an important constituent of skin, and it builds the matrix material in bone upon which the mineral constituents precipitate.

(a) Relaxed state

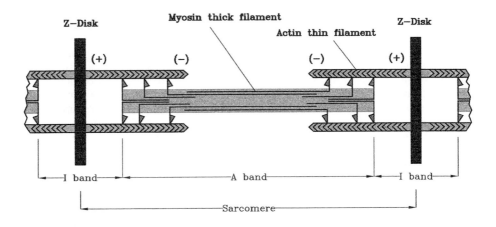

(b) Contracted state

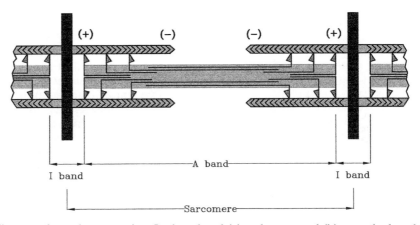

Fig. 9. Schematic diagram of muscle contraction. In the relaxed (a) and contracted (b) state the length of the A band (thick myosin filament) remains constant. Contraction occurs when myosin heads push actin filaments towards the center of the sarcomere.

The extracellular space of animal tissues is filled with ground substance, also called extracellular matrix, which is composed of an interlocking mesh of heteropolysaccharides and fibrous proteins including collagen. At least 19 different collagens and 10 proteins with collagen-like domains are known (PRO-KOP and KIVIRIKKO, 1995). In accordance to their polymeric structure and different features, the collagen superfamily can be divided into several classes:

- fibril forming collagens (types I, II, III, V, and XI) assemble into cross-striated fibrils. Type I collagen is the major component of most connective tissues and is accompanied by small quantities of the minor component type V. In cartilage and vitreous humor, type II collagen is the major collagen type besides small quantities of type XI. Extensible connective tissues such as skin, lung, and the vascular system contain large proportions of type III collagen;
- network forming collagens (type IV family, and type XII and type X). The type IV collagen family is found in basement membranes that separate endothelial and epithelial cells from the underlying tissue. Type VIII and type X collagen are constituents of the Descemet membrane, separating the corneal endothelial cells from stroma. Type X collagen was also found in the deep calcifying zone of cartilage;
- collagens found on the surface of collagen fibrils (fibril-associated collagens type IX, XII, XIV, XVI, and XIX);
- collagen that forms beaded filaments (type VI);
- collagen that forms anchoring fibrils to basement membranes (type VII) is responsible for attachment to the underlying extracellular matrix;
- collagens with transmembrane domains (type XIII and XVII) have been discovered recently and are probably not secreted into the extracellular matrix;
- newly discovered collagens (type XV and type XVIII) have only been partially characterized.

The overall function of collagens is to provide stability, strength, and rigidity within connective tissues. In the extracellular matrix collagen participates in holding cells of a tissue together. Cells are not directly associated with collagen. The association is mediated through intergrin, a plasma membrane protein located in cell membranes, and fibronectin, an extracellular matrix protein which has binding sites for collagen and intergrin.

In addition to the above mentioned collagens and other connective tissue components, yellow connective tissue, e.g., the walls of the large blood vessels or elastic ligaments contain elastin, a protein with rubber-like properties. Elastin fibers can stretch to several times their normal length, and they are also present in small amounts in most connective tissues, e.g., joints, skin, and tendons, mediating elasticity.

α-Keratin, the structural protein of hair, wool, feathers, nails, and much of the outer layer of the skin has evolved for strength.

To provide strength and elasticity is the unique characteristic of another group of extracellular fibrous proteins that are synthesized by insects and arachnids (spiders). Silk, produced by the domestic silkworm, has been used for several centuries as a high-quality material to produce textiles.

Based on their molecular structure, particularly elastin and spider silk have the potential to expand. For example, the spider major (dragline) silk has impressive characteristics of strength and elasticity (Xu and Lewis, 1990). Since the expression of this protein has recently been established (Lewis et al. 1996), the notable mechanical properties of dragline spider silk will lead to a broad area of application in the future.

3.5 Nutrient and Storage Proteins

Nutrient and storage proteins are distributed among animals and plants. The primary function of plant nutrient proteins is to provide germinating seeds with proteins required for growth. Plant nutrient proteins are also important in the nutrition of humans and animals.

Im mammals β-casein, the major milk protein, guarantees the essential supply of the infant with protein. The highly phosphorylated β-casein also functions as a transport protein for Ca^{2+}, an essential nutrient factor.

The iron storage protein ferritin also has dual functions: it is involved in iron detoxification and serves as iron reserve. Ferritin is ubiquitously distributed among living species including bacteria, and the three-dimensional structure is highly conserved (Harrison and Arosio, 1996). 24 protein subunits are arranged as a hollow shell with a cavity with a diameter of 80 Å, capable of storing up to 4,500 Fe(III) atoms as an inorganic complex.

3.6 Defense Proteins

In higher organisms, protection from injury and defense against invasion by other species is provided by defense proteins within the circulatory system. Defense proteins are also present in plants and bacteria, e.g., ricin or bacteriotoxins.

The self-healing properties of the circulatory system in vertebrates are based on a process known as hemostasis that prevents blood loss from even the smallest injury. The initial event and the cause for blood clotting is the injury of a blood vessel. The damaged vessel stimulates platelets (unpigmented enucleated blood cells) to adhere to the injured vessels as well as to each other forming a plug. This mechanism can stop minor bleedings. The process is mediated by the von Willebrandt factor, a multimeric plasma glycoprotein that binds to a specific receptor on the platelet membrane and to collagen and other subendothelial components which are exposed to the plasma during vascular injury. Once ag-

gregated, the platelets release several substances, such as serotonin and thromboxane A2, which stimulate vasoconstriction and, therefore, reduce the blood flow in the injured vessel. After these initial events, the aggregated platelets and the damaged tissue initiate blood clotting (coagulation), the major defense against blood loss. The blood clotting cascade represents a classical example for the enzymatic activation of slightly longer precursor proteins (zymogens). Nearly 20 different, mainly liver-synthesized glycoproteins, participate in this second step by forming a blood clot through a cascade of proteolytic reactions. These so-called clotting factors are, with the exception of fibrin, serine proteases. They are able to activate other factors of the cascade by proteolytic attack. Seven of the clotting factors are zymogens of serine proteases (inactive forms) that are activated by serine proteases further up in the cascade. The cascade of proteolytic activation may be initiated by blood exposure at damaged tissues (intrinsic pathway, Fig. 10) or by internal

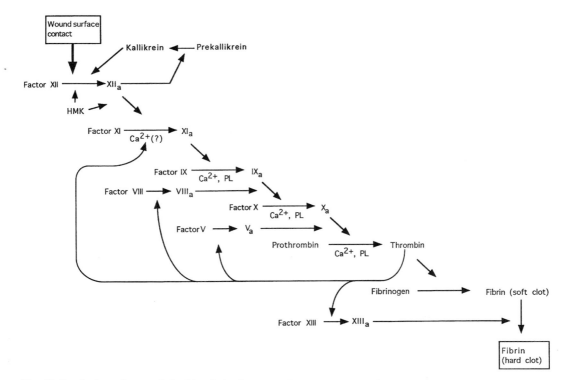

Fig. 10. Intrinsic pathway of the blood clotting cascade.

damage of blood vessels (extrinsic pathway). The last steps in both pathways are the same.

The blood clot itself consists of an insoluble fibrous network, formed of arrays of cross-linked fibrin. The fibrin polymer originates from the soluble plasma protein fibrinogen (factor I). This is polymerized by the proteolytic removal of the fibrinopeptides A and B catalyzed by the serine protease thrombin. A further modification of the initially formed "soft clot" is a covalent cross-linking of neighboring fibrin molecules within the fibrin network. The reaction is catalyzed by a transamidase, the fibrin stabilizing factor (FSF or factor XIII) which is itself activated by thrombin. The resulting "hard clot" contains isopeptide bonds between a glutamine residue of one side chain and the lysine residue of another. As shown in Fig. 10, the blood clotting cascade depends on other factors such as Ca^{2+} and phospholipid. Since most participating enzymes are serine proteases, the presence of Ca^{2+} is necessary in nearly all reactions of the cascade.

Since the multilevel cascade permits enormous amplification of the triggering signals, a variety of factors such as heparin (a glycosaminoglycan), antithrombin, and thrombomodulin limits clot growth, thus preventing uncontrolled clot expansion and avoiding occlusion of the damaged vessel.

The primary aim of the blood clotting cascade is the closure of "leaks" in damaged vessels to avoid blood loss. In addition, rapid sealing prevents the penetration of a large number of disease causing microorganisms and viruses (pathogens). Once these invaders have penetrated, either via traumatized vessels or by breaching the physical barrier of the skin or mucous membranes, they are inactivated by a second class of defense proteins called antibodies. Antibodies represent the effector molecules of a protective mechanism known as the humoral immune response.

The immune system of higher vertebrates generally consists of two facets of immune response once a foreign substance, a virus, a bacterium or even a foreign protein, invades the tissues. In the cellular immune response a special class of lymphatic cells (T lymphocytes) recognize and kill foreign cells. In the humoral immune response another class of lymphatic cells, so-called B lymphocytes synthesize specific immunoglobulin molecules. These proteins (antibodies) are able to recognize and bind to the invading substance. In consequence, the foreign molecules are precipitated (Fig. 11) or marked for destruction by other cells of the immune system, the macrophages. The substance that elicits an immune response is called an antigen and if the invador is large, like a virus or protein, many different antibodies may be elicited. Each different antibody possesses a specific binding site for a given antigenic determinant (or epitope) at the surface of the particle. Groups of amino acids on the surface of a given protein or sugar residues in a carbohydrate are examples for such antigenic determinants.

The immune response is the first line of defense against infection and probably against cancer cells. It has some remarkable features:

- It is incredibly versatile, because it is able to recognize an enormous number of different foreign substances ranging from small synthetic molecules to cells of another individual. The exposure to the same antigen at a later date results in a rapid and massive production of the specific immunoglobulin. Therefore, the immune response is said to have a memory.
- With their immense diversity in specific recognition and in their amino acid sequence, the antibody arsenal is based on clonal selection of those B cells that produce antibodies against substances regarded as foreign.
- The human genome alone does not include enough information to encode the genes for each of the millions of different antibodies. The recombination of exons is the major source of antibody diversity as well as the somatic mutation of antibody generating cells.

The various functions in the immune system are carried out by 5 different classes of antibody molecules. Each monomer antibody molecule consists of 4 chains, 2 heavy chains and 2 light chains, which are covalently linked by disulfide bonds forming a Y-shaped molecule. There are constant domains in each antibody chain which are identical in a given class of immunoglobulins listed in Tab. 10. Besides antibody monomers, di-, tri-, and pentamers are known (IgA and IgM), linked together by a joining chain (J chain). The variable domains in the Y-like fork of all antibody molecules contribute to the multiplicity of specific reactions to different epitopes. Antibodies are divalent; the presence of two binding sites in each antibody molecule leads to precipitation of the antigen if different individual antibodies against different epitopes at the surface of these antigens exist (Fig. 11).

Both, heavy and light chains contribute to the antigen-binding site at the extreme end of the variable domains. Differences in the amino acid sequence in these regions give rise to variable secondary and tertiary structures and hence define close contact surfaces to different antigens or antigenic determinants.

The constant regions in the base of the Y-shaped molecule have different functions:

- They serve to hold the chains together.
- They signal macrophages to eliminate those particles which are labeled with the antibody.
- Differences in these regions are responsible for the identification of immunoglobulin types for delivery to different tissues or secretion.

Antibodies only serve to identify foreign antigens. The inactivation and disposal of the invaders is carried out by another biological

Antigenic determinants

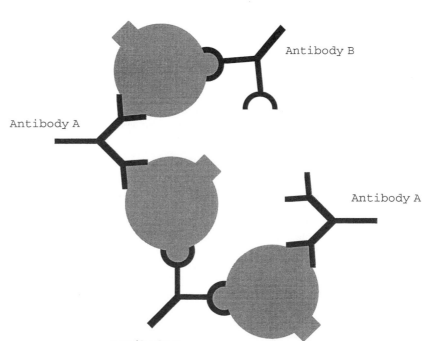

Fig. 11. Antibody–antigen complex. An antigen with different epitopes is recognized by different antibodies finally leading to precipitation of the complex.

Tab. 10. The 5 Classes of Immunoglobulins

IgA (mono-, di- or trimer) Heavy chain α Light chain κ or λ Subunits structure $(\alpha_2\kappa_2)_n J^a$ or $(\alpha_2\lambda_2)_n J^a$	– predominantly in the intestinal tract and secretions such as saliva, sweat, and tears – mw range from 360 to 720 kp depending on the subunit structure – held together by a J chain (joining chain) – block the attachment of invading pathogens to outer (epithelial) surfaces by binding their antigenic sites – major antibody of milk and colostrum (primary milk after pregnancy) – protects nursing infants from pathogens invading via the gastrointestinal tract
IgD (monomer) Heavy chain δ Light chain κ or λ Subunit structure $\delta_2\kappa_2$ or $\delta_2\lambda_2$	– small amounts in the blood on the surface of B cells – Y-shaped monomer (mw 160 kD) – function unknown
IgE (monomer) Heavy chain ε Light chain κ or λ Subunit structure $\varepsilon_2\kappa_2$ or $e_2\lambda_2$	– present in blood in minute concentrations – consists of a single Y-shaped unit (mw 190 kD) – associated with allergic responses – constant regions bind tightly to mast cells
IgG (monomer)[b] Heavy chain γ Light chain κ or λ Subunit structure $\gamma_2\kappa_2$ or $\gamma_2\lambda_2$	– most abundant antibody of the circulatory system and interstitial fluid – consists of a single Y-shaped unit (mw 150 kD) – production begins 2–3 d after first IgM appears – traverses blood vessels and is able to cross the placenta to provide the fetus with immunity – triggers complement system
IgM (pentamer) Heavy chain μ Light chain κ or λ Subunit structure $(\mu_2\kappa_2)J$ or $(\mu_2\lambda_2)J$	– restricted to the circulatory system – largest immunoglobulin (mw 950 kD) – contains 5 Y-shaped units with 2 light and 2 heavy chains each – held together by a J chain (joining chain) – the first immunoglobulin produced in response to an antigen – most effective against invading microorganisms

[a] $n = 1$, 2 or 3.
[b] IgG has 4 subclasses, IgG1-IgG4, which differ in their γ chains.

system, the complement system. This essential defense system "complements" the function of antibodies by eliminating foreign substances via a complex series of interacting plasma proteins. Most of the approximately 20 complement system proteins are serine proteases which act similar to the blood clotting pathway through sequential activation of precursor proteins (zymogens) (HALKIER, 1991). According to two related series of reactions two pathways can be distinguished, the classical pathway (antibody-dependent) and the alternative pathway (antibody-independent).

For example, the classical pathway, which follows the immune response, is triggered by antibody–antigen complexes. During the first step the recognition unit C1, consisting of a loosely bound complex of three complement proteins C1q, C1r and C1s, specifically binds to a target cell surface antigen-bound antibody through its recognition region (constant region of the Y-shaped molecule, Fc) of IgM and several subclasses of IgG. The binding is mediated through protein C1q of the recognition unit. This is followed by the sequential activation of the participating serine proteases C1r and C1s in the same complex.

In the second step of the classical pathway, the activation unit, consisting of three serine proteases (C2, C3, and C4), amplifies the recognition event through a proteolytic cascade initiated by the proteolytic activation through C1s of the recognition unit. The proteolytic cascade finally leads to the covalent linkage of the proteases C3 and C4 to the cell membrane.

The third and last step of the antibody-dependent classical pathway is called membrane attack complex (MAC) and consists of five proteins (C5, C6, C7, C8, and C9). Once activated by a proteolytic reaction, these proteins assemble to form a transmembrane pore (MAC) in the target cell plasma membrane. The MAC forms an aqueous channel (30–100 Å diameter) and, therefore, increases permeability to small molecules. In consequence, water is osmotically drawn into the cell, which finally leads to swelling and bursting of the target cell. MACs are very efficient; probably only a single one is able to lyse a cell.

The alternative pathway is antibody-independent, but uses many of the components of the classical pathway. Many of these components are not able to discriminate between foreign and host cells. Consequently, the complement is strictly regulated in order to prevent host cell destruction, which is the case in many autoimmune diseases. Regulation occurs in three ways:

(1) through spontaneous decay of activated complement components,
(2) through degradation of the complement component by specific proteases,
(3) through the association of activated complement components with specific binding proteins.

These three regulatory mechanisms minimize host cell damage while amplifying the targeting of foreign invaders.

3.7 Regulatory Proteins

In multicellular organisms cells do not live in isolation. An elaborate network coordinates growth, differentiation, and metabolism of tissues and organs. Over short distances, e.g., within a small group of cells, communica-tion often functions by direct contact via gap junctions (see Sect. 3.2). Over longer distances extracellular polypeptides, synthesized and released by signaling cells, facilitate communication by inducing a specific response only in those target cells provided with specific receptors for these signaling molecules. Communication by extracellular signals involves several steps. The signalling molecule is synthesized and released by the signaling cell and transported (e.g., via blood circulation) to the target cell. Specific cell surface receptors recognize and bind the signaling molecule. This is followed by a change in metabolism, mediated by the receptor-signaling molecule complex. Finally, the signal is terminated by different mechanisms.

Water soluble signaling molecules are not able to pass the plasma membrane. These include peptide hormones and growth factors. Some mammalian hormones and growth factors that interact with cell surface receptors are listed in Tab. 11.

The duration of the effects of receptor-bound hormones and growth factors are usually variable. Hormones initiate very short reaction times, the target cell reacts in milliseconds or a few seconds after the formation of the complex. The effects of growth factor binding often extend over days, and the changes in gene expression are often induced in a few minutes.

Five different types of cell surface receptors triggering different types of cellular responses are known. Binding of the signal molecule induces

(1) conformational changes in the receptor opening a specific ion channel in the transmembrane receptor molecule itself, resulting in a change of electric potential across the cell membrane. This type is mainly found in the nervous system, the signaling molecules are small (neurotransmitter), e.g., the acetylcholine receptor at the nerve–muscle junctions (RAFFERTY et al., 1980);
(2) protein kinase activity located at the intracellular domain of the transmembrane receptor molecule. In consequence, intracellular substrate proteins are phosphorylated, thus altering the activity of that

Tab. 11. Some Mammalian Hormones and Growth Factors which Interact with Cell Surface Receptors

Hormone or Growth Factor	Cells of Origin	Major Effects
Insulin	β cells of pancreas	stimulation of – glucose uptake into fat and muscle cells – carbohydrate metabolism – lipid synthesis by adipose tissue – protein synthesis and proliferation
Glucagon	α cells of pancreas	stimulation of – glucose synthesis and glycogen degradation in liver – lipid hydrolysis in adipose tissue
Secretin	small intestine	secretion of digestive enzymes
Gastrin	intestine	secretion of HCl and pepsin by stomach mucosa
ACTH (adrenocortico-tropic hormone)	anterior pituitary	stimulation of – lipid hydrolysis from adipose tissue – production of cortisol and aldoste-rol by adrenal cortex
FSH (follicle stimulating hormone)	anterior pituitary	stimulation of – oocyte and ovarian follicle growth – estrogen synthesis by follicles
LH (luteinizing hormone)	anterior pituitary	maturation of the oocyte stimulates estrogen and progesterone secretion by ovarian follicles
TSH (thyroid stimulating hormone)	anterior pituitary	release of thyroxine by thyroid cells
Parathyroid hormone	parathyroid	increases Ca^{2+} in blood, decreases phosphate; dissolution of bone calcium phosphate, resorption from kidney filtrate: Ca^{2+} is increased, PO_4^{2-} is decreased
TRH (THS releasing hormone)	hypothalamus	induces secretion of thyroid stimulating hormone by anterior pituitary
LHRH (LH releasing hormone)	hypothalamus, neurons	induces secretion of luteinizing hormone by anterior pituitary
Somatotropin (growth hormone)	anterior pituitary	stimulation of – amino acid uptake by many cells – IGF1 production by liver; IGF1 causes growth of bone and muscles
Erythropoietin	kidney	differentiation of erythrocyte stem cells
EGF (epidermal growth factor)	salivary and other glands	growth of epidermal and other body cells
Interleukin 2	T cells and macrophages	growth of T cells of the immune system
NGF (nerve growth factor)	all tissues innervated by sympathetic neurons	growth and differentiation of sensory and sympathetic neurons

protein. Examples for this type of receptor are the insulin receptor and the epidermal growth factor (EGF) receptor (FANTL et al., 1993);

(3) tyrosine phosphatase activity at the intracellular domain of the receptor. A phosphate residue attached to a tyrosine side chain of a certain intracellular substrate protein is removed and, therefore, activity of the substrate protein is altered. Leukocyte CD45 protein belongs to this type of receptor (WALTON and DIXON, 1993);

(4) direct synthesis of 3'-5' cyclic GMP, a second messenger molecule that triggers rapid alteration in enzyme activity or in non-catalytic proteins within the target cell, by initiating guanylate cyclase activity at the intracellular domain of the receptor (natriuretic peptide receptor, KOLLER et al., 1993);

(5) activation of a G protein (guanine nucleotide-binding regulatory protein). G proteins bind and activate two different kinds of enzymes. One group of G proteins stimulates (G_s protein) or inhibits (G_i protein) an adenylate (or guanylate) cyclase leading to an increase or decrease of the intracellular second messengers cAMP or cGMP. These second messenger molecules in turn stimulate protein kinases which alter target protein activity by phosphorylation. In addition, cAMP opens Ca^{2+}-channels. Another group of G proteins is coupled with the phosphoinositide signal system (STRYER and BOURNE, 1986; STRADER et al., 1994).

The signal transduction pathways summarized above are a major research field in biochemistry and the reader is be referred to relevant literature.

Another example of regulatory proteins are transcription factors (PABO and SAUER, 1992). These proteins have been implicated in regulating transcriptional initiation and thus modulating the rate of mRNA synthesis during gene expression. One of the best understood examples is the stimulation of gene expression by steroid hormones (TSAI and O'MALLY, 1994). These lipophilic molecules are able to penetrate the target cell by diffusion. Once in the nucleus, the hormone binds to a protein called glucocorticoid receptor. The hormone–receptor complex now activates transcription by binding to a specific DNA region. This principle of transcription control is known right down to prokaryotes (e.g., the *lac* repressor protein).

4 Conclusions

In the last sections, proteins were assigned different functions. In many cases proteins may fulfill several functions. As illustrated in a few examples above, lipocalins are classified as proteins with transport functions, but many of them have been implicated in the regulation of cell homeostasis and hence they possess a regulatory function (FLOWER, 1994). Furthermore, many of the proteins of the blood clotting cascade or single components of a complement system are enzymes, however, as a collective they have a defense function. Besides the main classes of protein function there are several proteins with rare exotic functions, e.g., the anti-freeze protein which prevents the growth of ice cristals in the blood plasma of arctic fishes (KNIGHT et al., 1984).

Modern molecular biological techniques have revealed a large number of hypothetical proteins with unknown function. In addition, many immunochemical markers for different cell types have been discovered (e.g., antigens of T lymphocytes), but the function of these proteins is not known in all cases. Even among the proteins with known function it can be expected that several other functions will be discovered in the future due to the functional variability of proteins.

5 References

BALDWIN, S. A., BALDWIN, J. M., LIENHARD, G. E. (1982), Monosaccharide transporter of the human erythrocyte. Characterization of an improved preparation, *Biochemistry* **21**, 3836–3842.

BEINERT, H. (1990), Recent developments in the field of iron–sulfur proteins, *FASEB J.* **4**, 2483–2491.

BRAUER, D., BRÖCKER, M., KELLERMANN, C., WINNACKER, E.-L. (1995), Biosafety in rDNA research and production, in: *Biotechnology* 2nd Edn., Vol. 12 (REHM, H.-J., REED, G., Eds.), pp. 63–116. Weinheim: VCH.

CLARKE, S. (1992), Protein isoprenylation and methylation at carboxyterminal cysteine residues, *Annu. Rev. Biochem.* **61**, 355–386.

CROSS, G. A. M. (1990), Glycolipid anchoring of plasma membrane proteins, *Annu. Rev. Cell. Biol.* **6**, 1–39.

DARNELL, J., LODISH, H., BALTIMORE, D. (1990), *Molecular Cell Biology,* 2nd Edn. New York: Scientific Books.

DAUTRY-VARSAT, A., CIECHANOVER, A., LODISH, H. F. (1983), pH and the recycling of transferrin during receptor-mediated endocytosis, *Proc. Natl. Acad. Sci. USA* **80**, 2258–2262.

DE SOUSA, M., BREEDVELT, F., DYNESIUS-TRENTHAM, R., TRENTHAM, D., LUM, J. (1988), Iron, iron-binding proteins and immune system cells, *Ann. N. Y. Acad. Sci.* **526**, 310–322.

ELBEIN, A. D. (1987), Inhibitors of the biosynthesis and processing of N-linked oligosaccharide side chains, *Annu. Rev. Biochem.* **56**, 497–534.

ENGLUND, P. T. (1993), The structure and biosynthesis of glycosyl-phosphatidyl-inositol protein anchors, *Annu. Rev. Biochem.* **62**, 121–138.

FANTL, W. J., JOHNSON, D. E., WILLIAMS, L. T. (1993), Signaling by receptor tyrosine kinases, *Annu. Rev. Biochem.* **62**, 453–481.

FASMAN, G. D. (Ed.) (1976), *Handbook of Biochemistry and Molecular Biology,* 3rd Edn. Vol. II *Proteins,* pp. 91–172. Boca Raton, FL: CRC Press.

FINDLAY, J. B. C. (1990), Purification of membrane proteins, in: *Protein Purification Applications – A Practical Approach* (HARRIS, E. L. V., ANGAL, S., Eds.). pp. 59–82. Oxford: IRL Press.

FLOWER, D. R. (1994), The lipocalin protein family: a role in cell regulation, *FEBS Lett.* **354**, 7–11.

GESTELAND, R. F., ATKINS, J. F. (Eds.) (1993), *The RNA World.* Cold Spring Harbor, NY: Cold Spring Harbor Laboratory Press.

GHISLA, S., MASSEY, V. (1989), Mechanisms of flavoprotein-catalyzed reactions, *Eur. J. Biochem.* **181**, 1–17.

GORSKI, J. P. (1992), Acidic phospho proteins from bone matrix: a structural rationalization of their role in biomineralization, *Calcif. Tissue Int.* **50**, 391–396.

GROS, P., CROOP, J., HOUSMAN, D. E. (1986), Mammalian drug resistance gene: complete cDNA sequence indicates strong homology to bacterial transport proteins, *Cell* **47**, 371–380.

HAEFLIGER, J. A., PEITSCH, M. C., JENNE, D. E., TSCHOPP, J. (1991), Structural and functional characterization of complement C8 gamma, a member of the lipocalin protein family, *Mol. Immunol.* **28**, 123–131.

HALKIER, T. (1991), *Mechanisms in Blood Coagulation, Fibrinolysis and the Complement System.* Cambridge: Cambridge University Press.

HANOVER, J. A. (1992), The nuclear pore: at the crossroads, *FASEB J.* **6**, 228–2295.

HARRISON, P. M., AROSIO, P. (1996), The ferritins: molecular properties, iron storage function and cellular regulation, *Biochim. Biophys. Acta* **1275**, 161–203.

KNIGHT, C. A., DEVRIES, A. L., OOLMAN, L. D. (1984), Fish antifreeze protein and the freezing and recrystallization of ice, *Nature* **308**, 295–296.

KOLLER, K. J., LIPARI, M. T., GOEDDEL, D. V. (1993), Proper glycosylation and phosphorylation of the A type natriuretic peptide receptor are required for hormone-stimulated guanylyl cyclase activity, *J. Biol. Chem.* **268**, 5997–6003.

KORNFIELD, R., KORNFIELD, S. (1985), Assembly of asparagine-linked oligosaccharides, *Annu. Rev. Biochem.* **54**, 633–664.

KRAGH-HANSEN, U. (1990), Structure and ligand binding properties of human serum albumin, *Dan. Med. Bull.* **37**, 57–84.

KRITCHEVSKY, D. (1986), Atherosclerosis and nutrition, *Nutr. Int.* **2**, 290–297.

LASH, A., SALEEM, A. (1995), Iron metabolism and its regulation. A review, *Ann. Clin. Lab. Sci.* **2555**, 20–30.

LAZCANO, A., MILLER, S. L. (1996), The origin and early evolution of life: Prebiotic chemistry, the pre-RNA world, and time, *Cell* **85**, 793–798.

LEWIS, R. V., HINMAN, M., KOTHAKOTA, S., FOURNIER, M. J. (1996), Expression and purification of a spider silk protein: A new strategy for producing repetitive proteins, *Prot. Exp. Purif.* **7**, 400–406.

LOW, M. G. (1987), Biochemistry of the glycosyl–phosphatidylinositol membrane protein anchors, *Biochem. J.* **244**, 1–13 (review article).

MARKELLO, T., ZLOTNIK, A., EVERETT, J., TENNYSON, J., HOLLOWAY, P. W. (1985), Determination of the topography of cytochrome b$_5$ in lipid vesicles by fluorescence quenching, *Biochemistry* **24**, 2895–2901.

MCCORKLE, G. M., ALTMANN, S. (1987), RNA as catalysts: a new class of enzymes, *J. Chem. Educ.* **64**, 221–226.

MONTREUIL, J. (1995), The history of glycoprotein research, a personal view, in: *Glycoproteins* (MONTREUIL, J., VLIEGENHART, J. F. G., SCHACHTER, H., Eds.), pp. 1–10. Amsterdam: Elsevier.

PABO, C. O., SAUER, R. (1992), Transcription factors: Structural families and principles of DNA recognition, *Annu. Rev. Biochem.* **61**, 1053–1095.

PEARSON, C. K., CUNNINGHAM, C. (1993), Multidrug resistance during cancer chemotherapy – biotechnological solutions to a clinical problem, *Trends Biotechnol.* **11**, 511–516.

PENEFSKY, H. S., TZAGOLOFF, A. (1971), Extraction of water-soluble enzymes and proteins from membranes, *Methods Enzymol.* **22**, 204–219.

PIKULA, S., HAYDEN, J. B., AWASTHI, S., AWASTHI, Y. C., ZIMNIAK, P. (1994a), Organic anion-transporting ATPase of rat liver. I. Purification, photoaffinity labeling, and regulation by phosphorylation, *J. Biol. Chem.* **269**, 27566–27573.

PIKULA, S., HAYDEN, J. B., AWASTHI, S., AWASTHI, Y. C., ZIMNIAK, P. (1994b), Organic anion-transporting ATPase of rat liver. II. Functional reconstitution of active transport and regulation by phosphorylation, *J. Biol. Chem.* **269**, 27574–27579.

PRICE, P. A., RICE, J. S., WILLIAMSON, M. K. (1994), Conserved phosphorylation of serines in the Ser-X-Glu/Ser(P) sequences of the vitamin K-dependent matrix Gla protein from shark, lamb, rat, cow, and human, *Protein Sci.* **3**, 822–830.

PROKOP, D. J., KIVIRIKKO, K. I. (1995), Collagens: Molecular biology, diseases, and potentials of therapy, *Ann. Rev. Biochem.* **64**, 403–434.

RAFFERTY, M. A., HUNKAPILLER, M. W., STRADER, C. D., HOOD, L. E. (1980), Acetylcholine receptor: Complex of homologous subunits, *Science* **208**, 1454–1457.

REASON, A. J., MORRIS, H. R., PANICO, M., MARAIS, R., TREISMAN, R. H., HALTIWANGER, R. S., HART, G. W., KELLY, W. G., DELL, A. (1992), Localization of O-GlcNAc modification on the serum response transcription factor, *J. Biol. Chem.* **267**, 16911–16921.

REITHMEIER, R. A. F. (1996), Assembly of proteins into membranes, in: *Biochemistry of Lipids, Lipoproteins and Membranes* (VANCE, D. E., VANCE, J., Eds.), pp. 425–471. Amsterdam: Elsevier.

ROBINSON, P. J., MILLRAIN, M., ANTONIOU, J., SIMPSON, E., MELLOR, A. L. (1989), A glycophospholipid anchor is required for Qa-2-mediated T cell activation, *Nature* **342**, 85–87.

ROTH, J., WANG, Y., ECKHARDT, A. E., HILL, R. L. (1994), Subcellular localization of the UDP-N-acetyl-D-galactosamine: polypeptide N-acetyl-galactosaminyl transferase-mediated O-glycosylation reaction in the submaxillary gland, *Proc. Natl. Acad. Sci. USA* **91**, 8935–8939.

SCHAUER, R. (1985), Sialic acids and their role as biological masks, *Trends Biochem. Sci.* **10**, 357–360.

SCHOPF, J. W. (1993), Microfossils of the early archean apex chert: new evidence of the antiquity of life, *Science* **260**, 640–646.

SCHWABE, J. R. W., KLUG, A. (1994), Zinc mining for protein domains, *Nature Struct. Biol.* **1**, 345–349.

SEGEL, I. H. (1975), *Enzyme Kinetics*. New York: John Wiley & Sons.

SINGER, S. J., NICHOLSON, G. L. (1972), The fluid mosaic model of the structure of cell membranes, *Science* **175**, 720–731.

STRADER, C. D., MING FONG, T., TOTA, M. R., UNDERWOOD, D., DIXON, R. A. F. (1994), Structure and function of G-protein coupled receptors, *Annu. Rev. Biochem.* **63**, 101–132.

STRYER, L., BOURNE, H. R. (1986), G Proteins: A family of signal transducers, *Ann. Rev. Cell Biol.* **2**, 391–419.

SÜDHOF, T. C., GOLDSTEIN, J. L., BROWN, M. S., RUSSELL, D. W. (1985), The LDL receptor gene: A mosaic of exons shared with different proteins, *Science* **228**, 815–822.

TSAI, M.-J., O'MALLY, B. W. (1994), Molecular mechanisms of action of steroid/thyroid receptor superfamily members, *Annu. Rev. Biochem.* **63**, 451–486.

ULRICH, A., BELL, J. R., CHEN, E. Y., HERRERA, R., PETRUZZELLI, R., DULL, T. J., GRAY, A., COUSSENS, L., LIAO, Y. C., TSUBOKAWA, M., MASON, A., SEEBERG, P. M., GREENFIELD, C., ROSEN, O. M., RAMACHANDRAN, J. (1985), Human insulin receptor and its relationship to the tyrosine kinase family of oncogens, *Nature* **313**, 756.

VANCE, D. E., VANCE, J. E. (Eds.) (1985), *Biochemistry of Lipids and Membranes*. Menlo Park, CA: The Benjamin Cummings Publishing Company.

VERBERT, A., CACAN, R., CECCHELLI, R. (1987), Membrane transport of sugar donors to the glycosylation sites, *Biochimie* **69**, 91–99.

WALTON, K. M., DIXON, J. E. (1993), Protein tyrosine phosphatases, *Annu. Rev. Biochem.* **62**, 101–120.

WU, H. C., TOKUNAGA, M. (1986), Biogenesis of lipoproteins in bacteria, *Curr. Top. Microbiol. Immunol.* **125**, 127–154.

XU, M., LEWIS, R. V. (1990), Structure of a protein superfiber: spider dragline silk, *Proc. Natl. Acad. Sci. USA* **87**, 7120–7124.

2 Sequence and Structure of Proteins

FRANK EISENHABER
PEER BORK

Heidelberg and Berlin, Germany

1 Introduction 47
2 Hierarchical Description of Protein Structure 47
3 Primary Structure of Proteins 48
  3.1 Chemical Structure of Proteins 49
  3.2 Protein Primary Structure as a Result of Transcription and Translation of Genetic Information 49
  3.3 Investigation of Primary Structural Features – Protein Sequence Analysis 50
    3.3.1 Quality of Sequence Databases 53
    3.3.2 Knowledgc-Based Prediction of Protein Structure and Function Using Protein Sequence Analysis 55
      3.3.2.1 Standard Procedure of Database-Aided Homology Search for a Query Sequence 57
      3.3.2.2 Significance of Weak Homologies 58
      3.3.2.3 Search for Internal Repeats 58
      3.3.2.4 Complex Regression of Protein Sequence–Structure Relationships 59
4 Secondary Structural Features of Proteins 59
  4.1 Types of Secondary Structures 59
    4.1.1 The α-Helix and the 3_{10}-Helix 61
    4.1.2 Extended Structures in Protein Structures 61
    4.1.3 Loop Segments 62
  4.2 Automatic Assignment of Secondary Structural Types in Three-Dimensional Structures 63
  4.3 The Concept of Secondary Structural Class 63
  4.4 Prediction of Secondary Structural Features 65
    4.4.1 Secondary Structural Class and Content Prediction 66
    4.4.2 Traditional Secondary Structure Prediction 66
    4.4.3 Prediction of Transmembrane Regions, Coiled-Coil Segments, and Antigenic Sites 67

5 Tertiary Protein Structures   68
  5.1 Phenomenology of Tertiary Protein Structures   68
      5.1.1 Construction Principle No. 1 – Close Packing   69
      5.1.2 Construction Principle No. 2 – Hydrophobic Interior and Hydrophilic Exterior   70
      5.1.3 Protein Structure Comparison and Structural Families   71
  5.2 Prediction and Modeling of Tertiary Structure   73
      5.2.1 Computation of Protein Structures Based on Fundamental Physical Principles (*ab initio* Approach)   73
      5.2.2 Threading Amino Acid Sequences through Structural Motifs   74
      5.2.3 Homology Modeling   75
      5.2.4 Prediction of Solvent Accessibility   76
6 Quarternary Structures of Proteins   77
  6.1 Phenomenology of Quarternary Structural Features   77
  6.2 Prediction of Protein–Protein Docking   77
7 Concluding Remark   78
8 References   78

List of Abbreviations

ASC	computer program for the analysis of surface properties of proteins
BLAST	tool for protein or DNA sequence homology search leading to optimal local alignment
BLIP	inhibitor of TEM-1β-lactamase
BLITZ	tool for finding homologous sequences in databases
BLOCKS	motif database
BRCA1	breast cancer gene
CAP	BLAST output parser
CATH	classification of protein structures
CD	circular dichroism
COMBI	protein secondary structure prediction method
DEFINE	computer program for objectivation of secondary structure assignment
DSSP	computer program for objectivation of secondary structure assignment
ε-helix	extended polypeptide conformation outside β-sheet
EST	expressed sequence tag
FASTA	tool for finding homologous sequences in databases
FLyBase	molecular database for *Drosophila*
FSSP	classification of protein structures
FTP	file transfer protocol
GENEQUIZ	tool for functional assignment of proteins by sequence homology
GOR/GORIII	secondary structure prediction methods
HMG-box	high mobility group (HMG) domain in several families of nuclear proteins
HMM	Hidden Markov Model
MoST	motif search tool
NP-complete	problem which cannot be solved with a polynominal algorithm (*non-polynominal complete problem*)
NSC	computer program for the analysis of surface properties of proteins
ORF	open reading frame
PATSCAN	WWW server for motif and profile searchers
PBE	Poisson-Boltzmann differential equation
P-CURVES	computer program for objectivation of secondary structure assignment
PDB	Brookhaven Protein Data Bank
PHD	engine for secondary structure prediction
PIMA	motif database
PPI-helix	extended conformation of *cis*-polyproline
PPII-helix	extended conformation of *trans*-polyproline
PREDATOR	engine for secondary structure prediction
PRINTS	motif database
PRODOM	motif database
PROFILE	WWW server for motif and profile searchers
PROSITE	motif database
r.m.s.d.	root mean square deviation
SCOP	classification of protein structures
SCR	structurally conserved regions
SOM	tool for pattern search
SOPM	program for secondary structure prediction
SRS	sequence retrieval system, query tool for biomolecular databases
SSCP	WWW program for secondary structure content prediction
SSPRED	engine for secondary structure prediction
STRIDE	modification of DSSP
SVR	structurally variable regions

SWISS-PROT	major annotated protein sequence database	TTOP	program for prediction of topology of transmembrane proteins
TMAP	program for prediction of transmembrane regions	URL	unique resource location (in the WWW)
TREMBL	protein sequences obtained from translation of nucleotide sequences in the EMBL database	UV	ultraviolet
		WWW	world wide web
		Yeast YPD	molecular database for yeast

1 Introduction

For biotechnological applications, it is very important to determine which type of specific structural information is necessary to understand or to modify the protein function in the desired manner. In this review, we will give an overview on protein structural features classified with increasing complexity and consider appropriate methodological approaches and theoretical concepts for their investigation. Special emphasis will be given to techniques applicable to predicting protein structural properties by relying only on the protein sequence (EISENHABER et al., 1995b). It is expected that the reader is familiar with basic knowledge on protein biochemistry and biophysics as given in standard university textbooks.

2 Hierarchical Description of Protein Structure

Whereas nucleic acids fulfill mainly the tasks of storage and transfer of genetic information in living organisms, the proteins form a complicated cellular machinery for realization of this genetic program dependent on and in response to changing environmental conditions. Some proteins are catalysts and enable chemical reactions which would otherwise not occur at temperatures (e.g., 37 °C) and pH (e.g., pH 7) values typical for living organisms. Others are involved in storage and transport of particles ranging from electrons to macromolecules. Proteins mediate signal transmission between cells, tissues, and organisms; they control the passage of molecules through membranes surrounding cellular compartments. Yet other proteins are involved in the mechanochemistry of motion as in muscles or serve a structural purpose in the filamentous architecture of cells and tissues.

The functionality of different protein molecules is tightly connected with their structural and dynamic properties. Amino acid sequences for linear polypeptides forming proteins are directly encoded in genes. At the same time, the three-dimensional protein architecture represents the ultimate in molecular information, and from it springs a variety of significant scientific results: the understanding of protein folding and structural stability, interactions between subunits, receptors, ligands, substrates, and the like, enzymatic catalysis, the understanding of molecular evolution, the ability to engineer and design proteins through synthesis and mutation, the creation of drugs and the utilization of protein-based processes to confront human disease and suffering. Inspite of many years of intensive research, the complete description of structure and dynamics of a protein molecule based on fundamental physical and chemical principles is still an unsolved scientific task. Therefore, theoretical protein science must settle for lesser goals as well as for the quest, which is described in this article.

The main vehicle to organize the current knowledge of protein structure is the so-called hierarchical description (Fig. 1). A protein molecule in solution is a very complex system with a huge number of degrees of freedom, albeit many of them are of little importance for biological function. Traditionally, primary, secondary, tertiary, and quarternary structural levels are considered which correspond to biochemical events of coding, synthesis, and function of proteins as well as to physico-chemical properties of isolated polypeptides in solution. Supersecondary structure, protein fold, topology, and structural domain are historically younger terms. Short definitions are given below:

1. The lowest level of structural organization, the *primary structure,* is identical with the amino acid sequence. The order of amino acids as genetically encoded may be changed posttranscriptionally through splicing, during translation as a result of recoding mechanisms, and posttranslationally due to chemical modifications (backbone scission, side chain modifications). Generally, the chemical identity of a residue is of critical importance only for a few sequence positions. At most sequence positions, many amino acid exchanges have

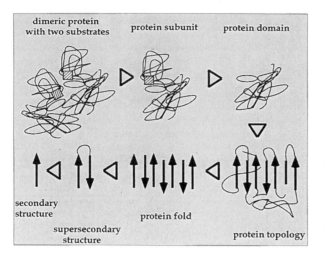

Fig. 1. Hierarchy of structural description of proteins. The flowchart presents the structural analysis of a dimeric β-sheet protein (at the upper left) with 2 ligands (squares) up to the level of secondary structural elements (β-strands represented as arrows, at the lower left). Successively, subunit, domain, topology, fold, and supersecondary structure are illustrated.

only a slight effect on three-dimensional structure and function.

2. The terms of *secondary structure* are used to describe preferred relative backbone locations of sequentially near residues mainly due to local interactions.

3. *Supersecondary structural motifs* are typical ways of packing between secondary structural units. A *protein fold* is commonly defined as the scaffold of secondary structural elements with repetitive backbone structure; i.e., α-helices and β-sheets. The term *"protein topology"* is more general and includes the spatial arrangement of all secondary structural elements including loop segments. The *tertiary structure* of a protein is described with the relative position of atoms in all residues of a polypeptide chain, both in regular secondary structural elements as well as in loops connecting them. The protein tertiary structure may also be represented in terms of pairwise distances between atoms of various residues. A particular feature of globular proteins is the existence of contacts between residues with large sequence separation (non-local interactions). As a rule, a tertiary structure has a typical densely packed hydrophobic core shielded from interaction with a solvent. A tertiary structure can comprise several *domains* which are distinguished structurally due to autonomous hydrophobic cores and, possibly

thermodynamically as melting independently. A domain is often considered an autonomous folding unit. At the same time, a structural domain does not need to be contiguous in amino acid sequence.

4. The *quarternary structure* is composed of several subunits, each being a polypeptide with its own tertiary structure. The situation is similar as with domains but the chemical (peptide) link between subunits is missing.

Protein structural features are greatly influenced not only by the type of the cellular compartment or the solution conditions (e.g., temperature, water content, ionic strength, inclusion into membranes of organelles or cells, extracellular space), but also by binding of cofactors and other ligands.

3 Primary Structure of Proteins

The primary structural level, the lowest in the hierarchical description of protein structure, corresponds essentially to the chemical structure. After consideration of the known chemical possibilities of protein diversity, we discuss modern aspects of protein sequence analysis.

3.1 Chemical Structure of Proteins

Proteins are biomacromolecules. Their main constituents are linear polypeptide chains. The monomeric units are α-L-amino acids or -imino acids, interconnected by peptide bonds. The succession of monomer types in the linear polypeptide is called the protein sequence. How do proteins achieve the impressing functional diversity?

1. No other biomacromolecule has such a variety of monomer types. In addition to the 20 traditional amino acid types with different side chains (plus selenocysteine, see Sect. 3.2), monomers already included into the polypeptide chain may also be chemically modified after translation (posttranslational modification). Both amino acid composition and the order of amino acid types in the sequence characterize an individual protein.
2. Proteins can contain small organic (prosthetic groups) or inorganic compounds as an integral part of their structure. For example, the type of cofactor can influence the reaction specificity of an enzyme.
3. The main source of functional diversity of proteins is the variety of three-dimensional domain structures each of which is preferred by a certain class of amino acid sequences. A crude classification distinguishes between globular and fibrillar domains, the latter being mainly in structural molecules.
4. Proteins vary widely in size. The range in the number of monomers is from a few dozen amino acids for small proteins like crambin up to the enormous value of 27,000 residues in the muscle protein titin. An eukaryotic domain consists typically of ~125 amino acid residues (~150 for prokaryota) as seen from sequence length distributions and the occurrence of leading methionines (BERMAN et al., 1994; KOLKER and TRIFONOV, 1995). Larger proteins are normally composed of many domains (BORK et al., 1996).

The experimental techniques for determining the sizes and the amino acid sequences of proteins have reached a very high level of maturity. As a result of standardization and automation, the productivity in genomic and protein sequencing has steadily increased. The size of sequence databases exploded during the last years. The number of entries in TREMBL (protein sequences obtained from translation of nucleotide sequences in the EMBL database, ftp://embl -ebi.ac.uk/pub/databases/trembl/) is larger than 200,000 and SWISS-PROT (http://expasy.hcuge.ch/) contains more than 60,000 annotated sequences. Even whole genomes become available (FLEISCHMANN et al., 1995).

3.2 Protein Primary Structure as a Result of Transcription and Translation of Genetic Information

The relation of genetic information and protein primary structure is very complex. Primarily, the succession of residue types in the polypeptide is genetically encoded in the triplet sequence of genes. The transcriptional events which may include synthesis of a transcript from one or several DNA segments and splicing (removal of intron segments and ligation of extrons) result in a messenger RNA (mRNA). Depending on cellular conditions, RNA splicing may even follow alternative pathways ("alternative splicing"). In the ribosomal machinery, this triplet code is translated into a polypeptide sequence. In dependence on the organism and the organelle type, different standard genetic translation tables apply. Additionally, probably in all organisms, a minority of genes relies on recoding of the canonical genetic table (RNA editing) for translation of their mRNAs (BÖRNER and PÄÄBO, 1996; GESTELAND and ATKINS, 1996; STADTMAN, 1996):

- Frameshifting at a particular site allows the expression of a protein from overlapping reading frames (possibly with several translation products in the case of non-100% frameshift efficiency).

- The meaning of code triplets may be altered. Stop codons can be redirected as tryptophane, glutamine, or even selenocysteine, the 21st translationally incorporated type of amino acid. The existence of UGA translation into a cysteine-type amino acid has already been predicted from symmetry considerations of the genetic code (SHCHERBAK, 1988, 1989). Also the editing of glycine codons into asparagine has been observed.
- Ribosomes may translate over coding gaps in mRNAs if the stop codon is hidden in some mRNA secondary structure (translational bypassing). If a stop codon is missing, the ribosome may use a piece of ribosomal RNA for completing the polypeptide (transtranslation).

RNA editing is one aspect of a more general phenomenon, namely the continuing evolution of the genetic code (OSAWA et al., 1992).

Finally, the polypeptide is subject to posttranslational modifications, necessary for a wide variety of reasons, e.g., protection against proteolysis, direction of transport, genetic regulation, membrane anchoring, and regulation of enzymatic activation or of degradation (CREIGHTON, 1992; HAN and MARTINAGE, 1992; RESH, 1994). In the simplest case, single amino acids are chemically altered. The modifications can be N-terminal (e.g., acetylation, myristoylation and pyroglutaminylation), C-terminal (e.g., amidation, isoprenylation and farnesylation), or affect the side chains (e.g., glycosylation, phosphorylation, hydroxylation, and disulphide bond formation). Another type of posttranslational chemical modification consists in cuts of peptide bonds with or without dissociation of the two resulting chains. The latter is the activation mechanism of trypsinogen and chymotrypsinogen. In the case of α-lytic protease, a segment of the precursor polypeptide chain acts as catalyst for achieving the native fold (BAKER et al., 1992a). Polypeptide chain scission may occur during functioning of a protein. For example, serpins are efficient inhibitors of plasma proteases. In the unbound state, serpins are in a metastable kinetically trapped state with a 5-stranded β-sheet and a largely unstructured reactive loop (MOTTONEN et al., 1992). Upon tight binding with serine proteases, the conformation rearranges to the 6-stranded latent form. During dissociation of the serpin molecules from the complex, they are slowly cleaved and inactivated as inhibitors (GOLDSMITH and MOTTONEN, 1994; HUANG et al., 1994; CARREL et al., 1994).

Multiple polypeptide chain scissions occur during the so-called protein splicing, the most recently discovered variant of posttranslational modification (COOPER and STEVENS, 1995). It involves the precise and autocatalytic excision of one or several intervening protein sequences from a precursor protein, coupled to the ligation (peptide bond formation) of the remaining sequence domains, which results in a spliced protein product. In full analogy to RNA splicing, the two types of segments of the precursor protein are named exteins (expressed part) and inteins (intervening part). In several cases, it was difficult to isolate the full-length precursor since inteins can actually splice while the C-terminal extein is still being translated (COOPER and STEVENS, 1995). Intein assignment was an important step in the genome analysis of *Methanococcus jannaschii* (BULT et al., 1996).

The consideration of RNA recoding and posttranslational chemical modification in an automatic manner is currently impossible (or, at least, very unreliable), and this adds another moment of uncertainty to the correctness of amino acid sequences derived from nucleotide sequence data canonically translated by computer programs.

3.3 Investigation of Primary Structural Features – Protein Sequence Analysis

Whereas in the early times of molecular biology, protein research concentrated mainly on the physico-chemical and functional characterization of selected proteins and only a few protein sequences were available, the situation has now reversed. In a typical case, the researcher has the protein sequence (mainly derived from corresponding cDNA sequences), but does know only little about its structure and function. The 28th of July, 1995, marks the beginning of a new era. The complete genome of *Haemophilus influenzae*, a bacterium, has been obtained as a result of world wide cooperation in large-scale sequencing projects (FLEISCHMANN et al., 1995). Other genomes followed (several eubacteria and archebacteria as well as bakers' yeast) or will be available in a foreseeable future (for nematode *Caenorhabditis elegans*, it will be available in 1998 and, for the human genome, it is expected in 2003, see Fig. 2). The genetic make-up (the complete DNA of an organism) should contain all information necessary for cells to mature, to reproduce and to interact with the environment as open, homeostatic system until their preprogramed death. Thus, knowledge about virtually all proteins in a living organism is available, albeit, at the moment, the experimental facts are primarily limited to the nucleotide se-

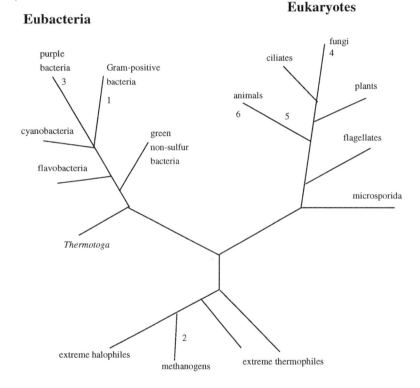

Fig. 2. Phylogeny of species and the size of genomes. A phylogenetic tree based on RNA data containing groups of organisms from all major phyla is shown.

Species	Genome Size [Mb]	Number of Genes (Proteins)
1 *M. genitalium*	0.6	470
2 *E. coli*	4.7	4000
3 *M. jannaschii*	1.7	1760
4 Yeast	13.5	6000
5 *C. elegans*	100	13000
6 Human	3000	70000

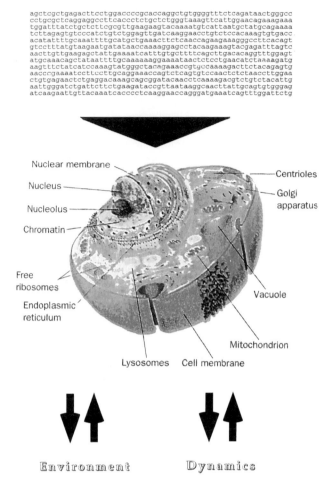

```
agctcgctgagacttcctggaccccgcaccaggctgtggggtttctcagataactgggcc
cctgcgctcaggaggccttcaccctctgctctgggtaaagttcattggaacagaaagaaa
tggatttatctgctcttcgcgttgaagaagtacaaaatgtcattaatgctatgcagaaaa
tcttagagtgtcccatctgtctggagttgatcaaggaacctgtctccacaaagtgtgacc
acatattttgcaaattttgcatgctgaaacttctcaaccagaagaaagggccttcacagt
gtcctttatgtaagaatgatataaccaaaaggagcctacaagaaagtacgagatttagtc
aacttgttgaagagctattgaaaatcatttgtgctttcagcttgacacaggtttggagt
atgcaaacagctataattttgcaaaaaaggaaaataactctcctgaacatctaaaagatg
aagtttctatcatccaaagtatgggctacagaaaccgtgccaaaagacttctacagagtg
aacccgaaaatccttccttgcaggaaaccagtctcagtgtccaactctctaaccttggaa
ctgtgagaactctgaggacaaagcagcggatacaacctcaaaagacgtctgtctacattg
aattgggatctgattcttctgaagataccgttaataaggcaacttattgcagtgtgggag
atcaagaattgttacaaatcaccctcaaggaaccagggatgaaatcagtttggattctg
```

Nuclear membrane

Nucleus

Nucleolus

Chromatin

Free ribosomes

Endoplasmic reticulum

Centrioles

Golgi apparatus

Vacuole

Mitochondrion

Lysosomes   Cell membrane

Environment   Dynamics

Fig. 3. Impact of the genetic make-up for a cell. In the upper part, a piece of the gene sequence of the breast cancer gene *BRCA*1 is presented (a few hundred bases). For comparison, the human genome comprises about $3 \cdot 10^9$ bases. The genetic information is sufficient to code for all cellular functions (lower part). It should be emphasized that living organisms composed of cells are not static devices but have different levels of ontogenetic development, exchange information, energy, and metabolic products with their environment, and finally react on impulses from outside.

quence of their genes (Fig. 3). The decoding of this enormous amount of data is an important goal and biological science is only starting to approach it, having in mind revolutionary applications in biotechnology, drug development, and in the treatment of diseases.

The main source of hypothetical information about such unknown proteins is amino acid sequence analysis and comparison (Fig. 4). This approach is knowledge-based and inductive. The query sequence or significant parts of it are compared with sequences or sequence motifs in databases. The annotated information (source organism, structural and functional data) of proteins or protein domains with similar sequences and the biochemical and molecular-biological context of

the query protein are used to extrapolate possible structural and functional features (see, e.g., TATUSOV et al., 1996). To obtain finally scientific data, experimental verification of these conclusions need to follow.

Protein function (Tab. 1) requires also a multilevel description similar to protein structure (see Sect. 2). Primarily, a protein has a molecular function. It may catalyze a specific reaction or transmit a signal. A set of many co-operating proteins can fulfill a physiological function. In the simplest case, this is a metabolic pathway. At the next level, protein function determines phenotypic properties (phenotypic function or disfunction). Protein activity might be limited to certain cellular compartments, the extracellular space, and

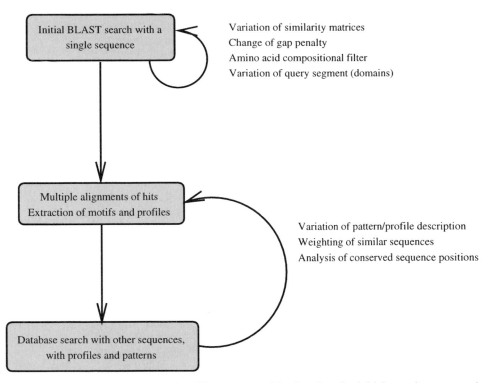

Fig. 4. Flowchart for iterative database searches. The sequence hits found at the initial searches are used for obtaining multiple alignments. They are analyzed and sequence motifs and/or profiles are extracted. With their help, a more sensitive database search is restarted.

types of cells (localization), to periods of the development (time restriction) or may depend on posttranslational modifications. Often, the corresponding genes are only expressed in a few phases of the ontogenesis and in some tissues (expression pattern).

In Sects. 3.3.1 and 3.3.2, we will consider both the issue of sequence databases and methods for knowledge-based sequence analysis. Protein property predictions relying on structural databases will be considered in Sect. 5.

3.3.1 Quality of Sequence Databases

The large amount of data on protein sequences calls for automation; also in sequence handling and analysis. Accurate storage and updating mechanisms as well as user-friendly retrieval software are required. Although database teams are aware of such demands and continuously improve the quality of data entries and computer software, databases have grown historically and are far from perfect. Thus, working with sequence databases requires knowledge about their pitfalls which can have a considerable influence on the interpretation of the data. We have already discussed in Sect. 3.2 possible modifications of translational decoding of mRNA sequences (organelle specific translation tables, RNA editing) and pre- (RNA splicing) and posttranslational changes, both resulting in a chemically altered polypeptide compared with the original DNA sequence. But there are also other, much more profane circumstances that result in uncertainty of sequence data. Errors can occur at each step of the multilevel procedure the result of which are protein sequences.

Tab. 1. What is Protein Function?

Molecular function	glycerol kinase
	contains an ATP-binding site
Physiological function	responsible for wing development
	nuclear transport
Phenotypic function	suppressor of SPT3 mutations
	involved in nucleotide metabolism
Dysfunction	deletion causes diabetis
	knockout lethal
Cellular localization	N-terminal myristilation
	transmembrane protein
Expression pattern	only in brain during embryogenesis
	activated by Gal4
Posttranslational modification	phosphotyrosine
	glycosylation

Many levels of protein function. Protein function is determined in the context of activity of other constituents of an organism. The primary molecular function of a protein may consist in catalyzing a specific reaction or in transmitting a signal. A set of many co-operating proteins accomplish a physiological function, e.g., a metabolic pathway. At the next level, proteins determine phenotype properties (phenotypic function or dysfunction). As a rule, protein activity is limited in space and time. The protein can be confined in certain cellular compartments, types of cells, or be located in the extracellular space. Posttranslational modifications often regulate protein activity. The genes coding proteins may be expressed in a few phases of the ontogenese and in some tissues only (expression pattern).

At the beginning, pieces of genomes are sequenced. Experimentally, the DNA sequences are obtained by reading-out electrophorese gels. Errors influence apparently 0.1% of all nucleotides (FLEISCHMANN et al., 1995) if DNA segments between 300 and 600 bases long are sequenced with 3–10 fold redundancy. This might affect 5–6% of all protein sequences (BIRNEY et al., 1996; TATUSOV et al., 1996). The rate is several orders of magnitude higher for so-called ESTs (expressed sequence tags), single gel reads of random cDNA. The comparison of 3,000 human proteins in SWISS-PROT that have been published more than once revealed differences in 0.3% of all amino acids (BORK and BAIROCH, 1996). This error rate seems to be a lower limit since sequences in different publications are often not independent, and many corrections have already been made. Anyway, mostly only erroneous stop codons and frameshift errors can be detected unambiguously. Single residue exchanges are hard to verify as strain differences or other types of natural polymorphism cannot be excluded.

The rate appears small but the error may accumulate in the sequence considered. This can lead to functional misinterpretations (BORK, 1996). Another serious problem is the contamination of cDNA libraries with material from hosts (usually fungal, bacterial, or viral DNA). A prominent example is the "human" EST library of Genethon having a surprisingly high rate of matches with the yeast genome. Gene sequences of annexin I and insulin from a sponge (their "existence" in lower eukaryotes is very unlikely) were closely related to mammalian homologs (BORK and BAIROCH, 1996). It turned out that the biological sample had been contaminated by DNA of a rodent species.

As a next step, the processing of raw DNA data includes identification of genes, open reading frames (ORFs), and the exon–intron structure. In addition to sequence signals for exon–intron junctions, recent exon recognition algorithms integrate similarity searches in existing sequence databases since homology hits are a strong indicator for a coding sequence. Other properties used for exon–intron classification are: (1) codon usage, (2) hexanucleotide frequencies, (3) local complexity (information content), (4) poly(A) ranges, regulatory sequences such as ribosome-binding segments (Shine-Delgarno segments) or promoters. The challenge is to improve the corresponding identification algorithms further since widely used programs for gene (exon) prediction in eukaryotes have an accuracy below 50% (BURSET and GUIGO, 1996). This rate drops further if the DNA sequences considered contain wrong nucleotide positions. The gene identification problem is avoided if mature mRNA is sequenced.

Finally, sequences are annotated by human beings or by their computer programs (BORK and BAIROCH, 1996). Hence, errors ranging from simple spelling ambiguities to semantic mistakes are common. For example, *SCD*25, the supressor gene of CDC25 in yeast, was so often misquoted as *SDC*25 that it has become an accepted synonym. Database queries are hindered by differences between spelling variants [e.g., hemoglobin (US) and haemoglobin (UK), upper and lower case in *H* (Hairless) and *h* (hairy) in the *Drosophila* genetic nomenclature], representation of non-English characters (e.g., 'ü' in *Krüppel, Krueppel,* or *Kruppel*). The same gene may have several synonymes (e.g., *TUP*1, *AER*2, *SFL*2, *CYC*9, *UMR*7, *AAR*1, *AAM*1, and *FLK*1 are the same gene in yeast and *hns, bns*A, *drd*X, *osm*Z, *bgl*Y, *msy*A, *cur, pil*G, and *top*S are the same gene in *Escherichia coli*). Similar multiplicity exists on the level of protein names (e.g., annexin V was also called lipocortin V, endonexin II, calphobindin I, placental anticoagulant protein I, thromboplastin inhibitor, vascular anticoagulant α, and anchorin CII). The opposite is also frequent – different genes or proteins having the same name (e.g., cyclin is the name for a variety of cell cycle components or *MRF*1 is the gene name in mitochondria of yeast both for the peptide chain releasing factor 1 and for the respiratory function protein 1).

A major problem is the functional description of genes and proteins (BORK and BAIROCH, 1996). For example, the ORGanelle division in the EMBL, Genebank nomenclature should be used only for sequences in the genome of mitochondria and plastids. Often

entries with nuclear-encoded genes for proteins targeted at organelles are wrongly entered in the same division. Since the number of sequences is large, functionalities are assigned automatically by the computer software based on sequence similarities. In this way, a single erroneous entry will lead to a whole sequence family with artificial functions [e.g., whether nifr3 has a role in nitrogen fixation, remains unclear, but it has already been assigned to several previously unannotated proteins (CASARI et al., 1994)].

Yet another layer of uncertainty consists of scope-related problems. Often, the retrieval system does not access all known sequence data but only a subset (due to license reasons or software limitations); sometimes even just a single sequence representative for a family. Many communities study particular classes of proteins and the information in such specialized databases is not pointed to from the general genetic databases but there is hope that links will appear in a near future (such as pointers between SWISS-PROT and the two databases FlyBase and yeast YPD within the SRS retrieval system, see Tab. 2, for WWW-links).

3.3.2 Knowledge-Based Prediction of Protein Structure and Function Using Protein Sequence Analysis

Generally, any property of the sequence representing an unknown protein can be compared with other sequences or families of sequences in databases. In cases of coincidence, the query protein is considered similar and the information on proteins which gave the hit is considered relevant also for the unknown protein. From the logical point of view, the discriminative power of this approach should be limited since it is not *a priori* clear whether a given sequence property is characteristic also for the sequence under consideration. But in practice, this way of thinking is unexpectedly often successful. The structure and function of an unknown protein can be tested experimentally if it is supposed to be related to a protein family with well-

Tab. 2. WWW Pointers for Important Programs and Databases for Similarity Searches in Protein Sequences are Presented
The list it not complete but the links here have been tested by the authors. A more detailed and updated list is available from http://www.embl-heidelberg.de/~bork/pattern.html

FTP Sites for Software Resources

Barton's flexible patterns	ftp://geoff.biop.ox.ac.uk
Propat (property pattern)	ftp://ftp.mdc.berlin.de/pub/makpat
SOM (neutral network)	ftp://ftp.mdc-berlin.de/pub/neural
SearchWise	http://www.ocms.ox.ac.uk/~birney/wise/topwise.html
PROFILE	ftp://ftp.ebi.ac.uk/pub/software/unix
MoST (motif search tool)	ftp://ncbi.nlm.nih.gov/pub/koonin/most
CAP (BLAST output parser)	ftp://ncbi.nlm.nih.gov/pub/koonin/cap

Searchable Motif Databases

PROSITE	http:/expasy.hcuge.ch/sprot/prosite.html
Motif search (ICR)	http://genome.ad.jp/SIT/MOTIF.html
Profile Scan (ISREC)	http://ulrec3.unil.ch/software/PFSCAN form.html
BLOCKS	http://www.blocks.fhcrc.org
PRINTS	http://www.biochem.ucl.ac.uk/~attwood/PRINTS/PRINTS.html
PIMA	http://dot.imgen.bcm.tmc.edu:9331/seq-search/protein-search.html
PRODOM	http://protein.toulouse.inra.fr/prodom.html

WWW Servers for Motif and Profile Searches

Regular expressions	http://ibc.wustl.edu/fpat/
PROFILE	http://sgbcd.weizmann.ac.il/Bic/ExecAppl.html
PATSCAN	hppt://www.mcs.anl.gov/home/papka/ROSS/patscam.html
PatternFind (ISREC)	http://ulrec3.unil.ch/software/PATFND_mailform.html
Pmotif	http://alces.med.umn.edu/pmotif.html
HMM	http://genome.wustl.edu/eddy/hmm.html
Discover	http://hertz.njit.edu/~jason/help.html

studied members. At present, sequence analysis alone cannot give final knowledge.

1st, the meaning of terms denoting sequence properties needs to be clarified. Often, similarities are more obvious if only a relevant part of the sequence information is used for comparison.

- *Sequence composition* is traditionally the proportion of amino acid residues in the sequence. About 40% of all sequences in the SWISS-PROT database have a pronounced compositional bias (WOOTTON, 1994). The *dipeptide composition* is a much stronger criterion for sequence comparison (VAN HEEL, 1992). Already the frequent occurrence or absence of certain amino acid types in a protein is sometimes indicative

for structure or function. Many transmembrane regions contain almost exclusively hydrophobic residues. A similar reduction of the amino acid alphabet is accompanied in coiled-coil regions with a position-dependent frequency of leucine and similar residue types (LUPAS et al., 1991; LUPAS, 1996). A glycine content of one third and a frequent occurrence of (hydroxy-)prolines with glycines mostly at every third sequence position are characteristic for a tropocollagen structure.

- A *motif* is a small conserved sequence region within larger entities. Sometimes, motifs are characteristic for structural and functional features (such as posttranslational glycosylation sites or SH3-binding sites) that develop independently from the surrounding sequence. For these (relatively rare) motifs, the concept of sequence homology is irrelevant.

- The terms "alignment block" and "pattern" are more technical compared with "motif" and are used to deal with difficulties associated with gaps (insertions or deletions) in sequence comparisons. The *alignment block* refers to conserved parts of multiple alignments containing no gaps. A *pattern* consists of one or several alignment blocks and can also contain gaps.
- A *profile* implies a description of a sequence or an alignment in other terms than just the letters denoting amino acids. Usually, conserved physical properties among different residues are used to characterize an alignment position and to derive position-dependent weights and penalties for all amino acid types or a gap.

There is no contradiction between "profile" and "motif" since profiles may be restricted to smaller regions and patterns can be also described in amino acid property terms.

Thus, given a single query sequence, database-aided protein sequence analysis can be based on

(1) comparison of general sequence properties such as amino acid composition,
(2) motif searches (sequence pieces), or
(3) full sequence comparison.

All three variants have been attempted both as restricted to analysis of strings composed of letters denoting amino acids ("linguistic" analysis) or as profile-based. Approaches (1) and (2) are in some contrast to techniques of type (3). Whereas the latter make an effort to utilize the complete sequence information to maximize the overall signal, the former generalize from the noise in variable regions and concentrate on key features. The applicability of any of the approaches depends solely on the specific sequence studied and the resources. Generally, compositional and motif analysis are more suitable for large database searches since they are harnessed to fast word look-up algorithms which require only little computational resources, whereas full sequence comparisons,

especially in profile representations, rely on exhaustive but slow dynamic programming algorithms and are applicable preferably to smaller subsets of sequences (BORK et al., 1995). Examples for motif and profile databases and searching software are listed in Tab. 2. BORK and GIBSON (1996) have evaluated different search algorithms in great detail.

3.3.2.1 Standard Procedure of Database-Aided Homology Search for a Query Sequence

The standard method in sequence analysis is to find distantly related proteins (grey zone homologies). Usually the only attempts undertaken are fast homology searches with one of the BLAST programs via World Wide Web (WWW) servers (ALTSCHUL et al., 1994), albeit FASTA or BLITZ are also in use. However, these techniques do not reveal many weakly homologous sequences. They omit proteins which are similar only in parts of their sequences (due to their multidomain structure) and, on the other hand, may find numerous similar proteins with very different functions if the sequence families are large. Sometimes, the application of special amino acid substitution matrices as offered in standard programs is an alternative if no significant hits are found (BORK and GIBSON, 1996). The probability to get a BLAST hit for a query sequence depends on the species: It is 70–90% for bacterial or yeast sequences but only ~50% for human sequences since the corresponding subset of the sequence databases is much less annotated. In the case of the genome analysis of *Haemophilus influenzae* and its comparison with the genome of *Escherichia coli* (TATUSOV et al., 1996), database homologies were found for about 90% of the genes. For about 80%, functional characterization of BLAST hits were available, but structural information was found only for 15%.

Given the concerns raised in Sect. 3.3.1, results of similarity searches in databases should be always considered with caution. Given the pressure on sequencing groups not to overlook interesting homologies, the methodology is stretched and, with small thresholds, spurious hits are taken as real; consequently, misinterpretations cannot be avoided. Sometimes, sequence homology is just a result of structural constraints (in coiled-coil regions of muscle and other structural proteins) or the similarity is limited to a single domain. Therefore, regions with compositional bias should be identified in advance and removed from the query sequence before the BLAST search. Biological diversity appears unlimited and simple schemata do not work in all cases.

The BLAST results provide usually only a starting point for motif and profile searches. The next step is the extraction of patterns or profiles characterizing the alignment of similar sequences. A weighting scheme is useful if some type of sequences is overrepresented in the multiple alignment (HIGGINS et al., 1996). Local, conserved motifs are often the only observable markers of structural or functional regions. Iterative searches with such motifs and profiles in the sequence databases are often successful and increase the original BLAST success rate by an additional 10–20%. Sensitive searches including many cycles of analysis take several weeks and are possible only for a few sequences (Fig. 4). The analysis of thousands of sequences from large-scale sequencing projects within a few days requires automated evaluation of database searches such as with the GENEQUIZ program for gene function prediction (CASARI et al., 1994).

3.3.2.2 Significance of Weak Homologies

In assessing the significance of weak homologies, purely statistical methods are currently only of little help (BORK and GIBSON, 1996). At the beginning, formal properties of the alignment should be checked: possibility of frameshift errors, sequence weighting in multiple alignments for motif extraction, appropriate handling of gaps and amino acid substitutions given the protein family and sequence length, as well as completeness of database searches including novel sequences. The second level of checks includes structural constraints which must apply between sequences as a consequence of homology between them. Such conclusions are often possible even without knowing the three-dimensional structure of a single protein in the multiple alignment. For example, Cys patterns are expected to match in Cys-rich proteins (conservation of disulphide bonds). Gly and Pro are unlikely at positions where all other sequences have different amino acids. Insertions/deletions in highly conserved regions are suspicious. Similar logics proved successful in retrieving GAL4 and SH_2 domains (BORK and GIBSON, 1996).

The knowledge of complete genomes opens new indirect ways for functional predictions. If the enzymes belonging to a metabolic pathway are well known and a few members are found in the given organism, there is a chance to find also the others or to draw conclusions on a pathway modification (KOONIN et al., 1996). The order of genes in the genome also gives information about common regulatory blocks such as operons which might help in functional assignments (TATUSOV et al., 1996).

3.3.2.3 Search for Internal Repeats

In addition to this standard way of sequence analysis, special techniques have been developed for specific applications. A major concern are intrinsic sequence repeats that indicate duplication of structural elements and, possibly, a preceding gene duplication. Fourier spectral analysis and autocorrelation techniques have proven successful only for special classes of sequences (MCLACHLAN, 1983; MAKEEV and TUMANYAN, 1996) since insertions between repeats can be of greatly different length. Direct alignments of sequences with themselves combined with graph-theoretical methods are a more general and very efficient approach (HERINGA and ARGOS, 1993).

3.3.2.4 Complex Regression of Protein Sequence–Structure Relationships

The utilization of neural network techniques (FRISHMAN and ARGOS, 1992; HANKE et al., 1996) or hidden Markov models (KROGH et al., 1994; BAIROCH and BUCHER, 1994) was attempted for the derivation of a complex regression function between pairs of protein sequence–structure relationships since the underlying physics between sequence and three-dimensional structural properties are not well known. Due to the heuristic nature of these approaches, the real impact is difficult to assess and, probably, currently overestimated. Not only the whole sequence but also typical sequence properties have been taken as input information. For example, DUBCHAK et al. (1993) use the amino acid composition as input for neural networks trained to recognize 4 helix bundles, parallel $(\alpha\beta)_8$ barrels, nucleotide binding folds, and immunoglobulin folds. The matrix of size $20 \cdot 20$ containing dipeptide frequencies in a query sequence was used as input for neural networks for checking the relatedness to 45 folding classes and 4 folding types (RECZKO et al., 1994; RECZKO and BOHR, 1994).

4 Secondary Structural Features of Proteins

After a detailed consideration of different secondary structural elements, we consider the problem of an objective definition which could be the basis for a computer program to assign secondary structural states in protein tertiary structures resolved with X-ray crystallography or NMR techniques. The concept of secondary structural class, based mainly on secondary structural content, was the first attempt of a structural classification of proteins. Finally, we analyze methods for prediction of secondary structure from protein sequence alone.

4.1 Types of Secondary Structures

The primary structure, the sequence of amino acids in the polypeptide, is the basis for the formation of complex spatial structures of proteins. Historically, regularities in the spatial location of sequentially near residues are named secondary structure. More traditionally, the term secondary structure is even confined to repetitive local conformations along the polypeptide chain and includes mainly α-helices and β-sheets. In contrast, tertiary structural description puts emphasis on residue–residue contacts which are distant in sequence. Before entering the typology of secondary structural features, the physical framework of the description of protein structure needs clarification.

In the molecular-mechanical approximation, protein atoms are considered mass points and centers of spherical potential functions. A *conformation* is defined as a set of all relative positions of atomic centers. Whereas bond lengths and valence angles allow only moderate changes, rotations around single bonds have relatively small energetic thresholds and constitute the main source of conformational variability of polypeptides (Fig. 5). A torsional angle is defined as the angle between the 2 planes spanned by the first 3 atoms and the last 3 atoms respectively of a quadruple of chemically connected atoms. The *cis* position (all 4 atoms in one plane and both marginal atoms in the same halfplane with respect to the bond between the 2nd and the 3rd atom) is defined as zero degrees. Among the 3 backbone torsional angles φ (dihedral angle $C'_{i-1}N_iC\alpha_iC'_i$), ψ (dihedral angel $N_iC\alpha_iC'_i N_{i+1}$), and ω (dihedral angle $C\alpha_{i-1}C'_{i-1}N_iC\alpha_i$), the last one is constrained to values near $180°$ or $0°$ due to the resonance effects in an almost planar peptide group for *trans* and *cis* peptide bonds respectively. The results of studying possible atomic clashes in a monomer surrounded by peptide-groups can be summarized in a Ramachandran φ vs. ψ-plot (see, e.g., CREIGHTON, 1992, p. 183). There are generally 2 accessible regions: a large area with $-180° < \varphi < 0°$ and $100° < \psi < 190°$ (region β) as well as a smaller region α (or α_R) with $-100° < \varphi < -50°$ and $-70° < \psi < -30°$. For residues with small side

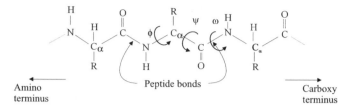

Amino terminus

Peptide bonds

Carboxy terminus

Fig. 5. The polypeptide backbone. The division of the polypeptide backbone into peptide units is a suitable methodological approach for describing the conformational properties of polypeptides.

chains, φ/ψ-values near 60°/40° are also possible (region α_L). In the case of proline, the φ is fixed to discrete values as a result of the ring structure including the backbone N-Cα bond. The conformational limitations described above restrict the possible forms of repetitive backbone structures. Repeated φ/ψ-combinations from region β result in extended chains, whereas many monomers with φ/ψ-values in the α-region form helical structures.

The analysis of local backbone conformations in crystallographic protein structure is in agreement with the molecular-mechanical treatment. Most of the residues belong to the allowed regions of the Ramachandran plot, although a few outsiders exist even in protein structures resolved with high resolution (KARPLUS, 1996). These residues are a sign of localized strain in the global protein structure which is compensated by other interactions and may indicate the non-optimality of the amino acid sequence for the given structure. In some cases, this strain has a functional role. Strain has also been observed in side

chain conformations (SCHRAUBER et al., 1993).

Secondary structures observed in experimentally determined protein structures are characterized by two properties: (1) whether they are composed of conformationally (nearly) identical monomer units and (2) whether direct interactions between sequentially near residues exist. The first property distinguishes between different types of helices (including extended structures, see Tab. 3) and segments changing the direction of the polypeptide chain (loops and turns). With the second classification, α-helices and 3_{10}-helices are united with most tight turns as they include backbone hydrogen bonds between residues with small sequence separation. The extended structures [β-strands (both in parallel and antiparallel sheets or the poly-Gly-I helix), PPI-helix, PPII-helix (ε-helix)] and larger loops form the other group.

Generally, the secondary structural preferences of oligopeptide fragments describe only the ease or difficulty with which a specific se-

Tab. 3. Types of Repetitive Secondary Structures
Hydrogen bonds are denoted relative to a residue with sequence position i if applicable. Whereas most helices consist of repetitive *trans*-peptide bonds, poly-Pro-I is a polypeptide conformation with *cis*-peptide bonds

Type of Helix	Torsion Angle [°]			Hydrogen Bond	Twist [°]	Pitch [Å]
	φ	ψ	ω			
α-Helix	− 57	−47	180	i...i+4	100	1.50
3_{10}-Helix	− 49	−29	180	i...i+3	120	2.00
β-Strand (parallel sheet)	−119	113	180	n.a.	180	3.2
β-Strand (antiparallel)	−139	135	182	n.a.	180	3.4
Poly-Gly-II	− 80	150	180	n.a.	120	3.1
Poly-Pro-I	− 83	158	0	n.a.	108	1.9
Poly-ProII	− 78	149	180	n.a.	120	3.12

quence adopts the conformation in a given tertiary structure. Possible strain at this level can also be sacrificed for a tertiary topology to be achieved. The lability of preferences for secondary structural types by some amino acid sequence is obviously demonstrated by the ease of large-scale conformational changes of serpins and prion proteins. In the case of serpins, a metastable kinetically trapped 5-stranded β-sheet conformation and a largely unstructured reactive loop were observed which only slowly rearrange to the native 6-stranded sheet (MOTTONEN et al., 1992; GOLDSMITH and MOTTONEN, 1994; HUANG et al., 1994; CARREL et al., 1994). Prion proteins may trigger into a completely different α-structure followed by massive aggregation of such conformationally changed proteins (RIEK et al., 1994; CERPA et al., 1996).

4.1.1 The α-Helix and the 3_{10}-Helix

The right-handed α-helix is the most widely studied form of secondary structures. The detailed geometry may deviate from that given in Tab. 2 with respect to amino acid types constituting the helix and the helix environment in the tertiary structure. Helices have a specific sequence pattern of hydrophobic and hydrophilic residues depending on their environment (BLUNDELL and ZHU, 1995); for example, helices at the surface to solvent are amphiphatic with hydrophilic residues clustered at the side contacting solvent and hydrophobic residues directed to the center of the protein. Often, the carbonyl groups tend to point outwards for interaction with solvent or other donors. Only a fraction of the helices is truely linear, most exhibit some type of curvature or are even kinked (BARLOW and THORNTON, 1988).

The ends of α-helices are special regions since the corresponding clusters of NH and CO groups are not saturated by helical hydrogen bonds. In context also with the helix dipole originating from the dipoles of peptide bonds and accumulated with helix length, α-helix ends are a perfect place for specific substrate and ligand binding [see, for example,

DORAN and CAREY (1996)]. At the termini, an α-helix may change into a 3_{10}-helix. Isolated 3_{10}-helices are rare in proteins but may be observed for small peptides (MILLHAUSER, 1995). Probably, 3_{10}-helices have a role in α-helix folding. The amino acid types have different propensities whether they are located in an α-helix or not and also depending on the specific locations, in the central region or at the N- or C-termini, the so-called capping boxes (PRESTA and ROSE, 1988; SERRANO et al., 1992; DOIG and BALDWIN, 1995; GONG et al., 1995; QIAN and CHAN, 1996). Generally, propensities vary among peptide host systems and are also different from those calculated from protein tertiary structures (BRYSON et al., 1995). Many helical segments observed in crystal structures of proteins are much less helical in the form of isolated peptides in solution.

4.1.2 Extended Structures in Protein Structures

The β-sheet is the second well characterized type of secondary structure. It consists of β-strands, a repetitive extended helical polypeptide conformation (Tab. 3). Poly-glycine I is essentially the same conformation. The polar NH and CO groups of the backbone are saturated with hydrogen bonds formed with peptide groups of neighboring strands. The helical parameters and the φ/ψ-values depend on the relative directionality of the strands inside the sheet, the amino acid composition, and the tertiary context in the protein structure. Peptides forming stable β-strands outside the protein context are difficult to design (MAYO et al., 1996).

The original models of β-sheets were planar and flat, but the structures observed in real proteins have a right-handed twist. Large sheets may even form a barrel. This is a consequence of tight packing of side chains on the surface of a sheet (MURZIN et al., 1994a; 1994b; VTYURIN, 1993; VTYURIN and PANOV, 1995). Few residues (generally 2) may not fit into the general β-sheet pattern, pucker out from an extended substructure between consecutive β-type hydrogen bonds joining adjacent strands and form bulges (CHAN et al., 1993). Bending of β-strands is another structural distortion (DAFFNER et al., 1994). As a rule, parallel β-sheets are buried inside the protein and hydrophobic residues dominate. Antiparallel sheets have often one side exposed to solvent, resulting in alternation of hydrophobic and hydrophilic residues. This strict periodicity may be broken by a β-bulge where an extra residue is accomodated in edge β-strands. Sheets may contain also both parallel and antiparallel strands.

The left-handed poly-proline-II-helix (PPII-helix, form of poly-proline in water, acetic acid or benzyl alcohol) or ε-helix is essentially also an extended structure as in β-strands but without grouping in sheets and the formation of hydrogen bond networks. It was demonstrated that this type of secondary structure is common in globular proteins (ADZHUBEI and STERNBERG, 1993) and conserved in homologous structures (ADZHUBEI and STERNBERG, 1994). It is an important feature in structural motifs such as the HMG boxes for DNA-binding (ADZHUBEI et al., 1995).

4.1.3 Loop Segments

Polypeptide segments without a repetitive backbone structure are called loops or turns (short loops) and have been long considered together with PPII-helical fragments as "random coil" structures. They connect helical and extended segments and make changes in backbone directionality possible. Long loop regions are the most flexible parts in protein structures for the accommodation of insertions and deletions (PASCARELLA and ARGOS, 1992).

A sharp reversal in chain direction of about 180° within only 4 residues is possible with a β-turn. Typically, they occur in β-hairpins and, generally, antiparallel sheets. β-turns as observed in crystallographic protein structures cluster in discrete regions with respect to the backbone torsion angles of the central residues i+1 and i+2 (Tab. 4). Most of the tight turns are characterized by a main-chain hydrogen bond CO_i-NH_{i+3}. Sometimes, main-chain side-chain hydrogen bonds are observed in the case of serine or aspartate residues. These interactions give rise to amino acid type preferences, so that hydrogen bond acceptors such as aspartate, serine, or asparagine are usually the first residues in type I β-turns, where they can hydrogen bond to the NH group of the central peptide group (WILMOT and THORNTON, 1990; SIBANDA and THORNTON, 1993).

The complete classification of loops is an unsolved scientific task since the role of loops in protein folding and the energetic contribu-

Tab. 4. Types of Tight 4 Residue Turns (β-Turns)
The conformational characteristics of residues i+1 and i+2 are listed. The classification is given in accordance with WILMOT and THORNTON (1990). "Y" denotes the existence of the hydrogen bond CO_i-NH_{i+3}. The regions α_R and α_L are the right- and left-handed α-region respectively, β stands for β-regions. The regions γ_L and ε are located on the Ramachandran plot for glycine-like residues for positive φ angles, γ_L is similar to α_L and ε is the part for highly negative ψ-angles.

Type of β-Turn	Positon i+1			Positon i+1			Hydrogen Bond
	φ [°]	ψ [°]	Region	φ [°]	ψ [°]	Region	
I	− 60	− 30	α_R	− 90	0	α_R	Y
I′	60	30	α_L	90	0	γ_L	Y
II	− 60	120	β	80	0	γ_L	Y
II′	60	−120	ε	− 80	0	α_R	Y
VIa	− 60	120	β	− 90	0	α_R	Y
VIb	−120	120	β	− 60	0	α_R	Y
VIII	− 60	− 30	α_R	−120	120	β	N

tion to their stabilization are not known and purely geometric principles work the less, the longer the loops are. There are also examples that the identity of loops is not important for folding of a small protein at all (BRUNET et al., 1993). 5-Membered π-turns (RAJASHANKAR and RAMAKUMAR, 1996) and loops consisting of 3–8 residues (KWASIGROCH et al., 1996) have been exhaustively studied. Long loops (≥ 10 residues) have been found to connect also mainly locally adjacent secondary structural elements ("long-closed loops"). Only 5% of the long loops (MARTIN et al., 1995) are between distant secondary structural units ("long-open loops"). Since they contain a larger percentage of proline residues the long-open loops probably contain some part of the PPII-helix.

4.2 Automatic Assignment of Secondary Structural Types in Three-Dimensional Structures

The definitions of secondary structural elements as described above are visual as derived from geometric models and are not quantitative. The secondary structure assignments given in the Brookhaven Protein Databank entries (ABOLA et al., 1987) by crystallographers and NMR spectroscopists are often subjective; therefore, a computer algorithm is necessary. The most widely used program is DSSP (KABSCH and SANDER, 1983), probably, since the corresponding software is widely available. DEFINE (RICHARDS and KUNDROT, 1988) or P-CURVES (SKLENAR et al., 1989) can also be utilized for objectification.

All 3 methods have been critically reviewed (COLLOCH et al., 1993). It was found that the assignments coincide only in 63% of the residues. This can be explained by the particularities of each method. The DSSP approach considers hydrogen bond patterns, while the P-CURVE algorithm finds regularities along the helicoidal axis and the DEFINE technique measures distances between Cα atoms. Therefore, when evaluating predictions, the "standard-of-truth" might vary depending on which property was used for the secondary

structure assignment. In addition, the most widely used DSSP algorithm produces many α-helices comprising 4 or fewer residues and β-strands consisting of 2 or even only one monomer. STRIDE (FRISHMAN and ARGOS, 1995) is a recent modification of DSSP reported to give secondary structure assignments somewhat more coinciding with subjective assignments of crystallographers than DSSP, but not much.

An interesting alternative for finding regularities in backbone structures is the algorithm of ADZHUBEI and STERNBERG (1993) relying only on Cα coordinates.

4.3 The Concept of Secondary Structural Class

Early in 1976, when only about 40 crystallographic structures of proteins were known, LEVITT and CHOTHIA (1976) studied the succession of secondary structural elements along the amino acid sequence. Intuitively, they grouped the proteins into 4 structural classes (or folding types):

- all-α proteins having *only* α-helix secondary structural elements (more than 60% of the residues adopt helical conformation, no residues in β-strands);
- all-β proteins consisting *mainly* of (often antiparallel) β-strands;
- α + β proteins having *independent clusters* of α-helices and (often antiparallel) β-strands in the sequence; and
- α/β proteins with mixed (often *alternating*) segments of α-helix and (mostly parallel) β-strands.

Many more protein structures are known today, and, for an increasing number, it is not easy to classify them in accordance with the definitions of LEVITT and CHOTHIA (1976). A variety of class definitions in terms of secondary structural content has been presented (Fig. 6). At the same time, there are no clear clusters any more in the α-content vs. β-content plot for large sets of protein structures as available in the PDB. Also the notion of α + β and α/β proteins is not longer applicable. For

Definitions of Folding Type
(Structural Class)

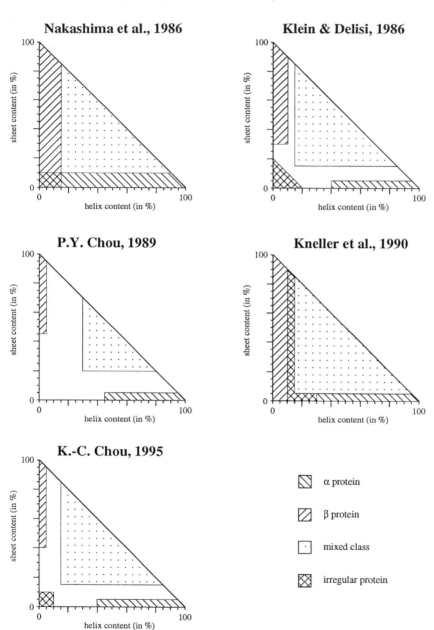

Fig. 6. Secondary structural class (folding type) definitions. The definitions of NAKASHIMA et al. (1986), KLEIN and DELISI (1986), CHOU (1989), KNELLER et al. (1990), and CHOU (1995) are illustrated in form of α vs. β content trigonal plots. The white regions correspond to proteins which are not assignable to any secondary structural class with the definition of the given authors.

example, both the acylphosphatase, PDB entry 1APS (PASTORE et al., 1992), and the B-chain of the regulatory domain of the aspartate carbamoyltransferase, PDB entry 8ATC (STEVENS et al., 1990), have a 2-layered structure consisting of an antiparallel β-sheet and two parallel α-helices. The existence of an antiparallel sheet is characteristic for α+β structures. But a more detailed investigation of the structures reveals a high degree of secondary structural alternation and βαββαβ and a doublet of the βαβ motif, both observations pointing to class α/β. It was argued by EISENHABER et al. (1996b) that the term "secondary structural class" may be applied today, if at all, only as a classification in α, β, mixed and irregular proteins. The best definition in thresholds of secondary structural content is that of NAKASHIMA et al. (1986) since

(1) it is applicable to all proteins,
(2) it is compatible with intuitive definitions of irregular and mixed proteins, and
(3) the secondary structural content thresholds are at minima of the occurrence vs. secondary structure contents plot for large selections of PDB structures (EISENHABER et al., 1996b).

Nevertheless, the concept of structural class, based on the secondary structural content of the protein and the directionality of β-strands, is useful from the experimental as well as the theoretical point of view. The folding type of a protein can be directly determined by relatively simple spectroscopic methods. With a sufficient quantity of the protein available, circular dichroism (CD) spectroscopy in the UV absorption range can be used to obtain reliable measures of secondary structure content, especially for α-helices, but also for parallel and antiparallel β-strands (JOHNSON, Jr., 1990; PERCZEL et al., 1991; SREERAMA and WOODY, 1994).

Secondary structural class restrictions have a high impact for secondary and tertiary structure prediction (EISENHABER et al., 1995b). The accuracy of secondary structure prediction from the amino acid sequence with methods designed for all-α proteins is larger than 80% compared with maximally ~70%

in the general case (KNELLER et al., 1990; MUGGLETON et al., 1992, 1993). The effect of knowledge of structural class alone in improving secondary structure prediction is comparable with the use of the extra information contained in multiple alignments of homologous sequences (LEVIN et al., 1993; ROST and SANDER, 1993b). The secondary structural class is related to various properties of a protein such as its location in extra- or intracellular compartments, biological function (being an enzyme or not), or the existence of disulfide bonds (NISHIKAWA and OOI, 1982; NISHIKAWA et al., 1983a, b).

4.4 Prediction of Secondary Structural Features

All secondary structure prediction methods currently in use are knowledge-based. A learning set of protein structures from the PDB is utilized to derive a prediction rule; e.g., for the calculation of propensities (the relation of the frequency of a given amino acid type in a certain secondary structure with the frequency of any residue to be observed in the same secondary structural state). The prediction rule is applied to protein sequences from a test set of PDB structures to estimate the expected rate of successful prediction.

Such an approach has two types of problems. Difficulties of the first type are of technical nature. Researchers often invent prediction functions with such a large number of parameters that the number of structurally nonhomologous proteins in the learning set is not sufficient to determine all parameters unambiguously. Also, the protein structure in the test set should not be contained in the learning set to exclude the information to be predicted from the rule. The latter difficulty can be avoided with the so-called "jackknife"-scheme: learning is done on N-1 proteins of a set and prediction is obtained for the Nth protein. This procedure is repeated for all proteins in the data set (EISENHABER et al., 1996a).

The second class of difficulties has its roots in the current state of the protein Data Bank.

Generally, these prediction techniques will work only for sequences representing globular proteins. All sequences with compositional bias constituting about 40% of SWISS-PROT (WOOTTON, 1994) cannot be well considered since an insufficient number of the structures of such proteins is known. Additionally, it is not clear whether the learning sets of proteins used now explore the available sequence and structure space for proteins comprehensively. Therefore, protein structures available in the PDB in future years will probably result in a decrease of the prediction power of today's algorithms and the success rates discussed in this work should be considered upper estimates.

4.4.1 Secondary Structural Class and Content Prediction

In the hierarchy of prediction methods, secondary structural class prediction [secondary structure above or below a threshold with classification into folding types all-α, all-β, mixed types (sometimes subclassified in $\alpha + \beta$ α/β) and irregular forms] corresponds to a lower level and appears a simpler task compared with secondary structural content prediction (fraction of residues in the 3 states helix, sheet, and coil) or even traditional secondary structure prediction (state of every residue among the 3 alternatives helix, sheet, and coil). Since the discovery of NISHIKAWA and OOI (1982) that the amino acid composition is tightly related to secondary structural class, both analytical distance criteria in the amino acid composition space and neural network methods have been applied for the jury decision between 3–5 folding types. A detailed review has been given by EISENHABER et al. (1995b, 1996b). Especially after 3 recent publications (CHOU, 1995; ZHANG and CHOU, 1995; CHOU and ZHANG, 1995), the paradoxical situation emerged that folding type prediction appears solved (reported prediction accuracies up to 100%) whereas secondary structure prediction even with multiple alignments approaches only about 70% accuracy, and the success rate of its class prediction is only near 75% (ROST and SANDER,

1993a; LEVIN et al., 1993; ROST and SANDER, 1994b). This paradox has now been solved (EISENHABER et al., 1996b). It was shown (1) that certain structural class definitions leave many proteins with intermediate secondary structural content without assignment to any class (predictions are made only for proteins with extreme contents); (2) that various analytical distance-based jury decision methods yield only prediction accuracies up to 55% for representative test sets even of extreme proteins; and (3) that the real impact of amino acid composition on secondary structural class is only about 60%. The amino acid composition determines the secondary structural content with an error of about 13% (EISENHABER et al., 1996a). Secondary structural content and class prediction based on the knowledge of only amino acid composition of the query protein is available on the WWW (program SSCP with the URL http://www.embl-heidelberg.de/~eisenhab/).

4.4.2 Traditional Secondary Structure Prediction

The prediction whether a residue in a protein sequence is in helix, sheet, or coil state is a classical problem in protein structure prediction. It was found early that different amino acid types have various preponderance for particular secondary structural environment. Consequently, methods of the first generation were propensity-based. The principal achievments are represented by the Chou-Fasman (CHOU and FASMAN, 1974a, b; YANG, 1996), GORIII (GARNIER et al., 1978) and COMBI (GIBRAT et al., 1987) methods [for detailed review see EISENHABER et al. (1995b) pp. 10–18]. The GORIII method relies not only on single residue propensities but also on statistically significant pairwise residue interactions. The prediction accuracy achieved was 60–63% (GIBRAT et al., 1987; GARNIER and LEVIN, 1991). A further improvement of about 4% (BIOU et al., 1988) was attained by combining the GORIII method with 2 other prediction schemes: one based on hydrophobicity patterns which are often observed in regular secondary structures (bit pattern method), and the other using structural similarity between short, sequentially homologous peptides (LEVIN and GARNIER, 1988). As was shown by GIBRAT et al. (1991), the predictive power of methods relying only on sequentially local structure information is limited to about 65%. With today's databases, the estimate would be even lower. The inability to find properly helical and strand segments is an even more important deficiency of the classical methods (ZHU, 1995).

Further development in secondary structure prediction occurred in two directions (EISENHABER et al., 1995b):

(1) It led to the involvement of multiple alignments of database sequences with the query sequence giving information about the mutability of sequence positions. The prediction accuracy increased to values of about 70% (LEVIN et al., 1993; MEHTA et al., 1995). Combined use of evolutionary information and capping rules was described by WAKO and BLUNDELL (1994a).
(2) Neural networks were applied to find a complex regression between diverse types of input information (e.g., data from sequence windows with respect to single sequences and multiple alignments, amino acid compositions and sequence length of the query sequence, etc.) and the secondary structural state of a residue in the center of the sequence window. The most prominent algorithm of this type is PHDIII (ROST and SANDER, 1994b).

Direct inclusion of non-local interactions proved to increase the prediction success significantly even for single sequence predictions. FRISHMAN and ARGOS (1996) used the statistics of hydrogen bonds for pairs of amino acid types in α-helices and β-strands and achieved correct predictions of about 68%.

Several engines for secondary structure prediction are available at http://www.embl-heidelberg.de [among which are PHD (ROST and SANDER, 1994b), SSPRED (MEHTA et al., 1995), PREDATOR (FRISHMAN and ARGOS, 1996)]. The server at http://www.gene-bee.msu.su provides a modification of the GOR algorithm implemented by BRODSKY and coworkers. Secondary structure prediction service is also provided by the servers at

- http://kiwi.imgen.bcm.tmc.edu [nearest neighbor method of SALAMOV et al., 1995),
- http://www.ibcp.fr/predict.html [SOPM method of GEOURJON and DELEAGE (1994)], and
- http://www.cmpharm.ucsf.edu [a neural network technique of KNELLER et al. (1990)].

4.4.3 Prediction of Transmembrane Regions, Coiled-Coil Segments, and Antigenic Sites

Until recently, *transmembrane segments* were known only as α-helices consisting of stretches of 21 hydrophobic residues. A sequence database analysis of putative transmembrane segments without any assumption of secondary structure has indeed shown that maximal sequence correlation is observed at a periodicity of 3.6 residues characteristic for α-helices (SAMATEY et al., 1995). Other types of transmembrane segments have been found only recently. The porins (WEISS and SCHULZ, 1992) have a build-up consisting of 16 β-sheets arranged as a complete transmembrane barrel giving rise to a big central hole. Recent three-dimensional structures also show further variants. The helices can be tilted against the perpendicular plane, like the case in a light harvesting complex (KÜHLBRANDT et al., 1994). Presence of helices parallel to the membrane plane has also been shown (KÜHLBRANDT et al., 1994; PICOT et al., 1994). There are also indications that the membrane-spanning segments can consist of single extended β-strand like structures, thus making it possible to span the membrane with fewer residues than in the case of an α-helix (HUCHO et al., 1994). All this implies that prediction of membrane-spanning regions might be a more difficult task than has hitherto been anticipated.

Nevertheless, recent prediction algorithms are still tuned to find only α-helical transmembrane segments. The algorithms rely on hydrophobicity scales and information from multiple alignments (EISENHABER et al., 1995b). The prediction function is sometimes found in form of a neural network (DOMBI and LAWRENCE, 1994; ROST et al., 1995). In other cases, explicit expressions are optimized for a given learning set (PERSSON and ARGOS, 1994). Profile techniques have also been employed (EFREMOV and VERGOTEN, 1996a, b). The topology prediction, the derivation of the intra- and extracellular sides of the helices (PERSSON and ARGOS, 1996; ROST et al., 1996), is based on the so-called "positive-inside rule" of VON HEIJNE (1986, 1995). It was found that internal loops between transmembrane segments are richer in positively charged residues. Although the authors usually claim very high success rates for prediction with their technique, the results should be considered with caution given the small number of known transmembrane segments. WWW servers for the prediction of transmembrane regions are available from http://www.embl-heidelberg.de [TMAP and TTOP (PERSSON and ARGOS, 1994, 1996), PredictProtein (ROST et al., 1995, 1996)].

Coiled-coils are intertwined α-helices. Due to their docking under a small angle, the residues at the docking side need to be large and hydrophobic; for example, such as leucine (WALTHER et al., 1996). It is this position-dependent pattern which is analyzed and searched for by the algorithm of LUPAS et al. (1991, 1996).

Transmembrane and coiled-coil regions are special examples of compositional bias. General software is available to diagnose sequence segments with low complexity and information content (WOOTTON, 1994). The analysis of compositional bias is an important initial step in protein sequence analysis.

An important step in the biochemical characterization of a protein is the detection of *antigenic sites*, responsible for specific antibody binding. Since epitopes are usually located in loop structures, this issue is discussed here as a secondary structural feature. Prediction algorithms attempt to locate antigenic sites indirectly as hydrophilic and malleable loops at the protein surface. Antigenic epitopes are also known as mutation hotspots. All these properties have been employed in sophisticated prediction algorithms, albeit with limited success. A detailed review has been given elsewhere (see pp. 8–9 of EISENHABER et al., 1995b).

5 Tertiary Protein Structures

This section is dedicated to the principles of tertiary structure construction and methods for prediction and modeling of tertiary structural features. The analysis and classification of known tertiary structures have helped in the derivation of rules followed by nature in the design of proteins. At the same time, we do not understand the structure formation sufficiently well as demonstrated by the small success of tertiary structure prediction just from amino acid sequence.

5.1 Phenomenology of Tertiary Protein Structures

Today, the three-dimensional structures of several thousand proteins are known from X-ray crystallographic or NMR studies, albeit many of the proteins have similar amino acid sequences [e.g., the PDB contains about 250 mutants of T4 lysozyme (ABOLA et al., 1987)]. Analysis and comparison of this structural information is and will continue to be our main source on protein tertiary structures since they are very complex. This view on tertiary structures has also its drawbacks: The protein is mainly seen as static entity with a fixed structure whereas other experimental techniques but with lower resolution (as well as the comparative analysis of the same protein in different crystallographic environments) deliver much information on small and large scale conformational fluctuations and transitions.

5.1.1 Construction Principle No. 1 – Close Packing

As known from statistical physics, hetero-polymers with primarily attractive forces between monomers form globular structures with locally confined conformational fluctuations. This is the case also for tertiary structures of small proteins. Their main characteristic is the close packing of atoms (RICHARDS and LIM, 1994) inside a volume of generally spherical shape with an irregular surface. The dense packing is achieved by contacts between residues with large sequence separation. Larger proteins appear to consist of several domains, each having its own densely packed core. Most likely, domains are connected with a single segment of the polypeptide chain and each domain consists of a single stretch. In some cases, protein domains are not so heavily segregated. For example, in pyruvat kinase, phosphofructokinase, and arabinose-binding protein, there are 2 or 3 links between domains. Much effort has been concentrated on elaborating objective criteria and automatic algorithms for domain recognition in large proteins (HOLM and SANDER, 1994a; NICHOLS et al., 1995; ISLAM et al., 1995; SWINDELLS, 1995a; SOWDHAMINI and BLUNDELL, 1995; SIDDIQUI and BARTON, 1995).

Valuable information can be obtained from studying the close packing of secondary structural elements in so-called supersecondary structures. Just the condition of packing optimization was sufficient to obtain a rigorous mathematical model and to derive equations for the parameters describing α-helix-α-helix docking (WALTHER et al., 1996). The authors showed:

(1) the existence of 3 different packing cell systems resulting in 5 types of docking angles,
(2) the dependence of the packing cells on helix radius and, therefore, on amino acid composition of the helix, and
(3) the hierarchy of optimal and suboptimal "knobs-into-holes" and "knobs-onto-knobs" packing schemes.

Similar studies obtained global conformational constraints for β-shects (MURZIN et al., 1994a, b; VTYURIN, 1993; VTYURIN and PANOV, 1995). Typical supersecondary structures with β-strands are the greek key and the meander. The $\beta\alpha\beta$-motif is a common mixed supersecondary structure where an α-helix is packed against 2 β-strands forming a parallel sheet (CREIGHTON, 1992).

The search of dense clusters of residues in a protein tertiary structure is yet another perspective for studying close packing. HERINGA and ARGOS (1993) have developed a strategy for locating such groups of residues. This condition has also been used to identify core regions in proteins (SWINDELLS, 1995b). Close packing is often only possible if some substructures accept strained conformation; for example, side chains may adopt non-rotameric torsion angle combinations (SCHRAUBER et al., 1993).

Often, packing is not optimal and does not exhaust the internal space of a protein. Cavities are wide-spread in protein structures. They are usually more readily tolerated than actively favored and can sometimes, be considered as a type of conformational strain since many cavities destabilize proteins (HUBBARD et al., 1994; HUBBARD and ARGOS, 1995).

The compactization of a protein is thought to proceed co-translationally; i.e., some part of the polypeptide folds whereas the remaining chain is still synthesized. Therefore, it is difficult for topological knots to appear in tertiary structures. Nevertheless, knots do happen. In the case of carbonic anhydrase (MANSFIELD, 1994), a C-terminal knot exists. (S)-adenosylmethionine synthetase has a real N-terminal knot (TAKUSAGAWA and KAMITORI, 1996). LIANG and MISLOW (1994a, b) and MAO (1993) have described topological chiralities formed through disulfide bonds, hydrogen bonds and coordination bonds.

5.1.2 Construction Principle No. 2 – Hydrophobic Interior and Hydrophilic Exterior

The uneven distribution of hydrophilic and hydrophobic residues between the interior and the surface is the second basic property of globular proteins. This is the result of interaction with water and ions of the solvent (hydrophobic effect). As a result, a protein core region with low dielectric permissiveness is embedded in an environment with high dielectric permittivity. Special restraints apply to atomic groups capable of hydrogen bonding. At the protein surface, they can have contact with water molecules. All hydrogen bond donors and acceptors buried inside must also have a hydrogen partner supplied by the protein itself. Unsatisfied hydrogen bonding abilities in the protein core are extremely destabilizing for tertiary structures.

For the analysis of surface properties of proteins, specialized and fast software tools are required. Traditionally, the van der Waals, the solvent-accessible (LEE and RICHARDS, 1971), and the molecular (CONNOLLY,

1983) surface are studied (Fig. 7). Recently, efficient techniques for computing both van der Waals and solvent-accessible surfaces of proteins have been published (EISENHABER and ARGOS, 1993; EISENHABER et al., 1995a). Programm ASC and NSC are available via WWW (http://www.embl-heidelberg.de/ ~eisenhab/).

Regions of the protein surface formed by hydrophobic atoms are unable to interact with surrounding water molecules in a similarly strong manner as polar atomic groups. Therefore, exposed hydrophobic surface patches are generally destabilizing for the protein and are often sites of binding subunits or other ligands. Hence, it is desirable to have objective criteria for locating hydrophobic surface patches. But paradoxically, the solvent-accessible surface as defined by LEE and RICHARDS (1971) which is traditionally used for the analysis of solvation properties of proteins is not informative for the determination of hydrophobic surface clusters. The hydrophobic part of the solvent-accessible surface of a typical monomeric globular protein consists of a single and large interconnected region formed from faces of apolar atoms and constituting about 60% of the solvent-accessi-

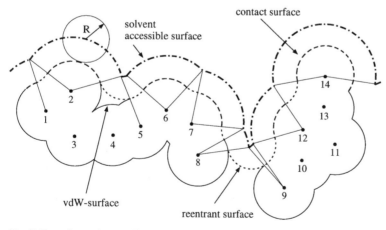

Fig. 7. Protein surfaces. The van der Waals surface (– – –) and the solvent-accessible surface (–·–·–) consist of pieces of spheres centred at atomic positions that are not occluded by neighboring spheres. In the first case, the radii are the so-called van der Waals radii, in the second case, they are incremented by the radius of a probe sphere modeling the solvent (usually the probe radius is 1.4 Å for water). The molecular surface is only in part identical with the van der Waals surface; the reentrant surface (– – – –) is formed by probe spheres at invaginations which are too small for allowing the probe to enter. The molecular surface surrounds the solvent-excluded volume.

ble surface area. Therefore, the direct delineation of hydrophobic surface patches on an atom-wise basis is impossible. Experimental data indicate that, in a 2-state hydration model, a protein can be considered as unified with its first hydration shell in its interaction with bulk water. It has been shown (EISENHABER and ARGOS, 1996) that, if the surface area occupied by water molecules bound at polar protein atoms is removed, only about two thirds of the hydrophobic part of the protein surface remain accessible to bulk solvent. Moreover, the organization of the hydrophobic part of the solvent-accessible surface experiences a drastic change such that the single interconnected hydrophobic region disintegrates into many smaller patches; i.e., the physical definition of a hydrophobic surface region as non-occupied by 1st hydration shell water molecules distinguishes between hydrophobic surface clusters and small interconnecting channels. The formation of hydrophobic surface regions owing to the structure of the first hydration shell can be computationally simulated by a small radial increment of solvent-accessible polar atoms (0.35–0.5 Å), followed by calculation of the remaining exposed hydrophobic patches.

Based on the area distribution of hydrophobic surface regions, a surface energy value of 18 ± 2 cal mol^{-1} Å^{-2} was obtained which compares favorably with the parameters for carbon obtained by other authors who use the crystal geometry of succinic acid or energies of transfer from hydrophobic solvent to water for small organic compounds (EISENHABER, 1996). Thus, the transferability of atomic solvation parameters for hydrophobic atoms to macromolecules has been directly demonstrated.

The solvation energy of a protein in an aqueous environment includes several components:

- the hydrogen bond formation with polar groups of the macromolecule,
- the entropy change of water molecules due to binding with polar groups or their release into bulk water,
- the cavity formation due to the solute excluded volume,

- non-valent interaction of non-hydrogen bonded water molecules with protein atoms at the surface, and
- the polarization of bulk water (and of the volume inside the macromolecule) and the changes of salt density.

Whereas the first 4 components can, in a first approximation, be linearly related to surface properties of the protein (short-range part), the last contribution is of electrostatic nature (long-range part) and requires, in principle, an integration over the whole volume of solute and solvent which is the equivalent solution of the Poisson–Boltzmann differential equation (PBE). It is not possible to reduce the whole solvation energy to a simple surface term (JUFFER et al., 1995, 1996).

5.1.3 Protein Structure Comparison and Structural Families

Protein tertiary structure comparison is necessary to elucidate topologically equivalent regions, to determine structural differences in space and to find insertions/deletions in one structure relative to others. To date, superimposing 3D protein structures provide the most sensitive technique for recognizing very distant relationships between amino acid sequences with low residue identities up to only 2% (HOLM and SANDER, 1996). It was one of the surprising news after a larger number of protein structures has been resolved with X-ray crystallography that many protein domains were visually similar above the regularities due to secondary structural constraints. The need for objective clarification of this similarity was felt by many researchers, and a variety of definitions of structural similarity and algorithms for protein structure comparison was proposed. The methods differ in the following:

1. Is the main emphasis put on global, overall coincidence or on the search of only local structural similarities?
2. Does the algorithm require that all structural elements occur in the same sequential

order or is it sensitive to sequential interchanges?

3. Are the structures compared up to atomic detail or is the algorithm oriented on a comparison of higher order structural blocks?

As shown by GODZIK (1996), the quantitative results of structural comparisons depend highly on the criterion applied. Indeed, the three best known classifications of protein structures

- SCOP (http://scop.mrc-lmb.cam.ac.uk/scop/index.html),
- CATH (http://www.biochem.ucl.ac.uk/bsm/cath/CATHintro.html), and
- FSSP (http://swift.embl-heidelberg.de/fssp/)

differ drastically even at higher hierarchical levels. This is just a reflection of the fact that the principles of tertiary structure formation are not well understood, and subjective choices such as mutual $C\alpha$ distance metrics or scaffolds of secondary structural elements are used as basis of the classifications. The situation is similar to that of Linné when he started to introduce systematics into botany based on the number of anthers in flowers.

Comparison techniques have been reviewed in detail by EISENHABER et al. (1995b). Here, we give a short outline of the variety of algorithms. The classical measure of distance between two structures is the global r.m.s.d. of the distance between equivalent $C\alpha$ atoms after spatial superposition. Early comparison techniques involve rigid body superpositions with the subjective assignment of initial equivalences. For largely divergent structures, window comparison techniques relying on successive oligopeptide superpositions along the sequence have been applied to locate similar substructures. Structure alignment methods based on dynamic programing have been developed to handle rigorously variable length gaps in the alignment (TAYLOR and ORENGO, 1989; ZUKER and SOMORJAI, 1989; SALI and BLUNDELL, 1990). The sensitivity of the technique has been enhanced by including hydrogen bonding, solvent exposure, torsional angles, secondary structural assignments and the like in addition to α-carbon distances. Dynamic programing cannot directly incorporate sequentially nonlocal effects, consequently, multiple levels of dynamic programing (TAYLOR and ORENGO, 1989; ORENGO and TAYLOR, 1990), self-consistency tests for suboptimal alignments (LUO et al., 1993), or stochastic optimization via genetic algorithms (MAY and JOHNSON, 1994) have been used.

Algorithms relying on two-dimensional plots of pairwise $C\alpha$ distances (or hydrogen bonds or main chain dihedral angle matches) do not depend on the sequential order of structural blocks. Parts of these plots can be compared to search for similar substructures or patterns (BARTON and STERNBERG, 1988; RICHARDS and KUNDROT, 1988; VRIEND and SANDER, 1991; HOLM and SANDER, 1993). The direct comparison of substructures (e.g., hexapeptides) and the search for the langest common homologous domain is a variant of this approach (BACHAR et al., 1993; ALEXANDROV and GO, 1994).

On the other hand, emphasis can be put on overall topological equivalence of secondary structural blocks rather than on detailed atomic correspondence (EISENHABER et al., 1995b). Characteristic patterns of secondary structure are much more robust to structural changes than individual amino acid positions, and even mutations often do not destroy the overall topology of the main chain. Simplified representations of building blocks in the form of vectors, e.g., along the axes of secondary structural elements, can be compared. In graph theoretic approaches, protein structural elements and their relations are coded in the form of nodes and edges.

Analyses of both known three-dimensional protein structures and amino acid sequences revealed that proteins are clustered into families whose members may have evolved from a common ancestor, share a characteristic fold and, sometimes, have a similar function (PASCARELLA and ARGOS, 1992; HOLM et al., 1992; YEE and DILL, 1993; ORENGO et al., 1993; HOLM and SANDER, 1994b; LESSEL and SCHOMBURG, 1994; SOWDHAMINI et al., 1996). Some authors even think that the total number of different folds may be in the range of a few thousand (for critical review, see EI-

SENHABER et al., 1995b). It must be emphasized that the number of structural families depends critically on the value of the homology threshold applied in the routine comparison of protein structures, and structural similarity is not structural identity. For pairs of distantly related proteins (residue identity ~20%), the regions with the same general fold comprise less than half of each molecule and the r.m.s. deviation between equivalent main chain atoms is 1.8–2.5 Å (CHOTHIA and LESK, 1986). The distance error can be in the range of up to 5 Å for $C\alpha$ atoms of proteins with about 25% residue identity (CHELVA-NAYAGAM et al., 1994). The equivalent secondary structural units may be shifted relative to each other by as much as 7 Å with rotations up to 30° (LESK and CHOTHIA, 1980; CHELVANAYAGAM et al., 1994). Pairwise residue–residue contacts may be conserved to only 12%, solvent accessibilities and secondary structures can be maintained to less than 40% (RUSSEL and BARTON, 1994). Thus, even if the scaffold of secondary structures is similar, the physical nature of stabilizing forces can be entirely different.

5.2. Prediction and Modeling of Tertiary Structure

Theoretical approaches to tertiary structure modeling are classified into 3 main streams.

(1) Attempts to predict the protein topology directly from sequence data alone based on the knowledge of intramolecular interactions are unified under the name *ab initio* approach. Since these techniques have not yet proven successful, other methods involving additional structural data on similar proteins have been developed.
(2) Threading tries to identify a suitable fold for the query sequence among those already known.
(3) Homology modeling includes techniques for fitting a new sequence into the known structure of another protein.

Solvent accessibility of residues is an important characteristic for the location of resi-dues in the tertiary structure. Methods developed for secondary structure prediction have also found application for this purpose.

5.2.1 Computation of Protein Structures Based on Fundamental Physical Principles (*ab initio* Approach)

The fundamental physical approach to the protein folding problem (predicting the tertiary fold from the amino acid sequence alone) relies on the hypothesis on the native protein structure as a minimum of free energy; i.e., the native protein structure corresponds to a system at thermodynamic equilibrium with a minimum of free energy. *In vitro* renaturation experiments strongly support this view (ANFINSEN, 1973; CREIGHTON, 1992) since they imply that the complete information necessary for protein folding is comprised in the amino acid sequence. Thus, it would be sufficient to compute an ensemble of conformations representative for the state of lowest free energy. The conformational invariants of this ensemble (e.g., the densely packed protein core) are the characteristics of the native structure. In a more simplified approach, a unique conformation with the lowest sum of intramolecular potential energy, conformational entropy term and solvation free energy is considered to represent the native state. This view is not unchallenged.

The computational problem of finding the lowest energy conformation of a polypeptide chain from an energy function containing pairwise terms and possibly other expressions is NP-complete (NGO and MARKS, 1992; UNGER and MOULT, 1994; FRAENKEL, 1993). The contention of LEVINTHAL (1968) that proteins search only a tiny fraction of the conformational space and move into the lowest kinetically accessible free energy minimum appears much more likely in this context. First experimental evidence in support of this view has been provided recently. The α-lytic protease was shown to exist in two forms: an inactive, metastable intermediate and an active native structure. Both conformations are separated by a barrier with activation energy of about 27 kcal mol^{-1}. A catalyst which is normally covalently attached to the protein is necessary to complete folding of the intermediate "molten globule", a less compact state compared with the native conformation (BAKER et al., 1992a, b). In the case of serpins, a metastable kinetically trapped 5-stranded β-sheet conformation was found which only slowly rearranges to the native 6-stranded form (MOTTONEN et al., 1992). During the last years, the role of biological factors such as the peptidyl–prolyl–isomerase, disulfide isomerase and molecular chaperonins in controlling the kinetics of protein folding and subunit assembly has been discovered (GETHING and SAMBROOK, 1992; HARTL, 1994). Often, the protein tertiary structure is incorporated into a system of a higher organizational level, and requirements of the latter may shift the conformation from a low-energy state.

In the "weak thermodynamic" hypothesis of UNGER and MOULT (1994), evolutionary arguments are taken into consideration. An originally functional structure of a protein corresponding to a local minimum will drift towards the global free energy minimum due to the combined effect of random mutations and the constant selective pressure of evolution. Hiding the native structure behind a large energy barrier may also be a sophisticated variant of enzyme activity regulation. Probably, only those sequences fold into unique conformations, fold fast and fold *via* the all-or-none transition (with the release of substantial latent heat) that have a pronounced energy minimum which is sufficiently distinguished from all other conformational states in the energy spectrum (SALI et al., 1994).

Hence, the search for low-energy conformations of polypeptides is still a promising plan for prediction of 3D structure and function. Two prerequisites for this approach are necessary:

- an energy function for discriminating the weight of different conformations in the native ensemble and
- a procedure for efficient searching of the conformational space.

A detailed review of both aspects has been given by EISENHABER et al. (1995b).

Although 3 decades of enormous scientific efforts have been concentrated on the *ab initio* folding problem, a solution to compute the structure from sequence has not yet been developed. The hardest problem waiting for solution is an appropriate energy function for the discrimination of native and non-native conformations. Electrostatic interactions in aqueous solutions, the solvation energy, hydrogen bonding, and entropic effects are not sufficiently well described. The identification of low-energy conformations in the highly dimensional conformation space is also not a trivial task, although different techniques for conformational searches such as Monte Carlo, genetic algorithms, and molecular dynamics have achieved a high level of maturity.

The main applications of the *ab initio* approach are conformational searches in combination with experimental restraints. Packing requirements and covalent strain are sufficiently well modeled by existing energy functions. Together with experimental data about pairwise distances (from NMR, cross-linking studies, and the like) or with X-ray diffraction data, the search techniques are applied for structure refinement and the generation of conformations satisfying restraints.

5.2.2 Threading Amino Acid Sequences Through Structural Motifs

The "threading" approach makes the far-reaching assumption that the query sequences under study might accept one of the protein folds already studied by X-ray crystallography or NMR. The structure prediction problem is thus greatly simplified since the allowable conformational space is reduced to about 100–300 unique protein topologies presently known. The primary goal of a "threading" method is to establish relationships between amino acid sequences and folding patterns, i.e., to select the most probable fold for a given sequence or to recognize suitable sequences that might fold into a given structure.

This approach has been stimulated by 3 observations:

(1) The number of different folds in the PDB grows more slowly than the number of new protein structures (notion of structural families).
(2) Distant relationships between sequences may be found by alignment to property profiles (see Sect. 3.3.2).
(3) Empirical potential functions for estimating solvation can distinguish incorrect folds.

As introduced by BRYANT and LAWRENCE (1993), "threading" a sequence through a fold implies a specific alignment between the amino acids of the sequence under consideration and the residue positions of the folding motif. The known structure establishes a set of possible amino acid positions in the three-dimensional space (the tertiary template) characterized physically by solvent accessibility, types and number of residue–residue contacts, backbone conformation and the like. The query sequence is made similar to the structure by placing its amino acids into their aligned positions and by taking into account the propensity of different amino acid types, oligopeptide fragments or residue pairs for a given physical environment. The recognition of sequence and structure is mediated by a suitable score or potential function for the evaluation of each alignment. The methods described in the literature vary

(1) in the derivation of the score function and
(2) in the alignment procedure for a single sequence with a single structure.

The technical details have been reviewed (EISENHABER et al., 1995b).

Threading has seen only a few real cases of competent application. Its efficiency in recognizing new distantly related homologues is low. Standard multiple sequence alignment methods or profile analysis are computationally cheaper and most often have the same predictive power. Threading methods will probably fail if the evolutionary divergence has removed most of the sequence similarity, if parts of the backbone have significantly moved and if secondary structural elements are inserted or deleted despite preservation of a similar basic fold pattern. This happens even within protein families (RUSSEL and BARTON, 1994). Such relative ineffectiveness is unexpected since the consideration of additional structural information should favor threading compared with a simple sequence pattern search. The crude formulation of the potential function compared even with that used in *ab initio* techniques and the NP-completeness of the alignment procedure appear responsible for this result. If threading is considered a problem of statistical hypothesis testing, it can be shown that the parameters currently used for structure description such as pairwise potentials differ if the learning set of tertiary structures is varied (S. SUNYAEV and F. EISENHABER, unpublished results).

5.2.3 Homology Modeling

The so-called modeling by homology can be applied if a protein with a given amino acid sequence is known (or supposed) to have a three-dimensional structure very similar to that of other proteins from the structural database. The unknown tertiary structure is produced by copying conserved parts of the structure (usually secondary structural elements) and fitting loop regions relying on the construction principles of close packing and hydrophobic-hydrophilic discrimination at the protein surface. The algorithm of homology modeling involves the following principal steps:

(1) Structurally conserved regions (SCR, the "tertiary template") are found on the basis of 3D structural comparisons and/or multiple sequence alignments within the protein family.
(2) The tertiary template (set of spatial positions of residues) must be aligned with the amino acid sequence that is a putative member of the same family. This step represents usually a multiple alignment with other sequences of the family or a profile analysis.

(3) Given the 3D-1D alignment, the new backbone of the protein being modeled is constructed. *Ab initio* modeling techniques are applied for constructing the loops which are usually the structurally variable regions (SVR).

(4) The conformations of the side chains anchored at the new backbone are placed with *ab initio* or knowledge-based methods.

(5) The structure proposal is finally subjected to several cycles of energy refinement and the check of verification criteria (control of strain, packing, and solvation energy based on stereochemical knowledge and tertiary structure database statistics).

The simple procedures in the early years requiring repeated human intervention have been gradually replaced by more sophisticated and generally automated techniques. In fact, there is not so much to predict. Since the fold is known, only local structural details need be tuned to comply with energetic and/or database criteria.

Another, conceptionally different approach is based on distance geometry and related algorithms. In this perspective, the tertiary template restrictions are translated into distance restraints which are used as input for distance geometry programs (HAVEL and SNOW, 1991; TAYLOR, 1993; SALI and BLUNDELL, 1993). This technique allows to integrate a variety of experimental information that can be used to formulate conformational restraints (SALI, 1995). The number of distance restraints sufficient for reproducing the protein structure correctly was also studied thoroughly (YCAS, 1990; OSHIRO et al., 1991; SIBBALD, 1995).

Technical details of homology modeling have been reviewed in great detail by JOHNSON et al. (1994) and EISENHABER et al. (1995b). The modeling error depends largely on the sequence identity between the query sequence and the known protein. If it is 40% or more, about 90% of the backbone atoms can be expected at a root-meansquare deviation of ≈ 1 Å. Side chain placement is usually worse. Below 40% sequence identity, misalignments with the target sequence become a major problem as well as the positioning of

large structurally variable regions. As a result, the error rate increases drastically.

Site directed mutagenesis aimed at changing physical and chemical properties of proteins (e.g., engineering enhanced thermostability) is a specific application for homology modeling methods since, as a rule, only a few amino acids are changed. The conformational space to be searched is, therefore, not very large and enumeration techniques can be applied. With a well refined, closely homologous structure as a starting point, the model can achieve accuracies in the range of 1 Å (VRIEND and EIJSINK, 1993; DE FILLIPIS et al., 1994).

5.2.4 Prediction of Solvent Accessibility

The solvent accessibility of residues is a major tertiary property. The knowledge which residues form the protein core and which residues are located at the surface significantly reduces the possible conformations accessible for a query sequence. YCAS (1990) has estimated the distance of an amino acid residue from the protein midpoint from its hydrophobicity.

Methods similar to those developed for secondary structure prediction and based on multiple sequence alignments have been applied to this problem. Neural network approaches to the prediction of amino acid accessibility have been described (HOLBROOK et al., 1990; ROST and SANDER, 1994a). Another method is based on environment specific amino acid substitution tables (WAKO and BLUNDELL, 1994b). Because of the lower conservation of accessibility within protein families (RUSSEL and BARTON, 1994), the improvement in prediction accuracy from the use of multiply aligned sequences is not as large as in the case of secondary structure prediction, and the correlation between the predicted and observed accessibilities is only in the range of 0.36–0.77 for different sets of sequences (ROST and SANDER, 1994a).

6 Quarternary Structures of Proteins

6.1 Phenomenology of Quarternary Structural Features

Many proteins exist in form of complexes of several polypeptide chains (called subunits or protomers), since the regulation of their function probably requires many types of interactions and binding sites. The subunits may be identical or different in sequence. The protein may be dimeric, trimeric or even a higher order aggregate, though dimers and tetramers are the most frequent combinations (JONES and THORNTON, 1996). Generally, each subunit is expected to fold into an apparently independent tertiary structure and to have its own hydrophobic core. This is not always the case; e.g., two polypeptide chains are intimately intertwined in the dimeric *trp*-repressor and in the *met*-aporepressor. The quarternary structure as a higher organizational level introduces its own requirements on the tertiary structures which may again result in conformational strain. For example, metabolic energy was found to be necessary for accurate folding, for correct disulphide bond formation and for maintaining influenza hemagglutinin in its oligomerization-competent state (BRAAKMAN et al., 1992). Sometimes, the quarternary structure is not even unique and depends on pH and salt concentration (HUANG et al., 1996).

It is known that a significant part of the subunit surface in multimeric proteins and complexes is shielded from contact with the solvent (ARGOS, 1988; JANIN et al., 1988; MILLER, 1989; JANIN and CHOTHIA, 1990). The typical surface buried by one partner in a subunit contact is about 600–1000 Å^2 with 55–70% non-polar (JANIN and CHOTHIA, 1990). The interfaces are generally more similar to the interior of proteins than to water-exposed surfaces and involve often large hydrophobic surface regions.

Contacting surfaces between subunits show a high degree of geometrical complementarity (on the level of van der Waals or molecular surfaces) and are closely packed, although intersubunit (and interdomain) cavities (packing defects) are commonly larger than inside single-domain proteins (HUBBARD and ARGOS, 1994). In some cases, these packing defects have a functional role and allow relative motions of domains and, probably, also subunits (HUBBARD and ARGOS, 1996). The types of interactions at protein–protein interfaces have been a subject of detailed investigations (JONES and THORNTON, 1996), and a database of protein interfaces is also available (TSAI et al., 1996). Complementarity between docking partners has also been observed with respect to the electrostatic potential energy (HONIG and YANG, 1995).

The hydration shell structure changes during ligand association. Bound water molecules have to be removed from the interface before the macromolecular contact can happen. This effect is responsible for repulsion at a distance of about 10 Å between the docking partners (LECKBAND et al., 1994).

6.2 Prediction of Protein–Protein Docking

The prediction of protein–protein docking is one aspect of the general problem of ligand binding by proteins. The various docking algorithms proposed in the literature try to utilize the properties of docking complexes described in the previous section and can be classified as follows:

(1) shape complementarity based techniques,
(2) approaches using solvation properties of interfaces, and
(3) methods developed for *ab initio* structure simulation.

The first group of algorithms puts major emphasis on close packing at the subunit interface. Both molecules are considered as rigid bodies and the level of complementarity at different mutual orientations is computed in a systematic or heuristic manner (LASKOWSKI et al., 1996; SOBOLEV et al., 1996). Alternatively, sections of the surface with pronounced shape complementarity can be searched (CONNOLLY, 1992).

Since the hydrophobic effect may be considered a driving force in protein association (DILL, 1990), it is desirable to elaborate techniques for the consideration of solvation in protein aggregation (KORN and BURNETT, 1991; YOUNG et al., 1994; COVELL et al., 1994; JACKSON and STERNBERG, 1995). Besides purely surface oriented algorithms aiming at burying as many hydrophobic patches as possible and making polar groups accessible to solvent (NAUCHITEL et al., 1995), electrostatic energy calculations have been applied (JACKSON and STERNBERG, 1995; WENG et al., 1996). Since the evaluation of the Poisson-Boltzmann equation is computationally time-consuming, effective charges for each of the macromolecules are proposed. The volume of the docking partner having also a low dielectric permittivity is ignored at the initial calculation (GABDOULLINE and WADE, 1996).

Whereas *ab initio* simulation techniques with full atomic detail and variable side chain conformations are extremely computer-time consuming if applied to docking problems (TOTROV and ABAGYAN, 1994), low-resolution studies relying even only on Cα-atom positions are already sufficient to predict the correct side of contact between subunits (VAKSER, 1996a, b).

The crystal structure of the complex of TEM-1 β-lactamase (262 residues) with one of its inhibitors (BLIP, 165 residues) was used as a large-scale test for various docking algorithms (STRYNADKA et al., 1996). The structures of both individual molecules were made known to the researchers. It is remarkable that all algorithms produced a solution with the correct overall mode of BLIP binding to the TEM-1 β-lactamase (association at the active site of the enzyme). At the same time, even a search with full atomic detail and an energy function of *ab initio* techniques as attempted by TOTROV and ABAGYAN (1994) as well as more simple approaches were not able to predict details of the interface like side chain rearrangements, correct residue–residue contacts or hydrogen bonds. Thus, the gross matching of molecular shapes is sufficient to yield an approximate docking solu-

tion. Further details are outside the scope of recently developed methods mainly due to the weakness of the energy function which has to discriminate between correct and wrong types of docking complexes.

7 Concluding Remark

The field of protein structure analysis and prediction has received an exciting development during the last three decades. The major challenge, cracking the protein folding puzzle, is still unsolved. Nevertheless, a wealth of valuable information in form of sequences and structures of proteins has been accumulated in databases and many algorithms able to predict structural and functional features of proteins have been developed. The necessity to rely on prediction techniques will even grow in the future with the successful realization of genome projects since many proteins will be known only in form of amino acid sequences. A wide field of activity has opened for applied research of practitioners who attempt the selection and modification of proteins for medical or biotechnological applications.

Acknowledgement
The authors thank JOACHIM SELBIG (GMD St. Augustin-Bonn) for critical reading of the manuscript.

8 References

ABOLA, E. E., BERNSTEIN, F. C., BRYANT, S. H., KOETZLE, T. F., WENG, J. (1987), Protein data bank, in: *Crystallographic Databases – Information Content, Software Systems, Scientific Applications* (ALLEN, F. H., BERGERHOFF, G., SIEVERS, R., Eds.), pp. 107–132. Bonn, Cambridge, Chester: Data Commission of the International Union of Crystallography.

ADZHUBEI, A. A., STERNBERG, M. J. E. (1993), Left-handed polyproline II helices commonly occur in globular protein, *J. Mol. Biol.* **229**, 472–493.

ADZHUBEI, A. A., STERNBERG, M. J. E. (1994), Conservation of polyproline II helices in homologous proteins: implications for structure prediction by model building, *Protein Sci.* **3**, 2395–2410.

ADZHUBEI, A. A., LAUGHTON, C. A., NEIDLE, S. (1995), An approach to protein homology modelling based on an ensemble of NMR structures: application to the Sox-5 HMG-box protein, *Protein Eng.* **8**, 615–625.

ALEXANDROV, N. N., GO, N. (1994), Biological meaning, statistical significance, and classification of local spatial similarities in non-homologous proteins, *Protein Sci.* **3**, 866–875.

ALTSCHUL, S., BOGUSKI, M., GISH, W., WOOTTON, J. C. (1994), Issues in searching molecular sequence databases, *Nature Genetics* **6**, 119–129.

ANFINSEN, C. B. (1973), Principles that govern the folding of protein chains, *Science* **181**, 223–230.

ARGOS, P. (1988), An investigation of protein subunit and domain interfaces, *Protein Eng.* **2**, 101–113.

BACHAR, O., FISCHER, D., NUSSINOV, R., WOLFSON, H. (1993), A computer vision based technique for 3-D sequence-independent structural comparison of proteins, *Protein Eng.* **6**, 279–288.

BAIROCH, A., BUCHER, P. (1994), PROSITE: recent developments, *Nucleic Acids Res.* **22**, 3583–3859.

BAKER, D., SOHL, J. L., AGARD, D. A. (1992a), A protein-folding reaction under kinetic control, *Nature* **356**, 263–265.

BAKER, D., SOHL, J. L., AGARD, D. A. (1992b), Protease Pro region required for folding is a potent inhibitor of the mature enzyme, *Proteins* **12**, 339–344.

BARLOW, D. J., THORNTON, J. M. (1988), Helix geometry in proteins, *J. Mol. Biol.* **201**, 601–619.

BARTON, G. J., STERNBERG, M. J. E. (1988), LOPAL and SCAMP: techniques for the comparison and display of protein structures, *J. Mol. Graph.* **6**, 190–196.

BERMAN, A. L., KOLKER, E., TRIFONOV, E. N. (1994), Underlying order in protein sequence organization, *Proc. Natl. Acad. Sci. USA* **91**, 4044–4047.

BIOU, V., GIBRAT, J. F., LEVIN, J. M., ROBSON, B., GARNIER, J. (1988), Secondary structure prediction: combination of three different methods, *Protein Eng.* **2**, 185–191.

BIRNEY, E., THOMPSON, J. D., GIBSON, T. J. (1996), Pairwise and Searchwise: finding the optimal alignment in a simultaneous comparison of a protein profile against all DNA translation frames, *Nucleic Acids Res.* **24**, 2730–2739.

BLUNDELL, T. L., ZHU, Z.-Y. (1995), The α-helix as seen from the protein tertiary structure: a 3-D structural classification, *Biophys. Chem.* **55**, 167–184.

BORK, P. (1996), Sperm-egg binding protein or proto-oncogene? *Science* **271**, 1431–1432.

BORK, P., BAIROCH, A. (1996), Go hunting in sequence databases but watch out for the traps, *Trends Genet.* **12**, 425–427.

BORK, P., GIBSON, T. J. (1996), Applying motif and profile searches, *Methods Enzymol.* **266**, 162–184.

BORK, P., GELLERICH, J., GROTH, H., HOOFT, R., MARTIN, F. (1995), Divergent evolution of a β/α-barrel subclass: detection of numerous phosphate-binding sites by motif search, *Protein Sci.* **4**, 268–274.

BORK, P., DOWNING, A. K., KIEFER, B., CAMPBELL, I. D. (1996), Structure and distribution of modules in extracellular proteins, *Q. Rev. Biophys.* **29**, 119–167.

BÖRNER, G. V., PÄÄBO, S. (1996), Evolutionary fixation of RNA editing, *Nature* **383**, 225.

BRAAKMAN, I., HELENIUS, J., HELENIUS, A. (1992), Role of ATP and disulphide bonds during protein folding in the endoplasmic reticulum, *Nature* **356**, 260–262.

BRUNET, A. P., HUANG, E. S., HUFFINE, M. E., LOEB, J. E., WELTMAN, R. J., HECHT, M. H. (1993), The role of turns in the structure of an α-helical protein, *Nature* **364**, 355–358.

BRYANT, S. H., LAWRENCE, C. E. (1993), An empirical energy function for threading protein sequence through the folding motif, *Proteins* **16**, 92–112.

BRYSON, J. W., BETZ, S. F., LU, H. S., SUICH, D. J., ZHOU, H. X., O'NEIL, K. T., DEGRADO, W. F. (1995), Protein design: a hierarchic approach, *Science* **270**, 935–941.

BULT, C. J., WHITE, O., OLSON, G. J., ZHOU, L., FLEISCHMANN, R. D., SUTTON, G. G., BLAKE, J. A., FITZGERALD, L. M., CLAYTON, R. A., GOCAYNE, J. D., KERVELAGE, A. R., DOUGHERTY, B. A., TOMB, J.-F., ADAMS, M. D., REICH, C. I., OVERBEEK, R., KIRKNESS, E. F., WEINSTOCK, K. G., MERRICK, J. M., GLODEK, A., SCOTT, J. L., GEOGHAGEN, N. S. M., WEIDMAN, J. F., FUHRMANN, J. L., NGYUEN, D., UTTERBACK, T. R., KELLEY, J. M., PETERSON, J. D., SADOW, P. W., HANNA, M. C., COTTON, M. D., ROBERTS, K. M., HURST, M. A., KAINE, B. P., BORODOWSKY, M., KLENK, H.-P., FRASER, C. M., SMITH, H. O., WOESE, C. R., VENTER, J. C. (1996), Complete genome sequence of the methanogenic archeon, *Methanococcus jannaschii*, *Science* **273**, 1058–1073.

BURSET, M., GUIGO, R. (1996), Evaluation of gene structure prediction programs, *Genomics* **34**, 353–367.

CARREL, R. W., STEIN, P. E., FERMI, G., WARDELL, M. R. (1994), Biological implications of a 3 Å structure of dimeric antithrombin, *Structure* **2**, 257–270.

CASARI, G., ANDRADE, M. A., BORK, P., BOYLE, J., DARUVAR, A., OUZOUNIS, C., SCHNEIDER, R., TAMAMES, J., VALENCIA, A., SANDER, C. (1994), Challenging times for bioinformatics, *Nature* **376**, 647–648.

CERPA, R., COHEN, F. E., KUNTZ, I. D. (1996), Conformational switching in designed peptides: the helix/sheet transition, *Folding Design* **1**, 91–101.

CHAN, A. W. E., HUTCHINSON, E. G., HARRIS, D., THORNTON, J. M. (1993), Identification, classification, and analysis of beta-bulges in proteins, *Protein Sci.* **2**, 1574–1590.

CHELVANAYAGAM, G., ROY, G., ARGOS, P. (1994), Easy adaptation of protein structure to sequence, *Protein Eng.* **7**, 173–184.

CHOTHIA, C., LESK, A. M. (1986), The relation between the divergence of sequence and structure in proteins, *EMBO J.* **5**, 823–826.

CHOU, P. Y. (1989), Prediction of protein structural classes from amino acid composition, in: *Prediction of Protein Structure* (FASMAN, G. D., Ed.), pp. 549–586. New York: Plenum Press.

CHOU, K.-C. (1995), A novel approach to predicting protein structural classes in a (20-1)-D amino acid composition space, *Proteins* **21**, 319–344.

CHOU, P. Y., FASMAN, G. (1974a), Conformational parameters for amino acids in helical, beta-sheet, and random coil regions calculated from proteins, *Biochemistry* **13**, 211–222.

CHOU, P. Y., FASMAN, G. (1974b), Prediction of protein conformation, *Biochemistry* **13**, 222–245.

CHOU, K.-C., ZHANG, C.-T. (1995), Prediction of protein structural classes, *CRC Crit. Rev. Biochem. Mol. Biol.* **30**, 275–349.

COLLOCH, N., ETCHEBEST, C., THOREAU, E., HENRISSAT, B., MORNON, J.-P. (1993), Comparison of three algorithms for the assignment of secondary structure in proteins: the advantage of a consensus assignment, *Protein Eng.* **6**, 377–382.

CONNOLLY, M. L. (1983), Analytical molecular surface calculation, *J. Appl. Cryst.* **16**, 548–558.

CONNOLLY, M. L. (1992), Shape distributions of protein topography, *Biopolymers* **32**, 1215–1236.

COOPER, A. A., STEVENS, T. H. (1995), Protein splicing: self-splicing of genetically mobile elements at the protein level, *Trends Biochem. Sci.* **20**, 351–356.

COVELL, D. G., SMYTHERS, G. W., GRONENBORN, A. M., CLORE, M. G. (1994), Analysis of hydrophobocity in the a and b chemokine families and its relevance to dimerization, *Protein Sci.* **3**, 2064–2072.

CREIGHTON, T. E. (1992), Protein Folding. New York: Freeman.

DAFFNER, C., CHELVANAYAGAM, G., ARGOS, P. (1994), Structural characteristics and stabilizing principles of bent beta-strands in protein tertiary structures, *Protein Sci.* **3**, 876–882.

DE FILLIPIS, V., SANDER, C., VRIEND, G. (1994), Predicting local structural changes that result from point mutations, *Protein Eng.* **7**, 1203–1208.

DILL, K. A. (1990), Dominant forces in protein folding, *Biochemistry* **29**, 7133–7155.

DOIG, A. J., BALDWIN, R. L. (1995), N- and C-capping preferences for all 20 amino acids in α-helical peptides, *Protein Sci.* **4**, 1325–1336.

DOMBI, G. W., LAWRENCE, J. (1994), Analysis of protein transmembrane helical regions by a neural network, *Protein Sci.* **3**, 557–566.

DORAN, J. D., CAREY, P. R. (1996), α-helix dipoles and catalysis: absorption and raman spectroscopic studies of acyl cysteine proteases, *Biochemistry* **35**, 12495–12502.

DUBCHAK, I., HOLBROOK, S. R., KIM, S.-H. (1993), Prediction of protein folding class from amino acid composition, *Proteins* **16**, 79–91.

EFREMOV, R. G., VERGOTEN, G. (1996a), Recognition of transmembrane α-helical segments with environmental profiles, *Protein Eng.* **9**, 253–263.

EFREMOV, R. G., VERGOTEN, G. (1996b), Hydrophobic organization of α-helix membrane bundle in bacteriorhodopsin, *J. Protein Chem.* **15**, 63–76.

EISENHABER, F. (1996), Hydrophobic regions on protein surfaces. Derivation of the solvation energy from their area distribution in crystallographic protein structures, *Protein Sci.* **5**, 1676–1686.

EISENHABER, F., ARGOS, P. (1993), Improved strategy in analytic surface calculation for molecular systems: handling of singularities and computational efficiency, *J. Comp. Chem.* **14**, 1272–1280.

EISENHABER, F., ARGOS, P. (1996), Hydrophobic regions on protein surfaces. Definition based on hydration shell structure and a quick method for region computation, *Protein Eng.* **9**, 1121–1133.

EISENHABER, F., LIJNZAAD, P., ARGOS, P., SANDER, C., SCHARF, M. (1995a), The double cubic lattice method: efficient approaches to numerical integration of surface area and volume and for generating dot surfaces of molecular assemblies, *J. Comp. Chem.* **16**, 273–284.

EISENHABER, F., PERSSON, B., ARGOS, P. (1995b), Protein structure prediction: recognition of primary, secondary, and tertiary structural features from amino acid sequence, *CRC Crit. Rev. Biochem. Mol. Biol.* **30**, 1–94.

EISENHABER, F., IMPERIALE, F., ARGOS, P., FRÖMMEL, C. (1996a), Prediction of secondary structural content of proteins from their amino acid composition alone. I. New vector decomposition methods, *Proteins* **25**, 157–168.

EISENHABER, F., FRÖMMEL, C., ARGOS, P. (1996b), Prediction of secondary structural content of proteins from their amino acid composition alone. II. The paradox with secondary structural class, *Proteins* **25**, 169–179.

FLEISCHMANN, R. D., ADAMS, M. D., WHITE, O., CLAYTON, R., KIRKNESS, E. F., KERLAGE, A. R., BULT, C. J., TOMB, J.-F., DOUGHERTY, B. A., MERRICK, J. M., McKENNEY, K., SUTTON, G., FITZHUGH, W., FIELDS, C., GOCAYNE, J. F., SCOTT, J., SHIRLEY, R., LIU, L.-I., GLODEK, A., KELLEY, J. M., WEIDMAN, J. F., PHILLIPS, C. A., SPRIGGS, T., HEDBLOM, E., COTTON, M. D., UTTERBACK, T. R., HANNA, M. C., NGUYEN, D. T., SAUDEK, D. M., BRANDON, R. C., FINE, L. D., FRITCHMAN, J. L., FUHRMANN, J. L., GEOGHAGEN, N. S. M., GNEHM, C. L., McDONALD, L. A., SMALL, K. V., FRASER, C. M., SMITH, H. O., VENTER, J. C. (1995), Whole-genome random sequencing and assembly of *Haemophilus influenzae* Rd, *Nature* **269**, 496–512.

FRAENKEL, A. S. (1993), Complexity of protein folding, *Bull. Math. Biol.* **55**, 1199–1210.

FRISHMAN, D. I., ARGOS, P. (1992), Recognition of distantly related protein sequences using conserved motifs and neural networks, *J. Mol. Biol.* **228**, 951–962.

FRISHMAN, D., ARGOS, P. (1995), Knowledge-based protein secondary structure assignment, *Proteins* **23**, 566–579.

FRISHMAN, D., ARGOS, P. (1996), Incorporation of non-local interactions in protein secondary structure prediction from the amino acid sequence, *Protein Eng.* **9**, 133–142.

GABDOULLINE, R. R., WADE, R. C. (1996), Effective charges for macromolecules in solvent, *J. Phys. Chem.* **100**, 3868–3878.

GARNIER, J., LEVIN, J. M. (1991), The protein structure code: what is its present status? *Comput. Appl. Biosci.* **7**, 133–142.

GARNIER, J., OSGUTHORPE, D. J., ROBSON, B. (1978), Analysis of the accuracy and implications of simple methods for predicting the secondary structure of globular proteins, *J. Mol. Biol.* **120**, 97–120.

GEOURJON, C., DELÉAGE, G. (1994), SOPM: a self-optimized method for protein secondary structure prediction, *Protein Eng.* **7**, 157–164.

GESTELAND, R. F., ATKINS, J. F. (1996), Recording: dynamic reprogramming of translation, *Annu. Rev. Biochem.* **65**, 741–768.

GETHING, M. J., SAMBROOK, J. (1992), Protein folding in the cell, *Nature* **355**, 33–45.

GIBRAT, J.-F., GARNIER, J., ROBSON, B. (1987), Further developments of protein secondary structure prediction using information theory. New parameters and consideration of residue pairs, *J. Mol. Biol.* **198**, 425–443.

GIBRAT, J.-F., ROBSON, B., GARNIER, J. (1991), Influence of the local amino acid sequence upon the zones of the torsional angles phi and psi adopted by residues in proteins, *Biochemistry* **30**, 1578–1586.

GODZIK, A. (1996), The structural alignment between two proteins: is there a unique answer? *Protein Sci.* **5**, 1325–1338.

GOLDSMITH, E. J., MOTTONEN, J. (1994), Serpins: the uncut version, *Structure* **2**, 241–244.

GONG, Y., ZHOU, H. X., GUO, M., KALLENBACH, N. R. (1995), Structural analysis of the N- and C-termini in a peptide with consensus sequence, *Protein Sci.* **4**, 1446–1456.

HAN, K.-K., MARTINAGE, A. (1992), Possible relationship between coding recognition of amino acid sequence motif or residue(s) and posttranslational chemical modification of proteins, *Int. J. Biochem.* **24**, 1349–1363.

HANKE, J., BECKMANN, G., BORK, P., REICH, J. G. (1996), Self-organizing hierarchic networks for pattern recognition in protein sequence, *Protein Sci.* **5**, 72–82.

HARTL, F. U. (1994), Secrets of a double-doughnut, *Nature* **371**, 557–559.

HAVEL, T., SNOW, M. E. (1991), A new method for building protein conformations from sequence alignments with homologues of known structure, *J. Mol. Biol.* **217**, 1–7.

HERINGA, J., ARGOS, P. (1993), A method to recognize distant repeats in protein sequences, *Proteins* **17**, 391–411.

HIGGINS, D., THOMPSON, J. D., GIBSON, T. J. (1996), Using CLUSTAL for multiple sequence alignment, *Methods Enzymol.* **266**, 383–402.

HOLBROOK, S. R., MUSKAL, S. M., KIM, S.-H. (1990), Predicting surface exposure of amino acids from protein sequence, *Protein Eng.* **3**, 659–665.

HOLM, L., SANDER, C. (1993), Protein structure comparison by alignment of distance matrices, *J. Mol. Biol.* **233**, 123–138.

HOLM, L., SANDER, C. (1994a), Searching protein structure databases has come of age, *Proteins* **19**, 165–173.

HOLM, L., SANDER, C. (1994b), The FSSP database of structurally aligned protein fold families, *Nucleic Acids Res.* **22**, 3600–3609.

HOLM, L., SANDER, C. (1996), Mapping the protein universe, *Science* **273**, 595–602.

HOLM, L., OUZOUNIS, C., SANDER, C., TUPAREV, G., VRIEND, G. (1992), A database of protein structure families with common folding motifs, *Protein Sci.* **1**, 1691–1698.

HONIG, B., YANG, A.-S. (1995), Free energy balance in protein folding, *Adv. Protein Chem.* **46**, 27–58.

HUANG, K., STRYNADKA, N. C. J., BERNARD, V. D., PEANASKY, R. J., JAMES, M. N. G. (1994), The molecular structure of the complex of *Ascaris* chymotrypsin/elastase inhibitor with porcine elastase, *Structure* **2**, 679–689.

HUANG, D.-B., AINSWORTH, C. F., STEVENS, F. J., SCHIFFER, M. (1996), Three quarternary structures for a single protein, *Proc. Natl. Acad. Sci. USA* **93**, 7017–7021.

HUBBARD, S. J., ARGOS, P. (1994), Cavities and packing at protein interfaces, *Protein Sci.* **3**, 2194–2206.

HUBBARD, S. J., ARGOS, P. (1995), Detection of internal cavities in globular proteins, *Protein Eng.* **8**, 1011–1015.

HUBBARD, S. J., ARGOS, P. (1996), A functional role for protein cavities in domain:domain motions, *J. Mol. Biol.* **261**, 289–300.

HUBBARD, S. J., GROSS, K.-H., ARGOS, P. (1994), Intramolecular cavities in globular proteins, *Protein Eng.* **7**, 613–626.

HUCHO, F., GÖRNE-TSCHELNOKOW, U., STRECKER, A. (1994), β-structure in the membrane-spanning part of the nicotinic acetylcholine receptor (or how helical are transmembrane helices?), *Trends Biochem. Sci.* **19**, 383–387.

ISLAM, S. A., LUO, J., STERNBERG, M. J. E. (1995), Identification and analysis of domains in proteins, *Protein Eng.* **8**, 513–525.

JACKSON, R. M., STERNBERG, M. J. E. (1995), A continuum model for protein-protein interactions: application to the docking problem, *J. Mol. Biol.* **250**, 258–275.

JANIN, J., CHOTHIA, C. (1990), The structure of protein-protein recognition site, *J. Biol. Chem.* **265**, 1627–1630.

JANIN, J., MILLER, S., CHOTHIA, C. (1988), Surface, subunit interfaces and interior of oligomeric proteins, *J. Mol. Biol.* **204**, 155–164.

JOHNSON, W. C., JR. (1990), Protein secondary structure and circular dichroism: a practical guide, *Proteins* **7**, 205–214.

JOHNSON, M. S., SRINIVASAN, N., SOWDHAMINI, R., BLUNDELL, T. L. (1994), Knowledge-based protein modelling, *CRC Crit. Rev. Biochem. Mol. Biol.* **29**, 1–68.

JONES, S., THORNTON, J. M. (1996), Principles of protein-protein interactions, *Proc. Natl. Acad. Sci. USA,* **93**, 13–20.

JUFFER, A., EISENHABER, F., HUBBARD, S. J., WALTHER, D., ARGOS, P. (1995), Comparison of atomic solvation parameter sets: applicability and limitations in protein folding and binding, *Protein Sci.* **4**, 2499–2509.

JUFFER, A. H., EISENHABER, F., HUBBARD, S. J., WALTHER, D., ARGOS, P. (1996), Erratum: Comparison of atomic solvation parameter sets: applicability and limitations in protein folding and binding, *Protein Sci.* **5**, 1748–1749.

KABSCH, W., SANDER, C. (1983), Dictionary of protein secondary structures: pattern recognition of hydrogen-bonded and geometrical features, *Biopolymers* **22**, 2577–2637.

KARPLUS, P. A. (1996), Experimentally observed conformation-dependent geometry and hidden strain in proteins, *Protein Sci.* **5**, 1406–1420.

KLEIN, P., DELISI, C. (1986), Prediction of protein structural class from the amino acid sequence, *Biopolymers* **25**, 1659–1672.

KNELLER, D. G., COHEN, F. E., LANGRIDGE, R. (1990), Improvements in protein secondary structure prediction by enhanced neural networks, *J. Mol. Biol.* **214**, 171–182.

KOLKER, E., TRIFONOV, E. N. (1995), Periodic recurrence of methionines: fossil of gene fusion? *Proc. Natl. Acad. Sci. USA* **92**, 557–560.

KOONIN, E. V., MUSHEGIAN, A. R., BORK, P. (1996), Non-orthologous gene displacement, *Trends Genet.* **12**, 334–336.

KORN, A. P., BURNETT, R. M. (1991), Distribution and complementarity of hydropathy in multisubunit proteins, *Proteins* **9**, 37–55.

KROGH, A., BROWN, M., MIAN, I. S., SJÖLANDER, K., HAUSSLER, D. (1994), Hidden Markov models in computational biology, *J. Mol. Biol.* **235**, 1501–1531.

KÜHLBRANDT, W., WANG, D. N., FUJIYOSHI, Y. (1994), Atomic model of plant light-harvesting complex by electron crystallography, *Nature* **367**, 614–621.

KWASIGROCH, J.-M., CHOMILIER, J., MORNON, J.-P. (1996), A global taxonomy of loops in globular proteins, *J. Mol. Biol.* **259**, 855–872.

LASKOWSKI, R. A., THORNTON, J. M., HUMBLET, C., SINGH, J. (1996), X-SITE: use of empirically derived atomic packing preferences to identify favourable interaction regions in the binding sites of proteins, *J. Mol. Biol.* **259**, 175–201.

LECKBAND, D. E., SCHMITT, F.-J., ISRAELACHVILI, J. N., KNOLL, W. (1994), Direct force measurements of specific and nonspecific protein interactions, *Biochemistry* **33**, 4611–4624.

LEE, B., RICHARDS, F. M. (1971), The interpretation of protein structures: estimation of static accessibility, *J. Mol. Biol.* **55**, 379–400.

LESK, A. M., CHOTHIA, C. (1980), How different amino acid sequences determine similar protein structures: the structure and evolutionary dynamics of the globins, *J. Mol. Biol.* **136**, 225–270.

LESSEL, U., SCHOMBURG, D. (1994), Similarities between protein 3-D structures, *Protein Eng.* **7**, 1175–1187.

LEVIN, J. M., GARNIER, J. (1988), Improvements in a secondary structure prediction method based on a search for local sequence homologies and its use as a model building tool, *Biochim. Biophys. Acta* **955**, 283–295.

LEVIN, J. M., PASCARELLA, S., ARGOS, P., GARNIER, J. (1993), Quantification of secondary structure prediction improvement using multiple alignment, *Protein Eng.* **6**, 849–854.

LEVINTHAL, C. (1968), Are there pathways for protein folding? *J. Chem. Phys.* **65**, 44–45.

LEVITT, M., CHOTHIA, C. (1976), Structural patterns in globular proteins, *Nature* **261**, 552–558.

LIANG, C., MISLOW, K. (1994a), Topological chirality of proteins, *J. Am. Chem. Soc.* **116**, 3588–3592.

LIANG, C., MISLOW, K. (1994b), Topological features of chorionic gonadotropin, *Biopolymers* **35**, 343–345.

LUO, Y., LAI, L., XU, X., TANG, Y. (1993), Defining topological equivalences in protein structures by means of a dynamic programming algorithm, *Protein Eng.* **6**, 373–376.

LUPAS, A. (1996), Coiled coils: new structures and new functions, *Trends Biochem. Sci.* **21**, 375–382.

LUPAS, A., VAN DYKE, M., STOCK, J. (1991), Predicting coiled coils from protein sequence, *Science* **252**, 1162–1164.

MAKEEV, V. J., TUMANYAN, V. G. (1996), Search of periodicities in primary structure of biopolymers: a general Fourier approach, *Comput. Appl. Biosci.* **12**, 49–54.

MANSFIELD, M. L. (1994), Are there knots in proteins? *Nature Struct. Biol.* **1**, 213–214.

MAO, B. (1993), Topological chirality of proteins, *Protein Sci.* **2**, 1057-1059.

MARTIN, A. C. R., TODA, K., STIRK, H. J., THORNTON, J. M. (1995), Long loops in proteins, *Protein Eng.* **8**, 1093–1101.

MAY, A. C. W., JOHNSON, M. S. (1994), Protein structure comparisons using a combination of a genetic algorithm, dynamic programming and least-squares minimization, *Protein Eng.* **7**, 475–485.

MAYO, K. H., ILYINA, E., PARK, H. (1996), A recipe for designing water-soluble, β-sheet-forming peptides, *Protein Sci.* **5**, 1301–1315.

MCLACHLAN, A. D. (1983), Gene duplications in the structural evolution of chymotrypsin, *J. Mol. Biol.* **128**, 49–79.

MEHTA, K. P., HERINGE, J., ARGOS, P. (1995), A simple and fast approach to prediction of protein secondary structure from multiply aligned sequences with accuracy above 70%, *Protein Sci.* **4**, 2517–2525.

MILLER, S. (1989), The structure of interfaces between subunits of dimeric and tetrameric proteins, *Protein Eng.* **3**, 77–83.

MILLHAUSER, G. L. (1995), Views of helical peptides: a proposal for the position of 3_{10}-helix along the thermodynamic folding pathway, *Biochemistry* **34**, 3873–3877.

MOTTONEN, J., STRAND, A., SYMERSKI, J., SWEET, R. M., DANLEY, R. E., GEOGHEGEN, K. F., GERARD, R. D., GOLDSMITH, E. J. (1992), Structural basis of latency in plasminogen activator inhibitor-1, *Nature* **355**, 270–273.

MUGGLETON, S., KING, R. D., STERNBERG, M. J. E. (1992), Protein secondary structure prediction using logic-based machine learning, *Protein Eng.* **5**, 647–657.

MUGGLETON, S., KING, R. D., STERNBERG, M. J. E. (1993), Corrigenda: protein secondary structure prediction using logic-based machine learning, *Protein Eng.* **6**, 549.

MURZIN, A. G., LESK, A. M., CHOTHIA, C. (1994a), Principles determining the structure of β-sheet barrels in proteins. I. A theoretical analysis, *J. Mol. Biol.* **236**, 1369–1381.

MURZIN, A. G., LESK, A. M., CHOTHIA, C. (1994b), Principles determining the structure of β-sheet barrels in proteins. II. The observed structures, *J. Mol. Biol.* **236**, 1382–1400.

NAKASHIMA, H., NISHIKAWA, K., OOI, T. (1986), The folding type of a protein is relevant to the amino acid composition, *J. Biochem.* **99**, 153–162.

NAUCHITEL, V., VILLAVERDE, M. C., SUSSMAN, F. (1995), Solvent accessibility as a predictive tool for the free energy of inhibitor binding to the HIV-1 protease, *Protein Sci.* **4**, 1356–1364.

NGO, T. J., MARKS, J. (1992), Computational complexity of a problem in molecular structure prediction, *Protein Eng.* **5**, 313–321.

NICHOLS, W. L., ROSE, G. D., TEN EYCK, L. F., ZIMM, B. H. (1995), Rigid domains in proteins: an algorithmic approach to their identification, *Proteins* **23**, 38–48.

NISHIKAWA, K., OOI, T. (1982), Correlation of the amino acid composition of a protein to its structural and biological characters, *J. Biochem.* **91**, 1821–1824.

NISHIKAWA, K., KUBOTA, Y., OOI, T. (1983a), Classification of proteins into groups based on amino acid composition and other characters. II. Grouping into four types, *J. Biochem.* **94**, 997–1007.

NISHIKAWA, K., KUBOTA, Y., OOI, T. (1983b), Classification of proteins into groups based on amino acid composition and other characters. I. Angular distribution, *J. Biochem.* **94**, 981–995.

ORENGO, C. A., TAYLOR, W. R. (1990), A rapid method of protein structure alignment, *J. Theor. Biol.* **147**, 517–551.

ORENGO, C. A., FLORES, T. P., TAYLOR, W. R., THORNTON, J. M. (1993), Identification and classification of protein fold families, *Protein Eng.* **6**, 485–500.

OSAWA, S., JUKES, T. H., WATANABE, K., MUTO, A. (1992), Recent evidence for evolution of the genetic code, *Microbiol. Rev.* **56**, 229–264.

OSHIRO, C. M., THOMASON, J., KUNTZ, I. D. (1991), The effects of limited input distance constraints upon the distance geometry algorithm, *Biopolymers* **31**, 1049–1064.

PASCARELLA, S., ARGOS, P. (1992), A data bank merging related protein structures and sequences, *Protein Eng.* **5**, 121–137.

PASTORE, A., SAUDEK, V., RAMPONI, G., WILLIAMS, R. J. P. (1992), Three-dimensional structure of acylphosphatase, *J. Mol. Biol.* **224**, 427–440.

PERCZEL, A., HOLLÓSI, M., TUSNÁDY, G., FASMAN, G. D. (1991), Convex constraint analysis: a natural deconvolution of circular dichroism curves of proteins, *Protein Eng.* **4**, 669–679.

PERSSON, B., ARGOS, P. (1994), Prediction of transmembrane regions in proteins utilising multiple sequence alignments, *J. Mol. Biol.* **237**, 182–192.

PERSSON, B., ARGOS, P. (1996), Topology prediction of membrane proteins, *Protein Sci.* **5**, 363–371.

PICOT, D., LOLL, P. J., GARAVITO, M. (1994), The X-ray crystal structure of the membrane protein prostaglandin H_2 synthase-1, *Nature* **367**, 243–249.

PRESTA, L. G., ROSE, G. D. (1988), Helix signals in proteins, *Science* **240**, 1632–1641.

QIAN, H., CHAN, S. I. (1996), Interactions between a helical residue and tertiary structures: helix propensities in small peptides and in native proteins, *J. Mol. Biol.* **261**, 279–288.

RAJASHANKAR, K. R., RAMAKUMAR, S. (1996), p-turns in proteins and peptides: classification, conformation, occurrence, hydration and sequence, *Protein Sci.* **5**, 932–946.

RECZKO, M., BOHR, H. (1994), The DEF data base of sequence based protein fold class predictions, *Nucleic Acids Res.* **22**, 3616–3619.

RECZKO, M., BOHR, H., SUBRAMANIAM, S., PAMIGIGHANTAM, S., HATZIGEORGIOU, A. (1994), Fold-class prediction by neural networks, in: *Protein Structure by Distance Analysis* (BOHR,

H., BRUNAK, S., Eds.), pp. 277–286. Amsterdam, Tokyo: IOS Press, Ohmsha.

RESH, M. D. (1994), Myristylation and palmitylation of Src family members: the fats of the matter, *Cell* **76**, 411–413.

RICHARDS, F. M., KUNDROT, C. E. (1988), Identification of structural motifs from protein coordinate data: secondary and first-level supersecondary structure, *Proteins* **3**, 71–84.

RICHARDS, F. M., LIM, W. A. (1994), An analysis of packing in the protein folding problem, *Q. Rev. Biophys.* **26**, 423–498.

RIEK, R., HORNEMANN, S., WIDER, G., BILLETER, M., GLOCKSHUBER, R., WÜTHRICH, K. (1994), NMR structure of the mouse prion protein domain PrP (121–231), *Nature* **382**, 180–182.

ROST, B., SANDER, C. (1993a), Secondary structure prediction of all-helical proteins in two states, *Protein Eng.* **6**, 831–836.

ROST, B., SANDER, C. (1993b), Prediction of protein secondary structure at better than 70% accuracy, *J. Mol. Biol.* **232**, 584–599.

ROST, B., SANDER, C. (1994a), Combining evolutionary information and neural networks to predict protein secondary structure, *Proteins* **19**, 55–72.

ROST, B., SANDER, C. (1994b), Conservation and prediction of solvent accessibility in protein families, *Proteins* **20**, 216–226.

ROST, B., CASADIO, R., FARISELLI, P., SANDER, C. (1995), Transmembrane helices predicted at 95% accuracy, *Protein Sci.* **4**, 521–533.

ROST, B., FARISELLI, P., CASADIO, R. (1996), Topology prediction for helical transmembrane proteins at 86% accuracy, *Protein Sci.* **5**, 1704–1718.

RUSSEL, R. B., BARTON, G. J. (1994), Structural features can be unconserved in proteins with similar folds. An analysis of side-chain to side-chain contacts, secondary structure and accessibility, *J. Mol. Biol.* **244**, 332–350.

SALAMOV, A. A., SOLOVYER, V. V. (1995), Prediction of protein secondary structure by combining nearest neighbor algorithms and multiple sequence alignments, *J. Mol. Biol.* **247**, 11–15.

SALI, A. (1995), Modelling mutations and homologous proteins, *Curr. Opin. Struct. Biol.* **6**, 437–451.

SALI, A., BLUNDELL, T. L. (1990), Definition of general topological equivalence in protein structures. A procedure involving comparison of properties and relationships through simulated annealing and dynamic programming, *J. Mol. Biol.* **212**, 403–428.

SALI, A., BLUNDELL, T. L. (1993), Comparative protein modelling by satisfaction of spatial restraints, *J. Mol. Biol.* **234**, 779–815.

SALI, A., SHAKHNOVICH, E., KARPLUS, M. (1994), Kinetics of protein folding. A lattice model study of the requirements for folding to the native state, *J. Mol. Biol.* **235**, 1614–1636.

SAMATEY, F. A., XU, C., POPOT, J.-L. (1995), On the distribution of amino acid residues in transmembrane α-helix bundles, *Proc. Natl. Acad. Sci. USA* **92**, 4577–4581.

SCHRAUBER, H., EISENHABER, F., ARGOS, P. (1993), Rotamers: to be or not to be? An analysis of amino acid side-chain conformations in globular proteins, *J. Mol. Biol.* **230**, 592–612.

SERRANO, L., SANCHO, J., HIRSHBERG, M., FERSHT, A. R. (1992), α-Helix stability in proteins. I. Empirical correlations concerning substitutions of side chains at the N and C-caps and the replacement of alanine by glycine or serine at solvent-exposed surfaces, *J. Mol. Biol.* **227**, 544–559.

SHCHERBAK, V. I. (1988), To co-operative symmetry of the genetic code, *J. Theor. Biol.* **132**, 121–124.

SHCHERBAK, V. I. (1989), The "START" and the "STOP" of the genetic code: why exactly ATG and TAG. TAA? *J. Theor. Biol.* **139**, 283–286.

SIBANDA, B. L., THORNTON, J. M. (1993), Accomodating sequence changes in b-hairpins in proteins, *J. Mol. Biol.* **229**, 428–447.

SIBBALD, P. R. (1995), Deducing protein structures using logic programming: exploiting minimum data of diverse types, *J. Theor. Biol.* **173**, 361–375.

SIDDIQUI, A. S., BARTON, G. J. (1995), Continuous and discontinuous domains: an algorithm for the automatic generation of reliable domain definitions, *Protein Sci.* **4**, 872–884.

SKLENAR, H., ETCHEBEST, C., LAVERY, R. (1989), Describing protein structure: a general algorithm yielding complete helicoidal parameters and a unique overall axis, *Proteins* **6**, 46–60.

SOBOLEV, V., WADE, R. C., VRIEND, G., EDELMAN, M. (1996), Molecular docking using surface complementarity, *Proteins* **25**, 120–129.

SOWDHAMINI, R., BLUNDELL, T. L. (1995), An automatic method involving cluster analysis of secondary structures for the identification of domains in proteins, *Protein Sci.* **4**, 506–520.

SOWDHAMINI, R., RUFINO, S. D., BLUNDELL, T. L. (1996), A database of globular protein structural domains: clustering of representative family members into similar folds, *Folding Design* **1**, 209–220.

SREERAMA, N., WOODY, R. W. (1994), Protein secondary structure from circular dichroism spectroscopy, *J. Mol. Biol.* **242**, 497–507.

STADTMAN, T. C. (1996), Selenocysteine, *Annu. Rev. Biochem.* **65**, 83–100.

STEVENS, R. C., GOUAUX, J. E., LIPSCOMB, W. N. (1990), Structural consequences of effector binding to the T state of aspartate carbamoyltransferase: crystal structure of the unligated and ATP- and CTP-complexed enzymes at 2.6-Å resolution, *Biochemistry* **29**, 7691–7701.

STRYNADKA, N. C. J., EISENSTEIN, M., KATCHALSKI-KATZIR, E., SHOICHET, B. K., KUNTZ, I. D., ABAGYAN, R., TOTROV, M., JANIN, J., CHERFILS, J., ZIMMERMANN, F., OLSON, A., DUNCAN, B., RAO, M., JACKSON, R., STERNBERG, M., JAMES, M. N. G. (1996), Molecular docking programs successfully predict the binding of a β-lactamase inhibitory protein to TEM-1β-lactamase, *Nature Struct. Biol.* **3**, 233–239.

SWINDELLS, M. B. (1995a), A procedure for the automatic determination of hydrophobic cores in protein structures, *Protein Sci.* **4**, 93–102.

SWINDELLS, M. B. (1995b), A procedure for detecting structural domains in proteins, *Protein Sci.* **4**, 103–112.

TAKUSAGAWA, F., KAMITORI, S. (1996), A real knot in protein, *J. Am. Chem. Soc.* **118**, 8945–8946.

TATUSOV, R. L., MUSHEGIAN, A. R., BORK, P., BROWN, N. P., HAYES, W. S., BORODOVSKY, M., RUDD, K. E., KOONIN, E. V. (1996), Metabolism and evolution of *Haemophilus influenzae* deduced from a whole-genome comparison with *Escherichia coli*, *Curr. Biol.* **6**, 279–291.

TAYLOR, W. R. (1993), Protein fold refinement: building models from idealized folds using motif constraints and multiple sequence data, *Protein Eng.* **6**, 593–604.

TAYLOR, W. R., ORENGO, C. A. (1989), Protein structure alignment, *J. Mol. Biol.* **208**, 1–22.

TOTROV, M., ABAGYAN, R. (1994), Detailed *ab initio* prediction of the lysozyme-antibody complex with 1.6 Å accuracy, *Nature Struct. Biol.* **1**, 259–263.

TSAI, C.-J., LIN, S. L., WOLFSON, H. J., NUSSINOV, R. (1996), Protein-protein interfaces: architectures and interactions in protein-protein interfaces and in protein cores. Their similarity and differences, *CRC Crit. Rev. Biochem. Mol. Biol.* **31**, 127–152.

UNGER, R., MOULT, J. (1994), Finding the lowest free energy conformation is an NP-hard problem: proof and implications, *Bull. Math. Biol.* **55**, 1183–1198.

VAKSER, I. A. (1996a), Main-chain complementarity in protein-protein recognition, *Protein Eng.* **9**, 741–744.

VAKSER, I. A. (1996b), Low-resolution docking: prediction of complexes for underdetermined structures, *Biopolymers* **39**, 455–464.

VAN HEEL, M. (1992), A new family of powerful multivariate statistical sequence analysis (MSSA) techniques, *J. Mol. Biol.* **216**, 877–887.

VON HEIJNE, G. (1986), The distribution of positively charged residues in bacterial inner membrane proteins correlates with transmembrane topology, *EMBO J.* **5**, 3021–3027.

VON HEIJNE, G. (1995), Membrane protein assembly: rules of the game, *Bioessays* **17**, 25–30.

VRIEND, G., EIJSINK, V. (1993), Prediction and analysis of structure, stability and unfolding of thermolysin-like proteases, *J. Comput. Aided Mol. Des.* **7**, 367–396.

VRIEND, G., SANDER, C. (1991), Detection of common three-dimensional substructures in proteins, *Proteins* **11**, 52–58.

VTYURIN, N. (1993), The role of tight packing of hydrophobic groups in β-structure, *Proteins* **15**, 62–70.

VTYURIN, N., PANOV, V. (1995), Packing constraints of hydrophobic side chains in $(\alpha/\beta)_8$ barrels, *Proteins* **21**, 256–260.

WAKO, H., BLUNDELL, T. L. (1994a), Use of amino acid environment-dependent substitution tables and conformational propensities in structure prediction from aligned sequences of homologues proteins. I. Solvent accessibility classes, *J. Mol. Biol.* **238**, 682–692.

WAKO, H., BLUNDELL, T. L. (1994b), Use of amino acid environment-dependent substitution tables and conformational propensities in structure prediction from aligned sequences of homologous proteins. II. Secondary structures, *J. Mol. Biol.* **238**, 693–708.

WALTHER, D., EISENHABER, F., ARGOS, P. (1996), Principles of helix-helix packing in proteins: the helical lattice superposition model, *J. Mol. Biol.* **255**, 536–553.

WEISS, M. S., SCHULZ, G. E. (1992), Structure of porin refined at 1.8 Å resolutiom, *J. Mol. Biol.* **227**, 493–509.

WENG, Z., VAJDA, S., DELISI, C. (1996), Prediction of protein complexes using empirical free energy functions, *Protein Sci.* **5**, 614–626.

WILMOT, C. M., THORNTON, J. M. (1990), β-Turns and their distortions: a proposed new nomenclature, *Protein Eng.* **3**, 479–493.

WOOTTON, J. C. (1994), Sequences with 'unusual' amino acid compositions, *Curr. Opin. Struct. Biol.* **4**, 413–421.

YANG, J. T. (1996), Prediction of protein secondary structure from amino acid sequence, *J. Protein Chem.* **15**, 185–191.

YCAS, M. (1990), Computing tertiary structures of proteins, *J. Protein Chem.* **9**, 177–200.

YEE, D. P., DILL, K. A. (1993), Families and the structural relatedness among globular proteins, *Protein Sci.* **2**, 884–899.

YOUNG, L., JERNIGAN, R. L., COVELL, D. G. (1994), A role for surface hydrophobicity in protein-protein recognition, *Protein Sci.* **3**, 717–729.

ZHANG, C.-T., CHOU, K.-C. (1995), An eigenvalue-eigenvector approach to predicting protein folding types, *J. Protein Chem.* **14**, 309–326.

ZHU, Z.-Y. (1995), A new approach to the evaluation of protein secondary structure predictions at the level of elements of secondary structure, *Protein Eng.* **8**, 103–108.

ZUKER, M., SOMORJAI, R. L. (1989), The alignment of protein structures in three dimensions, *Bull. Math. Biol.* **51**, 55–78.

3 Protein Interactions

MICHAEL MEYER

Jena, Germany

DIETMAR SCHOMBURG

Köln, Germany

1 Introduction 88
2 Biomolecular Interactions 88
3 Experimental Methods 90
  3.1 Calorimetry 90
  3.2 Analytical Ultracentrifugation 91
  3.3 Surface–Plasmon Resonance 92
  3.4 Protein Crystallography 92
  3.5 NMR Spectroscopy 92
4 Theoretical Methods 93
  4.1 Force Field Methods 93
  4.2 Continuum Models for Protein Interactions 94
  4.3 Protein Docking 94
      4.3.1 Protein–(Small)Ligand Docking and Ligand Design 95
      4.3.2 Docking of Proteins with Macromolecules 95
5 Protein Complexes 96
  5.1 Protein–(Small)Ligand Complexes 96
  5.2 Protein–Carbohydrate Interactions 98
  5.3 Protein–Steroid Complexes 98
  5.4 Protein–Protein Complexes 99
  5.5 Protein–Nucleic Acid Complexes 100
6 References 101

1 Introduction

As mentioned in Chapter 1, this volume, proteins are responsible for a large variety of tasks in biological systems, such as enzymatic catalysis of chemical reactions in metabolism, defence against infection, regulation and transport. For all these tasks highly specific or, in other cases, more general interactions with other molecules are necessary. Proteins achieve these by different surface motifs and a multitude of structural arrangements in the molecular interface.

The elucidation of biomolecular interactions is of steadily increasing importance. Exact knowledge of the principles governing the strengths and formation of molecular interactions is of highest importance for a huge number of applications in widely different areas. Only a small subset of these should be mentioned here:

The experimental determination, evaluation and/or prediction of protein interactions with other proteins, small molecules, sugars, or DNA is a requirement, e.g.,

– for the design of new drugs,
– for the design of proteins stable in different environments,
– for the understanding of cross-reactions of antibodies used in medical diagnosis or medical treatment,
– for the prediction of possible side-effects of drugs interacting with more than one protein,
– for the improvement of our understanding in diseases like BSE and Alzheimer's disease,
– for the understanding and regulation of biocatalytic activity (enzyme substrate specificity and inhibition),
– in the design of new materials based on proteins,
– understanding cell–cell communication and cell differentiation, etc.

Knowledge of the character of the interactions of a protein involves either the experimental determination of the full three-dimensional structure of the protein–second molecule complex plus the determination of thermodynamical association/dissociation constants. As an alternative of these experiments theoretical methods are currently developed that will allow the prediction of the structure of a protein complex and the strength of the interaction. This theoretical treatment – representing a highly active field of research in the last years – is an essential requirement for the prediction of the behavior of designed molecules before synthesis.

In the following section an overview of the principles of protein interactions is given and experimental and theoretical methods for their investigation are discussed. Some examples of protein complexes are presented in Sect. 5. To cope with the wide topic, we have concentrated on basic principles and supplemented them with results reported recently. Furthermore, reviews covering special aspects in detail are mentioned in the text.

2 Biomolecular Interactions

As is true for the description of most molecular properties a rigorous quantum mechanical treatment would be necessary in principle. As this is far beyond the possibilities of present computers several different approaches are used for a quantitative description of the strengths of protein interactions. The strengths, reflected in the association constant, depend on the difference in free energy between the interacting and the free solvated molecules. According to presently most popular force field approaches, the non-bonding energy E_{nb} for a complex consisting of a protein and another molecule may be written as a sum of terms for the Coulomb E_{Cb} and van der Waals energy E_{vdW}. Hydrogen bonds contribute also to the non-bonding energy.

$$E_{nb} = E_{Cb} + E_{vdW} = \sum_{i<j} \frac{q_i q_j}{4\pi\varepsilon\varepsilon_0 r_{ij}}$$
$$+ \sum_{i<j} E_{ij}\left(\frac{d_{ij}}{r_{ij}^{12}} - 2\frac{d_{ij}}{r_{ij}^{6}}\right) \quad (1)$$

The **long-range electrostatic energy** E_{Cb} is given by Coulomb's law, where q_i and q_j are the partial charges of the atoms i and j separated by the distance r_{ij}. In Eq. (1) ε_0 is the vacuum permittivity, and the dielectric constant ε describes the influence of a homogenous medium on the electrostatic energy. The dielectric constant ε is 1 for a vacuum, approximately 4 in a typical protein and 78.5 for water. The sum runs over all pairs of atoms in the general case and for the interaction between two non-bonded molecules over all pairs on the different molecules. This is an area of much discussion in the literature (BAYLY et al., 1993) as the right choice of these constants may greatly influence the results of the calculation.

Short-range non-bonded attraction and repulsion energies between molecules are summarized as van der Waals energy, which is frequently computed with a Lennard–Jones 6, 12 potential. The negative attraction term has its origin in the interaction between dipoles. It can be shown that the energies for dipole–dipole interaction between stationary dipole moments, the interaction between stationary and induced dipole moments and the interaction between induced dipole moments are proportional to the inverse 6th power of the distance between them. The latter interaction is also called London or dispersion energy. The repulsion from the electron shells rises drastically with decreasing distance. It can be calculated with an expression proportional to the inverse 12th power of the distance. E_{ij} is the depth of the van der Waals energy minimum at the distance d_{ij} between two atoms.

Eq. (1) is generally sufficient for the description of **hydrogen bonds,** but special expressions have been developed to improve the quality of the calculation. The hydrogen bond is mainly an electrostatic interaction between a partially positive charged hydrogen atom and a partially negative charged acceptor. Hydrogen bonds $D-H \cdots A$ occur when a hydrogen atom H is bound to an electronegative donor D and when it is in the vicinity of an electronegative acceptor A. Hydrogen bonds called salt bridges are formed between two oppositely charged residues, and so their energetical contribution is especially strong. Typical criteria for hydrogen bonds are a maximum distance $r(H \cdots A)$ of 2.5 Å and angles $\sphericalangle(D,H,A)$, $\sphericalangle(D,A,AA)$ and $\sphericalangle(H,A,AA)$ greater than 90° (BAKER and HUBBARD, 1984). AA symbolizes other atoms bound to the acceptor. Only a small number of oxygen or nitrogen atoms in proteins is not involved in hydrogen bonds (MCDONALD and THORNTON, 1994). This indicates the importance of hydrogen bonds for the secondary structure and formation of protein complexes (MEYER et al., 1996). In addition to hydrogen bonds with oxygen and nitrogen atoms, other types of hydrogen bonding have been proposed with a probably smaller energetic contribution. Examples involve hydrogen–sulfur interaction or aromatic rings as acceptors (LEVITT and PERUTZ, 1988) and $CH \cdots O$ hydrogen bonds (DEREWENDA et al., 1995).

Hydrophobic interactions cause an association of non-polar groups or molecules which arises from the tendency of water to exclude non-polar molecules. When a non-polar molecule is transferred into an aqueous medium the net of hydrogen bonds between the water molecules is partially destroyed, a reorientation of the water molecules at the surface of the solute takes place, and new hydrogen bonds are formed. The Gibbs free energy ΔG for a transfer of a molecule from the gas state into water may be used as a definition to describe the energetical aspects of these interactions associated with hydration (PRIVALOV and MAKHATADZE, 1993), but other reference states like the pure liquid state have also been proposed. Hydrophobic interactions play an important role for molecular recognition of host–guest complexes (BLOKZIJL and ENGBERTS, 1993) and hydrophobic interactions probably present the strongest driving force for protein interactions. An analysis of the hydrophobicity of amino acids at protein surfaces revealed that a hydrophobic cluster appears predominantly in the contact region of protein–protein complexes (YOUNG et al., 1994), and the free energy of complex formation has been related to the buried surface area (CHOTIA and JANIN, 1975; EISENBERG and MCLACHLAN, 1986). There is indeed a good correlation between measured free energies of association and the computed decrease in accessible surface area if structural changes associated with complex formation are small, but for those cases with an induced fit the correlation is rather poor (HORTON and LEWIS, 1992) which may be due to energy consumption needed for the rearrangement of the protein during complex formation.

3 Experimental Methods

Binding constants and thermodynamical data for protein complexes provide information about the affinity of proteins and their ligands, e.g., different new designed enzyme ligands can be compared in a quantitative way or specific and non-specific protein–DNA-binding can be distinguished. X-ray diffraction and NMR spectroscopy supplement thermodynamic studies, and molecular mechanisms can be explained with these methods at the atomic level. Several examples of protein complexes are discussed in Sect. 5 to illustrate applications of these methods.

3.1 Calorimetry

Calorimetric methods provide information about thermodynamic parameters and binding constants of a complex. The equilibrium between a protein P and a ligand L and the corresponding protein–ligand complex PL may be written as $P + L \rightleftharpoons PL$ for a 1:1 stoichiometry. The binding constant $K_b = [P][L]/[PL]$ ($[P]$ = concentration of P, etc.), and its inverse, the dissociation constant $K_d = 1/K_b$, are connected with the free energy of binding ΔG.

$$\Delta G = \Delta H - T\Delta S = -RT \ln K_b \qquad (2)$$

R denotes the gas constant and T the absolute temperature. A negative ΔG is required for spontaneous association. The enthalpy ΔH of binding can be evaluated in an indirect way from the temperature dependence of the binding constant. According to the van't Hoff

equation (Eq. 3) a plot of $\ln K_b$ vs. $1/T$ is a straight line with a slope of $\Delta H/R$.

$$\left(\frac{\partial \ln K_b}{\partial(1/T)} \right)_p = -\frac{\Delta H}{R} \qquad (3)$$

However, NAGHIBI et al. (1995) pointed out, that this method for the calculation of ΔH can lead to substantially different results compared to the direct calorimetric measurement of ΔH.

Binding reactions are often studied by isothermal titration calorimetry (FISCHER and SINGH, 1995). Small amounts of a solved protein are mixed with a solution of the reactant and the resulting temperature difference is measured with respect to a reference cell. The electrical energy required to compensate for the temperature difference is measured as a function of time to give the amount of heat absorbed or evolved in the binding reaction. An analysis of the titration curve yields the stoichiometry and the enthalpy of binding, the equilibrium-binding constant and the free energy of binding, which are model-dependent, except for the enthalpy. The heat q

$$q = V\Delta H [PL] \qquad (4)$$

measured in a calorimeter is proportional to the concentration of the bound ligand $[PL]$, which can be written as

$$[PL] = [P]_t \frac{K_b[L]}{1 + K_b[L]} \qquad (5)$$

where $[P]_t = [P] + [PL]$ is the total protein concentration. The binding enthalpy can then be derived with the cell volume V from Eq. (4).

The change of the heat capacity Δc_p follows from Eq. (6), if ΔH is constant over a small temperature range $T - T_0$.

$$\Delta H_T = \Delta H_{T_0} + \Delta c_p(T - T_0) \qquad (6)$$

For protein–nucleic acid complexes Δc_p depends on the interaction type, which can be specific or non-specific. In the first case the protein binds to a particular DNA sequence, in the latter case the binding is independent of the sequence. For example, TAKEDA et al.

(1992) determined $\Delta G = -67.4$ kJmol^{-1} (-16.1 kcal mol^{-1}), $\Delta H = 3.3$ kJmol^{-1} (0.8 kcal mol^{-1}), $\Delta S = 0.25$ kJmol^{-1} K^{-1} (59 cal mol^{-1} K^{-1}) and $\Delta c_p = -1.51$ kJmol^{-1} K^{-1} (-360 cal mol^{-1} K^{-1}) for the specific association of Cro protein with the high-affinity operator DNA OR3. For an unspecific interaction $\Delta G = -41$ kJmol^{-1} (-9.7 kcal mol^{-1}), $\Delta H = 18$ kJmol^{-1} (4.4 kcal mol^{-1}), $\Delta S = 0.21$ kJmol^{-1} K^{-1} (49 cal mol^{-1} K^{-1}) and $\Delta c_p \simeq 0$ Jmol^{-1} K^{-1} were determined. These associations are entropy driven since ΔH is positive, and only the entropic term $T\Delta S$ contributes to the negative change of the free energy. A Δc_p close to zero for non-specific interactions and a large negative value for specific interactions have also been derived for other cases of protein–nucleic acid interactions.

On the other hand the studies of antibody–lysozyme complexes (HIBITTS et al., 1994; TELL et al., 1993) show that this type of association is enthalpy driven, opposite to some enzyme–inhibitor complexes of trypsin, α-chymotrypsin and subtilisin (HIBITTS et al., 1994). Δc_p of association adopts always a negative value for these examples. A large negative change of the heat capacity has been interpreted sometimes as a consequence of a large apolar surface burial as a consequence of complex formation, but a recent comparison has shown a poor correlation between this interpretation of thermochemical data and known experimental structures of protein complexes, which is probably a consequence of the neglect of reorganization phenomena of water molecules (SINGHA et al., 1996).

3.2 Analytical Ultracentrifugation

Ultracentrifugation allows direct measurements of the mass M, the diffusion coefficient D and the sedimentation coefficient s of macromolecules. With an appropriate data analysis the determination of binding constants and ΔG is possible. An ultracentrifuge can be used for two types of experiments, sedimentation velocity measurements at high rotor speed and sedimentation equilibrium experiments at low speed.

Basically, the macromolecules at a radial distance r move to the bottom of the centrifugal cell under the influence of a centrifugal force $F = M\omega r^2$ proportional to the angular velocity ω, and the molecules build up a concentration gradient $\partial c / \partial r$, which is zero prior to the centrifugation.

The sedimentation velocity v is given by the Svedberg relation (Eq. 7).

$$v = s\omega r^2 = \frac{M(1 - \bar{v}\varrho)}{Nf} \omega r^2 \qquad (7)$$

The sedimentation velocity depends on the molecular mass M, the molecular size and shape via the friction coefficient f and the density of the solvent ϱ. The buoyant force of the displaced solvent compensates a fraction of the centrifugal force. In Eq. (8) f_s is the macromolecular flux caused by the centrifugal force,

$$f_s = vc = s\omega r^2 c \qquad (8)$$

and the flux f_D caused by diffusion in the opposite direction of f_s may be written as

$$f_D = -D\frac{\partial c}{\partial r} \qquad (9)$$

For a sedimentation equilibrium both fluxes are of identical magnitude and the molecules do not move, except for self-diffusion. The concentration as a function of the radius may be written apart from a baseline term as

$$c(r) = c(a)\exp[\sigma]$$
$$\sigma = M(1 - \bar{v}\varrho)\omega^2(r^2 - a^2)/2R \qquad (10)$$

for a single component system, where $c(r)$ and $c(a)$ are the concentrations at radius r and at the meniscus a. Eq. (10) gives information about the molecular mass M. For a dimerization reaction $P + P \rightleftharpoons PP$ Eq. (10) may be generalized

$$c(r) = c_P(a)\exp[\sigma_P] + c_{PP}(a)\exp[\sigma_{PP}] \qquad (11)$$

and with the binding constant $K_b = [P][P]/[PP]$. Eq. (11) may be rewritten

$$c(r) = c_P(a) \exp[\sigma_P] + \frac{c_P^2(a)}{K_b} \exp[\sigma_{PP}] \qquad (12)$$

to determine the binding constant with an appropriate data analysis (BROOKS et al., 1994).

3.3 Surface–Plasmon Resonance

Surface–plasmon resonance provides information about the kinetics and the binding constants of protein complexes. The experimental setup consists basically of a small flow cell mounted on a sensor chip with an immobilized reactant. A solution of the other reactant flows along the cell and upon formation of a complex the change of the refractive index is monitored. The binding constant K_b can be evaluated from corresponding rate constants of binding and dissociation, but it has been shown recently, that the heterogeneity of the immobilized site may lead to deviations from the expected pseudo-first-order kinetics for 1:1 stoichiometry (O'SHANNESSY and WINZOR, 1996). The method has been used to compare the binding affinities of cholera toxin with a series of gangliosides, which have been assembled on the sensor chip giving an artificial surface similar to a cell membrane (KUZIEMKO et al., 1996).

3.4 Protein Crystallography

The fact that X rays are scattered by electrons facilitates the computation of an electron density map and a subsequent model building of the molecular structure. For the measurement single crystals are required because the scattering of an individual protein is much too weak to be observed. Protein single crystals grow from a supersaturated solution and the determination of optimal crystallization conditions is one of the difficult steps of an X ray determination of protein complexes. It is often required to test empirically a wide range of conditions, such as pH, sample concentration and added precipitants. Protein crystals contain a high portion of water so that small ligands can diffuse through solvent channels into existing crystals, but large ligands can only be co-crystallized. This is also the only feasible method if major structural changes are necessary for complex formation (DUCRUIX and GIEGÉ, 1992). Also the formation of comparably weakly associated molecules can be observed by protein crystallography.

From an initial analysis of diffraction images information about the unit cell dimensions and space group can be derived. The measured intensity of the scattered X rays and a detailed assignment of the diffraction pattern are required to determine the phases by isomorphous replacement or molecular replacement techniques, since phase information cannot be determined by experiment. Then an electron density map can be computed by means of Fourier synthesis, and the interpretation of this map leads to an initial model of the structure, which has to be refined successively. Fourier maps based on the amplitude difference between reflections of the protein and the protein–ligand complex are very helpful for the investigation of structural changes upon binding of a small ligand and the locating of the binding site difference. A detailed description of crystallographic structure determination procedures is available in the textbooks of MCREE (1993), DRENTH (1994), and GLUSKER et al. (1994).

Some thousands of protein structures determined by protein crystallography and NMR spectroscopy have been deposited at the Brookhaven Protein Database (PDB), which can be accessed on the internet at *http://www.pdb.bnl.gov/*. Similarly, more than 180,000 small molecule structures, some of which are relevant as protein ligands and as reference molecules for theoretical methods, have been deposited at the Cambridge Structural Database (*http://www.ccdc.cam.ac.uk/*).

3.5 NMR Spectroscopy

NMR spectroscopy can be used for noncrystallizable biological macromolecules, and the solution conditions can be varied in a broad range, which may coincide with physiological conditions. On the other hand protein crystallography is the method of choice for large complexes, which cannot be investigated by NMR techniques.

Structural studies with NMR techniques require the presence of atoms with nuclear magnetic moments, such as ^{1}H, ^{13}C, ^{15}N, and ^{31}P. Basically, the shielding of nuclei by different chemical surroundings leads to absorptions at different energies in NMR experiments relative to a reference (chemical shift). The resonance lines may be split independently of an exterior magnetic field, if magnetically different atoms interact with each other through 3 or more bonds. A special case of such coupling, the vicinal amide–proton spin–spin coupling with Cα protons (^{3}J$_{HN\alpha}$-coupling), is torsional-angle dependent and gives information about α-helices in proteins. The nuclear Overhauser effect (NOE), a dipolar through space interaction observable in multidimensional NMR experiments, is of particular importance for protein structure determination. The NOE is proportional to the inverse 6th power of the distance of the interacting ^{1}H nuclei, and neighboring hydrogen atoms can thus be detected. Apart from 3D or 4D heteronuclear triple-resonance experiments with ^{13}C and ^{15}N labeled proteins, NOEs in combination with ^{3}J$_{HN\alpha}$-coupling form the basis of a sequence-specific assignment of small peptide fragments within a protein, which is achieved by comparison with the known amino acid sequence. Furthermore, close hydrogen atom pairs belonging to amino acids with a large distance in sequence can be identified and distance constraints can be set up, from which the structures are built-up (WÜTHRICH, 1995). NOEs are also a tool for the identification of interacting groups in a complex. In a complex the typical NOEs of the two components are found plus additional signals that originate from hydrogen atoms close in space but belonging to different molecules. The final step is the structural refinement, refined by using distance geometry or simulated annealing techniques.

4 Theoretical Methods

NMR spectroscopy and protein crystallography provide the basis of structural biology.

Theoretical methods use data from these methods and go one step further. They make the prediction of protein interactions of yet non-existing molecules possible or allow the guided setup of experiments on protein association. The structure-based design of new protein ligands has emerged as a powerful tool to support the development of new pharmaceutical products. An overview of structure-based theoretical methods is given in Sects. 4.1–4.3.

4.1 Force Field Methods

Energies of large biomolecules are often computed using empirical force fields like AMBER (CORNELL et al., 1995), CHARMM (BROOKS et al., 1983), MMFF94 (HALGREN, 1996) or MM3 (LI and ALLINGER, 1991) with parabolic energy terms for bond stretching and angle bending, a Fourier series for the torsional energy and the non-bonding energy E_{nb} (Eq. 1).

$$E = \sum_{bonds} \frac{1}{2} K_b (r - r_0)^2 + \sum_{angles} \frac{1}{2} K_\theta (\theta - \theta_0)^2$$
$$+ \sum_{torsions} \frac{1}{2} K_T [1 + \cos(n\tau - \delta)] + E_{nb} \qquad (13)$$

In practice force field calculations are carried out at $\varepsilon = 1$ with explicit solvent molecules or with a distant-dependent electric $\varepsilon = nr$ (n often between 1 and 4) to take solvent screening into account. The calculation of E_{nb} is time-consuming for large protein complexes since the summation scales with the square of the number of atoms. For computational efficiency contributions to the Coulomb energy are usually neglected outside a certain cut-off radius. A faster and more accurate variant for large complexes is the cell multipole approach (MATHINOWETZ et al., 1994), which divides the space into a hierarchy of cells. A pairwise calculation of the electrostatic energy is carried out for all atoms within a cell and its neighbors, but the electrostatic interaction between atoms in distant cells is based on a multipole expansion.

Force-field type energy calculations are the basis of molecular dynamics (MD) simulations (VAN GUNSTEREN and BERENDSEN, 1990), which are carried out to derive the macroscopic behavior of substances from their microscopic interactions. Potential biochemical applications include the docking of a ligand into a known binding pocket, calculations of differences in free binding energies for related ligands using thermodynamic cycles and conformational investigations, e.g., a refinement of approximate complex structures derived from docking methods based on rigid models (see Sect. 4.2). Another application is the refinement of NMR structures. Eq. (13) is supplemented with additional interatomic distance constraints to find low energy conformations satisfying the measured NOEs (WÜTHRICH, 1995).

4.2 Continuum Models for Protein Interactions

A rigorous force field calculation for the estimation of protein interactions in solution is very demanding, since it requires a high number of explicit solvent molecules in addition to the protein complex. Continuum models based on the Poisson–Boltzmann equation

$$
\begin{aligned}
&-\nabla \cdot [\varepsilon(r) \nabla \cdot \phi(r)] \\
&+ \varepsilon(r) \kappa(r)^2 \sinh[\phi(r)] = 4\pi\varrho(r)/kT
\end{aligned} \tag{14}
$$

are an interesting alternative. The electrostatic potential $\phi(r)$ is given in units of kT/q, where k is the Boltzmann constant and q the proton charge. The fixed charge density of the protein $\varrho(r)$ is represented by point charges of the atomic nuclei, and the dielectric $\varepsilon(r)$ is of the order of 4 inside of a protein, whereas outside a high dielectric for the solvent is used. The Debye–Hückel parameter $\kappa(r)$ is inversely proportional to the Debye length, which depends on the ionic strength of the bulk solution. The second term on the left side vanishes, if no additional ions are in the solvent, and the term $\sinh[\phi(r)]$ can be linearized to give $\phi(r)$, if $\phi(r)$ is small compared to kT. The Poisson–Boltzmann equation can be solved with fast numerical methods, which are even faster than standard methods for the linearized equation (HOLST et al., 1994).

Classical electrostatics has found a variety of applications. An important example is the visualization of electrostatic potentials of proteins, because the surfaces of interacting macromolecules often show a complementary electrostatic potential. In addition to functional recognition, the influence of the electrostatic potential on a charged substrate of an enzyme has given qualitative explanations for the observed kinetics (RIPOLL et al., 1993), and recently the method has been applied for protein–protein docking (JACKSON and STERNBERG, 1995).

4.3 Protein Docking

Docking algorithms predict the structure of a protein complex from known structures of the components. The prediction of complex structures is a formidable problem and only in the last few years has a series of promising methods been developed. Some programs are more suitable for protein–ligand complexes and others are designed for protein–protein complexes. *De novo* ligand design programs build up ligands from small molecular fragments at a receptor site with a known structure.

In general the docking simulation consists of 3 steps. In a first step the docking region of the protein has to be determined. Then potential structures are generated and finally some type of scoring is carried out for an assessment of the individual proposed complex structures. With the simplifying assumption of rigid partners the computational search can be reduced to an optimization of 3 rotational and 3 translational parameters. However, the computational fast rigid models have failed in several cases, and with increasing computer power the oversimplified models have been supplemented by some type of flexibility in the protein–ligand docking process. The rigid models are presently in use only for protein–protein docking and fast search methods, partly because the docking simulation in this case is much more demanding than for small molecules, partly because flexibility in proteins is in most cases confined to local positional changes, whereas in small molecules a rotation around one bond may change the surface completely on the side pointing to the protein.

The best scoring function is probably a computation of the free energy of binding, ΔG, but this is not practicable for a series of different relative orientations or a fast screening with potential ligands from a structural database. Therefore, simplifications have been introduced for practical applications like simple empirical scoring functions based on surface complementarity, hydrophobicity, electrostatic interaction energy and empirical free energy estimations. The methods are still open to discussion. Recently, a number of reviews (BLANEY and DIXON, 1993; WHITTLE and BLUNDELL, 1994; KUNTZ et al., 1994; ROSENFELD et al., 1995) covered several aspects of protein docking and described established programs.

4.3.1 Protein–(Small)Ligand Docking and Ligand Design

Binding sites for small ligands at the protein surface have often the shape of a groove or cavity. This enables a purely geometric approach for the identification of these areas based on the α-shape algorithm for a triangulation of the protein surface at different levels of resolution (PETERS et al., 1996). Besides geometric methods, probe molecules representing the essential features of a ligand can be used to find docking sites. These probe molecules are placed at the protein surface and minima of the interaction energy represent sites at which the probe molecules interact favorably with the protein. A complete functionality map of the protein can be computed using different types of probe molecules (GOODFORT, 1985; MIRANKER and KARPLUS, 1991).

SOBOLEV et al. (1996) generated a series of starting orientations of a ligand at a known binding pocket and carried out a maximization of a complementary function to optimize the orientation of the rigid ligand in a first step. The complementary function is based on a classification of the contact atoms as hydrophobic/hydrophilic, (neutral) donor/acceptor or aromatic. In a following step a refinement

of hydrogen bonding distances between the protein and the ligand was carried out, and a simple side chain flexibility model has been included. When an overlap between the ligand and a protein side chain occurs, the overlap contribution to the complementary function can be ignored, if a different side chain orientation without this overlap exists. In another approach the protein was assumed to be rigid and a conformational optimization of flexible ligands was carried out. For this type of optimization genetic algorithms have become popular (JUDSON et al., 1995; OSHIRO et al., 1995), because they provide good solutions for combinatorial optimization problems. Pattern recognition techniques can also be used for the placement of small molecules or fragments of larger molecules in the active site of an acceptor (RAREY et al., 1996). If it is necessary to dock a large number of ligands to an active site, atom-based and surface-based geometric shape descriptors can be used to reduce complex molecular structures to a small set of parameters in order to speed up the database screening (GOOD et al., 1995).

A series of programs have been written for the *de novo* design of enzyme inhibitors. Basically, these programs place small molecular fragments in such a way at the active site that hydrogen bonds can be formed and hydrophobic interactions are favorable according to a given set of rules. In a second step these fragments are linked together to form a complete molecule. The latest methods include also detailed analysis tools to design the new ligand similar or complementary to a given target (CLARK et al., 1995) or take into account conformational flexibility and synthetic accessibility (BÖHM, 1996).

4.3.2 Docking of Proteins with Macromolecules

Up to now only limited information on the performance of protein–protein docking algorithms is available, when structural changes occur (induced fit), because the algorithms have often been tested only for co-crystallized complexes. Recently, a series of docking algorithms have been able to predict successfully the structure of the β-lactamase inhibitory protein to TEM-1 β-lactamase (STRYNADKA et al., 1996) using non-co-crystallized structures.

The approach of JACKSON and STERNBERG (1995) for protein–protein docking is based on a rigid protein model. But their detailed analysis of the different contributions to free energy of binding leads to promising results. They linearized the Poisson–Boltzmann equation to compute the total electrostatic free energy and included also hydrophobic free energy and loss in side chain conformational energy to differentiate between false and correct docking orientations. A flexible protein–protein docking algorithm is the method of choice for protein–protein docking with an induced fit. Often the interior part and the backbone of a protein are quite rigid whereas the side chains located at the surface show major conformational changes upon complex formation. The ICM method (ABAGYAN et al., 1994) is based on this observation. It allows to fix arbitrary internal coordinates while others are optimized using a Monte Carlo variant for the search of the global energy minimum of protein complexes. The energy is evaluated from a detailed force field supplemented with solvation energy terms.

A highly reliable and very fast docking method has been developed by MEYER et al. (1996) by the combination of hydrogen bond geometry based prealignment of the proteins followed by an optimization using correlation analysis of surface complementarity. The method gives results very close to experimental results in all 95% of the tested cases when proteins do not lead to structure rearrangement.

For protein–DNA docking a binding site at the DNA, e.g., the major groove, is often known in advance. In this case a global search is not required and a local refinement with more time-consuming flexible docking models can be carried out. KNETGEL et al. (1994) have proposed a Monte Carlo technique-based program, which can handle flexible protein side chains at the docking site. The flexibility of the DNA is accounted for by unwinding or overwinding the DNA via slight changes of the helix-pitch. Then nucleotides having no contact with the protein are shifted towards the protein surface to bend the DNA along the protein prior to relaxing the protein structure.

5 Protein Complexes

Proteins form an enormous number of complexes, and for each purpose a specific recognition type has been developed. In Sects. 5.1–5.5 some complexes are discussed to illustrate how proteins achieve the necessary specific or non-specific recognition of their targets, and to show for selected examples how experimental and theoretical methods can·be used to extract aspects of protein function.

5.1 Protein–(Low Molecular Weight)Ligand Complexes

Enzyme–inhibitor complexes are an example for the widespread group of highly specific protein–ligand complexes. An important target enzyme for ligand design is HIV protease (HIV PR), because the enzyme is required for HIV maturation. NMR spectroscopy and protein crystallography have been powerful tools to show the interaction between inhibitors and the enzyme and to develop more specific substances. Quantum chemical studies of reduced active site model systems (LEE et al., 1996; SILVA et al., 1996) and a combined quantum mechanics/classical molecular dynamics study (LIU et al., 1996) based on the enzyme structure have been carried out to elucidate the mechanism of bond cleavage in the enzymatic process.

HIV protease is a dimer of two identical subunits, each subunit forming part of one active site which is C_2 symmetric (Fig. 1, see color plates, p. 105). The active site triad Asp25, Thr26 and Gly27 is located in a loop which is stabilized by a network of hydrogen bonds. The Asp carboxylate groups of both chains are in an almost coplanar orientation in the uncomplexed enzyme. The active site is cov-

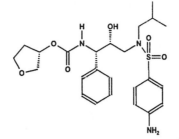

VX-478

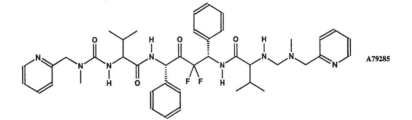

A79285

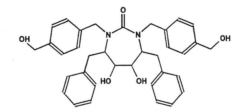

DMP323

SB203238

Fig. 2. Examples of HIV protease inhibitors.

ered by two flaps which are probably important for product release and stabilization of the substrate. A recent NMR structure of an HIV protease–inhibitor complex in solution shows that the tips of these flaps are indeed flexible (YAMAZAKI et al., 1996). In a high number of enzyme–inhibitor complexes a single water molecule is attached to the Ile50A and Ile50B amide groups. This water molecule is not accessible to the solvent. It has been proposed that this molecule is functionally important and the inclusion of a functional mimic of water at the corresponding orientation might lead to novel potent inhibitors. In the complex of HIV protease (Fig. 1) with the inhibitor VX-478 (Fig. 2) two hydrogen bonds between the inhibitor and this particular water molecule are formed. Another set of hydrogen bonds is formed between the inhibitor and the oxygen atoms of the catalytically important Asp25A and Asp25B (KIM et al., 1995).

The inhibitors designed for HIV protease may be divided into three groups, asymmetric and symmetric peptidic inhibitors and non-peptidic inhibitors. Some examples are shown in Fig. 2. For the design of HIV–protease inhibitors peptide substrate analogs have often been designed with the scissile peptide bond replaced by a non-hydrolyzable isostere with a tetrahedral geometry replacing the sp^2 carbonyl carbon atom (WLODAWER and ERICKSON, 1993). Reduced amides, hydroxyethylene or hydroxymethylamine derivatives have been proposed. Furthermore, hydrophobic blocking groups and non-naturally occurring side chains have been added. The experimental observation of a 2-fold symmetry of the HIV protease active site has led to the design of inhibitors having the same type of symmetry. Finally, database search techniques have been applied to search for non-peptidic compounds complementary to the HIV protease active site surface.

5.2 Protein–Carbohydrate Interactions

Interactions between carbohydrates and proteins are often required for protein function. Apart from enzymes acting on sugars, carbohydrate-specific antibodies, membrane transport proteins, bacterial periplasmic binding proteins, and lectins are known to interact with carbohydrates. The interactions between carbohydrates and proteins can be divided roughly in two groups. The first group binds the carbohydrates with high affinity in deep clefts so that little or no part of the carbohydrates is solvent accessible. The second group, proteins like, e.g., lectins bind carbohydrates at the surface in a shallow cleft with

low affinity. Lectins like bacterial or plant toxins and animal receptors for the uptake of glycoconjugates, are located on the cell surface to recognize sugars. As an example, methyl-α-D-mannopyranosid is shown in Fig. 3 (see color plates, p. 105) with the binding pocket located at the surface of concavalin A (NAISMITH et al., 1994). The sugar is fixed by hydrogen bonds at O2, O3, O4 and O6 to the protein. The stacking of a Tyr against the hydrophobic part of the sugar surface is a typical feature of protein–carbohydrate interactions. Similarly, lactose is fixed by hydrogen bonds between O3-O6 and peanut lectin (BANERJEE et al., 1996), and some of the residues involved in the hydrogen bonding and a Tyr can be superimposed closely with the corresponding ones of other legume lectins like pea or *Erythrina corallodendron* lectin. A packing of apolar regions of sugars is also a common feature of protein–carbohydrate interactions. Contrary to the low-affinity binding lectins high-affinity binding involves hydrogen binding with all polar sugar atoms, even the ring oxygen may participate. Aromatic residues appear at the active site of proteins with high affinity to sugars and a deep binding pocket as, e.g., in glucoseoxidase (HECHT et al., 1993).

5.3 Protein–Steroid Complexes

Steroid hormones bind to special protein receptors mediating their biological activity. Furthermore, carrier proteins distribute the hydrophobic molecules and protect them from degradation. Steroids play an important role for reproduction, behavior and cell differentiation. Disturbances in their metabolism and receptor interactions contribute to a series of diseases. Synthetic steroids have been used for fertility control and as immunosuppressants. The crystal structures of two complexes with steroid-binding enzymes have been reported, cholesteroloxidase with dehydroisoandrosterone (LII et al., 1993) and $3\alpha,20\beta$-hydroxysteroid dehydrogenase with carbenoxolan.

Cholesteroloxidase from *Brevibacterium sterolicum* is a monomeric oxidase containing one FAD coenzyme per protein molecule.

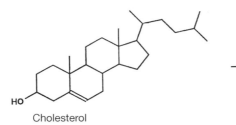 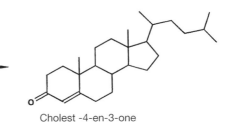

Cholesterol   Cholest -4-en-3-one

Fig. 4. The reaction catalyzed by cholesteroloxidase.

The enzyme catalyzes the oxidation of Δ^5-ene-3β-hydroxysteroids to a Δ^5-3-ketosteroid and the isomerization to the Δ^4-3-ketosteroid (Fig. 4). The enzyme can be reoxidized with molecular oxygen via the formation of hydrogen peroxide. The binding site is in an internal cavity of the enzyme, which is completely closed when a complex with dehydroisoandrosterone has been formed, even though the cavity is probably not large enough to also include the carbon chain of the natural substrate cholesterol, which is absent in dehydroisoandrosterone. Significant changes in the main chain coordinates upon steroid-binding are only observed for those residues that form a loop at the cavity entrance whereas side chain conformational changes are necessary for some of the hydrophobic side chains in the steroid binding region. The cavity is filled with 13 water molecules in the absence of the substrate. A single water molecule remains at the active site upon ligand binding (Fig. 5, see color plates, p. 106). This molecule forms a hydrogen bond with O1 of the substrate, and it takes probably part in the steroid oxidation. In a hydride transfer mechanism a proton from the substrate O1 is probably transferred to the water molecule, and a proton from the water molecule is transferred to His447. The hydrogen atom at the steroid C3 may be transferred as a hydride to N5 of the FAD coenzyme.

5.4 Protein–Protein Complexes

Interactions between proteins may be divided roughly into three groups. The first consists of proteins with a functional recognition, e.g., protease inhibitors and antigen–antibody complexes, the other groups are proteins with a quaternary structure (i.e., protein oligomers) plus protein–protein crystal contacts in the solid state. A common property of protease inhibitor and antigen antibody is the pairwise buried protein surface area in the range from 1,250–1,750 $Å^2$ upon complex formation. The complexes are stabilized by 7–13 hydrogen bonds (JANIN and RODIER, 1995). The pairwise buried surface area at protein–crystal contacts is generally lower (200–1,200 $Å^2$). These small interface contacts are similar to the magnitude of random contacts generated by computer experiments. Interface areas of oligomers are very variable (JANIN et al., 1988). For the extreme cases superoxide dismutase and catalase interface areas of 670 and 10,570 $Å^2$ per subunit have been determined corresponding to 9 and 40% of the subunit surfaces, respectively. The oligomeric interfaces are enriched in hydrophobic side chains. Arg and Leu contribute most to the surface area, and on the average there is one intermolecular hydrogen bond per 200 $Å^2$ surface area, even though there are examples which have none. Often charged residues are involved in hydrogen bonds: 22% of the hydrogen bonds are salt bridges and 35% of the hydrogen bonds are formed between a charged group and a neutral one.

Antibody–antigen complexes share a series of common properties. The light and the heavy chain of the antibodies have contact with the protein antigens. Often the contacts with the heavy chain predominate. The different binding sites and the specificity of binding are generated by the complementary determining regions of the variable domains of the light and heavy chains. In addition to hydrogen bonds between both proteins and water-mediated hydrogen bonds, aromatic residues of the complementary determining regions contribute to the binding of the antigen.

The high resolution structure of the Fv fragment of the anti-hen egg white lysozyme antibody D1.3 in free and complexed forms shows the details of the hydrogen bonding system for this type of protein–protein complexes (BHAT et al., 1994). A comparison of the water molecules located at the surface of the isolated proteins with the water molecules embedded in the interface of the complex shows that a few water molecules are displaced upon binding, whereas several water molecules retain their positions when the antigen is bound. The cavity between the $V_L - V_H$ interface is completely filled with water, and even additional water molecules are included in the hydrogen bonding network between both proteins upon binding (Fig. 6, see color plates, p. 106). Small conformational changes relative to the isolated protein bring antibody-bound water molecules in contact with the antigen.

Docking simulations of antibody–antigen complexes based on purely geometric criteria and rigid models have encountered difficulties because of the relative small complementary interface area at the epitope compared to larger interfaces at incorrect sites, small structural changes upon complex formation, and water-mediated hydrogen bonds.

5.5 Protein–Nucleic Acid Complexes

Protein–nucleic acid complexes are of high biochemical importance being often involved in gene regulation, transcription and translation. Three types of DNA are presently known. In B-DNA the base pairs are approximately perpendicular to the helix axis, the helix pitch is 34 Å. Nucleotide-specific recognition is difficult in double stranded DNA as the nucleotides are in the interior of the duplex and the surface is largely composed of the negatively charged phosphate and sugar backbone. Thus the B-DNA minor and major grooves with widths of 5 and 10 Å are important for recognition. In the coil-like A-DNA the major groove is narrower and deeper than

the one found in B-DNA. The minor groove of A-DNA is much flatter and more open than the major groove in B-DNA. In the left handed Z-DNA the minor groove is considerably deeper than in the right handed A and B-DNA. An interaction of proteins with DNA grooves requires interacting groups to protrude substantially from the residual molecular surface.

Several different binding motifs have been reported to date including the helix-turn-helix (hth) motif, different types of zinc-binding motifs, the basic leucine zipper and a β-ribbon. The first recognition type is important for repressor operator-binding and the hth-motif can be found in the Cro (TAKEDA et al., 1986), λ (BEAMER and PABO, 1992) and *trp* repressor (LAWSON et al., 1988), and the catabolic gene activator (CAP) protein (MONDRAGON and HARRISON, 1991).

In a complex one of the helices, the recognition helix, is located in the DNA major groove allowing extensive contacts between the side chains and the DNA bases. The proteins are dimers, and the distance between both recognition helices is 34 Å corresponding to the separation of the major grooves. The (approximate) 2-fold symmetry of the proteins has its counterpart in the palindromic DNA sequences generating two symmetrical sets of interactions. A sequence specificity of interaction can be achieved with hydrogen bond interactions between functional groups located at nucleotide edges in the major groove and amino acid side chains. No simple assignment between nucleotides and amino acid residue is possible since many contacts including water-mediated bonds contribute to specific recognition. In addition to protein–DNA interactions between the recognition helix and base pairs in the Cro/O_R1 complex (Fig. 7, see color plates, p. 107), the complex is additionally stabilized by hydrogen bonds to phosphates of the DNA backbone. In Fig. 8 (see color plates, p. 107) the close contact between the protein surface and the DNA is shown. It has been observed that the Cro–protein structures show no major differences upon DNA-binding, whereas the DNA is somewhat bent in the complex.

Zinc-binding domains consist of folded protein loops stabilized by zinc ions. In a classical zinc finger a zinc ion is ligated by two cysteine and two histidine residues. Additionally three hydrophobic residues are conserved. Multiple zinc fingers appear in proteins to attain a sequence-specific DNA recognition. They wrap around the DNA, and each zinc finger has a recognition helix that fits into the DNA major groove and makes contact to three bases.

In the class II of zinc-binding motifs two zinc ions are complexed by four cysteines and in class III two zinc ions are ligated by six cysteines. The recognized sequences of class II and III zinc-binding motifs are palindromic sites whereas zinc fingers of class I do not make use of symmetry. The human estrogen receptor shown in Fig. 9 (see color plates, p. 108) has two class II zinc-binding domains that bind to adjacent major grooves (SCHWABE et al., 1993). It has been shown by mutagenesis that three amino acids at the N-terminus of the recognition helix (reading head) are responsible for recognition. The protein is a monomer in solution, and the cooperative dimerization upon DNA-binding increases the specificity because binding depends on the spacing and helical repeat of the target sequence.

One of the structurally characterized examples for protein–RNA interaction is the complex of the RNA-binding domain of U1A spliceosomal protein with its target (OUBRIDGE et al., 1994). The binding domain of the protein consists of a four-stranded anti-parallel β-sheet (Fig. 10, see color plates, p. 108) with two additional helices. A loop between the β-strands protrudes between the RNA hairpin loop preventing the bases from pairing. The ten nucleotides of this hairpin are pointing away from the hairpin center, seven of these nucleotides have contact to the protein via direct or indirect hydrogen bonds via a water molecule. The complex formation requires probably an induced fit since hairpin loops are believed to be largely unstructured.

6 References

ABAGYAN, R., TOTROV, M., KUZNETSOV, D. (1994), ICM- a new method for protein modelling and design: applications to docking and structure prediction from distorted native conformation, *J. Comput. Chem.* **15**, 488–506.

BAKER, E. N., HUBBARD, R. E. (1984), Hydrogen bonding in globular proteins, *Prog. Biophys. Mol. Biol.* **44**, 97–179.

BANERJEE, R., DAS, K., RAVISHANAR, R., SUGUNA, K., SUROLIA, A., VIJAYAN, M. (1996), Conformation, protein–carbohydrate interactions and a novel subunit association in the refined structure of peanut lectin–lactose complex, *J. Mol. Biol.* **259**, 281–296.

BAYLY, C. I., CIEPLAK, P., CORNELL, W. D., KOLLMAN, P. A. (1993), A well-behaved electrostatic potential based method using charge restraints for deriving atomic charges: the RESP model, *J. Phys. Chem.* **97**, 10269–10280.

BEAMER, L. J., PABO, C. O. (1992), Refined 1.8 Å crystal structure of the λ repressor–operator complex, *J. Mol. Biol.* **227**, 177–196.

BHAT, T. N., BENTLEY, G. A., BOULOT, G., GREENE, M. I., TELLO, D. et al. (1994), Bound water molecules and conformational stabilization help mediate an antigen–antibody association, *Proc. Natl. Acad. Sci. USA* **91**, 1089–1093.

BLANEY, J. M., DIXON, J. S. (1993), A good ligand is hard to find: automated docking methods, *Perspect. Drug Discov. Design* **1**, 301–319.

BLOKZIJL, W., ENGBERTS, J. B. F. N. (1993), Hydrophobic effects: opinion and fact, *Angew. Chem.* **105**, 1610–1648, *Angew. Chem. (Int. Edn.)* **32**, 1545.

BÖHM, H.-J. (1996), Towards the automatic design of synthetically accessible protein ligands: peptides, amides and peptidomimetics, *J. Comput. Aided Mol. Des.* **10**, 265–272.

BROOKS, B. R., BRUCCOLERI, R. E., OLAFSON, B. D., STATES, D. J., SWAMINATHAN, S. et al. (1983), CHARMM: a program for macromolecular energy, minimization, and dynamics calculations, *J. Comput. Chem.* **4**, 187–217.

BROOKS, I., WETZEL, R., CHAN, W., LEE, G., WATTS, D. G. et al. (1994), Association of REI immunoglobulin light chain Vl domains: the functional linearity of parameters in equilibrium analytical ultracentrifuge models for-self-associating systems, in: *Modem Analytical Ultracentrifugation: Acquisition and Interpretation of Data for Biological and Synthetic Polymer Systems* (SCHUSTER, T. M., LAUE, T. M., Eds.) pp 15–36, Boston IMA: Birkhäuser.

CHOTIA, C., JANIN, J. (1975), Principles of protein–protein recognition, *Nature* **256**, 705–708.

CLARK, D. E., FRENKEL, D., LEVY, S. A., LI, J., MURRAY, C. W. et al. (1995), PRO_LIGAND: an approach to *de novo* molecular design, *J. Comput. Aided Mol. Des.* **9**, 13–32.

CORNELL, W. D., CIEPLAK, P., BAYLY, C. I., GOULD, I. R., MERZ, K. M. et al. (1995), A second generation force field for the simulation of proteins, nucleic acids, and organic molecules, *J. Am. Chem. Soc.* **117**, 5179–5197.

DEREWENDA, Z. S., LEE, L., DEREWENDA, U. (1995), The occurrence of C—H···O hydrogen bonds in proteins, *J. Mol. Biol.* **252**, 248–262.

DRENTH, J. (1994), *Principles of Protein X Ray Crystallography.* New York: Springer-Verlag.

DUCRUIX, A., GIEGÉ, R. (Eds.) (1992), *Crystallization of Nucleic Acids and Proteins, a Practical Approach.* Oxford: Oxford University Press.

EISENBERG, D., MCLACHLAN, A. D. (1986), Solvation energy in protein folding and binding, *Nature* **319**, 199–203.

FISCHER, H. F., SING, N. (1995), Caloric methods for interpreting protein–ligand interactions, *Methods Enzymol.* **259**, 194–221.

GLUSKER, J. P., LEWIS, M., ROSSI, M. (1994), *Crystal Structure Analysis for Chemists and Biologists.* New York: VCH.

GOOD, A. C., EWING, T. J. A., GSCHWEND, D. A., KUNTZ, I. D. (1995), New molecular shape descriptors: application in database screening, *J. Comput. Aided Mol. Des.* **9**, 1–12.

GOODFORT, P. J. (1985), A computational approach for determining energetically favorable binding sites on biologically important macromolecules (GRID), *J. Med. Chem.* **28**, 849–857.

HALGREN, T. A. (1996), The Merck molecular force field 1. Basis, form, scope, parametrization and performance of MMFF94, *J. Comput. Chem.* **17**, 490–501.

HECHT, H. J., KALISZ, H. M., HENDLE, J., SCHMID, R. D., SCHOMBURG, D. (1993), Crystal structure of glucose oxidase from *Aspergillus niger* at 2.3 Å resolution, *J. Mol. Biol.* **229**, 153–172.

HIBBITS, K. A., GILL, D. S., WILSON, R. C. (1994), Isothermal titration calorimetric study of the association of hen egg lysozyme and the anti-lysozyme antibody HyHEL-5, *Biochemistry* **33**, 3584–3590.

HOLST, M., KOZAK, R. E., SAIED, F., SUBRAMANIAM, S. (1994), Protein electrostatics: rapid multigrid algorithm for solution of the full nonlinear Poisson–Boltzmann equation, *J. Biomol. Struct. Dyn.* **11**, 1437–1445.

HORTON, N., LEWIS, M. (1992), Calculation of free energy of association for protein complexes, *Protein Sci.* **1**, 169–181.

JACKSON, R. M., STERNBERG, M. J. E. (1995), A continuum model for protein–protein interactions: application to the docking problem, *J. Mol. Biol.* **250**, 258–275.

JANIN, J., RODIER, F. (1995), Protein–protein interactions at crystal contacts, *Proteins* **23**, 580–587.

JANIN, J., MILLER, S., CHOTIA, C. (1988), Surface, subunit interface and interior of oligomeric proteins, *J. Mol. Biol.* **304**, 155–164.

JUDSON, R. S., TAN, Y. T., MORI, E., MELIUS, C., JAEGER, E. P. et al. (1995), Docking flexible molecules: a case study of three molecules, *J. Comput. Chem.* **16**, 1405–1419.

KIM, E. E., BAKER, C. T., DWYER, M. D., MURCKO, M. A., RAO, B. G. et al. (1995), Crystal structure of HIV-1 protease in complex with VX-478, a potent and orally bioavailable inhibitor of the enzyme, *J. Am. Chem. Soc.* **117**, 1181–1182.

KNEGTEL, R. M. A., BOELENS, R., KAPTEIN, R. (1994), Monte Carlo docking of protein–DNA complexes: incorporation of flexibility and experimental data, *Protein Eng.* **7**, 761–767.

KUNTZ, I. D., MENG, E. C., SHOICHET, B. K. (1994), Structure-based molecular design, *Accounts Chem. Res.* **27**, 117–123.

KUZIEMKO, G. M., STROH, M., STEVENS, R. C. (1996), Cholera toxin binding affinity and specificity for gangliosides determined by surface plasmon resonance, *Biochemistry* **35**, 6375–6384.

LAWSON, C. L., ZHANG, R., SCHEVITZ, R. W., OTWINOWSKI, Z., JOACHIMIAK, A., SIGLER, P. B. (1988), Flexibility of the DNA-binding domains of *trp* repressor, *Proteins* **3**, 18–31.

LEVITT, M., PERUTZ, M. F. (1988), Aromatic rings act as hydrogen bond acceptors, *J. Mol. Biol.* **201**, 751–754.

LEE, H., DARDEN, T. A., PEDERSEN, L. G. (1996), An *ab initio* quantum mechanical model for the catalytic mechanism of HIV-1 protease, *J. Am. Chem. Soc.* **118**, 3946–3950.

LI, J., VRIELINK, A., BRICK, P., BLOW, D. M. (1993), Crystal structure of cholesterol oxidase complexed with a steroid substrate: implications for flavin adenine dinucleotide dependent alcohol oxidases, *Biochemistry* **32**, 11507–11515.

LII, J.-H., ALLINGER, N. L. (1991), The MM3 force field for amides, polypeptides and proteins, *J. Comput. Chem.* **12**, 186–199.

LIU, H., MÜLLER-PLATHE, F., VAN GUNSTEREN, W. F. (1996), A combined quantum/classical molecular dynamics study of the catalytic mechanism of HIV protease, *J. Mol. Biol.* **261**, 454–469.

MATHINOWETZ, A. M., JAIN, A., KARASAWA, N., GODDARD, W. A. (1994), Protein simulations using techniques suitable for very large systems: the cell multipole method for nonbonded interactions and the Newton–Euler inverse mass operator method for internal coordinate dynamics, *Proteins* **20**, 227–247.

MCDONALD, T., THORNTON, J. M. (1994), Satisfying hydrogen bonding potential in proteins, *J. Mol. Biol.* **238**, 777–793.

MCREE, D. E. (1993), *Practical protein crystallography*. San Diego, CA: Academic Press.

MEYER, M., WILSON, P., SCHOMBURG, D. (1996), Hydrogen bonding and molecular surface shape complementary as a basis for protein docking, *J. Mol. Biol.* **264**, 199–210.

MIRANKER, A., KARPLUS, M. (1991), Functionality maps of binding sites: a multiple copy simultaneous search method (MCSS), *Proteins* **11**, 29–34.

MONDRAGON, A., HARRISON, S. C. (1991), The phage 434 Cro/O_R1 complex at 2.5 Å resolution, *J. Mol. Biol.* **219**, 321–334.

NAGHIBI, H., TAMARA, A., STURTEVANT, J. M. (1995), Significant discrepancies between van't Hoff and calorimetric enthalpies, *Proc. Natl. Acad. Sci. USA* **92**, 5597–5599.

NAISMITH, J. H., EMMERICH, C., HABASH, J., HARROP, S. J., HELLIWELL, J. R. et al. (1994), Refined structure of concavalin A complexed with methyl-α-D-mannopyranosid at 2.0 Å resolution and comparison with saccharide-free structure, *Acta Crystallogr. D* **50**, 847–858.

O'SHANNESSY, D. J., WINZOR, D. J. (1996), Interpretation of deviations from pseudo-first-order kinetic behavior in the characterization of ligand binding by biosensor technology, *Anal. Biochem.* **236**, 275–283.

OSHIRO, C. M., KUNTZ, I. D., DIXON, J. S. (1995), Flexible ligand docking using a genetic algorithm, *J. Comput. Aided Mol. Des.* **9**, 113–130.

OUBRIDGE, C., ITO, N., EVANS, P. R., TEO, C.-H., NAGI, K. (1994), Crystal structure at 1.92 Å resolution of the RNA-binding domain of the U1A spliceosomal protein complex with an RNA hairpin, *Nature* **372**, 432–438.

PETERS, K. P., FAUCK, J., FRÖMMEL, C. (1996), The automatic search for ligand binding sites in proteins of known three-dimensional structure using only geometric criteria, *J. Mol. Biol.* **256**, 201–213.

PIVALOV, P. L., MAKHATADZE, G. I. (1993), Contribution of hydration to protein folding thermodynamics, *J. Mol. Biol.* **232**, 660–679.

RAREY, M., WEFING, S., LENGAUER, T. (1996), Placement of medium-sized molecular fragments into active sites of proteins, *J. Comput. Aided Mol. Des.* **10**, 41–54.

RIPOLL, D. R., FAERMAN, C. H., AXELSEN, P. H., SILMAN, I., SUSSMAN, J. L. (1993), An electrostatic mechanism for substrate guidance down the aromatic gorge of acetylcholinesterase, *Proc. Natl. Acad. Sci. USA* **90**, 5128–5132.

ROSENFELD, R., VAJDA, S., DE LISI, C. (1995), Flexible docking and design, *Annu. Rev. Biophys. Biomol. Struct.* **24**, 677–700.

SCHWABE, J. W. R., CHAPMAN, L., FINCH, J. T., RHODES, D. (1993), The crystal structure of the estrogen receptor DNA-binding domain bound to DNA: how receptors discriminate between their response elements, *Cell* **75**, 567–587.

SILVA, A. M., CACHAU, R. E., SHAM, H. L., ERICKSON, Y. W. (1996), Inhibition and catalytic mechanism of HIV-1 aspartic protease, *J. Mol. Biol.* **255**, 321–346.

SINGHA, N., SUROLIA, N., SUROLIA, A. (1996), On the relationship of the thermodynamic parameters with the buried surface area in protein ligand-complex formation, *Biosci. Rep.* **16**, 1–10.

SOBOLEV, V., WADE, R., VRIENT, G., EDELMANN, M. (1996), Molecular docking using surface complementary, *Proteins* **25**, 120–129.

STRYNADKA, N. C. J., EISENSTEIN, M., KATCHALSKI-KATZIR, E., SHOICHET, B. K., KUNTZ, I. D. et al. (1996), Molecular docking programs successfully predict the binding of β-lactamase inhibitory protein to TEM-1 β-lactamase, *Nat. Struct. Biol.* **3**, 233–239.

TAKEDA, Y., KIM, J. G., CADAY, C. G., STEERS, E., OHLENDORF, D. H. et al. (1986), Different interactions used by Cro repressor in specific and nonspecific DNA binding, *J. Biol. Chem.* **261**, 8608–8616.

TAKEDA, Y., ROSS, P. D., MUDD, C. P. (1992), Thermodynamics of Cro protein-DNA interactions, *Proc. Natl. Acad. Sci. USA* **89**, 8180–8184.

TELLO, D., GOLDBAUM, F. A., MARIUZZA, R. A., YSERN, X., SCHWARZ, F. P., POLJAK, R. J. (1993), Immunoglobulin superfamily interactions, *Biochem. Soc Trans.* **21**, 943–946.

VAN GUNSTEREN, W. F., BERENDSEN, H. J. C. (1990), Computer simulation of molecular dynamics: methodology, applications, and perspectives, *Angew. Chem.* **102**, 1020–1055, *Angew. Chem. (Int. Edn.)* **29**, 992–1092.

WHITTLE, P. J., BLUNDELL, T. L. (1994), Protein structure-based drug design, *Annu. Rev. Biophys. Biomol. Struct.* **23**, 349–375.

WLODAWER, A., ERICKSON, J. W. (1993), Structure-based inhibitors of HIV-1 protease, *Annu. Rev. Biochem.* **62**, 543–585.

WÜTHRICH, K. (1995), NMR – this other method for protein and nucleic acid structure determination, *Acta Crystallogr.* D **51**, 249–270.

YAMAZAKI, T., HINCK, A. P., WANG, Y.-X., NICHOLSON, L. K., TORCHIA, D. A. et al. (1996), Three-dimensional solution structure of the HIV-1 protease complexed with DMP323, a novel cyclic urea-type inhibitor, determined by nuclear magnetic resonance spectroscopy, *Protein Sci.* **5**, 495–506.

YOUNG, L., YERNIGAN, R. L., COVELL, D. G. (1994), A role for surface hydrophobicity in protein–protein recognition, *Protein Sci.* **3**, 717–729.

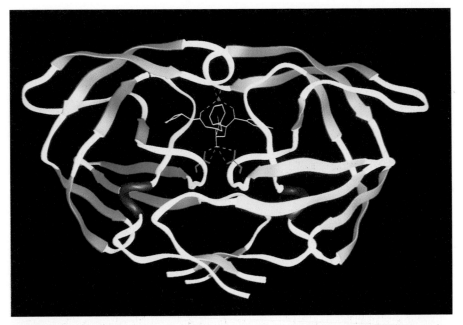

Fig. 1. Structure of HIV protease with the inhibitor VX-478. Hydrogen bonds formed by the inhibitor with Asp25A and Asp25B and with a water molecule are indicated by dashed lines. Hydrogen bonds exist also between the water molecule and Ile50A and Ile50B. The secondary structure indicates the C_2 symmetry of HIV protease (PDB entry 1hpv).

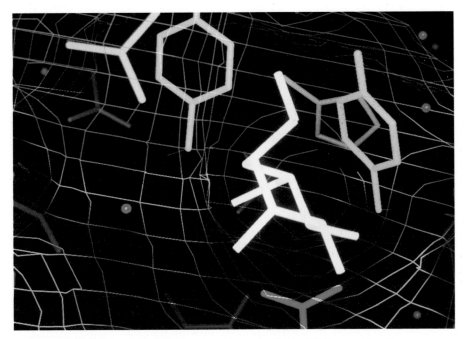

Fig. 3. Methyl-α-D-mannopyranosid at the binding pocket of concavalin A (PDB entry 5cna).

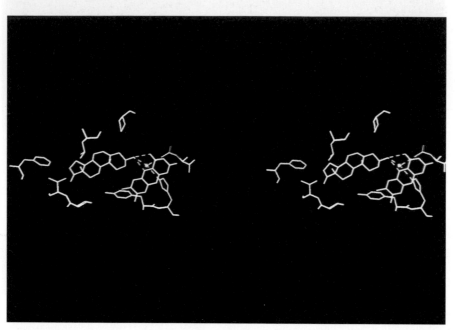

Fig. 5. Complex of cholesterol oxidase with dehydroandrosterone. The hydrophobic steroid is in a binding pocket containing hydrophobic and aromatic amino acids. The hydrogen bond with a water molecule relevant for the mechanism is shown.

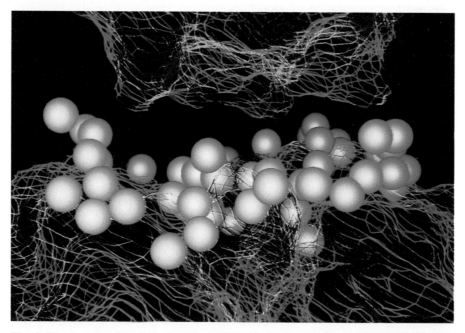

Fig. 6. Water molecules at the interface of the Fv fragment of monoclonal antibody D1.3 and lysozyme (PDB entry 1vfb). The antigen has been moved away from the interface in perpendicular direction. The small cavity between the V_L-V_H interface of the antibody is filled with water molecules behind the wire-frame (right from the middle).

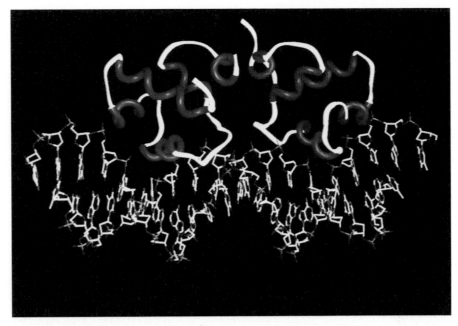

Fig. 7. Cro/O$_R$1 complex showing the helix-turn-helix recognition motif (PDB entry 3cro).

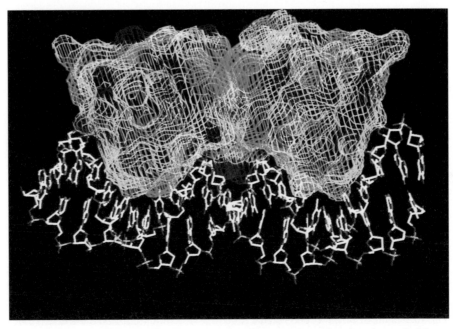

Fig. 8. The close protein–DNA contact of the Cro/O$_R$1 complex. The protein surface is color coded from red (hydrophobic) to blue (hydrophilic).

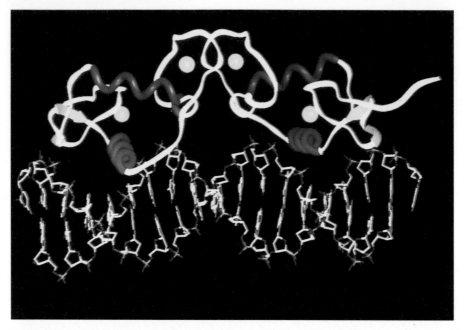

Fig. 9. The human estrogen receptor with two class II zinc motifs for DNA-binding. The zinc ions are shown in white (PDB entry 1hcq).

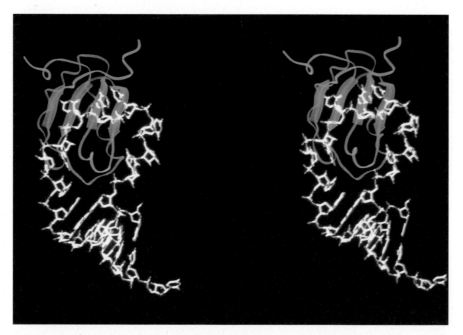

Fig. 10. Fragment of U1A spliceosomal protein bound to hairpin RNA (PDB entry 1urn).

Recombinant Proteins

4 *In vitro* Folding of Inclusion Body Proteins on an Industrial Scale

RAINER RUDOLPH
HAUKE LILIE
ELISABETH SCHWARZ

Halle, Germany

1 Introduction   112
2 Inclusion Body Formation   113
3 Cell Lysis and Inclusion Body Isolation   113
4 Solubilization of the Inclusion Bodies   114
5 *In vitro* Folding   115
   5.1 Disulfide Bond Formation   115
   5.2 Improving *in vitro* Folding by Low Molecular Weight Additives   116
   5.3 Optimizing the Transfer of the Unfolded Protein in Refolding Buffer   117
   5.4 Improved Refolding by Using Fusion Constructs with Hydrophilic Proteins or Peptides   119
   5.5 Improved Refolding by Transient Binding of the Polypeptide to Solid Supports   119
   5.6 Improving *in vitro* Folding by Molecular Chaperones or Folding Catalysts   120
6 Prevention of Inclusion Body Formation by Coexpression of Chaperones or Folding Catalysts   121
7 Conclusions   121
8 References   122

List of Abbreviations

DTE   dithioerythritol
DTT   dithiothreitol
Gdm/HCl   guanidinium chloride
IMAC   immobilized metal affinity
           chromatography
RS   reduced thiol reagent
RSSR   oxidized thiol reagent

1 Introduction

Proteins are extremely important in therapy, diagnostics, and other industrial applications. For example, in medical applications, efficient therapeutic intervention is possible by substituting deficient or disfunctional proteins, such as insulin, human growth hormone, or blood clotting factors, with the purified functional protein. The drawback, however, is that proteinaceous therapeutics in their purified forms are generally difficult to produce, characterize and formulate. Before the advent of recombinant DNA technology a few therapeutic proteins were obtained from human body fluids, animal or plant tissue, or from microorganisms. However, limited availability, the potential of viral contamination or antigenicity limited the use of natural proteins in therapy. Thus, the application of proteins from natural sources was primarily restricted to diagnostics and industrial applications.

The successful expression of chemically synthesized genes for somatostatin and human insulin in *Escherichia coli* opened a new era for the large scale production of recombinant proteins (ITAKURA et al., 1977; GOEDDEL et al., 1979). It was assumed that recombinant DNA technology would allow low cost synthesis of any protein in unlimited amounts. Soon, however, the negative aspect of recombinant protein expression in the cytosol of *E. coli* became apparent: Although protein production was feasable in most cases, many proteins did not fold to their native, functional structure. As described in an excellent review article, many recombinant proteins form insoluble, inactive aggregates (inclusion bodies) in the cytosol of the bacterial host cell (MARSTON, 1986). Because of this problem, other host systems for recombinant protein synthesis were tested. Though proving very efficient in some cases, none of these host systems offered the ultimate solution for recombinant protein production. Incorrect posttranslational modification (e.g., glycosylation) in yeast and insect cells, long development times and high production costs in mammalian cell culture and transgenic animal expression systems limited the general applicability of these alternatives to bacterial expression.

Since prokaryotes like *E. coli* are easy to manipulate by standard DNA technology and produce large amounts of recombinant proteins using simple, low-cost fermentation technology, many attempts were made to resurrect inclusion body aggregates by *in vitro* folding. Initially, this appeared to be a relatively easy and straightforward approach as by that time so much knowledge had already been gathered about *in vitro* folding. In the early experiments on protein folding, simple model proteins such as bovine pancreatic ribonuclease were unfolded using a high concentration of urea with concomitant reduction of disulfide bonds by β-mercaptoethanol (SELA et al., 1957; EPSTEIN et al., 1963). Upon transfer of the reduced, unfolded polypeptide chain into "native" buffer conditions, RNase spontaneously regained the native structure. Many experimental and theoretical attempts were undertaken to solve the code of protein folding. In the course of these studies, mechanistic aspects of the folding process, such as the nature of rate-determining steps and the structure of folding intermediates, were analyzed in great detail (JAENICKE, 1980; CREIGTHON, 1992). All this information helped to design folding strategies for inclusion body proteins. However, in many cases direct application of these standard folding protocols was not successful or inefficient for the *in vitro* folding of complex (large, multiple disulfide bonded, oligomeric) inclusion body proteins.

In order to achieve efficient refolding of these proteins at reasonable production costs, new techniques for *in vitro* folding were developed (RUDOLPH, 1990; RUDOLPH and LILIE, 1996). Thus, although being stigmatized as esoteric in the beginning, *in vitro* refolding of inclusion bodies is now well established in industrial production processes (RUDOLPH, 1996). For example, therapeutic proteins with a high market potential, such as human granulocyte colony stimulating factor, human growth hormone or human insulin, are produced by *in vitro* folding. Furthermore, target proteins for structure analysis or screening systems are now routinely produced by *in vitro* folding.

2 Inclusion Body Formation

Upon high expression of recombinant proteins in the cytosol of *E. coli,* inclusion body formation is frequently observed. Misfolding predominates when disulfide-bonded proteins are produced in the reducing cytosolic environment. Oxidative folding of secreted eukaryotic proteins, which usually contain disulfide bonds (FAHEY et al., 1977), occurs in the endoplasmic reticulum. This cell compartment offers the proper oxidizing milieu and harbors thiol–disulfide isomerases for correct and efficient disulfide bond formation. Upon overexpression of eukaryotic proteins containing multiple disulfide bonds in the reducing cytosolic environment, incorrect folding and, as a consequence, aggregate formation is to be expected.

Inclusion body formation, however, is also observed upon overexpression of proteins which do not contain disulfide bonds. Overexpression by itself seems to be sufficient to direct recombinant proteins into inclusion bodies, as indicated by aggregate formation observed at high level overproduction of endogenous *E. coli* proteins in their native cell compartment (GRIBSKOV and BURGESS, 1983). Upon production of a large amount of protein per time, higher order side reactions leading to aggregation presumably predominate first order folding reactions (KIEFHABER et al., 1991). Furthermore, essential components of the chaperone machinery may be titrated out upon massive overproduction of a recombinant protein. In these cases, inclusion body formation may be reduced or even prevented by decreasing the amount of recombinant protein produced per time unit. This can be achieved by fermentation at low temperatures or by limited induction (KOPETZKI et al., 1989). In some cases, inclusion body formation is suppressed by a co-overexpression of molecular chaperones (see Sect. 5.6). However, if *in vitro* folding can be achieved by a simple and efficient process step, recovery of the native protein by *in vitro* folding should be seriously considered as the method of choice.

3 Cell Lysis and Inclusion Body Isolation

Provided that overexpression is sufficiently high, *E. coli* inclusion bodies can be isolated by the application of a generic, straightforward cell lysis and subsequent solid–liquid separation protocol (RUDOLPH et al., 1997a, b). The first step of this isolation procedure is lysis of the *E. coli* cell walls by lysozyme treatment, subsequent high pressure dispersion and, finally, incubation with a detergent and high salt concentrations. This combination of cell disruption techniques guarantees the complete disintegration of all particulate matter (membrane fragments, cell wall debris, etc.) in such a manner that the material sedimented by the subsequent centrifugation step consists nearly exclusively of inclusion bodies. Following this procedure it is essential not to introduce extra centrifugation steps between the subsequent lysis steps. After maximum cell disruption, inclusion bodies are harvested by centrifugation (Fig. 1). Once pelleted, inclusion bodies exhibit a dense, rubbery consistency. In this form, impurities are difficult to remove by washing steps. However, buffer components and detergents, which may interfere with the subsequent solubilization and folding steps, can be removed at this stage by an additional washing and centrifugation cycle (RUDOLPH and LILIE, 1996).

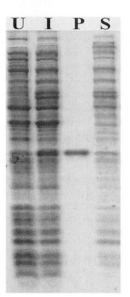

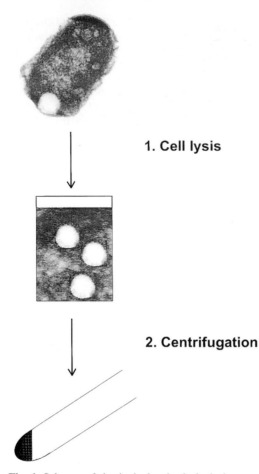

1. Cell lysis

2. Centrifugation

Fig. 2. SDS-PAGE of a representative inclusion body isolation according to protocol 3 in RUDOLPH et al. (1997b). U: total cell protein before induction; I: total cell protein after induction; P: inclusion body isolate; S: soluble protein fraction (supernatant after inclusion body sedimentation).

Fig. 1. Scheme of the inclusion body isolation process consisting of maximum cell lysis by combining lysozyme treatment, high pressure cell dispersion and finally detergent/high salt treatment. In the second step, the inclusion bodies are harvested by centrifugation.

As shown in Fig. 2, the recombinant protein in the inclusion body isolate is rather homogeneous in many cases, despite the fact that purification simply consisted of a solid–liqid separation. The protein is, however, in an aggregated, inactive and insoluble form. From this material, active protein is generated in the subsequent solubilization and folding steps.

4 Solubilization of the Inclusion Bodies

Inclusion body protein is usually solubilized in a buffer containing a chaotroph [urea or guanidinium chloride (Gdm/HCl)] at a high concentration and a thiol reductant (e.g., DTT, DTE or β-mercaptoethanol). Reductive solubilization is performed at alkaline pH values ($\geq$ pH 8.0), since reduction of disulfide bonds by thiol–disulfide interchange proceeds via the thiolate anion. If possible, Gdm/HCl should be given preference to urea as a denaturant. First, Gdm/HCl is a stronger denaturant than urea and thus guarantees faster and more efficient solubilization. Second, cyanate formed upon urea decomposition (especially at high pH values) may cause protein modification. The thiol component is included to reduce intra- and interchain disulfide bonds which are often found in inclusion body isolates. Although addition of this reductant can be omitted for a number of inclusion body isolates, it is absolutely essential for complete solubilization of those inclusion body preparations stabilized by interchain disulfide bonds. Cysteine oxidation will occur predominantly during the preparation procedure. Thus, longer process times for inclusion body isolation upon scaling up from the laboratory to the production scale often increase the extent of oxidation. In this case, the thiol reductant should be included in the solubilization buffer to ensure complete monomerization.

5 *In vitro* Folding

In vitro folding of small, one-domain proteins is often quantitative, i.e., all unfolded polypeptide chains fold back to their native form. For larger, multi-domain proteins the yield of *in vitro* folding is, however, often much lower, since unproductive side reactions (especially aggregation) compete with proper folding (Fig. 3). In this case, the yield can be improved by speeding up rate-determining folding steps, decelerating aggregate formation and/or destabilizing off-pathway products. This can be achieved by optimizing the folding conditions with respect to buffer composition, ionic strength, pH, folding time, temperature, protein concentration, cofactors, and, in the case of disulfide bonded proteins, additives which promote direct disulfide bond formation (see Sect. 5.1). Despite these optimization attempts, the yield of correctly folded species may still be zero under stand-

ard folding conditions. One of these examples is human tissue type plasminogen activator (tPA) which contains 527 amino acid residues, 17 disulfide bonds and one extra cysteine. The protein is organized in 5 structural domains and has a relatively low solubility even in its native form. Folding intermediates with exposed hydrophobic side chains of this protein should have an even lower solubility. Because of these structural characteristics it was assumed that tPA cannot be obtained by *in vitro* folding. Other publications claimed that this protein cannot be produced on an industrial scale because of exorbitant production costs (DATAR et al., 1993). In the following sections, techniques for *in vitro* folding in solution are described which were mainly developed upon investigating structure formation of tPA as a model protein. Technical application of these protocols renders production of tPA from inclusion body protein possible on an industrial scale (RUDOPLH, 1996).

5.1 Disulfide Bond Formation

In vitro disulfide bond formation should be extremely difficult for proteins containing multiple disulfide bonds. In the case of tPA, which contains 35 cysteines, the number of possible combinations of these cysteines to form 17 disulfide bonds exceeds $2.2 \cdot 10^{20}$. Therefore, upon random oxidation of 221,643 t of reduced protein only about 1 ng of correctly disulfide bonded protein would form. As demonstrated in the landmark work by AHMED et al. (1975) disulfide bond formation *in vitro* can be improved by including thiol–disulfide shuffling systems consisting of low molecular weight thiols in reduced and oxidized form in the refolding buffer. Glutathione, cysteine, cysteamine or β-mercaptoethanol in reduced (RS^-) and oxidized form ($RSSR$) can be used as low molecular weight thiols. As shown in Eq. (1), in the first reaction step, the reduced protein forms mixed disulfides with the low molecular weight thiol component. In a second reaction step, the disulfide bond within the polypeptide chain is formed by intrachain thiol–disulfide exchange.

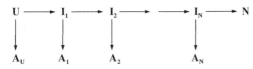

Fig. 3. Kinetic partitioning between folding and aggregation. On the correct folding pathway, unfolded protein (U), folds to the native state (N), via early (I_1, I_2) or late, native-like (I_N) folding intermediates. Unproductive aggregation reactions (A_U, A_1, A_2, A_N) compete with folding.

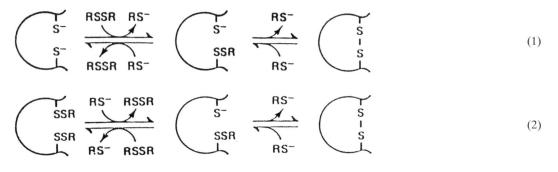

(1)

(2)

If this disulfide bond is not correct and is not stabilized by the conformational stability of the native protein, it is rapidly reduced by an excess of reduced low molecular weight thiol present in the refolding buffer. The reaction goes rapidly back-and-forth until all polypeptide chains are locked in the native disulfide pattern. With tPA, the yield of correct disulfide bond formation could be improved tremendously by starting refolding with the mixed disulfide derivative (Eq. 2). According to this protocol, the cysteine residues of the unfolded protein are first converted into mixed disulfides with glutathione. For the subsequent refolding, the oxidized but still denatured protein is diluted into renaturation buffer containing reduced glutathione to initiate thiol–disulfide interchange. The reason for the improved yield upon starting with the mixed disulfide form of the protein is, that a large number of charged residues are introduced via the glutathione moiety. These charged groups prevent hydrophobic precipitation during early stages of folding.

Since disulfide exchange reactions proceed via the thiolate anion, the folding buffer should have a slightly alkaline pH value. As shown in Fig. 4, folding with concomitant disulfide bond formation of tPA reaches a maximum around pH 8.5. The decrease in yield observed at higher pH values is due to structural destabilization and/or chemical modification of the protein at these high pH values.

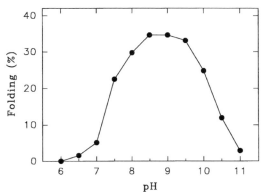

Fig. 4. pH dependence of the refolding of tPA (OPITZ, 1988). For refolding the unfolded mixed disulfide derivative of tPA with glutathione (for preparation of mixed disulfides, compare protocol 11 in RUDOLPH et al., 1997b), the protein was diluted to a final concentration of 3.3 μg mL^{-1} in 0.5 M L-arginine, 0.1 M Tris/HCl, 1 mM EDTA, 1 mg mL^{-1} BSA and 2 mM GSH at the indicated pH value. The final yield of reactivation relative to the total concentration of the unfolded protein was determined after 17 h of incubation at 25 °C.

5.2 Improving *in vitro* Folding by Low Molecular Weight Additives

Upon *in vitro* folding of disulfide bonded as well as non-disulfide bonded proteins the yield of correct folding can be improved by supplementing the refolding buffer with low molecular weight additives. In early examples, denaturants such as Gdm/HCl or urea at non-denaturing concentrations were found to improve the yield of correctly folded protein (ORSINI and GOLDBERG, 1978). As shown by an elaborate analysis of correct folding and

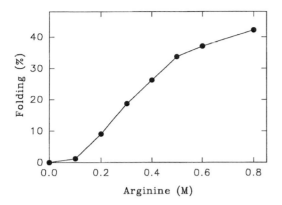

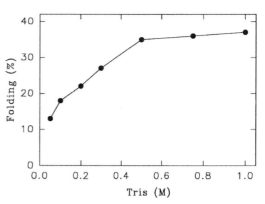

Fig. 5. Effect of L-arginine on the refolding of tPA (OPITZ, 1988). The mixed disulfide derivative of tPA with glutathione was refolded by dilution in 0.1 M Tris/HCl, 1 mM EDTA, 1 mg mL^{-1} BSA, 2 mM GSH, pH 8.5 at a protein concentration of 3 μg mL^{-1}. The folding yield as percentage of the total amount of protein was determined after 17 h of renaturation at 25°C.

Fig. 6. Effect of Tris buffer on the refolding of antibody Fab fragments (AMBROSIUS and RUDOLPH, 1997). Refolding was performed by a 1:100 dilution of the unfolded protein (in 5 M Gdm/HCl, 0.1 M Tris, 2 mM EDTA, 0.3 M DTE, pH 8.5 at a protein concentration of 5 mg mL^{-1}) in Tris buffer pH 7.5 of increasing molarity containing 2 mM EDTA and 6 mM GSSG. The folding yield expressed as the percentage relative to the total amount of protein was determined after 80 h of folding at 10°C.

aggregate formation using the oxidative folding of lysozyme as a model protein, Gdm/HCl decelerates aggregate formation more strongly than folding (HEVEHAN and DE BERNADEZ CLARK, 1997). Improving refolding by the addition of non-denaturing concentrations of denaturants is, however, only possible if the native state of the respective protein is sufficiently stable under these conditions.

As shown for numerous proteins, the yield of correct folding can be improved tremendously by adding L-arginine to the refolding buffer at relatively high molar concentrations (Fig. 5). Although containing a guanidino group, L-arginine has only a minor effect on protein stability. On the other hand this additive strongly enhances the solubility of folding intermediates. The increase in solubility of folding intermediates without significant destabilization of the final native structure results in an improved refolding of many different proteins.

Other low molecular weight additives such as Tris, carbonic acid amides, cyclodextrins, polyethylene-glycol, detergents, or mixed micelles were shown to improve *in vitro* folding (RUDOPLH, 1996). As illustrated in Fig. 6, Tris buffer leads to an increase in the refolding of antibody chains.

In some rare cases, additives which are strongly stabilizing native protein structures are essential for successful folding. Human placental alkaline phosphatase, e.g., could only be refolded *in vitro* in the presence of stabilizers such as sulfate or carbohydrates (Fig. 7).

5.3 Optimizing the Transfer of the Unfolded Protein in Refolding Buffer

Refolding is initiated by transfer of the protein from denaturing into the refolding conditions. Spontaconditions neous *in vitro* folding then starts when the concentration of denaturant is sufficiently low in the refolding buffer to allow structure formation. The critical denaturant concentration below which folding can occur depends on the structural stability of the respective protein. While some proteins are perfectly stable in 6 M urea, others unfold at urea concentrations as low as 0.5 M.

The yield of folding depends on the procedure chosen for transfer of the denatured

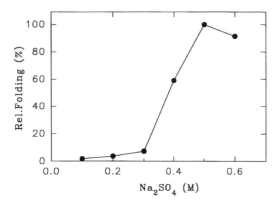

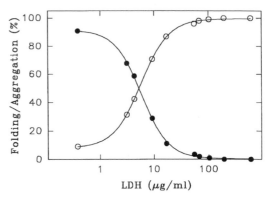

Fig. 7. Effect of Na₂SO₄ on the yield of folding of human placental alkaline phosphatase (MICHAELIS et al., 1995). The reduced, solubilized protein was refolded by rapid dilution in 0.2 M Tris/HCl, 10 mM MgCl₂, 0.1 mM ZnCl₂, 4 mM GSH, 2 mM GSSG and increasing amounts of Na₂SO₄ at a protein concentration of 10 μg mL^{-1}. After 40 h of folding at 20°C, the yields related to the maximum value obtained with these experimental series were calculated.

Fig. 8. Effect of protein concentration on the extent of reactivation and aggregation of lactic dehydrogenase (ZETTMEISSL et al., 1979). After acid denaturation of the tetrameric protein which does not contain disulfide bonds, refolding was achieved by rapid dilution in 0.1 M phosphate buffer, pH 7.0, containing 1 mM EDTA and 1 mM DTE. After 192 h of incubation at 20°C, folding (filled circles) was determined by activity measurements and aggregation (open circles) by light scattering.

protein from denaturing to folding conditions. This is commonly achieved by dialysis or dilution. During dialysis the denaturant is gradually removed. For some proteins the yield of refolding is higher upon gradual denaturant removal than upon rapid dilution into the refolding buffer. Many proteins, however, form structured intermediates which are prone to irreversible precipitation at intermediate denaturant concentrations. In these cases, rapid removal of the denaturant by dilution limits prolonged incubation at intermediate denaturant concentrations and, as a consequence, loss by aggregate formation. Therefore, transfer via rapid dilution usually results in higher yields of refolding than transfer by dialysis. Furthermore, dialysis procedures developed on the laboratory scale are difficult to scale up.

Upon protein folding *in vitro,* the yield critically depends on the initial concentration out of the denatured polypeptide chains, i.e., during dialysis or after diluting out the denaturant. As illustrated in Fig. 3, the final yield of correctly folded protein (N) depends on the kinetic partitioning between correct folding steps and unproductive side reactions such as

aggregation. While folding generally occurs as a first order reaction, aggregate formation has an apparent reaction order between 2 and 3 (ZETTMEISSL et al., 1979). As a consequence, the rate of aggregate formation increases upon increasing the initial concentration of unfolded protein. Above a critical level, which depends on the respective protein, the yield of correct folding decreases and aggregate formation increases (Fig. 8). For many proteins, the concentration up to which high yield refolding is still possible is in the range of 5–20 μg mL^{-1}. Accordingly, the production of large quantities of protein by *in vitro* folding would necessitate enormously large refolding tanks. If, e.g., the yield decreases above a concentration of denatured protein of 10 μg mL^{-1}, refolding containers of 100 m³ would be required for refolding 1 kg inclusion body protein. These excessive and cost-intensive volumes can be easily prevented by stepwise addition of the unfolded protein to much smaller refolding vessels (RUDOLPH and FISCHER, 1990). By this "pulse renaturation" technique, low concentrations of the unfolded protein are added to the refolding buffer. When the protein is either folded to a structured intermediate which is no longer susceptible to aggregation or even to the final native state, fresh denatured protein is added to the same refolding vessel and the procedure is repeated until a high concentration of correctly refolded protein is obtained. The time intervals between the addition of the unfolded polypeptides depend on the rate of folding of the given protein. In this context one has to remember that while some proteins fold very fast, others (as those where disulfide bond formation must occur during *in vitro* folding) sometimes require hours for native structure formation.

Besides the "fed-batch" mode described above, high yields of correctly folded protein at high final protein concentrations can also be obtained by a continuous addition of the unfolded protein in a "chemostat" mode. In industrial processes, where batches have to be defined properly for further characterization, the stepwise addition of the unfolded protein is, however, given preference.

In some recent publications successful *in vitro* folding by gel filtration chromatography has been reported (WERNER et al., 1994). In these procedures, the unfolded protein solubilized in the presence of denaturants such as Gdm/HCl is loaded on a gel filtration column equilibrated with the refolding buffer. Because of their high molecular weight, polypeptides escape the denaturant front during chromatography. Proteins then either fold to the native state and continue to elute or precipitate and stop migrating. These aggregates are subsequently resurrected by the following denaturant front which gives them a second chance for escape and folding. *In vitro* refolding by gel filtration chromatography should, however, only work if folding is much faster than the gel filtration process time.

5.4 Improved Refolding by Using Fusion Constructs with Hydrophilic Proteins or Peptides

If the yield of *in vitro* folding is limited by the low solubility of folding intermediates, it can be improved by fusing the protein of interest to hydrophilic partner proteins or peptides. If proteins or polypeptide domains are used as fusion partners to improve the folding of the desired protein, these proteins, besides being highly soluble, should refold fast and quantitatively. Furthermore, the fusion partner proteins should not interfere with the folding process of the target protein. These requirements are fulfilled by, e.g., the IgG-binding domain of staphylococcal protein A: When fused to insulin-like growth factor I, oxidative folding of the growth factor improved substantially (SAMUELSSON et al., 1991).

Similar effects could be observed using small hydrophilic peptide extensions (AMBROSIUS et al., 1996). As an example, the yield of oxidative folding of human granulocyte colony stimulating factor increased when small hydrophilic peptides were fused to its N-terminus. The drawback of fusion constructs is, however, that especially for therapeutic applications the extra proteins or peptides have to be removed by limited proteolysis.

5.5 Improved Refolding by Transient Binding of the Polypeptide to Solid Supports

It has been reported that *in vitro* folding can be improved by immobilization of the folding polypeptide on ion exchange matrices (CREIGHTON, 1990). This finding can be explained by transient binding of the unfolded polypeptides or folding intermediates to the ion exchange material, in analogy to chaperone function *in vivo*. The transient interaction of the folding polypeptides with the solid support reduces the concentration of folding polypeptides free in solution. This decrease in concentration of non-native polypeptides results in a decrease in aggregate formation. Furthermore, structure formation can occur while the protein is bound to the ion exchange matrix.

The disadvantage of the matrix-assisted folding using common chromatography supports is that it is difficult to generalize. If non-native proteins interact too strongly with the matrix, structure formation is completely blocked. Since folding intermediates are often very hydrophobic, strong hydrophobic interactions (especially at high ionic strength) also prevent structure formation. If, on the other hand, the interactions with the matrix are too weak, unproductive side reactions such as aggregation are not inhibited.

Specific binding of the unfolded polypeptide to the solid support while guaranteeing maximum freedom for structure formation can be achieved by using fusion constructs. As an example, histidine tags can be used for the immobilization of the unfolded material on metal affinity chromatography (IMAC) material (HOLZINGER et al., 1996; NEGRO et al., 1997). Upon buffer exchange to folding conditions, renaturation occurs. For the renaturation of disulfide bonded proteins, the IMAC material may, however, interfere with the redox buffer necessary for optimum disulfide bond formation.

As an alternative to histidine tags, polyionic sequences have been used for transient immobilization of the folding polypeptides (STEMPFER et al., 1996). In this study, the model protein, α-glucosidase, contained a polyarginine sequence fused to its N- or C-terminus. This polyionic sequence allowed a reversible immobilization of the unfolded protein on solid supports containing polyionic counter ions (Heparin-Sepharose®). After removal of the denaturant and protein contaminants, high yields of refolded species were observed at a high gel loading (Fig. 9). Folding is, however, only possible at moderate ionic strengths. At low salt concentrations, additional ionic interactions of the non-native polypeptide with the ion exchange matrix prevent structure formation. At high ionic strength, on the other hand, hydrophobic interactions are too strong to permit folding. Furthermore, one has to keep in mind that for large scale production low cost matrices other than Heparin-Sepharose^R are necessary.

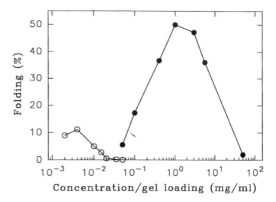

Fig. 9. Improved folding by transient immobilization (STEMPFER et al., 1996). Comparison of the refolding of an α-glucosidase fusion protein containing an extra hexa-arginine sequence at the C-terminus, free in the solution (open circles) and immobilized on Heparin-Sepharose® (filled circles). After denaturation in 8 M urea, refolding was performed at a given protein concentration in 0.01 M potassium phosphate, 30 mM NaCl, 8% ethylene glycol, pH 7.6. Folding was determined after 48 h of incubation at 10 °C.

5.6 Improving *in vitro* Folding by Molecular Chaperones or Folding Catalysts

It is now well established that protein folding *in vitro* can be improved considerably in the presence of molecular chaperones or folding catalysts (HARTL, 1996). While molecular chaperones reduce unproductive side reactions by transient interaction with the folding polypeptide, folding catalysts such as prolyl-peptidyl isomerases and protein disulfide isomerases speed up rate determining folding steps. Despite their efficiency in improving *in vitro* structure formation, these folding enhancers are not yet used in industrial production processes. The main reason to avoid folding catalysts in industrial processes is increasing production costs. In order to reduce increasing costs associated with folding enhancers, attempts were undertaken to use these proteins in an immobilized form (BUCHNER et al., 1992; ALTAMIRANO et al., 1997).

Although acting promiscuously molecular chaperones are to a certain extent specialized with respect to the nature of their substrates. Thus, despite *in vitro* folding of eukaryotic proteins may be substantially improved by using prokaryotic chaperones and *vice versa,* the different members of a chaperone family recognize different folding intermediates (e.g., unfolded polypeptides, folding intermediates with exposed hydrophobic patches, etc.). Because of this substrate specificity, it is not yet possible to pick out the right chaperone on a rational basis for a specific folding problem. As for folding catalysts, prolylpeptidyl isomerases and thiol–disulfide isomerases only improve the yield of *in vitro* folding when the rate limiting folding step competing with aggregation is determined by these isomerization steps. Thus, the application of molecular chaperones and folding catalysts for industrial protein folding remains to be established. As, however, many folding problems can be solved by using the above mentioned low molecular weight additives or by optimizing the process parameters. Expanding our knowledge of the mechanism by which folding enhancers stimulate structure formation will help to establish a future technical application for these molecules in production processes.

6 Prevention of Inclusion Body Formation by Coexpression of Chaperones or Folding Catalysts

Many attempts have been made to circumvent inclusion body formation by the coexpression of chaperones or foldases (GEORGIOU and VALAX, 1996). Upon cytosolic expression of non-disulfide bonded proteins correct folding can be improved considerably by the coexpression of molecular chaperones. There is, however, no explanation to date, why coexpression works in certain cases, while it fails in others.

In the case of proteins which contain disulfide bonds in their native form, correct folding is usually not possible in the reducing cytosol. It can, however, be improved by a modulation of the cytosolic redox conditions (DERMAN et al., 1993). On the other hand, these manipulations reduce the vitality and viability of the *E. coli* cells. As an alternative, correct disulfide bond formation can be achieved upon secretion of the recombinant protein into the periplasmic space. In this compartment, disulfide bonds are introduced into endogenous, periplasmic proteins which only contain few disulfide bonds. Correct disulfide bond formation of eukaryotic proteins with multiple disulfide bonds is, however, often very inefficient in the periplasm of *E. coli.*

7 Conclusions

Although quite exotic in the 1980s, inclusion body renaturation is now firmly established in industrial processes. It is applied in the large scale production of therapeutic and diagnostic proteins, as well as in the laboratory scale isolation of target proteins for establishing screening systems or structure analysis. As for other processes in biotechnology, the optimum conditions for *in vitro* folding have to be established on a case-by-case basis. The first round of the development of a folding process comprises a crude variation of the folding conditions such as various additives, protein concentration, pH, temperature, time, and ionic strength. In order to reduce production costs, expensive chemicals should then be decreased in concentration or exchanged by cheaper ones. The second round of process development involves the fine-tuning of the solvent conditions. After careful optimization, *in vitro* processes are competitive with alternative methods for recombinant protein production with respect to production costs, process development time, product quality and batch to batch consistency.

Acknowledgement

This work was supported by the Bundesministerium für Bildung, Wissenschaft, Forschung und Technologie (BMBF).

8 References

AHMED, A. K., SCHAFFER, S. W., WETLAUFER, D. B. (1975), Nonenzymatic reactivation of reduced bovine pancreatic ribonuclease by air oxidation and by glutathione oxidoreduction buffers, *J. Biol. Chem.* **250**, 8477–8482.

ALTAMIRANO, M. M., GOLBIK, R., ZAHN, R., BUCKLE, A. M., FERSHT A. R. (1997), Refolding chromatography with immobilized minichaperones, *Proc. Natl. Acad. Sci. USA* **94**, 3576–3578.

AMROSIUS, D., RUDOLPH, R. (1997), *U. S. Patent* 5618927.

AMBROSIUS, D., DONY, C., RUDOLPH, R. (1996), *U. S. Patent* 5578710.

BUCHNER, J., PASTAN, I., BRINKMANN, U. (1992), Renaturation of a single-chain immunotoxin facilitated by chaperones and protein disulfide isomerase, *Biotechnology* **10**, 682–685.

CREIGHTON, T. E. (1990), *U.S. Patent* 4977248.

CREIGHTON, T. E. (1992), *Protein Folding.* Freeman: New York.

DATAR, R. V., CARTWRIGHT, T., ROSEN, C. G. (1993), Process economics of animal cell and bacterial fermentations: a case study analysis of tissue plasminogen activator, *Bio/Technology* **11**, 349–357.

DERMAN, A. I., PRINZ, W. A., BELIN, D., BECKWITH, J. (1993), Mutations that allow disulfide bond formation in the cytoplasm of *Escherichia coli, Science* **262**, 1744–1747.

EPSTEIN, C. J., GOLDBERGER, R. F., ANFINSEN, C. B. (1963), The genetic control of tertiary protein structure: Studies with model systems. *Cold Spring Harbor Symposia* **28**, p. 439, New York: Cold Spring Harbor Press.

FAHEY, R. C., HUNT, J. S., WINDHAM, G. C. (1977), On the cysteine and cystine content of proteins. Differences between intracellular and extracellular proteins, *J. Mol. Evol.* **10**, 155–163.

GEORGIOU, G., VALAX, P. (1996), Expression of correctly folded proteins in *Escherichia coli, Curr. Opin. Biotechnol.* **7**, 190–197.

GOEDDEL, D. V., KLEID, D. G., BOLIVAR, F., HEYNEKER, H. L., YANSURA, D. G. et al. (1979), Expression in *Escherichia coli* of chemically synthesized genes for human insulin, *Proc. Natl. Acad. Sci. USA* **76**, 106–110.

GRIBSKOV, M., BURGESS, R. R. (1983), Overexpression and purification of the sigma subunit of *Escherichia coli* RNA polymerase, *Gene* **26**, 109–118.

HARTL, F. U. (1996), Molecular chaperones in cellular protein folding, *Nature* **381**, 571–579.

HEVEHAN, D., DE BERNARDEZ CLARK, E. (1997), Oxidative renaturation of lysozyme at high concentrations, *Biotechnol. Bioeng.* **54**, 221–230.

HOLZINGER, A., PHILLIPS, K. S., WEAVER, T. E. (1996), Single-step purification/solubilization of recombinant proteins: application to surfactant protein B, *Biotechniques* **20**, 804–806.

ITAKURA, K., HIROSE, T., CREA, R., RIGGS, A. D., HEYNECKER, H. L. et al. (1977), Expression in *Escherichia coli* of a chemically synthesized gene for the hormone somatostatin, *Science* **198**, 1056–1063.

JAENICKE, R. (1980), *Protein Folding.* Amsterdam, New York: Elsevier.

KIEFHABER, T., RUDOLPH, R., KOHLER, H.-H., BUCHNER, J. (1991), Protein aggregation *in vitro* and *in vivo*: a quantitative model of the kinetic competition between folding and aggregation, *Bio/Technology* **9**, 825–829.

KOPETZKI, E., SCHUMACHER, G., BUCKEL, P. (1989), Control of formation of active soluble or inactive insoluble bakers' yeast alpha-glucosidase PI in *Escherichia coli* by induction and growth conditions, *Mol. Gen. Genet.* **216**, 149–155.

MARSTON, F. A. O. (1986), The purification of eukaryotic polypeptides synthesized in *Escherichia coli, Biochem. J.* **240**, 1–12.

MICHAELIS, U., RUDOLPH, R., JARSCH, M., KOPETZKI, E., BURTSCHER, H., SCHUMACHER, G. (1995), *U. S. Patent* 5434067.

NEGRO, A., ONISTO, M., GRASSATO, L., CAENAZZO, C., GARBISA, S. (1997), Recombinant human TIMP-3 from *Escherichia coli*: synthesis, refolding, physico-chemical and functional insights, *Protein Eng.* **10**, 593–599.

OPITZ, U. (1988), Reoxidation, Reinigung und Charakterisierung von rekombinanten tissuetype Plasminogenaktivator aus *E. coli, Thesis,* University of Regensburg, Germany.

ORSINI, G., GOLDBERG, M. E. (1978), The renaturation of reduced chymotrypsinogen A in guanidine HCl. Refolding versus aggregation, *J. Biol. Chem.* **253**, 34–39.

RUDOLPH, R. (1990), Renaturation of recombinant, disulfide-bonded proteins from inclusion bodies, in: *Modern Methods Protein-and Nucleic*

Acid Research (TSCHESCHE, H., Ed.), pp. 149–172. New York, Berlin: deGruyter.

RUDOLPH, R. (1996), Successful protein folding on an industrial scale, in: *Protein Engineering: Principles and Practice* (CLELAND, J. L., CRAIK, C. S., Eds.), pp. 283–289, New York: John Wiley & Sons.

RUDOLPH, R., FISCHER, S. (1990), *U. S. Patent* 4 933 434.

RUDOLPH, R., LILIE, H. (1996), *In vitro* folding of inclusion body proteins, *FASEB J.* **10**, 49–56.

RUDOLPH, R., FISCHER, S., MATTES, R. (1997a), *U. S. Patent* 5 593 865.

RUDOLPH, R., BÖHM, G., LILIE, H., JAENICKE, R. (1997b), Folding proteins, in: *Protein Function* (CREIGTHON, T. E., Ed.), pp. 57–99. Oxford, New York, Tokyo: IRL Press, Oxford University Press.

SAMUELSSON E., WADENSTEN, H., HARTMANIS, M., MOKS, T., UHLEN, M. (1991), Facilitated *in vitro* refolding of human recombinant insulin-like growth factor 1 using a solubilizing fusion partner, *Bio/Technology* **9**, 363–366.

SELA, M., WHITE, F. H., ANFINSEN, C. B. (1957), Reductive cleavage of disulfide bridges in ribonuclease, *Science* **16**, 691–692.

STEMPFER, G., HÖLL-NEUGEBAUER, B., RUDOLPH, R. (1996), Improved refolding of an immobilized fusion protein, *Nature Biotechnology* **14**, 329–334.

WERNER, M. H., CLORE, G. M., GRONENBORN, A. M., KONDOH, A., FISHER, R. J. (1994), Refolding proteins by gel filtration chromatography, *FEBS Lett.* **345**, 125–130.

ZETTLMEISSL, G., RUDOLPH, R., JAENICKE, R. (1979), Reconstitution of lactic dehydrogenase. Noncovalent aggregation vs. reactivation. 1. Physical properties and kinetics of aggregation, *Biochemistry* **18**, 5567–5571.

5 Medical Applications of Recombinant Proteins in Humans and Animals

GAYLE DELMONTE WETZEL

Berkeley, CA, USA

1 Introduction   129
2 Human Medical Applications of Recombinant Proteins   129
  2.1 Specific Diseases and Applications   129
    2.1.1 Major Non-Immune Based Diseases   129
      2.1.1.1 Cardiovascular and Cerebrovascular Thrombembolic Disorders   129
      2.1.1.2 Stroke   130
      2.1.1.3 Restenosis   131
      2.1.1.4 Heart Failure   132
      2.1.1.5 Fibrosis   132
      2.1.1.6 Osteoporosis   133
      2.1.1.7 Angiogenesis   133
      2.1.1.8 Insulin Resistance and Non-Insulin-Dependent (Type II) Diabetes mellitus (NIDDM)   134
      2.1.1.9 Sepsis   134
    2.1.2 Immunoenhancement   135
      2.1.2.1 Tumor therapy   135
      2.1.2.2 Vaccines   137
    2.1.3 Immunosuppression   141
      2.1.3.1 Costimulation   141
      2.1.3.2 Antigen Presentation   141
      2.1.3.3 Immune Deviation   142
      2.1.3.4 Multiple Sclerosis   142
      2.1.3.5 Psoriasis   143
      2.1.3.6 Arthritis   143
      2.1.3.7 Insulin-Dependent Diabetes mellitus   144

2.1.3.8 Inflammatory Bowel Disease (IBD)   144
2.1.3.9 Allergy and Asthma   145
2.1.3.10 Allotransplantation   145
2.1.3.11 Xenotransplantation   146
2.2 Specific Recombinant Proteins and their Applications   146
2.2.1 Cytokines   146
2.2.1.1 Interferons   147
2.2.1.2 Interleukins   148
2.2.1.3 Growth and Colony Stimulating Factors   155
2.2.1.4 Nerve Growth Factors   158
2.2.1.5 Other Growth Factors   160
2.2.2 Hormones   163
2.2.2.1 Insulin   163
2.2.2.2 Growth Hormone and Insulin-Like Growth Factors   163
2.2.2.3 Other Hormones   164
2.2.3 Anti-Inflammatory Proteins   165
2.2.3.1 Chemokines   165
2.2.3.2 Complement   166
2.2.4 Recombinant Enzymes and Enzyme Inhibitors   167
2.2.4.1 For Cystic Fibrosis   167
2.2.4.2 For Inflammation   167
2.2.4.3 For Cancer   168
2.2.4.4 For Gaucher's Disease   168
2.2.4.5 Other Enzymes   168
2.2.5 Clotting Factors and Related Proteins   168
2.2.6 Fusion Proteins   169
2.2.7 Miscellaneous Proteins   170
3 Applications in Animals   170
4 Concluding Remarks   172
5 References   173

List of Abbreviations

AIDS	acquired immunodeficiency syndrome
ALS	amyotrophic lateral sclerosis disease
AML	acute myelogenous leukemia
ARC	AIDS related complex
ARDS	acute respiratory distress syndrome
AT	anti-thrombin
BCG	tuberculosis vaccine
BDNF	brain-derived neurotrophic factor
bFGF	basic FGF
BMP	bone morphogenetic protein
BMP-7	osteogenic protein 1, OP-1
BNP	brain natriuretic peptide
BPI	bacterial permeability increasing protein peptide
C3a	complement component
C4a	complement component
C5a	complement component
CCR3	chemokine receptor
CD14	LPS receptor
CD2	T cell surface protein
CD28	T cell surface molecule
CD4	T cell surface molecule
CD48	CD2 ligand
CD59	complement inactivator
CD8	T cell surface molecule
CD95	T cell surface molecule
CEA	carcinoembryonic antigen
CG	chorionic gonadotropin
CGRP	calcitonin gene related peptide
CHF	congestive heart failure
CKR	chemokine receptor
CML	chronic myelogenous leukemia
CMV	cytomegalovirus
CNTF	ciliary neurotrophic factor
CR	complement receptor
CR1	receptor for C3
CR2	receptor for C3
CR3	receptor for C3
CRH	corticotropin releasing hormone
CSF	colony stimulating factor
CTGF	connective tissue growth factor
CTL	cytotoxic T lymphocyte
DAF	decay accelerating factor
EAE	experimental autoimmune encephalitis
EDF	erythroid differentiation factor
EGF	epidermal growth factor
EGR	chemically inactivated factor Xa
EPO	erythropoietin
ER	endoplasmatic reticulum
ET-1	endothelin-a
Fas	apoptosis inducing ligand receptor, CD95
Fc	constant region of immunoglobulins
FGF	fibroblast growth factor
FSH	follicle stimulating hormone
G-CSF	granulocyte colony stimulating factor
GDF	growth and differentiation factor
GDNF	glial-derived neurotrophic factor
GGF	glial cell growth factor
GH	growth hormone
GHRF	growth hormone releasing factor
GHRP	growth hormone releasing peptide
GM-CSF	granulocyte macrophage colony stimulating factor
GMF	glial maturation factor
gp	glycoproteins of HIV
GRO	proinflammatory cytokine family member
HA	hemagglutinin protein
HAR	hyperacute rejection
hCG	human β-chorionic gonadotropin
HgB	hemoglobin B
HGF	hepatocyte growth factor
HIV	human immunodeficiency virus
HPV	human papilloma virus
HSV	herpes simplex virus
IBD	inflammatory bowel disease
ICAM	intercellular adhesion molecule
ICE	interleukin converting enzyme
ICP	immediate/early protein
IFN	interferon
Ig	immunoglobulin
IGF	insulin-like growth factor
IGF-BP	IGF-binding protein
IL	interleukin
IL-1ra	IL-1 receptor antagonist
IND	introduction of a new drug
IP-10	proinflammatory cytokine family member
LAK	lymphokine activated killer cells

LBP	LPS-binding protein	PECAM	platelet endothelial cell adhesion molecule
LeIF	*Leishmania*-derived protein		
LFA	leukocyte functional antigen	PEG	polyethylene glycol
LH	luteinizing hormone	PF	platelet factor
LIF	leukemia inhibitory factor	PMN	polymorphonuclear neutrophils
LPS	lipopolysaccharide	PTCA	percutaneous transluminal coronary angioplasty
mAb	monclonal antibody		
MAdCAM	Ig-family adhesion receptor, preferentially expressed by mucosal venula endothelial cells	PTH	parathyroid hormone
		RA	rheumatoid arthritis
		ra	receptor antagonist
MART-1	melanoma antigen recognized by T cells	RANTES	regulated upon activation, normal T cell expressed and secreted
MCP	monocyte chemoattractant protein	RBC	red blood cell(s)
MCP-1	membrane cofactor protein 1	RGD	arginine, glycine, aspartic acid
M-CSF	macrophage colony stimulating factor	rh	recombinant human
		rhEPO	recombinant human EPO
MGDF	megakaryocyte growth and development factor	r-ProUK	recombinant prourokinase
		RSV	respiratory syncytial virus
MGF	megakaryocyte growth factor	SCF	stem cell factor
MGSA	megakaryocyte growth stimulating activity	sIg	surface Ig
		SLE	systemic lupus erythematosus
MHC	major histocompatibility complex	SMC	smooth muscle cell
		SOD	superoxide dismutase
MI	myocardial infarction	SRC	SRC VB vector: commercial producer of erythropoietin
MIP	macrophage inhibitory protein		
MMP	matrix metalloproteinase	TAP	transporter associated proteins
MP-52	recombinant BMP from Hoechst	TcR	T cell receptor
		t-PA	Alteplase (tissue plasminogen activator), pharmaca
MS	multiple sclerosis		
Muc-1	mucine core peptide	TF	tissue factor
NEP	neutral endopeptidase	TGF	transforming growth factor
NGF	nerve growth factor	Th	T helper cell
NIDDM	non-insulin-dependent diabetes mellitus	TNF	tumor necrosis factor
		TNFR	tumor necrosis factor receptor
NIH	National Institutes of Health	TNK-tPA	t-PA, tissue plasminogen activator, pharmaca
NK cell	natural killer cell		
nPA	neutral plasminogen activator	TP	tissue plasminogen activator
NPY	neuropeptide Y	tPA	tissue plasminogen activator
NT	neurotrophin	TPO	thrombopoietin
OCT43	mutant IL-1β	TSA	tumor-specific antigen
OP	osteogenic protein	TSP	thrombospondin
OP-1	osteogenic protein 1, BMP-7	VEGF	vascular endothelial cell growth factor
OSM	Oncostatin M		
PAF	platelet activating factor	VLA	very late antigen
PDGF	platelet-derived growth factor	WBC	white blood cell

1 Introduction

The medical uses of recombinant proteins will be the focus of this chapter with concentration on the therapeutic uses. Although the diagnostic uses are many, these and the veterinary applications will receive only the briefest of mention. Furthermore, the use of recombinant receptors, or ligands, as tools for the development of non-peptide therapeutic agents is not considered here.

The breadth of human therapeutic applications could easily command an entire book for adequate coverage. According to an article in a recent *Genetic Engineering News* (STERLING, 1996), there are 284 medically related, biotechnology-derived candidates in some stage of development. Of these, 28 are for gene therapy and 62 are vaccines for various infectious diseases, cancers and other indications. Of the remaining 194, some 14 or so represent cell-based therapies and over 60 are various monoclonal antibodies or their fragments. This leaves in excess of 100 recombinant proteins in development as therapeutic agents, not to mention candidates for diagnostic purposes. As each effort is significant and would require some detail for complete understanding of the technical aspects of development, a thorough accounting of the development of recombinant proteins might easily represent an effort of epic proportion. For this reason, some principles in disease mechanisms and cytokine/recombinant protein function will be concentrated on, as understood at the time of writing. These principles hopefully will remain relevant for a significant time period subsequent to the printing of this chapter.

The previous paragraph notwithstanding, any attempt to count products in various stages of development runs a high risk of being both inaccurate and outdated by the time it is read. Therefore, only some of the more important candidate products in development and some of the underlying reasons for such efforts will be presented. It is also important to remember that development of some recombinant proteins can be halted at any point, due to disappointing clinical results or changing market concerns, only to be re-sumed at a later point for perhaps another indication or for other reasons. Therefore, information will also be included on some proteins which have not yielded data as promising as originally hoped but whose development may yet be renewed. Some of the categorization that follows may appear rather arbitrary since several proteins can be used for individual diseases and several diseases may be treated with each individual protein. Since the intent is to address each focus area in some detail, a small amount of overlap is included.

2 Human Medical Applications of Recombinant Proteins

2.1 Specific Diseases and Applications

2.1.1 Major Non-Immune-Based Diseases

2.1.1.1 Cardiovascular and Cerebrovascular Thrombembolic Disorders

Arterial and venous occlusion subsequent to thrombus formation is a life threatening situation and is responsible for significant mortality and morbidity. Plasminogen activators are being explored, and have provided some therapeutic effect, for patients of acute myocardial infarction and ischemic stroke. For a review on the biology of plasminogen activator see CAMERER et al. (1996). Fibrinolysis due to plasmin can result in restored blood flow in coronary arteries when recombinant tissue plasminogen activator (t-PA) is given shortly after MI, usually within 6 h. Although one study using i.v. *Alteplase* (t-PA) was able to demonstrate some effectiveness in improving functional and neurological measures, significantly higher rates of intracranial hemorrhage were a risk. Two other studies, however, showed that despite this increased risk, in ischemic stroke patients given *Alteplase* within the first 3 h the 3 month clinical outcome was improved with no increase in mortality (FRÖHLICH, 1996) [for data from other relevant clinical trials see GOLDHABER et al. (1996); SASAHARA et al. (1995); FALANGA et al. (1996); ROMEO et al. (1995), SMALLING et al. (1995) and CAMERER et al. (1996)]. Much work now concentrates on improved 2nd generation, plasminogen activators. *Reteplase,* from Boehringer Mannheim GmbH, has a 3–4 fold longer half-life than t-PA and appears to be at least as efficacious. Compared to streptokinase, it was as effective at reducing mortality but had a lower associated incidence of heart failure (MARTIN, 1996b). Another variant t-PA called TNK-tPA demonstrated a prolonged half-life in humans in a phase I trial. Staphylokinase has also been examined in MI patients and succeeded in quickly recanalizing coronary arteries. This protein is more fibrin specific than t-PA but a drawback to repeated use is its ability to induce neutralizing antibodies (VANDERSCHUEREN, 1996). For a comparison of tPA and staphylokinase see VANDERSCHUEREN et al. (1995). Bristol-Myers Squibb has produced a recombinant t-PA lacking a glycosylation site, a fibronectin domain and an epidermal growth factor domain. These variations provide a longer plasma half-life while demonstrating clot lysis equivalent to *Alteplase* (CHEW, 1996). Schering has a novel approach using a naturally occurring plasminogen activator from the vampire bat, *Desmodus rotundus* [for description see GOHLKE et al. (1995); SCHLEUNING (1996)]. One of 4 isolated bat plasminogen activators demonstrated a remarkable specificity for fibrin. When activity in the presence of fibrin versus in the presence of fibrinogen is measured, it is far superior to t-PA. This suggests that this protein might provide plasminogen activation solely on the surface of clots and not lead to fibrinogen consumption. This might be a safer thrombolytic than those currently used.

Another xenogeneic plasminogen activator is being investigated. *Calin* is derived from a leech and has the ability to inhibit collagen-induced platelet aggregation. As such, it can inhibit clot formation in certain settings. Biopharm and Merck are developing this agent for use in several surgical procedures including cardiovascular surgery.

The cascade of events leading to clot formation provides several potential targets for therapeutic intervention. Similar to the situation with restenosis, platelet aggregation is an initiating event whose final common pathway involves the platelet integrin $\alpha_{IIIb}\beta_3$, regardless of the activation mechanism. Blockade of the integrin receptor has been attempted with antagonists such as *ReoPro* (mAb) (see Chapter 16, this volume), *Integrelin, Triofiban* and *Lamifiban*. Although bleeding may be a complication (AGUIRRE et al., 1995), some clinical benefit could be seen for a period of 6 months, perhaps due to early benefit and longer-term prevention of restenosis (TOPOL et al., 1994).

Anti-coagulants targeting thrombin and factor Xa can provide protection against thrombosis in several settings. Deep vein thrombosis subsequent to orthopedic or other surgery can be treated with non-peptide molecules like heparin. However, recombinant hirudin (*Revasc*) has been shown to provide a superior benefit to risk ratio, compared to unfractionated hirudin and to low molecular weight heparin, in a study of patients undergoing total hip replacement (ERIKSSON, 1996). For some results of some clinical trials with hirudin see ERIKSSON et al. (1996) and COFRANCESCO et al. (1996). Endpoints in such studies are reduced mortality and reduced pulmonary embolism rates. On the other hand, the incidence of hemorrhage is a trade-off which must be monitored. For a recent review on anti-thrombotics see BEIJERIN et al. (1996).

2.1.1.2 Stroke

Cerebral ischemia and subsequent reperfusion provoke many biochemical and cellular responses. Immediate early genes are rapidly induced which provide transcription factors. Later gene expression leads to production of neurotrophic factors and neurotransmitters. Inflammatory cell influx and elaboration of mediators like TNFα and IL-1 ensue. There is evidence indicating these cytokines function in ischemic damage and contribute to necrotic and apoptotic cell death. For recent reviews on stroke see MORLEY et al. (1996) and BRICKNER (1996). One potentially fruitful approach might be to employ anti-cytokine treatments in managing stroke since there is ample evidence of local cytokine production in this disease (WANG et al., 1994). An alternate approach is to use basic fibroblast growth factor (bFGF). This cytokine has been shown to be a potent neuroprotective protein in vitro. Animal models employing cerebral artery occlusion have shown that bFGF administration can reduce infarct size by as much as 50% (KLINGBEIL and EPSTEIN, 1994). A clinical trial in humans has been initiated by Scios, Inc (GROSSBARD, 1996).

A recent study by the NINDS Stroke Study Group showed that early administration of t-PA reduced injury so that treated patients were 40–80% more likely to have minimal or no disability. Another thrombolytic, recombinant prourokinase (*r-ProUK*) developed by Abbott Labs, showed a 3–4 fold increase in recanalization of middle cerebral arteries after stroke-induced ischemia. Reduced 90 d mortality rates were also observed (ZARICH et al., 1995; SASAHARA et al., 1995; BARKER, 1996).

Some brain injury following infarction appears to be mediated by neutrophils. These cells use ICAM-1 (intercellular adhesion molecule-1) for adhesion to endothelium and migration into the brain parenchyma. ICAM-1 also appears to be involved in activation of neutrophil cytotoxicity. Boehringer Ingelheim Pharmaceuticals, Inc, has developed a murine anti-ICAM-1 mAb in phase III clinical development (POLMAR, 1996). A similar potential approach without the complications of a foreign antibody protein might be to employ recombinant truncated ICAM-1 as an antagonist.

2.1.1.3 Restenosis

Restenosis is a secondary closure of blood vessels subsequent to opening by surgical mechanical means and the use of stents. For oc-cluded coronary arteries the angioplasty is often referred to as PTCA (percutaneous transluminal coronary angioplasty), however, this technique is also used for vascular access grafts for hemodialysis and for vascular grafts in treating peripheral artery disease. Although the high economic and medical costs are clear, and despite intensive research on its pathophysiology, the process of restenosis remains poorly understood. For recent reviews see DANGAS and FUSTER (1996) and STOUFFER et al. (1996).

Among the many contributing factors is early platelet deposition, activation and release of vasoactive mediators through degranulation. This acute response to vessel injury leads to thrombotic complications in transition to the later chronic response involving remodeling, matrix deposition and smooth muscle cell (SMC) proliferation. One hypothesis is that SMC proliferation leads to hypercellularity due to deregulated apoptosis of these cells. Endothelin-a (ET-1) is a potent vasoconstrictor peptide which is mitogenic for smooth muscle cells. Subsequent to interaction with its receptors on SMCs, several growth factors are released. These properties have led to the hypothesis that ET-1 may function in restenosis. A receptor antagonist protein has been developed by Abbott Labs which blocks the biological properties of ET-1 and has been used in animal models of balloon-induced vascular injury and neointima formation (BURKE, 1996). A similar approach taken by Prizm Pharmaceuticals involves the use of a recombinant fibroblast growth factor-2-saporin mitotoxin conjugate (LAPPI et al., 1991) to inhibit SMC proliferation (AUKERMAN, 1996). SMC migration is induced by osteopontin and vitronectin. Protein agents inhibiting interactions of these proteins with their receptors would have to compete with small molecules which appear to be effective, although their modes of inhibition may be different. Early thrombotic events appear related to the thickness of the subsequently formed neointima. Therefore, one approach might be to develop inhibitors of thrombin, factor Xa and GPIIb/IIIa. A competing principle from Centocor is a monoclonal antibody to specific integrin receptors on platelets ($\alpha_{IIb}\beta_3$) and SMC (SMC $_a\omega_{b3}$), referred to in

Sect. 2.1.1.1 as *ReoPro* [see WOODY (1996) for a related publication]. Clinical efficacy has been demonstrated with this agent and it is approved for use in humans. There is, however, significant room for improvement in success rates.

2.1.1.4 Heart Failure

Heart failure has a remarkable prevalence in developed countries. About 1% of individuals aged 50–59 are affected and this figure doubles with each succeeding decade of life. Angiotensin II now is being appreciated as a major inducer of both pressure-dependent and pressure-independent mechanisms. Endothelin also functions to modulate myocardial contraction. Levels of this protein increase with congestive heart failure development. It may also play an important role in the pathology of pulmonary hypertension secondary to congestive heart failure (CHF). For a recent review of endothelins see MICHAEL and MARKEWITZ (1996); for a study examining effects of tPA and endothelin antagonists in animals see UMEMURA et al. (1995). Hence, targeting the endothelins with protein inhibitors may be a viable route of therapy for CHF. To date, most efforts to inhibit the activities of these proteins have focused on converting enzyme or receptor blockade.

Progressive dilatation is a feature of the failing heart. This is brought about by the function of metalloproteinases involved in degradation of interstitial extracellular matrix proteins. As such, these enzymes are also potential targets for therapy of CHF. Neutral endopeptidase is a membrane bound member of the MMPs (matrix metalloproteinases) and it degrades natriuretic peptides and possibly bradykinin. Inhibition of this enzyme has been shown to reverse renal hyporesponsiveness to natriuretic peptides following CHF (YASUHARA et al., 1994). Little work in inhibiting these enzymes has been done with recombinant proteins. However, some treatments are designed to be restorative and employ administration of the α-human atrial natriuretic peptide itself (UKAI et al., 1995).

2.1.1.5 Fibrosis

Tissue undergoes a normal turnover. When an individual suffers from persistent infections or is subject to chronic inflammation, abnormalities in tissue turnover can occur. Excess collagen deposition characterized by extensive scarring can result. This then disrupts the three-dimensional architecture of an organ. The aberrant environment is either conducive to, or is associated with, further cellular metabolism abnormalities leading to organ failure. Such fibrotic events occurring in the liver and lung are often fatal. During vascular repair, SMCs proliferate and migrate to sites of injury. They then synthesize extracellular matrix proteins as a normal part of the repair process. Agents which stimulate these processes in SMCs include PDGF, bFGF, TGFβ as well as α-thrombin. Excessive production of these agents may well lead to a fibrotic state. For reviews of pulmonary fibrosis, one of the more fatal forms, see MAPEL et al. (1996) and MARTINET et al. (1996). Although the lung is a major target of fibrosis, a treatment for fibrosis would also be beneficial for fibroses of the kidney and liver and for systemic fibrosis in scleroderma.

TGFβ has been identified as a potent stimulator of collagen synthesis, matrix production, matrix metalloproteinase inhibitor production, fibroblast mitogenesis and inhibition of matrix degradation. Decorin is a potent inhibitor of TGFβ (YAMAGUCHI et al., 1990; HAUSSER et al., 1994). Neutralization of TGFβ by antibodies or decorin is effective in blocking collagen deposition and fibrosis in animal models. Hence, it may be quite effective as an anti-fibrotic agent. Telios is investigating the use of recombinant decorin in fibrotic diseases of the kidney, lung and liver as well as the skin.

There are several other proteins involved in fibrosis which may serve as targets for therapy. TGFβ has also been shown to induce connective tissue growth factor (CTGF) production in fibroblasts (IGARASHI et al., 1993). Several studies have shown that blocking the production or action of CTGF can also inhibit fibrotic events (GROTENDORST, 1996; MARTIN 1996a). Hence, inhibitors of CTGF may serve as therapeutic proteins in regulating fibrotic events. Prolyl hydroxylase C initiates collagen precipitation and is required for proper synthesis of matrix collagen. Inhibition of this enzyme also inhibits fibrosis.

Relaxin is a protein, elevated during pregnancy, which can increase skin elasticity. This occurs presumably since it inhibits collagen and fibronectin synthesis while at the same time increasing collagenase production. Systemic administration is capable of blocking pulmonary fibrosis (UNEMORI et al., 1993). For these reasons, relaxin is currently in clinical evaluation as a therapeutic agent for treatment of the systemic fibrotic disease scleroderma (AMENTO, 1996a).

2.1.1.6 Osteoporosis

Loss of bone mass and strength is a major medical problem with older people resulting in bone fractures and poor healing. For a review of the nature of osteoporosis and current approaches towards treatment see BELLANTONI (1996) and MARCUS (1996).

Biotechnology has focused on members of the IL-6 family of cytokines. These proteins have potent effects on bone metabolism through regulating osteoblast and osteoclast development and function. It is known that sex steroid hormones are powerful regulators of IL-6 gene expression as well as expression of genes for both the gp80 and gp130 chains of the IL-6 receptors. Hence, decreased levels of these hormones may release IL-6 from inhibition of expression and increase sensitivity of osteocytic lineage cells to this cytokine. These observations make IL-6 a likely target for antagonism in designing a therapy for osteoporosis.

On the other hand, the insulin like growth factors (IGFs) are very important in development and maintenance of bone mass in early adulthood. They are also important mediators of the normal bone remodeling processes. Delivery of a bound form of IGF-1 complexed to its most prevalent circulating binding proteins is a powerful and apparently safe method for introducing the effects of this hormone (ROSEN, 1997).

A series of bone morphogenetic proteins has been described. Members stimulate growth and function of osteoblasts. Some therapeutic approaches to treating osteoporosis involve these anabolic proteins. For example, studies are being done with BMP-7/OP-1 (osteogenic protein-1) (KUBER Sampath, 1997) in fracture repair and other settings. Creative Biomolecules has established a relationship with Biogen to develop this protein for several therapeutic uses (CRAIG, 1996).

A related hormone, parathyroid hormone (PTH) and derivative peptides from the protein itself are being tested since they can apparently induce increases in bone mass and strength in some animal models (VICKERY, 1997). Allelix is investigating this molecule.

2.1.1.7 Angiogenesis

Organogenesis, tissue repair and tumor growth are dependent on a common process known as angiogenesis. This process is under strict regulation by several agonist and antagonist cytokines and other mediators. Several animal models, including rabbit and mouse models of corneal neovascularization, a rabbit skin patch model and healing of induced wounds, have revealed the angiogenic properties of basic fibroblast growth factor (bFGF) and vascular endothelial cell growth factor (VEGF). Naturally occurring antagonists include TSP1 and TSP2, 2 members of the family of 5 thrombospondins, and platelet factor 4 (PF4) a component of platelet α granules. The process of angiogenesis is primarily dependent on the $\alpha_v\beta_3$ receptor (although the $\alpha_v\beta_5$ receptor may also be involved) on endothelial cells interacting with a variety of extracellular matrix components [for a recent review on cell adhesion molecules see POLVERINI (1996)]. Antagonizing this interaction has been shown to block new blood vessel formation as well as causing regression of some human tumors (see below). Lack of supportive interaction with these receptors leads to apoptosis of endothelial cells. These processes, then, provide targets for therapeutic intervention.

Aside from small molecules, and mAbs to integrins, and VEGF, recombinant soluble receptor for VEGF has been efficacious in an animal model of neovascularization (AIELLO et al., 1995). Recombinant PF4 has also been used in the clinic and had some activity in inhibiting angiogenesis (MAIONE et al., 1991; DEHMER et al., 1995; MAIONE, 1996). Two peptides, angiostatin an endostatin, have been found to inhibit endothelial cell proliferation. Endostatin is a fragment of one of the less abundant collagen proteins, collagen XVIII. Systemic administration of both angiostatin and endostatin has been shown to prevent tumor metastasis growth and reduce primary tumor growth in animal models. EntreMed Inc and others are pursuing these findings (SIM, 1997). Following the proof-of-principle demonstration that targeting the VEGF receptor with a monoclonal antibody blocked neovascularization (FERRARA, 1996), Merck has found a soluble truncated form of the VEGF receptor. It binds to cellular VEGF receptors and blocks transactivation from oligomerization subsequent to ligand binding (THOMAS, 1997). A surprising aside comes from a study with mAb to TNF in rheumatoid arthritis. The pannus surrounding the inflamed joints is highly vascular and serum VEGF levels are readily detectable. Treatment with anti-TNF mAb has been shown to reduce these serum VEGF levels (WOODY, 1996). This may suggest that treatments with other recombinant proteins reducing TNF would prevent angiogenesis.

Osteopontin is an RGD-containing adhesin glycoprotein implicated in chronic inflammatory and fibrotic disease states including atherosclerosis and restenosis. It interacts with multiple receptors including the $\alpha_v\beta_3$ receptor to stimulate smooth muscle cell migration and proliferation thereby functioning in neointima formation. Cyclic peptides antagonizing the $\alpha_v\beta_3$ receptor can disrupt angiogenic events in the chick chorioallantoic membrane model and integrin antagonists have been shown to cause tumor regression in animal models (BROOKS et al., 1994).

2.1.1.8 Insulin Resistance and Non-Insulin-Dependent (Type II) Diabetes mellitus (NIDDM)

Obesity and NIDDM syndromes share some common features including resistance to the metabolic effects of insulin, impaired glucose tolerance and hyperinsulinemia. Hyperglycemia, most likely, is a major initiator of vascular and neurological complications. To date the causative molecular mechanisms responsible for induction of insulin resistance remain incompletely understood. However, treatment of adipocytes with TNFα leads to a breakdown in lipid homeostasis and changes in gene expression which may be related to development of insulin resistance. On the other hand, it is unclear how useful therapies involving TNFα antagonists, including antibodies, might be as treatments for this chronic condition.

Recent breakthroughs involving the discovery of a novel protein, called leptin, are illuminating previously unknown relationships between obesity, adipocyte metabolism and insulin sensitivity (SEGAL et al., 1996; HAFFNER et al., 1996). Leptin is a circulating protein whose blood levels correspond closely to amounts of fat stored. It acts on a receptor in the hypothalamus to participate in inhibition of neuropeptide Y (NPY) synthesis and release, thereby influencing lipid metabolism and obesity (ERIKSON et al., 1996). Although leptin has been proposed as a treatment for obesity, mice transgenic for this protein may develop kidney disease. Its association with diabetes has been suggested by the observation that insulin resistance in rodents is linked to circulating leptin levels (SEGAL et al., 1996). No similar correlation has yet been demonstrated in humans however (HAFFNER et al., 1996).

Human insulin like growth factor I (IFG-I) regulates carbohydrate metabolism and acts to lower blood glucose levels without causing weight gain. Some success in inhibiting progression of pathology has been achieved using IFG-I in patients with various stages of insulin resistance and type II diabetes mellitus (CHEETHAM et al., 1995; MOSES, 1996; ZYSOW, 1996). For these reasons, therapeutic use of this protein is being pursued by Genentech, Inc.

2.1.1.9 Sepsis

Infection following major burns or trauma is life threatening and a major cause of death despite modern therapy and antibiotics. Mortality is estimated to be between 20% and 40%. Endotoxin release is a major mediator of morbidity and of pathology. LPS (lipopolysaccharide) binding protein (LBP) catalyzes the transfer of LPS to the LPS receptor CD14. CD14 can exist on cell membranes and in solution where it can still interact with cell membranes, when charged with LPS. A cascade of events and prodigious production of the inflammatory cytokines TNFα, IL-1 IL-8 and IL-6 is triggered. Not only gram-negative bacteria, but gram-positive bacteria, viruses, and deep trauma are capable of initiating the syndrome of shock and the sequellae. Cytokine production can appear to be cyclical since these mediators bind to cellular receptors, inducing both inflammatory damage and initiating regulatory compensating mechanisms. Nitric oxide is one of several induced mediators resulting in organ damage. For recent reviews see THIJS et al. (1996); BLACK-WELL and CHRISTMAN (1996); BONE (1996) and AYALA and CHAUDRY (1996).

Several approaches targeting intermediates in the cascade of events have been tried. IL-1 receptor antagonist was very promising in early clinical trials, as were early trials with antibodies to TNF. To date, however, no successful clinical trials have been completed which demonstrate statistically significant efficacy of the agent being tested, perhaps due to small sample sizes or flawed theoretical approaches (for clinical trial results see references given with each of the individual cytokines/antagonists in Sect. 2.2.1). Synergy between IL-1 and TNF in inducing symptoms of septic shock make these targets particularly difficult. However, a recent phase II trial (Hoffmann-LaRoche) using a fusion protein of the p55 TNF receptor fused to an antibody Fc region to provide bivalency and high anti-TNF affinity interactions (LEIGHTON, 1996; ABRAHAM, 1996) has shown promise. Other approaches target CD14, TNF and IL-1 signaling pathway molecules and enzymes responsible for releasing IL-1 and TNF. Use of recombinant soluble receptors for TNF in treating sepsis is being explored by Yeda and Serono. Immunex has had some experience in this area, as well. Others with experience developing TNF receptors and TNF receptor fu-

sion proteins include Genentech, Amgen/Synergen and Hoechst. Recombinant IL-11 has been used to maintain gut epithelial membrane integrity with some success although this approach does not target the initiating mediators (OPAL, 1996).

2.1.2 Immunoenhancement

2.1.2.1 Tumor Therapy

Surgery can be used to remove the bulk of solid tumors but metastases and leukemias are unaffected. Treatment to kill rapidly dividing cells, such as irradiation and chemotherapy, are useful in eradicating this type of tumor cell. In the biotechnology arena where recombinant proteins are some of the weapons of choice for battling tumors, it is hoped that monoclonal antibodies, immunotoxin–conjugates, new vaccines and immunoadjuvants will also be effective.

Effective cancer vaccines are now potentially in sight due to recent advances in identification of tumor associated antigens. For example, idiotype vaccines are a possibility for B cell lymphomas; since these are of clonal origin and, therefore, express only a single combining site specificity on their surface immunoglobulins (KOBRIN, 1996). One disadvantage to this type of therapy is the ability of lymphoma cells to modulate expression of surface Ig (sIg) and thereby escape therapeutic modalities directed at the sIg or components thereof. Other target candidates include point-mutated ras oncogene products, carcinoembryonic antigen (colon, lung, breast and pancreatic carcinoma), Muc-1 [mucin core peptide, an antigen recognized by cytotoxic T lymphocytes (CTL) in patients with adenocarcinomas of the breast, ovaries and pancreas], melanoma antigen recognized by T cells (MART-1) and prostate specific antigen, for example (SCHLOM, 1996). Since most tumor antigens are weakly immunogenic, the goal is to enhance the immune response to these antigens through use of either improved adjuvants, such as a recombinant *Leishmania*-derived protein inducing potent Th1 responses (REED, 1996, 1997), or other biological response modifiers like recombinant IL-4 or IL-2 or TNFα. The interleukins are hoped to enhance host immune responses to tumors. IL-2 has been used for renal cell carcinoma, melanoma and myeloid leukemia. Until recently, high doses have appeared to be required for adequate activation of cytotoxic T lymphocytes, lymphokine activated killer cells, natural killer cells and tumor infiltrating lymphocytes. The toxic side effects of high doses of IL-2 can be limiting to this form of tumor therapy. Some have sought to circumvent this toxicity by expanding tumor specific killers, and non-specific killers, ex vivo by culture with IL-2. IL-4 has also been used in an attempt to amplify CTL responses. More specific information on the use of these cytokines is given in Sect. 2.2.1. To date, only limited efficacy has been achieved with either cytokine alone.

An approach to minimizing the toxicity problem of native IL-2 is to create a fusion protein where IL-2 is linked to an antibody molecule specific for a tumor associated antigen. This approach proved useful in stimulating cytotoxic T cells capable of eliminating melanoma metastases in mice (BECKER et al., 1996; REISFELD, 1996).

Some efforts have focused on making the weakly immunogenic tumor targets more immunogenic. IFNγ and IL-4, for example, have been used to increase antigen presentation by increasing major histocompatibility complex antigens bearing tumor antigen-associated immunogenic peptides.

Parallel approaches involve gene transduction into either tumor cells, making them more immunogenic, or into T cells making them more effective killers. Some have transfected or transduced into tumor cells costimulatory molecules such as B7. B7 engages the T cell surface molecule CD28 and enhances T cell activation by antigen. This strategy is more effective in inducing T cell cytotoxic responses. So far, most efforts along these lines have been performed only in animals. As demonstrated by investigators associated with Bristol-Myers Squibb (HELLSTROM, 1996), cotransfection of B7 plus CD48, a CD2 ligand, (LI et al., 1996) or CD80 (CHEN et al., 1994) has conferred immunogenicity on a normally non-immunogenic tumor.

A complementary approach is to identify universal tumor specific antigens (TSA) and then isolate high affinity human T cell receptors (TcR) specific to these TSAs. Introduction of these TcRs into host lymphocytes might then render them cytotoxic to host-derived tumors. This technique seeks to bypass the need for accessory molecule signaling of cytotoxic T cells due to the superior strength of signaling resulting from high affinity TcR ligand interactions. Proof-of-principle experiments have been performed using the p53 tumor suppressor gene (THEOBALD et al., 1995). Introducing chimeric antigen receptors composed of an antibody-like extracellular antigen-recognizing domain fused with an intracellular domain capable of interacting with T cell receptor triggering components has also been examined. This strategy provides antigen recognition which is not MHC restricted and, therefore, is universally applicable to tumor patients. Weak antigen presentation and processing by some tumor cells is a problem still remaining and to be overcome with these genetic transduction efforts. Other generic problems to be solved pertain to definition of appropriate antigens for recognition on tumors and to relevant delivery of these chimeric antigen recognition arming devices.

There is some evidence that certain proteins can be directly tumoricidal or at least tumoristatic. Indeed, TNF was one of the first agents to be used for tumor therapy, as its name might suggest. Now it is recognized that both TNFα and TNFβ are cytotoxic. Despite the significant side-effects with these proteins, Dainippon, Asahi Chemical and Genentech have continued to develop recombinant TNF for anti-tumor therapy.

Other recombinant proteins under exploration for direct anti-tumor activity exist. Sheffield Medical Technologies is developing a recombinant tumor suppressor protein. A specific urogenital sinus-derived growth inhibitory factor may be useful in treating various cancers especially prostate cancers. Finally, platelet factor 4 has been produced in recombinant form by RepliGen. It inhibits angiogenesis and so may be useful in treating cancer metastasis (MAIONE et al., 1991). Results of clinical testing are reported by DEHMER et al. (1995).

As a by-product of non-specific cytocidal therapies, patients are often left debilitated and vulnerable to infections due to lowered blood cell levels. Platelets are perhaps even more critical when so drastically depleted following chemotherapy. The inabilites to control bleeding and to combat normally benign infectious agents render chemotherapy itself life threatening. The use of recombinant proteins to bolster immune function and recombinant growth factors to restore normal levels of blood cells is a very exciting prospect for the treatment of cancer patients following anti-tumor regimens. Growth factors specific for various blood cells are now being developed in several cancer therapy settings. These include the colony stimulating factors (CSF), G-CSF, GM-CSF, erythropoietin (EPO), M-CSF, TPO as well as IL-6, IL-11, stem cell factor (SCF) and IL-3. For recent reviews on hematopoietic growth factors with regards to cancer therapy see BOCIEK and ARMITAGE (1996); ROWE (1996) and MORSTYN et al. (1996). Toxicities aside (RYFFEL, 1996), some efficacy has been achieved but the results are far from optimal.

Some investigators supplement these cytokines with autologous (or matched) stem cells. Growth of these stem cells *ex vivo* is aided by exposure to recombinant cytokine growth factors. Of these, stem cell factor, IL-3 and leukemia inhibitory factor (LIF) are a few examples. One problem to this approach is that the culture conditions and growth factors used so far tend to result in differentiated cells, to a greater or lesser degree, and hence some of the pluripotency of primitive stem cells is lost.

2.1.2.2 Vaccines

Recombinant proteins are also being used as vaccines for prevention of several diseases (Tab. 1). Most are being developed or used for viral diseases although some target bacterial and parasitic pathogens. Although the current vaccines, for the most part, are efficacious and relatively few problems have resulted from them, there are some problems with residual infectivity and contamination.

Recombinant proteins would not have these problems associated with them.

There remain some other considerations, however. A currently emerging paradigm in immunology is that an antigen must persist for immunological memory to last. With attenuated live vaccines, this is a real possibility. Living vectors may reside in occlusion for long periods, resurfacing periodically to give the necessary boost to an individual's immune system. This capacity for persistence is much less pronounced when one uses recombinant proteins. Certainly propagation of the proteins within a host is impossible and their disappearance should be rather rapid. These considerations may help explain the need for repeated boosts with some of the current recombinant vaccines and the sometimes weak responses which result. One possibility is that antigen can persist for extended times in the form of immune complexes which are antibody–antigen aggregates. These can reside on certain reticular cells for long periods. As immunity wanes, some of these aggregates may dissociate releasing antigen to boost the immune system again. Another emerging paradigm maintains that soluble antigens, in the absence of costimulatory signals, are tolerogenic rather than immunogenic. In this context, one important way of enhancing antigen persistence, and immunogenicity is formulating the antigen within an adjuvant. Development of better adjuvants which cause protracted antigen exposure, persistence and costimulatory signals without unwanted toxicities will certainly benefit vaccine development efforts. For example, some have tried using recombinant IL-2 to boost the efficacy of hepatitis B vaccination in previous vaccine non-responders (JUNGERS et al., 1994). Corixa has also developed a recombinant *Leishmania*-derived protein, LeIF, with potent adjuvant properties. LeIF induces IL-12 and B7 costimulator expression from human macrophages and dendritic cells and thereby is effective in enhancing induction of human CTL activity (REED, 1996, 1997).

Viral Vaccines

There are few good reviews of the entire field of vaccines but for some relevant read-

Tab. 1. Some Recombinant Proteins Used as Vaccines Currently Under Development

Recombinant Protein	Company	Indication	Status
gp160Mn	Immuno AG	HIV	Phase II
HIV-1 peptide	NCI	HIV	Phase I
gp120	Biocene/Chiron	HIV	Phase I/II
gp120	Genen Vax	HIV	Phase II
HIV-1	Immune Response	HIV	Phase III
gp160	MicroGeneSys	HIV	Phase II/III
Gastrin G17	Aphton	Cancers, ulcers	Phase I/II
hCG	Aphton	Contraception	Phase II
Borrelia proteins	Connaught Labs	Lyme disease	Phase III
	SmithKline Beecham	Lyme disease	Phase III
			Phase I
BCG: *Borrelia* protein	MedImmune	Lyme disease	Phase I
Pertussis subunit	Biocine/Chiron	Pertussis	Phase III
p53 + RAS	NCI	Cancers	Phase I
gp100 adenovirus	NCI	Melanoma	Phase I
HPV I6, E6, E7	NCI	Cervical cancer	Phase I
MART-1	NCI	Melanoma	Phase I
M24	Targeted Genetics	Melanoma	Phase I
CEA	MicroGeneSys	Cancers	Phase I
CEA	NCI	Cancers	Phase I
TcR peptides	Immune Response	MS	Phase I
TcR peptides	Connective Therapeutics	MS	Phase I
MHC:peptide	Anergen	MS	Phase I
TcR peptides	Immune Response	RA	Post Phase II
MHC:peptide	Anergen	RA	Phase I
TcR peptides	Immune Response	Psoriasis	Post Phase II (f)
Hepatitis B	SmithKline Beecham	Hepatitis B	Phase II
HSV	SmithKline Beecham	Herpes simplex	Phase II
HSV	Biocene/Chiron	Herpes simplex	Phase III
RSV subunit	Wyeth-Ayerst	RSV	Phase II
CMV subunit	Biocine/Chiron	CMV	Phase II

ing see ZIMMERMAN et al. (1996); KIMMEL et al. (1996) and MOLLNES and HARBOE (1996). Microgenesys, Inc, has expressed recombinant influenza hemagglutinin proteins (HA) which have been used in human clinical studies. These proteins induce neutralizing antibodies and are well tolerated (WILKINSON, 1996). Results of some clinical trials are given by POWERS et al. (1995). Protection is seen as lower infection rates in immunized individuals in a winter when influenza is epidemic. They are also developing vaccines against parainfluenza and respiratory syncytial virus for immunization of children.

AIDS is probably the single largest area of vaccine development. Vaccines are being de-veloped both for prevention and for therapeutic treatment. The envelope glycoproteins gp160, and its processed offspring gp120, are major protein vaccine candidates. The core antigen p24 is also a potential target. The goal would be to develop antibodies against envelope proteins to prevent infection and virus spread. CTL development against cells already infected and expressing core antigens would also kill infected cells and hinder virus spread. Companies actively involved in, or who have been actively involved in, vaccine research and development for AIDS include MicroGeneSys (gp160), SmithKline Beecham (gp120), Ajinomoto, British Biotech, Chiron (gp120, env2–3), Bristol-Myers Squibb, Genentech (gp120), Immuno (gp160) MedImmune (gag), Pasteur Merieux (gp160), Repligen (gp120) and others. For the most part, the vaccines are well tolerated and tend to elicit neutralizing antibodies. The mutability of the virus is a hurdle to complete efficacy of these vaccines. Recent progress in other areas of treatment, particularly protease inhibitors and nucleoside analogs is very exciting. The battle against AIDS, however, is not over and significant medical benefit in a worldwide setting may be realized with these vaccine development efforts. For results of some clinical trials with recombinant HIV protein vaccines see SALMON-CERON et al. (1995) and GRAHAM et al. (1966).

Herpes simplex vaccines are a potential aid for a common and highly contagious class of viruses for which no current vaccines exist. A major strategy is to prevent infection by inducing antiviral antibodies. This prophylactic approach may be more effective, since as discussed below, herpes simplex encodes early genes which shut off a part of the antigen processing system of most cells involved with presenting peptides derived from viruses infecting the cell. Chiron is using cloned and expressed HSV2 surface glycoprotein B for vaccine development. SmithKline Beecham is using the gD protein in similar efforts. Some clinical results are available (LANGENBERG et al., 1995).

Hepatitis B, C, δ and E viruses are major worldwide pathogens. As such the market for a treatment is enormous. Vaccine development is underway in both the prophylactic and therapeutic arenas. Surface antigens of hepatitis B have been developed as vaccines used by Biogen, Chemo, Chiron, Genentech, SmithKline Beecham, Genelabs, Pasteur Merieux, Research Development Corp, Biotechnology General, American Biogenetic Sciences, Medeva, Johnson and Johnson and most likely others. Some clinical results can be seen in LEE et al. (1995). The interferons are presently being used for therapy, but dose limiting toxicities make the effectiveness of this approach incomplete to date. Induction of host immunity to the pathogen could prove more beneficial. Most hosts mount some form of immune response to these viruses, but for unknown reasons these responses are ineffec-

tive or inappropriate. Cytel has taken the lead in attaching immunostimulating peptides to viral sequence peptides. The intent is for these novel fusion proteins to stimulate the appropriate form of immune response, e.g., a CTL response as opposed to a humoral response, where it is needed (SETTE, 1996, 1997). Trials with this approach have not been overwhelmingly positive, and this approach may need further refinement.

Other work has been done on viral vaccine development for Epstein-Barr virus (SPRINGS et al., 1996). The gp340 and gp220 recombinant envelope proteins have been used and the aim is to protect against glandular fever. This work has been done by 2 companies, CRC technology and BioResearch Ireland.

Human papilloma virus (HPV) causes genital warts and is thought to lead to cervical cancer as a possible sequella. MedImmune is developing a recombinant vaccine approach where viral particles are derived from transfected insect cells. Cantab and CRC Technology are using HPV proteins in vaccine development as well. For other work on these vaccines see SCHILLER and OKUN (1996).

Bacterial Vaccines

Malaria is certainly a major problem in the underdeveloped world, and a vaccine would have enormous impact. So far, a number of recombinant protein vaccines against the offending *Plasmodium* species are under development. Since a number of *Plasmodium* organisms are disease causing, yet they have some proteins in common, it is hoped that using proteins shared between several species may afford a greater range of protection. Companies developing malarial vaccines using recombinant proteins include Roche and SmithKline Beecham. Some clinical results are given in GORDON et al. (1995a).

Ajinomoto has cloned several mycobacterial protein antigens. These may be useful in vaccine development for leprosy. As adjuvants, they may also be useful immunostimulants for vaccines against AIDS, malaria and other disease causing pathogens.

Orovax is investigating *Helicobacter pylori* which has recently been shown to be causative for peptic ulcers. Vaccine development efforts against this bacterium are at an early stage but might provide enormous medical benefit worldwide.

Connaught Laboratories is using recombinant forms of outer surface protein A in vaccine development efforts against the spirochete *Borrelia burgdorferi*, which causes Lyme disease. MedImmune is also developing a Lyme disease vaccine using the adjuvant properties of recombinant BCG. Others developing recombinant protein-based vaccines to *Borrelia* include New York Medical College, SmithKline Beecham and Yale University. Clinical trial results are given in SCHOEN et al. (1995).

Otitis media can be caused by several bacteria including species of *Haemophilus*, *Streptococcus* and *Branhamella*. These pathogens employ selectins for adherence to a substratum. Taking advantage of this fact, Micro-Carb and Pasteur Merieux have employed recombinant bacterial selectins or adhesins for immunization efforts in vaccine development.

Tuberculosis is rapidly developing antibiotic resistance in humans and may well present a danger similar in magnitude to that prior to antibiotic development. Pasteur Merieux has expressed the mycobacterial protein invasin used by the bacteria as a pathogenetic factor. Vaccine development is underway using this protein.

Immunosuppressive or Immune Deviation-Inducing Vaccines

Allergy and autoimmune disease result from immune responses gone awry. The resulting secretion of inflammatory cytokines is in large part responsible for the pathology seen in these diseases. One approach to reducing disease or pathology, therefore, is to substitute one form of immune response for another in which inflammatory cytokine production is reduced. TGFβ can profoundly alter immune responses. It can be produced in large quantities by T cells found in the mucosae, particularly of the gut, subsequent to oral antigen administration. This approach can lead to the establishment of large numbers of antigen-reactive T cells which secrete TGFβ upon antigen exposure. The TGFβ down-regulates other cells secreting inflammatory cytokines. In this way one form of immune response modulates another, the end result being a subsiding of apparent symptomology. Another approach to altering responses employs antagonists of certain cytokines. For example, IL-4 is necessary for a switch to IgE production and can also function in generating more T cells capable of secreting IL-4. Since IgE is central to many allergic phenomena, an IL-4 antagonist such as a soluble receptor might subvert responses leading to allergic states. A similar approach involves cytokines or agents which directly induce differentiation along different pathways not leading to IL-4 secretion, rather than inhibiting the mediators which drive this process.

Autoimmune Inc is a leader in the development of products for oral tolerance induction (see below). They are examining the use of a recombinant form of myelin basic protein for treatment of multiple sclerosis (MS), collagen as a potential therapeutic for rheumatoid arthritis, and a recombinant human S-antigen for treatment of autoimmune uveitis. With Lilly, Autoimmune Inc is also using a recombinant protein, presumably of pancreatic origin, for diabetes treatment. ImmuLogic is attempting to induce T cell anergy in MS by administering relevant antigen-based peptides in the absence of adjuvants (HAPP, 1997). As discussed above, lymphocyte encounter of antigen in the absence of secondary signals has been proposed as a mechanism leading to unresponsiveness.

T Cell Sciences and Astra are developing vaccines consisting of recombinant peptides derived from specific T cell receptors. The T cell receptors from which the peptides derive appear to be very commonly used by T cells mediating autoimmune phenomena. This is true in some animal models of autoimmmune disease, especially very early in the onset of autoimmunity. Hence, an immune response to these proteins would represent an autoimmune response to a self antigen involved in mediating rejection of self, so to speak. Procept is also following this avenue of development for potential treatment of diabetes.

Other Vaccines

One particularly novel vaccine approach is that taken by the Center for Molecular Immunology. Recombinant EGF is used to immunize. The resulting immune response deprives EGF-dependent tumors of their EGF supply and it is hoped the tumor will then be more easily cleared. Development to date is in animals only.

Chlamydia infections are prevalent and cause particular problems in pregnant women MicroCarb is developing a recombinant vaccine against *Chlamydia trachomatis*. Their efforts are at an early stage.

SRC VB Vector is developing a recombinant vaccine against encephalitis-inducing organisms carried by ticks.

Chiron is developing a pertussis toxin vaccine. The protein is recombinant and lacks enzymatic activity. It is marketed in other parts of the world and is in late stage trials elsewhere.

Mycobacterial heat shock proteins can serve as adjuvants to stimulate the immune system. Connaught Laboratories with Stree-Gen and Pasteur Merieux are using recombinant forms of these proteins as immunostimulants agents for vaccines against various pathogens.

The use of recombinant human reproductive hormones as vaccines is in practice and being developed further for contraception (JONES, 1996). Initially, complete human β-chorionic gonadotropin (hCG) was used successfully. This hormone is required for implantation of a fertilized egg in the uterus. Antagonism results in failure of pregnancy. Aphton is one company with experience developing this hormone for commercial use. This concept is being developed further with various forms of reproductive hormones coupled to various carriers.

2.1.3 Immunosuppression

Organ transplantation and autoimmune diseases are two major areas where an individual's own defense systems give rise to undesired conditions. Immunosuppression achieved through vaccination is one strategy

to induce immunosuppression, as discussed above. In addition, agents which could act to suppress the normally occurring immune responses, or to deviate them to a different form of response, might be beneficial and are being investigated.

2.1.3.1 Costimulation

One general immunosuppressive strategy is being implemented by Procept, which is to produce a soluble form of the T cell surface protein CD2. This molecule is a costimulatory receptor involved in some forms of T cell activation signaling. Inhibition of its interaction with ligands might then inhibit activation and subsequent immune responses. Potential areas of use would include the T cell mediated autoimmune diseases and transplanted organ rejection.

2.1.3.2 Antigen Presentation

The major histocompatibility complex (MHC) encodes a series of genes, most of which fall into two classes. The function of these genes is to present peptides derived from proteins within cells and from ingested extracellular proteins to the immune system. If the immune system recognizes these processed and presented peptides as foreign or novel, an immune response ensues. In many autoimmune diseases, some incident, generally unknown, results in aggressive responses to self peptides which are normally presented but do not induce a pathological response. Several teams are targeting the antigen presentation process as a way to down-regulate autoimmune responses. One potential avenue is to antagonize MHC-self peptide interaction by loading all MHC molecules with a non-immunogenic peptide or peptidomimetic. Before MHC molecules are loaded with peptides, they are occupied with an invariant protein which acts to help assembly and transport of the two chain MHC molecules. Hence, the invariant protein is universally accepted by one class of MHC molecules and is a possible candidate for the approach just outlined. This approach is being followed by Antigen

Express, Inc (HUMPHREYS, 1996). Hoffmann-LaRoche is examining down-regulation of MHC class II expression to render these molecules non-immunogenic using antibodies to the binding site of these molecules (NAGY, 1996). Another related approach towards immunosuppression takes advantage of the fact that T cells recognizing MHC class II peptide complexes in the absence of costimulatory molecule engagement are rendered anergic or are induced to apoptose. Anergen, Inc (NAG, 1996) has developed a treatment composed of exposing individuals to soluble class II molecules charged with peptide. This therapy is being examined in multiple sclerosis and rheumatoid arthritis but, if successful, would provide a platform for treatment of other autoimmune phenomena.

The other side of disease mechanisms involving antigen presentation is seen in certain viral diseases and with some tumors. These agents code for and express molecules which down-regulate MHC expression or interfere with antigen presentation. For example, herpes simplex encodes an immediate/early protein (ICP47) shutting off peptide transport into the ER (by binding to TAP proteins). Therefore, the viral peptides do not associate with MHC molecules and are not presented to the immune system. The task in enhancing immunity to tumor cells and viruses is to circumvent their ability to preempt normal host processes. Few recombinant proteins have been developed which are fully capable of doing this, to date.

2.1.3.3 Immune Deviation

With the concept that T cell helper responses can sometimes be classified in two ways, considerable research went into characterizing the class of response in autoimmune diseases and other chronic disease states. Now, at least 4 distinct phenotypes of CD4 T cells are recognized and 2–3 CD8 T cell subsets. The distinction between CD4 and CD8 positivity and association with helper function is also becoming blurred. Nevertheless, the simplicity of a bipolar model in which all mature T helper cells are either Th1 (characterized as responsible for cellular immunity) or

Th2 (characterized as responsible for humoral immunity) is attractive and has become rather firmly entrenched in the minds of many, perhaps because of the simplicity. Although it may be more useful to think in terms of IL-4 associated cytokines as opposed to interferon-γ associated cytokines, it is likely that most simple paradigms will ultimately be proven too simplistic. Philosophy aside, an early finding is that T cells producing IL-2 and IFNγ are common in lesions in multiple sclerosis and diabetes. In animal models of both MS (experimental autoimmune encephalitis EAE) and diabetes, injection of IL-4 has been shown to alleviate pathology. Since IL-4 can direct T cell development away from the IFNγ/IL-2 secreting T cell phenotype and direct it more towards a phenotype of IL-4/IL-5 secretion, a conceptual framework has arisen where deviation from one extreme T cell phenotype to another can be therapeutically useful. Multiple approaches to effect this immune deviation are being examined in several settings.

2.1.3.4 Multiple Sclerosis

Autoreactive lymphocyte-mediated demyelination in the central nervous system leads to a progressive, debilitating and chronic disease state. Although new evidence for a role of pathogenic autoantibodies is emerging (GENAIN et al., 1996), animal models and the finding of autoreactive T cells in lesions have conditioned most thinking to regard this disease as primarily T cell-mediated. For reviews of pathogenic mechanisms in multiple sclerosis see BROD et al. (1996) and STEINMAN (1966).

Treatment of MS has been generally disappointing. Presently, type 1 interferons, particularly interferon β are being used for treatment of MS. For a discussion of this approach see KELLEY (1996); ARNASON (1996) and LUBLIN et al. (1996). Novel approaches are being developed. In animal models, there is clear evidence for a restricted usage of T cell receptor (TcR) genes, especially early in the development of disease. It is hoped that a similar restriction might be seen in humans. With this hypothesis, several companies have launched programs to identify peptides specifically derived from these limited TcRs. The intent is to use these peptides to vaccinate and reduce the usage of these TcRs by the immune system, thereby inhibiting disease initiation, maintenance or progression. For a discussion see VANDENBARK et al. (1996). Both The Immune Response Corporation (BROSTOFF, 1996a) and Connective Therapeutics, Inc (AMENTO, 1996b), have pursued this approach. As an alternate, Anergen Inc (SPACK, 1996) is trying to capitalize on the fact that soluble MHC class II peptide complexes induce T cell unresponsiveness. Identification of the specific peptides, in conjunction with the appropriate MHC class II presentation molecules, is emphasized. Delivery of these complexes might then prove to reduce specific autoaggression and thereby alleviate disease. Other potential approaches involve delivering apoptosis-inducing antigens. Alexion Pharmaceuticals has developed a fusion protein of myelin basic protein and proteolipid protein for this purpose and some clinical benefit has been seen (MATIS, 1996). Induction of oral tolerance is being pursued by Autoimmune Inc (BISHOP, 1996a). Delivery of peptides related to the immunogenic peptides but which are antagonists of T cell activation is an alternative approach presently being undertaken by Neurocrine Biosciences (GAUR, 1996). The complexity of this disease and the late stage at which treatments often begin suggest that simple and limited approaches may be inadequate and combined therapies may be more efficacious.

2.1.3.5 Psoriasis

Psoriasis is a condition where T cells infiltrate the skin and induce excessive epidermal proliferation and differentiation through cellular interactions and cytokine release. For recent reviews describing psoriasis see PHILLIPS (1996) and GRIFFITHS and VOORHEES (1996). The T cells responsible for this state are usually characterized as Th1 cells, that is they release proinflammatory cytokines of which IFNγ and TNFα are important examples. Although the antigens responsible for activation of the T cells in the skin are not well characterized at present, some research-

ers believe that only a small set of antigens is responsible. This implies that T cell receptor usage is restricted in this disease. If true, vaccination with peptides related to these TcRs may be useful in inhibiting the activation of the offending cells. This approach is being used by both The Immune Response Corporation (BROSTOFF, 1996c), although it recently failed in a phase II trial (SEACHRIST, 1997) and Connective Therapeutics, Inc (AMENTO, 1996c). Activated T cells also express IL-2 receptors. Therefore, Seragen has developed an IL-2 toxin fusion protein and is examining the ability to kill T cells and thereby reduce pathology in psoriasis (SCHMALBACH, 1996). Other approaches might involve immunomodulation with monoclonal antibodies or soluble ligands to lymphocytes and their stimulatory receptors, or immune deviation with cytokines redirecting T cell cytokine production and, therefore, the class of response.

2.1.3.6 Arthritis

Joint inflammation has several causes and can also develop into a progressively debilitating disease. Destruction of tissue architecture and aberrant cellular metabolism develop. Normal remodeling processes appear to become uncontrolled leading ultimately to joint destruction. Initiating events are not presently well-defined but it appears that once normally cryptic tissue components are exposed in the context of inflammatory cytokines, induction of an autoimmune response can occur. This response can maintain and exacerbate the ongoing process. In addition to destruction by inflammatory cytokines and degradative enzymes, nitric oxide and oxygen radicals are thought to participate in pathogenesis. For recent reviews of rheumatoid arthritis see POPE (1996) and FELDMANN et al. (1996).

Several potential treatment approaches offer themselves. Anabolic agents which hasten chondrocyte proliferation and tissue repair might be delivered, either systemically or locally. BMP-7/OP-1, also being developed for osteoporosis, has the ability to stimulate extracellular matrix synthesis and deposition by chondrocytes. Creative Biomolecules, Inc is one company developing this agent to accelerate repair in arthritic indications (KUBER Sampath, 1996). Others are attempting to block ongoing destruction to allow normal restorative processes to begin. In this effort, enzyme inhibitors are being examined, although only a limited set of possibilities exists in this specific arena for use of recombinant proteins. Approaches similar to those mentioned above for psoriasis and multiple sclerosis are also relevant to this autoimmune response. Therefore, vaccination with peptides from relevant TcRs, or induction of tolerance with peptides is being examined by the same companies, i.e., Autoimmune Inc (BISHOP, 1996b), Connective Therapeutics (ALLEGRETTA, 1996) and The Immune Response Corporation (BROSTOFF, 1996b). The intent of vaccination is to reduce autoreactivity resulting in tissue destruction. Recent results of a phase II trial by Immune Response were favorable (SEACHRIST, 1997). Targeting inflammatory cytokines, including IL-1, TNFα, and the newly described IL-17 (LOTZ, 1996), is also being tried. Companies involved in this line of investigation include, Roche, Immunex, Amgen and others. These companies are using proteins other than monoclonal antibodies. Since anti-C5 antibodies have been shown to alleviate joint inflammation (WANG, 1996), another possibility is targeting the complement cascade with recombinant versions of naturally occurring or other specific inhibitors of C5. Immune deviation where a shift in the character of an immune response away from an inflammatory type response might be another approach. Success in this approach has been observed using IL-4 in animal models of diabetes and encephalitis, where a shift from Th1 like to Th2 like T cells is thought to occur, although few data are available for attempts to treat arthritides.

2.1.3.7 Insulin-Dependent Diabetes mellitus

Type 1 diabetes is an autoimmune disease thought to be mediated by IFNγ and TNFα-producing autoreactive Th1 cells. For recent reviews of diabetes see TISCH and McDEVITT (1996) and ROEP (1996). Induction of anergy by use of soluble complexes of MHC II peptides is being attempted by Anergen Inc. The peptides used are not derived from TcRs but from antigens exposed on or by pancreatic beta cells. Anergen is using a similar MHC II peptide approach for treatment of myasthenia gravis. For treatment of diabetes, immunmodulation or immune deviation is also possible since the mechanisms of pathology share similarities with those for other autoimmune diseases. Others are targeting adhesion molecules (YANG et al., 1996) or using recombinant IGF-I as a therapeutic (FROESCH et al., 1996; SAVAGE and DUNGER, 1996).

2.1.3.8 Inflammatory Bowel Disease (IBD)

These are complex diseases with undefined genetic components and an underlying autoimmune pathology resulting from an imbalance of immune regulation processes [for recent reviews see ELSON (1996) and HAMILTON (1996)]. Use of an anti-TNF monoclonal antibody has established a pathogenic role for TNF in Crohn's disease, as shown by Centocor, Inc (SCHAIBLE, 1996), and in a primate model of ulcerative colitis, as shown by Celltech Therapeutics, Ltd (HEATH, 1996). As in other inflammatory states, Interleukin-1 has also been implicated in IBD. Interleukins 10 and 11 are potent anti-inflammatory cytokines and so are potentially clinically beneficial in these diseases. IL-10 is being developed by Schering Plough Research Institute (LEBEAUT, 1996). It has shown some efficacy in animal models and has recently been administered to humans with steroid resistant Crohn's disease to test safety and tolerance. Recombinant IL-11 also has been shown to inhibit TNFα. It provides some benefit in animal models of inflammatory bowel disease and is also being evaluated in a clinical trial in humans by Genetics Institute, Inc (KEITH, 1996). Parke-Davis has shown that EGF may act both as a growth factor and as an anti-inflammatory regulatory protein in animal models of IBD (GUGLIETTA, 1996). For a review on the potential use of cytokines in IBD see NASSIF et al. (1996).

Adhesion molecules are important targets for some of the events in many pathological processes. While PECAM-1 (platelet endothelial cell adhesion molecule) functions in transendothelial migration rather than in leukocyte rolling or adherence, MAdCAM-1 is another Ig-family adhesion receptor which is preferentially expressed by mucosal venule endothelial cells. It mediates lymphocyte extravasation into mucosal lymphoid tissues and the lamina propria. Antagonists of this molecule may find a role in inflammatory bowel disease.

2.1.3.9 Allergy and Asthma

Immune responses under the influence of IL-4 can give rise to secreted IgE specific for the offending antigens. Binding of IgE to granulocytic cells arms them for release of many vasoactive and destructive proteins as a degranulation response to a secondary exposure to the recall antigens. These antigens are now allergens and the responses provoked involve many mediators. Among these are the leukotrienes, which are potent bronchoconstrictors, and kinins involved in neurogenic inflammation and bronchoconstriction. They release other mediators and cytokines eliciting secretion of mucus, edema, tissue destruction and leukocyte recruitment. A result of mast cell, basophil and eosinophil degranulation in response to antigen exposure is release of many Th2 type cytokines involved in further granulocyte recruitment and amplification of IL-4 axis cytokines. The allergic response is very complex, involving dozens of mediators, and consisting of early, intermediate and late stages sometimes lasting several weeks. Given the complexity of events, a host of potential targets is available for biotechnologically-derived medical intervention. For recent reviews of T cells in allergy see DASER et al. (1995) and ROMAGNANI (1995).

Reduction of immediate inflammatory processes can be achieved, in most cases, with steroid intervention. When this is not effective, longer range therapies are necessary. IL-5 is a chemoattractant for eosinophils, major mediators of tissue destruction. Targeting IL-5 with a monoclonal antibody is an approach taken by Schering-Plough Research Institute (GARLISI, 1996). Leukosite, Inc, has also targeted the CCR3 eosinophil eotaxin receptor (NEWMAN, 1996). The goal of a longer-term approach using anti-IgE antibody is to prevent binding of newly synthesized allergen specific IgE antibodies to their arming Fc receptors on granulocytes. This is the approach taken by Genentech and Tanox (FICK, 1996; BRAUN, 1996). Use of non-antibody recombinant protein approaches are also available. Muteins of IL-5 (perhaps a bioinactive monomer) or Fc fragments of IgE might antagonize the activity of the wild-type proteins. A mutant IL-4 protein which blocks the agonistic properties of wild-type IL-4 has been described (KRUSE et al., 1992; AVERSA et al., 1993). Efforts to redistribute the balance of cells secreting IL-4 axis proteins compared to IFNγ axis proteins are underway. The concept here is that Th2 cells mediate allergy/asthma and Th1 cells would be regulatory and non-allergic. IL-12 is being considered by Hoffmann-LaRoche for tilting the Th balance (RENZETTI, 1996). Alternatively, ImmuLogic Pharmaceutical Corp has developed peptide vaccines for cat and ragweed allergies. The principle here is to anergize the offending allergen responsive T cells (WALLNER, 1996).

2.1.3.10 Allotransplantation

In the absence of effective immunosuppression, T lymphocytes quickly infiltrate an allograft through use of a set of cell surface adhesion molecules. One specific adhesion molecule on T cells is the integrin VLA-4 (very late antigen 4). Agents which inhibit either induction or function of this adhesin might prevent T cell homing to the graft. Interaction of LFA (leukocyte functional antigen) and ICAM-1 is also involved in homing to foreign grafts. Most attempts to block these homing receptors use monoclonal antibodies and are addressed in chapter 12, this volume. One mechanism for providing a particular anatomical site or organ, privilege from the immune system is expression of CD95 ligand, or FAS ligand, on the potential target tissue. This has been demonstrated for the testis and the eye, for example. T cells bear the receptor for CD95 ligand and interaction, and engagement of this receptor initiates the process of apoptotic death in the lymphocytes involved. Alteration of some graft tissue, through genetic means, to express CD95 ligand is one approach to transplantation. Use of recombinant CD95 ligand protein locally has not been extensively examined. Once graft-reactive T cells are eliminated, however, and inflammation subsequent to surgery is reduced, newly developing T cells have a good chance of accepting the grafted organ as self. Activation of T cells may involve oligomerization of surface triggering molecules. One such molecule is thought to be CD4. Specific peptide stretches in the CD4 molecule are hypothesized to function in this putative oligomerization process. Use of peptides containing these amino acid sequences has been shown to inhibit some T cell activation processes *in vitro* (KORNGOLD, 1997). A treatment for organ transplantation using local delivery of similar peptides for human T cells could be envisioned. To date, the most advanced approaches to transplantation with recombinant proteins involve antibodies. However, it is clear other proteins might afford some efficacy in this area of medical intervention.

2.1.3.11 Xenotransplantation

One successful treatment for end stage organ failure is organ transplantation. The demand for human organs, however, is such that the present supply is insufficient. An alternative is the use of xenogeneic grafts, primarily using the pig as a source organism. The major obstacle to successful xenotransplantation is the phenomenon known as hyperacute rejection (HAR). This involves reactivity and deposition of host antibodies, primarily against a carbohydrate Galα(1,3)Gal xenoepitope, onto the graft. Complement is then activated and the terminal components destroy endothelial cells leading to graft rejection. If xenoantibodies are removed and/or complement components are blocked, a phenomenon called delayed xenograft rejection occurs, a reaction in which platelets aggregate, thereby resulting in initiation of coagulation. A

subsequent infiltration of activated monocytes and natural killer cells then causes complete failure of the graft. To date, most approaches to this problem have involved monoclonal antibodies to complement components (ROLLINS, 1996b), to lymphocytes (SIKES, 1996) or genetic manipulations to make the graft organ appear more human, either by expressing human sugars (FODOR, 1996) or anticomplementary surface proteins (LOGAN, 1996). Companies taking these approaches are Alexion Pharmaceuticals Inc and Nextran. The use of other recombinant proteins as therapeutic interventions is not yet as well developed. However, potential benefit might arise from use of soluble recombinant inhibitors of complement activation like decay accelerating factor or CD59. For a review of the approach using complement inhibition for xenotransplantation see WHITE (1996).

2.2 Specific Recombinant Proteins and their Applications

2.2.1 Cytokines

The term cytokines can mean many things. In the following text the word is used in its broadest sense to refer to proteins secreted by cells which act upon those or other cells. As such, interleukins, growth factors, hormones, interferons and even some members of the complement cascades are included. Enzymes and various clotting factors are discussed separately.

Often treatment with cytokines has been fraught with unwanted toxic side-effects of these proteins. One approach to circumvent these side-effects would be to design forms of these immunoenhancers reactive only with lymphocytes which would then be selective in their activities. A model system for selective action is provided in the natural case of the IL-13/IL-4 system. The activities of these cytokines overlap extensively in humans except for T cells which lack receptors for IL-13 and are, therefore, selectively responsive to IL-4. The class I interferons appear also to have target cell selectivity in that some cells respond better to some isoforms than to other isoforms. The IL-2/IL-15 system may also provide some selectivity, since some receptor components are shared and others are unique. Further research on receptor structures and interactions with cognate ligands is necessary, but there is reason to believe that knowledge provided by such research could lead to rational design of selective agonists (or antagonists) with the desired spectra of activities. For a recent review see COHEN and COHEN (1996).

2.2.1.1 Interferons

The interferons were among the earliest cytokines to be developed for therapeutic use. Broadly speaking, there are two classes of interferons, type I and type II. There are several subclasses of type I interferons including α, β, ω and perhaps others. Of the α interferons, there are at least 23 variants. Type I interferons are produced by many cell types including fibroblasts and lymphoid cells. The type II interferon consists of a single member, IFNγ, produced mainly by T lymphocytes and NK cells. Many texts have been written about the interferons. For a brief overview see KALVAKOLANU and BORDEN (1996).

Type I Interferons

The interferon α proteins have been developed by several companies and for several uses, as shown in Tab. 2. Primary indications are for treatment of viral diseases like hepatitis and AIDS. Other indications include boosting immune function to improve anti-tumor defenses. IFNβ is also being used as a treatment for multiple sclerosis. Although some benefit has been seen with IFNα, or type I IFNs, particularly in anti-viral therapy, dose-limiting toxicity is a major drawback to achieving improved efficacy. For results of

Tab. 2. Type I Interferons, Indications and Suppliers

Company	Product	Some Indications
Biogen	IFN α2B	Hairy cell leukemia, multiple myeloma
Schering Plough	IFN α2B	Kaposi's sarcoma, melanoma, papillomas
Yamanouchi	IFN α2B	Warts, chronic myelogenous leukemia
Enzon	IFN α2B	Hepatitis C and B, multiple sclerosis, renal cell carcinoma, psoriasis, other
Boehringer Ingelheim	IFN α2C	Hairy cell leukemia, multiple myeloma papillomas, breast cancer
Cheil	IFN α2	Similar to above
Sidus	IFN α2	Similar to above
Dong-A	IFN α2	Similar to above
SCR VC Vector	IFN α2	Similar to above
Amgen	IFN α consensus	Similar to above
Gentech	IFNα2A	Similar to above
Roche	IFNα2A	
Interferon Sciences	IFNαn3	ARC, AIDS, cervical dysplasia, HPV, hepatitis C
Ciba-Geigy	IFNα	HIV, similar to above
Chiron	IFNβ1b	Multiple sclerosis, similar to above
Berlex	IFNβ1b	Multiple sclerosis, similar to above
Biogen	IFNβ	Multiple sclerosis, hepatitis
Berlex	IFNβ	Similar to above
Serono	IFNβ	Similar to above

some clinical trials with type I interferons see REID et al. (1991); PICHERT et al. (1991); KRUIT et al. (1995); IOFFE et al. (1996) and VILLA et al. (1996). For results of IFN treatment of multiple sclerosis in the clinic see DURELLI et al. (1996) and KHAN et al. (1996).

Type II Interferon (Tab. 3)

Human IFNγ is isolated in a 20 kD form glycosylated at a single site and as a 25 kD form glycosylated at both consensus N-glycosylation sites. There are two alleles of this protein described so far with a single amino acid difference at position 137. The mature protein circulates as an apparent dimer. The receptor consists of a 90 kD glycoprotein and at least one additional chain, currently poorly characterized of which a good candidate has been identified.

Several groups have developed IFNγ for therapeutic use primarily in chronic viral infections and in cancer treatment, similar to type I interferons (Tab. 3). Since IFNγ is particularly effective at activating macrophage cytocidal activity, it is very useful in diseases with intracellular pathogens. Some examples include listeriosis and tuberculosis. It is possible that with the rise in drug-resistant variants of mycobacteria, IFNγ will have a larger medical impact. This protein also has anti-protozoal and immunomodulatory activities. The induction of a dioxygenase appears to account for its anti-protozoal activity. IFNγ tends to suppress production of Th2-derived cytokines like IL-4 and, therefore, to promote Th1 development and has been used with IL-2 as a cancer therapy. For results of a preliminary study see BAARS et al. (1991); WEINER et al. (1991); PUJADE-LAURAINE et al. (1996) and LUMMEN et al. (1996). Others are investigating treatment of systemic sclerosis with IFNγ (POLISSON et al., 1996). For results of treatment of chronic granulomatous disease see WEENING et al. (1995). Treatment of osteopetrosis has been investigated by KEY et al. (1995). To date however, successful therapy with IFNγ has been limited and its therapeutic potential remains to be developed.

2.2.1.2 Interleukins

Interleukins are a set of proteins divided into several families based upon structural similarities and cognate receptors to which they bind. As the name implies, they were first described to be produced by white cells and to act upon white cells. It now appears that many cells can produce these proteins and many cell types can respond to them. Each of the interleukins has multiple activities on several different target cells. While some of these effects may be beneficial, many appear to be counter productive and result in dose-limiting toxicities. Often the therapeutic dose is the same as that inducing toxicity, thus limiting efficacy (Tab. 4).

Tab. 3. Type II Interferons, Indications and Suppliers

Company	Product	Some Indications
SRC VB Vector Sidus	IFNγ	Chronic viral infections, cancer Parasitic infections
Connective Therapeutics	IFNγ	Atopic dermatitis
Genentech Biogen Daiichi Boehringer Ingelheim	IFNγ1b	Viral infections, cancer, mycobacterial infections
Suntory Schering Plough	IFNγ1a	Skin cancer

Tab. 4. Some Therapeutic Recombinant Interleukins and Related Proteins Currently Under Development

Protein	Company	Indication	Status
IL-1	Roche/Immunex	Hematopoiesis	Preclinical
IL-1R	Immunex	Asthma	Phase I
IL-1RA	Amgen	Inflammations	Phase II
IL-2	Biomira	Renal cell carcinoma	Phase II
IL-2-PEG	Chiron	HIV	Phase II
IL-2 toxin	Seragen	Lympomas, T-cell dependent abnormalities	Phase I/II/III
IL-3 mutein	Searle	Hematopoiesis induction	Phase II
IL-4	Schering-Plough	Lung cancers	Phase II
IL-4R	Immunex	Asthma	Phase I
IL-4 toxin	Seragen	Tumors	
IL-6 mutein	Imclone Systems	Thrombocytopenia	Phase I
IL-6	Sandoz	Platelet restoration	Phase I/II
IL-6	Serono	Thrombocytopenia, cancer	Phase I/II
IL-6 toxin	Seragen	Tumors	
IL-8RA	Repligen	Anti-inflammatory	Preclinical
IL-10	Schering-Plough	Tumors, autoimmunity	Clinical
IL-11	Genetics Institute Schering-Plough	Thrombocytopenia	Phase III
IL-3	Sandoz	Stem cell transplantation	Phase III
IL-12	Genetics Institute Wyeth-Ayerst	HIV, cancer	Phase I/II
IL-15	Immunex	Mucositis, infections	Preclinical

Interleukin-1

Interleukins-1α and -1β, distinct proteins from distinct genes, activate T cells and provide support to hematopoietic progenitors. They interact with each of 2 cell surface receptors. The receptors are distributed differently on various cell types. One receptor (type 2) may not transduce a signal, serving as a decoy receptor. It is a potential source of a soluble cleavage product, which might function as an IL-1 carrier or antagonist. The type 2 receptor is found mainly on B cells and polymorphonuclear neutrophils (PMN), whereas the type 1 receptor is on many other cell types. Generally speaking, the type I receptor binds IL-1α more strongly and the type II receptor binds IL-1β more strongly. Each IL-1 protein is synthesized as a 31 kD precursor and is cleaved by an interleukin converting enzyme (ICE) to mature 17 kD forms. They share only 25% amino acid identity and are produced by a wide variety of cells including monocytes, macrophages, Kupffer cells, os-

teoblasts, keratinocytes, hepatocytes, some epithelial cells, fibroblasts, glial cells and others. The IL-1 proteins have a wide variety of biological activities and are involved in the acute phase response, inflammation, fever, immune function, bone metabolism, hematopoiesis and many others. For a review of IL-1 see DINARELLO (1996).

The therapeutic potential of using the IL-1 isoforms in treating aplastic anemia, neutropenias and myeloid disorders is being explored. Dainippon is developing IL-1α to promote lymphocyte activity to minimize myeloid dysfunction after cancer therapy. Although flu like side effects were observed, some acceleration in recovery of platelet activity and numbers was observed. Immunex produced IL-1α, and now Roche is developing it as a radioprotective agent and for its ability to restore levels of neutrophils and platelets. Immunex also developed IL-1β, and Roche is pursuing its development for blood cell restoration following chemotherapy and also for wound healing.

A mutant IL-1β, OCT43 has been developed in Japan for use in treating some fungal infections. For results of clinical trials using IL-1 see CROWN et al. (1991); WEISDORF et al. (1994); NEMUNAITIS et al. (1994); JANIK et al. (1996) and KOPP et al. (1996). As suggested above, use of IL-1 is not yet widespread generally due to associated toxicities.

Interleukin-1 Antagonist

IL-1 receptor antagonist (IL-1ra) is a naturally occurring soluble form of an IL-1 molecule which binds to both IL-1 receptors but induces no biological effects. As such, it antagonizes the natural activities of both IL-1 variants. IL-1ra was discovered in the urine of monocytic leukemia patients and is a molecule of 25 kD with approximately 20% amino acid identity with either of the IL-1 isoforms. Nevertheless, it binds to the type 1 IL-1 receptor with affinity of the same order as seen with the IL-1 agonists (~200 pM). IL-1ra is produced by monocytes, neutrophils, macrophages and fibroblasts. Its production is enhanced by IL-4, IL-13, IFNγ and other cytokines. IL-1ra is believed to function as a natural regulator of IL-1 activity in vivo. For a review of IL-1ra and complement receptor antagonists see MANTOVANI et al. (1996).

The recombinant form of the molecule was developed by Synergen, Inc, for treatment of inflammatory processes and sepsis, in particular. Early clinical trials were very promising but later and larger trials were unconvincing with regard to efficacy in reducing mortality in sepsis. Muteins of the antagonist have been developed by Dompe. These have apparent higher binding capabilities and show better potency both *in vitro* and *in vivo* at blocking IL-1-mediated effects. Amgen is investigating potential use of IL-1ra in several indications including rheumatoid arthritis, asthma, psoriasis, diabetes, chronic myelogenous leukemia (CML) and others.

IL-1 receptors have been cloned by Immunex and Affymax. Each is developing either soluble IL-1 receptor type 1 (Immunex) or types 1 and 2 for possible treatment of rheumatoid arthritis, allergy, asthma, systemic lupus erythematosus (SLE), IBD, septic shock, osteoarthritis and other inflammatory disorders. Development in the US has been slower than in other parts of the world. Collaborators in development include Glaxo Wellcome and Hoechst. For data on some clinical trials using IL-1ra and IL-1 receptor see FISHER et al. (1994a, b); BOERMEESTER et al. (1995); CAMPION et al. (1996) and DREVLOW et al. (1996).

Interleukin-2

IL-2 was initially named T cell growth factor to reflect one of the first known functions of this protein. As more data were generated, multiple biological activities were described. IL-2 is a strong immunostimulant and is one of the primary growth factors for lymphocytes. It is produced primarily by T cells as a 15–18 kD glycoprotein. The IL-2 receptor is complex and can consist of 3 different polypeptide chains. The individual chains have various functions and some are shared with receptors for other interleukins, including IL-4, IL-7, IL-9, IL-13, and IL-15. This suggests that some overlap in the activities of these interleukins exists, and indeed this has been found. These interleukins and their receptor interactions are the subjects of intense investigation presently. IL-2 is predicted to have a 4 helix-bundle conformation like other members in the family which include IL-4, IL-7, IL-9 and IL-13. The medical interest lies primarily in its immunostimulatory properties. For a review of IL-2 see COHEN and COHEN (1996).

Interleukin-2 is available from several sources now and is primarily indicated for treatment of renal cell carcinoma and hemangioendothelioma. The anti-cancer value lies in the ability of IL-2 to stimulate lymphoid cells to proliferate, secrete cytokines and activate other effector functions including cytolysis. For a review of the use of IL-2 for leukemias see FOA (1996). Use of IL-2 is limited by unwanted toxicities, as has been seen with virtually all cytokine therapies. Currently, lower doses are being used in repeated, staggered administrations in attempts to restore immune T cell function in individuals with HIV. IL-2 has been used for treatment of renal cell carcinoma and a variety of other tumors, malignant pleurisy and for expansion of autologous lymphocytes *ex vivo*.

IL-2 is sold by Chiron under the name *Aldesleukin* for renal cell carcinoma. *Teceleukin* is the name of the product from Biogen for use in Japan. Other commercial names are *Proleukin* and *Leuferon*. IL-2 is also being developed for use in treatments for hepatitis B, melanoma, colorectal carcinoma, acute myeloid leukemia, non-Hodgkin's lymphoma, and bladder cancer. Additional efforts are underway to use IL-2 in combination with IFNα for cancer therapies. Companies involved in the development of IL-2 include Chiron, Ajinomoto, Roche, Hoechst, Sanofi, Immunex and Biogen. For data on some clinical trials with IL-2, and associated toxicities, see KLIMAS et al. (1994); CONANT et al. (1989); MEMOLI et al. (1995); KORETZ et al. (1991) and CLAYMAN et al. (1992). For results using PEG-IL-2 see YANG et al. (1995) and RAMACHANDRAN et al. (1996).

Interleukin-3

Interleukin-3, or multi-colony stimulating factor, is pleiotropic but stimulates proliferation and differentiation of hematopoietic stem cells and committed progenitors. It can act in synergy with IL-1, IL-6, SCF and G-CSF to support proliferation of multipotent progenitors. Although IL-3 is produced mainly by activated T cells, it can also be induced from eosinophils, keratinocytes, neutrophils and other cell types. The mature protein exists as a 28 kD glycoprotein and requires a disulfide bond for biological activity. To date, only a single class of high affinity receptor has been described for human IL-3. There are at least two different proteins which comprise the receptor. The degree of multimerization of these chains is subject to debate. The larger subunit, a 120 kD β chain, is shared with receptors for IL-5 and GM-CSF. The 70 kD α chain also has homology to α chains in receptors for IL-5 and GM-CSF. In addition to supporting hematopoiesis, IL-3 induces IL-8 production by monocytes, endothelial cell adhesion and chemotaxis of eosinophils. For

a review on IL-3 biology see OSTER et al. (1991).

Since IL-3 supports hematopoiesis, it is being developed as a blood cell restoration therapy in cancer patients following chemotherapy or irradiation. Genetics Institute cloned the gene for the human protein and this is being developed by Sandoz. The primary indication is in cancer patients with myelosuppression. Clinical trials have been performed with bone marrow transplant patients, those with ovarian and breast cancers, as well as in patients with Hodgkin's disease. Combination therapy with GM-CSF and other colony stimulating factors is also being examined. Searle/Monsanto has developed a mutant IL-3, called *Synthokine,* with higher receptor binding affinity. The rationale is that this mutein will be clinically effective at doses low enough to avoid triggering undesired side-effects. Some benefit in restoring platelet function has been observed with this synthetic. For results of clinical trials using IL-3 see SEIPELT et al. (1994); RAEMAEKERS et al. (1993); OTTMANN et al. (1990); LINDEMANN et al. (1991) and GANSER et al. (1992).

Interleukin-4

IL-4, a pleiotropic cytokine, was originally described as a B cell growth factor. It is now known to act on a wide variety of cells including those of hematopoietic and mesenchymal origin. IL-4 itself has powerful anti-inflammatory properties, enhancing secretion of soluble IL-1 receptors, for example, and can influence T cell differentiation towards secretion of anti-inflammatory cytokines as well. It is secreted as a 14–18 kD glycoprotein with multiple glycosylation sites. The receptors for IL-4 are still being elucidated. One form is well known and consists of a specific binding chain and a chain shared with the IL-2 receptor. IL-13 has many of the properties of IL-4 and may even use at least one of the IL-4 receptors although the ligands share only 25% amino acid identity. The 140 kD binding chain of the type 1 IL-4 receptor can exist as a truncated soluble protein in biological fluids and acts to transport IL-4 at a stoichiometry of 1:1. IL-4 appears to be crucial in the development of an IgE antibody response. With IL-11, it enhances proliferation of primitive and committed blood cell progenitors. Somewhat surprisingly, IL-4 is also capable of enhancing development of cytotoxic T cells. For a recent review see COHEN and COHEN (1996).

Interleukin-4 was first explored clinically as a T cell activator for tumor therapy. DNAX of Schering Plough, Inc, cloned and expressed the human gene for clinical use in treatment of infectious diseases, since it is a B cell stimulating factor and can enhance some forms of antibody production. The tumor therapy efforts were based on reports of IL-4-mediated stimulation of CTL function. Currently, IL-4 is viewed as a regulator of T cells considered to mediate chronic inflammation. In addition to inhibiting the production of proinflammatory cytokines like IL-1, TNFα, IL-6 and IL-8, IL-4 can direct T cell development towards a non-inflammatory phenotype. It is this activity which is being explored as a treatment for diabetes and multiple sclerosis. For data regarding IL-4 *in vivo* see GILLEECE et al. (1992); RUBIN et al. (1992); MILES et al. (1994); STADLER et al. (1995); ATKINS et al. (1992) and WHITEHEAD et al. (1995).

Interleukin-4 Antagonist

Another role of IL-4 is manifested in its central function in allergic and asthmatic phenomena. IL-4 is crucial for IgE production, and IgE is a central mediator of many allergic events. Hence, inhibiting the function of IL-4 may be beneficial in some clinical settings. A naturally occurring soluble receptor chain with IL-4 binding properties has been shown to inhibit some IL-4 functions in vitro as well as in vivo [for a recent review see FERNANDEZ-BOTRAN et al. (1996)]. Similarly, a mutant IL-4 which binds to the IL-4 receptor binding chain but which apparently fails to transduce a signal in most IL-4 responsive cells has been described (KRUSE et al., 1992). These agents target IL-4 interactions with its cell bound receptor and by disrupting this interaction prevent IL-4-mediated effects. Another approach is to down-regulate IL-4 production. This has been attempted *in vitro* by

exposure to agents which tend to polarize T cell development away from IL-4 axis cytokine production. Such agents include IL-12 and IFNγ. Development of these IL-4 inhibitors is still not far enough along to comment on clinical results.

Immunex has cloned a receptor binding chain specific for IL-4. Early trials in asthma are complete. Further clinical development may include the areas of intracellular infections by parasites and AIDS.

Interleukin-6

IL-6 is multifunctional and is involved in acute phase reactions, immune responses and hematopoiesis. It is produced by a wide variety of cells including lymphocytes, monocytes, macrophages, fibroblasts, hepatocytes, astrocytes, keratinocytes, osteoblasts, endothelial cells and many more. Expression of IL-6 in monocytes is inhibited by IL-4 and IL-13. The human glycoprotein is approximately 26 kD. IL-6 is predicted to exist as a 4 helix bundle type protein like other members in the family including IL-11, EPO, G-CSF, OSM, CNTF, LIF, MGF, prolactin and others (see below). The receptor consists of an 80 kD binding protein and a 130 kD signal transducing chain. Signal transduction may be generated by a multimeric ligand–receptor complex. A hexamer consisting of 2 each of the α, β receptor subunits and IL-6 ligand is the currently favored model of a signal transducing complex. Soluble IL-6R, arising from alternate splicing of the gp80 message and also perhaps from proteolysis of the membrane form, can act as an IL-6 carrier and initiate high affinity receptor-mediated signaling on cells. The gp130 subunit is shared with receptors for IL-11, OSM, LIF and CNTF. A soluble form of gp130 has also been found. It is capable of inhibiting gp130-mediated signaling. Obviously the biology of IL-6 is complex. For a recent review see SEYMOUR and KURZROCK (1996).

The hematopoietic ability of IL-6, especially thrombopoiesis, is being exploited in treatment of cancer patients after ablative therapy (chemotherapy and irradiation) to restore normal platelet levels. Genetics Institute has cloned and expressed human IL-6 for development in humans, as has Ajinomoto. Phase III trials in humans have shown some enhancement of platelet and neutrophil recovery when used in combination with G-CSF, although side effects were common. Some of the developers of IL-6 include Ajinomoto, Sandoz and Yeda. Imclone has cloned and expressed a mutein of IL-6. IL-6 is also claimed to have some anti-viral activity, particularly with respect to HSV-2, although most development efforts are not focusing on this characteristic. The major role played by IL-6 in inflammation may limit its use as a therapeutic inducer of hematopoiesis. For results of clinical trials using IL-6 and associated toxicities see ATKINS et al. (1995), GORDON et al. (1995b), VAN GAMEREN et al. (1994), D'HONDT et al. (1995) and STOUTHARD et al. (1996).

Interleukin-7

IL-7 is a 20–25 kD glycoprotein produced by bone marrow-derived stromal cells, thymic epithelial cells and perhaps other cell types. It has growth factor activity on pre-B cells and T cells and is being examined for its ability to restore lymphocyte levels in lymphopenic patients. It promotes induction of CTL and LAK function and, therefore, also might be a candidate for cancer therapies. For a recent review see KOMSCHLIES et al. (1995).

Interleukin-8 Receptor Antagonist

IL-8 is involved in several inflammatory events and is a powerful neutrophil chemoattractant and activator. It is produced by many cell types including monocytes, macrophages, T cells, NK cells, endothelial cells, keratinocytes, hepatocytes, astrocytes and chondrocytes. IL-8 is a member of a proinflammatory cytokine family, which includes GROα/MGSA, platelet factor 4, β thromboglobulin and IP-10, and is known as the α subfamily of chemokines, or the CXC chemokines. The β or CC chemokine subfamily includes

RANTES, MIP-1α, MIP-1β, MCP-1 and others. The CXC chemokines activate neutrophils whereas the CC chemokines chemoattract and activate monocytes and T cells. IL-8 is a dimer of approximately 20 kD. Two different receptors for human IL-8 have been identified with 77% amino acid identity. They are G-protein coupled 7 transmembrane domain receptors. A third receptor has also recently been identified. Since IL-8 is a powerful inflammatory protein and high levels of IL-8 have been reported in several inflammatory diseases, development of an IL-8 antagonist might be therapeutically useful. A soluble receptor could act as an antagonist. Repligen has cloned and expressed such a protein. Development for use in chronic inflammatory states such as arthritis, acute respiratory distress syndrome (ARDS), IBD or SLE, however, is progressing slowly.

Interleukin-10

IL-10 exists in solution as a non-covalently linked homodimer of approximately 30–35 kD. In humans, it is produced by many different T cell subsets and activated monocytes as well as other cells. The receptor contains a 110 kD chain which shares some characteristics of the IFNγ binding chain. IL-10 is pleiotropic and a potent modulator of monocyte/macrophage function. As such, it inhibits production of proinflammatory cytokines like TNFα, IL-1, IL-6, IL-8 as well as prostaglandins. It also inhibits macrophage antigen presentation. These activities make IL-10 a candidate for treatment of rheumatoid arthritis and possibly other diseases associated with chronic inflammation. It is being developed by DNAX Research Institute of Schering Plough. For data regarding clinical use of IL-10 see HUHN et al. (1995, 1996).The immunoregulatory properties of IL-10 have suggested its use in treating diabetes. This is addressed by BALASA and SARVETNICK (1996).

Investigators at the NIH have found HIV replication within macrophages, monocytes and some T cell lines to be inhibited by IL-10. Although this is intriguing, clinical development in this area is not proceeding rapidly.

Interleukin-11

Human IL-11 is an approximately 23 kD glycoprotein produced by stromal and fetal lung fibroblasts, trophoblast cells, articular chondrocytes and synoviocytes. It can inhibit adipogenesis in addition to supporting hematopoiesis, lymphopoiesis and induction of acute phase proteins. IL-11 can induce synthesis of tissue inhibitor of metalloproteinase and thereby functions in cartilage homeostasis and tissue remodeling. Additional functions include neuronal differentiation and osteoclast development. The receptor is related to the IL-6 receptor in structure and shares the gp130 common chain. For reviews of the biology and clinical applications of IL-11 see GORDON (1995) and TERAMURA et al. (1996).

Interleukin-11 supports thrombopoiesis similarly to IL-6 but must act in concert with another hematopoietic growth factor for this function. Production of megakaryocytes and platelets was stimulated in phase II trials with cancer patients. Therefore, IL-11 has the ability to reduce the number of platelet transfusions in cancer patients after chemotherapy. It is being developed by Genetics Institute, Inc, Schering Plough, Wyeth Ayerst/American Home Products and Yamanouchi. It is also being evaluated as an anti-inflammatory agent in Crohn's disease and some forms of mucositis. For results of clinical trials using IL-11 see GORDON et al. (1996) and TEPLER et al. (1996).

Interleukin-12

IL-12 is a 75 kD heterodimeric glycoprotein consisting of p40 and p35 subunits. While the p35 subunit is homologous to IL-6 and GM-CSF, the p40 subunit is homologous to IL-6Rα. Monocytes, macrophages, B cells and some mast cells are known to produce IL-12. It is multifunctional in that it promotes cytolytic activity of NK cells and T lymphocytes. Indeed, IL-12 was initially identified as NK cell stimulatory factor. It enhances certain forms of humoral immunity, and is a strong activator of T cell secretion of proinflammatory cytokines such as IFNγ and TNFα. The ability to couple cellular and humoral immunity, as well as strongly stimulating non-specific (NK) and specific (T cell) forms of immunity, make IL-12 a key cytokine. There is some evidence IL-12 can act with other growth factors to enhance development of myeloid cells from primitive progenitors. For recent reviews of the biology of IL-12 see TRINCHIERI and GEROSA (1996) and STERN et al. (1996).

The multifunctionality of IL-12 suggests therapeutic potential for a spectrum of disease states. For one, it is being developed as a potential anti-tumor therapy. Enhancement of both NK and CTL activity would be of definite benefit in this application. The tendency to polarize T cells to secretion of inflammatory cytokines also may be beneficial in anti-tumor defenses (BRUNDA et al., 1996). Genetics Institute with Wyeth Ayerst have examined the effects of IL-12 in a phase II trial for advanced kidney cancer. Initial adverse reactions were noted but collaborations with Roche are now underway for renal cell carcinoma treatment. As an anti-tumor therapy it can be as efficacious, or more so than IL-2 and IFNα. As an adjuvant, or immune deviation mediator IL-12 promotes Th1 immunity and can suppress Th2 type immunity. These abilities suggest that it may be useful for treatment of allergies and asthma. Another potential therapeutic application takes advantage of its adjuvant properties in the management of intracellular pathogens. IL-12 is also being examined for treatment of HIV and hepatitis B and C.

Interleukin-15

Interleukin-15 is similar, but not identical, to IL-2 in its target cell range and activities. Part of this similarity may be due to the fact that 2 of the 3 chains of the receptor are shared with the IL-2 receptor. IL-15 nevertheless does have some specific features. Its development as a therapeutic might focus on the T cell activation properties of IL-15 and with tumors and viral immunity as potential targets. Immunex, which discovered IL-15, is examining the molecule's ability to provide therapeutic advantage in treatment of chemotherapy-induced mucositis, as well as in treating infectious diseases. For recent work on IL-15 see KANEGANE and TOSATO (1996) and KUMAKI et al. (1996).

Interleukin-16

Interleukin-16 has been described as a chemotactin for T cells. As such, it is possible to envisage therapeutic uses of such a protein to attract T cells to a site where an immune response is desired, perhaps in cancer therapy. On the other hand, antagonizing the influx of T cells to a site of inflammation might also be of some benefit and so an antagonist of IL-16 might be beneficial. However, the exact molecular nature of this protein is in dispute. It is unclear whether or not IL-16 is secreted and the exact size of the potentially secreted protein is unknown. Resolution of these questions will aid the development of therapeutic recombinant proteins involving IL-16.

2.2.1.3 Growth and Colony Stimulating Factors

The term growth factor is a functional term and derives from historic usage. Several of the interleukins were originally called growth factors. Their names have since changed as more pleiotropy in their activities was discovered. Some proteins, however, retain their earlier names as colony stimulating factors or growth factors. This division into interleukins, colony stimulating factors, hormones and growth factors is more one of convenience than anything else. Tab. 5 lists some therapeutic recombinant growth factors currently under development.

The first recombinant growth factors were introduced to the market in 1988. Erythropoietin (EPO), granulocyte colony stimulating factor (G-CSF) and granulocyte-macrophage colony stimulating factor (GM-CSF) were among the first. Others still in development include thrombopoietin (TPO), macrophage colony stimulating factor (M-CSF), transforming growth factors (TGF) α and β, expidermal growth factor (EGF), fibroblast

growth factor (FGF), platelet-derived growth factor (PDGF), nerve growth factor (NGF), ciliary neurotrophic factor (CNTF) and brain-derived neurotrophic factor (BDNF).

Erythropoietin, thrombopoietin and the colony stimulating factors induce proliferation of hematopoietic precursors and are useful for restoring levels of blood cells, including red blood cells, megakaryocytes, platelets, granulocytes, monocytes and macrophages, particularly after ablative chemotherapy. EPO, in particular, stimulates the development of erythrocytes and so has been used in diseases associated with anemia. For example, renal disease and dialysis associated anemia, chemotherapy associated anemias and those associated with surgery or rheumatoid arthritis. IL-3 (multi-CSF), GM-CSF, M-CSF and G-CSF are active on the myeloid pathway of differentiation and are being used for chemotherapy associated leukopenia in cancer patients.

Erythropoietin

Circulating EPO exists as a 165 amino acid protein with an apparent molecular weight of 30 kD. It is produced primarily by endothelial cells of the kidney but also by hepatocytes. It stimulates erythropoiesis but thrombopoiesis as well. The receptor consists of a 66 kD glycoprotein but an additional 95 kD accessory chain has also been implicated.

Amgen first introduced recombinant human EPO and markets it for anemias associated with chemotherapy, dialysis and other conditions. Recombinant human EPO has also been cloned and expressed by Genetics Institute. These companies have formed alliances with a number of other companies, including Pharmacia and Upjohn, Boehringer Mannheim, and Chugai for marketing in various parts of the world. A PEG-derivatized form of EPO has been generated by Cangene with a longer half-life than conventional rhEPO. Elanex Pharmaceuticals also has produced a rhEPO derived from mammalian cell expression systems and has launched this product in several countries. Other producers of EPO include Serono, Snow Brand and SRC VB Vector. EPO is certainly one of the

Tab. 5. Some Therapeutic Recombinant Growth Factors Currently Under Development

Protein	Company	Indication	Status
GM-CSF	Immunex	Neutropenias	Pending approval
G-CSF	Amgen	Infectious diseases etc.	Phase II/III
		Myeloid disorders	Phase II/III
SCF	Amgen	Hematopoietic restoration	Post phase III
Pixykine	Immunex	Myeloid restoration	Post phase II
Synthokine	Searl	Myeloid restoration	Clinical
Erythropoietin	Ortho Biotech	Transfusion reduction	Approved
Erythropoietin	Amgen	Anemias	Approved
Thrombopoietin	Genentech	Thrombocytopenias	Phase I
LIF	AMRAD	Thrombocytopenias	Phase I
PDGF-BB	Chiron/RW Johnson	Wound healing	Phase III
EGF-toxin	Seragen	Cancers	Phase I/II
FGF	Scios	Stroke	Post Phase I/II (f)
IGF-1	Cephalon	ALS	Phase III
		Peripheral neuropathies	Phase II
IGF-1	GeneMedicine Inc	Motor neuropathy	Phase I
IGF-1	Genentech	Diabetes	Phase II/III
NGF	Genentech	Peripheral neuropathies	Phase II
BDNF	Amgen/Regeneron	ALS	Post Phase III (f)
Neutrophin-3	Amgen/Regeneron	Peripheral neuropathies	Phase I/II
CNTF	Regeneron	ALS	Post Phase III (f)
F-spondin	Cambridge Neurosciences	Spinal cord injury	Preclinical
GDF-1	Camb Neuro	MS	Preclinial
GGF-2	Camb Neuro	MS	Preclinical
GMF	Regeneron		Preclinical
NT-3	Regeneron	Motor neuropathies	Preclinical
TGFβ	Genzyme	Skin ulcers/MS	Phase II/I
BMP-2	Genetics Institute	Bone and cartilage repair	Clinical trials
BMP-7/OP-1	Biogen/Creative Bio	Bone and cartilage repair	Preclinical
BMP/MP-52	Hoechst	Bone and cartilage repair	

efficacious recombinant human proteins and has set a standard for success in biotechnology. For results of some clinical trials with EPO see NEGRIN et al. (1996); SCHREIBER et al. (1996) and BROWN and SHAPIRO (1996).

Erythroid Differentiation Factor

EDF is a protein capable of inducing differentiation of erythrocyte progenitors. One obvious possible application is anemia in combination with EPO to hasten restoration of red blood cells. As a treatment for chemotherapy patients, it may also be of some benefit. A less obvious application relates to anti-cancer efforts. Some forms of cancer are preerythrocytic in nature. EDF has the ability to differentiate these cells and thereby inhibit their uncontrolled proliferation. Ajinomoto is following up on this possibility.

Granulocyte Colony Stimulating Factor

G-CSF is produced by monocytes in addition to fibroblasts, endothelial cells, astrocytes, osteoblasts and bone marrow stromal cells, and acts primarily on neutrophilic granulocytes. It can also stimulate stem cell proliferation and has been found to be chemotactic for mesenchymal cells. Since cells of the placenta, decidua and endometrium express G-CSF receptors, it may function during pregnancy. G-CSF is glycosylated and migrates at approximately 20–25 kD. The receptor is closely related to gp130, and like the growth hormone (GH) receptor, homodimerization apparently results in signal transduction. A soluble form of the G-CSF receptor has been described.

G-CSF is marketed by Amgen and Roche under the trade name *Neupogen,* among others which include *Lenograstim* and *Filgrastim.* The principal uses are prevention of neutropenia following chemotherapy and ablative treatment for bone marrow transplantation. The factor is also useful for mobilization of blood cell progenitors into the peripheral blood prior to autologous (or heterologous) transplantation. G-CSF is also in early stage development by Dong-A. A mutant form of this protein is sold by Kyowa Hakko for enhancing granulopoiesis. Chugai has a recombinant form of G-CSF marketed as *Neutrogin.* The applications are similar to the uses of other G-CSFs. For results of some clinical trials using G-CSF see GANSER et al. (1993); CORTELAZZO et al. (1995); WEXLER et al. (1996) and NEGRIN et al. (1996). G-CSF may also be beneficial in sepsis treatment (NELSON and BAGBY, 1996).

Granulocyte Macrophage Colony Stimulating Factor

GM-CSF is a glycoprotein of 127 amino acid residues, which stimulates myeloid progenitor proliferation and differentiation. For this reason it has been used for treatment of neutropenia (WELTE et al., 1996; BERNINI, 1996). In addition, GM-CSF stimulates megakaryocyte and erythroid precursors, and can stimulate proliferation by endothelial cells as well as dendritic cells. GM-CSF has some toxicities associated with its ability to induce secretion of proinflammatory cytokines like IL-8, IL-1 and TNFα.

GM-CSF has been examined by Amgen for treatment of immune deficiencies and for infectious diseases. It also has been used for chemotherapy patients to restore myeloid cell function. This protein may also have some benefit for burn patients who are usually immunosuppressed and subject to severe infections. Significant toxicities have been a hurdle for development of this growth factor. GM-

CSF is also known as *Leucotropin* and *Sargrastim.*

Genetics Institute, Sandoz and Schering Plough are developing rhGM-CSF as a therapeutic for granulocytopenia in several settings.

Immunex markets a recombinant GM-CSF known as *Leukine.* It is used for replacing the myeloid cell compartments after several depletion regimes for various forms of cancer. For results of some clinical trials using GM-CSF see ZHANG et al. (1990); FAY et al. (1994); BRETTI et al. (1996) and O'SHAUGHNESSY et al. (1996).

Macrophage Colony Stimulating Factor

M-CSF is both a secreted and cell surface cytokine. The biologically active form of this glycoprotein is a dimer. It is synthesized by several cell types including fibroblasts, bone marrow stromal cells, astrocytes, osteoblasts, keratinocytes, macrophages, B cells, T cells and endothelial cells. M-CSF promotes macrophage progenitor proliferation, differentiation, and survival. It also has similar activities on osteoclasts and microglia. In mice, only macrophages involved in organogenesis and remodeling are dependent on M-CSF. These include alveolar, splenic and synovial macrophages, Kupffer cells and microglia, but not macrophages involved in immune and inflammatory responses. M-CSF may also play a role in pregnancy, like the other CSFs described above.

M-CSF has been used therapeutically in treatment of acute monocytic type myeloid leukemia and hypoplastic leukemia. It is also being examined for a possible role in treatment of other cancers and fungal diseases. For results of a clinical trial see MINASIAN et al. (1995).

Thrombopoietin – Megakaryocyte Growth and Development Factor

MGDF is being developed by Amgen and Kirin Brewery for treatment of thrombocytopenia. It stimulates the differentiation and proliferation of platelet progenitors and so is potentially useful in recovery from chemotherapy. For a recent review see KAUSHANSKY (1996).

Thrombopoietin itself was cloned independently by several investigators. It is being independently developed by Genentech, Novo Nordisk and Amgen/Kirin brewery for thrombocytopenia subsequent to chemotherapy.

Stem Cell Factor

SCF exists in two forms with molecular weights of 45 kD and 32 kD resulting from alternative splicing. Both are cell surface molecules which are cleaved to yield soluble molecules of 31 kD and 23 kD. SCF is produced by fibroblasts, hepatocytes, endothelial cells, neurons, macrophages, Schwann cells, epithelial cells and some other cell types. It synergizes with other growth factors to induce proliferation of the progenitors of myeloid and erythroid cell lineages. For recent reviews see GLASPY et al. (1996) and HASSAN and ZANDER (1996).

Soluble human SCF has been isolated, cloned and expressed by Sandoz and Systemix. Its primary use has been for expansion of hematopoietic stem cells ex vivo prior to reintroduction into a patient. British Biotech is also investigating the use of stem cell factor as an anti-cancer therapy. Bone marrow stem cell mobilization appears to be a major function of this protein and this would be useful in bone marrow transplants. For some clinical results with SCF see ORAZI et al. (1995) and GRICHNIK et al. (1995). Amgen has recently completed a clinical phase III trial of SCF for treatment of chemotherapy related to breast cancer. SCF together with *Neupogen* was more effective than *Neupogen* alone (SEACHRIST and YOFFEE, 1997).

Leukemia Inhibitory Factor

This protein is related to IL-11 and IL-6 and, like these cytokines, supports thrombopoiesis. AMRAD Corp has examined this protein as a treatment for thrombocytopenia and observed some success. Mature human

LIF is heavily glycosylated and has an apparent molecular weight of 38–67 kD. It is produced by T cells, monocytes, thymic epithelial cells, embryonic stem cells, bone marrow stromal cells, astrocytes mast cells, astrocytes and other cell types. Since the receptor for LIF consists of a specific chain and the gp130 chain common to the IL-6 family receptors, many of the activities of LIF overlap with those of these other family members. Not only can LIF stimulate hematopoietic progenitor proliferation, it stimulates acute phase protein synthesis, myoblast proliferation, osteoblast function and can maintain embryonic stem cells in an undifferentiated state. Other activities on neuroglia and adipocytes have been noted.

2.2.1.4 Nerve Growth Factors

Several growth factors have been identified which are neurotrophic. They act on one or several cell types of nervous tissue and promote growth, maturation and survival. For example, NGF, CNTF and BDNF all stimulate nerve growth and could be useful in neuropathies like Alzheimer's disease, peripheral neuropathies, stroke, Huntington's and Parkinson's disease. Another potentially large indication is as a treatment for chronic pain. For recent reviews on neurotrophins see TOMLINSON et al. (1996); LINDSAY (1996); PUSCHEL (1996) and KNUSEL and GAO (1996).

Nerve Growth Factor

NGF is a 26 kD, nonglycosylated polypeptide normally circulating as a homodimer. It is synthesized by some epithelial cells, neurons, astrocytes, Schwann cells, smooth muscle cells, fibroblasts, mast cells, T cells and macrophages, among others. Two receptors have been identified which bind NGF. NGF acts on various cells including Schwann cells, bone marrow fibroblasts, keratinocytes, lymphocytes, neurons and some epithcial cells. It is highly pleiotropic and rescues or blocks cell death of sensory and sympathetic neurons in several situations. Its functions outside the nervous system are now being examined but among other activities, NGF acts with other CSFs in generating hematopoietic colony formation, is chemotactic for neutrophils and can influence immunoglobulin isotypes which are secreted. Use in treatment of Alzheimer's disease has been proposed. A recombinant human NGF has been developed by Genentech for treatment and repair of nerve damage. It has shown some protection of neurons in experimental animal models. Amgen and Roche are developing an NGF for Alzheimer's disease.

Ciliary Neurotrophic Factor

CNTF is a 23 kD protein with no apparent glycosylation. It is derived from Schwann cells and type 1 astrocytes, and can act on sensory ganglia neurons, non-autonomic motor neurons and sympathetic motor neurons. CNTF promotes survival and differentiation of these cells. It has been suggested to be a rescue factor for damaged neurons and a differentiation inducing factor for developing neurons and glial cells. In addition, CNTF can maintain the undifferentiated state of pluripotent embryonic stem cells (a property shared with LIF), and can induce fibrinogen and acute phase proteins. Consistent with these properties is the finding that CNTF shares the gp130 common chain of the IL-6 family receptors. Other members of this family sharing neurotrophic properties include LIF and *Oncostatin M* (OSM). The ability to promote neuron survival has been observed in animal models where CNTF reduces pathology associated with progressive motor neuropathies.

Regeneron is developing a form of CNTF for potential treatment of amyotrophic lateral sclerosing disease (ALS). Fidia and Dompe are developing CNTF, as well, for treatment of acute myelogenous leukemia (AML) and peripheral neuron disease. Roche is also looking at CNTF, formerly with Synergen, for treatment of progressive bilbar palsy, primary lateral sclerosis, progressive muscular atrophy, spinal muscular atrophies and possibly for kidney disease and AIDS. For results of a recent study see ALS-CNTF Study a (1996);

ALS-CNTF Study b (1995) and ALS-CNTF Study c (1995). A phase III trial was stopped for lack of efficacy (SEACHRIST and YOFFEE, 1997). Regeneron is also developing a 2nd generation protein related CNTF called *Axokine*.

Brain-Derived Neurotrophic Factor

BDNF acts to promote neuron survival and differentiation of dopaminergic neurons. It has shown promise in promoting reinnervation when infused into rats and so might be useful in treating Parkinson's disease (YUREK et al., 1996). BDNF is being developed with CNTF by Amgen for possible treatment of diseases with neuronal malfunction, degeneration or dysfunction. In other animal models of motor neuron disease, synergy between BDNF and CNTF was observed in reducing disease progression. The potential indications for BDNF are being expanded. In addition to those listed above, stroke, Alzheimer's disease and peripheral neuropathies are being considered. In phase I/II clinical trials BDNF was well-tolerated and safe. Some reduction in the rate of breathing capacity loss was observed, as well. Improvements in delivery and pharmacokinetics, problems presented by many recombinant proteins as therapeutics, should enhance the possibility of significant gains in disease treatment with these growth factors. Recently the factor failed to show efficacy compared to controls, when administered s.c., in a phase III trial. Amgen apparently is still pursuing delivery directly into the spinal cord (SEACHRIST and YOFFEE, 1997).

Glial-Derived Neurotrophic Factor

GDNF is a disulfide linked homodimer, a member of the TGFβ superfamily, originally discovered by scientists at Synergen (LIN et al., 1993). It is capable of inhibiting degeneration of dopaminergic neurons of the midbrain. As such, GDNF represents a potential therapeutic for Parkinson's disease and is being investigated by Amgen. In animal models it promotes neuron survival and enhances embryonic neuron maturation. It has been shown to act synergistically with BDNF and CNTF (ZURN et al., 1996). Therefore, a combination of these proteins may be more effective in a therapeutic setting than either alone, as is possibly suggested by the recent failures of BDNF and CNTF alone in human trials (see below).

Glial Cell Growth Factor 2

GGF2 is a protein which promotes growth and survival of nerve cells and is indirectly involved in neurite outgrowth (MAHANTHAPPA et al., 1996). It may be useful in chronic neurological disorders such as multiple sclerosis. GGF2 is a neuregulin family member, and might function in treatment for muscle disorders and retinal cell survival.

Neurotrophin-3

NT-3, a brain-derived neurotrophic factor with high homology to NGF, is being developed by Amgen and Regeneron. The indications are similar as for other neurotrophic factors, i.e., Parkinson's disease, spinal cord injury and motor neuron disease. NT-3 plays an important role in nervous tissue development (CHALAZONITIS, 1996), and in animal models it promotes neuron regeneration and prevents cell death subsequent to ischemia.

Growth and Differentiation Factor-1

GDF-1 is a member of a family of growth and differentiation factors in the TGFβ superfamily (LEE, 1990). It is another neurotrophic factor being investigated by Cambridge Neurosciences. GDF-1 also functions in nerve cell regeneration and may be useful in multiple sclerosis therapy, or other indications where repair of nervous tissue would be useful.

Noggin

Noggin is a neurotrophic factor involved in embryonic neural tissue formation (ZIMMER-MAN et al., 1996a) and is expressed in specific regions of adult mammalian brain (VALENZUELA et al., 1995). Should noggin be capable of enhancing neural tissue development or regeneration in adult tissues, it might serve as a useful therapeutic in several neurological diseases, assuming appropriate delivery.

F-Spondin

F-spondin is a novel nerve growth factor which may be useful in promoting axon growth following spinal cord injury or in damage to other nerve tissue. Cambridge Neurosciences is developing this principle.

Glial Maturation Factor

GMF promotes maturation of both glial and nerve cells. It is being investigated by Regeneron.

2.2.1.5 Other Growth Factors

The members of the TGF family stimulate fibroblast responses and may be useful in wound healing. They also have antiproliferative effects on some lymphoid cells and may be useful in diseases such as rheumatoid arthritis, psoriasis, leukemia and others where lymphoproliferation contributes to pathology. EGF stimulates epidermal cell growth and may be useful in healing skin wounds as well as ulcers and may be of use in eye surgery. FGF and PDGF stimulate angiogenesis and fibroblast growth and may also be of use in wound healing applications.

Transforming Growth Factor-β

A family of related proteins consists of 5 TGFβ proteins and the more distantly related bone morphogenetic proteins, activins and inhibins, which share some 30–40% amino acid sequence identity. Remarkably, nearly every cell type has receptors for TGFβ-1 suggesting its central roles. This is paralleled by sequence conservation for TGFβ-1 of nearly

100% across many mammalian species. Given these observations, it is not surprising that the role of TGFβ depends largely on the context in which it is perceived, i.e., the activation state and cytokine milieu of the responding cell. In general, TGFβ stimulates mesenchymal cell responses while inhibiting those of epithelial and neuroectodermal cells. TGFβ is synthesized as a large precursor which is cleaved. The products remain associated and are secreted in a biologically inactive form. Activation can occur through acidification, which releases mature, disulfide-linked, dimeric TGFβ of approximately 25 kD. Cell surface receptors remain incompletely described but at least 3 classes of binding proteins have been identified.

The TGFβ-1, -2 and -3 proteins appear to have similar functions but potent abilities of promoting fibroblast proliferation and extracellular matrix deposition make TGFβ an obvious candidate for treatment of dermal ulcers. This is being pursued although TGFβ has not performed well in trials for treatment of retinal macular holes. TGFβ2 has been investigated by Celtrix in collaboration with others.

As mentioned above, TGFβ also has the ability to profoundly alter immunological responses, being generally immunosuppressive. The factor is found in the anterior chamber of the eye and has been suggested to participate in what has been called the "immunologically privileged" status of this site. It is also produced in large quantities by T cells found in the mucosae particularly of the gut. TGFβ appears to be responsible, at least in part, for the phenomenon of oral tolerance, an immunological state in which inflammatory responses are reduced subsequent to oral administration of an antigen [for recent reviews see KAGNOFF (1996), WHITACRE et al. (1996) and FRIEDMAN (1996)]. Although significant effort has been expended with this approach, success has been sporadic. This outcome may relate to a recent observation that autoimmunity rather than tolerance can be induced by oral administration of autoantigen (BLANAS et al., 1996). Animal experiments have shown some efficacy of oral administration of antigen in alleviating pathology observed in murine models of multiple sclerosis and diabetes. Several phenomena have been invoked to explain this effect, and all may participate to varying degrees. One is the establishment of large numbers of antigen-reactive T cells which secrete TGFβ, upon antigen exposure. Since TGFβ down-regulates other T cell responses which result in inflammatory cytokine secretion, the end result is a decrease in apparent symptomology. Apparently in consideration of these results, TGFβ is being promoted as a potential therapy for multiple sclerosis by Celtrix.

Oncogene Science is examining a TGFβ-3 precursor protein for treatment of chronic skin wounds, and is collaborating with Ciba-Geigy and Pfizer for development in additional indications including cancer therapy induced oral mucositis.

TGFβ-1 is being developed by Mitsubishi for mucositis, and is being examined for potential use in rheumatoid arthritis and other inflammatory diseases where the anti-inflammatory activities of TGFβ might be beneficial.

Platelet-Derived Growth Factor

PDGF stimulates mitogenesis in many cell types including fibroblasts, and consists of 2 chains, an α and a β chain. Although PDGF must exist as a dimer for activity, any combination of two chains appears to have biological activity. PDGF is part of a larger family of growth factors including vascular endothelial growth factor, which is analogous to PDGF-BB, and placental growth factor, a PDGF-AA analog. The dimeric proteins have apparent molecular weights of 30 kD although some alternative splicing of the A chain can give rise to slightly shorter variants with full bioactivity. Among the activities of PDGF isoforms are mitogenesis for many connective tissue cells, dermal fibroblasts, glial cells, smooth muscle cells and some endothelial and epithelial cells. The factor stimulates collagen synthesis and collagenase secretion, and is chemotactic for neutrophils, fibroblasts and smooth muscle cells. It also has some immunomodulatory activities. This multifunctional protein also participates in tissue remodeling and chemotaxis. Two distinct receptors exist with varying distributions among different cell types. The demonstration of accelerated wound healing when applied topically stimulated interest in these proteins as therapeutic agents.

PDGF is being examined as a treatment for skin ulcers and wound healing by Chiron and Johnson and Johnson. Creative Biomolecules is developing rhPDGF for treatment of gastrointestinal ulcers and perhaps periodontal tissue repair. Novo Nordisk is developing PDGF-BB for treatment of dermal wounds. Early clinical trials with foot ulcers of diabetic patients did not encourage further development in this indication, however. Treatment of other wounds remains a possibility particularly as PDGF-BB is known to stimulate fibrosis. Results of preliminary clinical trials can be found in PIERCE et al. (1995) and STEED (1995).

Epidermal Growth Factor

EGF exists both as a soluble and cell surface protein growth factor. The precursor protein is 130 kD consisting of 9 EGF domains. The mature protein exists as a single domain of 6–10 kD. The soluble pro-EGF protein can bind to heparin binding protein, and this complex may prolong the existence of EGF in body fluids. Human EGF has been isolated from urine and has been called urogastrone due to its gastric secretion inhibitory functions. The receptor consists of a single chain which signals upon dimerization. This protein is often amplified in several tumors. Like other cytokines and growth factors, EGF is highly pleiotropic. It is a mitogen for mesenchymal and epithelial cells. It stimulates endothelial cell proliferation and angiogenesis.

EGF can promote growth of cells in the epidermis and therefore might be useful in treating skin ulcers, wounds and perhaps burn patients. Johnson and Johnson is developing EGF, after Chiron discontinued its efforts, and the primary focus is a form of ulcers. Other clinical studies have examined the ability of EGF to promote wound healing. Some results of clinical trials with EGF are published by HAEDO et al. (1996).

Fibroblast Growth Factor

Fibroblast growth factors represent one of the more complicated families of proteins. There are 9 FGFs described so far with some 30–50% amino acid identity between members. Acidic and basic FGF refer to FGF-1 and FGF-2, respectively. Keratinocyte growth factor is FGF-7, androgen-induced growth factor is FGF-8 and glia-activating factor is FGF-9. The FGFs are synthesized by various cells. FGF-2 ranges in size from 18–24 kD while FGF-1 is more homogenous at 16–18 kD. The family members are active on various cell types including those of ectodermal, endodermal and mesodermal origin. Functions of FGFs include inducing mitogenesis, migration and changes in morphology as well as other cellular functions. FGFs bind to heparin and heparan sulfate proteoglycans. The FGFs are mostly sequestered on cell surfaces and extracellular matrices in a bound state. At least five distinct FGF receptors have been discovered. Among the FGF activities of potentially therapeutic importance are angiogenesis, wound healing, tissue repair and hematopoiesis.

Scios Nova has cloned and expressed recombinant human FGF for use in repair of soft tissue and bone. The angiogenic properties of FGF also indicate potential use in treatment of cardiovascular disease and ulcers. Diabetics are particularly prone to development of sores and ulcers and represent a target population for therapeutic use of FGF. Angiogenesis enhancement would benefit repair in stroke victims and phase I/II trials have been started for this indication. Both Kaken Pharmaceuticals and Merck have formed alliances with Scios for development of this principle. Ramot, on the other hand, is an Israeli concern with the intention of developing FGF as a treatment for regeneration of both myocardium and corneal epithelium. Some clinical trial results are given by RICHARD et al. (1995).

Bone Morphogenetic Proteins

Genetics Institute has cloned and expressed a series of 9 proteins within a family which may contain more related proteins. These are involved in growth of osteoblasts and function in bone formation. Their obvious applications are in the areas of large fractures, osteoporosis and osteoarthritis. A related indication is in repair of bone damage caused by periodontal disease. Following up on efficacy of BMP-2 in humans and an animal bone repair model (KIRKER-HEAD et al., 1995; BOSTROM et al., 1996), Genetics Institute and Yamanouchi are developing BMP-2 for bone and cartilage repair. However, it is clear that several BMPs may be useful in several clinical settings. Development of these proteins is an exciting area and progress is rapid. The BMPs function on cells other than those strictly involved in bone formation. They can apparently act on several different mesenchymal cells. BMP-7, for example, has been shown to be useful for repair of kidney epithelium. Creative Biomolecules has explored the possibility of using BMP-7 in tissue repair subsequent to damage from periodontal disease. After demonstration of efficacy in primates (COOK et al., 1995) they are also developing BMP-7 (also known as osteogenic protein 1 or OP-1) for non-union bone fracture treatment.

Hoechst is developing a recombinant BMP called MP-52. It has the ability to promote chondrocyte and osteoblast maturation. As such, it could possibly be a treatment for osteoporosis or perhaps other bone disorders.

Hepatocyte Growth Factor

HGF, also known as scatter factor, stimulates angiogenesis and mitogenesis of several cell types including hepatocytes. It is derived from fibroblasts. Its wide spectrum of target cell types suggest that it may be useful in tissue repair including regeneration of liver, kidney, and lung. For a recent review of HGF and its multipotency see MATSUMOTO and NAKAMURA (1996).

2.2.2 Hormones

The difference between the term hormone and cytokine or interleukin is subtle. Although all are proteins secreted by cells which act on other cells, some may also act on the cells secreting them. As more is learned, the differences between these terms become less obvious. It is now more fashionable to name newly discovered bioactive cytotropic proteins cytokines or interleukins rather than hormones. Thus the terms are more historic than anything else. Tab. 6 lists some therapeutic recombinant hormones currently under development.

2.2.2.1 Insulin

Insulin treatment for diabetics has progressed now to include the use of a number of recombinant human insulin preparations. These replace porcine insulin and are both less expensive and less immunogenic. Producers of this recombinant protein include Chiron, BiobrasLilly and Novo Nordisk, in collaboration with others. Novo Nordisk has developed a recombinant form of glucagon for use in cases of insulin-induced hypoglycemia. Trials of insulin versus an insulin analog are reported by NIELSEN et al. (1995).

2.2.2.2 Growth Hormone and Insulin-Like Growth Factors

IGFs are two non-glycosylated single chain polypeptides with approximately 76% amino acid identity. IGF-I is produced by a variety of cells including hepatocytes, neurons, astrocytes, macrophages, fibroblasts, several muscle cell types and neutrophils, among others. IGF-II, on the other hand, is produced by osteoblasts, type II lung alveolar epithelial cells, chondrocytes, endothelial cells and Kupffer cells. Six forms of IGF binding proteins ranging in size from 24–45 kD have been described. These act as carriers for the IGFs and control interaction of the ligands with receptors. For recent reviews on growth hormone see BAILEY et al. (1996); HOLMES et al. (1996) and KELLEY et al. (1996).

Human growth hormone and IGFs are being developed as human growth hormones for treatment of several disorders. Primary in-

Tab. 6. Some Therapeutic Recombinant Hormones Currently Under Development

Protein	Company	Indication	Status
Testosterone	Bio-Technol Gen	Hypogonadism	IND
GH	Genentech/Lilly	Turner's Syndrome	Approved
		Abnormal growth	Approved
GH	Novo Nordisk	Abnormal growth	Approved
GH	Alkermes	Abnormal growth	Phase I
GH	Serono Labs	Abnormal growth	Pending approval
		Cachexia in AIDS	Pending approval
Atrial natriuretic peptide	Genentech/Scios	Renal failure	Phase III
LH	Serono	Infertility	Phase III
PTH	Allelix Biopharma	Osteoporosis	Phase II
Insulin	Lilly	Diabetes	Pending approval
Glucagon	Novo Nordisk	Hypoglycemia, etc.	Phase III
Prolactin	Genzyme	Adjuvant	
FSH	Serono Labs	Infertility	Pending approval
CRH	Neurobiological Technologies	RA, edemas	Phase II

dications include dwarfism resulting from inadequate pituitary function, muscle atrophy, fractured bones, Turner's syndrome, wasting, bone disorders and wound healing although there is some reason to believe further indications may be possible. Results of clinical trials for treatment of Turner's syndrome and pituitary deficiency can be found in MASSA et al. (1995); TAKANO et al. (1995) and CHIPMAN et al. (1995). Genentech's version of recombinant human growth hormone, *Nutropin,* recently was approved for treatment of Turner's syndrome by the FDA (CRAIG, 1997). Atherosclerosis treatment may be possible due to the ability of IGF-1 to lower lipoprotein A levels (OSCARSSON et al., 1995). A link to neurotrophic factor turnover may exist and open the way to use of IGF-1 in ALS disease. Further indications include diabetes, kidney failure, osteoporosis, possibly cachexia and inflammatory bowel disease (LUND et al., 1996). These recombinant human growth hormones are produced both without and with an N-terminal methionyl group. Developers of IGF-1 or somatomedin-1 include Fujisawa, Biogen, Sumitomo, Chiron, Cephalon, Genentech, Pharmacia, Kabi, Lilly, Sanofi, Dong-A, Bio-Technology, CSL, Novo Nordisk, Serono and others.

Another approach involves recombinant IGF binding proteins. Companies which have investigated developing IGF-BPs for treatment of various diseases include Celtrix Pharmaceuticals and Amgen. These proteins may be useful in treating muscle wasting in chronic pulmonary disease and several other diseases, as indicated above. For results of some recent clinical trials with growth hormones see SAGGESE et al. (1995); WIT et al. (1995); GOETERS et al. (1995); LEINSKOLD et al. (1995); MAXWELL and REES (1996); SARNA et al. (1996); FINE et al. (1996) and MACGILLIVRAY et al. (1996).

2.2.2.3 Other Hormones

Several hormones are being developed for infertility treatment. Follicle stimulating hormone (FSH) is being developed by several companies including Serono, Cell Genesys/Theriak, Organon and Unigene. For some early clinical trial results see HOMBURG et al. (1995) and BEN-CHETRIT et al. (1996). Serono is exploring development of FSH, recombinant luteinizing hormone (LH) and human chorionic gonadotropin (CG) for infertility. Inhibin is a naturally occurring protein which prevents synthesis and secretion of FSH in both men and women. It is being developed by Biotech Australia and has been shown to block both sperm production and ovulation. LH has the capacity to become a reversible contraceptive for both men and women.

Calcitonin is being developed for musculoskeletal disorders. Suntory and Unigene are developing recombinant calcitonin products for treatment of hypercalcemia, osteoporosis, Paget's disease, bone pain and perhaps fractures. Unigene is also developing a calcitonin gene related peptide (CGRP) for possible use in treating hypertension since it has demonstrated some ability to cause vasodilation. Others developing calcitonins include Chugai which is investigating a hybrid form of calcitonin resulting from sequences of both human and salmon calcitonin.

Several hormones are being investigated for possible treatment of osteoporosis. One is parathyroid hormone which Allelix has cloned, expressed and is developing. Cambridge Biotech is considering development of growth hormone releasing factor (GHRF) for this indication.

Amgen is developing the product of the obesity gene, leptin, as a potential treatment for obesity. It controls appetite as well as interacts with the central nervous system to influence lipid metabolism. Recent work indicates a role in puberty induction, as well.

Other hormones being developed include activin for induction of erythropoiesis. Thyroid stimulating hormone is being developed by Genzyme for use in treatment of cancers of the thyroid. An atrial natriuretic factor peptide derivative is being developed by Scios Nova with Genentech for treatment of dialysis patients. Some efficacy has been observed in reducing morbidity in a cohort of patients. Brain natriuretic peptide is also being developed for treatment of congestive heart failure. Suntory is developing a different electrolyte absorption agonistic peptide called atriopeptin being targeted for use with patients suffering acute congestive heart failure. Advanced Polymer Sciences is developing recombinant melanin as a sun protection product. Melatonin has also been tested in humans for treatment against non-small cell lung cancer by LISSONI et al. (1992, 1994).

2.2.3 Anti-Inflammatory Proteins

Several general approaches to reducing inflammation and consequent damage have been examined. Some have been discussed above with reference to specific disease indications. One can attempt to reduce inflammatory cell influx to sites of injury by antagonizing chemokines attracting cells to the injury locus. Blocking receptors for these chemokines is one avenue, as is blocking other receptors, like cell adhesion molecules used in normal and disease related trafficking. Cells at the site of injury can be blocked from releasing mediators, although this is a difficult process, if the goal is to employ recombinant proteins. One potentially successful approach would be to block costimulation of lymphocytes and render them anergic or apoptotic. One potential route for this might be specific delivery of Fas ligand or a toxin to the targeted cells, thus inducing specific apoptosis. Antibody fusion proteins are well suited for this and are dealt with in chapter 7, this volume. Another approach employs the fact that activated cells often express receptors for interleukins which are not expressed on resting cells. Hence, interleukin toxin fusion proteins may be useful. After mediator release, some antagonists might be useful. Among the potentially useful antagonists are IL-1ra (receptor antagonist) or other muteins blocking receptor activation. Antagonists for activated complement components would be particularly useful, as these components are powerful mediators of inflammation. Specific approaches will be given within the discussion of the individual recombinant proteins in the following sections.

2.2.3.1 Chemokines

Trafficking of leukocytes is mediated by many molecules but among them is a diverse family of proteins called chemokines. They are cationic and heparin-binding chemoattractants synthesized by several cell types including those in endothelium, epithelium and some hematopoietic cells. These strongly proinflammatory proteins are generally about

10 kD in size. They can induce degranulation of granulocytes in addition to their chemotactic properties. Chemokines may also function as growth factors for various cell types including fibroblasts and contribute to wound healing and angiogenesis. Some members of this family of proteins are IL-8, RANTES (regulated upon activation, normal T cell expressed and secreted), MCP-1 (monocyte chemoattractant protein-1), MIP-1α and MIP-1β (macrophage inhibitory proteins). These proteins fall into two classes identified as CC and CXC depending on whether or not two cysteine residues are separated by another amino acid in their primary structure, as mentioned in more detail in Sect. 2.2.1.2. The list of chemokines is growing rapidly and 10 or more proteins fall into this category at present. On the other hand, only two chemokine receptors, CKR-1 and CKR-2, mediate the activities of most of the chemokines. A third receptor, CKR-3, appears to be specifically distributed on eosinophils, although it binds the chemokines eotaxin, RANTES and MCP-3. For a recent review on the chemokines see HOWARD et al. (1996) and for a review on the pathophysiology of chemokines see ROT et al. (1996).

Some therapeutic efforts now are focused on small molecular weight antagonists of receptors, or on anti-IL-8 antibodies, since IL-8 binds to two receptors. However, a recombinant truncated form of MCP-1 has been generated (HANDEL, 1996). This may be important since mice whose MCP-1 genes have been deleted have demonstrated that MCP-1 specifically is necessary for accumulation of monocytes at sites of injury (ROLLINS, 1996a). The clinical usefulness of this approach remains in doubt until further work is done.

A different approach takes advantage of the ability of several chemokines to mobilize primitive hematopoietic precursors into the blood. British Biotech Pharmaceuticals Ltd. has developed an engineered form of MIP-1α which does not polymerize as readily as the wild type protein and hence lacks some of the inflammatory aspects of this protein. It does mobilize monocytes but tests in humans have so far failed to reveal mobilization of primitive hematopoietic cells (EDWARDS, 1996).

Other chemokines have recently been discovered with mobilization properties which lack chemoattractant properties. Further work may reveal an advantage of using these for mobilization in cancer patients.

Dompe has cloned several chemokine proteins but its development efforts are at present preliminary.

2.2.3.2 Complement

Activated complement components, particularly C3a, C5a and C4a, initiate anaphylactic reactions with inflammatory consequences. They stimulate cellular degranulation, respiratory burst, smooth muscle contraction, cytokine release and chemotaxis. They have been shown to play a role in the pathology of many models of disease states including Alzheimer's disease, multiple sclerosis, meningitis and neurotrauma. For recent reviews on complement and therapeutic intervention see DAVIES (1996) and MORGAN (1996). There are at least 3 receptors for the 3rd component of complement C3, named CR1, CR2 and CR3. Soluble CR1 is capable of inhibiting both the alternate and classical pathways of complement activation. It has been used in animal models of autoimmune disease as well as transplant rejection, ischemia–reperfusion injury and inflammation with some efficacy in reducing pathology. Phase I safety trials have been performed in humans at risk of ARDS and in 1st time myocardial infarction patients scheduled for PTCA. T Cell Sciences has shown that the protein has been well tolerated, has a half-life of over 30 h and can inhibit complement activation for longer than 1 d (RYAN, 1996a, b).

Several inhibitors of complement activation are being developed. CytoMed and Chiron are among the companies trying to capitalize on this effort. Proteins acting to inhibit the convertases of C3 and C5 are being examined as is membrane cofactor protein. Application of these inhibitors is presently targeted to trauma but infections subsequent to burns might be a likely indication for this kind of approach. Treatment of SLE, transplant rejection, sepsis, stroke, myocardial infarction and others may be amenable to therapy by complement activation inhibition. DNX is developing naturally occurring complement inactivators known as CD59, decay accelerating factor (DAF) and membrane cofactor protein. CD59 and DAF especially have been targeted for prevention of hyperacute rejection of transplanted xenogeneic organs. T Cell Sciences and Yamanouchi are developing a soluble complement receptor type 1 for treatment of ARDS. Stroke, trauma and transplantation are other potential indications. T Cell Sciences apparently has other recombinant complement inhibitory proteins under development, as well.

2.2.4 Recombinant Enzymes and Enzyme Inhibitors

Enzymes are an important class of proteins involved in many biological processes. In cases where there is a particular enzyme deficiency, either acquired or genetic, restoration with a recombinant form represents one kind of treatment.

2.2.4.1 For Cystic Fibrosis

Exocrine gland dysfunction leading to chronic pulmonary disease is characteristic of this disease. Pulmonary involvement results from excessive production of mucous in the respiratory tract. Other problems are also associated with the disease but treatment of the respiratory occlusion is a primary aim. Genentech has developed a recombinant human enzyme which degrades DNA. It is marketed for treatment of cystic fibrosis in several countries under the name *Pulmozyme*. It also has been investigated for benefit in chronic obstructive pulmonary disease, but development in this arena is not progressing due to either weak or nonexistent efficacy, at present. Several companies are developing lipolytic and other enzymes for treatment of this disease. One such company is Jouveinal, which is investigating the use of a recombinant gastric lipase. Biogen is also investigating a human plasma-derived protein called *Gelsolin* which can cleave actin filaments.

Genzyme is examining the development of a recombinant human cystic fibrosis transmembrane regulator as a replacement therapy for treatment of cystic fibrosis. Excepting the use of DNAses, development of the other candidate treatments is at an early stage. Results of clinical trials with DNAse for cystic fibrosis can be found in SHAH et al. (1996).

Another recombinant DNAse protein is being developed by Applied Genetics. Although Genentech already markets a similar product for cystic fibrosis, another potential application might be xeroderma pigmentosum. For results of some clinical trials with recombinant DNAse proteins see WILMOTT et al. (1996) and LAUBE et al. (1996).

2.2.4.2 For Inflammation

As outlined above, inflammation is an integral process of many disease states. Treatments targeting enzymes involved in this process are under examination and may provide some important benefit in several settings. For example, Genentech has cloned and produced a recombinant neutral endopeptidase (NEP). Khepri is developing this protein for potential use as an anti-inflammatory. Indications may include asthma, chronic obstructive pulmonary disease and perhaps some cancers. Lexin Pharmaceutical has produced a recombinant chymotrypsin inhibitor. Its potential therapeutic value lies in treatment of pancreatitis and reperfusion injury. ICOS, with others, is developing an enzyme which degrades platelet activating factor (PAF). Potential benefit lies in treatment of inflammation. ICOS is also developing a phosphodiesterase inhibitor with Glaxo Wellcome. This class of protein is anti-inflammatory and of potential use in cardiovascular disease treatment, as mentioned above. Superoxide dismutase (SOD) may have some value in treating inflammatory diseases and other conditions. It is being developed by several companies including IXIS, Yeda, Chiron, Cephalon, Bio-Technology General and perhaps others. Enzon is developing a PEG-derivatized form of this enzyme. Some clinical results with SOD are documented in ROSENFELD et al. (1996).

2.2.4.3 For Cancer

A recombinant arginine deiminase has been investigated by Japan Energy. A recombinant form of cobra venom factor, which activates the complement cascade, has been produced by Knoll. Although these recombinant enzymes may be useful as anti-tumor agents, development of each is at an extremely early stage.

2.2.4.4 For Gaucher's Disease

Deficient glucoceramidase activity can result in an accumulation of lipids in storage cells in the liver, spleen, lymph nodes, alveolar capillaries and bone marrow. This lipidosis disease is presently being treated with alglucerase. Treatment, however, results in increased platelet and hemoglobin levels. Genzyme has developed a new recombinant glucocerebrosidase which replaces alglucerase. It presently is marketed in the US. Enzon also produces a recombinant glucocerebrosidase with PEG-derivatization for the same indication. For results of clinical trials see GRABOWSKI et al. (1995) and ZIMRAN et al. (1995).

2.2.4.5 Other Enzymes

Carboxypeptidase-G2 is under experimental investigation by the National Cancer Institute. It may prove useful in treating *Methotrexate* poisoning resulting from cancer and other chemotherapies.

2.2.5 Clotting Factors and Related Proteins

Treatment of genetic deficiencies in clotting is perhaps one of the big success stories of recombinant proteins and biotechnology, so far. Replacement of plasma derived protein therapies carries with it significantly fewer risks as well as greater access to therapeutic material. Factors VIII and IX are now available as recombinant proteins and are being supplied by Bayer, Baxter, Genetics In-

stitute and others. Factor VIIa is used as a treatment for hemophilia A by Novo Nordisk.

A recombinant form of factor Xa inhibitor, which inhibits the activity of prothrombinase, is under development by Cor Therapeutics. Indications include treatment for prevention of pulmonary emboli. Mochida is developing a recombinant protein with a Kunitz-type inhibitor domain from ulinastatin. This protein is capable of inhibiting the enzymatic activity of several proteins including elastase, chymotrypsin as well as factor Xa and others. Development remains at an early stage.

Plasminogen activators to dissolve clots are also being marketed and developed. Abbot Labs has been investigating prourokinase for treatment in stroke cases. Results of early clinical trials with a recombinant prourokinase are given by SASAHARA et al. (1995). Merck has formed an agreement with Biopharm and is developing recombinant hementin. This enzyme derived from leeches is a fibrinolytic agent. As mentioned above, a bat (*Desmodus*)-derived plasminogen activator is being developed by Schering AG for potential use in coronary thrombosis. Genentech is also developing improved plasminogen activators. Others working along these lines include Bristol-Myers Squibb, Genetics Institute, Suntory, Eisai, Mitsubishi, Boehringer Ingelheim, Kyowa Hakko, Sumitomo, Mitsui, Mochida and others. Boehringer Mannheim is developing a thrombolytic called *Reteplase*. A similar agent from Gruenenthal is called *Saruplase*. Ciba-Geigy has investigated a chimeric protein consisting of domains derived from both tissue plasminogen activator and from prourokinase. Potential areas of treatment include thrombosis, arteriosclerosis, infarcts and other vasculopathies. Results of some clinical trials are given in SMALLING et al. (1995); ROMEO et al. (1995); GOLDHABER et al. (1996) and FALANGA et al. (1996).

Biotech Australia has a recombinant plasminogen activator inhibitor type 2. It inhibits urokinase type serine proteases which are involved in remodeling as well as cellular migration phenomena.

Genzyme has produced recombinant human anti-thrombin III (AT-III) in transgenic goats. They are developing it with Sumitomo for treatment of blood clot disorders in patients with AT-III deficiency. Lexin Pharmaceutical has developed a thrombin inhibitor *LEX026* for coagulation disorders and reperfusion injury. Japan Energy has a recombinant hirudin analog under development as an anticoagulant. One potential use of this thrombin inhibitor is in angioplasty. Ciba-Geigy also is developing a hirudin analog called *Revasc* for prevention of restenosis. Hoechst has a similar approach in developing recombinant hirudin itself. Knoll has produced a PEG conjugated recombinant hirudin as a thrombin inhibitor. The PEG-derivatization is designed to provide a longer serum half-life than the naked protein. Corvas has produced NAP-5 which is a recombinant factor Xa inhibitor derived from nematodes. Monsanto and Chiron have a recombinant form of a natural inhibitor of tissue factor and the extrinsic coagulation pathway. Indications are microvascular surgery and potentially sepsis.

Asahi has produced a recombinant form of soluble thrombomodulin. It is being developed for treatment of patients with disseminated intravascular clotting.

A recombinant aprotonin has been developed. Its use in coronary bypass operations has been tested and reported by GREEN et al. (1995).

A recombinant thromboplastin has been studied for management of patients on anticoagulant therapy (BARCELLONA et al., 1996).

2.2.6 Fusion Proteins

Recombinant DNA technology has the ability to construct chimeric proteins. Functional domains from different proteins can be joined at the coding level to generate proteins with many different functions. The technologist can select functions, virtually at will, and transplant them onto a protein backbone. This allows a single protein to perform the functions of many. A clear example involves joining the binding specificity of immunoglobulins with the toxicity of various toxins. This results in changing the binding domain on the toxin to a specificity desired by the designer. This approach, as mentioned elsewhere in this volume, is being pursued for cancer treatments. Another approach is to use non-immunoglobulin protein domains for specific binding, or other functions.

Seragen and Lilly have constructed a protein fusing epidermal growth factor domains to the toxic moiety of diphtheria toxin. This results in a protein which may be specifically toxic for tumors with elevated levels of EGF receptors. Since elevated EGF receptor expression is observed with tumors of the esophagus, colon, lung, ovary, prostate, pancreas and several other tissues, this approach may be very broadly applicable and efficacious. Furthermore, the protein may be useful in treating non-malignant tissues which express elevated EGF receptor levels where the function of these cells is undesirable. Some examples might be restenosis or psoriasis where excessive proliferation of smooth muscle cells or epidermal cells is pathologic. A similar approach has been developed by Merck and Aronex. They have fused TGFα sequences, which bind to EGF receptors, with *Pseudomonas* exotoxin. Aronex has fused erbB-2 receptor binding domain sequences with a modified *Pseudomonas* exotoxin for treatment of adenocarcinomas of the lung, breast, stomach, ovary, etc.

Immunex has created a novel cytokine consisting of components of IL-3 and GM-CSF, named *Pixykine*, which exhibits activity due to each of the components. For example, it promotes bone marrow cell growth 10fold better than a combination of the individual cytokines *in vitro*. *Pixykine* represents a novel, and potentially more powerful, treatment for neutrophil and platelet restoration, after cancer treatments. For results of a clinical trial using *Pixykine* see VADHAN-RAJ et al. (1995).

Pharmacia/Upjohn has been investigating the use of a human CD4-*Pseudomonas* exotoxin fusion protein for AIDS treatment. The HIV-binding region of CD4 targets the toxin to HIV infected cells, and free virus as well. Seragen has also investigated a similar CD4-toxin fusion protein. Some indication of inhibition of smooth muscle cell proliferation was observed which might indicate a possible use in preventing restenosis and other fibrotic events.

Enzon is investigating the anti-tumor potential of TNFα. Results of a phase I study administering TNF and IL-2 have been published (KRIGEL et al., 1995; THOMAS et al., 1995). Although other efforts along these lines have produced undesired side effects, a possible advantage might be achieved by coupling TNF to monoclonal antibodies to tumor antigens. In this way, TNFα could be acting as a toxin and act specifically on cells bearing the target antigen.

Seragen and Lilly have produced a fusion protein of IL-2 and domains of the diphtheria toxin. The conjugate is being developed for several indications including dermatological problems mediated by T cells, Hodgkin's disease and certain lymphomas. Takeda has also produced a fusion protein of IL-2 and herpes simplex virus glycoprotein D for treatment of HSV-2 infections. Some results of clinical trials with IL-2 fusion proteins are seen in MORELAND et al. (1995) and FOSS et al. (1994).

Seragen has produced a fusion toxin of IL-4. Patients with some leukemias, allergy and perhaps autoimmune diseases are potential recipients of this form of protein. An IL-6 toxin fusion protein for treatment of myelomas and Kaposi's sarcoma has also been developed by Seragen. A fusion of melanocyte stimulating hormone with the diphtheria toxin has been produced for potential melanoma treatment by Seragen.

2.2.7 Miscellaneous Proteins

Recombinant albumin for use in plasma volume replacement is being developed by Ohmeda, Genentech, Mitsubishi and perhaps others.

A soluble form of truncated ICAM-1 is being developed for treatment of the common cold by Boehringer Ingelheim. This is the major receptor for a class of rhinoviruses. A similar protein is being developed by Bayer for treatment of human rhinovirus infections in children.

Xoma is developing bacterial permeability increasing protein, and fragments of it, for treatment of infections subsequent to surgery or traumas. Results of some clinical trials are published by VON DER MOHLEN et al. (1995) and BAUER et al. (1996). Incyte Pharmaceuticals has produced a fusion protein of BPI and LPS binding protein for potential treatment of septic shock, which is claimed to have a longer half-life than BPI itself.

Lactoferrin is a protein which transports iron but has some natural antimicrobial activity as well. Agennix is developing 2 recombinant forms of this protein. One is for use as an antibacterial agent while the other form, which is saturated with iron, is being developed for its antidiarrheal properties.

Recombinant prolactin is being developed by Genzyme. Its immunostimulating properties include stimulating mitogenesis, phagocytosis and NK cell activity. It may be useful as an adjuvant.

Various anti-inflammatory proteins, enzymes and enzyme inhibitors, clotting factors and related proteins, and miscellaneous proteins are listed in Tab. 7.

3 Applications in Animals

The use of recombinant proteins in a therapeutic mode for veterinary medicine is rather limited, at present. This is due to the relative expense of these products and to the species-specificity demonstrated by several medically relevant cytokines. The use of gene therapy may be less expensive, however, and is becoming more commonly used in animals. Not surprisingly, veterinary use of recombinant proteins falls into two broad classes. One is the treatment of commercially valuable livestock for improvement of animal size or mass, offspring number or size, production of commercially valuable animal products and preservation of livestock through medical intervention and vaccination. The other broad use is improving the health, or rescuing from death, of domesticated animals or pets in particular. Use of vaccination for prevention of zoonoses has not yet received widespread popular acceptance or even recognition as a tool for human health improvement.

Tab. 7. Some Other Therapeutic Recombinant Proteins Currently Under Development

Protein	Company	Indication	Status
Factor VIIa	Novo Nordisk	Hemophilias A + B	Phase III
Factor IX	Genetics Institute	Hemophilia B	Pending approval
Factor VIII	Pharmacia/Upjohn	Hemophilia	Phase III
GHRP	Wyeth-Ayerst	Abnormal growth	Phase III/I
ICAM-1	Boehringer Ingelheim	Rhinovirus infection	Phase I
TNFR	Hoffmann-LaRoche	Septic shock, sepsis	Phase III
		Rheumatoid arthritis, MS	Phase II
TNFR	Immunex	Rheumatoid arthritis	Phase III
	Wyeth-Ayerst		
TP10	T Cell Sciences	ARDS/reperfusion	Phase II/I
t-PA	Genentech	Ischemic stroke	Pending approval
nPA	Bristol-Myers Squibb	Myocardial infarct	Phase II
t-PA	Boehringer Ingelheim	Myocardial infarct	Pending approval
t-PA mutein	Genentech	Cardiovascular disease	Phase II
TF inhibitor	Chiron	Microsurgeries	Phase II
DNAse	Genentech	Cystic fibrosis/nephritis	Approved/Phase I
Atrial natriuretic peptide	Genentech/Scios	Renal failure	
Relaxin	Connective	Scleroderma	Phase III
	Therapeutics		Phase I/II
CTLA4IG	Bristol-Myers Squibb	Immunosuppression	
EGR-Xa	Cor Therapeutics	Surgeries	Phase I
CRH	Neurobiological	RA, edemas	Phase I
	Technologies		Phase II
CD4-toxin	Upjohn/Pharmacia	HIV	
BPI	Xoma	Trauma, infections	Phase I/II
Lactoferrin	Agennix	Infections	
Albumin	Genentech/others	Volume replacement	
BNP	Scios	Congestive heart failure	Phase II
PAF	ICOS	ARDS, etc.	Phase I
HgB	Eli Lilly	O_2 transport	Phase II
HgB	Somatogen	RBC formation	Phase I
CR1	T Cell Sciences	ARDS	Phase II
Neutral endopeptidase	Khepri	Chronic inflammation	Clinical
Cerebrosidase-PEG	Enzon	Gaucher's disease	Phase I
Cerebrosidase	Genzyme	Gaucher's disease	Approved
Carboxypeptidase-G2	NCI	Methotrexate poisoning	Phase I
Urate oxidase	Sanofi	Hyperuricemia	Phase II
Superoxide dismutase	Bio-Technology	O_2 toxicity in pre-	Phase III
	General	mature infants	

One particular use which lends itself well to animal medicine is vaccination with DNA or recombinant organisms (MARTINEZ and WEISS, 1993). Nevertheless, recombinant papilloma virus L1 protein has been used to successfully vaccinate beagles (SUZICH et al., 1995). Immunization of cattle with recombinant *Babesia bovis* merozoite surface antigen-1 has been reported (HINES et al., 1995). Foot-and-mouth disease vaccines are also un-

der development (MCKENNA et al., 1996; HINES et al., 1995).

Surprisingly, IL-2 has been examined as a tumor therapy in animals, not just as an animal model in development of proteins for human use. Recombinant human IL-2 has been given to 5 cows with bovine ocular squamous cell carcinoma. Total tumor regression was seen in 3 of the 5 cases (RUTTEN et al., 1989). IL-2 has also been injected into papillomas or carcinomas of the bovine vulva. In one study, 19 of 23 animals showed reduction of tumor burden and 3 animals showed complete regression (HILL et al., 1994). Furthermore, recombinant human TNF has been used to treat equine sarcoid tumours. Regression was seen in almost every tumor (OTTEN et al., 1994).

Several instances where recombinant proteins have been used in animals for treatment of infectious diseases have been reported. Bovine IL-2 has also been used as adjunct therapy in antibiotic treatment of *Staphylococcus aureus* mediated mastitis (REDDY et al., 1992). IFNγ has also been used in this disease and in combination with IL-2 (QUIROGA et al., 1993). Interferons have been used in cattle to reduce morbidity and mortality resulting from bovine respiratory disease and mastitis (BABIUK et al., 1991).

Recombinant proteins have been used to treat diseases of a non-infectious nature. In one study, recombinant bovine somatotropin was given to cows prior to calving to reduce fat cow syndrome (MAISEY et al., 1993). When plasma urea concentrations were taken as a measure of disease, some efficacy was observed. An anemic cat suffering from chronic renal failure apparently was treated successfully with recombinant human erythropoietin (HENRY, 1994). Recombinant canine G-CSF has been used to treat hematopoietic deficiencies in Grey collies and other neutropenic dogs (MISHU et al., 1992). This breed demonstrates cyclic fluctuations in blood cell counts caused by an autosomal-recessive regulatory defect of hematopoietic stem cells (PRATT et at., 1990). The combination of recombinant canine stem cell factor and canine G-CSF appeared to have some efficacy in restoring white blood cell (WBC) counts to normal in Grey collies (DALE et al., 1995). Recombinant bovine growth hormone has been used for treatment of iron-deficient veal calves (CEPPI and BLUM, 1994).

Increasing body mass, offspring and production of certain animal-derived products represents a commercially important opportunity. The use of recombinant porcine somatotropin has been approved in Australia. The protein is given to swine to improve muscle mass at the sacrifice of fat. (The implications for human use give one pause). Somatotropin has been examined in swine for its effect on lactation performance of first-litter sows (TONER et al., 1996). Although this appeared to have some efficacy, a larger market may exist for cattle, swine, goats, sheep and poultry transgenic for growth hormone. Such an approach in livestock for human consumption remains in its early stages but certainly has a strong motivation behind it.

4 Concluding Remarks

Despite a sincere effort to relay basic principles and approaches, any attempt to be complete in dealing with the medical uses of recombinant proteins is beyond the scope of this monograph and may not be possible. Furthermore, the delays between manuscript preparation and publication render much of what is contained in this kind of effort outdated by the time of publication. Indeed, during the writing of the manuscript, results of several clinical trials were announced necessitating an update. It is unclear whether this reflects the rapid pace of research in the field, or the pace of preparing this chapter. Nevertheless, this chapter may be useful as a reference and for choosing future directions of research. In some ways the chapter is like a snapshot in a family photo album, serving to document events of the past, however fondly. The principles enumerated should be valid for quite some time and provide new avenues for exploration and exploitation of recombinant proteins, both in medicine and other commercial applications.

Acknowledgements

In most cases, the relevant scientific literature has been cited. Review articles have been suggested, as providing reference for each point made is not within the scope of this chapter. It should be noted that some of the information included in this document was obtained from the documentation of Pharmaceutical Research and Manufacturers of America 1995 and 1996 (see Tabs. 1–7), on-line sources from individual biotechnology companies, *Genetic Engineering News* and *BioWorld Today* (BWT). In particular, information in the tables was compiled and amended from *Genetic Engineering News* V16, No16 29-34, *BioWorld Today* as well as several other individual publications.

I wish to thank Professors PETER STADLER and MOSHE STERNBERG for the opportunity to provide this chapter. I also wish to thank Drs. KUN HUANG and KEN LEMBACH for critical comments on the manuscript.

5 References

ABRAHAM, E. (1996), Status report of soluble TNF receptors in the treatment of septic shock. *Abstract* (IBC Sepsis Advances in Treatment and Drug Development Meeting).

AGUIRRE, F. V., TOPOL, E. J., FERGUSON, J. J., ANDERSON, K., BLANKENSHIP, J. C., HEUSER, R. R., SIGMON, K., TAYLOR, M., GOTTLIEB, R., HANOVICH, G. et al. (1995), Bleeding complications with the chimeric antibody to platelet glycoprotein IIb/IIIa integrin in patients undergoing percutaneous coronary intervention. EPIC Investigators, *Circulation* **91**, 2882–2890.

AIELLO, L. P., PIERCE, E. A., FOLEY, E. D., TAKAGI, H., CHEN, H., RIDDLE, L., FERRARA, N., KING, G. L., SMITH, L. E. (1995), Suppression of retinal neovascularization *in vivo* by inhibition of vascular endothelial growth factor (VEGF) using soluble VEGF-receptor chimeric proteins, *Proc. Natl. Acad. Sci. USA* **92**, 10457–10461.

ALLEGRETTA, M. (1996), Prospects for T-cell receptor peptide vaccination in rheumatoid arthritis. *Abstract* (IBC New Therapeutic Advances in Arthritis Meeting).

ALS-CNTF Study a (1996), The amyotrophic lateral sclerosis functional rating scale. Assessment of activities of daily living in patients with amylotrophic lateral sclerosis. The ALS CNTF Treatment Study (ACTS) Phase I–II Study Group, *Arch. Neurol.* **53**, 141–147.

ALS-CNTF Study b (1995), A phase I study of recombinant human ciliary neurotrophic factor (rHCNTF) in patients with amyotrophic lateral sclerosis. The ALS CNTF Treatment Study (ACTS) Phase I–II Study Group, *Clin. Neuropharmacol.* **18**, 515–532.

ALS-CNTF Study c (1995), The pharmacokinetics of subcutaneously administered recombinant human ciliary neurotrophic factor (rHCNTF) in patients with amyotrophic lateral sclerosis: relation to parameters of the acute-phase response. The ALS CNTF Treatment Study (ACTS) Phase I–II Study Group. *Clin. Neuropharmacol.* **18**, 500–514.

AMENTO, E. P. (1996a), Potential therapeutic role of relaxin in fibrotic disease. *Abstract* (IBC Fibrosis Identification and Development of Novel Anti-Fibrotic Agents Meeting).

AMENTO, E. P. (1996b), TCR peptides as therapeutics. *Abstract* (IBC Advances in the Understanding and Treatment of Multiple Sclerosis Meeting.)

AMENTO, E. P. (1996c), Immune regulation by TCR self-recognition: mechanism of action. *Abstract* (IBC Psoriasis Latest Advances in Understanding and Therapeutic Development Meeting).

ARNASON, B. G. (1996), Interferon beta in multiple sclerosis, *Clin. Immunol. Immunopathol.* **81**, 1–11.

ATKINS, M. B., VACHINO, G., TILG, H. J., KARP, D. D., ROBERT, N. J., KAPPLER, K., MIER, J. W. (1992), Phase I evaluation of thrice-daily intravenous bolus interleukin-4 in patients with refractory malignancy, *J. Clin. Oncol.* **10**, 1802–1809.

ATKINS, M. B., KAPPLER, K., MIER, J. W., ISAACS, R. E., BERKMAN, E. M. (1995), Interleukin-6-associated anemia: determination of the underlying mechanism, *Blood* **86**, 1288–1291.

AUKERMAN, S. L. (1996), Inhibition of vascular smooth muscle cell hyperplasia by a recombinant fibroblast growth factor-2-saporin mitotoxin. *Abstract* (IBC Restenosis Advances in Therapeutics and Drug Development Meeting).

AVERSA, G., PUNNONEN, J., COCKS, B. G., WAAL MALEFYT, R. DE, VEGA, F. JR., ZURAWSKI, S. M., ZURAWSKI, G., VRIES, J. E. DE (1993), An interleukin 4 (IL-4) mutant protein inhibits both IL-4 or IL-13-induced human immunoglobulin G4 (IgG4) and IgE synthesis and B cell proliferation: support for a common component shared by IL-4 and IL-13 receptors, *J. Exp. Med.* **178**, 2213–2218.

AYALA, A., CHAUDRY, I. H. (1996), Platelet activating factor and its role in trauma, shock, and sepsis, *New Horiz.* **4**, 265–275.

BAARS, J. W., WAGSTAFF, J., BOVEN, E., VERMORKEN, J. B., GROENINGEN, C. J. VAN, SCHEPER, R. J., FONK, J. C., NIJMAN, H. W., FRANKS, C. R., DAMSMA, O., et al. (1991), Phase I study on the sequential administration of recombinant human interferon-gamma and recombinant human interleukin-2 in patients with metastatic solid tumors, *J. Natl. Cancer Inst.* **83**, 1408–1410.

BABIUK, L. A., SORDILLO, L. M., CAMPOS, M., HUGHES, H. P., ROSSI-CAMPOS, A., HARLAND, R. (1991), Application of interferons in the control of infectious diseases of cattle, *J. Dairy Sci.* **74**, 4385–4398.

BAILEY, D. A., FAULKNER, R. A., McKAY, H. A. (1996), Growth, physical activity, and bone mineral acquisition, *Exerc. Sport Sci. Rev.* **24**, 233–266.

BALASA, B., SARVETNICK, N. (1996), The paradoxical effects of interleukin 10 in the immunoregulation of autoimmune diabetes, *J. Autoimmun.* **9**, 283–286.

BARCELLONA, D., BIONDI, G., VANNINI, M. L., MARONGIU, V. F. (1996), Comparison between recombinant and rabbit thromboplastin in the management of patients on oral anticoagulant therapy, *Thromb. Haemost.* **75**, 488–490.

BARKER, W. B. (1996), PROACT 1, a pilot study using r-prourokinase for the treatment of acute ischemic stroke. *Abstract* (IBC Recent Advances in the Understanding, Therapy and Diagnosis of Ischemic Stroke Meeting).

BAUER, R. J., WHITE, M. L., WEDEL, N., NELSON, B. J., FRIEDMANN, N., COHEN, A., HUSTINX, W. N., KUNG, A. H. (1996), A phase I safety and pharmacokinetic study of a recombinant amino terminal fragment of bactericidal/permeability-increasing protein in healthy male volunteers, *Shock* **5**, 91–96.

BECKER, J. C., PANCOOK, J. D., GILLIES, S. D., MENDELSOHN, J., REISFELD, R. A. (1996), Eradication of human hepatic and pulmonary melanoma metastases in SCID mice by antibody-interleukin 2 fusion proteins, *Proc. Natl. Acad. Sci. USA* **93**, 2702–2707.

BEIJERING, R. J., CATE, H. TEN, CATE, J. W. TEN (1996), Clinical applications of new antithrombotic agents, *Ann. Hematol.* **72**, 177–183.

BELLANTONI, M. F. (1996), Osteoporosis prevention and treatment, *Am. Fam. Physician* **54**, 986–992, 995–996.

BEN-CHETRIT, A., GOTLIEB, L., WONG, P. Y., CASPER, R. F. (1996), Ovarian response to recombinant human follicle-stimulating hormone in luteinizing hormone-depleted women: examination of the two cell, two gonadotropin theory, *Fertil. Steril.* **65**, 711–717.

BERNINI, J. C. (1996), Diagnosis and management of chronic neutropenia during childhood, *Clin. North. Am.* **43**, 773–792.

BISHOP, R. C. (1996a), Clinical updates of myloral. *Abstract* (IBC Advances in the Understanding and Treatment of Multiple Sclerosis Meeting).

BISHOP, R. (1996b), Clinical development of colloral for rheumatoid arthritis. *Abstract* (IBC New Therapeutic Advances in Arthritis Meeting).

BLACKWELL, T. S., CHRISTMAN, J. W. (1996), Sepsis and cytokines: current status, *Br. J. Anaesth.* **77**, 110–117.

BLANAS, E., CARBONE, F. R., ALISON, J., MILLER, J. F. A. P. (1996), Hearh WR Induction of autoimmune diabetes by oral administration of autoantigen, *Science* **274**, 1707–1709.

BOCIEK, R. G., ARMITAGE, J. O. (1996), Hematopoietic growth factors [see comments], *CA Cancer J. Clin.* **46**, 165–184.

BOERMEESTER, M. A., LEEUWEN P. A. VAN, COYLE, S. M., WOLBINK, G. J., HACK, C. E. (1995), Lowry SF Interleukin-1 blockade attenuates mediator release and dysregulation of the hemostatic mechanism during human sepsis, *Arch. Surg.* **130**, 739–748.

BONE, R. C. (1996), Why sepsis trials fail, *JAMA* **276**, 565–566.

BOSTROM, M., LANE, J. M., TOMIN, E., BROWNE, M., BERBERIAN, W., TUREK, T., SMITH, J., WOZNEY, J., SCHILDHAUER, T. (1996), Use of bone morphogenetic protein-2 in the rabbit ulnar nonunion model, *Clin. Orthop.* **327**, 272–282.

BRAUN, D. G. (1996), Anti-IgE antibody a drug for therapy of allergies? *Abstract* (IBC Asthma and Allergy Advances in Understanding and Novel Therapeutic Approaches Meeting).

BRETTI, S., GILLEECE, M. H., KAMTHAN, A., FITZSIMMONS, L., HICKS, F., ROWLANDS, M., BISHOP, P., PICARDO, A. M., DEXTER, T. M., SCARFFE, J. H. (1996), An open phase I study to assess the biological effects of a continuous intravenous infusion of interleukin-3 followed by granulocyte macrophage-colony stimulating factor, *Eur. J. Cancer* **32A**, 1171–1181.

BRICKNER, M. E. (1996), Cardioembolic stroke, *Am. J. Med.* **100**, 465–474.

BROD, S. A., LINDSEY, J. W., WOLINSKY, J. S. (1996), Multiple sclerosis: clinical presentation, diagnosis and treatment, *Am. Fam. Physician* **54**, 1301–1306, 1309–1311.

BROOKS, P. C., MONTGOMERY, A. M., ROSENFELD, M., REISFELD, R. A., HU, T., KLIER, G., CHERESH, D. A. (1994), Integrin alpha v. beta 3 antagonists promote tumor regression by inducing apoptosis of angiogenic blood vessels, *Cell* **79**, 1157–1164.

BROSTOFF, S. (1996a), T cell receptor peptide vaccination as immunotherapy for multiple sclerosis. *Abstract* (IBC Advances in the Understanding and Treatment of Multiple Sclerosis Meeting).

BROSTOFF, S. (1996b), Phase II rheumatoid arthritis clinical trials using T-cell receptor peptides as immunotherapy. *Abstract* (IBC New Therapeutic Advances in Arthritis Meeting).

BROSTOFF, S. (1996c), T cell receptor peptide vaccinates as immunotherapy for psoriasis. *Abstract* (IBC Psoriasis Latest Advances in Understanding and Therapeutic Development Meeting).

BROWN, M. S., SHAPIRO, H. (1996), Effect of protein intake on erythropoiesis during erythropoietin treatment of anemia of prematurity, *J. Pediatr.* **128**, 512–517.

BRUNDA, M. J., LUISTRO, L., RUMENNIK, L., WRIGHT, R. B., DVOROZNIAK, M., AGLIONE, A., WIGGINTON, J. M., WILTROUT, R. H., HENDRZAK, J.A., PALLERONI, A.V. (1996), Antitumor activity of interleukin 12 in preclinical models, *Cancer Chemother. Pharmacol.* **38** (Suppl.), 16–21.

BURKE, S. E. (1996), Selective antagonism of endothelin-A receptors as therapeutic agents approach to prevention of restenosis. *Abstract* (IBC Restenosis Advances in Therapeutics and Drug Development Meeting).

CAMERER, E., KOLSTO, A. B., PRYDZ, H., KOLST, A. B. (1996), Cell biology of tissue factor, the principal initiator of blood coagulation, *Thromb. Res.* **81**, 1–41.

CAMPION, G. V., LEBSACK, M. E., LOOKABAUGH, J., GORDON, G., CATALANO, M. (1996), Dose-range and dose-frequency study of recombinant human interleukin-1 receptor antagonist in patients with rheumatoid arthritis. The IL-1Ra Arthritis Study Group, *Arthritis Rheum.* **39**, 1092–1101.

CEPPI, A., BLUM, J. W. (1994), Effects of growth hormone on growth performance, hematology, metabolites and hormones in iron-deficient veal calves, *Zentralbl. Veterinaermed. [A]* **41**, 443–458.

CHALAZONITIS, A. (1996), Neurotrophin-3 as an essential signal for the developing nervous system, *Mol. Neurobiol.* **12**, 39–53.

CHEETHAM, T. D., HOLLY, J. M., CLAYTON, K., CWYFAN-HUGHES, S., DUNGER, D. B. (1995), The effects of repeated daily recombinant human insulin-like growth factor I administration in adolescents with type 1 diabetes, *Diabet. Med.* **12**, 885–892.

CHEN, L., MCGOWAN, P., ASHE, S., JOHNSTON, J. V., HELLSTROM, K. E. (1994), B7-1/CD80-transduced tumor cells elicit better systemic immunity than wild-type tumor cells admixed with *Corynebacterium parvum*, *Cancer Res.* **54**, 5420–5423.

CHEW, P. H. (1996), BMS-200980: a novel plasminogen activator. *Abstract* (IBC Advances in Anticoagulant Antithrombolytic and Thrombolytic Drugs Meeting).

CHIPMAN, J. J., HICKS, J. R., HOLCOMBE, J. H., DRAPER, M. W. (1995), Approaching final height in children treated for growth hormone deficiency, *Horm. Res.* **43**, 129–131.

CLAYMAN, G. L., YOUNG, G., TAYLOR, D. L., SAVAGE, H. E., LAVEDAN, P., SCHANTZ, S. P. (1992), Detection of regulatory factors of lymphokine-activated killer cell activity in head and neck cancer patients treated with interleukin-2 and interferon alpha, *Ann. Otol. Rhinol. Laryngol.* **101**, 909–915.

COFRANCESCO, E., CORTELLARO, M., LEONARDI, P., CORRADI, A., RAVASI, F., BERTOCCHI, F. (1996), Markers of hemostatic system activation during thromboprophylaxis with recombinant hirudin in total hip replacement, *Thromb. Haemost.* **75**, 407–411.

COHEN, M. C., COHEN, S. (1996), Cytokine function: a study in biologic diversity, *Am. J. Clin. Pathol.* **105**, 589–598

CONANT, E. F., FOX, K. R., MILLER, W. T. (1989), Pulmonary edema as a complication of interleukin-2 therapy, *AJR Am. J. Roentgenol.* **152**, 749–752.

COOK, S. D., WOLFE, M. W., SALKELD, S. L., RUEGER, D. C. (1995), Effect of recombinant human osteogenic protein-1 on healing of segmental defects in non-human primates, *J. Bone Joint Surg. Am.* **77**, 734–750.

CORTELAZZO, S., VIERO, P., BELLAVITA, P., ROSSI, A., BUELLI, M., BORLERI, G. M., MARZIALI, S., BASSAN, R., COMOTTI, B., RAMBALDI, A. et al. (1995), Granulocyte colony-stimulating factor following peripheral-blood progenitor-cell transplant in non-Hodgkin's lymphoma [see comments], *J. Clin. Oncol.* **13**, 935–1941.

CRAIG, C. (1996), Creative biomolecules in $122M deal with Biogen, in: *Bioworld Today,* Vol. 7, No. 240, pp. 1–5. Atlanta, G. A.: D. R. Johnson Publisher.

CRAIG, C. (1997), FDA approves Genentech's Nutropin for Turner's syndrome, in: *Bioworld Today,* Atlanta, G. A.: Vol. 8, No. 1, pp. 1–3. D. R. Johnson Publisher.

CROWN, J., JAKUBOWSKI, A., KEMENY, N., GORDON, M., GASPARETTO, C., WONG, G., SHERIDAN, C., TORNER, G., MEISENBERG, B., BOTET, J. et al. (1991), A phase I trial of recombinant human interleukin-1 beta alone and in combination with myelosuppressive doses of 5-fluorouracil in patients with gastrointestinal cancer, *Blood* **78**, 1420–1427.

DALE, D. C., RODGER, E., CEBON, J., RAMESH, N., HAMMOND, W. P., ZSEBO, K. M. (1995), Long-term treatment of canine cyclic hematopoiesis with recombinant canine stem cell factor, *Blood* **85**, 74–79.

DANGAS, G., FUSTER, V. (1996), Management of restenosis after coronary intervention, *Am. Heart J.* **132**, 428–436.

DASER, A., MEISSNER, N., HERZ, U., RENZ, H. (1995), Role and modulation of T-cell cytokines in allergy, *Curr. Opin. Immunol.* **7**, 762–770.

DAVIES, K. A. (1996), Michael Mason Prize Essay 1995. Complement, immune complexes and systemic lupus erythematosus, *Br. J. Rheumatol.* **35**, 5–23.

DEHMER, G. J., FISHER, M., TATE, D. A., TEO, S., BONNEM, E. M. (1995), Reversal of heparin anticoagulation by recombinant platelet factor 4 in humans, *Circulation* **91**, 2188–2194.

D'HONDT, V., HUMBLET, Y., GUILLAUME, T., BAATOUT, S., CHATELAIN, C., BERLIERE, M., LONGUEVILLE, J., FEYENS, A. M., GREVE J. DE, VAN OOSTEROM, A. et al. (1995), Thrombopoietic effects and toxicity of interleukin-6 in patients with ovarian cancer before and after chemotherapy: a multicentric placebo-controlled, randomized phase Ib study, *Blood* **85**, 2347–2355.

DINARELLO, C. A. (1996), Biologic basis for interleukin-1 in disease, *Blood* **87**, 2095–2147.

DREVLOW, B. E., LOVIS, R., HAAG, M. A., SINACORE, J. M., JACOBS, C., BLOSCHE, C., LANDAY, A., MORELAND, L. W., POPE, R. M. (1996), Recombinant human interleukin-1 receptor type I in the treatment of patients with active rheumatoid arthritis, *Arthritis Rheum.* **39**, 257–265.

DURELLI, L., BONGIOANNI, M. R., FERRERO, B., FERRI, R., IMPERIALE, D., BRADAC, G. B., BERGUI, M., GEUNA, M., BERGAMINI, L., BERGAMASCO, B. (1996), Interferon alpha-2a treatment of relapsing-remitting multiple sclerosis: disease activity resumes after stopping treatment, *Neurology* **47**, 123–129.

EDWARDS, M. (1996), Clinical experience with BB-10010, an engineered form of MIP-1α with reduced self-association properties. *Abstract* (IBC Chemokines *in vivo* function, Signaling Pathways, Drug Development, Clinical Applications Meeting).

ELSON, C. O. (1996), The basis of current and future therapy for inflammatory bowel disease, *Am. J. Med.* **100**, 656–662.

ERIKSON, J. C., HOLLOPETER, G., PALMITER, R. D. (1996), Attenuation of the obesity syndrome of ob/ob mice by the loss of neuropeptide Y, *Science* **274**, 1704–1707.

ERIKSSON, E. (1996), Direct inhibitors of thrombin with recombinant hirudin, a new method for the prophylaxis of deep vein thrombosis after orthopedic surgery. *Abstract* (IBC Advances in Anticoagulant Antithrombolytic and Thrombolytic Drugs Meeting).

ERIKSSON, B. I., EKMAN, S., KALEBO, P., ZACHRISSON, B., BACH, D., CLOSE, P. (1996), Prevention of deep-vein thrombosis after total hip replacement: direct thrombin inhibition with recombinant hirudin, CGP 39393 [see comments], *Lancet* **347**, 635–639.

FALANGA, V., CARSON, P., GREENBERG, A., HASAN, A., NICHOLS, E., MCPHERSON, J. (1996), Topically applied recombinant tissue plasminogen activator for the treatment of venous ulcers. Preliminary report, *Dermatol. Surg.* **22**, 643–644.

FAY, J. W., LAZARUS, H., HERZIG, R., SAEZ, R., STEVENS, D. A., COLLINS, R. H. JR., PINEIRO, L. A., COOPER, B. W., DiCESARE, J., CAMPION, M. et al. (1994), Sequential administration of recombinant human interleukin-3 and granulocyte–macrophage colony-stimulating factor after autologous bone marrow transplantation for malignant lymphoma: a phase I/II multicenter study, *Blood* **84**, 2151–2157.

FELDMANN, M., BRENNAN, F. M., MAINI, R. N. (1996), Rheumatoid arthritis, *Cell* **85**, 307–310.

FERNANDEZ-BOTRAN, R., CHILTON, P. M., MA, Y., WINDSOR, J. L., STREET, N. E. (1996), Control of the production of soluble interleukin-4 receptors: implications in immunoregulation, *J. Leukoc. Biol.* **59**, 499–504.

FERRARA, A. (1996), Therapeutic applications of a humanized anti-VEGF monoclonal antibody. *Abstract* (IBC Angiogenesis Inhibitors and Stimulators Meeting).

FICK, R. B. JR. (1996), Recombinant humanized anti-IgE: update of safety and efficacy. *Abstract* (IBC Asthma and Allergy Meeting).

FINE, R. N., KOHAUT, E., BROWN, D., KUNTZE, J., ATTIE, K. M. (1996), Long-term treatment of growth retarded children with chronic renal insufficiency, with recombinant human growth hormone, *Kidney Int.* **49**, 781–785.

FISHER, C. J. JR., SLOTMAN, G. J., OPAL, S. M., PRIBBLE, J. P., BONE, R. C., EMMANUEL, G., NG, D., BLOEDOW, D. C., CATALANO, M. A. (1994), Initial evaluation of human recombinant interleukin-1 receptor antagonist in the treatment of sepsis syndrome: a randomized, open-label, placebo-controlled multicenter trial. The IL-1RA Sepsis Syndrome Study Group [see comments], *Crit. Care Med.* **22**, 12–21.

FISHER, C. J. JR., DHAINAUT, J. F., OPAL, S. M., PRIBBLE, J. P., BALK, R. A., SLOTMAN, G. J., IBERTI, T. J., RACKOW, E. C., SHAPIRO, M. J., GREENMAN, R. L. et al. (1994b), Recombinant human interleukin 1 receptor antagonist in the treatment of patients with sepsis syndrome. Results from a randomized, double-blind, placebo-controlled trial. Phase III rhIL-1ra Sepsis Syndrome Study Group [see comments], *JAMA* **271**, 1836–1843.

FOA, R. (1996), Interleukin 2 in the management of acute leukaemia, *Br. J. Haematol.* **92**, 1–8.

FODOR, W. L. (1996), Enzymatic remodeling of cell surface carbohydrates on xenogeneic cells. *Abstract* (IBC Xenotransplantation Meeting).

FOSS, F. M., BORKOWSKI, T. A., GILLIOM, M., STETLER-STEVENSON, M., JAFFE, E. S., FIGG, W. D., TOMPKINS, A., BASTIAN, A., NYLEN, P., WOODWORTH, T. et al. (1994), Chimeric fusion protein toxin DAB486IL-2 in advanced mycosis fungoides and the Sezary syndrome: correlation of activity and interleukin-2 receptor expression in a phase II study, *Blood* **84**, 1765–1774.

FRIEDMAN, A. (1996), Induction of anergy in Th1 lymphocytes by oral tolerance. Importance of antigen dosage and frequency of feeding, *Ann. N. Y. Acad. Sci.* **778**, 103–110.

FROEHLICH, J. K. (1996), Alteplase in acute ischemic stroke: I.V. administration. *Abstract* (IBC Advances in Anticoagulant Antithrombolytic and Thrombolytic Drugs Meeting).

FROESCH, E. R., BIANDA, T., HUSSAIN, M. A. (1996), Insulin-like growth factor-I in the therapy of non-insulin-dependent diabetes mellitus and insulin resistance, *Diabetes Metab. Rev.* **22**, 261–267.

GANSER, A., LINDEMANN, A., OTTMANN, O. G., SEIPELT, G., HESS, U., GEISSLER, G., KANZ, L., FRISCH, J., SCHULZ, G., HERRMANN, F. et al. (1992), Sequential *in vivo* treatment with two recombinant human hematopoietic growth factors (interleukin-3 and granulocyte–macrophage colony-stimulating factor) as a new therapeutic modality to stimulate hematopoiesis: results of a phase I study, *Blood* **79**, 2583–2591.

GANSER, A., HEIL, G., KOLBE, K., MASCHMEYER, G., FISCHER, J. T., BERGMANN, L., MITROU, P. S., HEIT, W., HEIMPEL, H., HUBER, C. et al. (1993), Aggressive chemotherapy combined with G-CSF and maintenance therapy with interleukin-2 for patients with advanced myelodysplastic syndrome, subacute or secondary acute myeloid leukemia-initial results, *Ann. Hematol.* **66**, 123–125.

GARLISI, C. G. (1996), Antibody-mediated inhibition of interleukin-5 as a therapy for allergy and asthma. *Abstract* (IBC Asthma and Allergy Advances in Understanding and Novel Therapeutic Approaches Meeting).

GAUR, A. (1996), Therapeutic effect of MBP peptide analogs in a murine model of multiple sclerosis. *Abstract* (IBC Advances in the Understanding and Treatment of Multiple Sclerosis Meeting).

GENAIN, C. P., ABEL, K., BELMAR, N., VILLINGER, F., ROSENBERG, C. P., LININGTON, C., RAINSE, S., HAUSER, S. L. (1996), Late complications of immune deviation therapy in a nonhuman primate, *Science* **274**, 2054–2057.

GILLEECE, M. H., SCARFFE, J. H., GHOSH, A., HEYWORTH, C. M., BONNEM, E., TESTA, N., STERN, P., DEXTER, T. M. (1992), Recombinant human interleukin 4 (IL-4) given as daily subcutaneous injections – a phase I dose toxicity trial, *Br. J. Cancer* **66**, 204–210.

GLASPY, J., DAVIS, M. W., PARKER, W. R., FOOTE, M., MCNIECE, I. (1996), Biology and clinical potential of stem-cell factor, *Cancer Chemother. Pharmacol.* **38** (Suppl.), S53–157.

GOETERS, C., MERTES, N., TACKE, J., BOLDER, U., KUHMANN, M., LAWIN, P., LOHLEIN, D. (1995), Repeated administration of recombinant human insulin-like growth factor-I in patients after gastric surgery. Effect on metabolic and hormonal patterns, *Ann. Surg.* **222**, 646–653.

GOHLKE, M., BAUDE, G., NUCK, R., GRUNOW, D., KANNICHT, C., BRINGMANN, P., DONNER, P., REUTTER, W. (1996), O-linked L-fucose is present in *Desmodus rotundus* salivary plasminogen activator, *J. Biol. Chem.* **271**, 7381–7386.

GOLDHABER, S. Z., HIRSCH, D. R., MACDOUGALL, R. C., POLAK, J. F., CREAGER, M. A. (1996), Bolus recombinant urokinase versus heparin in deep venous thrombosis: a randomized controlled trial, *Am. Heart J.* **132**, 314–318.

GORDON, M. S. (1995), Interleukin-11: a new thrombopoietic agent. *Meeting Abstract* **31** (3rd International Conference: Clinical Applications of Cytokines and Growth Factors in Hematology and Oncology, June 8–10, 1995, Atlanta, GA).

GORDON, D. M., MCGOVERN, T. W., KRZYCH, U., COHEN, J. C., SCHNEIDER, I., LACHANCE, R., HEPPNER, D. G., YUAN, G., HOLLINGDALE, M., SLAOUI M. et al. (1995a), Safety, immunogenicity, and efficacy of a recombinantly produced *Plasmodium falciparum* circumsporozoite protein-hepatitis B surface antigen subunit vaccine, *J. Infect Dis.* **171**, 1576–1585.

GORDON, M. S., NEMUNAITIS, J., HOFFMAN, R., PAQUETTE, R. L., ROSENFELD, C., MANFREDA, S., ISAACS, R., NIMER, S. D. (1995b), A phase I trial of recombinant human interleukin-6 in patients with myelodysplastic syndromes and thrombocytopenia, *Blood* **85**, 3066–3076.

GORDON, M. S., MCCASKILL-STEVENS, W. J., BATTIATO, L. A., LOEWRY, J., LOESCH, D., BREEDEN, E., HOFFMAN, R., BEACH, K. J., KUCA, B., KAYE, J., SLEDGE, G. W. JR. (1996), A phase I trial of recombinant human interleukin-11 (neumega rhIL-11 growth fractor) in women with breast cancer receiving chemotherapy, *Blood* **87**, 3615–3624.

GRABOWSKI, G. A., BARTON, N. W., PASTORES, G., DAMBROSIA, J. M., BANERJEE, T. K., MCKEE, M. A., PARKER, C., SCHIFFMANN, R., HILL, S. C., BRADY, R. O. (1995), Enzyme therapy in type 1 Gaucher disease: comparative efficacy of mannose-terminated glucocerebrosidase from natural and recombinant sources, *Ann. Intern. Med.* **122**, 33–39.

GRAHAM, B. S., KEEFER, M. C., MCELRATH, M. J., GORSE, G. J., SCHWARTZ, D. H., WEINHOLD, K., MATTHEWS, T. J., ESTERLITZ, J. R., SINANGIL, F., FAST, P. E. (1996), Safety and immunogenicity of a candidate HIV-1 vaccine in healthy adults: recombinant glycoprotein (rgp) 120. A randomized, double-blind trial. NIAID AIDS Vaccine Evaluation Group, *Ann. Intern. Med.* **125**, 270–279.

GREEN, D., SANDERS, J., EIKEN, M., WONG, C. A., FREDERIKSEN, J., JOOB, A., PALMER, A., TROWBRIDGE, A., WOODRUFF, B., MOERCH, M. et al. (1995), Recombinant aprotinin in coronary artery bypass graft operations, *J. Thorac. Cardiovasc. Surg.* **110**, 963–970.

GRICHNIK, J. M., CRAWFORD, J., JIMENEZ, F., KURTZBERG, J., BUCHANAN, M., BLACKWELL, S., CLARK, R. E., HITCHCOCK, M. G. (1995), Human recombinant stem-cell factor induces melanocytic hyperplasia in susceptible patients, *J. Am. Acad. Dermatol.* **33**, 577–583.

GRIFFITHS, C. E., VOORHEES, J. J. (1996), Psoriasis, T cells and autoimmunity, *J. R. Soc. Med.* **89**, 315–319.

GROSSBARD, E. B. (1996), Basic fibroblast growth factor in stroke. *Abstract* (IBC Recent Advances in the Understanding, Therapy and Diagnosis of Ischemic Stroke Meeting).

GROTENDORST, G. R. (1996), Connective tissue growth factor as a target for blockade of TGF-β induced fibrosis. *Abstract* (IBC Fibrosis Identification and Development of Novel Anti-Fibrotic Agents Meeting).

GUGLIETTA, A. (1996), Activity of growth factors in animal models of inflammatory bowel disease. *Abstract* (IBC Novel Anti-Inflammatory Therapeutics for Inflammatory Bowel Disease Meeting).

HAEDO, W., GONZALEZ, T., MAS, J. A., FRANCO, S., GRA, B., SOTO, G., ALONSO, A., LOPEZ-SAURA, P. (1996), Oral human recombinant epidermal growth factor in the treatment of patients with duodenal ulcer, *Rev. Esp. Enferm. Dig.* **88**, 409–418.

HAFFNER, S. M., STERN, M. P., MIETTINEN, H., WEI, M., GINGERICH, R. L. (1996), Leptin concentrations in diabetic and nondiabetic Mexican-Americans, *Diabetes* **45**, 822–824.

HAMILTON, J. (1966), Signposts to therapy: recent advances in inflammatory bowel disease research, *Can. Med. Assoc. J.* **154**, 1513–1516.

HANDEL, T. (1996), Structure of the macrophage chemoattractant protein-1 and characterization of the oligomerization state of truncation mutants. *Abstract* (IBC Chemokines *in vivo* function, Signaling Pathways, Drug Development, Clinial Applications Meeting).

HAPP, M. P. (1997), Peptide vaccines for the treatment of multiple sclerosis. *Abstract* (IBC Vaccines New Advances in Technologies and Applications Meeting).

HASSAN, H. T., ZANDER, A. (1996), Stem cell factor as a survival and growth factor in human normal and malignant hematopoiesis, *Acta Haematol.* **95**, 257–262.

HAUSSER, H., GRONING, A., HASILIK, A., SCHONHERR, E., KRESSE, H. (1994), Selective inactivity of TGF-beta/decorin complexes, *FEBS Lett.* **353**, 243–245.

HEATH, P. K. (1996), Early clinical experience with CDP571, an engineered human anti-TNFa antibody, in ulcerative colitis and Crohn's disease. *Abstract* (IBC Novel Anti-Inflammatory Therapeutics for Inflammatory Bowel Disease Meeting).

HELLSTROM, K. E. (1996), Co-stimulator molecules. *Abstract* (IBC Immunotherapeutic Strategies for Cancer Meeting).

HENRY, P. A. (1994), Human recombinant erythropoietin used to treat a cat with anemia caused by chronic renal failure, *Can. Vet. J.* **35**, 375.

HILL, F. W., KLEIN, W. R., HOYER, M. J., RUTTEN, V. P., KOCK, N., KOTEN, J. W., STEERENBERG, P. A., RUITENBERG, E. J., DEN OTTER, W. (1994), Antitumor effect of locally injected low doses of recombinant human interleukin-2 in bovine vulval papilloma and carcinoma, *Vet. Immunol. Immunopathol.* **41**, 19–29.

HINES, S. A., PALMER, G. H., JASMER, D. P., GOFF, W. L., MCELWAIN, T. F. (1995), Immunization of cattle with recombinant *Babesia bovis* merozoite surface antigen-1, *Infect. Immun.* **63**, 349–352.

HOLMES, S. J., SHALET, S. M. (1996), Role of growth hormone and sex steroids in achieving and maintaining normal bone mass, *Horm. Res.* **45**, 86–93.

HOMBURG, R., LEVY, T., BEN-RAFAEL, Z. (1995), A comparative prospective study of conventional regimen with chronic low-dose administration of follicle-stimulating hormone for anovulation associated with polycystic ovary syndrome, *Fertil. Steril.* **63**, 729–733.

HOWARD, O. M., BEN-BARUCH, A., OPPENHEIM, J. J. (1996), Chemokines: progress toward identifying molecular targets for therapeutic agents, *Trends Biotechnol.* **14**, 46–51.

HUHN, R. D., O'CONNELL, S. M., RADWANSKI, E., CUTLER, D. L., STURGILL, M., CLARKE, L., SEIBOLD, J. R. (1995), Phase I pharmacodynamic study of intravenous (iv) recombinant human interleukin-10 (RHIL-10) in normal volunteers

(Meeting abstract), *Clin. Pharmacol. Ther.* **57**, 162.

HUHN, R. D., RADWANSKI, E., O'CONNELL, S. M., STURGILL, M. G., CLARKE, L., CODY, R. P., AFFRIME, M. B., CUTLER, D. L. (1996), Pharmacokinetics and immunomodulatory properties of intravenously administered recombinant human interleukin-10 in healthy volunteers, *Blood* **87**, 699–705.

HUMPHREYS, R. E. (1996), MHC class II "Antigen supercharging" therapeutics. *Abstract* (IBC Antigen Processing and Presentation Novel Therapeutics Development Meeting).

IGARASHI, A., OKOCHI, H., BRADHAM, D. M., GROTENDORST, G. R. (1993), Regulation of connective tissue growth factor gene expression in human skin fibroblasts and during wound repair, *Mol. Biol. Cell.* **4**, 637–645.

JANIK, J. E., MILLER, L. L., LONGO, D. L., POWERS, G. C., URBA, W. J., KOPP, W. C., GAUSE, B. L., CURTI, B. D., FENTON, R. G., OPPENHEIM, J. J., CONLON, K. C., HOLMLUND, J. T., SZNOL, M., SHARFMAN, W. H., STEIS, R. G., CREEKMORE, S. P., ALVORD, W. G., BEAUCHAMP, A. E., SMITH, J. W. (1996), 2nd phase II trial of interleukin 1 alpha and indomethacin in treatment of metastatic melanoma, *J. Natl. Cancer. Inst.* **88**, 44–49.

JOFFE, J. K., BANKS, R. E., FORBES, M. A., HALLAM, S., JENKINS, A., PATEL, P. M., HALL, G. D., VELIKOVA, G., ADAMS, J., CROSSLEY, A., JOHNSON, P. W., WHICHER, J. T., SELBY, P. J. (1996), A phase II study of interferon-alpha, interleukin-2 and 5-fluorouracil in advanced renal carcinoma: clinical data and laboratory evidence of protease activation, *Br. J. Urol.* **77**, 638–649.

JONES, W. R. (1996), Contraceptive vaccines, *Baillieres Clin. Obstet. Gynaecol.* **10**, 69–86.

JUNGERS, P., DEVILLIER, P., SALOMON, H., CERISIER, J. E., COUROUCE, A. M. (1994), Randomised placebo-controlled trial of recombinant interleukin-2 in chronic uraemic patients who are non-responders to hepatitis B vaccine [see comments], *Lancet* **344**, 856–1857.

KAGNOFF, M. F. (1996), Oral tolerance: mechanisms and possible role in inflammatory joint diseases, *Baillieres Clin. Rheumatol.* **10**, 41–54.

KALVAKOLANU, D. V., BORDEN, E. C. (1996), An overview of the interferon system: signal transduction and mechanisms of action, *Cancer Invest.* **14**, 25–53.

KANEGANE, H., TOSATO, G. (1996), Activation of naive and memory T cells by interleukin-15, *Blood* **88**, 230–235.

KAUSHANSKY, K. (1996), Thrombopoietin: biological and preclinical properties, *Leukemia* **10** (Suppl. 1), S46–S48.

KEITH, J. C. JR. (1996), Beneficial effects of interleukin-11 in animal models of Crohn's disease. *Abstract* (IBC Novel Anti-Inflammatory Therapeutics for Inflammatory Bowel Disease Meeting).

KELLEY, C. L. (1996), The role of interferons in the treatment of multiple sclerosis, *J. Neurosci. Nurs.* **28**, 114–120.

KELLEY, K. W., ARKINS, S., MINSHALL, C., LIU, Q., DANTZER, R. (1996), Growth hormone, growth factors and hematopoiesis, *Horm. Res.* **45**, 38–45.

KEY, L. L. JR., RODRIGUIZ, R. M., WILLI, S. M., WRIGHT, N. M., HATCHER, H. C., EYRE, D. R., CURE, J. K., GRIFFIN, P. P., RIES, W. L. (1995), Long-term treatment of osteoporosis with recombinant human interferon gamma [see comments], *N. Engl. J. Med.* **332**, 1594–1599.

KHAN, O. A., XIA, Q., BEVER, C. T. JR., JOHNSON, K. P., PANITCH, H. S., DHIB-JALBUT, S. S. (1996), Interferon beta-1b serum levels in multiple sclerosis patients following subcutaneous administration, *Neurology* **46**, 1539–1643.

KIMMEL, S. R., MADLON-KAY, D., BURNS, I. T., ADMIRE, J. B. (1996), Breaking the barriers to childhood immunization, *Am. Fam. Physician* **63**, 1648–1666.

KIRKER-HEAD, C. A., GERHART, T. N., SCHELLING, S. H., HENNIG, G. E., WANG, E., HOLTROP, M. E. (1995), Long-term healing of bone using recombinant human bone morphogenetic protein 2, *Clin. Orthop.* **318**, 222–230.

KLIMAS, N., PATARCA, R., WALLING, J., GARCIA, R., MAYER, V., MOODY, D., OKARMA, T., FLETCHER, M. A. (1994), Clinical and immunological changes in AIDS patients following adoptive therapy with activated autologous CD8 T cells and interleukin-2 infusion, *AIDS* **8**, 1073–1081.

KLINGBEIL, C., EPSTEIN, S. E. (1994), Basic fibroblast growth factor enhances myocardial collateral flow in a canine model, *Am. J. Physiol.* **266**, H1558–F1595.

KNUSEL, B., GAO, H. (1996), Neurotrophins and Alzheimer's disease: beyond the cholinergic neurons, *Life Sci.* **58**, 2019–2027.

KOBRIN, C. (1996), Application of the idiotype vaccine concept to traditional as well as novel settings. *Abstract* (IBC Immunotherapeutic Strategies for Cancer Meeting).

KOMSCHLIES, K. L., GRZEGORZEWSKI, K. J., WILTROUT, R. H. (1995), Diverse immunological and hematological effects of interleukin 7: implications for clinical application, *J. Leukoc. Biol.* **58**, 623–633.

KOPP, W. C., URBA, W. J., RAGER, H. C., ALVORD, W. G., OPPENHEIM, J. J., SMITH, J. W. II, LONGO, D. L. (1996), Induction of interleukin 1 receptor antagonist after interleukin 1 therapy in patients with cancer, *Clin. Cancer Res.* **2**, 501–506.

KORETZ, M. J., LAWSON, D. H., YORK, R. M., GRAHAM, S. D., MURRAY, D. R., GILLESPIE, T. M., LEVITT, D., SELL, K. M. (1991), Randomized study of interleukin 2 (IL-2) alone vs IL-2 plus lymphokine-activated killer cells for treatment of melanoma and renal cell cancer, *Arch. Surg.* **126**, 898–903.

KORNGOLD, R. (1997), Effect of CD4 peptide analogs on allotransplantation. *Abstract* (IBC Tolerance and Immunotherapy in Allotransplantation New Discoveries for Drug Development Meeting).

KRIGEL, R. L., PADAVIC-SHALLER, K., TOOMEY, C., COMIS, R. L., WEINER, L. M. (1995), Phase I study of sequentially administered recombinant tumor necrosis factor and recombinant interleukin-2, *J. Immunother. Emphasis Tumor Immunol.* **17**, 161–170.

KRUIT, W. H., GOEY, S. H., CALABRESI, F., LINDEMANN, A., STAHEL, R. A., POLIWODA, H., OSTERWALDER, B., STOTER, G. (1995), Final report of a phase II study of interleukin 2 and interferon alpha in patients with metastatic melanoma, *Br. J. Cancer* **71**, 1319–1321.

KRUSE, N., TONY, H. P., SEBALD, W. (1992), Conversion of human interleukin-4 into a high affinity antagonist by a single amino acid replacement, *EMBO J.* **11**, 3237–3244.

KUBER SAMPATH, T. (1996), Osteogenic protein-1 (OP-1/BMP-7) as an anabolic agent for cartilage repair and regeneration. *Abstract* (IBC New Therapeutic Advances in Arthritis Meeting).

KUBER SAMPATH, T. (1997), Use of osteogenic protein-1 (OP-1/BMP-7) as a basis to treat osteoporosis and related metabolic bone disorders. *Abstract* (IBC Emerging Therapies for Osteoporosis Meeting).

KUMAKI, S., ARMITAGE, R., AHDIEH, M., PARK, L., COSMAN, D. (1996), Interleukin-15 up-regulates interleukin-2 receptor alpha chain but down-regulates its own high-affinity binding sites on human T and B cells, *Eur. J. Immunol.* **26**, 1235–1239.

LANGENBERG, A. G., BURKE, R. L., ADAIR, S. F., SEKULOVICH, R., TIGGES, M., DEKKER, C. L., COREY, L. (1995), A recombinant glycoprotein vaccine for herpes simplex virus type 2: safety and immunogenicity [corrected] [published erratum appears in Ann. Intern. Med. (1995), **123**, 395], *Ann. Intern. Med.* **122**, 889–898.

LAPPI, D. A., MAHER, P. A., MARTINEAU, D., BAIRD, A. (1991), The basic fibroblast growth factor-saporin mitotoxin acts through the basic fibroblast growth factor receptor, *J. Cell. Physiol.* **147**, 17–26.

LAUBE, B. L., AUCI, R. M., SHIELDS, D. E., CHRISTIANSEN, D. H., LUCAS, M. K., FUCHS, H. J., ROSENSTEIN, B. J. (1996), Effect of rhDNase on airflow obstruction and mucociliary clearance in cystic fibrosis, *Am. J. Respir. Crit. Care Med.* **153**, 752–760.

LEBEAUT, A. (1996), Interleukin-10: a novel therapy for inflammatory bowel disease. *Abstract* (IBC Novel Anti-Inflammatory Therapeutics for Inflammatory Bowel Disease Meeting).

LEE, S. J. (1990), Identification of a novel member (GDF-1) of the transforming growth factor-beta superfamily, *Mol. Endocrinol.* **4**, 1034–1040.

LEE, P. I., LEE, C. Y., HUANG, L. M., CHANG, M. H. (1995), Long-term efficacy of recombinant hepatitis B vaccine and risk of natural infection in infants born to mothers with hepatitis B antigen, *J. Pediatr.* **126**, 716–721.

LEIGHTON, A. (1996), Consideration for phase III design of anti-TNF treatment. *Abstract* (IBC Sepsis Advances in Treatment and Drug Development Meeting).

LEINSKOLD, T., PERMERT, J., OLAISON, G., LARSSON, J. (1995), Effect of postoperative insulin-like growth factor I supplementation on protein metabolism in humans, *Br. J. Surg.* **82**, 921–925.

LI, Y., HELLSTROM, K. E., NEWBY, S. A., CHEN, L. (1996), Costimulation by CD48 and B7-1 induces immunity against poorly immunogenic tumors, *J. Exp. Med.* **183**, 639–644.

LIN, L.-F. H., DOHERTY, D. H., LILE, J. D., BEKTESH, S., COLLINS, F. (1993), GDNF: a glial cell line-derived neurotrophic factor for midbrain dopaminergic neurons, *Science* **260**, 1130–1132.

LINDEMANN, A., GANSER, A., HERRMANN, F., FRISCH, J., SEIPELT, G., SCHULZ, G., HOELZER, D., MERTELSMANN, R. (1991), Biologic effects of recombinant human interleukin-3 *in vivo*, *J. Clin. Oncol.* **9**, 2120–2127.

LINDSAY, R. M. (1996), Role of neurotrophins and trk receptors in the development and maintenance of sensory neurons: an overview, *Philos. Trans. R. Soc. Lond. B. Biol. Sci.* **351**, 365–373.

LISSONI, P., TISI, E., BARNI, S., ARDIZZOIA, A., ROVELLI, F., RESCALDANI, R., BALLABIO, D., BENENTI, C., ANGELI, M., TANCINI, G. et al. (1992), Biological and clinical results of a neuroimmunotherapy with interleukin-2 and the pineal hormone melatonin as a first line treatment in advanced non-small cell lung cancer, *Br. J. Cancer* **66**, 155–158.

LISSONI, P., BARNI, S., TANCINI, G., ARDIZZOIA, A., RICCI, G., ALDEGHI, R., BRIVIO, F., TISI, E., ROVELLI, F., RESCALDANI, R. et al. (1994), A randomised study with subcutaneous low-dose interleukin 2 alone vs interleukin 2 plus the pineal neurohormone melatonin in advanced solid neoplasms other than renal cancer and melanoma, *Br. J. Cancer* **69**, 196–199.

LOGAN, J. S. (1996), The potential use of genetically modified pigs as organ donors for transplantation into humans. *Abstract* (IBC Xenotransplantation Meeting).

LOTZ, M. (1996), IL-17, a novel cytokine with IL-1/TNF like activities on chondrocytes and synoviocytes. *Abstract* (IBC New Therapeutic Advances in Arthritis Meeting).

LUBLIN, F. D., WHITAKER, J. N., EIDELMAN, B. H., MILLER, A. E., ARNASON, B. G., BURKS, J. S. (1996), Management of patients receiving interferon beta-1b for multiple sclerosis: report of a consensus conference, *Neurology* **46**, 12–18.

LUMMEN, G., GOEPEL, M., MOLLHOFF, S., HINKE, A., OTTO, T., RUBBEN, H. (1996), Phase II study of interferon-gamma versus interleukin-2 and interferon-alpha 2b in metastatic renal cell carcinoma, *J. Urol.* **155**, 455–458.

LUND, P. K., ZIMMERMANN, E. M. (1996), Insulin-like growth factors and inflammatory bowel disease, *Baillieres Clin. Gastroenterol.* **10**, 83–96.

MacGILLIVRAY, M. H., BAPTISTA, J., JOHANSON, A. (1996), Outcome of a four-year randomized study of daily versus three times weekly somatropin treatment in prepubertal native growth hormone-deficient children. Genentech Study Group, *J. Clin. Endocrinol. Metab.* **81**, 1806–1809.

MAHANTHAPPA, N. K., ANTON, E. S., MATTHEW, W. D. (1996), Glial growth factor 2, a soluble neuregulin, directly increases Schwann cell motility and indirectly promotes neurite outgrowth, *J. Neurosci.* **16**, 4673–4683.

MAIONE, T. (1996), Natural anti-endothelial and neovascular targeting mechanisms of platelet factor 4 (PF4) angiostatic activity. *Abstract* (IBC Angiogenesis Inhibitors and Stimulators Meeting).

MAIONE, T. E., GRAY, G. S., HUNT, A. J., SHARPE, R. J. (1991), Inhibition of tumor growth in mice by an analog of platelet factor 4 that lacks affinity for heparin and retains potent angiostatic activity, *Cancer Res.* **51**, 2077–2083.

MAISEY, I., ANDREWS, A. H., LAVEN, R. A. (1993), Efficacy of recombinant bovine somatotropin in the treatment of fat cow syndrome, *Vet. Rec.* **133**, 293–296.

MANTOVANI, A., MUZIO, M., GHEZZI, P., COLOTTA, F., INTRONA, M. (1996), Negative regulators of the interleukin-1 system: receptor antagonists and a decay receptor, *Int. J. Clin. Lab. Res.* **26**, 7–14.

MAPEL, D. W., SAMET, J. M., COULTAS, D. B. (1996), Corticosteroids and the treatment of idiopathic pulmonary fibrosis. Past, present, and future, *Chest* **110**, 1058–1067.

MARCUS, R. (1996), Clinical review 76: the nature of osteoporosis, *J. Clin. Endocrinol. Metab.* **81**, 1–5.

MARTIN, G. R. (1996a), Current concepts in fibrosis. *Abstract* (IBC Fibrosis Identification and Development of Novel Anti-Fibrotic Agents Meeting).

MARTIN, U. (1996b), Reteplase: a bolus injectable plasminogen activator. *Abstract* (IBC Advances in Anticoagulant Antithrombolytic and Thrombolytic Drugs Meeting).

MARTINET, Y., MENARD, O., VAILLANT, P., VIGNAUD, J. M., MARTINET, N. (1996), Cytokines in human lung fibrosis, *Arch. Toxicol. Suppl.* **18**, 127–139.

MARTINEZ, M. L., WEISS, R. C. (1993), Applications of genetic engineering technology in feline medicine, *Vet. Clin. North Am. Small Anim Pract.* **23**, 213–226.

MASSA, G., OTTEN, B. J., MUINCK KEIZER-SCHRAMA SM DE, DELEMARRE-VAN DE WAAL, H. A., JANSEN, M., VULSMA, T., OOSTDIJK, W., WAELKENS, J. J., WIT, J. M. (1995), Treatment with two growth hormone regimens in girls with Turner syndrome: final height results. Dutch Growth Hormone Working Group, *Horm. Res.* **43**, 144–146.

MATIS, L. A. (1996), Apoptosis-inducing antigens. *Abstract* (IBC Advances in the Understanding and Treatment of Multiple Sclerosis Meeting).

MATSUMOTO, K., NAKAMURA, T. (1996), Emerging multipotent aspects of hepatocyte growth factor, *J. Biochem. (Tokyo)* **119**, 591–600.

MAXWELL, H., REES, L. (1996), Recombinant human growth hormone treatment in infants with chronic renal failure, *Arch. Dis. Child.* **74**, 40–143.

McKENNA, T. S., RIEDER, E., LUBROTH, J., BURRAGE, T., BAXT, B., MASON, P. W. (1996), Strategy for producing new foot-and-mouth disease vaccines that display complex epitopes, *J. Biotechnol.* **44**, 83–89.

MEMOLI, B., DE NICOLA, L., LIBETTA, C., SCIALO, A., PACCHIANO, G., ROMANO, P., PALMIERI, G., MORABITO, A., LAURIA, R., CONTE, G. et al. (1995), Interleukin-2-induced renal dysfunction in cancer patients is reversed by low-dose dopamine infusion, *Am. J. Kidney Dis.* **26**, 27–33.

MICHAEL, J. R., MARKEWITZ, B. A. (1996), Endothelins and the lung, *Am. J. Respir. Crit. Care Med.* **154**, 555–581.

MILES, S. A., MITSUYASU, R., LaFLEUR, F., RYBACK, M., KASDEN, P., SUCKOW, C., GROOPMAN, J., SCADDEN, D. (1994), Phase I/II trial of interleukin-4 in KS (ACTG 224). *Abstract* (Int. Conf. AIDS) **10**, 46.

MINASIAN, L. M., YAO, T. J., STEFFENS, T. A., SCHEINBERG, D. A., WILLIAMS, L., RIEDEL, E., HOUGHTON, A. N., CHAPMAN, P. B. (1995), A phase I study of anti-GD3 ganglioside monoclonal antibody R24 and recombinant human macrophage-colony stimulating factor in patients with metastatic melanoma, *Cancer* **75**, 2251–2257.

MISHU, L., CALLAHAN, G., ALLEBBAN, Z., MADDUX, J. M., BOONE, T. C., SOUZA, L. M., LOTHROP, C. D., JR. (1992), Effects of recombinant canine granulocyte colony-stimulating factor on white blood cell production in clinically normal and neutropenic dogs, *J. Am. Vet. Med. Assoc.* **200**, 1957–1964.

MOLLNES, T. E., HARBOE, M. (1996), Clinical immunology, *BMJ* **312**, 1465–1469.

MORELAND, L. W., SEWELL, K. L., TRENTHAM, D. E., BUCY, R. P., SULLIVAN. W. F., SCHROHENLOHER, R. E., SHMERLING, R. H., PARKER. K. C., SWARTZ, W. G., WOODWORTH, T. G. et al. (1995), Interleukin-2 diphtheria fusion protein (DAB486IL-2) in refractory rheumatoid arthritis. A double-blind, placebo-controlled trial with open-label extension, *Arthritis Rheum.* **38**, 1177–1186.

MORGAN, B. P. (1996), Intervention in the complement system: a therapeutic strategy in inflammation, *Biochem. Soc. Trans.* **24**, 224–229.

MORLEY, J., MARINCHAK, R., RIALS, S. J., KOWEY, P. (1996), Atrial fibrillation, anticoagulation, and stroke, *Am. J. Cardiol.* **77**, 38A–44A.

MORSTYN, G., FOOTE, M., LIESCHKE, G. J. (1996), Hematopoietic growth factors in cancer chemotherapy, *Cancer Chemother. Biol. Response Modif.* **16**, 295–314.

MOSES, A. C. (1996), Recombinant human insulin-like growth factor I (rhIGF-I) as a treatment for severe insulin resistance. *Abstract* (IBC Advances in the Understanding and Treatment of Insulin Resistance Novel Drug Development Strategies for Type II Diabetes and Obesity Meeting).

NAG, B. (1996), Antigen presenting by individual α and β chains of MHC class II molecules. *Abstract* (IBC Antigen Processing and Presentation Novel Therapeutics Development Meeting).

NAGY, Z. (1996), Tageting the MHC class II binding site for immunosuppression. *Abstract* (IBC Antigen Processing and Presentation Novel Therapeutics Development Meeting).

NASSIF, A., LONGO, W. E., MAZUSKI, J. E., VERNAVA, A. M., KAMINSKI, D. L. (1996), Role of cytokines and platelet-activating factor in inflammatory bowel disease. Implications for therapy, *Dis. Colon. Rectum.* **39**, 217–223.

NEGRIN, R. S., STEIN, R., DOHERTY, K., CORNWELL, J., VARDIMAN, J., KRANTZ, S., GREENBERG, P. L. (1996), Maintenance treatment of the anemia of myelodysplastic syndromes with recombinant human granulocyte colony-stimulating factor and erythropoietin evidence for in vivo synergy, *Blood* **87**, 4076–4081.

NELSON, S., BAGBY, G. J. (1996), Granulocyte colony-stimulating factor and modulation of inflammatory cells in sepsis, *Clin. Chest. Med.* **17**, 319–332.

NEMUNAITIS, J., APPELBAUM, F. R., LILLEBY, K., BUHLES, W. C., ROSENFELD, C., ZEIGLER, Z. R., SHADDUCK, R. K., SINGER, J. W., MEYER, W., BUCKNER, C. D. (1994), Phase I study of recombinant interleukin-1 beta in patients undergoing autologous bone marrow transplant for acute myelogenous leukemia, *Blood* **83**, 3473–3479.

NEWMAN, W. (1996), CCR3 eotaxin receptor as a target in the management of asthma. *Abstract* (IBC Asthma and Allergy Advances in Understanding and Novel Therapeutic Approaches Meeting).

NIELSEN, F. S., JORGENSEN, L. N., IPSEN, M., VOLDSGAARD, A. I., PARVING, H. H. (1995), Long-term comparison of human insulin analog B10Asp and soluble human insulin in IDDM patients on a basal/bolus insulin regimen, *Diabetologia* **38**, 592–598.

OPAL, S. M. (1996), The utility of recombinant human interleukin-11 (rhIL-11) in the prevention of sepsis. *Abstract* (IBC Sepsis Advances in Treatment and Drug Development Meeting).

ORAZI, A., GORDON, M. S., JOHN, K., SLEDGE, G. JR., NEIMAN, R. S., HOFFMAN, R. (1995), In vivo effects of recombinant human stem cell factor treatment. A morphologic and immunohistochemical study of bone marrow biopsies, *Am. J. Clin. Pathol.* **103**, 177–184.

OSCARSSON, J., LUNDSTAM, U., GUSTAFSSON, B., WILTON, P., EDEN, S., WIKLUND, O. (1995), Recombinant human insulin-like growth factor-I decreases serum lipoprotein(a) concentrations in normal adult men, *Clin. Endocrinol. (Oxf.)* **42**, 673–676.

O'SHAUGHNESSY, J. A., TOLCHER, A., RISEBERG, D., VENZON, D., ZUJEWSKI, J., NOONE, M., GOSSARD, M., DANFORTH, D., JACOBSON, J., CHANG, V., GOLDSPIEL, B., KEEGAN, P., GIUSTI, R., COWAN, K. H. (1996), Prospective, randomized trial of 5-fluorouracil, leucovorin, doxorubicin, and cyclophosphamide chemotherapy in combination with the interleukin-3/granulocyte-macrophage colony-stimulating factor (GM-CSF) fusion protein (PIXY321) versus GM-CSF in patients with advanced breast cancer, *Blood* **87**, 2205–2211.

OSTER, W., FRISCH, J., NICOLAY, U., SCHULZ, G. (1991), Interleukin-3. Biologic effects and clinical impact, *Cancer* **67** (Suppl.) 2712–2717.

OTTEN, N., MARTI, E., SODERSTROM, C., AMTMANN, E., BURGER, D., GERBER, H., LAZARY, S. (1994), Experimental treatment of equine sarcoid using a xanthate compound and recombinant human tumor necrosis factor alpha, *Zentralbl. Veterinärmed. [A]* **41**, 757–765.

OTTMANN, O. G., GANSER, A., SEIPELT, G., EDER, M., SCHULZ, G., HOELZER, D. (1990), Effects of recombinant human interleukin-3 on human hematopoietic progenitor and precursor cells *in vivo, Blood* **76**, 1494–1502.

PHILLIPS, T. J. (1996), Current treatment options in psoriasis, *Hosp. Pract. (Off. Ed.)* **31**, 155–157, 161–164, 166.

PICHERT, G., JOST, L. M., FIERZ, W., STAHEL, R. A. (1991), Clinical and immune modulatory effects of alternative weekly interleukin-2 and interferon α-2a in patients with advanced renal cell carcinoma and melanoma, *Br. J. Cancer* **63**, 287–292.

PIERCE, G. F., TARPLEY, J. E., TSENG, J., BREADY, J., CHANG, D., KENNEY, W. C., RUDOLPH, R., ROBSON, M. C., VANDE BERG, J., REID, P. et al. (1995), Detection of platelet-derived growth factor (PDGF)-AA in actively healing human wounds treated with recombinant PDGF-BB and absence of PDGF in chronic nonhealing wounds, *J. Clin. Invest.* **96**, 1336–1350.

POLISSON, R. P., GILKESON, G. S., PYUN, E. H., PISETSKY, D. S., SMITH, E. A., SIMON, L. S. (1996), A multicenter trial of recombinant human interferon gamma in patients with systemic sclerosis: effects on cutaneous fibrosis and interleukin 2 receptor levels, *J. Rheumatol.* **23**, 654–658.

POLMAR, A. H. (1996), Enlimomab (anti-ICAM-1-therapy for acute ischemic stroke. *Abstract* (IBC Recent Advances in the Understanding, Therapy and Diagnosis of ischemic Stroke Meeting).

POLVERINI, P. J. (1996), Cellular adhesion molecules. Newly identified mediators of angiogenesis, *Am. J. Pathol.* **148**, 1023–1029.

POPE, R. M. (1996), Rheumatoid arthritis: pathogenesis and early recognition, *Am. J. Med.* **100**(2A), 3S–9S.

POWERS, D. C., SMITH, G. E., ANDERSON, E. L., KENNEDY, D. J., HACKETT, C. S., WILKINSON, B. E., VOLVOVITZ, F., BELSHE, R. B., TREANOR, J. J. (1995), Influenza A virus vaccines containing purified recombinant H3 hemagglutinin are well tolerated and induce protective immune responses in healthy adults, *J. Infect. Dis.* **171**, 1595–1599.

PRATT, H. L., CARROLL, R. C., McCLENDON, S., SMATHERS, E. C., SOUZA, L. M., LOTHROP, C. D. JR. (1990), Effects of recombinant granulocyte colony-stimulating factor treatment on hematopoietic cycles and cellular defects associated with canine cyclic hematopoiesis, *Exp. Hematol.* **18**, 1199–1203.

PUJADE-LAURAINE, E., GUASTALLA, J. P., COLOMBO, N., DEVILLIER, P., FRANCOIS, E., FUMOLEAU, P., MONNIER, A., NOOY, M., MIGNOT, L., BUGAT, R., MARQUES, C., MOUSSEAU, M., NETTER, G., MALOISEL, F., LARBAOUI, S., BRANDELY, M. (1996), Intraperitoneal recombinant interferon gamma in ovarian cancer patients with residual disease at second-look laparotomy, *J. Clin. Oncol.* **14**, 343–350.

PUSCHEL, A. W. (1996), The semaphorins: a family of axonal guidance molecules? *Eur. J. Neurosci.* **8**, 1317–1321.

QUIROGA, G. H., SORDILLO, L. M., ADKINSON, R. W., NICKERSON, S. C. (1993), Cytologic responses of *Staphylococcus aureus*-infected mammary glands of heifers to interferon gamma and interleukin-2 treatment, *Am. J. Vet. Res.* **54**, 1894–1900.

RAEMAEKERS, J. M., IMHOFF, G. W. VAN, VERDONCK, L. F., HESSELS, J. A., FIBBE, W. E. (1993), The tolerability of continuous intravenous infusion of interleukin-3 after DHAP chemotherapy in patients with relapsed malignant lymphoma. A phase-I study, *Ann. Hematol.* **67**, 175–181.

RAMACHANDRAN, R., KATZENSTEIN, D. A., WINTERS, M. A., KUNDU, S. K., MERIGAN, T. C. (1996), Polyethylene glycol-modified interleukin-2 and thymosin alpha 1 in human immunodeficiency virus type 1 infection, *J. Infect. Dis.* **173**, 1005–1008.

REDDY, P. G., REDDY, D. N., PRUIETT, S. E., DALEY, M. J., SHIRLEY, J. E., CHENGAPPA, M. M., BLECHA, F. (1992), Interleukin 2 treatment of *Staphylococcus aureus* mastitis, *Cytokine* **4**, 227–231.

REED, S. G. (1996), A novel adjuvant for anti-tumor responses. *Abstract* (IBC Antigen Processing and Presentation Novel Therapeutics Development Meeting).

REED, S. G. (1997), A recombinant protein from *Leishmania* with potent adjuvant effects: cloning of *Mycobacterium tuberculosis* vaccines. *Abstract* (IBC Vaccines New Advances in Technologies and Applications Meeting).

REID, I., SHARPE, I., McDEVITT, J., MAXWELL, W., EMMONS, R., TANNER, W. A., MONSON, J. R. (1991), Thyroid dysfunction can predict response to immunotherapy with interleukin-2 and interferon-2 alpha, *Br. J. Cancer* **64**, 915–918.

REISFELD, R. A. (1996), Tumor therapy with antibody–cytokine fusion proteins. *Abstract* (IBC Immunotherapeutic Strategies for Cancer Meeting).

RENZETTI, L. M. (1996), Interleukin-12: a therapeutic for treatment of asthma? *Abstract* (IBC Asthma and Allergy Advances in Understanding and Novel Therapeutic Approaches Meeting).

RICHARD, J. L., PARER-RICHARD, C., DAURES, J. P., CLOUET, S., VANNEREAU, D., BRINGER, J., RODIER, M., JACOB, C., COMTE-BARDONNET, M. (1995), Effect of topical basic fibroblast growth factor on the healing of chronic diabetic neuropathic ulcer of the foot. A pilot, randomized, double-blind, placebo-controlled study, *Diabetes Care* **18**, 64–69.

ROEP, B. O. (1996), T-cell responses autoantigens in IDDM. The search for the Holy Grail, *Diabetes* **45**, 1147–1156.

ROLLINS, B. J. (1996a), MCP-1 activity *in vivo*. *Abstract* (IBC Chemokines in vivo function, Signaling Pathways, Drug Development, Clinical Applications Meeting).

ROLLINS, S. A. (1996b), Anti-C5 monoclonal antibodies serve as potent soluble inhibitors of complement-mediated hyperacute rejection. *Abstract* (IBC Xenotransplantation Meeting).

ROMAGNANI, S. (1995), Atopic allergy and other hypersensitivities editorial overview: technological advances and new insights into pathogenesis prelude novel therapeutic strategies [editorial], *Curr. Opin. Immunol.* **7**, 745–750.

ROMEO, F., ROSANO, G. M., MARTUSCELLI, E., DE LUCA, F., BIANCO, C., COLISTRA, C., COMITO, M., CARDONA, N., MICELI, F., ROSANO, V. et al. (1995), Concurrent nitroglycerin administration reduces the efficacy of recombinant tissue-type plasminogen activator in patients with acute anterior wall myocardial infarction, *Am. Heart J.* **130**, 692–697.

ROSEN, D. M. (1997), Potential use of the IGF-I/IGFBP-3 complex (SomatoKine) as an anabolic agent: initial clinical experiences. *Abstract* (IBC Emerging Therapies for Osteoporosis Meeting).

ROSENFELD, W. N., DAVIS, J. M., PARTON, L., RICHTER, S. E., PRICE, A., FLASTER, E., KASSEM, N. (1996), Safety and pharmacokinetics of recombinant human superoxide dismutate administered intratracheally to premature neonates with respiratory distress syndrome, *Pediatrics* **97**, 811–817.

ROT, A., HUB, E., MIDDLETON, J., PONS, F., RABECK, C., THIERER, K., WINTLE, J., WOLFF, B., ZSAK, M., DUKOR, P. (1996), Some aspects of IL-8 pathophysiology. III: Chemokine interaction with endothelial cells, *J. Leukoc. Biol.* **59**, 39–44.

ROWE, J. M. (1996), Use of growth factors during induction therapy for acute myeloid leukemia, *Leukemia* **10** (Suppl.), S40–S43.

RUBIN, J. T., LOTZE, M. T. (1992), Acute gastric mucosal injury associated with the systemic administration of interleukin-4, *Surgery* **111**, 274–280.

RUTTEN, V. P., KLEIN, W. R., DE JONG, W. A., MISDORP, W., DEN OTTER, W., STEERENBERG, P. A., DE JONG, W. H., RUITENBERG, E. J. (1989), Local interleukin-2 therapy in bovine ocular squamous cell carcinoma. A pilot study, *Cancer Immunol. Immunother.* **30**, 165–169.

RYAN, U. (1996a), Complement inhibition by sCR1 in clinical trials. *Abstract* (IBC Controlling the Complement System for Novel Drug Development Meeting).

RYAN, U. S. (1996b), Complement inhibitor therapeutics. *Abstract* (IBC Anti-Inflammatory Drug Discovery Novel Approaches for Therapeutic Development Meeting).

RYFFEL, B. (1996), Unanticipated human toxicology of recombinant proteins, *Arch. Toxicol. Suppl.* **18**, 333–341.

SAGGESE, G., PASQUINO, A. M., BERTELLONI, S., BARONCELLI, G. I., BATTINI, R., PUCARELLI, I., SEGNI, M., FRANCHI, G. (1995), Effect of combined treatment with gonadotropin releasing hormone analog and growth hormone in patients with central precocious puberty who had subnormal growth velocity and impaired height prognosis, *Acta Paediatr.* **84**, 299–304.

SALMON-CERON, D., EXCLER, J. L., SICARD, D., BRANCHE, P., FINKIELSTZJEN, L., GLUCKMAN, J. C., AUTRAN, B., MATTHEWS, T. J., MEIGNIER, B., KIENY, M. P. et al. (1995), Safety and immunogenicity of a recombinant HIV type 1 glycoprotein 160 boosted by a V3 synthetic peptide in HIV-negative volunteers, *AIDS Res. Hum. Retroviruses* **11**, 1479–1486.

SARNA, S., SIPILA, I., RONNHOLM, K., KOISTINEN, R., HOLMBERG, C. (1996), Recombinant human growth hormone improves growth in children receiving glucocorticoid treatment after liver transplantation, *J. Clin. Endocrinol. Metab.* **81**, 1476–1482.

SASAHARA, A. A., BARKER, W. M., WEAVER, W. D., HARTMANN, J., ANDERSON, J. L., REDDY, P. S., VILLIARD, E. M. (1995), Clinical studies with the new glycosylated recombinant prourokinase, *J. Vasc. Interv. Radiol.* **6**, 84S–93S.

SAVAGE, M. O., DUNGER, D. B. (1996), Recombinant IGF-I therapy in insulin-dependent diabetes mellitus, *Diabetes Metab.* **22**, 257–260.

SCHAIBLE, T. (1996), Use of chimeric anti-TNF antibody (cA2) in the successful treatment of severe Crohn's disease. *Abstract* (IBC Novel Anti-

Inflammatory Therapeutics for Inflammatory Bowel Disease Meeting).

SCHILLER, J. T., OKUN, M. M. (1996), Papillomavirus vaccines: current status and future prospects, *Adv. Dermatol.* **11**, 355–381.

SCHLEUNING, W.-D. (1996), *Desmodus rotundus* (vampire bat) salivary plasminogen activator DSPα1. *Abstract* (IBC Advances in Anticoagulant Antithrombolytic and Thrombolytic Drugs Meeting).

SCHLOM, J. (1996), Carcinoma-associated antigens as targets for immunotherapy. *Abstract* (IBC Immunotherapeutic Strategies for Cancer Meeting).

SCHMALBACH, T. (1996), DAB$_{389}$IL-2 and the treatment of psoriasis. *Abstract* (IBC Psoriasis Latest Advances in Understanding and Therapeutic Development Meeting).

SCHOEN, R. T., MEURICE, F., BRUNET, C. M., CRETELLA, S., KRAUSE, D. S., CRAFT, J. E., FIKRIG, E. (1995), Safety and immunogenicity of an outer surface protein A vaccine in subjects with previous Lyme disease, *J. Infect. Dis.* **172**, 1324–1329.

SCHREIBER, S., HOWALDT, S., SCHNOOR, M., NIKOLAUS, S., BAUDITZ, J., GASCHE, C., LOCHS, H., RAEDLER, A. (1996), Recombinant erythropoietin for the treatment of anemia in inflammatory bowel disease [see comments], *N. Engl. J. Med.* **334**, 619–623.

SEACHRIST, L. (1997), Immune response reports positive phase II results for rheumatoid arthritis drug, in: *Bioworld Today,* Vol. 8, No. 5, pp. 1–5. Atlanta, GA: D. R. Johnson Publishers.

SEACHRIST, L., YOFFEE, L. (1997), Regeneron's lead candidate BDNF fails in phase III, in: *Bioworld Today,* Vol. 8, No. 8, pp. 1–4. Atlanta, GA: D. R. Johnson Publisher.

SEGAL, K. R., LANDT, M., KLEIN, S. (1996), Relationship between insulin sensitivity and plasma leptin concentration in lean and obese men, *Diabetes* **45**, 988–991.

SEIPELT, G., GANSER, A., DURANCEYK, H., MAURER, A., OTTMANN, O. G., HOELZER, D. (1994), Induction of soluble IL-2 receptor in patients with myelodysplastic syndromes undergoing high-dose interleukin-3 treatment, *Ann. Hematol.* **68**, 167–170.

SETTE, A. (1996), HLA supermotifs and epitope immunodominance in development of peptide-based CTL vaccines. *Abstract* (IBC Antigen Processing and Presentation Novel Therapeutics Development Meeting).

SETTE, A. (1997), HLA supermotifs and epitope immunodominance in development of peptide-based CTL vaccines. *Abstract* (IBC Vaccines New Advances in Technologies and Applications Meeting).

SEYMOUR, J. F., KURZROCK, R. (1996), Interleukin-6: biologic properties and role in lymphoproliferative disorders. (231 Refs), *Cancer Treat. Res.* **84**, 167–206.

SHAH, P. L., SCOTT, S. F., KNIGHT, R. A., MARRIOTT, C., RANASINHA, C., HODSON, M. E. (1966), *In vivo* effects of recombinant human DNase I on sputum in patients with cystic fibrosis, *Thorax* **51**, 119–125.

SIKES, M. (1996), Xenogeneic tolerance. *Abstract* (IBC Xenotransplantation Meeting).

SIM, B. K. L. (1997), Recombinant human angiostatin protein: an anti-angiogenic protein with inhibitory activity against experimental metastatic cancer. *Abstract* (IBC Angiogenesis Novel Therapeutic Development Meeting).

SMALLING, R. W., BODE, C., KALBFLEISCH, J., SEN, S., LIMBOURG, P., FORYCKI, F., HABIB, G., FELDMAN, R., HOHNLOSER, S., SEALS, A. (1995), More rapid, complete, and stable coronary thrombolysis with bolus administration of reteplase compared with alteplase infusion in acute myocardial infarction. RAPID Investigators, *Circulation* **91**, 2725–2732.

SPACK, E. G. (1996), Inducing antigen-specific unresponsiveness with soluble MHC II:MBP peptide complexes. *Abstract* (IBC Advances in the Understanding and Treatment of Multiple Sclerosis Meeting).

SPRING, S. B., HASCALL, G., GRUBER, J. (1996), Issues related to development of Epstein-Barr virus vaccines, *J. Natl. Cancer Inst.* **88**, 1436–1441.

STADLER, W. M., RYBAK, M. E., VOGELZANG, N. J. (1995), A phase II study of subcutaneous recombinant human interleukin-4 in metastatic renal cell carcinoma, *Cancer* **76**, 1629–1633.

STEED, D. L. (1995), Clinical evaluation of recombinant human platelet-derived growth factor for the treatment of lower extremity diabetic ulcers. Diabetic Ulcer Study Group, *J. Vasc. Surg.* **21**, 71–81.

STEINMAN, L. (1996), Multiple sclerosis: a coordinated immunological attack against myelin in the central nervous system, *Cell* **85**, 299–302.

STERLING, J., (Ed.) (1996), Biotechnology, therapeutic medicines and vaccines under development, in: *Genetic Engineering News,* Vol. 16, No. 16, pp. 29–34. New York: Mary Ann Liebert Publishers.

STERN, A. S., MAGRAM, J., PRESKY, D. H. (1996), Interleukin-12 an integral cytokine in the immune response, *Life Sci.* **58**, 639–654.

STOUFFER, G. A., SCHMEDTJE, J. F., GULBA, D., HUBER, K., BODE, C., AARON, J., RUNGE, M. S. (1996), Restenosis following percutaneous revascularization – the potential role of thrombin and the thrombin receptor, *Ann. Hematol.* **73** (Suppl. 1), S39–S41.

STOUTHARD, J. M., GOEY, H., VRIES, E. G. DE, MULDER, P. H. DE, GROENEWEGEN, A., PRONK, L., STOTER, G., SAUERWEIN, H. P., BAKKER, D. J., VEENHOFF, C. H. (1996), Recombinant human interleukin 6 in metastatic renal cell cancer: a phase II trial, *Br. J. Cancer* **73**, 789–793.

SUZICH, J. A., GHIM, S. J., PALMER-HILL, F. J., WHITE, W. I., TAMURA, J. K., BELL, J. A., NEWSOME, J. A., JENSON, A. B., SCHLEGEL, R. (1995), Systemic immunization with papillomavirus L1 protein completely prevents the development of viral mucosal papillomas, *Proc. Natl. Acad. Sci. USA* **92**, 11553–11557.

TAKANO, K., SHIZUME, K., HIBI, I., OGAWA, M., OKADA, Y., SUWA, S., TANAKA, T., HIZUKA, N. (1995), Long-term effects of growth hormone treatment on height in Turner syndrome: results of a 6-year multicenter study in Japan. Committee for the Treatment of Turner Syndrome, *Horm. Res.* **43**, 141–143.

TEPLER, I., ELIAS, L., SMITH, J. W. 2ND, HUSSEIN, M., ROSEN, G., CHANG, A. Y., MOORE, J. O., GORDON, M. S., KUCA, B., BEACH, K. J., LOEWY, J. W., GARNICK, M. B., KAYE, J. A. (1996), A randomized placebo-controlled trial of recombinant human interleukin-11 in cancer patients with severe thrombocytopenia due to chemotherapy, *Blood* **87**, 3607–3614.

TERAMURA, M., KOBAYASHI, S., YOSHINAGA, K., IWABE, K., MIZOGUCHI, H. (1996), Effect of interleukin 11 on normal and pathological thrombopoiesis, *Cancer Chemother. Pharmacol.* **38** (Suppl.), S99–S102.

THEOBALD, M., BIGGS, J., DITTMER, D., LEVINE, A. J., SHERMAN, L. A. (1995), Targeting p53 as a general tumor antigen, *Proc. Natl. Acad. Sci. USA* **92**, 11993–11997.

THIJS, L. G., GROENEVELD, A. B., HACK, C. E. (1996), Multiple organ failure in septic shock, *Curr. Top. Microbiol. Immunol.* **216**, 209–237.

THOMAS, K. A. (1997), Inhibition of vascular endothelial cell mitogenesis and tumor growth by sFLT-1, a naturally expressed VEGF receptor extracellular fragment. *Abstract* (IBC Angiogenesis Novel Therapeutic Development Meeting).

THOMAS, P. S., YATES, D. H., BARNES, P. J. (1995), Tumor necrosis factor-alpha increases airway responsiveness and sputum neutrophilia in normal human subjects, *Am. J. Respir. Crit. Care Med.* **152**, 76–80.

TISCH, R., MCDEVITT, H. (1996), Insulin-dependent diabetes mellitus, *Cell* **85**, 291–297.

TOMLINSON, D. R., FERNYHOUGH, P., DIEMEL, L. T. (1996), Neurotrophins and peripheral neuropathy, *Philos. Trans. R. Soc. Lond. B. Biol. Sci.* **351**, 455–462.

TONER, M. S., KING, R. H., DUNSHEA, F. R., DOVE, H., ATWOOD, C. S. (1996), The effect of exogenous somatotropin on lactation performance of first-litter sows, *J. Anim. Sci.* **74**, 167–172.

TOPOL, E. J., CALIFF, R. M., WEISMAN, H. F., ELLIS, S. G., TCHENG, J. E., WORLEY, S., IVANHOE, R., GEORGE, B. S., FINTEL, D., WESTON, M. et al. (1994), Randomised trial of coronary intervention with antibody against platelet IIb/IIIa integrin for reduction of clinical restenosis: results at six months. The EPIC Investigators [see comments], *Lancet* **343**, 881–886.

TRINCHIERI, G., GEROSA, F. (1996), Immunoregulation by interleukin-12, *J. Leukoc. Biol.* **59**, 505–511.

UKAI, M., NISHINAKA, Y., SOBUE, T., MIYAHARA, T., YOKOTA, M. (1995), Improvement in exercise-induced left ventricular dysfunction by infusion of alpha-human atrial natriuretic peptide in coronary artery disease, *Am. J. Cardiol.* **75**, 449–454.

UMEMURA, K., ISHIYE, M., KOSUGE, K., NAKASHIMA, M. (1995), Effect of combination of a tissue-type plasminogen activator and an endothelin receptor antagonist, FRI39317, in the rat cerebral infarction model, *Eur. J. Pharmacol.* **275**, 17–21.

UNEMORI, E. N., BECK, L. S., LEE, W. P., XU, Y., SIEGEL, M., KELLER, G., LIGGITT, H. D., BAUER, E. A., AMENTO, E. P. (1993), Human relaxin decreases collagen accumulation *in vivo* in two rodent models of fibrosis, *J. Invest. Dermatol.* **101**, 280–285.

VADHAN-RAJ, S., BROXMEYER, H. E., ANDREEFF, M., BANDRES, J. C., BUESCHER, E. S., BENJAMIN, R. S., PAPADOPOULOS, N. E., BURGESS, A., PATEL, S., PLAGER, C. et al. (1995), *In vivo* biologic effects of PIXY321, a synthetic hybrid protein of recombinant human granulocyte–macrophage colony-stimulating factor and interleukin-3 in cancer patients with normal hematopoiesis: a phase I study, *Blood* **86**, 2098–2105.

VALENZUELA, D. M., ECONOMIDES, A. N., ROJAS, E., LAMB, T. M., NUNEZ, L., JONES, P., LP, N. Y. (1995), Espinosa R 3rd Brannan CI Gilbert DJ et al. Identification of mammalian noggin and its expression in the adult nervous system, *J. Neurosci.* **15**, 6077–6084.

VANDENBARK, A. A., HASHIM, G. A., OFFNER, H. (1996), T cell receptor peptides in treatment of autoimmune disease: rationale and potential, *J. Neurosci. Res.* **43**, 391–402.

VANDERSCHUEREN, S. (1996), Recombinant staphylokinase for thrombolytic therapy. *Abstract*

(IBC Advances in Anticoagulant Antithrombolytic and Thrombolytic Drugs Meeting).

VANDERSCHUEREN, S., BARRIOS, L., KERDSINCHAI, P., VAN DEN HEUVEL, P., HERMANS, L., VROLIX, M., DE MAN, F., BENIT, E., MUYLDERMANS, L., COLLEN, D. et al. (1995), A randomized trial of recombinant staphylokinase versus alteplase for coronary artery potency in acute myocardial infarction. The STAR Trial Group, *Circulation* **92**, 2044–2049.

VAN GAMEREN, M. M., WILLEMSE, P. H., MULDER, N. H., LIMBURG, P. C., GROEN, H. J., VELLENGA, E, VRIED, E. G. DE (1994), Effects of recombinant human interleukin-6 in cancer patients: a phase I–II study, *Blood* **84**, 1434–1441.

VICKERY, B. H. (1997), Bone formation can be uncoupled from resorption by members of a new class of hPTHr(1–34) analogs. *Abstract* (IBC Emerging Therapies for Osteoporosis Meeting).

VILLA, E., TRANDE, P., GROTTOLA, A., BUTTAFOCO, P., REBECCHI, A. M., STROFFOLINI, T., CALLEA, F., MERIGHI, A., CAMELLINI, L., ZOBOLI, P., COSENZA, R., MIGLIOLI, L., LORIA, P., IORI, R., CARULLI, N., MANENTI, F. (1996), Alpha but not beta interferon is useful in chronic active hepatitis due to hepatitis C virus. A prospective, double-blind, randomized study, *Dig. Dis. Sci.* **41**, 1241–1247.

VON DER MOHLEN, M. A., KIMMINGS, A. N., WEDEL, N. I., MEVISSEN, M. L., JANSEN, J., FRIEDMANN, N., LORENZ, T. J., NELSON, B. J., WHITE, M. L., BAUER, R. et al. (1995), Inhibition of endotoxin-induced cytokine release and neutrophil activation in humans by use of recombinant bactericidal/permeability-increasing protein, *J. Infect. Dis.* **172**, 144–151

WALLNER, B. P. (1996), Peptide treatment of allergy diseases. *Abstract* (IBC Asthma and Allergy Advances in Understanding and Novel Therapeutic Approaches Meeting).

WANG, Y. (1996), Subcutaneous administration of anti-C5 mAb induces systemic complement inhibition and ameloriates established joint inflammation. *Abstract* (IBC New Therapeutic Advances in Arthritis Meeting).

WANG, X., YUE, T. L., BARONE, F. C., WHITE, R. F., GAGNON, R. C., FEUERSTEIN, G. Z. (1994), Concomitant cortical expression of TNF-alpha and IL-1 beta mRNAs follows early response gene expression in transient focal ischemia, *Mol. Chem. Neuropathol.* **23**, 103–114.

WEENING, R. S., LEITZ, G. J., SEGER, R. A. (1995), Recombinant human interferon-gamma in patients with chronic granulomatous disease – European follow up study, *Eur. J. Pediatr.* **154**, 295–298.

WEINER, L. M., PADAVIC-SHALLER, K., KITSON, J., WATTS, P., KRIGEL, R. L., LITWIN, S. (1991), Phase I evaluation of combination therapy with interleukin 2 and gamma-interferon, *Cancer Res.* **51**, 3910–3918.

WEISDORF, D., KATSANIS, E., VERFAILLIE, C., RAMSAY, N. K., HAAKE, R., GARRISON, L., BLAZAR, B. R. (1994), Interleukin-1 alpha administered after autologous transplantation: a phase I/II clinical trial, *Blood* **84**, 2044.

WELTE, K., GABRILOVE, J., BRONCHUD, M. H., PLATZER, E., MORSTYN, G. (1996), Filgrastim (r-metHuG-CSF): the first 10 years, *Blood* **88**, 1907–1929.

WEXLER, L. H., WEAVER-MCCLURE, L., STEINBERG, S. M., JACOBSON, J., JAROSINSKI, P., AVILA, N., PIZZO, P. A., HOROWITZ, M. E. (1996), Randomized trial of recombinant human granulocyte-macrophage colony-stimulating factor in pediatric patients receiving intensive myelosuppressive chemotherapy, *J. Clin. Oncol.* **14**, 901–910.

WHITACRE, C. C., GIENAPP, I. E., MEYER, A., COX, K. L., JAVED, N. (1996), Treatment of autoimmune disease by oral tolerance to autoantigens, *Clin. Immunol. Immunopathol.* **80**, S31–S39.

WHITE, D. (1996), Alteration of complement activity: a strategy for xenotransplantation, *Trends Biotechnol.* **14**, 3–5.

WHITEHEAD, R. P., FRIEDMAN, K. D., CLARK, D. A., PAGANI, K., RAPP, L. (1995), Phase I trial of simultaneous administration of interleukin 2 and interleukin 4 subcutaneously, *Clin. Cancer Res.* **1**, 1145.

WILKINSON, B. E. (1996), A baculovirus expressed recombinant hemagglutinin influenza protein as a vaccine is immunogenic, well-tolerated and protective in young adults. *Abstract* (IBC Influenza and Other Respiratory Disorders Latest Therapeutic and Vaccine Developments Meeting).

WILMOTT, R. W., AMIN, R. S., COLIN, A. A., DEVAULT, A., DOZOR, A. J., EIGEN, H., JOHNSON, C., LESTER, L. A., MCCOY, K., MCKEAN, L. P., MOSS, R., NASH, M. L., JUE, C. P., REGELMANN, W., STOKES, D. C., FUCHS, H. J. (1996), Aerosolized recombinant human DNase in hospitalized cystic fibrosis patients with acute pulmonary exacerbations, *Am. J. Respir. Crit. Care Med.* **153**, 1914–1917.

WIT, J. M., BOERSMA, B., MUINCK KEIZER-SCHRAMA, S. M. DE, NIENHUIS, H. E., OOSTDIJK, W., OTTEN, B. J., DELEMARRE-VAN DE WAAL, H. A., REESER, M., WAELKENS, J. J., RIKKEN, B. et al. (1995), Long-term results of growth hormone therapy in children with short stature, subnormal growth rate and normal growth hormone response to secretagogues. Dutch Growth Hormone Working Group, *Clin. Endocrinol. (Oxf.)* **42**, 365–372.

WOODY, J. N. (1996), Reduction in VEGF levels in patients with severs rheumatoid arthritis treated with cA2, an anti-TNF monoclonal antibody. *Abstract* (IBC Angiogenesis Inhibitors and Stimulators Meeting).

YAMAGUCHI, Y., MANN, D. M., RUOSLAHTI, E. (1990), Negative regulation of transforming growth factor-beta by the proteoglycan decorin, *Nature* **346**, 281–284.

YANG, J. C., TOPALIAN, S. L., SCHWARTZENTRUBER, D. J., PARKINSON, D. R., MARINCOLA, F. M., WEBER, J. S., SEIPP, C. A., WHITE, D. E., ROSENBERG, S. A. (1995), The use of polyethylene glycol-modified interleukin-2 (PEG-IL-2) in the treatment of patients with metastatic renal cell carcinoma and melanoma. A phase I study and a randomized prospective study comparing IL-2 alone versus IL-2 combined with PEG-IL-2, *Cancer* **76**, 687–694.

YANG, X. D., MICHIE, S. A., MEBIUS, R. E., TISCH, R., WEISSMAN, I., McDEVITT, H. O. (1996), The role of cell adhesion molecules in the development of IDDM: implications for pathogenesis and therapy, *Diabetes* **45**, 705–710.

YASUHARA, M., YAMAGUCHI, M., SHIMIZU, H., HASHIMOTO, Y., HAMA, N., ITOH, H., NAKAO, K., HORI, R. (1994), Natriuretic peptide-potentiating actions of neutral endopeptidase inhibition in rats with experimental heart failure, *Pharm. Res.* **11**, 1726–1730.

YUREK, D. M., LU, W., HIPKENS, S., WIEGEND, S. J. (1996), BDNF enhances the functional reinnervation of the striatum by grafted fetal dopamine neurons, *Exp. Neurol.* **137**, 105–118.

ZARICH, S. W., KOWALCHUK, G. J., WEAVER, W. D., LOSCALZO, J., SASSOWER, M., MANZO, K., BYRNES, C., MULLER, J. E., GUREWICH, V. (1995), Sequential combination thrombolytic therapy for acute myocardial infarction: results of the pro-urokinase and t-PA enhancement of thrombolysis (PATENT) trial, *J. Am. Coll. Cardiol.* **26**, 374–379.

ZHANG, X. G., BATAILLE, R., JOURDAN, M., SAELAND, S., BANCHEREAU, J., MANNONI, P., KLEIN, B. (1990), Granulocyte–macrophage colony-stimulating factor synergizes with interleukin-6 in supporting the proliferation of human myeloma cells [see comments], *Blood* **76**, 2599–2605.

ZIMMERMAN, L. B., DE JESUS-ESCOBAR, J. M., HARLAND, R. M. (1996), The Spemann organizer signal noggin binds and inactivates bone morphogenetic protein 4, *Cell* **86**, 599–606.

ZIMMERMAN, R. K., KIMMEL, S. R., TRAUTH, J. M. (1996b), An update on vaccine safety, *Am. Fam. Physician* **54**, 185–193.

ZIMRAN, A., ELSTEIN, D., LEVY-LAHAD, E., ZEVIN, S., HADAS-HALPERN, I., BAR-ZIV, Y., FOLDES, J., SCHWARTZ, A. J., ABRAHAMOV, A. (1995), Replacement therapy with imiglucerase for type 1 Gaucher's disease [see comments], *Lancet* **345**, 1479–1480.

ZURN, A. D., WINKEL, L., MENOUD, A., DJABALI, K., AEBISCHER P. (1996), Combined effects of GDNF, BDNF, and CNTF on motoneuron differentiation *in vitro*, *J. Neurosci. Res.* **44**, 133–141.

ZYSOW, B. R. (1996), RhIGF-I: overview of clinical trials experience in NIDDM and IDDM. *Abstract* (IBC Advances in the Understanding and Treatment of Insulin Resistance Novel Drug Development Strategies for Type II Diabetes and Obesity Meeting).

6 Enzymes for Industrial Applications

WOLFGANG AEHLE
ONNO MISSET

Delft, The Netherlands

1 Introduction   191
2 α-Amylases   192
   2.1 α-Amylases in Starch Liquefaction   193
   2.2 α-Amylases in Textile Desizing   194
   2.3 α-Amylases for Paper Production   195
   2.4 α-Amylases for Detergent Applications   196
   2.5 Development of New α-Amylases   197
3 Lipases   199
   3.1 Lipases in Laundry Detergents   200
   3.2 New Developments in the Lipase Field   202
4 Cellulases   202
   4.1 Properties of Microbial Cellulases   202
   4.2 Industrial Applications of Recombinant Cellulases   204
      4.2.1 Detergent Industry   204
      4.2.2 Textile Industry   204
5 Proteases   205
   5.1 Introduction   205
   5.2 Classes of Proteases   206
      5.2.1 Serine Proteases   206
      5.2.2 Cysteine Proteases   207
      5.2.3 Aspartyl Proteases   207
      5.2.4 Metalloproteases   207
   5.3 Detergent Proteases   208
   5.4 Chymosin   209
6 Oxidoreductases   209
   6.1 Technical Applications of Oxidoreductases   209

7 Phytase   210
    7.1 Properties of *Aspergillus* Phytases   210
    7.2 Application of Recombinant Phytase in the Feed Industry   212
        7.2.1 Development of the Enzyme Product   212
        7.2.2 The Industrial Use of Natuphos®   212
8 Conclusions   213
9 References   214

1 Introduction

Recombinant technology has become the main technology for the development and production of industrial enzymes. Since in 1985 the Dutch company Gist-Brocades introduced the α-amylase from *Bacillus licheniformis* and Genencor Inc. (now International) introduced a protease from *Bacillus subtilis,* which are produced in a recombinant production system, more and more enzymes for industrial applications are produced in a homologous or heterologous expression system with genetically modified microorganisms. Most of these enzymes would not be available because they can not be produced in the source organism or the production would be too expensive to give a competitive product for industrial applications.

The Association of Manufacturers of Fermentation Enzyme Products (Avenue de Roodebeeklaan 30, B-1030 Brussels) summarized in 1995 the advantages of the usage of recombinant technology for enzyme production in the policy statement on modern biotechnology:

"The microbial enzymes produced by AMFEP members are used in a wide variety of industrial applications. The introduction of genetic modification can offer the following advantages in the production and/or quality of these enzymes:

- higher production efficiency and thus less use of energy and raw materials, and less waste;
- availability of enzyme products which for economic, occupational or environmental reasons would otherwise not be available, thus enabling new applications;
- technical improvement through higher specificity and purity of enzyme product."

Recombinant enzyme technology allows the production of highly pure enzyme products for a competitive price. But recombinant techniques also allow the production of a mixture of enzymes in one microorganism with constant composition by genetic modification of the organism as has been demonstrated for cellulases from *Trichoderma reesei* (SUOMINEN et al., 1992).

In addition to that, the application of recombinant technology during the improvement of existing enzyme products gives the developer the opportunity to modify enzymes specifically. This has been successfully demonstrated for the development of oxidation resistant detergent proteases (see Sect. 5.3).

Fig. 1a shows that the majority of technical enzymes is used in detergents, textile and

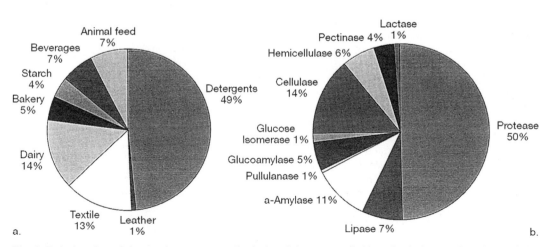

Fig. 1. Relative size of the market segments for industrial enzymes (left) and relative market share of the different enzyme functionalities in the industrial enzyme market (right) (based on sales 1995; approx. 1 billion US $).

dairy applications. Minor quantities are used in starch processing, paper production, animal nutrition and chemical processes like the production of the sweetener aspartame. In molecular biology, an important application of recombinant enzymes is the use of DNA polymerase, which is in fact used to construct new recombinant proteins.

From enzyme sales in terms of enzyme functionality (Fig. 1b) it can be deduced that the major part of the technical enzymes nowadays is of recombinant origin. Protease, α-amylase and lipase alone make up 67% of the enzyme sales. They are generally, except for some food applications, produced in recombinant production systems. From the food industry, the well-known example for the application of a recombinant enzyme is the use of recombinant chymosin and microbial rennet in the production of cheese.

In this chapter we shall present the main industrial enzymes, their application and new developments for improvement or new applications for the presented enzymes. In the description of the respective enzymes we always highlight and explain in detail one general aspect of enzyme usage in industrial applications. In Sect. 2 the different applications and application conditions and the necessary small scale test systems of α-amylase are described. In Sect. 4 it is described how the selection of one enzyme of the same functionality can influence the effect in the application. Even mixtures of enzymes of the same type are produced through recombinant technologies to achieve a certain effect in the application. Sect. 5 exemplifies the usage of protein engineering for the improvement of existing technical enzymes. Sect. 7 finally shows the necessity of modern recombinant technologies for the development and production of new technical enzymes.

2 α-Amylases

α-Amylases (1,4-α-D-glucan-glucanohydrolases, EC 3.2.1.1) catalyze the hydrolysis of the α-1,4-glucosidic linkages in starch. α-Amylase is an endoamylase that randomly cleaves the internal bonds in amylose and amylopectin, the two carbohydrate polymers of natural starch (Fig. 2).

α-Amylases are found in a wide variety of organisms as in mammals, plants and several different microorganisms. A summary of the properties of the most important commercial α-amylases is shown in Tab. 1. It can be seen that α-amylases are available for a wide temperature and pH range. There are α-amylases that act under alkaline conditions, and others can be used in acidic applications. The temperature where α-amylases are applicable can be higher than 100°C. Almost all of the technical α-amylases, however, need a certain amount of calcium ions in the application, because their thermostability depends on the presence of a structural calcium ion (see, e.g., CHIANG et al., 1979 for *B. licheniformis*). It has been shown recently that this calcium dependence of the *B. licheniformis* enzyme can be affected by protein engineering (VAN DER LAAN and AEHLE, 1995).

The range of technical applications for α-amylases, is very wide. α-Amylases are used for the liquefaction of cereal starch during the production of high fructose corn syrup or fuel ethanol. The paper and textile industry apply α-amylases in the sizing/desizing process. In household laundry and automatic dishwashing detergents, α-amylases remove starch containing stains from clothes and dishes.

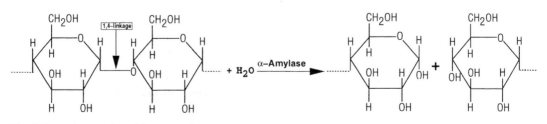

Fig. 2. Reaction catalyzed by α-amylases.

Tab. 1. Properties of Technical α-Amylases

Trivial Name	Source Organism	pH Range[a]	T Range[a] [°C]	Commercial Products[b]
(Conventional) Bacterial amylase	*B. amyloliquefaciens*	5.5–6	50–70	BAN (N)
	B. subtilis	5.5–6	43–70	Aquazyme® (N) Dezyme® (N) Dex-lo® (Gb) Desize® 40, 160 and 900 (GCI)
Thermostable bacterial amylase	*B. licheniformis*	6–9	76–100	Spezyme® AA 20 (GCI) Termamyl® (N) Liquozyme® 280 L (N) Purafect® HP Am L and OxAm G(GCI) Desize® HT and TEX (GCI)
	B. stearothermophilus	4.5–6	60–80	Maltogenase® (N)
Fungal amylase	*A. oryzae*	4.5–6	35–55	Fungamyl® (N) Mycolase® (GCI)
Acid amylase	*A. niger*	4–6	50–60	Hazyme® (Gb)

[a] Data range taken from VIHINEN and MÄNTSÄLÄ (1989).
[b] GCI: Genencor International, Gb: Gist-Brocades, N: Novo Nordisk.

2.1 α-Amylases in Starch Liquefaction

Starch liquefaction is the first step in the preparation of important raw materials from natural starch for the food and the fuel industry. Starch liquefaction itself is a 2-step process composed of a gelatinization step and a dextrinization step. Starch liquefaction is part of the production process for high fructose corn syrup (HFCS), which is shown schematically in Fig. 3.

A starch slurry from water and 30–40% dry solid starch is prepared. The pH of the slurry is adjusted to approximately 6 and, if necessary, calcium is added to a final concentration of 40–100 ppm. The α-amylase is added and, after transfer to a jet cooker, it is brought by steam injection to a temperature of about 108°C. The reaction mixture is kept at this temperature for about 20 min until gelatinization is finished. After flash cooling to 95°C, the slurry is allowed to react for 2 h, while cooling slowly down to 80°C, to finish the dextrinization step. As a result, a solution of maltodextrins of variable chain length is obtained. Fig. 4 shows that the viscosity of the starch slurry is drastically reduced during the liquefaction step and the number of dextrose equivalents, a measure for the number of free accessible sugars, increases.

For the saccharification step, the pH of the maltodextrin solution must be adjusted to 4.3 and the metal ions must be removed to enable the degradation of the maltodextrins into glucose monomers. After isomerization with glucose isomerase at pH 7.7, the high fructose corn syrup is obtained that is, e.g., used in huge quantities in the beverages industry as sweetener for soft drinks. The description of the process makes clear that a highly thermostable α-amylase is required for starch liquefaction. The most preferred products for starch liquefaction are thus the thermostable α-amylase from *B. licheniformis* and *B. stearothermophilus*. For HFCS production, the *B. licheniformis* enzyme is preferred, because

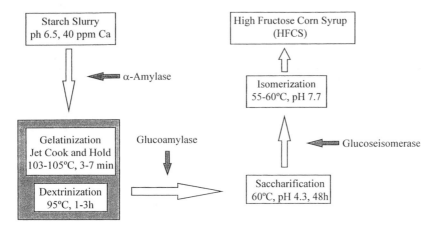

Fig. 3. Schematical representation of the enzymatical production of high fructose corn syrup (HFCS) from cereal starch.

the composition of the chain length of the maltodextrins in the product mixture is better for the subsequent treatment with glucoamylase. The *B. stearothermophilus* enzyme shows another composition of the maltodextrin mixture which is less suitable for the glucoamylase treatment. This enzyme is applied either in the HFCS starch liquefaction as a blend with other α-amylases or in starch liquefaction for the fermentation of fuel ethanol, where the maltodextrin solution is used as carbon source.

The properties of an ideal starch liquefaction α-amylase can also be deduced from the description of the HFCS process. The ideal α-amylase would act at an acidic pH and high temperature without being dependent on the presence of calcium ions. That would allow the starch liquefaction industry to avoid addition and removal of calcium before and after the liquefaction step, and it would be necessary to adjust the pH only once for the first two steps of the process.

2.2 α-Amylases in Textile Desizing

The modern production processes for textiles introduce a considerable strain on the warp during weaving. The yarn must, therefore, be prevented from breaking. For this purpose a removable protective layer is applied to the threads. The materials that are used for this size layer are quite different. From natural sources starch, gelatin or plant gums and in modified form water soluble methyl or carboxymethyl cellulose can be used. The very stable chemical polymers like polyvinyl alcohol are also very popular sizes. In most cases the sizes must be removed from the textile after weaving in the so-called desizing process, because a size has also an impregnating effect and prevents the absorption of the staining dyes into the fibers. An easily removable material is a preferred property of a good size. This explains the interest in the water soluble cellulose derivatives. Starch is a very attractive size, because it is cheap, easily available in most regions of the world, and it can be removed, despite its water insolubility, quite easily. A good desizing of starch sized textiles is achieved by the application of α-amylases. This is another advantage of starch as size, because α-amylases selectively remove the size and do not attack the fibers. α-Amylases randomly cleave the starch into dextrins that are water soluble and can consequently be removed by washing.

In general a desizing process is carried out in four stages which can be performed in batch and continuous processes.

The textiles are first prepared for desizing in the prewashing step. During prewashing the textile is saturated with water and the water soluble additives of the starch size are re-

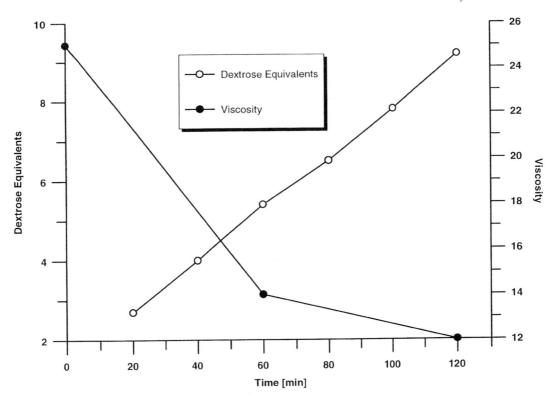

Fig. 4. The starch liquefaction process – increase in dextrose equivalents (○) and decrease in viscosity (●).

moved. This improves the access of the α-amylases to the starch.

The prewashed textiles are in the second step impregnated with an α-amylase containing liquor. The cloth is saturated with the enzyme liquor and prepared for the starch hydrolysis. The temperature is already close to the optimum temperature of the applied α-amylases (65–70 °C for conventional and 75–80 °C for thermostable amylases). The starch hydrolysis begins during this stage. The main step of textile desizing is of course the starch hydrolysis step. In batch processes this stage can take 2–16 h, depending on conditions like pH and temperature. During this time the cloth is on a roll and must be prevented from drying by wrapping it in plastic.

The continuous hydrolysis process is much shorter. It takes only 1–4 h with conventional amylases. With thermostable α-amylases the hydrolysis step can be performed within not more than 20–60 s at 90–110 °C. Drying is here no problem, because the hydrolysis step is either performed in a steam or a wet box. The whole desizing process finishes with the after wash step. The α-amylase hydrolysis products have to be removed as good as possible. The cloth is, therefore, washed in hot water in presence of detergents and sodium chloride. After thorough rinsing the cloth is ready for the subsequent process steps. A detailed description of the textile desizing process can be found in GODFREY (1996).

2.3 α-Amylases for Paper Production

Sizing of paper is not only performed to protect the paper against mechanical damage during processing, but it also improves the quality of the finished paper. The size enhances stiffness and strength in paper. It improves the erasibility and is a good coating for the final product.

Starch is, as for textiles, also a good sizing agent for the finishing of paper. The starch is added to the paper in a size press. The paper picks up the starch by passing through two rolls that transfer a starch slurry. The temperature of this process lies between 45 and 60°C, and a constant viscosity of the starch is required for a reproducible result. The viscosity of natural starch is much too high for paper sizing and is adjusted by partially degrading the carbohydrate polymers with α-amylases. This can be done in batch and continuous processes. The conditions depend on the source of the starch and the α-amylase that is used (TOLAN, 1996).

2.4 α-Amylases for Detergent Applications

All over the world the daily food contains as one of the major components starch or modified starches. Consequently stains on clothes or dishes generated during preparation or consumption of our meals contain notable amounts of starch. Especially stains derived from food ingredients like spaghetti sauce, porridge, mashed potatoes or chocolate are based on starch, which was modified during the preparation of the food. Removal of starch from cloth and dishes by detergent compositions is necessary for a good washing result. Especially in automatic dishwashing, where the mechanical treatment of conventional hand dishwashing is replaced by aggressive alkaline conditions, enzymes that degrade starch are an efficient aid for cleaning. Consequently, α-amylases are used in powder laundry detergents since 1975. Nowadays 90% of all liquid laundry detergents contain α-amylase (KOTTWITZ et al., 1994).

The demand for α-amylases for automatic dishwashing detergents is growing. The latter is seen for enzymes in automatic dishwashing detergents in general, because of safety and quality reasons. In general enzymes are used in detergents for two reasons. First, enzymes

attack selectively the key ingredients in the stains and not the material of the goods to be washed (but see the cellulases in Sect. 4.2.1). The main advantage of enzyme application in detergents is, however, the chance to go to much milder conditions than with enzyme free detergents. Enzymes permit lowering of washing temperatures and are one of the ingredients that allowed the development of the modern compact detergents. The early very aggressive automatic dishwashing detergents could cause injury when ingested and were not compatible with delicate china and wooden dishware. This forced the detergent industries to search for less aggressive, more effective solutions (VAN EE et al., 1992).

Determination of the performance of an enzyme is very important for enzyme producers and applicants for two reasons:

- It must be possible to determine the benefit of enzyme addition in reproducible full scale systems. This allows the detergent producer to determine the necessary enzyme dosage in all his formulations and to demonstrate the benefit of enzyme application.
- Predictive small scale tests are necessary for the evaluation of new enzymes or for the improvement of existing enzymes. These tests must allow the reliable discrimination between enzymes with good and poor performance in the application. They also must allow screening of a huge number of samples in a wild-type or variant screening system.

JENSEN and THELLERSEN (1993) described a typical full scale test system for α-amylase performance in automatic dishwashing applications. For this test different plates and glasses are soiled with 2% (weight) oatmeal porridge. The dishes are then dried for 16–22 h at 22°C and 50–60% relative humidity. After this treatment the dishes have a dull look due to the starch layer. The dishes are subsequently washed in a standard commercial dishwasher with a standard washing program. Starch removal by the washing procedure is evaluated through visual inspection by a trained panel. A second determination of the residual starch is performed by ranking in a scale from 1–5 after iodine coloring. This test is repeated for all changes in the detergent formulation or in the enzyme.

A small scale test system as described by TIERNY et al. (1995) is designed quite differently. Here rice flour is hydrated at room temperature and then cooked to obtain a consistent starch slurry. The starch slurry is coated onto frosted glass slides, which are dried in an oven. "Washing" of the coated class slides can be performed in Erlenmeyer flasks. Starch removal as a measure for cleaning performance is determined gravimetrically.

Examples, how the results of these tests can be used are shown in Sect. 2.5. A detailed description of performance tests for enzymes in automated dishwashing detergents can be found in the article of MAGG and WÄSCHENBACH (1997). Some of these tests are even standardized by several national organizations all over the world.

2.5 Development of New α-Amylases

Despite the fact that α-amylases have been used in a wide variety of technical applications for several years, there have not been many new developments. The available enzymes are good and fulfilled until recently the needs of the customers. However, since a few years the interest in new and improved α-amylase is growing, and consequently the research is intensified as well. From the description of the HFCS process it can be concluded that there is a need for an α-amylase that has the same thermostability as the *B. licheniformis* enzyme, but is also independent of the calcium concentration. Such an enzyme would fit much better into the process, because the addition and removal of the metal ions before and after the liquefaction step could be avoided. On the other hand, for the detergent industry a better compatibility with the detergent formulations is necessary. A detergent enzyme has to resist the oxidative potency of the bleach system in detergents, the low calcium concentration and the high concentration of surfactants in the wash liquor. Two potential weaknesses of the *B. licheni-*

formis α-amylase in technical applications have been the subject of protein engineering projects. The stability against oxidative agents in detergents has been addressed by two companies and the stability at low calcium concentrations by another. It can be expected that the protein engineering efforts for the improvement of this enzyme will increase since the recent determination of its X-ray structure will allow rational protein engineering (MACHIUS et al., 1995; SVENDSEN et al., 1996).

Nevertheless, the first protein engineering attempts were successful even without an X-ray 3D-structure. The problem of the calcium dependence was solved by using a hypothetical 3D-model, which was obtained by homology modeling using the X-ray structure of the α-amylase from *Aspergillus oryzae* (SWIFT et al., 1991) as a template (AEHLE et al., 1995a). Based on this model the region around the structural calcium-binding site was investigated for positions where a negative charge could be introduced or a positive charge could be removed by amino acid substitutions. The rationale behind this approach was to increase the negative potential around the calcium-binding site to make this area more attractive for positive ions like calcium. As a result the calcium ions should be bound more tightly in the binding site giving a more thermostable enzyme. Fig. 5 shows that the thermostability of the variant enzymes was indeed increased in several variants. The results of small scale application tests for textile desizing and starch liquefaction in Fig. 6 show that the increased stability under conditions with low calcium concentrations indeed supports the performance of the enzymes in these applications under practical conditions.

The stability against oxidants in household detergents was approached generally utilizing successful strategies followed with other enzymes such as proteases (see Sect. 5.3). Scientists from the two major detergent enzyme suppliers Novo Nordisk and Genencor International independently replaced the oxidation sensitive amino acids by other amino acids (TIERNY et al., 1995; SVENDSEN and BISGÅRD-FRANTZEN, 1994). It turned out that replacement of Met at position 197 by Leu in the *B. licheniformis* enzyme resulted in

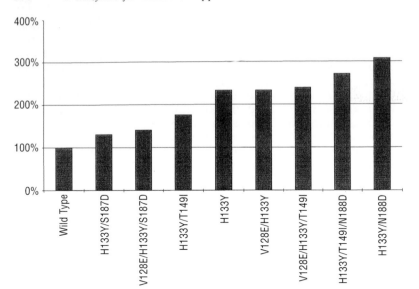

Fig. 5. Improvement of the thermostability of *B. licheniformis* α-amylase by site directed mutagenesis. Increase of the half-life at 93°C, 1 mM Ca^{2+} and pH 7.5 compared to the wild-type (100%).

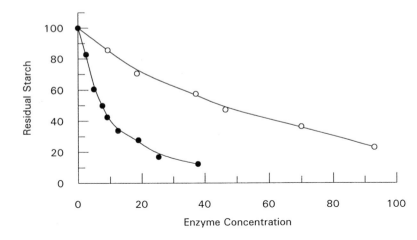

Fig. 6. Dose-response curve of α-amylases for textile desizing demonstrated in a lab scale experiment, wild-type (○), variant α-amylase (●).

an α-amylase with improved resistance against the oxidative compounds. The improved oxidation stability is beneficial for the usage of the enzyme for two reasons:

- The storage stability in bleach containing detergents of the variant enzyme is improved with respect to the wild-type enzyme (Fig. 7).
- The performance of the mutated enzyme in bleach containing detergent formulations is

much better than the performance of the wild-type enzyme (Fig. 8). This relation between stability of enzymes against detergent ingredients and their performance has been discussed by MISSET (1997).

Meanwhile, Genencor International and Novo Nordisk introduced their new products on the market under the trade names Purafect OxAm® and Duramyl®, respectively.

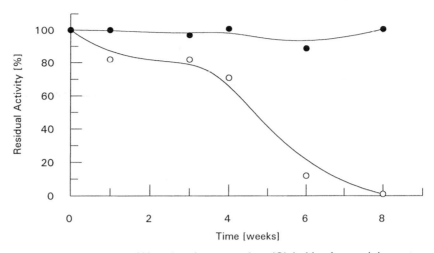

Fig. 7. Storage stability of wild-type α-amylase (O) and variant α-amylase (●) in bleach containing automatic dishwashing detergents at ambient temperature.

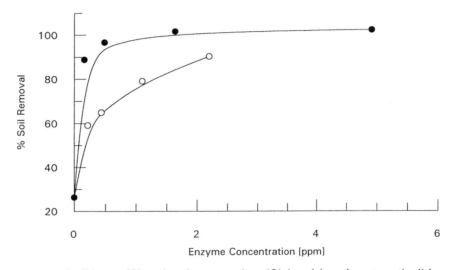

Fig. 8. Dose-response curve of wild-type (O) and variant α-amylase (●) in a lab scale automatic dishwashing experiment.

3 Lipases

Lipases (triacylglycerol esterases EC 3.1.1.3) hydrolyze triacylglycerols into free fatty acids and glycerol (Fig. 9). They are found in a growing number of different microorganisms. Until now lipases have been identified in bacteria, fungi, mammals and plants. Almost every week the discovery of a new lipase, often from new sources, is reported in the literature. The research on lipases received a major push by a research project that was funded by the European Community. During this project, several structures of lipases were solved by X-ray crystallography, and knowledge about the aspects of the molecular biology and the kine-

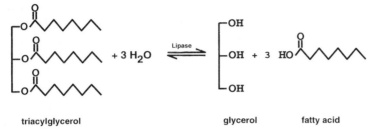

Fig. 9. Reaction catalyzed by lipases.

triacylglycerol   glycerol   fatty acid

tics of this class of enzymes has increased tremendously.

Lipases belong to the class of hydrolytic enzymes that are characterized by the so-called α/β-hydrolase fold (OLLIS et al., 1992). Similar to the serine proteases, the catalytic mechanism uses a catalytic triad consisting of Ser, Hist and Asp (BRADY et al., 1990). The latter is replaced in at least one case by Glu (SCHRAG et al., 1991). Lipases, however, are unusual enzymes with respect to their substrates and the interaction with these substrates. Triglycerides are amphiphilic substances which form aggregates at the air–water interface or micelles in the water phase. The catalysis of the triglyceride hydrolysis requires, therefore, an enzyme that can operate at these water lipid interfaces. This is not a solid–liquid interface as in the case of cellulases or starch degrading enzymes, but a liquid–liquid interface. Nature has equipped lipases with a special feature to deal with this peculiar situation. In solution, lipases are comparable to conventional water soluble enzymes; the hydrophilic amino acids are at the surface and the hydrophobic amino acids form the core of the enzyme. Lipases have a structural feature which allows them to adopt a completely different conformation thereby converting themselves into amphiphilic molecules. During this structural change a helical lid, which covers the active site in the hydrophilic state of the enzyme, moves away from its position and opens a huge hydrophobic region around the active site of the molecule, whereas the rest of the enzyme remains water soluble giving a perfect amphiphilic enzyme (BRADY et al., 1990; BRZOZOWSKI et al., 1991).

Lipases from recombinant production organisms are relatively new in the group of in-

dustrial enzymes whereas lipases from non-recombinant sources have been used already for a long time for flavor development during cheese ripening. The first recombinant lipase product contains the *Humicola lanuginosa* lipase, produced in *Aspergillus* for detergent applications. There is great interest from the scientific world in using lipases for the synthesis of organic compounds. The reviews of KAZLAUSKAS (1994) and CYGLER et al. (1995) discuss possible syntheses with lipases as well as the stereochemistry of lipase catalyzed reactions.

3.1 Lipases in Laundry Detergents

Lipases have great potency as cleaning aids in household laundry detergents. The majority of the stains that are recognized by the consumer as "problem" stains contain triglycerides as main components. These stains are derived from greasy food (tomato-based sauces, butter, edible oils and chocolate), cosmetics (lipstick and mascara) and different kinds of body soils (sebum and sweat on collars, cuffs and underarm areas). A triglyceride degrading enzyme such as a lipase would be a good aid for the removal of these stains. Most of the fat components are very hydrophobic and can not easily be dissolved into the washing liquor. Lipases catalyze the hydrolysis of the water insoluble triglycerides into the more water soluble mono- and diglycerides, glycerol and free fatty acids. These can be transferred much better by the surfactants into the wash liquor. Therefore, in the 1980s several patents described the benefit of lipases in laundry detergents. In 1983 the Japanese detergent manufacturer Lion filed a patent describing the use of a lipase in detergents in a practical form. Consequently, Lion introduced the first lipase containing detergent, Hi-Top, in 1987. The Danish enzyme manufacturer Novo Nordisk announced the first detergent lipase for the world market in the same year. The lipase with the brand name Lipolase® was isolated from the fungus *Humicola lanuginosa* and is produced from *Aspergillus oryzae*.

In 1994 the 2 *Pseudomonas*-derived lipases Lumafast® and Lipomax® were introduced into the detergent enzyme market by Genencor International and Gist-Brocades, respectively. Both are recombinant enzymes. Lumafast® originally from *Pseudomonas mendocina* is produced in a *Bacillus* species. Lipomax® is already a protein engineered variant of the wild-type enzyme to overcome its oxidation sensitivity. Also in 1994 Novo Nordisk introduced an improved variant of its detergent lipase called Lipolase® Ultra.

As expected lipases show a demonstrable effect in the removal of greasy stains, if they are added to detergents. Fig. 10 shows the lipase benefit in detergents on an artificial monitor stain. It depicts also an effect which is typical for lipases usage. The benefit of lipases usage becomes only visible after using a lipase containing detergent in several subsequent wash cycles. This strange behavior of lipases in detergent applications is often referred to as the multi-cycle effect. Fig. 11 shows an explanation for this behavior. Lipolase® is almost inactive on the wet textile. The lipase activity increases during the drying period and goes through an optimum at a relative humidity of 20%. After drying the released free fatty acids are still adsorbed to the fiber and still visible as stain. It remains on the cloth until the next wash. The practical performance of lipases in detergents has not fulfilled the expectations of the detergent industry. WOLFF and SHOWELL (1997) showed that available lipases aid stain removal but definitely need improvement to become the robust performers the industry needs. Research should address the surfactant compatibility, low temperature performance and a broad substrate specificity, to allow removal of all greasy stains by one lipase.

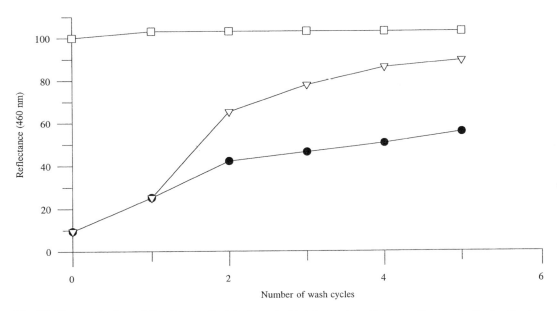

Fig. 10. Demonstration of the lipase effect in household detergent applications. Removal of a lard/sudan red monitor on polyester/cotton swatch with detergent alone (●), detergent plus Lipolase® (△) and a blank polyester/cotton swatch as reference (□) in a multi-cycle experiment (data taken from AASLYNG et al., 1991).

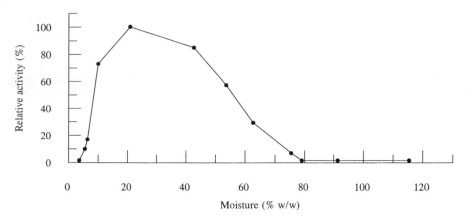

Fig. 11. Activity of Lipolase® during the drying period of the clothes (data taken from AASLYNG et al., 1991).

3.2 New Developments in the Lipase Field

The development of commercial lipases focuses on the main problems of these enzymes in detergent applications. The compatibility with surfactants – a problem that applies to all detergent enzymes – is addressed. The other major research item is the single-cycle behavior of lipases in laundry washing.

OKKELS et al. (1996) described an attempt to increase the low wash performance of Lipolase®. Their research aimed at an improvement of the washing performance in one- and multi-cycle wash. In a rational protein engineering experiment the negatively charged residues around the lipid contact zone of the *H. lanuginosa* lipase were replaced. This experiment was performed based on the hypothesis that repulsive interactions between the negatively charged fatty acids and surfactants bound to the lipid interface and the enzyme could be decreased. Experiments had shown that amphiphilic surfactants accumulate where the lipase should accumulate for good washing performance. One of the variants showed indeed improved washing performance in multi-cycle experiments. The improvement factor of 4–5 times, however, can not be seen in the first cycle. This lipase variant with the replacement of Asp 96 by Leu is meanwhile marketed under the tradename Lipolase® Ultra.

An attempt to improve the stability of the *Pseudomonas alcaligenes* lipase against the denaturing influence of the surfactants (AEHLE et al., 1995b) has been reported in the article of MISSET (1997). The purpose of this research was to stabilize Lipomax® against the denaturing activity of anionic surfactants in the detergent mixture. By introducing either space-filling amino acids into holes or removal of positive charges near cavities on the surface of the enzyme by site-directed mutagenesis the enzyme was stabilized against sodium laurylsulfate. The half-life of one variant was 10 times improved under certain conditions.

4 Cellulases

4.1 Properties of Microbial Cellulases

Cellulose is a linear polymer made up of glucose residues that are connected to each other through 1,4-β-glucosidic linkages; it contains no branches. The fundamental, constantly recurring unit is cellobiose (for reviews on cellulose and cellulases see KOTTWITZ and UPADEK (1997), HOSHINO and ITO (1997) and MAURER (1997) and references cited therein.) Cellulose differs from starch in the absence of branch points and the type of linkage which is 1,4-α-glucosidic in the case of starch. As a result of their molecular structure and physical and chemical properties, the glucose chains are arranged in parallel bundles and linked by hydrogen bonds resulting in almost water-free fibrillar structures (so called microfibrils). This type of interaction continues up to higher levels (macrofibrils) ultimately giving rise to crystalline aggregates that are water insoluble and very resistant to chemical and enzymatic attack. In addition, cellulose contains regions of less ordered structure also referred to as the amorphous regions; these are more easily degraded by cellulolytic enzyme systems.

Cellulose, together with hemicellulose, is the most abundant biopolymer on Earth. It is found in wood, cotton, and the walls of plant cells. It is, therefore, no surprise that (micro)organisms can be found in nature that try to utilize cellulose as a source of carbon and energy. However, before being able to utilize the glucose constituent as a nutritive source, the complex cellulose structures must be degraded.

Most microorganisms of bacterial and fungal origin produce a mixture of different cellulolytic enzymes that are capable of degrading cellulose in a synergistic mode. Analogous to the enzymes that degrade other biological macromolecules such as starch and protein, one can distinguish between cellulases with an endo- and with an exoactivity (Tab. 2). Endoglucanase (recommended name is cellulase) is the only enzyme capable of cleaving the glucose chains in an endofashion. Cellobiohydrolase exerts its exoactivity by liberating cellobiose moieties from either the reducing (e.g., CBH-1 from *Trichoderma longibrachiatum*) or the non-reducing end (e.g. CBH-II from *T. longibrachiatum*) (TEERI et al., 1995). β-Glucosidase, also called cellobiase, cleaves cellobiose into two molecules of glucose.

Cellulases that are used in industrial applications are often differentiated by their microbial origin. Fungal enzymes are produced by species such as the afore-mentioned *T. longibrachiatum* and *Humicola insolens*. Usually the cellulases are secreted as a complex mixture composed of several endoglucanases (up to 6), cellobiohydrolases (2–3) and usually one β-glucosidase. Fungal enzymes are in general active in the acidic (pH > 3) to mild alkaline region (pH = 7–8). Differences in pH optima between the various enzymes do occur and can be exploited in the different applications. Bacterial enzymes are mainly derived from *Bacillus* species and display their enzymatic activity at more alkaline pH than their fungal counterparts (HOSHINO and ITO, 1997). Bacteria produce either one or a limited number of only endoglucanases or they produce cellulosomes: multienzyme complexes containing several cellulolytic activities that are bound via short sequences (docking domains) to a large non-catalytic protein (scaffolding subunit).

Depending on the desired effect and on the application conditions, a single or a defined mixture of cellulolytic activities is required. In such a case, cloning and expression of the cel-

Tab. 2. Cellulolytic Activities

Type EC-Number	Enzyme	Substrate	Products
Endo EC 3.2.1.4	endoglucanase	cellulose	β-glucans
Exo EC 3.2.1.91	cellobiohydrolase	cellulose β-glucans	cellobiose
EC 3.2.1.21	β-glucosidase	cellobiose	glucose

lulase genes involved is a powerful tool to produce enzyme products with the desired characteristics. Until now, only cloned endoglucanases are used in detergent and textile applications as will be described in Sects. 4.2.1 and 4.2.2. Other applications such as feed and food still make use of the cellulase complexes as produced by the various microorganisms.

4.2 Industrial Applications of Recombinant Cellulases

4.2.1 Detergent Industry

The application and development of detergent cellulases have been described in great detail by KOTTWITZ and UPADEK (1997), HOSHINO and ITO (1997) and MAURER (1997). This paragraph will give only a short overview of the current state of the art.

Since 1987, cellulases have been used in detergents. Their use started in Japan with an endoglucanase from *Bacillus* that was developed by KAO. In subsequent years the use of cellulase expanded toward Europe (1991) and America. Also Novo Nordisk introduced a fungal cellulase complex from *Humicola insolens* for detergent application (tradename Celluzyme). In order to get a cost-effective enzyme product, researchers at the KAO company undertook efforts to create an overproducing strain by cloning the gene of the alkaline *Bacillus* endoglucanase and express it in a closely related host (self-cloning – see HOSHINO and ITO, 1997, for a detailed description of the work carried out at KAO).

Similarly, at Novo Nordisk 1 of the 5 endoglucanases from *H. insolens* was cloned and brought to expression in *Aspergillus oryzae* (heterologous expression) and further developed into a detergent enzyme product (Carezyme®). This endoglucanase was chosen because it possesses the highest pH optimum of all the endoglucanases from *H. insolens* (required for detergent applications and strong antipilling activities.)

In contrast to the other detergent enzymes (protease, lipase and amylase), cellulases do not degrade the cellulose component of a stain, but are active on the cellulose fibers of the cotton fabric. Several effects of cellulase are described in the literature:

- *stain removal:* according to KAO, the alkaline bacterial cellulase degrades the amorphous regions of the cotton fibers thereby liberating particulate soil and sebum which are preferentially immobilized on this part of the cellulose.
- *fabric softening:* for bacterial cellulase the softening effect is explained by their effect on particulate soil removal from the amorphous cellulose regions on the surface of the cotton fabric as well as between the fibers. For the fungal cellulase the softening effect is explained by its effect on the removal of fibrils thus smoothening the cotton surface thereby preventing the interlining of the fibers.
- *color care:* some cellulases have been shown to improve the appearance of fabrics by removing the grayish cast which arises on a used, colored garment.
- *depilling and pilling prevention:* these effects are closely related. Wash-and-wear of cotton gives rise to the formation of pills which deteriorate the appearance of the fabric (pills are composed of microfibrils that protrude from the surface together with cellulose fragments that are torn off). Removal of pills by cellulase, therefore, results in cleaning up and smoothening of the fabric. The smoothening effect of cellulase also prevents the easy formation of new protruding microfibrils, i.e., pilling prevention.
- *antiredepositing:* the elimination of microfibrils by cellulase also avoids redeposition of particulate soil that is already present in the wash water as a result of surfactant action. These protruding microfibrils can otherwise serve as ideal anchor points for soil particles.

4.2.2 Textile Industry

There are two different applications of cellulolytic enzymes in the textile industry: "stonewashing" and "biopolishing". The most prominent one is the one in denim processing where cellulase can replace the traditional pumice stones in the washing step. Cellulases replace the pumice stones and the extra space in the washing machines allows a higher load of textiles in the washing step. This results in lower energy consumption and lower damage of the washing machines. The purpose of this step is to remove (an excess of) indigo to give the denim the "right" look. While the stones achieve this through abrasion, cellulase enzymes remove cellulose fibrils from the cotton surface thereby liberating the attached indigo. Since the penetration of the enzyme into the cellulose fibers is limited, the complete removal of indigo is avoided. Although the cellulase complexes, both the acidic ones (*Trichoderma*) and the more neutral/alkaline (*Humicola insolens*) ones, are adequate for this process, cloned endoglucanase from *H. insolens* (Denimax® Ultra = Carezyme® for detergents) is now also available. Other possibilities are the use of genetically modified *Trichoderma* strains in which genes for undesired endoglucanases and/or cellobiohydrolases are deleted, favoring the expression of the desired enzymes, or in which the desired enzymes are overexpressed, thus creating mixtures of certain activity ratios (e.g., SUOMINEN et al., 1992). Such mixtures are meanwhile commercially used, e.g., in the IndiAge® product line of Genencor International for cotton and denim garment washing and cellulosic fabric finishing.

In the biopolishing application, the cellulase treatment of new cotton fabric results in a smooth surface. This prevents or at least slows down the formation of pills and, therefore, helps in maintaining the new look of clothes for a longer time (color maintenance).

5 Proteases

5.1 Introduction

Proteins are one of the most abundant classes of biological macromolecules in the living world. Protein molecules fulfil a large number of different roles which depend mainly on their structural properties. Common to all proteins is that they are linear polymers of amino acids that are connected to each other via peptide bonds (Fig. 12). In nature, 20 different amino acids occur which means that a large variation of amino acid sequences is possible. This is illustrated by the following calculation: The total number of different dipeptides is $20 \cdot 20 = 2^2 = 400$; in general, a polypeptide composed of n amino acids can occur in 20^n possible combinations. Consequently, for a protein of 300 amino acids (e.g., a small enzyme) there are $20^{300} = 10^{390}$ sequences possible. Let's suppose that of each possible sequence 1 molecule would be present, then the total weight would amount to $5 \cdot 10^{367}$ kg. The weight of the planet Earth is only $6 \cdot 10^{24}$ kg! Cross-links between peptide chains can exist through the formation of disulfide bridges between the SH-group of Cys residues or through a chemical bond between the side chains of Lys and Glu. These cross-

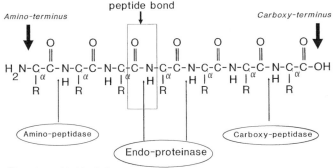

Fig. 12. Action of the different types of proteases on a polypeptide chain.

links can either confer stability to globular proteins such as enzymes or they can give rise to very large and complex protein molecules such as the gluten network in wheat dough.

For the enzymatic breakdown of the protein molecule, proteases and peptidases from different families and classes are available. Yet, they have in common that they hydrolyze only one type of chemical bond in the protein, i.e., the peptide bond which links the amino acids together. All other chemical bonds, including the cross-links described above, are left "unharmed". On the basis of where they "attack" their substrate, proteolytic enzymes can be divided into two groups: endopeptidases and exopeptidases (Tab. 3). Endopeptidases (also called proteases or proteinases) are enzymes that cleave somewhere in the middle of the substrate molecule, while exopeptidases start from either one or the other end. We discriminate between aminopeptidases that cleave 1, 2 or 3 amino acids from the N-terminus of the protein substrate and carboxypeptidases that cleave amino acids from the C-terminus (Fig. 12).

Proteolytic enzymes can also be differentiated on the basis of their substrate specificity, i.e., which of the 20 amino acids, or which sequence motif is recognized by the different enzymes. The pancreatic protease trypsin, for instance, only cleaves a peptide bond when the amino acid at the N-terminal side of the scissile bond is positively charged: Lys or Arg. Likewise, the pancreatic protease chymotrypsin and the bacterial (high) alkaline protease "subtilisin" prefer to cleave adjacent to more hydrophobic amino acids such as Phe, Tyr, Met and others. Similar substrate specificities exist for the exopeptidases.

The endopeptidases, now being referred to as proteases, are grouped on the basis of the molecular mechanism of the catalyzed reaction. Tab. 3 gives the names of groups as well as the corresponding EC-numbers.

5.2 Classes of Proteases

5.2.1 Serine Proteases

Serine proteases (EC 3.4.21) are characterized by the presence of the amino acid Ser in the active site of the enzyme, that plays a crucial role during catalysis. Together with two other amino acids (His and Asp), this Ser performs a nucleophilic attack on the peptide bond in the substrate. This step results in a covalent acyl-enzyme intermediate. The second step in the reaction mechanism is the hy-

Tab. 3. Classification of Peptidases

Peptidase	Enzyme Group	EC-Number
Exopeptidase		
Aminopeptidase	• aminopeptidases	EC 3.4.11
	• dipeptidases	EC 3.4.13
	• di- and tripeptidyl peptidases	EC 3.4.14
Carboxypeptidase	• peptidyl-dipeptidases	EC 3.4.15
	• serine type carboxypeptidases	EC 3.4.16
	• metallocarboxypeptidases	EC 3.4.17
	• cysteine type carboxypeptidases	EC 3.4.18
Endopeptidase		
Proteinase	• serine endopeptidases	EC 3.4.21
	• cysteine endopeptidases	EC 3.4.22
	• aspartic endopeptidases	EC 3.4.23
	• metalloendopeptidases	EC 3.4.24

drolysis of the acyl-enzyme by water. On the basis of their three-dimensional structure two families of the serine proteases are recognized: trypsin-like and the subtilases (SIEZEN et al., 1991; SIEZEN and LEUNISSEN, 1997). Well known examples of the first family are the pancreatic enzymes trypsin, chymotrypsin and elastase. To the subtilase family belong the subtilisins (alkaline bacterial proteases of various *Bacillus* species) but also fungal proteases such as proteinase K and thermitase and many others. As a result of the chemistry of the catalytic mechanism, serine proteases are active at pH 7–13; at a lower pH the His in the active site becomes protonated which makes the enzyme inactive. These proteases can be used at temperatures up to 60–70°C (depending on the particular protease).

The industrial use of serine proteases is widespread. Especially the bacterial subtilisins are abundantly found in household detergents all over the world in which they act to facilitate the removal of proteinaceous stains (see EGMOND, 1997; BOTT, 1997, for reviews). Most, if not all of these proteases are produced on an industrial scale using genetically modified bacteria that overexpress either the wild-type enzyme (homologous expression or self-cloning) or variants thereof which have been obtained through protein engineering. Other applications are in the food industry where certain subtilisins are used in the meat and fish processing industry.

5.2.2 Cysteine Proteases

Similar to serine proteases which contain a Ser residue in their active site, cysteine proteases (EC 3.4.22) contain the amino acid Cys as the crucial catalytic residue. They are active in the pH region 4–8 and at temperatures up to 60–70°C. The majority of industrial cysteine proteases is used in the food industry and is of vegetal origin: papain from *Carica papaya,* ficin from *Ficus glabrata* and bromelain from the pineapple (*Ananas comosus*). Papain is used in the beverage (beer) industry as an anti-haze enzyme and in the meat processing industry as a tenderizer.

Since at present, none of the commercial cysteine proteases originates from genetically modified (micro)organisms, this class of proteases falls out of the scope of this chapter and will not be discussed further.

5.2.3 Aspartyl Proteases

Aspartyl proteases (EC 3.4.23) contain two aspartic acid residues in the active site which "do the job". As a result of the involvement of two residues, the enzymes of this class display a so-called bell shaped pH optimum with the maximum activity in the acidic to neutral region. A very well known example is the gastric enzyme pepsin which can be found in all vertebrates including man and plays an indispensable role in the first stages of the digestive tract. Another well known example and one which is used on a worldwide basis in the dairy industry is chymosin (rennin), the protease found in the 4th stomach of calves (BROWN, 1993; WIGLEY, 1996). The enzyme is added in the cheese making process with the purpose to coagulate the casein fraction of milk by making a very specific cleavage in one of the subunits of casein (see Sect. 5.4). In addition to the calf enzyme chymosin, similar rennin-like proteases of fungal origin (e.g., *Mucor miehei*) are also used, in particular for the manufacturing of vegetarian cheese.

The calf chymosin is not only available in its original form – an extract of the calf stomach – but also as a product from genetically modified microorganisms in which the calf gene has been cloned (heterologous expression).

5.2.4 Metalloproteases

The class of metalloproteases (EC 3.4.24) contains a metal ion (usually Zn^{2+}) in the active site and this metal ion acts as a cofactor. This means that the enzyme is not active when the metal ion is absent [e.g., after removal by metal ion sequestering agents such as EDTA and (poly)phosphates]. These enzymes are also active in the neutral to alkaline pH region.

The best known example is thermolysin, a heat stable protease that is industrially used for the synthesis (reversed reaction!) of aspartame, the artificial dipeptide sweetener that is composed of Asp and Phe. Other examples of industrial use of metalloproteases are in the beverage industry (anti-haze) and baking industry (limited degradation of gluten).

5.3 Detergent Proteases

Proteases are applied in household detergents for several decades now (for reviews on the application and development of detergent proteases see EGMOND, 1997; and BOTT, 1997). Bacterial proteases from the subtilisin family have been shown to possess the best properties for this application: good activity and stability at neutral to (high) alkaline pH (from 7–12) and at elevated temperatures (up to 60 °C) combined with a good compatibility with detergent ingredients such as surfactants, bleaches, builders, etc. Their contribution to (proteinaceous) stain removal is regarded as essential and unique in the sense, that it cannot be achieved by other detergent ingredients.

Three different protease "backbones" are currently used: subtilisin Carlsberg from *Bacillus licheniformis,* subtilisin BPN' from *B. amyloliquefaciens* and a high alkaline subtilisin from *B. lentus* (or *B. alcalophilus*). The first two are applied in (liquid) detergents with a more neutral pH (7–8) wash water while the *B. lentus* subtilisin is used in laundry and automatic dishwash detergents of high-(er) pH (>10).

Perhaps with the exception of the Carlsberg subtilisin, all detergent proteases (both wild-type and the various variants) are produced by genetically modified microorganisms (GMOs). The major purpose to use GMOs is to get price reduction. For this purpose the enzyme manufacturers (Presently there are two major companies (Novo Nordisk and Genencor International), a few smaller ones (e.g., Showa Denko, KAO) and one captive producing company (Biozyme for Henkel) that produce detergent proteases. The detergent enzyme business of Gist-Bro-

cades and Solvay was acquired by Genencor International in 1995 and 1996 respectively; for details see HOUSTON, 1997). They have constructed overproducing strains (Bacilli) that secrete the protease into the fermentation broth up to very high levels, resulting in almost pure enzyme products (in terms of protein composition).

A second reason for using GMOs is to produce variant enzymes. Despite the fact that the subtilisins are already suitable candidates for detergent applications, there appears to be "room for improvement", especially since this could be carried out by using the protein engineering technology that was developed in the early 80ies. A first improvement that has been achieved was the oxidative stability of subtilisin BPN' and *B. lentus* subtilisin (oxidation and inactivation occur especially during storage of the enzymes in bleach containing powder detergents). Hence a Met residue in the protein molecule, adjacent to the active site Ser, was substituted by non-oxidizable amino acids. This indeed conferred oxidative stability but only a replacement by Ser appeared to result in a variant enzyme with equal wash performance when compared to the wild-type enzyme (BOTT, 1997). Maxapem®, the commercial enzyme product containing the engineered *B. lentus* subtilisin M222S, was the first variant enzyme to appear on the market for non-captive users and has now been used in bleach containing powder detergents for 9 years after its introduction in 1988 by Gist-Brocades. Later, Novo Nordisk launched a similar enzyme product (Durazyme®) in which the Met has been replaced by an Ala, yet, the wash performance of this variant is considered to be inferior when compared with Maxapem or the wild-type subtilisin.

Other mutations that are introduced into either subtilisin BPN' or the *B. lentus* subtilisin aimed at increasing the specific wash performance (same cleaning at lower enzyme dosage) or a higher wash performance at lower washing temperature (e.g., Properase® – see BOTT, 1997, for details). The mutations involved affect positions in the active site that interact with the substrate and, consequently, also with the protein constituents of the stain. It is expected that protein engineering of detergent proteases will continue in the future. Targets for improvement are stain specificity (broader), cold water performance and (anionic) surfactant compatibility.

5.4 Chymosin

Chymosin (EC 3.4.23.2) plays a crucial role in the manufacture of cheese. As a result of its very narrow substrate specificity, this proteolytic enzyme is able to cleave one peptide bond in the milk protein casein thereby initiating the coagulation of milk, i.e., the precipitation of the casein micelles (for reviews on the cheese manufacturing process and the enzymes used therein, see BROWN, 1993, and WIGLEY, 1996).

Milk contains approximately 2.5% (w/v) casein, comprising α-, β- and κ-casein subunits (DRIESSEN, 1994). These constituents aggregate to form complexes (micelles) which appear as spherically shaped particles. κ-Casein is located mainly at the exterior of the micelle and protrudes from the surface with a polar tail. This tail stabilizes the micelles and prevents them from coagulation. During cheese making, chymosin cleaves the peptide bond between a Phe and a Met in the κ-casein subunit which releases the polar tail from the micelle. Subsequently, the casein micelles coagulate, assisted by calcium phosphate clusters which are also part of the micelles.

Traditional coagulating enzyme products are animal rennets, extracted from the stomachs of calves, but also from other animals such as sheep and goat. In addition, microbial milk coagulants are used for special cheese varieties. The animal rennets contain chymosin as major enzyme although pepsin is also present at varying levels. Proteases with substrate specificities different from chymosin (e.g., pepsin and the microbial rennets) may have undesired side-effects in the sense that they produce peptide fragments during the clotting process and later in the cheese ripening stage, that can give rise to off-flavor and bitter taste.

Several companies have undertaken efforts to clone the calf chymosin gene in order to provoke its expression in microorganisms (heterologous expression) and to develop an economic production process. The advantage of cloning chymosin is that a protease pure rennet can be produced without "contamination" by non-specific proteases such as pepsin. Other advantaces compared to the classical animal rennets are the constant quality, supply and price.

6 Oxidoreductases

Oxidoreductases are a huge class of enzymes which catalyze oxidation–reduction reactions. Their classification is based on the nature of the hydrogen acceptor and donor. The reaction that is catalyzed needs always two reaction partners: one to be oxidized and one to be reduced. One of the two substrates is either a cofactor like a metal ion, NAD or a heme group or a particular cosubstrate that acts as hydrogen donor or acceptor respectively. The requirement of oxidoreductases for a cofactor or a cosubstrate makes the usage in technical processes a bit more difficult. Either a minimum amount of the cosubstrate must be present during the reaction or the cofactor must be as stable as the enzyme and it must be possible to regenerate the cofactor during the usage of the enzyme. It is also possible that the enzymes loose their cofactor during usage, which would be recognized by the user as "instability" of the enzyme. This might explain the observation that, although there are plenty of oxidoreductases described, only a few of them are used as technical enzymes.

6.1 Technical Applications of Oxidoreductases

Two oxidoreductases are used in industrial processes. Catalase (EC 1.11.1.6) catalyzes the decomposition of hydrogen peroxide into molecular oxygen and water (Fig. 13). It is used in processes where removal of hydrogen peroxide is necessary. In the textile industry catalase is used to remove the excess hydro-

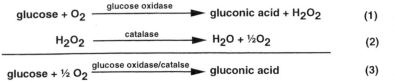

$$glucose + O_2 \xrightarrow{\text{glucose oxidase}} gluconic\ acid + H_2O_2 \qquad (1)$$

$$H_2O_2 \xrightarrow{\text{catalase}} H_2O + \tfrac{1}{2}O_2 \qquad (2)$$

$$glucose + \tfrac{1}{2}\ O_2 \xrightarrow{\text{glucose oxidase/catalse}} gluconic\ acid \qquad (3)$$

Fig. 13. Reaction catalyzed by glucose oxidase (1), catalase (2) and the net reaction of glucose oxidase and catalase (3).

gen peroxide from the bleaching process. In food applications catalase decomposes hydrogen peroxide that has been added to food for disinfection, e.g., for transport of milk from the farmer to the dairy.

Glucose oxidase (EC 1.1.3.4) catalyzes the oxidation of glucose to gluconic acid (Fig. 13). It is used in food applications where sugar has to be removed. A combination of glucose oxidase and catalase is used as deoxygenation system. The net reaction (Fig. 13) of the two enzymes shows that the combination of glucoseoxidase and catalase catalyzes the oxidation of glucose to gluconic acid by consuming oxygen. In air-tight glucose containing containers the excess oxygen can thus effectively be removed. This can be necessary for storage of food products. Oxidation causes, e.g., the rancidity in fats or fat containing preparations like mayonnaise.

7 Phytase

Animal feed is primarily of vegetal origin and contains much of phytic acid (for a review see GRAF, 1986). An average figure for phytate content in plant raw materials is 1–1.2% although large variations exist (COWAN, 1996). This compound cannot be degraded by the enzymes of the digestive track of livestock, in particular not by monogastric animals (pigs, poultry, fish). Therefore, the enzyme manufacturing and feed industries became interested in phytase since it was expected and subsequently confirmed that, when phytase was added to the feed, these monogastric animals could utilize the liberated phosphate which otherwise would end up as non-degraded phytic acid in the manure. The addition of phytase to feed then would have two advantages: firstly it avoids

the addition of supplementary inorganic phosphate to feed and secondly, it reduces considerably the phosphate load in the manure, thereby reducing environmental pollution and contributing to a cleaner world (VAN GORCOM et al., 1991, and references therein).

Phytases are *myo*-inositol hexaphosphate phosphohydrolases and comprise a family of enzymes that catalyze the stepwise removal of inorganic orthophosphate from phytic acid = *myo*-inositol 1,2,3,4,5,6-hexakisphosphate (for a review see WODZINSKI and ULLAH, 1996). Depending on their mode of action two different phytases are recognized: 3-phytase (EC 3.1.3.8) and 6-phytase (EC 3.1.3.26) that have the 3-phosphate and 6-phosphate group as their primary site of attack, respectively (according to the 1992 Recommendations of the Nomenclature Committee of the International Union of Biochemistry and Molecular Biology). In the following paragraphs, the enzymatic properties of phytase as well as the development and application of the industrial enzyme product will be described.

7.1 Properties of *Aspergillus* Phytases

Phytases are produced by a large variety of fungi, yeasts, and bacteria. However, it appears that only the fungal enzymes, particular those from *Aspergillus* species, have enzymological properties that allow their application as a feed additive. These properties comprise a high activity and stability in the acidic pH range (2–5), resistance against acidic proteases (from the animal's digestive tract, e.g., pepsin) and sufficient heat stability in order to survive the feed pelleting step.

Aspergillus niger strain NRRL 3135, formerly known as *Aspergillus ficuum* and recently designated by the Centraalbureau voor Schimmelcultures as *A. niger* van Tieghem (WODZINSKI and ULLAH, 1996), produces a few enzymes with phytase and/or acidic phosphatase activity that in the various (patent) literature are designated as follows:

(1) phytase (phyA);
(2) pH 2.5 acid phosphatase (phyB);
(3) pH 6.5 acid phosphatase.

When grown under phosphate starvation conditions in starch medium, approximately 50% of the excreted protein consists of these 3 enzymes: 5% phyA, 5% phyB and 40% pH 6.5 acid phosphatase (WODZINSKI and UL-LAH, 1996).

Phytase (phyA) is a 3-phytase (EC 3.1.3.8) and has all the right properties for feed application. The gene for the enzyme has been cloned and sequenced (Sect. 7.2) and encodes a polypeptide of 467 amino acids: a signal peptide of 19 amino acids and the mature phytase of 448 amino acids (GORCOM et al.,

1991; VAN HARTINGSVELDT et al., 1993). On the basis of the amino acid composition, the molecular weight (mw) can be calculated to be 48,851 D. However, after purification of the enzyme to homogeneity a mw on SDS-gels of 85 kD is observed (GORCOM et al., 1991). This higher mw is caused by glycosylation of the enzyme: in the amino acid sequence 10 putative N-linked glycosylation sites can be found (VAN HARTINGSVELDT et al., 1993) of which 9 are glycosylated (ULLAH and DISCHINGER, 1993). Deglycosylation results in a smaller mw of 48–56 kD which is in good agreement with the calculated value. The enzyme has a broad pH activity profile (pH 2–7) with optima at $pH = 2.5$ and $pH = 5.5$. In an elegant study, ULLAH and PHILIPPY (1994) determined the kinetic parameters of this enzyme together with the other two enzymes for several *myo*-inositol-phosphate substrates (Tab. 4).

In contrast to phyA that is active from pH 2–7, phyB is only active around pH 2–3 (UL-LAH and PHILIPPY, 1994) but nevertheless

Tab. 4. Substrate Specificities of *Aspergillus niger* NRRL3135 Phytase and Acid Phosphatases (Reproduced from ULLAH and PHILIPPY, 1994)

Substrate parameters	inositol phosphatase	Phytase EC 3.1.3.8 PhyA	pH 2.5 Acid Phosphatase EC 3.1.3.8 PhyB	pH 6.0 Acid Phosphatase EC 3.1.3.2
	Assay conditions:	37.5 mM sodium acetate pH=5 58°C, 3 min	37.5 mM glycine/HCl pH=2.5 58°C, 3 min	37.5 mM imidazole pH=6.0 58°C, 103 min
K_M (µM)	IP6	27	103	315
	IP5	161	330	1265
	IP4	1000	690	400
	IP3	200	2000	400
k_{cat} (s^{-1})	IP6	348	628	2
	IP5	477	1479	38
	IP4	1554	1967	10
	IP3	164	7395	2
k_{cat}/K_M $\cdot 10^{-3}$ M^{-1} s^{-1}	IP6	12900	6100	8
	IP5	2960	4480	30
	IP4	1550	2850	26
	IP3	820	3700	64

shows considerable phytase activity at its optimum pH (Tab. 4). However, the very restricted pH activity profile makes this enzyme much less suitable for feed applications. The pH 6.5 acid phosphatase does not reveal any significant phytase activity (Tab. 4) and will, therefore, not be discussed further.

7.2 Application of Recombinant Phytase in the Feed Industry

7.2.1 Development of the Enzyme Product

In 1991, Gist-Brocades in the Netherlands was the first company worldwide to introduce "recombinant phytase" on the market (VAN DIJCK and GEERSE, 1994). The commercial product is named Natuphos® and contains phytase (phyA) from *Aspergillus niger* van Tieghem.

Phytase is a nice illustration of how the industrial application of an interesting enzyme can be realized only through the use of recombinant DNA technology. Namely, to make Natuphos®' application economically possible, the gene encoding phyA had to be cloned and brought to overexpression in a suitable production strain. So, nowadays, the large scale production of phytase is achieved by the use of a selected strain of the same microorganism *Aspergillus niger* which contains several copies of the gene (i.e., an example of "self-cloning").

Prior to the introduction of Natuphos® in 1991, Alko offered an *Aspergillus* phytase product named Finase which was derived from a non-recombinant strain (*A. niger* var. *awamori*). However, the price of this product was such that it was not compatible with large scale application in animal feed. This can easily be explained by the low production level of phyA by the non-recombinant strain (Sect. 7.1–5% of the total protein secreted by the similar wild-type strain *A. niger*). Consequently, attempts were also made at Alko to clone the gene in order to make an overproducing strain and a cost-effective product. These efforts resulted in a (cloned) phytase

(gene) which is nearly identical with the phytase from Gist-Brocades; it contains only 12 amino acid substitutions (out of a total of 467 amino acids).

Gist-Brocades started their work on gene cloning and sequencing in 1988 (R. BEUDEKER, personal communication). The result was obtained but not without much difficulty. At that time ULLAH and GIBSON (1987) and ULLAH (1988) obtained some sequence information of N-terminal and internal peptides from a supposedly homogenous phytase sample. However, it turned out that this preparation was not pure and the sequence information was, therefore, questionable. At least 2 out of the 4 reported sequences turned out not to belong to the phyA but to the pH 6.5 acid phosphatase and another, still unidentified, protein. Apparently, these proteins all copurified with the phyA and contributed to the 85–100 kD protein band pattern on SDS-gels instead of the postulated heterogenous glycosylation of phyA (ULLAH and GIBSON, 1987). Furthermore, there was also some doubt about the sequence of the internal peptide obtained by CNBr cleavage since it started with 2 Met while CNBr cleaves at the C-terminal side of Met rather than at the N-terminal side. Therefore, the researchers at Gist-Brocades were urged to develop a new purification protocol in order to get rid off all the contaminating, copurifying proteins. Finally, they were successful and could identify the 85 kD protein as the only phyA with the desired properties for feed application. The final proof, however, was obtained after sequencing of the internal peptide fragments, followed by the construction of DNA probes, cloning and expression of the gene and measuring substantially increased phytase levels in the fermentation broth. These results are described in detail in the European and World Patent Applications which were filed in September 1989 (VAN GORCOM et al., 1991).

7.2.2 The Industrial Use of Natuphos®

Phytase can be applied in industrial processes that make use of phytate rich vegetal raw materials such as soy and corn. The most widespread application of phytase today is in compound feed although it can be used in liquid feeding as well. The efficacy of phytase (Natuphos®) has been demonstrated in a large number of animal trials and under a broad range of dietary fibers (JONGBLOED, 1992). From physiological studies it could be concluded that with optimal activity of Natuphos® at pH 2.5 and 5.5, a substantial hydrolysis of phytic acid occurs in the stomach and duodenum, whereas liberated phosphates are absorbed in the small intestine. Fig. 14 shows that phytase, when dosed to either a corn–soybean starter pig diet (diet A) or a diet which contained higher quantities of less digestible feed ingredients (diet B), enhanced the phosphorus digestibility from about 30% (no phytase added) up to a maximum of 60% (at 1800 phytase units per kg feed) (reproduced from VAN DIJCK and GEERSE, 1994). Similar results were obtained with laying hens. Other applications of phytase are recognized as well (VAN GORCOM et al., 1991)

such as in the industrial process for the production of starch from corn or wheat. Waste products from these processes comprising, e.g., corn gluten are sold as animal feed. Addition of phytase to the steeping process will result in waste products containing phosphate instead of phytate.

In soy processing, phytase may be used so as to reduce the high levels of phytate that otherwise act as an anti-nutritional factor which renders the protein source unsuitable for applications such as baby food and feed for fish, calves and other non-ruminants (VAN GORCOM et al., 1991).

8 Conclusions

Enzymes are used in an increasing number of technical applications. Since the conditions in these various applications are very different, enzymes of the same functionality may

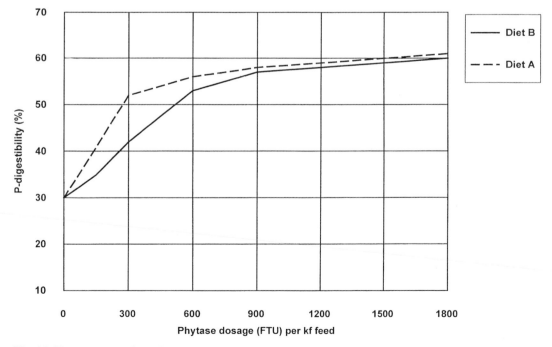

Fig. 14. The response of varying phytase dosages on the P-digestibility coefficient of two starter pig diets (data taken from DIJCK and GEERSE, 1994).

Tab. 5. Application of Recombinant Industrial Enzymes in Different Market Segments

| Enzyme | | Non-Food-Applications | | | | | Food Applications | | | |
Function	EC-Number	Deter-gents	Leath-er	Tex-tile	Paper	Feed	Dairy	Bakery	Starch	Bever-ages
α-Amylase	3.2.1.1	•		•	•	•		•	•	•
Catalase	1.11.1.6			•			•			
Cellulase	3.2.1.4	•		•		•				
Chymosin	3.4.23.4						•			
β-Glucanase	3.2.1.6					•			•	•
Glucose isomerase	5.3.1.5								•	•
Glucose oxidase	1.1.3.4						•	•		
Hemicellulase	div.							•		
Lipase	3.1.1.3	•	•		•		•	•		
Phytase	3.1.3.8					•			•	
Protease	3.4.2X.X	•	•	•	•	•	•	•	•	•
Pullulanase	3.2.1.41							•	•	•
Xylanase	3.2.1.8			•	•			•	•	•

be required but with different optima in order to match the operational conditions.

In the past 15 years, recombinant technology has permitted the industry to develop and produce enzymes for various conditions and applications. Most of these developments would not have been possible without recombinant technology: either the high costs of enzymes would prohibit their development and use, or the properties of the enzymes were not suited to survive the conditions of the application. Recombinant technology has made a considerable contribution to the solution of these problems: first, overexpression of the genes coding for the enzymes leads to higher production yields and lower costs and second, protein engineering allows the adaptation of the enzymes to conditions of their use.

An interesting observation can be made from Tab. 5. The majority of technical enzymes consists of hydrolyzing enzymes. Only a few exceptions such as catalase, glucose oxidase and glucose isomerase do not belong to the group of hydrolases. It is thus obvious that industry is using enzymes mostly for degradation purposes and the degradable materials are in almost all cases natural polymers like starch or proteins. The reason for this might be the prejudice that enzymes as natural products are only capable of operating under natural conditions on natural substrates. We have, however, shown in this chapter that

enzymes are used at temperatures even above the boiling point of water, at extreme pH values and that they can act in chemical syntheses in organic solvents on non-natural substrates.

Taking into account that enzymes can be manufactured today in all quantities at a competitive price and can also be modified to match the majority of technical applications, it can be predicted that enzymes will become available for many more applications than currently used.

9 References

AASLYNG, D., GORMSEN, E., MALMOS, H. (1991), Mechanistic studies of proteases and lipases for the detergent industry, *J. Chem. Tech. Biotechnol.* **50**, 321–330.

AEHLE, W., ARFMAN, N., KOENHEN, E., SCHOLTE, M., LAAN VAN DER, J. M. (1995a), Improvement of a *Bacillus* α-amylase for industrial applications by protein engineering, *Am. Oil Chem. Soc. 86th Annual Meeting*, San Antonio.

AEHLE, W., GERRITSE, G., LENTING, H. B. M. (1995b), *PCT Patent Publication* WO 95/30744.

BOTT, R. (1997), Development of new proteases for detergents, in: *Enzymes in Detergency* (EE VAN, J. H., BAAS, E.-J., MISSET, O., Eds.), pp. 75–91. New York: Marcel Dekker.

BRADY, L., BRZOZOWSKI, A. M., DEREWENDA, Z. S., DODSON, E. et al. (1990), A serine protease triad forms the catalytic centre of a triacylglycerol lipase, *Nature* **343**, 767–770.

BROWN, R. J. (1993), Dairy products, in: *Enzymes in Food Processing* (NAGODAWITHANA, T., REED, G., Eds.), pp. 347–357. New York: Academic Press.

BRZOZOWSKI, A. M., DEREWENDA, U., DEREWENDA, Z. S., DODSON, G. G. et al. (1991), A model for interfacial activation in lipases from the structure of a fungal lipase-inhibitor complex, *Nature* **351**, 491–494.

CHIANG, J. P., ALTER, J. E., STERNBERG, M. (1979), Purification and characterization of a thermostable alpha-amylase from *Bacillus licheniformis, Starch/Stärke* **31**, 86–92.

COWAN, W. D. (1996), Animal feed, in: *Industrial Enzymology* (GOODFREY, T., WEST, S., Eds.), 2nd Edn., pp. 69–86. New York: Stockton Press.

CYGLER, M., GROCHULSKY, P., SCHRAG, J. D. (1995), Structural determinants defining common stereoselectivity of lipases towards secondary alcohols, *Can. J. Microbiol.* **41** (Suppl. 1), 289–296.

DIJCK VAN, P. W. M., GEERSE, C. (1994), Specific examples of enzyme products for the compound feed industry, in: *Enzymes: Tools for Success,* pp. 6.1–6.26. Delft: Gist-Brocades.

DRIESSEN, M. (1994), Enzymes essential for the dairy industry, in: *Enzymes: Tools for Success,* pp. 4.1–4.18. Delft: Gist-Brocades.

EGMOND, M. R. (1997), Application of proteases in detergents, in: *Enzymes in Detergency* (EE VAN, J. H., BAAS, E.-J., MISSET, O., Eds.), pp. 61–74. New York: Marcel Dekker.

GODFREY, T. (1996), Textiles, in: *Industrial Enzymology* (GODFREY, T., West, S., Eds.), 2nd Edn., pp. 359–371. New York: Stockton Press, London: Macmillan Press Ltd.

GRAF, E. (1986), Chemistry and application of phytic acid: an overview, in: *Phytic Acid Chemistry and Applications* (GRAF, E., Ed.). Minneapolis: Pilatus Press.

HARDINGSVELDT VAN, W., ZEIJL VAN, C. M., HARTEVELD, G. M., GOUKA, R. J., SUYKERBUYK, M. et al. (1993), Cloning, characterization and overexpression of the phytase-encoding gene (*phy*A) of *Aspergillus niger, Gene* **127**, 87–94.

HOSHINO, E., ITO, S. (1997), Application of alkaline cellulases that contribute to soil removal in detergents, in: *Enzymes in Detergency* (EE VAN, J. H., BAAS, E.-J., MISSET, O., Eds.), pp. 149–175. New York: Marcel Dekker.

HOUSTON, J. H. (1997), Detergent enzymes market, in: *Enzymes in Detergency* (EE VAN, J. H.,

BAAS, E.-J., MISSET, O., Eds.), pp. 11–22. New York: Marcel Dekker.

JENSEN, G., THELLERSEN, M. (1993), Enzyme für Maschinen-Geschirrspülmittel, *SÖFW-J.* **119**, 37–41.

JONGBLOED, A. W., MROZ, Z., KEMME, P. A. (1992), The effect of supplementary *Aspergillus niger* phytase in diets for pigs on concentration and apparent digestibility of dry-matter, total phosphorus, and phytic acid in different sections of the alimentary tract, *J. Anim. Sci.* **70**, 1159–1168.

KAZLAUSKAS, R. J. (1994), Elucidating structure-mechanism relationships in lipases: prospects for predicting and engineering catalytic properties, *TIBTECH* **12**, 464–472.

KOTTWITZ, B., UPADEK, H. (1997), Application of cellulases that contribute to color revival and softening in detergents, in: *Enzymes in Detergency* (EE VAN, J. H., BAAS, E.-J., MISSET, O., Eds.). pp. 133–148. New York: Marcel Dekker.

KOTTWITZ, B., UPADEK, H., CARRER, G. (1994), Application and benefits of enzymes in detergents, *Chim. Oggi* **12**, 21–24.

MACHIUS, M., WIEGAND, G., HUBER, R. (1995), Crystal structure of calcium-depleted *Bacillus licheniformis* α-amylase at 2.2 Å resolution, *J. Mol. Biol.* **246**, 545–559.

MAGG, H., WÄSCHENBACH, G. (1997), Application of enzymes in automatic dishwashing detergents, in: *Enzymes in Detergency* (EE, VAN, J. H., BAAS, E. J., MISSET, O., Eds.), pp. 231–249. New York: Marcel Dekker.

MAURER, K.-H. (1997), Development of new cellulases, in: *Enzymes in Detergency* (EE VAN, J. H., BAAS, E.-J., MISSET, O., Eds.), pp. 175–202. New York: Marcel Dekker.

MISSET, O. (1997), Development of new lipases, in: *Enzymes in Detergency* (EE VAN, J. H., BAAS, E. J., MISSET, O., Eds.). pp. 107–131. New York: Marcel Dekker.

OKKELS, J. S., SVENDSEN, A., PATKAR, S. A., BORCH, K. (1996), Protein engineering of microbial lipases with industrial interest, in: *Engineering of/with Lipases* (MALCATA, F. X., Ed.), pp. 203–217. Dordrecht: Kluwer Academic Publishers.

OLLIS, D. L., CHEAH, E., CYGLER, M., DIJKSTRA, B., FROLOW, F. et al. (1992), The α/β-hydrolase fold, *Protein Eng.* **5**, 197–211.

SCHRAG, J. D., LI, Y., WU, S., CYGLER, M. (1991), Ser-His-Glu triad forms the catalytic site of the lipase from *Geotrichum candidum, Nature* **351**, 761–764.

SIEZEN, R. J., LEUNISSEN, J. A. M. (1997), Subtilases: the superfamily of subtilisin-like serine protease, *Protein Sci.* **6**, 501–523.

SIEZEN, R. J., VOS DE, W. M., LEUNISSEN, J. A. M,, DIJKSTRA, B. W. (1991), Homology modelling and protein engineering strategy of subtilases, the family of subtilisin-like proteinases, *Protein Eng.* **4**, 719–737.

SUOMINEN, P., MÄNTYLÄ, A., KARHUNEEN, T., PARKKINEN, E., NEVALAINEN, H. (1992), Genetic engineering of *Trichoderma reesei* to produce novel enzyme combinations for different applications, *3rd Nordic Conference on Protein Engineering*, Korpilampi.

SVENDSEN, A., BISGÅRD-FRANZEN, H. (1994), *PCT Patent Publication* WO 94/02597.

SVENDSEN, A., BISGÅRD-FRANTZEN, H., BORCHERT, T. (1996), *PCT Patent Publication* WO 96/23874.

SWIFT, H. J., BRADY, L., DEREWENDA, Z. S., DODSON, E. J., DODSON, G. G. et al. (1991), Structure and molecular model refinement of *Aspergillus oryzae* (TAKA) α-amylase: an application of the simulated-annealing method, *Acta Cryst.* **B47**, 535–544.

TEERI, T., KOIOULA, A., LINDER, M., REINIKAINEN, T., RUOHONEN, L. et al. (1995), Modes of action of two *Trichoderma reesei* cellobiohydrolases, *Carbohydr. Bioeng.* **10**, 211–224.

TIERNY, L., DANKO, S., DAUBERMAN, J., VAHA-VAHE, P., WINETZKY, D. (1995), Performance advantages of novel α-amylases in automatic dishwashing, *Am. Oil Chem. Soc. 86th Annual Meeting*, San Antonio.

TOLAN, J. S. (1996), Pulp and paper, in: *Industrial Enzymology* (GODFREY, T., WEST, S., Eds.), 2nd Edn., pp. 327–338. New York: Stockton Press.

ULLAH, A. H. J. (1988), Production, rapid purification and catalytic characterization of extracellular phytase from *Aspergillus ficuum*, *Prep-Biochem.* **18**, 459–471.

ULLAH, A. H. J., DISCHINGER, H. C. (1993), *Aspergillus ficuum* phytase: complete primary structure elucidation by chemical sequencing, *Biochem. Biophys. Res. Commun.* **192**, 747–753.

ULLAH, A. H. J., GIBSON, D. M. (1987), Extracellular phytase from *Aspergillus ficuum* NRRL3135: purification and characterisation, *Prep. Biochem.* **17**, 63–91.

ULLAH, A. H. J., PHILIPPY, B. Q. (1994), Substrate selectivity in *Aspergillus ficuum* phytase and acid phosphatases using *myo*-inositol phosphates, *J. Agric. Food Chem.* **42**, 423–425.

VAN DER LAAN, J. M., AEHLE, W. (1995), *PCT Patent Publication* WO 95/35382.

VAN EE, J. H., RIJSWIJK VAN, W. C., BOLLIER, M. (1992), Enzymatic automated dishwash detergents, *Chim. Oggi* **10**, 21–24.

VAN GORCOM, R. F., VAN HARTINGSVELDT, W., PARIDON, P. A., VEENSTRA, A. E., LUITEN, R. G. M., SELTEN, G. C. M. (1991), *Eur. Patent Application* 0 420 358 A1.

VIHINEN, M., MÄNTSÄLÄ, P. (1989), Microbial amylolytic enzymes, *Crit. Rev. Biochem. Mol. Biol.* **24**, 329–418.

WIGLEY, R. C. (1996), Cheese and whey, in: *Industrial Enzymology* (GODFREY, T., WEST, S., Eds.), 2nd Edn., pp. 135–153. New York: Stockton Press.

WODZINSKI, R. J., ULLAH, A. B. J. (1996), Phytase, *Adv. Appl. Microbiol.* **42**, 263–302.

WOLFF, A. M., SHOWELL, M. S. (1997), Application of lipases in detergents, in: *Enzymes in Detergency* (VAN EE, J. H., BAAS, E. J., MISSET, O., Eds.), pp. 93–106. New York: Marcel Dekker.

Monoclonal Antibodies

7 Antibody Engineering and Expression

JOHN R. ADAIR

Cambridge, UK

1 Introduction   221
2 Antibody-Binding Sites   221
  2.1 Antigen-Binding Sites – Structures and Generation   221
  2.2 Sources of Antibody-Binding Sites   226
    2.2.1 Murine and Humanized Binding Sites   226
      2.2.1.1 Antibody Humanization   227
    2.2.2 Primatized Antibodies   228
    2.2.3 Human mAbs   228
      2.2.3.1 Transgenic Animals Producing Human Antibody Repertoires   228
      2.2.3.2 *In vitro* Production of Human mAbs   229
    2.2.4 Low Molecular Weight Antibody-Based Binding Sites   229
  2.3 Binding Site Formats   229
  2.4 Number of Binding Sites per Product   230
3 Effector Mechanism(s)   230
  3.1 Second Binding Site (Bispecific Antibodies)   231
  3.2 Antibody Fc Functions (Immune System Recruitment)   231
  3.3 Antibody–Effector Fusions and Conjugates   232
    3.3.1 Cytokines   232
    3.3.2 Enzymes   232
      3.3.2.1 Immunotoxins   232
      3.3.2.2 ADEPT/ADAPT   233
      3.3.2.3 Thrombolytic and Coagulation Enzymes   233
    3.3.3 Drugs   233
    3.3.4 Radioisotopes   234
    3.3.5 DNA (Antibody Targeted Gene Therapy)   234
4 Linking the Binding Site to the Effector Mechanism   234
5 Manipulating the Half-Life of Antibody Products   234
6 Production of Recombinant Antibodies   235

6.1 Mammalian Cell Culture   236
  6.1.1 Antibody Glycosylation and Mammalian Cell Expression   236
6.2 *E. coli* Fermentation   236
6.3 Alternative Production Systems   236
  6.3.1 Transgenic Animals   237
  6.3.2 Transgenic Plants   237
  6.3.3 Insect Cells   237
  6.3.4 GRAS (Generally Regarded as Safe) Organisms   237
7 Conclusions   237
8 References   238

List of Abbreviations

ADAPT	antibody-dependent abzyme prodrug therapy
ADCC	antibody-dependent cellular cytotoxicity
ADCMC	antibody-dependent complement-mediated cytolysis
ADEPT	antibody directed enzyme prodrug therapy
CDR	complementarity determining region
CEA	carcinoembryonic antigen
CpG2	carboxypeptidase G2
DT	diphtheria toxin
Fab	fragment antigen binding
Fc	fragment crystallizable
GRAS	generally regarded as safe
HAMA	human anti-murine antibody
Mbp	megabase pairs
PCR	polymerase chain reaction
PE	*Pseudomonas* exotoxin
RES	reticuloendothelial system
RSV	respiratory syncytical virus
scFv	single chain Fv
TNF	tumor necrosis factor

1 Introduction

Antibodies have many properties which make them attractive as the starting point for novel biopharmaceutical products. They occur naturally in serum and mucosal surfaces at high concentrations and have long intrinsic half-lives. They combine high affinity and specificity for ligand (antigen) with innate effector elements. These functions are located in separate protein domains within the antibody which can be isolated by genetic engineering techniques and can be rearranged into novel combinations with other non-antibody elements. These novel molecules can be produced in a range of expression systems.

There are now several licensed monoclonal antibody (mAb) therapeutics and diagnostic products (Tab. 1) and a much larger number of murine, human and recombinant antibody candidates in research, preclinical and clinical development phases. It is anticipated that the rate of product licensing will continue to increase.

This chapter summarizes recent progress in the development of antibody-based products. The product concept and targets for antibody-based products are outlined and basic antibody structure, and the underlying genetic organization which allows easy antibody gene manipulation, and the isolation of novel antibody binding sites are then described. Features considered in the design and construction of antibody-based products are then summarized and finally, recent developments in options for production are noted.

2 Antibody-Binding Sites

Recombinant DNA technologies allow a modular approach to the design of antibody-based products. The product can be viewed as comprising one or more binding or targeting domains, one or more effector domains or elements, and suitable linking element(s), as shown schematically in Fig. 1. The composition and derivation of these elements are discussed in more detail later. Basic antibody protein structure is summarized in Fig. 2.

This section summarizes how the antibody-binding site structure, which provides the affinity and specificity suitable for its role as the binding and targeting domain, is generated.

2.1 Antigen-Binding Sites – Structures and Generation

The binding site is formed by non-covalent association of the N-terminal "variable" domains of the heavy and light chains (V_H and V_L) respectively (see Fig. 2). Structural studies of antibodies complexes with antigen (WILSON and STANFIELD, 1994; MACCALLUM et al., 1996) show that the antigen-binding site is composed from the solvent exposed side chains of amino acids in 6 loops, 3 from V_H and 3 from V_L, which emerge from the underlying β-sandwich frameworks of the

Tab. 1. Monoclonal Antibody-Based Products

Disease Area	Licensed mAb Product or Development Stage	Product Type	Date of First Product Launch
Transplantation	Orthoclone-OKT®3 (muromonab-CD3)	murine mAb	1986
	Zenapax™ (daclizumab)	humanized mAb	1997
	Simulect® (basiliximab)	chimeric mAb	1998
Cancer	Panorex® (edrecolomab)	murine mAb	1995
	Rituxan™ (MabThera® (rituximab)	chimeric mAb	1997
	Herceptin® (trastuzumab)	humanized mAb	1998
	Bexxar™ (^{131}I Therapeutic) (tositumomab)	murine mAb	application late 1998
	OncoScint® CR/OV (^{111}In Diagnostic) (Satumomab Pendetide)	murine mAb	1992
	ProstaScint® (^{111}In Diagnostic) (Capromab Pendetide)	murine mAb	1996
	CEA-Scan® (^{99m}Tc Diagnostic) (arcitumomab)	murine Fab	1996
	Verluma™ (^{99m}Tc Diagnostic) (Nofetumomab Merpentan)	murine Fab	1996
Cardiovascular	MyoScint® (^{111}In Diagnostic) (Imciromab Pentetate)	murine mAb	1989
	ReoPro® (abciximab)	chimeric mouse-human Fab	1994
Infectious disease	Synagis™ (palivizumab)	humanized mAb	1998
Auto-immune disease	Remicade™ (infliximab)	chimeric mAb	1998
Allergy/asthma	phase III (RhuMab E25)	humanized mAb	1998

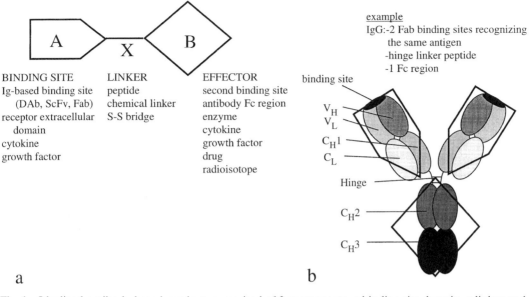

Fig. 1a. Idealized antibody-based product, comprised of 3 components, a binding site domain, a linker and an effector domain. These components may be derived from a variety of sources as noted in the figure and described in the text, **b** IgG antibody product, comprising 2 identical binding sites, an Ig hinge linker and one Fc region.

variable domains (Fig. 3). Of these loops, H3, which is most variable in sequence and length, and L3, occupy the center of the binding site and tend to make most of the key contacts with antigen. These 6 loop regions are substantially, but not exactly, similar in extent to the previously defined complementarity determining regions (CDRs) (WU and KABAT, 1970). The CDR were initially identified as regions within the variable domains which exhibited even greater between antibody sequence variability (hypervariability).

For 5 out of the 6 loops, but not yet H3, it has been possible to identify a small number of canonical conformations that each loop can adopt depending on sequence, length and the identity of key framework residues which contact the loops (CHOTHIA et al., 1989; CHOTHIA and LESK, 1987; TRAMONTANO et al., 1990). Using this information predictions can be made about the conformation for a newly derived antibody sequence. Some estimate of the likely structure of the binding site, from sequence analysis or calculation or structure determination, is important for

many antibody engineering projects (e.g., humanization or affinity modification).

The sequences of H3 and L3 are generated during the process of genomic recombination and somatic mutation which occurs during B cell development. The coding sequence of H3 is formed when one member from each of 3 genetic segments, the V_H minigenes, D_H (diversity) segments and J_H (joining) segments, is brought together during the maturation of the Ig heavy chain. The coding sequence for L3 is formed by recombination of one each from a set of V_L minigenes and J_L segments during the formation of the light chain. H1 and H2, and L1 and L2 are derived from the coding sequence in the V_H and V_L minigenes, respectively, but can also be altered later during the somatic mutation process (WABL and STEINBERG, 1996).

The number of these sequences in humans, primates and rodents, their organization into sequence subfamilies and their relative utilization in the immune response is of interest to antibody engineering projects. For example, the human immunoglobulin loci have

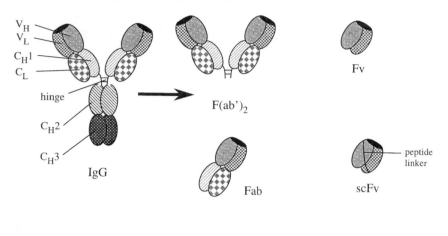

V_H
V_L
C_H1
C_L

hinge

C_H2

C_H3

IgG

F(ab')$_2$

Fab

Fv

scFv

peptide linker

a   b

Fig. 2a. Antibody structure. The basic immunoglobulin unit, as revealed by electron microscopy and X-ray crystallography experiments, adopts a Y or T shaped form. The binding site "arms" (termed Fab, for fragment antigen binding) can be isolated by digestion of IgG with enzymes. Papain digestion releases individual Fab fragments by cleavage on the N-terminal side of the interheavy chain disulfide bridges. Fab comprises the light chain and the Fd fragment, the 2 N-terminal domains of the heavy chain. The non-covalent interactions between Fd and light chain lead to a strong association ($k_D > 10^{10}$ M^{-1}), although the affinity of the individual interacting domains is much less (k_D approx. 10^{6-7} M^{-1}) (e.g., HORNE et al., 1982 and references therein). Pepsin digestion releases the F(ab')$_2$ fragment by digestion on the C-terminal side of the interheavy chain disulfide bridges. F(ab')$_2$, therefore, contains both light chains and 2 Fd hinge heavy chain fragments linked by disulfide bridges. The Fd hinge heavy chain fragment is known as the Fd'. The N-terminal variable domains together are termed Fv (fragment variable). In a small number of cases the Fv can be isolated from IgG with pepsin (GIVOL, 1991). The associating surface between the 2 domains of Fv derives from CDR and framework sequences and, therefore, the stability of Fvs varies from antibody to antibody. The other major fragment of IgG formed after pepsin digestion is termed the Fc (fragment crystallizable). The Fc contains the domains which interact with immunoglobulin receptors on cells and with the initial elements of the complement cascade. Pepsin sometimes also cleaves before the third constant domain (C_H3) of the heavy chain to give a large fragment F(abc) and a small fragment pFc'. These terms are also used for analogous regions of the other immunoglobulins. **b** Antibody fragments which derive from IgG and which can be produced by means of genetic manipulation are shown. The scFv comprises the 2 domains in Fv genetically linked with a peptide which links the C-terminus of one domain to the N-terminus of the other linker.

been substantially cloned and mapped. There are believed to be approximately 95 V_H segments, although only about half of these would be able to contribute to forming functional antibody; 30 (approx.) D_H and 6 J_H sequences over a 1.1 megabase pairs (Mbp) region of chromosome 14q (COOK and TOMLINSON, 1995; MATSUDA and HONJO, 1996). There are 2 light chain loci κ and λ. At the κ locus there are 76 Vκ, of which 32 are potentially functional, and 5 Jκ followed by a single Cκ gene in 2 Mbp of chromosome 2q (ZACHAU, 1993) (Fig. 4). The λ locus extends

over 1.15 Mbp on chromosome 22q (WILLIAMS et al., 1996). There are 52 (approx.) Vλ, an undefined number of these are pseudogenes. The Jλ and Cλ are linked and duplicated. There are 7 Jλ-Cλ miniloci although 3 are pseudogenes. V-J recombination, therefore, determines which Cλ will be used.

Heavy chain recombination occurs before light chain reorganization and the heavy variable domain appears on the surface of pre-B cells as membrane bound μH in association with a surrogate light chain. Later in B cell development κ locus rearrangement occurs,

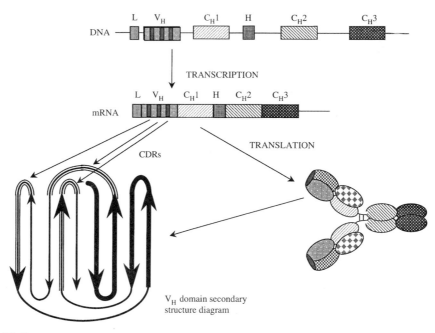

Fig. 3. V_H domain of IgGκ.

and if no productive κ chain is made at either of the κ alleles then λ rearrangement occurs until a functional light chain is formed (Fig. 5). A functional light chain then appears on the B cell surface with a heavy chain. Exposure to antigen and selection for B cells expressing binding sites with reasonable affinity for antigen then occurs.

Within a species the V minigene sequences can be grouped into a number of families according to amino acid or nucleotide sequence homology. In some cases interspecies homology between these families appear higher than intraspecies homology. The interspecies homologies can range from <40% to >80% between mouse and man, and are very high between primates and man. The sequence differences are sufficient to cause murine antibodies to be recognized as foreign by humans.

2.2 Sources of Antibody-Binding Sites

A number of sources of suitable antibody-binding sites is available. These are:

1. murine mAbs and humanized binding sites
2. primatized antibodies
3. human mAbs
   - transgenic animals producing human antibody repertoires
   - *in vitro* production of human mAbs
4. lower molecular weight antibody-based binding sites

2.2.1 Murine and Humanized Binding Sites

The variable domain coding sequences can be obtained from among the many thousands of murine mAbs which have been generated since 1975, including many with clinical applications. Methods for production of hybridomas are reviewed by DONOHUE et al. (1995).

Murine mAbs may provide binding sites which are suitable for single use, *in vivo* diagnostic or therapeutic purposes. However, in most but not all cases, sequence differences between murine and human antibodies lead

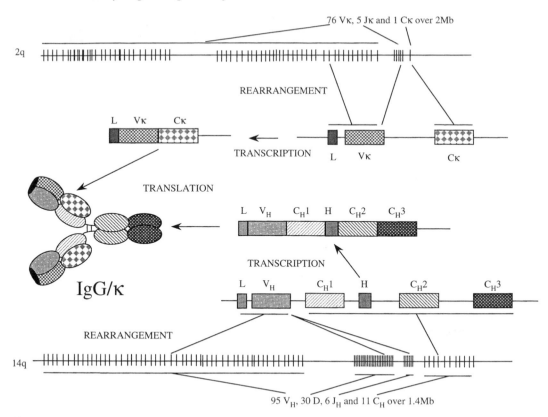

Fig. 4. IgGκ locus rearrangement.

to an immune response (human anti-murine antibody, HAMA) when murine mAbs are used as diagnostics or therapeutics, even in immunocompromised patients. This HAMA response is against both variable and constant regions so constructing a chimeric antibody (MORRISON et al., 1984; NEUBERGER et al., 1985; LOBUGLIO et al., 1989) by attaching the constant regions of a human antibody to the binding site domains of a murine antibody does not always remove the problem. Clinical data with chimeric antibodies are reviewed by SALEH et al. (1994). The HAMA response can, in principle, be avoided by various strategies, including coadministration of tolerizing anti-CD4 antibodies (MATHIESON et al., 1990), conjugation with polyethylene glycol ("pegylation") (NUCCI et al., 1991; INADA et al., 1995) or transferring the binding site from the murine mAb and transplanting it to a human mAb (humanization).

2.2.1.1 Antibody Humanization

Antibody humanization involves the substitution of sufficient residues from the variable domains of a non-human mAb into the variable domains from a human mAb so as to reconstitute in the human mAb the binding affinity and specificity of the non-human mAb. This notion was first raised at around that time chimeric antibodies were initially being described (MUNRO, 1984) because it was anticipated that chimeric antibodies would still be immunogenic in man due to their foreign variable domains. When humanization was suggested many of the structural features of antibody variable domains discussed in Sect. 2.1 above were not yet known, and the relative importance of the framework residues, including sequences introduced during somatic mutation, in determining binding site affinity and specificity was unknown. It was also

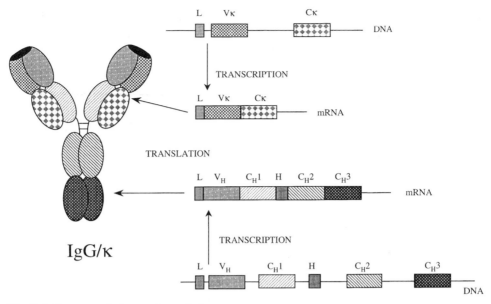

Fig. 5. IgGκ transcription and translation.

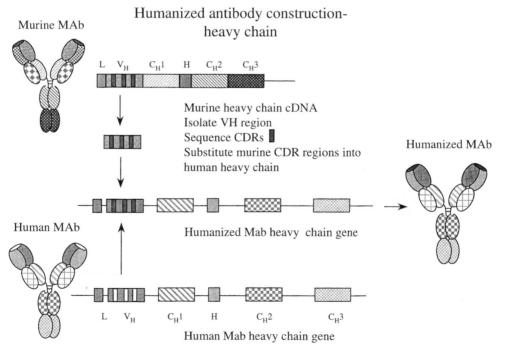

Fig. 6. Humanized antibody construction – heavy chain.

not clear to what extent any conformational adjustments of the CDRs after interaction with antigen, or solvent effects, contributed to net binding affinity, or, more importantly, how to take these factor into account during a humanization process (Fig. 6).

Nevertheless, the first successful experiments involving simple CDR transfers from a mouse mAb to a human mAb were soon described (JONES et al., 1986; VERHOEYEN et al., 1988) and were rapidly followed by the first humanization of a clinically relevant mAb, CAMPATH-1 (CD52), (RIECHMANN et al., 1988) The latter subsequently entered clinical investigation (HALE et al., 1988).

These and many other humanization experiments confirmed that certain murine framework residues were indeed necessary for restitution of significant binding activity. This has led to the development of a number of processes of humanization and descriptions of humanized mAb products (MOUNTAIN and ADAIR, 1992). Each in its own way seeks to identify those key amino acids (mainly buried within each variable domain) which interact with or affect, directly or indirectly, the conformation of the binding site. The success of these procedures can be measured by the fact that there are now well over a hundred disclosed examples of humanized mAbs, with a considerable number in clinical evaluation (ADAIR and BRIGHT, 1995).

2.2.2 Primatized Antibodies

As an alternative to murine systems, cynomolgus macaques have been used to develop mAbs to human antigens (NEWMAN et al., 1992). These primates are sufficiently related to humans to generate antibodies very similar in sequence and structure to humans, but are distinct enough to recognize many human antigens as foreign. Hybridomas can be generated and the binding site coding sequences extracted. First examples of their use have been macaque variable domain–human constant domain chimeric antibodies. One of these "primatized" antibodies, a CD4, chimeric mAb, is currently in human phase II clinical studies for rheumatoid arthritis, and a

number of others are in preclinical development.

2.2.3 Human mAbs

The human immune system is a further source of useful binding sites. Volunteers or vaccinees immunized with particular foreign antigens, or patients convalescing after an infection, or cancer patients can be the source of B cells from which human hybridomas expressing human mAbs to a range of antigens can be prepared (JAMES, 1994). A number of human antiviral and antitumor IgG mAbs from human hybridomas has been tested in the clinic. Alternatively the coding sequences for the antibody binding sites can be extracted for later genetic manipulation (LARRICK et al., 1992).

2.2.3.1 Transgenic Animals Producing Human Antibody Repertoires

An alternative approach to human mAb production is to transplant sufficient of the human Ig loci into the genome of another species and then to develop hybridomas after hyperimmunization with the antigen of interest. Mice are obvious candidates because of the established hybridoma technology. LONBERG (1994) and BRUGGEMANN and NEUBERGER (1996) and BRUGGEMANN and TAUSSIG (1997) have recently reviewed developments in the production of mAbs from transgenic mice. High affinity human mAbs with therapeutic potential have successfully been developed by this approach (FISHWILD et al., 1996; MENDEZ et al., 1997).

2.2.3.2 *In vitro* Production of Human mAbs

There have been many attempts to derive B cells expressing specific antibodies by immunizing *in vitro* which have met with little success. Primary or secondary *in vitro* immunization could be achieved but in each case an *in vivo* step was required. Recently, however, DUENAS et al. (1996) have described an *in vitro* system for generating B cells producing high affinity IgG.

Alternatively, it is possible to prepare large libraries of human antibody binding sites and present these to antigen *in vitro* and this approach has been thoroughly reviewed (WINTER and MILSTEIN, 1991; GALLOP et al., 1994; BURTON and BARBAS, 1994; WINTER et al., 1994; HOOGENBOOM, 1997b).

Human V_H and V_L can be isolated from B cell populations of "naive" (non-immunized) individuals, or from immunized or convalescent individuals, or can be prepared *in vitro* from germline V, D and J segments by using PCR technology. *In vitro* "affinity maturation" can be achieved (LOW et al., 1996). Very large libraries have been prepared (GRIFFITHS et al., 1994), and these libraries can be screened against a variety of antigens, including self-antigens. Several high affinity antibodies, some of potential clinical interest (e.g., anti-tumor necrosis factor α) have been developed from such libraries.

2.2.4 Low Molecular Weight Antibody-Based Binding Sites

Smaller binding sites based on antibody domains can be prepared which brings the binding site element closer to the realms of the pharmaceutical organic chemist (see, e.g., CUNNINGHAM and WELLS (1997) for a review of how protein engineering can reduce binding domain size to the scope of organic synthesis). WARD et al. (1989) showed that single domains of antibodies can retain significant binding affinity and HAMERS-CASTERMAN et al. (1993) have demonstrated that single domain binding sites occur naturally. These sequences provided the rationale for improving the solubility properties of human or mouse single domain binding sites (DAVIES and RIECHMANN, 1995). SIMON and RAJEWSKY (1992) showed that the binding site surface can be increased by incorporating an adjacent non-CDR loop into the binding site of a single domain. PESSI et al. (1993) have

developed a small V_H-based binding domain, which comprises 2 loops on a supporting β-sandwich miniframework. Potentially useful low molecular weight binding sites have already been isolated by this approach (MARTIN et al., 1994, 1996).

2.3 Binding Site Formats

To be useful the binding site must be relatively stable and, for most purposes, must be capable of being linked to an effector element. Currently the most useful and generally applicable formats for high affinity antibody binding sites are single chain Fv (scFv) and Fab (see Fig. 2b). The coding sequences of the binding sites are usually isolated by recombinant DNA procedures using oligonucleotide primers based on known and/or likely antibody sequences and usually, by some form of polymerase chain reaction (PCR) (SAIKI et al., 1985; JONES and BENDIG, 1991).

scFv are formed by joining the coding sequences for the 2 variable domains with a sequence coding for a linker peptide which may be of variable sequence and length (HUSTON et al., 1993). This linkage maintains the association of the domains even at the low concentrations that would be observed during *in vivo* dosing.

Fabs are stabilized by the extent of the intermolecular surfaces between the Fd (V_H plus C_H1 domains) and light chain, also in some cases by the retention of the light–heavy chain disulfide bridge. To generate Fab the variable region DNAs are joined to coding sequences for the C_H1 or $C\kappa(\lambda)$ domains. The joining may incorporate part or all of the V-C intron, usually depending on what type of expression system will be used later.

2.4 Number of Binding Sites per Product

The finished product will contain one or more binding domains. How many are needed will be determined by the affinity of the single binding site for antigen or ligand, and by any need to cross-link the ligand after binding. If the single site binding affinity is sufficient, then products based on monovalent binding sites are feasible. An example of such is the licensed product ReoPro™, an anti-gpIIb/IIIa chimeric Fab used to prevent blood clot formation (WEISMAN et al., 1995). Others include radioisotope labelled scFv for tumor imaging (and therapy) (BEGENT et al., 1996), and scFv or Fab fused to toxins, cytokine or growth factors (reviewed by PIETERSZ and McKENZIE, 1995).

In some of these blocking situations a Fab or scFv may not be suitable because of their short *in vivo* half-life (see Sect. 5). Monovalent Fab can be attached to the Fc region of an IgG either by chemical conjugation (STEVENSON et al.. 1993) or by genetic fusion (ROUTLEDGE et al., 1991). These methods have the advantage that cross-linking of the ligand cannot occur, which is of benefit when binding to certain cellular receptors. Alternatively an IgG could be used, which has the combined advantage of having 2 binding sites per molecule and a long serum half-life. Several humanized IgG antibodies now in late stage clinical development act as blocking agents (STEPHENS et al., 1995; TAYLOR et al., 1995).

Increasing the number of binding sites from 1 to 2, as in IgG or $F(ab')_2$, markedly increases the binding strength (avidity) of the molecule, often by more than an order of magnitude. However, further increasing the number of binding sites does not lead to ever increasing avidity. KING et al. (1994, 1995) have shown that trimeric Fab' molecules, constructed by chemically cross-linking antitumor Fabs, have higher avidity than the $F(ab')_2$ but tetrameric molecules show little further improvement in avidity (KING et al., 1992). STEVENSON et al. (1993) and SCHOTT et al. (1993) provide other similar examples of the properties of multivalent cross-linked Fab, and a humanized $F(ab')_3$ has been developed for clinical trials (KING et al., 1995). Similar results are also seen for scFv, when modified to include antibody hinge sequences at the C-terminus and then cross-linked to the dimeric and trimeric forms, $scFv_2$ and $scFv_3$ (KING et al., 1992, 1994).

In the examples noted above with 2 or more binding sites, the domains were linked either by disulfide bridges or by chemical cross-linking. Other options are possible, including replacing the hinge sequences with other associating sequences, e.g., *fos/jun* or amphipathic helical peptide sequences (PACK and PLÜCKTHUN, 1992). In addition, it is possible to genetically fuse the binding domains to form $scFv_2$ (MALLENDER and VOSS, 1994; JOST et al., 1996) or Fab' [termed $L-F(ab')_2$] (ZAPATA et al., 1995). Recent developments in producing multivalent products have been summarized by HOGGENBOOM (1997a).

3 Effector Mechanism(s)

In the preceding section a number of examples was noted where the product acts as a simple blocking agent and the binding site itself acts as the effector. A key point is that the blocking action should not lead to the functional consequences of the normal ligand–receptor interaction (e.g., receptor signaling). The product must be present at sufficient concentration and must persist for a sufficient time to ensure effective inhibition by maintaining a minimum threshold level during the critical therapeutic window. This is important in situations where transient interactions may allow the passing of a signal (e.g., activation, differentiation or homing) to a recipient cell. The therapeutic window may be short, e.g., for immunosuppression during or immediately after organ transplantation, or during a period of high risk of infection, but in other cases, e.g., chronic autoimmune diseases, continuous surveillance may be needed.

There are other opportunities for using the binding site as the effector. For example, a catalytic antibody, upon binding to its target, inactivates or alters it in an enzymatic process. JACOBSEN and SCHULTZ (1995) review recent progress in production of catalytic antibodies. Potential applications include detoxification after deliberate or inadvertant administration of poisons or drugs, e.g., to remove the remainder of a therapeutic low molecular weight drug after the desired effect is achieved.

Recently MAGLIANI et al. (1997) have described a scFv derived from an anti-idiotypic antibody which functionally resembles the original antigen, a killer toxin. The scFv exerts a cytocidal activity on target *Candida* cells.

Finally, antibody genes can be delivered intracellularly (to date only *in vitro*) and can inhibit various cellular functions, including viral replication and protein assembly and secretion, by blockade. The latter has implications for autocrine stimulation of tumor growth. Intracellular immunization is reviewed by BIOCCA and CATTANEO (1995), while HARRIS and LEMOINE (1996) have recently summarized progress with targeted gene therapy for various applications.

In most situations, however, the effector will be provided by attaching one or more other elements to the binding site domain(s) (e.g., NEUBERGER et al., 1984). The effector elements can be summarized as follows:

1. antigen binding site
   - blockade
   - catalytic antibodies
   - receptor stimulation
2. second antigen binding site (bispecific products)
3. antibody Fc domains (immune system recruitment)
4. cytokines (immune system recruitment)
5. enzymes
   - immunotoxins
   - ADEPT/ADAPT
   - thrombolytic and coagulation enzymes
6. superantigens (immune system recruitment)
7. drugs
8. radioisotopes
9. DNA (gene therapy)

The examples described in the following sections provide a brief rationale and describe recent developments.

3.1 Second Binding Site (Bispecific Antibodies)

Here the two different binding sites act as a bridge, e.g., to bring cells together by effec-

tively cross-linking receptors on their surfaces, or the second antigen binding site may act as a specific carrier for drugs or radioisotopes. For this approach to be sensible the single site binding affinity of each arm of the bispecific molecule must be quite high. Recent developments in the production of multi binding site complexes were noted earlier. Bispecific or multispecific entities of various sizes can be envisaged and several types have been constructed (summarized by HOOGENBOOM, 1997a). A number of examples of bispecific F(ab')$_2$, one arm of which binds to FcγR1 (CD64) on macrophages and other effector cells, are under clinical investigation as antitumor agents (VALONE et al., 1995).

3.2 Antibody Fc Functions (Immune System Recruitment)

The Fc regions of immunoglobulins are multifunctional and contribute not only to antibody stability and regulation of half-life, but also provide for interaction with various cells of the immune system and with complement. These latter attributes can be used as therapeutic effectors (summarized by CLARK, 1997).

These interactions can lead to phagocytosis, to antibody-dependent cellular cytotoxicity (ADCC) mediated by FcγRIII (CD16) NK cells, to IgE dependent mast cell and basophil degranulation and histamine release via FcϵR, and can also induce antibody-dependent complement-mediated cytolysis (ADCMC).

IgG$_1$ is the isotype often chosen in antibody engineering projects because of its ability to interact with all of the FcγR (and potentially to lead to ADCC) and with the initial components of complement cascade to initiate ADCMC. IgG$_1$ also has the advantage of a longer biological half-life compared to IgG$_3$ and IgM which are also effective, at least *in vitro*, at causing ADCC and ADCMC.

The rat CD52 mAb CAMPATH-1G and its humanized IgG$_1$ variant, CAMPATH-1H, are believed to act via ADCC in depleting CD52$^+$ cells (BRETT et al., 1996), while the chimeric anti-CD20 mAb Rituxan™ (IgG$_1$ isotype) (MALONEY et al., 1994) may act by a combination of ADCC and ADCMC to deplete CD20$^+$B cells in B cell lymphoma patients.

In some situations a long half-life antibody may be required which does not bind well to FcR or complement components, e.g., if recruitment of an immune cell will lead to activation of a target cell, or if uptake of an infectious organism into FcR$^+$ cells is to be avoided. This can be achieved using an "inert" isotype, e.g., IgG$_4$ which has low natural affinity for the FcγR and complement, and by further reducing the affinity of the binding motifs by point mutagenesis (ALEGRE et al., 1992). It is possible to generate novel isotypes which retain some but not all aspects of Fc function (DUNCAN et al., 1988; MORGAN et al., 1995). These novel antibodies are usually constructed by genetic fusions of the constant region exons to the desired variable region exons.

3.3 Antibody–Effector Fusions and Conjugates

Antibodies can also be used to target a non-antibody effector agent to specific tissues and cells. Examples of these include specific targeting of cytokines; enzymes, e.g., ADEPT/ADAPT systems (see Sects. 3.3.1–3.3.5) and immunotoxins; radioisotopes and drugs, and liposomes. Examples of the therapeutic applications of these potential products are described in Chapters 12–16 of this volume, as is the antibody targeting of gene therapy. This section serves as an introduction to these antibody products.

3.3.1 Cytokines

Specific targeting of a cytokine is of benefit when systemic administration would lead to toxic effects. Several antibody–cytokine fusions have been described (PIESTERSZ and McKENZIE, 1995). The cytokine will generally be used to recruit and stimulate specific effector lymphocytes. Increasingly the structures of cytokines are becoming available which should aid in more efficient design of genetic fusions. For example, the structural homologies between members of the hematopoietin family have been reviewed by WELLS and DE VOS (1996). No antibody–cytokine fusions are in clinical studies as yet.

3.3.2 Enzymes

Various enzymes can be considered as effector elements. Some require to be internalized and then act to disrupt cellular processes (e.g., the immunotoxins) while some require a sufficient concentration of antigen–antibody complex to remain externalized.

3.3.2.1 Immunotoxins

Several toxic proteins with enzymatic activity have been used as cytotoxic agents, particularly for cancer therapies (KREITMAN and PASTAN, 1997). The most extensively studied toxins, diphtheria toxin (DT), *Pseudomonas* exotoxin A (PE) and the plant toxin ricin, have distinct domains associated with cell-binding and entry, translocation to the cytoplasm, and catalysis. These toxins bind to a cell surface and are then internalized into endosomes. They translocate into the cytoplasm before exerting their cytotoxic effect, usually by interfering with aspects of protein synthesis. As they act catalytically only minute quantities are required to reach the cytoplasm for a cytocidal effect. The translocation and catalysis domains can be isolated and attached to an alternative binding site which confers a desired cell selectivity. Both chemical conjugation and genetic fusion have been used.

The toxin portion normally is immunogenic and it appears that the process of uptake, proteolysis and representation of toxin-derived peptides by antigen presenting cells avoids exposing their cytoplasm to the functional toxin sufficiently to lead to a strong immune response. Measures to reduce immunogenicity include pegylation, which also has the effect of extending the half-life of the toxin, concomitant immunosuppression regimes, or attempts to genetically remove immunodominant epitopes on the toxins.

Human enzymes may also be used if they can be delivered to the target cell and brought efficiently to the cytoplasm. Human serum ribonuclease A (angiogenin) cleaves cytoplasmic RNA and is very toxic when delivered by direct injection. Angiogenin has genetically been fused to transferrin and to Fab (RYBAK et al., 1992). Other enzymes which could act as cytosolic toxins include eosinophilic granular toxins (which are also ribonucleases), DNAses, and proteases (DEONARAIN and EPENETOS, 1994; GADINA et al., 1994).

3.3.2.2 ADEPT/ADAPT

In Antibody Directed Enzyme ProDrug Therapy (ADEPT) the binding site of the antibody–enzyme fusion recognizes a cell surface antigen which does not internalize after antibody-binding, or which does so very slowly. The chosen enzyme can cleave a protective side group from a specific chemical compound, thereby turning an inert proform into an active cytotoxic or cytostatic drug. Therefore, a high local, concentration of the active drug builds up at or near the binding site of the fusion, usually at a tumor (BENHAR and PASTAN, 1997).

Early studies focused on the use of bacterial enzymes, e.g., carboxypeptidase G2 (CpG2) and β-lactamase. Clinical responses have been seen when an anti-CEA–CpG2 conjugate was used to activate a benzoic acid mustard in colorectal cancer patients (summarized in BLAKEY et at., 1995). Recently, a human enzyme, β-glucuronidase, has now also been used, fused to the humanized Fab of an anti-carcinoembryonic antigen (CEA) antibody (BOSSLET et al., 1994).

An extension of the ADEPT principle is to use a catalytic antibody or an abzyme to cleave the prodrug (MIYASHITA et al., 1993). WENTWORTH et. al. (1996) discussed some of the desired catalytic features required for this ADAPT (antibody-dependent abzyme prodrug therapy) process.

3.3.2.3 Thrombolytic and Coagulation Enzymes

Fab fragments of antibodies which recognize fibrin but not fibrinogen have been fused or chemically conjugated to various forms of tissue type and urokinase type plasminogen activators (HABER, 1994; MURPHY, 1996; VERSTRAETE et al., 1995). This leads to a more specific activation of plasmin near to fibrin clots. Alternatively HUANG et al. (1997) have recently demonstrated in an *in vivo* model system that antibody–tissue factor hybrids can facilitate site-specific coagulation leading to occlusion of tumor blood vessels and reduction in tumor size.

3.3.3 Drugs

Several cytotoxic and cytostatic drugs have been chemically conjugated to mAbs to improve their selectivity for tumor cells over other rapidly dividing cells, thereby reducing dose-limiting toxicity and this has been reviewed by MOUNTAIN and ADAIR (1992) and YARRANTON (1997). A case study of the development of an antibody cytotoxic conjugate is described in Chapter 15, this volume.

Recently, antibody targeting has been combined with liposome drug delivery systems. Liposomes have been used as delivery systems for cytotoxic drugs for over 30 years (reviewed by GREGORIADIS, 1995). Their usefulness has been enhanced by attachment of polyethyleneglycol derivatives to form sterically stabilized ("stealth") liposomes (ALLEN, 1994). The antibody–liposome combination provides a long lasting method of delivering drug to target at a much higher ratio than in the case of linking drug directly to antibody. The Fab′ fragment of the humanized antibody Herceptin™ is being evaluated in this liposome format, to carry doxorubicin to breast tumors (PARK et al., 1995).

3.3.4 Radioisotopes

Radioisotopes are used not only as therapeutics, where the isotope has a cytotoxic effect, but also as diagnostics, where the location of the targeted isotope can be detected by means of external body imaging. The factors involved in the selection of isotopes for both imaging and therapy have been reviewed (ADAIR et al., 1993; GOLDENBERG et al., 1994; MONTAIN and ADAIR, 1992).

A number of ^{111}In and ^{99}mTc–mAb conjugates are now available as approved diagnostic products (Tab. 1), and RIGScan CR49 (^{125}I-labeled CC49 mAb), for use in radioimmuno guided surgery, is in phase II/III studies for detection of primary and metastatic colorectal cancers (ARNOLD et al., 1995).

Antibodies are also being assessed clinically for cancer therapy after conjugation with various β-emitters (GOLDENBERG et al., 1994). For example oncolym (^{131}I-labeled mAb Lym-1, an anti-MHC class II mAb) (DeNARDO et al., 1994), ^{131}I-labeled B1 (anti-CD20 mAb) (PRESS et al., 1995) and IDEC-Y2B8, (^{90}Y-labeled 2B8, an anti-CD20 mAb) (REFF et al., 1994) are mAbs in late stage clinical studies for treatment of non-Hodgkins lymphoma. Other mAbs, conjugated to ^{131}I, ^{90}Y, ^{177}Lu, ^{186}Re or ^{188}Re are being investigated in clinical studies for therapy of a wide variety of tumor types.

3.5 DNA (Antibody Targeted Gene) Therapy

Gene therapies require targeted gene expression. This can be achieved by a combination of targeting the vector or vehicle to the tissue or cells of interest and providing tissue-specific regulation of gene expression. Antibodies can provide the initial targeting element for viral, liposomal or DNA-based approaches, and these are being actively pursued (HARRIS and LEMOINE, 1995; PONCET et al., 1996; SCHNIERLE et al., 1996). To date there are no clinical studies reported using antibody targeted gene therapy.

4 Linking the Binding Site to the Effector Mechanism

For most of the protein-based effectors the linkage to the binding site can be achieved by direct genetic fusion using standard molecular biology procedures, an approach which provides a consistent expression product. During the early stages of assessment of a new product it may, however, prove more practical to chemically conjugate the binding sites to the effector, e.g., via disulfide bridges (STEVENSON et al., 1993).

The drug and radioisotope effector molecules require chemical coupling to the antibody binding domains. This coupling can be direct, often after modification of antibody side chains to derive reactive thiols, or by selective reduction of the hinge region, or indirectly, using a bifunctional linking agent. Often metallic radioisotopes are "caged" in macrocyclic chelators which are conjugated to the antibody surface (ADAIR et al., 1993).

5 Manipulating the Half-Life of Antibody Products

A key element in developing antibody products is the ability to predict and manipulate the pharmacokinetics, biodistribution and elimination of the product. The catabolic fate of naturally occurring immunoglobulin is not completely understood but involves proteolysis of serum immunoglobulin and also uptake by phagocytic cells of the reticuloendothelial system (RES).

It has been proposed that normally IgG is recognized by cellular receptors which bind to and protect Ig from proteolytic attack (BRAMBELL et al., 1964; summarized by RAVETCH, 1997 and references therein). This process may include recycling into and out of intracellular compartments, thereby physically partitioning the antibody away from proteolytic attack. According to this proposal antibody not bound by receptors is destined for catabolism by proteolysis. Breakdown products which are below the cut-off size for kidney filtration (<50 kDa) are excreted. Modifications to the Fc region which affect structure may perturb this interaction and allow increased levels of proteolysis. This view can guide design of antibody products to modulate biological half-life.

A general point which emerges from tests of antibody-based products in various animal models and in human clinical studies is that incorporation or retaining a native Fc region extends biological half-life. For example, generating chimeric or fully humanized antibodies from murine mAbs substantially extends half-life in humans compared to the murine mAb (ANASETTI et al., 1994; BROWN et al., 1991; DHAINAUT et al., 1995; HAKIMI et al., 1991; STEPHENS et al., 1995; TAYLOR et al., 1995). Murine mAbs tend to have half-lives of around 50 h in humans (reviewed by SALEH et al., 1994).

Conversely modifications which affect the structure of the Fc region reduce antibody half-life. These modifications include deletion of the C_H2 domain of IgG (MUELLER et al., 1990), making certain amino acid substitutions at the C_H2–C_H3 interface in IgG (KIM et al., 1994a, b, 1995) and modifications to, or removal of, the N-linked carbohydrate on the Fc region (WAWRZYNCZAK et al., 1992; WRIGHT and MORRISON, 1994). An effect on half-life of this latter modification is not universally accepted.

A very simple way to reduce biological half-life is to remove the Fc region entirely and generate antibody fragments, such as Fab or scFv (see Fig. 2b). For example, a human antitumor scFv, being developed as an imaging agent, has been shown to have an elimination half-life ($t_{1/2}\beta$) of 5.3 h (BEGENT et al., 1996). This is in good agreement with primate data for scFv showing half-lives of around 4 h (MILENIC et al., 1991).

Methods to promote or avoid uptake by cells of the RES, thereby influencing biological half-life, are also available. Antibodies modified by galactose conjugation are recognized by the asialoglycoprotein receptor in the liver, giving rapid clearance from the blood. A galactosylated anti-enzyme antibody has been used in a clinical study to achieve rapid clearance of a mAb–enzyme fusion (BAGSHAWE et al., 1992). In contrast, conjugation with polyethyleneglycol (PEDLEY et al., 1994) extends the half-life of an antibody, fragment or fusion, with the additional benefit of reduced immunogenicity. This appears to work by protecting the complex from uptake by cells of the RES.

Finally, the antigenic burden may affect half-life. A humanized IgG_1 antibody recognizing the CD33 differentiation marker showed a much longer $t_{1/2}\beta$ in primates, where the antigen is not present, than in humans (CARON et al., 1994). However, studies using a humanized IgG_1 anti-RSV antibody on healthy volunteers, where the target antigen is not present, showed a half-life of 23 d (TAYLOR et al., 1995). The antigen burden, and the use of doses which do not saturate all binding sites, may explain why many chimeric and humanized antibodies do not achieve the half-lives predicted from studies on bulk human IgG (WALDMANN and STROBER, 1969).

6 Production of Recombinant Antibodies

Features of large scale production of antibodies are dealt with in Chapter 8, this volume, but this section will outline the variety of expression systems available and highlight some recent developments.

Some options which exist for the expression of antibody-based products are:

- mammalian cells
- microorganisms (including GRAS)
- transgenic animals
- transgenic plants
- insect cells

Expression in mammalian cells and in the bacterium *E. coli* has been widely used for some time and these systems provide the means of production for all of the current antibody-based products (reviewed by MOUNTAIN and ADAIR, 1992; WARD and BEBBING-

TON, 1995; PLÜCKTHUN, 1994; WALL and PLÜCKTHUN, 1995; WEICKERT et al., 1996). The alternative systems each have their proponents and some may eventually be adopted for large scale production but these will have to overcome a large acceptance barrier to displace the "traditional" mammalian cells and *E. coli* routes.

6.1 Mammalian Cell Culture

Mammalian expression is an option for virtually all antibody products, exceptions being some immunotoxins.

Short-term, self-limiting transient expression systems have been described which are relatively low yielding (0.1–10 μg mL^{-1} with volumes generally not larger than 1L), but rapidly provide a means of producing analytical quantities of material. For production of larger amounts of material the antibody genes are usually stably integrated into the host cell genome. A key factor in cell line productivity is the status of the integration site, i.e., the chromatin organization around the integrated antibody genes must be conducive to high level and persistent transcription of the genes. Efforts to integrate the engineered antibody gene preferentially into such chromosomal loci by homologous recombination with flanking sequences, or to effect a reorganization of the chromatin for this purpose are underway (HOLLIS and MARK, 1995).

6.1.1 Antibody Glycosylation and Mammalian Cell Expression

Each of the heavy chains of an antibody bears a common core oligosaccharide at residue 297 in the C_H2 domain. However, the host cell type and growth conditions affect outer arm glycosylation, leading to a complex spectrum of isoforms in the IgG preparation. The consequences for antibody function of variable glycosylation have recently been reviewed (WRIGHT and MORRISON, 1997). The full significance of subtle differences in attached carbohydrate structure for pharmacokinetics and biodistribution in patients remains to be determined.

6.2 *E. coli* Fermentation

E. coli expression is an option for most of the non-glycosylated forms of antibody products, particularly for forms toxic to mammalian cells (e.g., immunotoxins). Assembly of multivalent (and/or multispecific) products using gene fusions [e.g., L(Fab')$_2$ (ZAPATA et al., 1995)] or self-assembling protein subdomains (PLÜCKTHUN, 1994) has recently extended the utility of *E. coli* expression systems.

Factors influencing protein folding, intracellular stability and export, including the role of chaperonins, are becoming reasonably well understood, if not yet all controllable or manipulable. Key factors in efficient large scale production are plasmid stability (including stability of the product genes in culture) and plasmid loss from cells in culture which are under environmental stress while producing the product. These are major concerns to the regulatory authorities.

Antibody products can be produced intracellularly or, more usually, by export to the periplasmic space. Expression at high levels tends to cause product to aggregate in the periplasm as an insoluble material, although in most cases the signal sequence is correctly removed. It is now known that solubility in the periplasm and the rate of correct folding of antibody variable domains are affected by the primary amino acid sequence (KNAPPICK and PLÜCKTHUN, 1995) and these can be manipulated to increase soluble product levels.

E. coli can be grown in very large volumes and at high cell density in simple media formulations and with tightly controlled transcriptional regulation. Under these conditions yields of antibody fragments can be in the range of gL^{-1}.

6.3 Alternative Production Systems

Alternatives to mammalian cells and *E. coli* are being developed as outlined in Sects. 6.3.1–6.3.2 and have recently been reviewed by PORTER et al. (1977).

6.3.1 Transgenic Animals

Since the first descriptions of the production of therapeutic proteins in the milk of transgenic animals (CARVER et al., 1993; ECHELARD, 1996; MAGA and MURRAY, 1995) there has been substantial interest and effort in using large mammals for the production of antibodies in milk. IgG production in goat milk at >4 g L^{-1} has been reported (DI-TULLIO et al., 1995). The long timescale of generation of the transgenic animal remains to be addressed, however, the recent demonstration of cloning of a sheep following transplantation of the nucleus from a cell of an adult animal to an enucleated oocyte may lead eventually to substantial numbers of genetically identical animals each producing large quantities of recombinant protein (WILMUT et al., 1997).

Glycosylation may be an issue and one study has suggested that interteron-γ produced in mouse milk differs from that produced by CHO cells (JAMES et al., 1995).

6.3.2 Transgenic Plants

Since the first description in 1989 a variety of antibodies and fragments have been produced in plant tissues (MA and HEIN, 1995). Recently it has been shown that scFv can survive intact in seeds (FIEDLER and CONRAD, 1995). This may reduce the need to harvest and then process the tissues immediately to avoid proteolysis. Plant-derived antibody is unlikely to have mammalian type glycosylation. This may affect the product by inducing immunogenicity, by reducing half-life due to uptake by mannose receptors on cells and by altering effector functions of the antibody. Plants may offer a viable alternative to the "standard" procedures above for production of deglycosylated antibodies and fragments.

6.3.3 Insect Cells

IgG, scFv and IgA have been produced by infection of insect cells (*Spodoptera frugiperda* and *Bombyx mori*) using recombinant expression vectors based on baculoviruses (CARAYANNOPOULOS et al., 1994; HASEMANN and CAPRA, 1990; NESBIT et al., 1992; REIS et al., 1992; ZU PULITZ et al., 1990). Expression levels of around 30 mg L^{-1} have been reported. There are concerns about the glycosylation status of insect cell-derived antibody. The glycosylation pattern seen from insect-derived material differs from mammalian cells, with mannose sugars being present. However, NESBIT et al. (1992) reported no difference in ADCC between a mammalian and insect-derived mAb. Recently, a clonal cell line of *S. frugiperda* has been described which produces glycoproteins with complex sugars similar to those of mammalian cells (OGONAH et at., 1996).

6.3.4 GRAS (Generally Regarded as Safe) Organisms

U.S. Department of Health and Human Services (1997)

Expression of antibody scFv at 5 mg L^{-1} has been reported in *Bacillus subtilis* (WU et al., 1993) and at >100 mg L^{-1} in the methylotrophic yeast *Pichia pastoris* (RIDDER et al., 1995).

Antibody Fab fragments have been expressed in the filamentous fungus, *Trichoderma reesei* (NYSSÖNEN et al., 1993). Yields increased from 1 to 40 mg L^{-1} when a Fab fragment was produced fused to the major cellulase, cellobiohydrolase I, compared to expression of the Fab alone.

Historically yeasts such as *Saccharomyces cerevisiae* have been candidate expression hosts for biotechnology products. Low level antibody expression in *S. cerevisiae* was demonstrated very early (WOOD et al., 1984) and was considered for some time (HORWITZ et al., 1988), but little real progress has been described recently.

7 Conclusions

Generation of antibody-based products and their testing in preclinical and clinical systems is accelerating. A recent survey by the Pharmaceutical Research and Manufacturers of

America described 284 biotechnology products in various states of clinical testing. Of these 80 are antibody-based products (the largest single product type) and a large proportion of these are in late stage studies. In the next year or so several of these will add to the currrently approved biotechnology products.

8 References

ADAIR, J. R., BRIGHT, S. (1995), Progress with humanized antibodied – an update, *Exp. Opin. Invest. Drugs* **4**, 863–870.

ADAIR, J. R., BAKER, T. S., BEBBINGTON, C. R., BEELEY, N. R. A., EMTAGE, J. S. et al. (1993), Engineering monoclonal antibody B72.3 for cancer therapy, in: *Protein Engineering of Antibody Molecules for Prophylactic and Therapeutic Applications in Man* (CLARK, M. R., Ed.), pp. 145–158. Nottingham: Academic Titles.

ALEGRE, M.-L., COLLINS, A.-M., PULITO, V. L., BROSIUS, R. A., OLSON, W. C. (1992), Effect of a single amino acid mutation on the activating and immonosuppressive properties of a "humanized" OKT3 monoclonal antibody, *J. Immunol.* **148**, 3461–3468.

ALLEN, T. M. (1994), Long-circulating (sterically stabilized) liposomes for targeted drug delivery, *Trends Pharmacol. Sci.* **15**, 215–220.

ANASETTI, C., HANSEN, J. A., WALDMANN, T. A., APPELBAUM, F. R., DAVIS, J. et al. (1994), Treatment of acute graft-versus-host disease with humanized anti-Tac: an antibody that binds to the interleukin-2 receptor, *Blood* **84**, 1320–1327.

ARNOLD, M. W., YOUNG, D. C., HITCHCOCK, C. L., SCHNEEBAUM, S., MARTIN, E. W., JR. (1995), Radioimmunoguided surgery in primary colorectal carcinoma: an intraoperative prognostic tool and adjuvant to traditional staging, *Am. J. Surg.* **170**, 315–318.

BAGSHAWE, K. D., SHARMA, S. K., SPRINGER, C. I., ANTINIW, P., ROGERS, G. T. et al. (1992), Antibody directed enzyme prodrug therapy (ADEPT) – report on a pilot clinical trial in progress, *Antibody Immunoconj. Radiopharm.* **5**, 133.

BEGENT, R. H. J., VERHAAR, M. J., CHESTER, K. A., CASEY, J. L., GREEN, A. J. et al. (1996), Clinical evidence of efficient tumor targeting based on single-chain Fv antibody selected from a combinatorial library, *Nature Medicine* **2**, 979–984.

BENHAR, I., PASTAN, I. H. (1997), Tumor targeting by antibody – drug conjugates, in: *Antibody Therapeutics* (HARRIS, W. J., ADAIR, J. R., Eds.), pp. 73–85. Boca Raton, FL: CRC Press.

BIOCCA, S., CATTANEO, S. (1995), Intracellular immunization: antibody targeting to subcellular compartments, *Trends Cell. Biol.* **5**, 248–252.

BLAKEY, D. C., BURKE, P. J., DAVIES, D. H., DOWELL, R. I., MELTON, R. G. et al. (1995), Antibody-directed enzyme prodrug therapy (ADEPT) for treatment of major solid tumor disease, *Biochem. Soc. Trans.* **23**, 1047–1050.

BOSSLET, K., CZECH, J., HOFFMANN, D. (1994), Tumor selective prodrug activation by fusion protein-mediated catalysis, *Cancer Res.* **54**, 2151–2159.

BRAMBELL, F. W. R., HEMMINGS, W. A., MORRIS, I. G. (1964), A theoretical model of γ-globulin catabolism, *Nature* **203**, 1352–1355.

BRETT, S., BAXTER, G., COOPER, H., JOHNSTON, J. M., TITE, J., RAPSON, N. (1996), Repopulation of blood lymphocyte sub-populations in rheumatoid arthritis patients treated with the depleting humanized monoclonal antibody, CAMPATH-1H, *Immunology* **88**, 13–19.

BROWN, P. S. JR., PARENTEAU, G. L., DIRBAS, F. M., GARSIA, R. J., GOLDMAN, C. K. et al. (1991), Anti-Tac-H, a humanized antibody to the interleukin 2 receptor, prolongs primate cardiac allograft survival, *Proc. Natl. Acad. Sci. USA* **88**, 263–2667.

BRUGGEMANN, M., NEUBERGER, M. (1996), Strategies for expressing human antibody repertoires in transgenic mice, *Immunol. Today* **17**, 391–397.

BRUGGEMANN, M., TAUSSIG, M. I. (1997), Production of human antibody repertoires in transgenic mice, *Curr. Opin. Biotechnol.* **8**, 455–458.

BURTON, D. R., BARBAS III, C. F. (1994), Human antibodies from combinatorial libraries, *Adv. Immunol.* **57**, 191–280.

CARAYANNOPOULOS, L., MAX, E. E., CAPRA, J. D. (1994), Recombinant human IgA expressed in insect cells, *Proc. Natl. Acad. Sci. USA* **91**, 8348–8352.

CARON, P. C., JURCIC, J. G., SCOTT, A. M., FINN, R. D., DIVGI, C. R. et al. (1994), A phase 1B trial of humanized monoclonal antibody M195 (anti-CD33) in myeloid leukemia: specific targeting without immunogenicity, *Blood* **83**, 1760–1768.

CARTER, P., KELLEY, R. F., RODRIGUES, M. L., SNEDCOR, B., COVARRUBIAS, M. et al. (1992), High-level *Escherichia coli* expression and production of a bivalent humanized antibody fragment, *Bio/Technology* **10**, 163–167.

CARVER, A. S., DALRYMPLE, M. A., WRIGHT, G., COTTAM, D. S., REEVES, D. B. et al. (1993), Transgenic livestock as bioreactors: stable expression of human alpha-1-antitrypsin by a flock of sheep, *Bio/Technology* **11**, 1263–1270.

CHOTHIA, C., LESK, A. M. (1987), Canonical structures for the hypervariable regions of immunoglobulins, *J. Mol. Biol.* **196**, 901–917.

CHOTHIA, C., LESK, A. M, TRAMONTANO, A., LEVITT, M., SMITH-GILL, S. J. et al. (1989), Conformations of immunoglobulin hypervariable regions, *Nature* **342**, 877–883.

CLARK, M. (1997), Unconjugated antibodies as therapeutics, in: *Antibody Therapeutics* (HARRIS, W. J., ADAIR, J. R., Eds.), pp. 3–31. Boca Raton, FL: CRC Press.

COOK, G. P., TOMLINSON, I. M. (1995), The immunoglobulin VH repertoire, *Immunol. Today* **16**, 237–242.

CUNNINGHAM, B. C., WELLS, I. A. (1997), Minimized proteins, *Curr. Opin. Struct. Biol.* **7**, 457–462.

DAVIES, J., RIECHMANN, L. (1995), Antibody VH domains as small recognition units, *Bio/Technology* **13**, 475–479.

DeNARDO, G. L., LEWIS, J. P., DeNARDO, S. J., O'GRADY, L. F. (1994), Effect of Lym-1 radioimmunoconjugate on refractory chronic lymphocytic leukaemia, *Cancer* **73**, 1425–1432.

DEONARAIN, M. P., EPENETOS, A. A. (1994), Targeting enzymes for cancer therapy: old enzymes in new roles, *Br. J. Cancer* **70**, 786–794.

DHAINAUT, J.-F. A., VINCENT, J.-L., RICHARD, C., LEJEUNE, P., MARTIN, C. et al. (1995), CDP571, a humanized antibody to human TNFα: safety, pharmacokinetics, immune response and influence on cytokine levels in patients with septic shock, *Crit. Care Med.* **23**, 1461–1469.

DiTULLIO, P., EBERT, K., POLLOCK, D., HARVEY, M., WILLIAMS, J. et al. (1995), High-level production of human monoclonal antibody in the milk of transgenic mice and a transgenic goat, *IBC Conference on Antibody Engineering,* La Jolla, CA, USA.

DONOHUE, P. J., MACARDLE, P. J., ZOLA, H. (1995), Making and using "conventional" mouse monoclonal antibodies, in: *Monoclonal Antibodies: The Second Generation* (ZOLA, H., Ed), pp. 15–42. Oxford: BIOS Scientific Publishers.

DUENAS, M., CHIN, L.-T., MALMBORG, A.-C., CASALVILLA, R., OHLIN, M., BORREBAECK, C. A. K. (1996), *In vitro* immunization of naive human B cells yields high affinity immunoglobulin G antibodies as illustrated by phage display, *Immunology* **89**, 1–7.

DUNCAN, A. R., WOOF, I. M., PARTRIDGE, L. J., BURTON, D. R., WINTER, G. (1988), Localization of the binding site for the human high affinity Fc receptor on IgG, *Nature* **332**, 563–564.

ECHELARD, Y. (1996), Recombinant protein production in transgenic animals, *Curr. Opin. Biotechnol.* **7**, 536–570.

FIEDLER, U., CONRAD, U. (1995), High-level production and long-term storage of engineered antibodies in transgenic tobacco seeds, *Bio/Technology* **13**, 1090–1093.

FISHWILD, D. M., O'DONNELL, S. L., BENGOECHEA, T., HUDSON, D. V., HARDING, F. et al. (1996), High avidity human IgGκ monoclonal antibodies from a novel strain of minilocus transgenic mice, *Nature Biotechnology* **14**, 845–851.

GALLOP, M. A., BARRETT, R. W., DOWER, W. J., FODOR, S. P. A., GORDON, E. M. (1994), Applications of combinatorial technologies to drug discovery. 1. Background and peptide combinatorial libraries, *J. Med. Chem.* **37**, 1233–1251.

GADINA, M., NEWTON, D. L., RYBAK, S. M., WU, Y.-N., YOULE, R. J. (1994), Humanized immunotoxins, *Ther. Immunol.* **1**, 59–64.

GIVOL, D. (1991), The minimal antigen-binding fragment of antibodies-Fv fragment, *Mol. Immunol.* **28**, 1379–1386.

GOLDENBERG, D. M., BLUMENTHAL, R. D., SHARKEY, R. M. (1994), Prospects for cancer imaging and therapy with radioimmunoconjugates, in: *The Pharmacology of Monoclonal Antibodies, Handbook of Experimental Pharmacology* (ROSENBERG, M., MOORE, G. P., Eds.), Vol. 113, pp. 347–367. Berlin: Springer-Verlag.

GREGORIADIS, G. (1995), Engineering liposomes for drug delivery: progress and problems, *Trends Biotechnol.* **13**, 527–537.

GRIFFITHS, A. D., WILLIAMS, S. C., HARTLEY, O., TOMLINSON, I. M., WATERHOUSE, P. et al. (1994), Isolation of high-affinity human antibodies directly from large synthetic repertoires, *EMBO J.* **13**, 3245–3260.

HABER, E. (1994), Antibody–enzyme fusion proteins and bispecific antibodies, in: *Monoclonal Antibodies, The Pharmagology of Handbook of Experimental Pharmacology* (ROSENBERG, M., MOORE, G. P., Eds.), Vol. 113, pp. 178–197. Berlin: Springer-Verlag.

HALE, G., DYER, M. J., CLARK, M. R., PHILLIPS, J. M., MARCUS, R. et al. (1988), Remission induction in non-Hodgkin lymphoma with reshaped human monoclonal antibody CAMPATH-1H, *Lancet* **ii**, 1394–1399.

HAKIMI, J., CHIZZONITE, R., LUKE, D. R., FAMILLETTI, P. C., BAILON, P. et al. (1991), Reduced immunogenicity and improved pharmacokinetics of humanized anti-Tac in cynomolgus monkeys, *J. Immunol.* **147**, 1352–1359.

HAMERS-CASTERMAN, C., ATARHOUCH, T., MUYLDERMANS, S., ROBINSON, G., HAMERS, C. et al. (1993), Naturally occurring antibodies devoid of light chains, *Nature* **363**, 446–448.

HARRIS, J. D., LEMOINE, N. R. (1996), Strategies for targeted gene therapy, *Trends Genet.* **12**, 400–405.

HASEMANN, C. A., CAPRA, J. D. (1990), High-level production of functional immunoglobulin heterodimer in a baculovirus expression system, *Proc. Natl. Acad. Sci. USA* **87**, 3942–3946.

HOLLIS, G. F., MARK, G. E. (1995), *Homologous Recombination Antibody Expression System for Murine Cell*, WO95/17516.

HOOGENBOOM, H. R. (1997a), Mix and match: building manifold binding sites, *Nature Biotechnology* **15**, 125–126.

HOOGENBOOM, H. R. (1997b), Designing and optimizing library selection strategies for generating high-affinity antibodies, *Trends Biotechnol.* **15**, 62–70.

HORNE, C., KLEIN, M., POLIDOULIS, I., DORRINGTON, K. J. (1982), Non-covalent association of heavy and light chains of human immunoglobulins, *J. Immunol.* **129**, 660–664.

HORWITZ, A. H., CHANG, C. P., BETTER, M., HELLSTROM, K. E., ROBINSON, R. R. (1988), Secretion of functional antibody and Fab fragment from yeast cells, *Proc. Natl. Acad. Sci. USA* **85**, 8678–8682.

HUANG, X., MOLEMA, G., KING, S., WATKINS, L., EDGINGTON, T. S., THORPE, P. E. (1997), Tumor infarction in mice by antibody-directed targeting of tissue factor to tumor vasculature, *Science* **275**, 547–550.

HUSTON, J. S., MCCARTNEY, J., TAI, M.-S., MOTTALA-HARTSHORN, C., JIN, D. et al. (1993), Medical applications of single-chain antibodies, *Int. Rev. Immunol.* **10**, 195–217.

INADA, Y., FURUKAWA, M., SASAKI, H., KODERA, Y., HIROTO, M. et al. (1995), Biomedical and biotechnological applications of PEG- and PM-modified proteins, *Trends Biotechnol.* **13**, 86–91.

JACOBSEN, J. R., SCHULTZ, P. G. (1995), The scope of antibody catalysis, *Curr. Opin. Struct. Biol.* **5**, 818–824.

JAMES, K. (1994), Human monoclonal antibody technology, in: *The Pharmacology of Monoclonal Antibodies, Handbook of Experimental Pharmacology* (ROSENBERG, M., MOORE, G. P., Eds.), Vol. 113, pp. 1–22. Berlin: Springer-Verlag.

JAMES, D. C., FREEDMAN, R. B., HOARE, M., OGONAH, D. W., ROONEY, B. C. et al. (1995), N-glycosylation of a recombinant human interferon-γ produced in different animal expression systems, *Bio/Technology* **13**, 592–596.

JONES, S. T., BENDIG, M. (1991), Rapid PCR cloning of full length mouse immunoglobulin variable regions, *Bio/Technology* **9**, 88–89.

JONES, P. T., DEAR, P. H., FOOTE, J., NEUBERGER, M. S., WINTER, G. (1986), Replacing the complementarity-determining regions in a human antibody with those from a mouse, *Nature* **321**, 522–525.

JOST, C. R., TITUS, J. A., KURUCZ, I., SEGAL, D. M. (1996), A single-chain bispecific Fv2 molecule produced in mammalian cells redirects lysis by activated CTL, *Mol. Immunol.* **33**, 211–219.

KIM, J.-K., TSEN, M.-F., GHETIE, V., WARD, E. S. (1994a), Identifying amino acid residues that influence plasma clearance of murine IgG$_1$ fragments by site-directed mutagenesis, *Eur. J. Immunol.* **24**, 542–548.

KIM, J.-K., TSEN, M.-F., GHETIE, V., WARD, E. S. (1994b), Catabolism of the murine IgG$_1$ molecule: Evidence that both C$_H$2-C$_H$3 domain interfaces are required for persistence of IgG$_1$ in the circulation of mice, *Scand. J. lmmunol.* **40**, 457–465.

KIM, J.-K., TSEN, M.-F., GHETIE, V., WARD, E. S. (1995), Evidence that the hinge region plays a role in maintaining serum levels of the murine IgG$_1$ molecule, *Mol. Immunol.* **32**, 467–475.

KING, D. J., TURNER, A., BEELEY, N. R. A., MILLICAN, T. A. (1992), *Tri- and Tetravalent Monospecific Antigen-Binding Proteins*, WO92/22583.

KING, D. J., TURNER, A., FARNSWORTH, A. P. H., ADAIR, J. R., OWENS, R. J. et al. (1994), Improved tumor targeting with chemically cross-linked recombinant antibody fragments, *Cancer Res.* **54**, 6176–6185.

KING, D. J., ANTONIW, P., OWENS, R. J., ADAIR, J. R., HAINES, A. M. R. et al. (1995), Preparation and preclinical evaluation of humanised A33 immunoconjugates for radioimmunotherapy, *Br. J. Cancer* **72**, 1364–1372.

KNAPPICK, A., PLÜCKTHUN, A. (1995), Engineered turns of a recombinant antibody improve its *in vivo* folding, *Protein Eng.* **8**, 81–89.

KREITMAN, R. J., PASTAN, I. (1997), Immunotoxins for treating cancer and autoimmune disease, in: *Antibody Therapeutics* (HARRIS, W. J., ADAIR, J. R., Eds.), pp. 33–51. Boca Raton, FL: CRC Press.

LARRICK, J. W., WALLACE, E. F., COLOMA, M. J., BRUDERER, U., LANG, A. B., FRY, K. E. (1992), Therapeutic human antibodies derived from PCR amplification of B cell variable regions, *Immunol. Rev.* **130**, 69–85.

LOBUGLIO, A. F., WHEELER, R. H., TRANG, J., HAYNES, A., ROGERS, K. et al. (1989), Mouse/human chimeric monoclonal antibody in man:

kinetics and immune response, *Proc. Natl. Acad. Sci. USA*. **86**, 4220–4224.

LONBERG, N. (1994), Transgenic approaches to human monoclonal antibodies, in: *The Pharmacology of Monoclonal Antibodies Handboock of Experimental Pharmacology* (ROSENBERG, M., MOORE, G. P., Eds.), Vol. 113, pp. 49–101. Berlin: Springer-Verlag.

LOW, N. M., HOLLIGER, P., WINTER, G. (1996), Mimicking somatic hypermutation: affinity maturation of antibodies displayed on bacteriophage using a bacterial mutator strain, *J. Mol. Biol.* **260**, 359–368.

MA, J. K.-C., HEIN, M. (1995), Immunotherapeutic potential of antibodies produced in plants, *Trends Biotechnol.* **13**, 522–527.

MACCALLUM, R. M., MARTIN, A. C. R., THORNTON, J. M. (1996), Antibody–antigen interactions: contact analysis and binding site topography, *J. Mol. Biol.* **262**, 732–745.

MAGA, E. A., MURRAY, J. D. (1995), Mammary gland expression of transgenes and the potential for altering the properties of milk, *Bio/Technology* **13**, 1452–1457.

MAGLIANI, W., CONTI, S., DE BERNARDIS, F., GERLONI, M., BERTOLOTTI, D. et al. (1997), Therapeutic potential of antiidiotypic single chain antibodies with yeast killer toxin activity, *Nature Biotechnology* **15**, 155–158.

MALLENDER, W. D., VOSS, E. W. (1994), Construction, expression, and activity of a bivalent bispecific single-chain antibody, *J. Biol. Chem.* **269**, 199–206.

MALONEY, D. G., LILES, T. M., CZERWINSKI, D. K., WALDICHUK, C., ROSENBERG, J. et al. (1994), Phase I clinical trial using escalating single-dose infusion of chimeric anti-CD20 monoclonal antibody (IDEC-C2B8) in patients with recurrent B cell lymphoma, *Blood* **84**, 2457–2466.

MARTIN, F., TONIATTI, C., SALVATI, A. L., VENTURINI, S., CILIBERTO, G. et al. (1994), The affinity selection of a minibody polypeptide inhibitor of human interleukin-6, *EMBO J.* **13**, 5303–5309.

MARTIN, F., TONIATTI, C., SALVATI, A. L., CILIBERTO, G., CORTESE, R., SOLLAZZO, M. (1996), Coupling protein design and *in vitro* selection strategies: improving specificity and affinity of a designed β-protein IL-6 antagonist, *J. Mol. Biol.* **255**, 86–97.

MATHIESON, P. W., COBBOLD, S. D., HALE, G., CLARK, M. R., OLIVERA, D. B. et al. (1990), Monoclonal antibody therapy in systemic vasculitis, *N. Engl. J. Med.* **323**, 250–254.

MATSUDA, F., HONJO, T. (1996), Organization of the human immunoglobulin heavy-chain locus, *Adv. Immunol.* **62**, 1–29.

MENDEZ, M. J., GREEN, L. L., CORVALAN, J. R. F., JIA, X.-C., MAYNARD-CURRIE, C. E. et al. (1997), Functional transplant of megabase human immunoglobulin loci recapitulates human antibody response in mice, *Nature Genetics* **15**, 146–156.

MILENIC, D. E., YOKOTA, T., FILPULA, D. R., FINKELMAN, M. A. J., DODD, S. W. et al. (1991), Construction, binding properties, metabolism, and tumor targeting of a single-chain Fv derived from the pancarcinoma monoclonal antibody CC49, *Cancer Res.* **51**, 6363–6371.

MIYASHITA, H., KARAKI, Y., KIKUCHI, M., FUJII, I. (1993), Prodrug activation via catalytic antibodies, *Proc. Natl. Acad. Sci. USA* **90**, 5337–5340.

MORGAN, A., JONES, N. D., NESBITT, A. M., CHAPLIN, L., BODMER, M. W., EMTAGE, J. S. (1995), The N-terminal end of the C_H2 domain of chimeric human IgG_1 anti-HLA-DR is necessary for C1q, FcγRI and FcγRIII binding, *Immunology* **86**, 319–324.

MORRISON, S. L., JOHNSON, M. J., HERZENBERG, L. A., OI, V. T. (1984), Chimeric human antibody molecules: mouse antigen-binding domains with human constant region domains, *Proc. Natl. Acad. Sci. USA* **81**, 6851–6855.

MOUNTAIN, A., ADAIR, J. R. (1992), Engineering antibodies for therapy, *Biotechnol. Genet. Eng. Rev.* **10**, 1–142.

MUELLER, B. M., REISFELD, R. A., GILLIES, S. D. (1990), Serum half-life and tumor localization of a chimeric antibody deleted of the C_H2 domain and directed against the disialoganglioside GD2, *Proc. Natl. Acad. Sci. USA* **87**, 5702–5705.

MUNRO, A (1984), Uses of chimaeric antibodies, *Nature* **312**, 597.

MURPHY, J. R. (1996), Protein engineering and design for drug delivery, *Curr. Opin. Struct. Biol.* **6**, 541–545.

NESBIT, M., FANG FU, Z., MCDONALD-SMITH, J., STEPLEWSKI, Z., CURTIS, P. J. (1992), Production of a functional monoclonal antibody recognizing human colorectal carcinoma cells from a baculovirus expression system, *J. Immunol. Methods* **151**, 201–208.

NEUBERGER, M. S., WILLIAMS, G. T., FOX, R. O. (1984), Recombinant antibodies possessing novel effector functions, *Nature* **312**, 604–608.

NEUBERGER, M. S., WILLIAMS, G. T., MITCHELL, E. B., JOUHAL, S. S., FLANAGAN, J. G., RABBITTS, T. H. (1985), A hapten-specific chimaeric IgE antibody with human physiological effector function, *Nature* **314**, 268–270.

NEWMAN, R., ALBERTS, J., PERSON, D., CARNER, K., HEARD, C. et al. (1992). "Primatization" of recombinant antibodies for immunotherapy of human disease: a macaque/human chimeric antibody against human CD4, *BioTechnology* **10**, 1455–1460.

NUCCI, M., SHORR, R., ABUCHOWSK, A. (1991), The therapeutic value of PEG-modified proteins, *Adv. Drug Delivery Rev.* **6**, 133–151.

NYYSSÖNEN, E., PENTTILÄ, M., HARKKI, A., SALOHEIMO, A., KNOWLES, J. K. C., KERÄNEN, S. (1993), Efficient production of antibody fragments by the filamentous fungus *Trichoderma reesei*, *Bio/Technology* **11**, 591–595.

OGONAH, O. W., FREEDMAN, R. B., JENKINS, N., PATEL, K., ROONEY, B. C. (1996), Isolation and characterisation of an insect cell line able to perform complex N-linked glycosylation on recombinant proteins, *Nature Biotechnology* **14**, 197–202.

PACK, P., PLÜCKTHUN, A. (1992), Miniantibodies: use of amphipathic helices to produce functional, flexibly linked dimeric Fv fragments with high avidity in *Escherichia coli*, *Biochemistry* **31**, 1579–1584.

PARK, J. W., HONG, K., CARTER, P., ASGARI, H., GUO, L. Y. et al. (1995), Development of anti-p185HER2 immunoliposomes for cancer therapy, *Proc. Natl. Acad. Sci. USA* **92**, 1327–1331.

PEDLEY, R. B., BODEN, J. A., BEGENT, R. H. J., TURNER, A., HAINES, A. M. R., KING, D. J. (1994), The potential for enhanced tumour localisation by poly(ethyleneglycol) modification of anti-CEA antibody, *Br. J. Cancer* **70**, 1126–1130.

PESSI, A., BIANCHI, E., CRAMERI, A., VENTURINI, S., TRAMONTANO, A., SOLLAZZO, M. (1993), A designed metal-binding protein with a novel fold, *Nature* **362**, 367–369.

PIETERSZ, G. A., McKENZIE, I. F. C. (1995), The genetic engineering of antibody constructs for diagnosis and therapy, in: *Monoclonal Antibodies: The Second Generation* (ZOLA, H., Ed.), pp. 93–117. Oxford: BIOS Scientific Publishers.

PLÜCKTHUN, A. (1994), Antibodies from *Escherichia coli*, in: *The Pharmacology of Monoclonal Antibodies, Handbook of Experimental Pharmacology* (ROSENBERG, M., MOORE, G. P., Eds., Vol. 113, pp. 269–315. Berlin: Springer-Verlag.

PONCET, P., PANCZAK, A., GOUPY, C., GUSTAFSSON, K., BLANPIED, C. et al. (1996), Antifection – an antibody-mediated method to introduce genes into lymphoid cells *in vitro* and *in vivo*, *Gene Ther.* **3**, 731–738.

PORTER, A. J. R., BENTLEY, K. J., CUPIT, P. M., WALLACE, T. P. (1997), Emerging production systems for antibody therapeutics, in: *Antibody Therapeutics* (HARRIS, W. J., ADAIR, J. R., Eds.), pp. 247–273. Boca Raton, FL: CRC Press.

PRESS, O. W., EARY, J. F., APPLEBAUM, F. R., MARTIN, P. J., NELP, W. B. et al. (1995), Phase II trial of ^{131}I-B1 (anti-CD20) antibody therapy with autologous stem cell transplantation for relapsed B cell lymphoma, *Lancet* **346**, 336–340.

RAVETCH, J. V. (1997), Fc receptors, *Curr. Opin. Immunol.* **9**, 121–125.

REFF, M. E., CARNER, K., CHAMBERS, K. S., CHINN, P. C., LEONARD, J. E. et al. (1994), Depletion of B cells *in vivo* by a chimeric mouse human monoclonal antibody to CD20, *Blood* **83**, 435–445.

REIS, U., BLUM, B., VON SPECHT, B. U., DOMDEY, H., COLLINS, J. (1992), Antibody production in silkworm cells and silkworm larvae infected with a dual recombinant *Bombyx mori* nuclear polyhedrosis virus, *Bio/Technology* **10**, 910–912.

RIDDER, R., SCHMITZ, R., LEGAY, F., GRAM, H. (1995), Generation of rabbit monoclonal antibody fragments from a combinatorial phage display library and their production in the yeast *Pichia pastoris*, *Bio/Technology* **13**, 255–260.

RIECHMANN, L., CLARK, M., WALDMANN, H., WINTER, G. (1988), Reshaping human antibodies for therapy, *Nature* **332**, 323–327.

ROUTLEDGE, E. G., LLOYD, I., GORMAN, S. D., CLARK, M., WALDMANN, H. (1991), A humanized monovalent CD3 antibody which can activate homologous complement, *Eur. J. Immunol.* **21**, 2717–2725.

RYBAK, S. M., HOOGENBOOM, H. R., MEADE, H. M., RAUS, J. C. M., SCHWARTZ, D., YOULE, R. J. (1992), Humanization of immunotoxins, *Proc. Natl. Acad. Sci. USA* **89**, 3165–3169.

SAIKI, R. K., SCHARF, S., FALOONA, F., MULLIS, K. B., HORN, G. T. et al. (1985), Enzymatic amplification of β-globin genomic sequences and restriction site analysis for diagnosis of sickle anemia, *Science* **230**, 1350–1354.

SALEH, M. N., CONRY, R. M., LoBUGLIO, A. F. (1994), Chiral experience with murine, human and genetically engineered monoclonal antibodies, in: *The Pharmagology of Monoclonal Antibodies Handbook of Experimental Pharmacology* (ROSENBERG, M., MOORE, G. P., Eds.), Vol. 113, pp. 369–386. Berlin: Springer-Verlag.

SCHNIERLE, B. S., MORITZ, D., JESCHKE, M., GRONER, B. (1996), Expression of chimeric envelope proteins in helper cell lines and integration into Moloney murine leukemia virus particles, *Gene Ther.* **3**, 334–342.

SCHOTT, M. E., FRAZIER, K. A., POLLOCK, D. K., VERBANAC, K. M. (1993), Preparation, characterization, and *in vivo* biodistribution properties of synthetically cross-linked multivalent antitumour antibody fragments, *Bioconjug. Chem.* **4**, 153–165.

SIMON, T., RAJEWSKY, K. (1992), A functional antibody mutant with an insertion in the framework region 3 loop of the V_H domain: implications for antibody engineering, *Protein Eng.* **5**, 229–234.

STEPHENS, S., EMTAGE, S., VETTERLEIN, O., CHAPLIN, L., BEBBINGTON, C. et al. (1995), Comprehensive pharmacokinetics of a humanized antibody and analysis of residual anti-idiotypic responses, *Immunology* **85**, 668–674.

STEVENSON, G. T., GLENNIE, M. J., KAN, K. S. (1993), Chemically engineered chimeric and multi-Fab antibodies, in: *Protein Engineering of Antibody Molecules for Prophylactic and Therapeutic Applications in Man* (CLARK, M., Ed.), pp. 127–141. Nottingham: Academic Titles.

TAYLOR, G., PORTER, T., DILLON, S., TRILL, J., GANGULY, S. et al. (1995), Anti-respiratory syncytial virus monoclonal antibodies show promise in the treatment and prophylaxis of viral disease, *Biochem. Soc. Trans.* **23**, 1063–1067.

TRAMONTANO, A., CHOTHIA, C., LESK, A. M. (1990), Framework residue 71 is a major determinant of the position and conformation of the second hypervariable region in the V_H domains of immunoglobulins, *J. Mol. Biol.* **215**, 175–182.

U.S. Department of Health and Human Services (1997), Proposed Rule *Fed. Reg.* **62**, 18937–18964.

VALONE, F. H., KAUFMAN, P. A., GUYRE, P. M., LEWIS, L. D., MEMOLI, V. et al. (1995), Phase Ia/Ib trial of bispecific antibody MDX210 in patients with advanced breast or ovarian cancer that express the proto-oncogene HER-2/*neu, J. Clin. Oncol.* **13**, 2281–2292.

VERHOEYEN, M., MILSTEIN, C., WINTER, G. (1988), Reshaping human antibodies: grafting an anti-lysozyme activity, *Science* **239**, 1534–1536.

VERSTRAETE, M., LIJNEN, H. R., COLLEN, D. (1995), Thrombolytic agents in development, *Drugs* **50**, 29–42.

WABL, M., STEINBERG, C. (1996), Affinity maturation and class switching, *Curr. Opin. Immunol.* **8**, 89–92.

WALDMANN, T. A., STROBER, W. (1969), Metabolism of immunoglobulins, *Prog. Allergy* **13**, 1–110.

WALL, J. G., PLÜCKTHUN, A. (1995), Effects of overexpressing folding modulators on the *in vivo* folding of heterologous proteins in *Escherichia coli, Curr. Opin. Biotechnol.* **6**, 507–516.

WARD, E. S., GUSSOW, D., GRIFFITHS, A. D., JONES, P. T., WINTER, G. (1989), Binding activities of a repertoire of single immunoglobulin domains secreted from *Escherichia coli, Nature* **341**, 544–546.

WARD, E. S., BEBBINGTON, C. R. (1995), Genetic manipulation and expression of antibodies, in: *Monoclonal Antibodies – Principles and Applications* (BIRCH, J. R., LENNOX, E. S., Eds.), pp. 137–185. New York: Wiley-Liss.

WAWRZYNCZAK, E. J., CUMBER, A. J., PAMELL, G. D., JONES, P. T., WINTER, G. (1992), Blood clearance in the rat of a recombinant mouse monoclonal antibody lacking the N-linked oligosaccharide side chains of the C_H2 domains, *Mol. Immunol.* **29**, 213–220.

WEICKERT, M. J., DOHERTY, D. H., BEST, E. A., OLINS, P. O. (1996), Optimization of heterologous protein production in *Escherichia coli, Curr. Opin. Biotechnol.* **7**, 494–499.

WEISMAN, H. F., SCHAIBLE, T. F., JORDAN, R. E., CABOT, C. F., ANDERSON, K. M. (1995), Antiplatelet monoclonal antibodies for the prevention of arterial thrombosis: experience with Reo-Pro, a monoclonal antibody directed against the platelet GPIIb/IIIa receptor, *Biochem. Soc. Trans.* **23**, 1051–1057.

WELLS, J. A., DE VOS, A. M. (1996), Hematopoietic receptor complexes, *Annu. Rev. Biochem.* **65**, 609–634.

WENTWORTH, P., DATTA, A., BLAKEY, D., BOYLE, T., PARTRIDGE, L. J., BLACKBURN, G. M. (1996), Toward antibody-directed "abzyme" prodrug therapy, ADAPT: carbamate prodrug activation by a catalytic antibody and its *in vitro* application to human tumor cell killing, *Proc. Natl. Acad. Sci. USA* **93**, 799–803.

WILLIAMS, S. C., FRIPPIAT, J.-P., TOMLINSON, I. M., IGNATOVICH, O., LEFRANC, M.-P., WINTER, G. (1996), Sequence and evolution of the human germline Vλ repertoire, *J. Mol. Biol.* **264**, 220–232.

WILMUT, I., SCHNIEKE, A. E., MCWHIR, J., KIND, A. J., CAMPBELL, K. H. S. (1997), Viable offspring derived from fetal and adult mammalian cells, *Nature* **385**, 810–813.

WILSON, I. A., STANFIELD, R. L. (1994), Antibody–antigen interactions: new structures and new conformational changes, *Curr. Opin. Struct. Biol.* **4**, 857–867.

WINTER, G., MILSTEIN, C. (1991), Man-made antibodies, *Nature* **349**, 293–299.

WINTER, G., GRIFFITHS, A. D., HAWKINS, R. E., HOOGENBOOM, H. R. (1994), Making antibodies by phage display technology, *Annu. Rev. Immunol.* **12**, 433–455.

WRIGHT, A., MORRISON, S. L. (1994), Effect of altered C_H2-associated carbohydrate structure on the functional properties and *in vivo* fate of chimeric mouse–human immunoglobulin $G_1, J. Exp. Med.* **180**, 1087–1096.

WRIGHT, A., MORRISON, S. L. (1997), Effect of glycosylation on antibody function: implications for genetic engineering, *Trends Biotechnol.* **15**, 26–32.

WU, T. T., KABAT, E. A. (1970), An analysis of the sequences of the variable regions of Bence Jones proteins and myeloma light chains and their implications for antibody complementarity, *J. Exp. Med.* **132**, 211–250.

WU, X.-C., NG, S.-C., NEAR, R. I., WONG, S.-L. (1993), Efficient production of a functional single-chain antidigoxin antibody via an engineered *Bacillus subtilis* expression–secretion system, *Bio/Technology* **11**, 71–76.

YARRANTON, G. T. (1997), Antibodies as carriers of drugs and radioisotopes, in: *Antibody Therapeutics* (HARRIS, W. J., ADAIR, J. R., Eds.), pp. 53–72. Boca Raton, FL: CRC Press.

ZACHAU, H. G. (1993), The immunoglobulin κ locus – or what has been learned from looking closely at one tenth of a percent of the human genome, *Gene* **135**, 167–173.

ZAPATA, G., RIDGWAY, J. B. B., MORDENTI, J., OSAKA, G., WONG, W. L. T. et al. (1995), Engineering linear F(ab')₂ fragments for efficient production in *Escherichia coli* and enhanced antiproliferative activity, *Protein Eng.* **8**, 1057–1062.

ZU PUTLITZ, J., KUBASEK, W. L., DUCHENE, M., MARGET, M., VON SPECHT, B. U., DOMDEY, H. (1990), Antibody production ih baculovirus infected insect cells, *Bio/Technology* **8**, 651–654.

8 Manufacture of Therapeutic Antibodies

ANDREW J. RACHER
JERRY M. TONG
JULIAN BONNERJEA

Slough, UK

1 Introduction 247
2 Cell Line 248
  2.1 Choice of Cell Line 248
  2.2 Gene Expression Systems for Antibodies 250
    2.2.1 DHFR Expression Systems 250
    2.2.2 GS Expression Systems 250
    2.2.3 High Expression by Gene Amplification 251
    2.2.4 High Expression from Low Copy Numbers 252
    2.2.5 Stability of Expression 254
    2.2.6 Summary 254
  2.3 Product Consistency 254
  2.4 Cloning and Selection 256
  2.5 Serum-Free and Suspension Adaptation 256
3 Cell Banking 257
4 Cell Line Stability 258
5 Production Systems 259
6 Process Definition 261
7 Primary Recovery 262
8 Antibody Purification 263
  8.1 Purity Requirement 264
    8.1.1 Protein Contaminants 264
    8.1.2 DNA 265
    8.1.3 Viruses 265
  8.2 Robustness 265

8.3 Economic Aspects   266
8.4 Purification Methods   267
 9 Good Manufacturing Practice   268
10 Conclusions   269
11 References   270

List of Abbreviations

ALF	airlift bioreactor
BHK cell line	baby hamster kidney cell line
BSA	bovine serum albumin
CDR	complementarity determining region
cGMP	current GMP
CHO cell line	Chinese hamster ovary cell line
CMV	cytomegalovirus
COS cell line	monkey kidney cell line
DEAE	diethylaminoethyl
DHFR	dihydrofolate reductase
EBV	Epstein-Barr virus
FDA	Food and Drug Administration
FLP	site-specific recombinase from yeast
FRT	FLP recognition targets
GMP	good manufacturing practice
GS	glutamine synthetase
HAMA	human anti-mouse antibody
Hams's F12	basal medium
HPLC	high performance liquid chromatography
MCB	master cell bank
MSX	methionine sulfoxamine
MTX	methotrexate
MWCB	manufacturer's WCB
NSO cell line	myeloma cell line
ppm	parts per million
QA	quality assurance
QC	quality control
RPMI 1640	basal medium
SDS	sodium dodecylsulfate
SDS-PAGE	SDS-polyacrylamide gel electrophoresis
Sp2/0 cell line	myeloma cell line
STR	stirred tank bioreactor
TAG-72	tumor associated antigen
tPA	tissue plasminogen activator
WCB	working cell bank

1 Introduction

The manufacture of antibodies for therapeutic use is more than simply the low cost production of large amounts of protein. The development of animal cell bioprocesses is focused upon a large number of issues. Firstly, there are the generation and selection of a cell line for use in a production process. The characteristics of a suitable cell line will ideally include: stable production of a product with consistent characteristics; the ability to grow in simple media after a rapid period of adaptation; suitable growth kinetics in culture; and high productivity. Secondly, there is the need to be able to produce sufficient quantities of material to meet the market demand. This requires optimization of the production process through: medium optimization to support high cell densities and product secretion rates; development of strategies for bioreactor operation; and design of bioreactors to support high cell densities. It is pointless to develop a cell culture process capable of producing large quantities of product if it is then impossible to recover the product from the spent culture broth. Therefore it must be possible to separate the product from cell debris and present it in a form suitable for purification. Finally, the antibody product must be purified to an acceptable level of purity. Ideally the last two activities should occur with minimum product losses.

The above are all important process considerations. Regulatory considerations are also important for the production of bulk purified antibodies for clinical use, and require definition of the complete manufacturing process. The process must consistently produce a product with minimal variation in biochemical characteristics, and this must be capable of being demonstrated to the relevant regulatory authorities. The cell line and cell bank must be characterized and the complete process must be shown to be free of adventitious agents.

This chapter presents an overview of the production of therapeutic antibodies for either clinical trials or to satisfy the market demands of a licensed product. It examines issues ranging from the choice of host cell line through to generation of the bulk purified product. Consideration of the steps involved in converting the bulk purified product to an injectable drug is beyond the scope of this chapter and is not discussed. This review is also limited to discussing mammalian cell systems, since these are currently the most widely used for production of therapeutic antibodies.

2 Cell Line

2.1 Choice of Cell Line

Murine hybridomas are an obvious route for the manufacture of antibodies. These are readily created by non-molecular methods and are used very successfully to produce antibodies for both therapeutic and diagnostic applications. However, the effective therapeutic use of mouse monoclonal antibodies in humans is limited partly by the development of a strong human anti-mouse antibody (HAMA) response by the recipient. Non-recombinant human monoclonal antibodies can be produced for therapeutic use. For example, anti-D antibody (used to prevent hemolytic disease of the newborn) is produced by EBV-transformed human lymphoblastoid cells (DAVIS et al., 1995). An alternative method of producing human monoclonal antibodies is to create human–human or human–mouse hybridomas. However, human cell lines appear to be used infrequently to produce non-recombinant monoclonal antibodies.

Another approach to alleviate the HAMA response is to use human–mouse chimeric antibodies or CDR-grafted antibodies, in which either the variable regions or CDR-regions of the mouse antibody are genetically engineered into a human antibody framework. Genetic engineering can also be used to express murine monoclonal antibodies in alternative cell lines to hybridomas, which often results in higher yields. Recombinant antibodies have been expressed in a variety of mammalian cell lines including Chinese hamster ovary (CHO), myeloma (NSO and

Sp2/0), monkey kidney (COS), and baby hamster kidney (BHK) cell lines.

The key issues affecting the choice of a cell line for use in a manufacturing process are: the capability to produce high product titers in the chosen production system; the ability to produce consistently a product of uniform characteristics; and the speed with which a high yielding cell line can be obtained. The availability of a suitable expression system and the importance of post-translational modifications of the recombinant antibodies may also affect this choice.

The time constraints of cell line selection programs undertaken in the pharmaceutical and biotechnology industries require the rapid evaluation of different antibody gene constructs and the production of milligram amounts of antibody for preliminary analysis. Transient expression systems using COS cells allow the rapid screening of multiple constructs within a few days (FOUSER et al., 1992; KETTLEBOROUGH et al., 1991). Product titers obtained using the COS cell system are variable, typical values ranging from 0.1–4 ng mL^{-1} (TRILL et al., 1995). This variation in expression level appears to be intrinsic to the antibody, as it was seen when different antibodies were expressed from the same vector. WHITTLE et al. (1987) used COS cells to express a full length version of the B72.3 antibody that recognizes an epitope on a tumor-associated antigen TAG-72. These COS cells produced sufficient product to allow physical and immunological study of the secreted antibody.

COS cells have been used to express a variety of antibody constructs in addition to whole immunoglobulin molecules. Fab' fragments, bifunctional chimeric antibodies, mono- and bi-specific single chain antibody derivatives have all been expressed in COS cells (TRILL et al., 1995).

There are disadvantages to using COS cells. The data obtained by transient expression in COS cells are not predictive and do not necessarily correspond to the levels of expression expected from stable expression systems (e.g., CHO or NSO). Additionally, the levels of expression can be too low for initial characterization so multiple transfections have to be undertaken to obtain sufficient material. There are alternatives to COS cell lines for the transient expression of antibodies, such as the human embryonic kidney cell line 293 that has been used to express single chain Fv proteins (DORAI et al., 1994).

Once an antibody construct has been assessed in a transient expression system, the host cell type for preparation of the stable cell line must be chosen. Both CHO and myeloma (e.g., NSO and Sp2/0) cells satisfy the criteria for cell line selection discussed above and are used extensively for the stable expression of recombinant antibodies.

Characteristics common to both CHO and myeloma cell types have favored their use. A range of high yielding expression systems is available for use with these two cell types and both are readily transfectable using a variety of methods. Recombinant derivatives of both cell types can grow in suspension culture and after a period of adaptation both recombinant CHO and myeloma cell lines can grow in serum-free, and even protein-free media. The product titers achievable are acceptable for use in an industrial process when these cell lines are grown in large scale bioreactors. After process optimization, product titers greater than 1 g L^{-1} can be obtained using both CHO and NSO cell lines. Biological activity, determined using *in vitro* systems, of an antibody produced in both CHO and NSO cell lines has been shown to be identical (PEAKMAN et al., 1994). RAY et al. (1997) demonstrated that a humanized antibody produced in both NSO and Sp2/0 cell lines at the 2,000 L scale was equivalent following analytical characterization.

CHO cells are widely used to produce recombinant antibodies using both the dihydrofolate reductase (DHFR) and glutamine synthetase (GS) expression systems (TRILL et al., 1995). The most commonly used CHO strains are DUKX-B11 and DG44, which are both *dhfr$^-$*, or CHO-K1 with the GS system. CHO cells have been used to express both full length antibodies (PAGE and SYDENHAM, 1991; FOUSER et al., 1992) and antibody fragments (DORAI et al., 1994). Recombinant CHO cell lines show efficient post-translational processing of complex proteins, while the glycosylation patterns of native and CHO-derived recombinant proteins are similar.

BEBBINGTON et al. (1992) reported the development, using the GS vector system, of a NSO cell line expressing the antibody cB72.3. The parental cell type of the NSO cell line is a differentiated B cell, which is inherently capable of high levels of immunoglobulin production. The genotype of NSO cells makes them particularly suited for use with the GS expression system. Unlike other cell types, NSO cells are obligate glutamine auxotrophs: glutamine independence can be conferred upon NSO cells following transfection with a functional GS gene. Spontaneous Sp2/0 and Y0 mutants can be generated that no longer require glutamine: this is not observed with NSO cells. The GS vector system in NSO cells has become an extremely important system for the manufacture of recombinant antibodies.

As has been discussed above, the majority of antibodies manufactured for therapeutic use are produced by mammalian cell culture. Alternative systems for the manufacture of antibodies are available. If the requirement for post-translational modification is minimal or if antibody fragments rather than multimeric immunoglobulins are required, then bacteria (CARTER, 1991), yeast (WOOD et al., 1985) or fungi (NYYSSÖNEN et al., 1993) can be used. Biologically active chimeric mouse–human antibodies have been expressed in a variety of insect cell lines (DERAMOUDT et al., 1995; KIRKPATRICK et al., 1995), although product titers obtained are low at about 5 µg mL^{-1}. There is increasing interest in the use of transgenic animals (JAKOBVITS, 1995) and plants (DE NEVE et al., 1993; MA et al., 1995) for the production of whole molecules or large multimeric immunoglobulin complexes, with both systems having the potential for large scale production. By using transgenic animals it is possible to produce fully human antibodies (JAKOBVITS, 1995). This abolishes any possibility of the recombinant antibody being immunogenic since the non-human sequences, found in even CDR-grafted antibodies, are eliminated.

Once the decision is made to produce a recombinant antibody from a mammalian cell line, a suitable expression system must be chosen.

2.2 Gene Expression Systems for Antibodies

The ability to express the product at a high level is the critical issue for any manufacturing process using recombinant cell lines. Consequently, expression vectors have been developed that combine suitable promoters and favorable RNA processing signals. These vectors also contain selectable marker genes to facilitate the selection of transfectants. Selectable marker genes that have been used in antibody expression vectors include GS (BEBBINGTON et al., 1992), DHFR (PAGE and SYDENHAM, 1991; FOUSER et al., 1992), and antibiotic resistance (KALUZA et al., 1991). These systems frequently require the use of selective media that are deficient in some essential nutrient or contain selective agents that are toxic to non-recombinant cells. If large quantities of recombinant product are required, gene expression can be increased using an amplifiable expression system. For some antibody expression systems, (notably those based upon DHFR) a dramatic increase in productivity is seen following amplification (PEAKMAN et al., 1994). However, high levels of amplification can have detrimental affects upon other cell line characteristics. In contrast, the GS system (BEBBINGTON et al., 1992) and some variants of the DHFR system (REFF, 1994) do not rely upon amplification to achieve high productivities. Instead these systems rely upon insertion of the antibody construct into a transcriptionally active region to achieve high productivities.

2.2.1 DHFR Expression Systems

DHFR expression systems rely upon the folate analog methotrexate (MTX) blocking the action of the essential metabolic enzyme DHFR. Provision of a DHFR gene in the expression vector enables transfected cells to overcome MTX poisoning while non-transfected cells are killed. DHFR expression vectors often contain an antibiotic resistance gene for selection of transfectants. The primary function of the DHFR gene is then to facilitate amplification of the vector sequence. It is better if the amplifiable and selectable marker genes are the same since transfected DNA is liable to undergo rearrangement and mutation. Consequently, not all clones selected for the presence of one marker will express the second marker. DHFR can be used for both selection and subsequent amplification, but this requires the use of a *dhfr$^-$* cell line. The DHFR gene is usually under the control of a weak promoter, such as one from SV40. The use of a weak promoter to regulate DHFR gene expression should reduce promoter interference (due to read-through from the upstream promoter inhibiting expression from the downstream promoter) thus increasing expression of the gene(s) of interest. Immunoglobulin heavy and light chain genes are inserted downstream of a strong promoter, e.g., from β-actin or CMV. The CMV promoter is highly efficient in most cell types including lymphocytes, CHO and COS cells, whereas the SV40 promoter functions poorly in lymphoid cells. The CMV promoter is also reported to be a stronger promoter than the known immunoglobulin myeloma promoter enhancers (BEBBINGTON et al., 1992).

2.2.2 GS Expression Systems

GS synthesizes glutamine from glutamate and ammonium. Since glutamine is an essential amino acid, transfection of cells that lack endogenous GS with the GS vector confers the ability to grow in glutamine free media. GS expression vectors contain the GS gene downstream of a SV40 promoter and the gene of interest downstream of a CMV promoter (BEBBINGTON et al., 1992). The use of a weak promoter to regulate GS gene expression offers similar benefits to those seen when a weak promoter is used to drive DHFR gene expression. To achieve high levels of immunoglobulin gene expression the CMV promoter is used upstream of both the heavy and light chain genes.

2.2.3 High Expression by Gene Amplification

Gene amplification is usually achieved by constructing the expression vector so that the immnunoglobulin genes are linked to an amplifiable gene (e.g., thymidine kinase, adenosine deaminase, GS or DHFR). The amplifiable gene may be the same as the selectable marker gene used to select for transfectants. GS and DHFR are the most commonly used genes in antibody expression vectors that can function as both amplifiable and selectable markers. By exposing the transfectants to successively higher concentrations of appropriate toxic drugs, the stringency of selection is increased. If the drug inhibits an enzyme essential for the survival of the cell, only cells that overproduce this enzyme will survive. The overproduction of the enzyme commonly results from increased levels of its particular mRNA. This can result from an increase in the gene copy number (i.e., amplification), or alternatively from more efficient transcription of the gene (GILLIES et al., 1989). Often more DNA than just the relevant gene is amplified, with regions of the chromosome as large as 1,000 kb amplified. Therefore, when the transfected genes are amplified, other sequences, including that of the gene(s) of interest, on the vector are also amplified.

MTX is used for both amplification and selection of antibody gene constructs based upon the DHFR vector system. The concentration of MTX needed to amplify the DHFR gene depends upon the efficiency with which the gene is expressed. Cells transfected with the DHFR gene under the control of a strong promoter are more resistant to MTX than cells where DHFR expression is controlled by a weak promoter. Assuming that the kinetics of the binding of MTX to the enzyme are unaltered, the resistance is increased because the same amount of MTX blocks the activity of a smaller fraction of the cellular DHFR pool. Thus cells where the DHFR gene is downstream of a strong promoter generally require higher levels of MTX to achieve the same level of amplification. Due to random integration of the expression vector into the genome, there is variability in the tolerance to

MTX between transfectants from the same transfection. After amplification, relatively high expression levels (80–110 pg cell^{-1} d^{-1}) have been obtained in CHO cells using this system (PAGE and SYDENHAM, 1991; FOUSER et al., 1992). These specific production rates can be up to 100-fold higher than the rate obtained in the non-amplified parent (PEAKMAN et al., 1994). However, large increases in gene copy number and concomitant increases in mRNA levels are not necessarily reflected in increases in product titers since other cellular functions, e.g., protein secretion or mRNA translation, may become limiting (PENDSE et al., 1992).

PAGE and SYDENHAM (1991) evaluated two selection/amplification procedures to generate high producing DHFR-CHO clones. Cells producing moderate product titers were obtained in a single step by imposing direct selection using high MTX concentrations upon the initial transformants. These cells were stable and could rapidly generate adequate amounts of antibody for research purposes. Higher titers could be achieved by continued amplification although this was accompanied by a decrease in the stability of product expression. Cell lines possessing both good productivity and good growth were generated in a more reliable fashion by a two-step process. This involved preliminary selection and establishment of basal transformants that were then amplified in a stepwise manner. Although this is a longer process, the behavioral properties of the cell lines so derived make them more suitable for manufacturing processes.

Use of the DHFR/MTX system is not restricted to CHO cell lines. DHFR-antibody gene constructs in Sp2/0 cell lines have been amplified using MTX (GILLIES et al., 1989; DORAI et al., 1994).

For cell lines transfected with antibody constructs based upon the GS selection system, amplification is achieved by applying increasing concentrations of the GS inhibitor methionine sulphoxamine (MSX). The weak promoter upstream of the GS gene should minimize the amount of MSX needed to select for gene amplification. BEBBINGTON et al. (1992) found that increasing the MSX concentration was accompanied by an increase in copy number of the vector from 1–4; similar results have been described by other groups (PEAKMAN et al., 1994; BIBILA et al., 1994). This is much lower than the levels observed with the DHFR-CHO system (PEAKMAN et al., 1994). In CHO cell lines the copy numbers of DHFR-linked genes can be as high as 1,000 whereas in NSO clones the copy number of GS-linked genes rarely exceeds 20. This variation in copy number may result in inherent differences in the stability of product expression (HASSELL et al., 1992).

The copy number of either the GS or immunoglobulin heavy chain genes may be the same in different NSO clones but it does not follow that mRNA levels will be the same (PEAKMAN et al., 1994). Amplification of a GS-CHO cell line increased the product titer from 110–250 mg L^{-1} and a titer of 560 mg L^{-1} was obtained when the same antibody was expressed in an amplified GS-NSO (HASSELL et al., 1992). Although amplification of DHFR-CHO and GS-NSO cell lines may result in markedly different copy numbers, the two cell lines may express approximately the same amount of antibody since mRNA levels are virtually identical (PEAKMAN et al., 1994). Therefore, for cell lines generated using the GS vector system, amplification of copy number is not necessarily as critical for generating high producing clones as with the DHFR expression system. With the GS expression system the position of integration of the transfected DNA is the important factor in determining whether the cell line will ultimately be a high producer (discussed below).

The high copy numbers of the antibody construct seen upon amplification, especially with the DHFR expression system, may increase the cell-specific productivity but it can also have a detrimental effect upon other properties of the cell. Amplification of the desired gene will frequently result in poor growth performance of the resulting cell population and may alter cellular metabolism. These effects have been seen in both GS-NSO and DHFR-CHO cell lines (PENDSE et al., 1992; BIBILA et al., 1994). GU et al. (1996) suggested that the poor growth seen upon amplification of the DHFR gene is not due to increased expression of a recombinant proteins. It is a consequence of the higher metabolic burden imposed upon the cell by the increase in specific DHFR activity. The problem of poor growth (low values for the maximum viable cell concentration and time integral of the viable cell concentration) can be alleviated by using the combination of good growth characteristics with a high cell-specific production rate as selection criteria. Amplification can also alter the stability of expression and often requires the continued presence of the amplification agent (discussed in Sect. 2.2.5).

Amplification of the number of copies of the immunoglobulin genes can lead to an increase in productivity of the cell line. This is not the only method for generating high yielding cell lines.

2.2.4 High Expression from Low Copy Numbers

Currently the generation of high yielding cell lines is characterized by screening large numbers of transfectants or a combination of amplification plus screening. There are problems with these approaches due to the time needed to obtain a cell line with an acceptable productivity, the potential instability of amplified cell lines, and the deleterious affect of amplification upon other characteristics of a cell line. High yielding cell lines are difficult to obtain because large regions of the genome are organized into heterochromatin, which is believed to be transcriptionally inactive. The level of mRNA expression from a vector that integrates into the heterochromatin will be low. Since there are only a few loci within the genome capable of expressing the selectable marker gene and the linked immunoglobulin genes at high levels, it follows that the probability of integration into such a transcriptionally active locus is low. Thus large numbers of transfectants normally have to be screened to isolate those few clones where the vector has integrated into transcriptionally active loci, with concomitant high product expression levels.

Two methods were recently developed to reduce the time needed to obtain high yielding clones. Both approaches exploit the importance of the chromosomal locus in determining the level of gene expression to increase the proportion of transfectants with the expression vector integrated into a transcriptionally active locus by up to 10-fold. Since these approaches generate a clone with only one or at most a few copies of the expression vector (BARNETT et al., 1995), the problems of reduced cell line stability and poor growth associated with high levels of amplification are eliminated.

The first method is the construction of high yielding cell lines by insertion, through site-specific recombination, of the gene of interest into a transcriptionally active locus. Expression vectors can be constructed that contain a specific targeting sequence that will direct the vector to integrate by homologous recombination into a particular active site. Such a sequence has been identified in the immunoglobulin locus of NSO cells (HOLLIS and MARK, 1995). Vectors containing this sequence are targeted to the immunoglobulin locus in more than 50% of high producing NSO clones. In a variation on this approach, a new host cell line is constructed in which a recombinase recognition sequence is introduced into a transcriptionally active locus. Upon transfection with an expression vector containing a sequence homologous to the recombinase recognition sequence, the vector containing the gene(s) of interest can then be specifically targeted to the same active locus. This strategy is useful when multiple products are expressed in the same cell line because the expression vector is always integrated into a transcriptionally active locus. Recombinase driven gene replacement has been demonstrated in BHK cells using the *Saccharomyces cerevisiae* FLP/FRT recombination system (KARREMAN et al., 1997). These approaches are still being evaluated, although both should rapidly produce a high yielding cell line with a low copy number for the expression vector.

A second approach is to transfect the cells with a conventional expression vector (i.e., randomly integrate the expression vector into the genome) but then bias the selection method so that only transfectants, where the vector integrated into a transcriptionally active site, are progressed. This can be done by using a selection system that only allows trans-

fectants producing high levels of the selectable marker gene product to proliferate. Expression systems using a selectable marker gene with either the weak SV40 promoter (BEBBINGTON et al., 1992) or an impaired Kozak sequence upstream of the marker gene (REFF, 1994) are included in this class of selection system.

The Kozak sequence is the consensus sequence for initiation of mRNA translation in higher eukaryotes. REFF (1994) postulated that if the dominant selectable marker gene (including *DHFR* and antibiotic resistance genes) was designed such that translation of that gene was impaired by alterations in the Kozak sequence, then only those cells that could overcome such impairment by overexpression of the dominant selectable gene would survive. This approach differs from the GS system where the naturally weak SV40 promoter, rather than a deliberately impaired RNA processing signal, is used with the selectable GS gene, although again only those cells overexpressing the selectable marker gene are allowed to survive.

When the expression vector contains a weak promoter or an impaired RNA processing signal upstream of the selectable marker gene, only transfectants with the vector integrated into transcriptionally active loci will survive selection. This is because these transfectants are the only ones that produce sufficient product from the selectable marker gene to survive. Linkage of the antibody construct to the selectable marker gene results in the overproduction of antibody as both genes are integrated into a transcriptionally active locus. The choice of selection conditions is extremely important for the success of this approach. For vectors based upon the impaired Kozak sequence system, selection is achieved by including low levels of the selective agent in the culture medium. Marked increases in gene expression can be achieved through use of impaired Kozak sequences: clones producing more than 1 g L^{-1} antibody can be obtained without amplification (REFF, 1994). For the GS system, transfected cells are simply grown in glutamine-free media. In the absence of MSX, the proportion of high producers (those having specific productivities of more than 10 pg cell^{-1} d^{-1}) is less than 2%.

Inclusion of 10 μM MSX at transfection increased the proportion to more than 20%. MSX is only needed during the transfection step (BEBBINGTON, unpublished data). This approach therefore generates stable cell lines and increases the likelihood of identifying a high yielding cell line. Since amplification is not necessary, the problems associated with amplification are avoided. But once high yielding clones are isolated, further amplification is not precluded.

Use of the vector systems described above obviates the problems associated with the conventional methods of obtaining high yielding cell lines. Since fewer transfectants have to be screened to obtain one with an acceptable productivity, the process is faster. As amplification is not necessary, this also reduces time but has additional benefits: stability is potentially improved; since the copy number of the expression vector in these cell lines is low, no metabolic burden (with its concomitant deleterious affects upon other cellular characteristics) is imposed upon the cell. The selective agent needed to maintain amplification is no longer required.

2.2.5 Stability of Expression

A critical issue in the manufacture of a recombinant biological is the ability to produce the recombinant product consistently, i.e., is the expression of the gene(s) of interest stable? Several factors have been reported to influence the stability of gene expression: the host cell line; the expression system; and the presence of a selective agent. The observation that stability of expression is influenced by a variety of factors indicates that stability is dependent upon the treatment of the cell and is not an inherent property of the cell.

Stability of product expression is cell line dependent. BROWN et al. (1992) reported data on the stability of product expression by CHO and NSO cell lines producing recombinant antibodies using the GS expression system. The CHO cell line was stable in the presence of MSX for at least 60 cell doublings, while in the absence of MSX productivity was lost by 25–30 cell doublings. It was presumed that endogenous GS activity in CHO cells

makes the need for vector-encoded GS redundant. Therefore, recombinant sequences are lost as there was no selective pressure to maintain them. The NSO cell lines were stable for at least 60 generations in both the presence and absence of MSX. This is presumably because NSO cells are obligate glutamine auxotrophs and require vector-encoded GS to survive in glutamine free medium. Thus, selection is maintained regardless of whether MSX is present or absent.

Production of antibody by amplified cell lines is not necessarily stable even in the presence of selective agents. A loss of productivity by recombinant CHO cell lines can be observed in the presence of selective drugs; RAPER et al. (1992) observed a decrease in antibody titer upon repeated subculture of an amplified DHFR-CHO cell line in the presence of MTX. A possible explanation is that a mutation in the expression vector reduced gene expression. It is unlikely that such mutants would be selected in the presence of selective agents. An alternative explanation is the selection of MTX transport mutants.

2.2.6 Summary

A variety of expression systems are available for the production of recombinant antibodies. High levels of immunoglobulin mRNA production are achieved by placing the genes downstream of strong promoters. High yielding cell lines can be generated by increasing the copy number of the expression vector by amplification but there are disadvantages with amplification and alternative methods of obtaining high yielding cell lines with low copy numbers are now available. Once created, a cell line must maintain stable gene expression and this stability is influenced by a number of factors.

2.3 Product Consistency

A structurally defined, or at least biochemically consistent, product helps to ensure reproducible and acceptable behavior upon administration to the patient. Therefore, a critical issue for the manufacture of any biological is the ability to produce consistently a product of uniform characteristics. Monoclonal and recombinant antibodies are glycoproteins with oligosaccharides attached to the C_H2 domain within the Fc region. These oligosaccharides are important for effector functions (phagocytosis, complement activation, etc.) while the composition of the carbohydrate moieties may influence the behavior of the antibody, with regard to its immunogenicity, pharmacokinetics or specific activity. *In vivo*, these moieties exhibit microheterogeneity due to stochastic elements in the synthetic process. The distribution of glycoforms is reproducible for a particular glycoprotein produced using mammalian cell culture but this may be different from the *in vivo* profile. The purification method can then further alter the glycoform distribution observed in the final product. Intentional or unintentional changes in the bioprocess can alter the glycoform distribution, which consequently may affect the *in vivo* behavior of the antibody. Therefore, the bioprocess must be reproducible to minimize variation in the glycoform distribution.

Various bioprocessing factors influence the composition of the carbohydrate moieties: these factors include choice of host cell species, cell type, and culture conditions. The choice of species and type for the host cell can have a marked effect upon glycosylation and, consequently, biological activity (LIFELY et al., 1995; JENKINS et al., 1996; KOPP et al., 1997). This appears to be related to the distribution of glycosyltransferases and glycosidases and their kinetic characteristics (GRAMER and GOOCHEE, 1994). The culture system itself can also affect the glycoform distribution and physiochemical properties of an antibody (GOOCHEE et al., 1991). Downstream processing also affects glycoform distribution in the purified product as oligosaccharide composition affects the behavior of glycoproteins in chromatographic systems based upon ion exchange and hydrophobic interaction. Since the host cell species and type, culture system and the purification methods will be defined before the process is used to supply either clinical trials or the market, the variation in culture conditions becomes the major factor affecting glycoform distribution during the production process.

A variety of culture conditions have been shown to influence glycosylation (GOOCHEE et al., 1991; JENKINS et al., 1996). These include glucose and ammonium concentrations, hormonal content of the media, pH, culture age, and serum. The importance of these conditions has been demonstrated for a number of glycoproteins including antibodies. Mechanisms that could explain these effects include: alteration of the cellular energy state; changes in substrate concentrations; disruption of the ER and Golgi; interference in vesicle trafficking; or modulation of enzyme activities.

For recombinant antibodies, unlike other recombinant glycoproteins, the influence of variation in oligosaccharide composition upon immunogenicity and clearance times of parenterally administered therapeutic antibodies may not be important (BERGWERFF et al., 1995). In IgG immunoglobulin molecules the N-glycans are enclosed by the peptide chain of the Fc region so the various epitopes generated by using different host cell lines may be masked from the immune system. The role of carbohydrates in the clearance of antibodies is still debatable. Currently the prevalent view is that the carbohydrate moieties may have only a limited effect upon circulatory half-life. Thus the changes in glycoform distribution caused by variation in the cultural conditions may only be of minor importance. However, this must be balanced by the observations that changes in glycoform distribution can affect biological activity (LIFELY et al., 1995).

Heterogeneity can be found in the polypeptide chains due to mutations in the immunoglobulin genes (HARRIS et al., 1993). This will generate a polyclonal cell population but, if the mutation is silent, the mutation may not cause product heterogeneity. Polypeptide clipping by extracellular protease activity can also change the product identity (FROUD et al., 1991). The level of clipping can be influenced by the cell type as well as culture conditions but, as the choice of cell line will have been made early in process development, variation in culture conditions becomes the main influence upon polypeptide clipping (FROUD et al., 1991).

2.4 Cloning and Selection

The primary function of preparative cloning is to ensure that all the cells in daughter cultures are descended from a single cell. Transfection results in the random integration of the expression vector into the genome. Although the selection system may allow only specific subpopulations of phenotypes to proliferate, these subpopulations will still be genetically heterogeneous as the integration loci will not be identical. In turn, different subpopulations will be selected as the culture conditions change. This may result in unpredictable variation as one subpopulation is favored over another, making the consistent production of a product with uniform characteristics difficult. By ensuring that all daughter cultures originate from the same cell, process consistency is improved. It should be noted that a cloned cell line can still change with time. This can be either a change in the genotype, as a result of the accumulation of mutations, or in the phenotype due to altered environmental conditions.

The selection system used following transfection may increase the proportion of high producing cells within the transfectant pool. Cloning permits the isolation of individual clones from this pool.

A variety of cloning techniques are available, although none are ideal. Limiting dilution cloning should be used with suspension cells or cells that are very mobile when attached. The problems with this method are due to the false assumption that every viable clone has the same random chance of monoclonality (UNDERWOOD and BEAN, 1988). The probability of monoclonality is maximized by using several rounds of limiting dilution cloning (COLLER and COLLER, 1986). This method is the most widely used during the generation of production cell lines. Capillary aided cell cloning (CLARKE and SPIER, 1980) has the advantage over limiting dilution cloning in that clonality is confirmed by visual inspection. Thus only a single round of cloning is required, decreasing the time requirements. Other alternative cloning methodologies use soft agar or cloning rings.

Typically several criteria are used to select the production cell line from a range of available clones. The criteria include: a high cell specific production rate; growth characteristics such as length of culture (this influences the occupation time of production facilities) and maximum cell concentration; harvest titer; cell line stability; relationship between growth and production kinetics; and product integrity. However, the weight applied to each of these criteria can vary, for example, depending upon whether the aim is to run a continuous or a batch process (RACHER et al., 1994). The importance of screening prior to cell line selection in a system that has relevance to the manufacturing process was demonstrated by BRAND et al. (1994). They found that there is poor correlation between productivity of recombinant myelomas in static culture (cloning plates and flasks) and agitated suspension culture.

2.5 Serum-Free and Suspension Adaptation

Serum in cell culture media is used to supply growth and attachment factors, matrix proteins, and transport proteins carrying hormones, lipids, minerals, etc., while serum albumin has protective and detoxification functions. These components are often required for cell proliferation and function. For some media, serum may also function as a source of amino acids, etc. However, there are disadvantages with using serum. Serum is a complex substance with a high protein content (including immunoglobulins). This makes it difficult to recover and purify products present at lower concentrations. As serum is of animal origin, concerns arise about the presence of adventitious agents, lot-to-lot variability in composition while cost is also a major issue. To circumvent the need for serum, various serum- and protein-free media that support cell growth and function have been developed. These media offer the advantage that they are free from adventitious agents, typically of lower cost and they are better defined chemically. Such media also tend to offer greater consistency in cell growth and product uniformity plus simplification of product recovery and purification. For the manufacture of antibodies, serum-free culture also eliminates a major contaminant (the serum-sourced immunoglobulins) of the purified product.

Typically, cloning and cell line selection is done using serum-containing media yet the manufacturing process uses serum-free media. The cell line must, therefore, be adapted to serum-free growth. Hybridomas and myelomas can be rapidly adapted (1–2 months) to serum-free media by a process of sequentially lowering the serum level until the cells are growing with a much reduced serum concentration or without serum. CHO cell lines can also be adapted using this approach but may require more intermediate steps over a longer period. The use of a nutritionally optimized medium, with minimal growth factor supplementation, may hasten serum-free adaptation.

Manufacture of antibodies is frequently undertaken in suspension systems in which the cells are not attached to a surface. Transfection and cloning are done in static culture in which the cells are attached, possibly only loosely, to a surface. The purpose of suspension adaptation is then to obtain a cell line that will grow unattached to a surface. Adaptation to suspension culture can be achieved by a variety of methods (MATHER, 1990) and can be done before, after or during adaptation to serum-free media. Some cell lines, e.g., those derived from some BHK strains or NSO cells, start growing in suspension within one or two passages using this approach. Other cell lines, including some CHO strains, can be more problematic.

The time needed to adapt a recombinant cell line to suspension culture is variable but can be reduced by preadapting the parent cells (SINACORE et al., 1996). These workers showed that preadaptation of CHO DUKX cells to serum-free suspension culture reduced the overall time from transfection back to suspension growth in serum-free media.

The process of adaptation to suspension culture in serum-free medium can dramatically reduce the number of cell lines from which the final production cell line is finally selected. This is because the productivity of an individual clone can change during adaptation. For example, ROBINSON et al. (1994) screened 179 transfectants and identified 17

(ca. 10%) as being high producers. Four (ca. 2%) remained as high producers after adaptation to growth in serum-free suspension culture.

3 Cell Banking

Cell banking is the first stage of the manufacturing process. A bank system for a cell line should provide a source of cells sufficient for the entire lifetime of the product. The cell banks are also one of the major stages of consideration for the regulatory authorities during the approval stage of a product. Consequently they must be produced under cGMP conditions (see Sect. 9).

The first important stage of a cell bank preparation is the derivation of the cells used to create the bank. Ideally there should be sufficient details to permit a complete history of the cells to be traced, right back to the original donor organism. If the cell line is of human origin or is derived from fusion with a human cell line, then details of the original human donor such as health and viral status (e.g., absence of hepatitis and HIV) are essential.

Preparation of the cell banks should begin with cells only a few generations from the origin of the specific cell line. If the cell line produces a recombinant product, full details of the construction of the vectors and transfection methods are required, together with subsequent subculturing, adaptation and selection procedures. Thus access to appropriate development records will be required to support the regulatory review of a cell bank. Similarly details of the fusion and subsequent selection of a hybridoma cell line are required. In either case, a prime concern is to ensure that the cell line is clonal before starting preparation of the cell bank.

All cell lines should be banked using the two tiered banking systems of a master cell bank (MCB) that is then used to create a working cell bank (WCB – sometimes termed a manufacturer's working cell bank, MWCB) (WIEBE and MAV, 1990). Each production batch is then derived from an ampule of the WCB. In each instance, the method of preparation should be a predefined and well-documented method within a GMP system. If one WCB is exhausted, subsequent banks can be prepared from the MCB so the MCB must be suitably sized. Often early development work may not require the creation of full cell banks but investment in cell banks at these early stages can be advantageous as these will be required further down the clinical development process. A example of how a cell bank system functions is shown in Fig. 1.

Once created the cell banks must be stored in a suitable manner. Cryopreservation using liquid nitrogen is the norm, with either liquid or vapor phase storage. The advantages of the liquid phase are that the temperature of ampules throughout a dewar is more consistent and the speed of equilibration to this temperature is faster (ONADIPE, unpublished data). In the vapor phase the risk of cross contamination of either microbial or viral contaminants between ampules is greatly reduced. In either case full documentation of the storage, labeling and control of issue of the correct ampule is required.

Characterization of a cell bank requires examination of a range of parameters. Firstly, each bank must be homogeneous. This can be ensured by using a single pool of cells for the aliquoting of the ampules of a cell bank. Once frozen the homogeneity of the banks can be assessed by reviving a sample of ampules across the bank and comparing the growth and viability characteristics of the cultures. The banks must also be shown to be free from microbiological contamination (bacteria and fungi) and from mycoplasma. The viral status of the banks should elucidated out by a range of tests determined by the species of origin of the cell line. An assessment of the clonality of the MCB is often required by regulatory authorities. Fuller details on the testing that should be performed are defined elsewhere (CBER, 1993; MCA, 1994). Testing of banks must be well documented and audited to be suitable for inclusion in regulatory submissions.

4 Cell Line Stability

As previously discussed, the ability to manufacture consistently a product with uniform characteristics is critical for any bioprocess. Product consistency is affected by a number of parameters including choice of cell line, culture system and culture conditions. The major source of variation during the manufacture of antibodies is variation in culture conditions, but as was discussed above, stability of expression is not guaranteed even in the presence of selective agents. Modern analytical techniques allow accurate monitoring of the impact of process changes upon the product, however product analysis does not eliminate the need to demonstrate cell line stability. Stability can be demonstrated by showing that the cell line produces the same product throughout a simulated or real production process (CASTILLO et al., 1994). Stability can be studied at the level of the phenotype, e.g., productivity of the cell line, or by looking for changes in DNA sequences, mRNA levels or gene copy numbers. Changes at the DNA or RNA level will often be reflected in the variation in productivity and growth kinetics of the cell line with increasing generation number.

Regardless of how cell line stability is assessed, a key parameter of any study is its duration. Duration is best described by the increase in generation number of the cells, which is the cumulative total of population doublings (calculated as described in MORRIS and WARBURTON, 1994) from an arbitrarily defined starting point. Often the starting point for counting generation numbers is either the last cloning performed during the creation of the cell line or the frozen cell stock from which the cell banks are prepared (see Sect. 3). In either case the generation number is arbitrarily set to zero at this point.

Stability of the cell line should be considered over a generation number range equivalent to a period from the initiation of the pro-

Fig. 1. Function of a cell bank system.

duction process to at least, and often beyond, its normal end. For most situations this generation number range will start at the WCB, with the end being the maximum generation number that would occur in the production system (often an additional safety factor of up to 50% is added onto this end point). In a batch culture production system the study period will be easy to define. However, for a continuous or perfusion culture process the end point may be less easy to define, and the elapsed generations that will need to be studied could be quite large.

The stability study should ideally be performed on the full production process, but this is often too complex and costly to implement. Therefore, a simplified laboratory version of the production process can be used, with the generation number increased by repeated subculture using a method similar to the process used to prepare inoculum cells from a cell bank. At regular intervals simple overgrow cultures in conditions similar to the final production can then be used to assess the various stability parameters discussed below.

Having defined the range of generation numbers over which to examine the cell line and the culture method to use, a formal stability study must be implemented to examine the stability parameters. A major parameter in any stability study is product formation. The cell line should be shown to produce the same product, both in terms of identity and quality, throughout any production process. Other stability parameters to examine include cell growth, maximum cell concentration and culture duration as well as product parameters such as product accumulation, specific production rate and product quality. The change in each parameter that is considered acceptable will depend on the parameter concerned.

Two parameters that should not change are product identity and product quality. The importance of product quality depends upon the nature of the product. The exact nature of the product and how changes in quality will affect its usability and efficacy will determine the degree of characterization during the stability study. An overall assessment of the stability of the cell line can be made once all these parameters have been analyzed, with some parameters having more importance than others depending on the nature of the production process. An alternative to a separate stability study is a planned, detailed examination of several production batches.

Such stability studies are a key part of regulatory submissions to obtain either clinical trial or market clearance. Thus the studies must be carried out in a controlled manner and all data fully reviewed. The regulatory guidelines for cell banks are covered by such documents as the *"Points to Consider in Characterization of Cell Lines"* (CBER, 1993).

Cell lines that use recombinant DNA technology, and especially those that employ amplification to enhance productivity, need careful consideration of their stability (PALLAVICINI et al., 1990; HASSELL et al., 1992). Recent regulatory updates have included consideration of genetic stability as a major factor in the stability of cell lines for the production of recombinant products (CBER, 1992; ICH, 1996). These guidelines cover four complementary features of genetic stability of a cell line:

1. vector DNA source, construction and description,
2. vector characterization in the master cell bank,
3. *in vitro* cell age limit validation,
4. end of production cell testing and/or product characterization.

The above issues need to be examined but do not require a continuous stability study to be set up as with the product formation study described above. The latter two issues need to be considered at either the pilot or manufacturing scale production stages.

5 Production Systems

A consequence of the stoichiometric interaction between antibody and antigen molecules is that many therapeutic applications of antibodies require high dosages (0.5 to more than 5 mg kg^{-1}). To supply in-market demands would require production of 10-100 kg per year. Therefore there is a need for large culture volumes of cells engineered with high expression constructs as the first stage in a highly productive process. A number of approaches exists to increase the product titer at harvest. The titer can be raised by increasing the specific production rate by manipulating either the genotype or the phenotype. As discussed above, cells with suitable genotypes can be obtained through careful choice of host cell line and expression vector, and by using the appropriate criteria for cell line selection. The phenotype can be altered to a high producing one by manipulation of the bioreactor environment. Conditions that stress the cells and reduce the growth rate are reported to increase product titers of hybridoma cultures through enhancement of the specific antibody production rate (SUZUKI and OLLIS, 1990; PARK and LEE, 1995). Gene expression can be manipulated more directly by inclusion of growth factors in the medium.

An alternative to increasing the specific productivity is to increase the cell space–time yield by extending the length of the culture or increasing the cell concentration (FIELD et al., 1991). This can be achieved through better bioreactor design, improved process control, better medium design and the use of feeds.

A variety of bioreactor types have been used to produce recombinant antibodies for use for either clinical trials or market supply. The two main types of bioreactors used commercially are airlift bioreactors (ALF) or stirred tank bioreactors (STR). ALFs up to 2,000 L (RHODES and BIRCH, 1988; RAY et al., 1997) are currently used for commercial production while STRs of 8,000 L (KEEN and RAPSON, 1995) or larger are also used. Hollow fiber bioreactors are often used to produce small quantities of antibody but are not suited for large scale operations.

The performance of both ALFs and STRs can be improved by design. For example, different aeration systems can have dissimilar gas transfer characteristics. This influences the maximum cell concentration that may be supported as a result of different oxygen supply or carbon dioxide removal rates. Bio-

reactor geometry affects mixing: poor mixing can lead to a heterogenous environment within the bioreactor due to the establishment of nutrient or waste metabolite gradients. In turn this can impact upon product yield if zones with conditions suboptimal for growth or productivity are established. Poor design can create regions within the bioreactor where the cells are exposed to large hydrodynamic forces, e.g., around the impeller in a STR, that can damage the cells. Better bioreactor design can improve yield by increasing the cell concentration that can be supported or alternatively by reducing cell death due to fluid mechanical damage.

An ideal bioreactor process would maintain cells in a physiological state that ensures high growth, viability and productivity, with a product of an acceptable quality. Improved process control therefore has the potential to enhance process yield and consistency. Realization of improved control requires such considerations as: on-line measurement and estimation of key process variables; a model to predict how the cells will respond to process changes; feedback control; and the hardware to implement control actions.

A bioreactor can be operated in a variety of modes including batch, fed-batch, continuous or perfusion. A batch culture is a closed system in which cells and media are added to the bioreactor at the start and the culture is harvested at an appropriate point. No further additions of media are made. In a fed-batch culture small volumes of nutrients or other substances are added to the culture during the growth cycle to enhance or sustain its growth and productivity. For continuous and perfused cultures, fresh medium is continuously added to the bioreactor and spent medium continuously removed. The medium and, if applied, the feed are therefore critical components in optimizing cell growth.

Basal media, e.g., RPMI 1640 or Ham's F12, provide many soluble, low molecular weight nutrients to the cell. However, most of these media were developed to support clonal growth of relatively quiescent cells, not the high densities (more than $2 \cdot 10^6$ cell mL^{-1}) of cells secreting high levels of proteins found in modern commercial cell cultures. Optimization of the basal medium can lead to dramatic improvements in product titer (KEEN and HALE, 1996). Key steps in such process development are identifying limiting nutrients and inhibitory by-products, determination of maximum tolerated levels of supplemented nutrients, and optimization of the mixture of nutrients in the feed mixture. The process is iterative, requiring several passes because changes in one nutrient may affect toxicities or requirements for other components.

In batch and fed-batch cultures, the final product concentration is the product of the cell-specific productivity and the time integral of the viable cell concentration. Most successful fed-batch processes aim at maximizing these two parameters. By identifying and responding to the limitation of specific nutrients, the time integral of the viable cell concentration can be increased. Periodic replacement of spent medium components (amino acids, lipids, proteins, glucose) improved the productivity of a NSO cell line producing a recombinant antibody from 80–100 mg L^{-1} to 850 mg L^{-1} (ROBINSON et al., 1994). A stoichiometric model describing the nutritional balances in a mammalian cell was used to design a feed for stoichiometric feeding in fed-batch cultures (XIE and WANG, 1994a, b). This approach has increased the product titer at harvest to 2.4 g L^{-1} (XIE and WANG, 1996), almost 50-fold higher than the original batch cultures. This increase resulted from an increase in the time integral of the viable cell concentration rather than an increase in the specific production rate. For some cell lines, addition of a feed can increase the cell specific productivity. However, sometimes, although there is an increase in the time integral, this is coupled to an overall decrease in cell specific productivity. Since transfection and amplification can markedly affect metabolism (BIBILA et al., 1994), feeding strategies and feed composition are generally cell line specific.

Once both the cell line and production system have been decided upon, along with any optimization of the medium or feeding strategy, the actual large scale manufacture of the product can be started.

6 Process Definition

Arguably the most critical phase of the development of a process for the manufacture of all biologicals is the transfer of the ideas and concepts of a process from the people who have been developing it to those who have to operate it at the manufacturing scale. To do this effectively, the process must be clearly defined so that nothing is omitted in the transfer. In this phase of work a consideration of the transfer between scales of operation of the process must be considered. A process that is highly effective in development situations but is problematic or significantly less effective at the manufacturing scale is undesirable. This concept requires that those undertaking process development must always keep in mind the scale and environment in which the final process will operate.

Process definition has to start by identifying the key parameters of a process. High on the list of these will be the viable cell concentrations expected at the various points of the process. Not only should ideal values be set for these types of parameters but also a working range should be examined to allow for flexibility and robustness. For example, the seeding viable cell concentration for the inoculum grow-up phase of a hybridoma process might have a value of $1.0 \cdot 10^5$ cell mL^{-1}. If, when seeded at $0.8 \cdot 10^5$ cell mL^{-1} the culture lags for half a day longer than normal, thus upsetting the subsequent subcultures and initiation of the fermentation process, the overall process would not be reliable since small variations are inevitable in routine manufacturing. Obviously a consideration of the accuracy normally attainable with any particular parameter is required before a suitable range can be set. Again considering the cell concentration example, a range of $\pm 0.5 \cdot 10^5$ cell mL^{-1} would normally be sensible. If $1.0 \cdot 10^5$ cell mL^{-1} was the desired cell concentration then tests of both 0.5 and $1.5 \cdot 10^5$ cell mL^{-1} should be examined to ensure that acceptable growth patterns for that stage of the process can be achieved. The parameters needing detailed consideration will depend upon the exact details of the process being developed. For example, the growth rate observed in the final production bioreactor might be a key concern in a stirred tank system but if a hollow fiber system was employed then such a parameter would be of less importance.

Parameters describing both the cells, such as concentration, growth rate and viability, and non-cell factors, need examining. Non-cell related parameters include the physical characteristics of the process such as temperature, pH, dissolved oxygen concentration and many others. For those characteristics that are actively controlled during the process a consideration of both alarm limits (when operators are alerted to a problem) and alert limits (past which an assessment of effect upon the product is required) must be made.

Many of these parameters can be examined at the development scale but caution should be exercised during such work to ensure that the process is operated as closely to the manufacturing scale as is possible. Sometimes minor alterations in practices can have major effects on the overall efficiency of a process. Pilot scale fermentations are an invaluable part of any process definition program. A pilot scale facility should, wherever possible replicate the manufacturing scale only at a smaller capacity. Again the exact nature of the process will dictate how similar the two scales can be. For example, with a hollow fibre system, the move to the manufacturing scale may be made by multiplexing the number of units used so a pilot scale of a single unit can exactly mimic the larger scale. In suspension fermentation, the increase in scale is a more complex process. The increase in volume, being the normal measure of scale, is not mimicked by similar changes in other significant parameters. These include mixing times, shear levels and gaseous conditions. Pilot scale runs also allow other features of a process to be examined in conditions similar to those to be used in manufacturing. An example of this would be the change to steam and clean in place procedures. Most development scale reactors are not steam in place and are manually cleaned but the use of steam and clean in place procedures is common at larger manufacturing scales. The usefulness of the pilot scale bioreactors will be determined by how well results generated from them can be

used to predict the behavior of a process at the manufacturing scale.

Once all these types of considerations have been made and successful pilot runs have occurred then the final process transfer can be made. This requires both the transfer of process details into the manufacturing plant but also the generation of the full documentation required to operate the process under GMP.

Once a process is in the final manufacturing plant the routine operations will be dictated by the GMP defined procedures and directions. Regulatory bodies require that the process used is consistent; so often the first stage of full scale manufacturing is to produce several consistency batches. During these batches the analysis of the running of the batch will be more detailed than would normally occur. Additional samples at all stages of the production process are required, together with an examination of stability of both the final product and the various intermediates that might be stored for short periods during manufacture. The data from such consistency lots often form a key part of the manufacturing details that support the licence application for a therapeutic antibody.

Routine operation for market supply needs less monitoring than the consistency lots but the performance of the process should be monitored and trended to ensure that no changes occur over time. Once the process is running at manufacturing scale there is often a desire to increase its productivity or efficiency. These can be implemented in a manufacturing process, but an assessment of the significance of the process change needs to be made. A minor alteration may not need major regulatory submissions beyond notification but more major changes may require repeats of consistency lots and demonstration of product equivalence between the old and new process.

7 Primary Recovery

Primary recovery involves the harvesting of the cell containing media for use in the purification process. It can be considered as part of downstream processing or as part of the harvest procedure but conventionally it is considered as one or more unit operations that need to integrate well with both the preceding and subsequent steps in the manufacturing process. The main activity of primary recovery is usually to remove cells (if present) and debris from the harvest stream. The harvest will often require some form of polishing or preparation before the first purification step. This may include concentration of the bulk product to reduce the volumes to be handled and filtration to remove particulates. Filtration can also be used to produce a low bio-burden intermediate that may be stored within a defined expiry date to enable more convenient scheduling of purification activities.

Cell and debris removal is conventionally achieved either by centrifugation or by some form of microfiltration. These options are discussed in detail in BERTHOLD and KEMPKEN (1994) and in WHITTINGTON (1990). Following clarification of the harvest, the subsequent steps can include sequential filtration through reducing pore size filters to achieve microbiological and submicron sized particle clearance. Concentration can be performed by ultrafiltration and other steps such as virus inactivation can be included during the primary recovery phase.

An alternative method of product recovery has recently been introduced that promises a much simpler overall process (THÖMMES et al., 1995). Expanded bed chromatography allows crude feedstocks which may contain cells and cell debris, to be applied to a chromatography column without pretreatment. Filtration or centrifugation are therefore not required, and if used in combination with an appropriate affinity ligand, expanded bed chromatography allows for a very simple and elegant combined recovery and purification process.

The optimum recovery method is dependent on a number of factors including the cell line being used and the type and scale of fermentation system employed. Care must be taken during scale-up to ensure that a consistent intermediate product is transferred for downstream processing irrespective of the scale of operation. Even when similar cell lines are used (such as hybridomas) in the same fermentation system significant differences in the characteristics of the harvest mean that generic methodologies for primary recovery require detailed examination for each process to ensure adequate performance.

8 Antibody Purification

Therapeutic antibodies, like all medicines, must meet high standards of safety, efficacy and potency. This translates into very high purity levels, with the suggested target of no more than 1 ppm of any contaminant present in the purified product, i.e., 1 ng of contaminant per mg of monoclonal antibody (CBER, 1994). Even higher specifications relate to DNA and viral contaminants. However, the definition of "contaminant" is not straightforward as "related" or "modified" forms of the antibody are always present in all cell culture supernatants and these molecules can be extremely difficult to remove. Analytical methods such as tryptic mapping and capillary electrophoresis can detect protein molecules modified through oxidation or deamidation of amino acid residues (CLARKE et al., 1992) or proteolytic or glycosylation variants (RAO and KROON, 1993). For example, highly purified antibody preparations appear as "pure" single-component preparations by many different analytical techniques, e.g., SDS-PAGE or gel permeation HPLC, but appear as a number of discrete bands by charge-based separations such as isoelectric focusing or capillary electrophoresis. Whether these molecules are considered contaminants is debatable. A pragmatic approach is that the purified antibody product must be equivalent to the product used in clinical trials and toxicology studies, and must be equivalent to the reference standard laid down early in the product development program.

The term antibody can be used to describe a wide variety of different molecules ranging from IgM molecules with a molecular weight (mw) of approximately 900,000 to small fragments of antibodies with mw of 25,000 or less. All these molecules are of clinical interest in various different indications (MOUNTAIN and ADAIR, 1992; LADYMAN and RITTER, 1995), and they can be produced in a number of different systems, e.g., microbial or mammalian cell culture, transgenic animals, plants, etc. Despite this wide diversity in molecular type and production system, all therapeutic antibodies need to be purified to similar standards. Although the exact specification for a therapeutic antibody can vary depending on dose and indication, a typical specification for a bulk purified product will contain a large number of tests that measure purity, integrity, identity and safety of the product. Typical purity specifications are shown in Tab. 1. A much greater range of analytical tests is used to characterize key batches, e.g., those used for clinical trials, for reference standards, or for process consistency batches.

In addition to the obvious requirement that the purified product meets the specification, there exist a number of other requirements, e.g., robustness, economy, scaleability, etc., that are also important in the design of a successful purification process.

8.1 Purity Requirement

8.1.1 Protein Contaminants

The 1994 draft, *"Points to Consider in the Manufacturing and Testing of Monoclonal Antibody Products for Human Use"* issued by the FDA (CBER, 1994) suggest that "wherever possible, contaminants should be below detectable levels using a sensitive assay capable of detecting 1 ppm expressed on a weight basis with respect to the mAb". In practice, this often requires immunological assays such as ELISAs developed specifically to quantify known contaminants. This approach is likely to be required to measure the amounts of any cell culture medium additives, e.g., bovine serum albumin, transferrin, methotrexate, or any process related contaminants such as protein A or G molecules leached from chromatography columns. Host cell proteins also need to be measured and the development of such assays is not a trivial task as potentially a very wide range of molecules could contami-

Tab. 1. Typical Purity Specification for a Bulk Purified Therapeutic Grade Monoclonal Antibody Prior to Formulation and Vialling

Test	Comments
General	
Appearance	Color of solution, presence or absence of any particulate or "extraneous" matter
Protein concentration	Determined using Bradford assay or A280nm
Purity	
SDS-PAGE with Coomassie blue and silver staining	Purity expressed as percentage on the basis of scanning densitometry and also compared to the reference standard
Gel permeation HPLC	Used to measure antibody aggregate content. Result expressed as percent monomer and/or percent aggregates
Specific contaminants from the fermentation medium, e.g., serum proteins, methotrexate, etc.	Results expressed as ng contaminant per mg antibody, i.e., ppm
Specific contaminants from the process, e.g., protein A or G residues	Results expressed as ng contaminant per mg antibody, i.e., ppm
DNA	Expressed as pg per dose or pg per mg (ppb)
Host cell protein content	Results expressed as ng contaminant per mg antibody, i.e., ppm
Endotoxin level	e.g. <1 Endotoxin unit mL^{-1}
Virus tests	Results of tests on in-process or purified bulk product included

nate the purified product (EATON, 1995). Fortunately, with highly productive cell lines grown in serum-free or protein-free media, the purity target is usually not too difficult to achieve. This is particularly true when affinity chromatography is used in conjunction with one or more ion exchange separations.

8.1.2 DNA

The DNA specification is tighter than that of other contaminants with a recommended limit of no more than 100 pg cellular DNA per dose (CBER, 1994). Often a target specification is set in the early stages of a product development program before the dose has been finalized, and in these cases a specification is set based on the weight of antibody, e.g., 1–10 pg DNA mg^{-1} antibody. It is not unusual for the specification to be set at the limit of quantification of the assay, as it is technically quite challenging to measure pg or sub-pg quantities of DNA in a concentrated protein solution. The level of DNA clearance required can be equivalent to 10 logs of clearance or more.

8.1.3 Viruses

Many mammalian cell lines are known to harbor viruses and there are cases where biopharmaceutical products have been contaminated by viruses which have infected patients (SHADLE et al., 1995). Therefore, viruses present a particular issue for the production of antibodies when mammalian cell lines are used (LEES and DARLING, 1995). In addition to a thorough characterization of the cell line used for antibody manufacture and the testing of in-process samples, there are guidelines issued by US and European regulatory authorities (CPMP, 1996; CBER, 1994) relating to the incorporation of virus removal or inactivation steps in the purification process. In general, it is recommended that purification schemes use at least one "robust" virus removal or inactivation step such as solvent/detergent or low pH treatment, or virus removal filtration steps. In addition, the chromatographic procedures used should also remove

or inactivate viruses. The overall level of virus removal or clearance required is set by the number of virus-like particles detected in the unpurified bulk supernatant, which can be of the order of 10^8 or 10^9 mL^{-1} (POILEY et al., 1994). A safety margin of 3–6 logs is suggested (CBER, 1994) but it has been proposed that the safety margin should be variable and be dependent on the consequences of accidental exposure to the virus and the likelihood that a single virus particle will be pathogenic (SHADLE et al., 1995). A similar strategy is to base the calculation of virus removal or clearance required on the probability that a single dose of product will contain a virus particle. The figure that has been used for tPA is a probability of less than 1 in 10^6, i.e., less than 1 virus particle in a million doses (LUBINIECKI et al., 1990). Typically, these types of calculations require a minimum of 15 logs of clearance/inactivation for the whole purification process.

8.2 Robustness

In addition to the quantifiable targets in terms of the clearance of various contaminants, a key aspect of the antibody purification process is its "robustness" or tolerance to small variations in processing conditions. Robustness is directly related to the failure rate, and with considerable time and money invested in the upstream production of a batch of antibody, the failure rate must be kept to an absolute minimum. It is essential to determine for each step of a purification process the key parameters that influence the performance of that step. For example, ion exchange operations are sensitive to the pH and conductivity of the various equilibration, wash and elution buffers used, and it is important to design the step such that minor variations in these parameters do not result in a decreased purification performance.

As an additional precaution, some redundancy is frequently built into a purification process so that should one step perform poorly, subsequent steps can make up the deficiency and the final bulk product will still meet specifications. Thus the penultimate or the final step may be a polishing step that can remove process contaminants but in most cases does not actually do so, as they have been removed by preceding steps.

Another aspect of robustness relates to the variable titer expected in the upstream cell culture operations. There is always some inherent variability in the growth and synthesis of product in any mammalian cell culture system, and this results in a range of product titer expected in the unpurified cell culture supernatant.

The first chromatography step in particular must be robust enough to cope with this variability in titer. The easiest mode of operation is to load the entire contents of the bioreactor on to the first chromatography column irrespective of the exact titer or product-to-contaminant ratio. However, this may require a large margin of safety to ensure that overloading the column does not occur even with the maximum expected titer. This has the disadvantage that for most batches of product the chromatography column is larger than necessary with greater plant capacity, larger tanks, etc., than actually required for the quantity of antibody produced. Alternatively, the product titer in the unpurified supernatant may be determined and the cell culture supernatant divided into a number of aliquots which are then loaded on to the column as successive cycles. This results in better utilization of the matrix and plant capacity but requires a variable production schedule depending on product titer, and may require the process to be interrupted awaiting assay results. Which mode of operation is selected will depend on many factors such as the stability of the molecule, the design of the production plant, the time available to purify one batch of antibody, etc. With both modes of operation, it is essential to test the process with a wide range of supernatants that cover the expected variations in product titer.

8.3 Economic Aspects

Increasing attention is being paid to the economic aspects of all biopharmaceutical production processes. Although there is not a great deal of published information on the contribution of downstream operations to overall costs in biopharmaceutical manufacturing and how it breaks down, one aspect of downstream processing economics is the high cost of make-up and storage tanks for the various solutions required (DATAR and ROSEN, 1990; DATAR et al., 1993). Therefore, it is important to keep the number of different solutions required to run a process to a minimum. For most chromatography operations at least four different solutions are required: an equilibration buffer, an elution buffer, a regeneration solution and a storage solution. For more complex chromatography steps, e.g., metal chelate chromatography, the number of different solutions required can be significantly higher, with additional steps for the loading and stripping of the metal ion. The scheduling of tank usage is therefore a key aspect of plant operation and can have a significant impact on annual throughput.

Another aspect of production economics is the reuse of chromatography matrices and ultrafiltration membranes (SEELY et al., 1994; BARRY and CHOJNACKI, 1994). While in most cases reuse of matrices and membranes is essential for a cost effective process, in some cases it may not be essential. For the production of one or a small number of batches, e.g., for early clinical trials, the cost of validating, cleaning and reuse of the matrices may outweigh the cost of purchasing new matrices. Also, some of the matrices may bind contaminants particularly tightly and if it is not possible to demonstrate recovery of that contaminant in one of the wash steps, it can be difficult to ensure that there is not a build up and subsequent leaching of the contaminant in successive cycles. Therefore, the cost of a validation study demonstrating the acceptability of matrix or membrane reuse must be balanced against the cost of new matrices.

Affinity matrices such as protein A or G resins, are often the most expensive single item on a list of direct materials used in antibody production, and their expense is often quoted as one key disadvantage of affinity methods. Fortunately, protein A is a very robust molecule and relatively harsh cleaning conditions are possible. Therefore, it is feasible to validate reuse for 50 cycles (BAKER et al., 1994) or more. In these cases the initial cost is spread over many batches, and the protein A cost per batch can become minor.

8.4 Purification Methods

The addition of affinity tails to protein products by genetic engineering techniques has been used as a means of facilitating purification (NYGREN et al., 1994). Fortunately, IgG antibodies have a built in affinity tail that interacts with a widely available and well characterized ligand, making affinity chromatography a particularly attractive option. Affinity techniques can also be useful for certain antibody fragments that do not have an Fc "tail" (PROUDFOOT et al., 1992). Although protein A affinity chromatography has its critics (DUFFY et al., 1989), most of its disadvantages (e.g., high cost and protein A leakage) can be overcome relatively easily by careful process design. The leakage of protein A residues from various matrices is well characterized (GODFREY et al., 1993; ROE, 1994), and validated assays are available that can reliably measure nanogram quantities of protein A (BLOOM et al., 1989). Protein A matrices are undoubtedly expensive, but there are now many published guidelines relating to matrix reuse validation (LEVINE et al., 1992; SEELY et al., 1994; BARRY and CHOJNACKI, 1994) that enable the high initial cost of the resins to be depreciated over many batches of the antibody product. A further advantage of protein A affinity chromatography is its tolerance to a very wide range of process conditions, making for a very robust process (KANG and RYU, 1991). For example, protein A is much less sensitive to conductivity, pH and loading variations than ion exchange or hydrophobic interaction chromatography (Pharmacia, 1993). Consequently, a large number of purification processes for therapeutic antibodies now utilize protein A or G matrices.

Another commonly used method for antibody purification is ion exchange chromatography (KENNEY, 1989). Affinity chromatography and ion exchange chromatography are complementary techniques as many processes use an affinity step first to obtain antibody purities of more than 95%, followed by one or more ion exchange steps to remove the remaining trace amounts of impurities, including leached affinity ligand. With good column washing procedures it is not unusual to obtain a very clean preparation after a protein A or protein G affinity step, with only antibody related bands visible on a SDS-PAGE gel even when visualized with a sensitive silver stain. This is generally possible even when cells are grown in protein containing cell culture media. When the cell culture medium contains bovine serum, it can be difficult to reach the 1 ppm suggested target specification for bovine polyclonal antibodies derived from the serum, since some of these antibodies in the polyclonal mixture will have similar chemical properties to the monoclonal antibody produced by the cell line. However, most cell culture media used for the production of therapeutic antibodies are serum-free, and, therefore, removal of unrelated antibodies should not be required. Nonetheless, if required, some fractionation of bovine antibodies away from the target antibody is possible by conventional chromatographic means such as ion exchange or hydrophobic interaction chromatography.

An anion exchange step is often used as one of the steps following protein A or G chromatography as anion exchangers bind DNA, BSA, endotoxin and certain viruses (NG and MITRA, 1994). For basic antibodies with a high isoelectric point, it is possible to select pH and ionic conditions using a DEAE or quaternary ammonium resin that results in the tight binding of DNA and other impurities but allows the antibody to flow through the column. This results in a simple, high capacity process as the resin is used to bind only trace quantities of impurities, while the purified product can be collected in the unbound fraction. A third chromatography step is frequently added, particularly if additional clearance of an impurity or of virus is required, or if the antibody tends to aggregate. Purified antibody preparations must have a very low content of antibody aggregates, as aggregates of even human antibodies are potentially immunogenic. Therefore, if the antibody of interest tends to aggregate either during the cell culture stage of the manufacturing process or during downstream processing, then a third chromatographic step of size exclusion is often included. However, size exclusion chromatography is a slow process requiring relatively large columns. Therefore, it is advisable wherever possible to investigate alternative strategies such as inhibiting the formation of aggregates in the first place.

However, for relatively small batches of therapeutic antibodies, e.g., up to a 100 g, it is possible to use a simple 3-step process that utilizes an affinity step, a ion exchange step and a size exclusion step:

Protein A/G affinity chromatography
↓
Ion exchange chromatography
↓
Size exclusion chromatography

For the purification of therapeutic antibodies from mammalian cell lines, one or more virus removal and inactivation steps may be introduced into the purification scheme. The bulk product purified by such a process is likely to meet the purity requirements for most therapeutic applications and will be ready for formulation and vialling.

9 Good Manufacturing Practice

Good manufacturing practice (GMP) is the mode of production operation that must be applied to the final manufacture of all biological products destined for either clinical trial or for market supply. A process compliant with current GMP (cGMP) ensures that the product is consistently produced and controlled to the standards required by the relevant regulator and is appropriate to the products intended final use. GMP cannot be considered as separate from both quality assurance (QA) and quality control (QC) considerations, both of which will have fundamental influences on the activities performed under GMP. QA and QC will not be discussed further here but they must be well integrated into a production environment to ensure cGMP.

GMP has a series of basic requirements: (taken from the Orange Guide, MCA. Medicines Control Agency 1993)

(1) All manufacturing processes are clearly defined, systematically reviewed in the light of experience and shown to be capable of consistently manufacturing medicinal products of the required quality and complying with their specifications.
(2) Critical steps of manufacturing processes and significant changes to the process are validated.
(3) All necessary facilities for GMP are provided including: appropriately qualified and trained personnel; adequate premises and space; suitable equipment and services; correct materials, containers and labels; approved procedures and instructions, suitable storage and transport.
(4) Instructions and procedures are written in an instructional form in clear and unambiguous language, specifically applicable to the facilities provided.
(5) Operators are trained to carry out procedures properly.
(6) Records are made, manually and/or by recording instruments, during manufacture which demonstrate that all steps required by the defined procedures and instructions were in fact taken and that the quantity and quality of the product was as expected. Any significant deviations are fully recorded and investigated.
(7) Records of manufacture including distribution, which enable the complete history of a batch to be traced, are retained in a comprehensible and accessible form.
(8) The distribution of the products minimizes any risk to their quality.
(9) A system is available to recall any batch of product, from sale or supply.
(10) Complaints about marketed products are examined, the causes of quality defects investigated and appropriate measures taken in respect of defective products and to prevent reoccurrence.

When the final product or a stage in the cGMP manufacturing process is sterile, special consideration must be given to minimizing the risks of microbiological, particulate and pyrogen contamination. The manufacture of biological products (including antibodies) also requires consideration of the issues of viral contamination, bioprocess containment and stability of recombinant DNA sequences.

This is not intended to be a complete description of the requirements of GMP manufacture. A fuller consideration can be found in regulations produced by the individual regulatory bodies [e.g., Commission Directive of 13 June 1991 laying down the principles and guidelines of good manufacturing practice for medicinal products for human use (91/356/ EEC; or Federal Drug Administration, 1991)]. *"The Rules and Guidance for Pharmaceutical Manufacturers"* (MCA. Medicines Control Agency, 1993) (referred to as *"The Orange Guide"*) provides a good working source of information on GMP compliance in the UK.

Regulatory authorities will inspect the GMP manufacturing facilities used to produce product destined for use in the geographical areas for which they are responsible, even if production occurs outside their area. There has been little direct regulatory definition as to when GMP operations should start in the lifetime of a biological product. However, general practice which is reflected in the latest EU directive on clinical trials, to be enacted later this year, is that all material for clinical use should be produced under GMP. Once a product reaches the licence application stage and during market supply, the relevant regulators will perform regular inspections of the facilities to ensure cGMP. This is in addition to ensuring compliance with any specific requirements of the particular regulatory authority. To satisfy these inspections, full records of all the procedures performed during manufacture must be kept. Thus the documentation and recording of information are a key part of the GMP system, together with a careful review and control by QA. The storage and maintenance of this information also needs careful consideration since regulators will require access to records from historical batches during inspections. Some of the complex interactions required to operate a GMP facility are shown in the Fig. 2.

Compliance to GMP will be required for all stages of production of a biological product: ranging from the creation of the cell banks, through fermentation and during purification. Creation of the original cell line is not required to be carried out under GMP but documentation and traceability of the process performed would be required to satisfy other regulatory requirements.

10 Conclusions

The production of bulk purified antibody for clinical use requires the definition of the complete manufacturing process, from characterization of the cell line and cell bank to demonstrating consistency of the production process to demonstrating the absence of adventitious agents in the final product. All stages of the production must be carried out in a manner that satisfies the relevant regulatory authorities for product to be used in either clinical trials or for market supply.

Fig. 2. Some interactions required to operate a GMP facility.

11 References

BAKER, R. M., BRADY, A.-M., COMBRIDGE, B. S., EJIM, L. J., KINGSLAND, S. L., LLOYD, D. A., ROBERTS, P. L. (1994), Validation of a primary capture process for production of human monoclonal antibodies, in: *Separations for Biotechnology* 3 (PYLE, D. L., Ed.), London: Royal Society of Chemistry.

BARNETT, R. S., LIMOLI, K. L., HUYNH, T. B., OPLE, E. A., REFF, M. E. (1995), Antibody production in Chinese hamster ovary cells using an impaired selectable maker, in: *Antibody Expression and Engineering* (WANG, H. Y., Ed.), pp. 27–40. Washington, DC: American Chemical Society.

BARRY, A. R., CHOJNACKI, R. (1994), Biotechnology product validation, part 8: chromatography media and column qualification, *BioPharm* **7**, 43–47.

BEBBINGTON, C. R., RENNER, G. L., THOMSON, S., KING, D., ABRAMS, D., YARRANTON, G. T. (1992), High-level expression of a recombinant antibody from myeloma cells using a glutamine synthetase gene as an amplifiable selectable marker, *Bio/Technology* **10**, 169–175.

BERGWERFF, A. A., STROOP, C. J. M., MURRAY, B., HOLTROF, A.-P., PLUSCHKE, G., VAN OOSTRUM, J., KAMERLING, J. P., VLIEGENTHART, J. F. G. (1995), Variation in N-linked carbohydrate chains in different batches of two chimeric monoclonal IgG1 antibodies produced by different murine Sp2/0 transfectoma cell subclones, *Glycoconj. J.* **12**, 318–330.

BERTHOLD, W., KEMPKEN, R. (1994), Interaction of cell culture with downstram purification: a case study, *Cytotechnology* **15**, 229–242.

BIBILA, T. A., RANUCCI, G., GLAZOMITSKY, K., BUCKLAND, B. C., AUNINS, J. G. (1994), Investigation of NSO cell metabolic behavior in monoclonal antibody producing clones, *Ann. NY Acad. Sci.* **745**, 277–284.

BLOOM, J. W., WONG, M. F., MITRA, G. (1989), Detection and reduction of protein A contamination in immobilized protein A purified monoclonal antibody preparations, *J. Immunol. Methods* **117**, 83–89.

BRAND, H. N., FROUD, S. J., METCALFE, H. K., ONADIPE, A. O., SHAW, A., WESTLAKE, A. J. (1994), Selection strategies for highly productive recombinant cell lines, in: *Animal Cell Technology: Products for Today, Prospects for Tomorrow* (SPIER, R. E., GRIFFITHS, J. B., BERTHOLD, W., Eds.), pp. 55–60. Oxford: Butterworth-Heinemann.

BROWN, M. E., RENNER, G.L., FIELD, R. P., HASSELL, T. E. (1992), Process development for the production of recombinant antibodies using the glutamine synthetase (GS) system, *Cytotechnology* **9**, 231–236.

CARTER, P. J. (1991), Expression in *Escherichia coli* of antibody fragments having at least a cysteine present as a free thiol, use for the production of bifunctional F(ab')₂ antibodies, *U.S. Patent* 91-762292910919.

CASTILLO, F. J., MULLEN, L. J., GRANT, B. C., DELEON, J., THRIFT, J. C., CHANG, L. W., IRVING, J. M., BURKE, D. J. (1994), Hybridoma stability, *Dev. Biol. Stand.* **83**, 55–64.

CBER (Center for Biologics Evaluation and Research) (1992), *Supplement to the Points to Consider in the Production and Testing of New Drugs and Biologicals Produced by Recombinant DNA Technology: Nucleic Acid Characterization and Genetic Stability.* Rockville: Food and Drug Administration.

CBER (Center for Biologics Evaluation and Research) (1993), *Points to Consider in Characterization of Cell Lines Used to Produce Biologicals.* Rockville: Food and Drug Administration.

CBER (Center for Biologics Evaluation and Research) (1994), *Points to Consider in the Manufacturing and Testing of Monoclonal Antibody Products for Human Use.* Rockville: Food and Drug Administration.

CLARKE, J. B., SPIER, R. E. (1980), Variation in the susceptibility of BHK populations and cloned cell lines to three strains of foot-and-mouth disease virus, *Arch. Virol.* **63**, 1–9.

CLARKE, S., STEPHENSON, R. C., LOWENSON, J. D. (1992), Lability of asparagine and aspartic acid residues in proteins and peptides: spontaneous deamidation and isomerization reactions, in: *Stability of Protein Pharmaceuticals: Part A* (AHERN, T. J., MANNING, M. C., Eds.), pp. 2–29. New York: Plenum Press.

COLLER, H. S., COLLER, B. S. (1986), Poisson statistical analysis of repetitive subcloning by the limiting dilution technique as a way of assessing hybridoma monoclonality, *Methods Enzymol.* **121**, 412–417.

CPMP Biotechnology Working Party (1996), *Note for Guidance on Virus Validation Studies: the Design, Contribution and Interpretation of Studies Validation the Inactivation and Removal of Viruses.* London: The European Agency for the Evaluation of Medicinal Products, Canary Wharf.

DATAR, R., ROSEN, C.-G. (1990), Downstream process economics, in: *Separation Processes in Biotechnology* (ASENJO, J. A., Ed.), pp. 741–793. New York: Marcel Dekker.

DATAR, R. V., CARTRIGHT, T., ROSEN, C.-G. (1993), Process economics of animal and bacterial fermentations: a case study analysis of tissue plasminogen activator, *Bio/Technology* **11**, 349–357.

DAVIS, J. M., LAVENDER, C. M., BOWES, K. J., HANAK, J. A. J., COMBRIDGE, B. S., KINGSLAND, S. L. (1995), Human therapeutic monoclonal antibody anti-D antibody produced in long-term hollow-fiber culture, in: *Animal Cell Technology; Developments towards the 21st Century* (BEUVERY, E. C., GRIFFITHS, J. B., ZEIJLEMAKER, W. P., Eds.), pp. 149–153. Dordrecht: Kluwer Academic Publishers.

DE NEVE, M., DE LOOSE, M., JACOBS, A., VAN HOUDT, H., KALUZA, B., WEIDLE, U. H., VAN MONTAGU, M., DEPICKER, A. (1993), Assembly of an antibody and its derived antibody fragment in *Nicotiana* and *Arabidopsis, Transgen. Res.* **2**, 227–237.

DERAMOUDT, F. X., CHAABIHI, H., POUL, M. A., MARGRITTE, C., CERUTTI, M., DEVAUCHELLE, G., BERNARD, A., LEFRANC, M. P., KACZOREK, M. (1995), Production of a biologically active chimeric mouse/human monoclonal antibody in insect cells, in: *Animal Cell Technology: Developments towards the 21st Century* (BEUVERY, E. C., GRIFFITHS, J. B., ZEIJLEMAKER, W. P., Eds.), pp. 469–473. Dordrecht: Kluwer Academic Publsihers.

DORAI, H., McCARTNEY, J. E., HUDZIAK, R. M., TAI, M.-S., LAMINET, A. A., HOUSTON, L. L., HUSTON, J. S., OPPERMANN, H. (1994), Mammalian cell expression of single-chain Fv (sFv) antibody proteins and their C-terminal fusions with interleukin-2 and other effector domains, *Bio/Technology* **12**, 890–897.

DUFFY, S. A., MOELLERING, B. J., PRIOR, G. M., DOYLE, K. R., PRIOR, C. P. (1989), Recovery of therapeutic-grade antibodies: protein A and ion-exchange chromatography, *BioPharm* **6**, 34–47.

EATON, L. C. (1995), Host cell contaminant protein assay development for recombinant biopharmaceuticals, *J. Chromatogr.* **A 705**, 105–114.

Federal Drug Administration (1991), *Code of Federal Regulations; 21. Parts 200, 600.* USA: FDA.

FIELD, R. P., BRAND, H. N., RENNER, G. L., ROBERTSON, H. A., BORASTON, R. C. (1991), Production of a chimeric antibody for tumor imaging and therapy from Chinese hamster ovary (CHO) and myeloma cells, in: *Production of Biologicals from Animal Cells in Culture* (SPIER, R. E., GRIFFITHS, J. B., MEIGNIER, B., Eds.), pp. 742–744. Oxford: Butterworth-Heinemann.

FOUSER, L. A., SWANBERG, S. L., LIN, B.-Y., BENEDICT, M., KELLEHER, K., CUMMING, D. A., RIEDEL, G. E. (1992), High level expression of a chimeric anti-ganglioside GD2 antibody: genomic kappa sequences improve expression in COS and CHO cells, *Bio/Technology* **10**, 1121–1127.

FROUD, S. J., CLEMENTS, G. J., DOYLE, M. E., HARRIS, E. L. V., LLOYD, C., MURRAY, P., PRENETA, A., STEPHENS, P. E., THOMPSON, S., YARRANTON, G. T. (1991), The development of a process for the production of HIV1 gp120 from recombinant cell lines, in: *Production of Biologicals from Animal Cells in Culture* (SPIER, R. E., GRIFFITHS, J. B., MEIGNIER, B., Eds.), pp. 110–115. Oxford: Butterworth-Heinemann.

GILLIES, S. D., DORAI, H., WESOLOWSKI, J., MAJEAU, G., YOUNG, D., GARDNER, J., JAMES, K. (1989), Expression of human anti-tetanus toxoid antibody in transfected murine myeloma cells, *Bio/Technology* **7**, 799–804.

GODFREY, M. A. J., KWASOWSKI, P., CLIFT, R., MARKS, V. (1993,) Assessment of the suitability of commercially available SpA affinity solid phases for the purification of murine monoclonal antibodies at process scale, *J. Immunol. Methods* **160**, 97–105.

GOOCHEE, C. F., GRAMER, M. J., ANDERSEN, D. C., BAHR, J. B., RASMUSSEN, J. R. (1991), The oligosaccharides of glycoproteins: bioprocess factors affecting oligosaccharide structure and their effect upon glycoprotein properties, *Bio/Technology* **9**, 1347–1355.

GRAMER, M. J., GOOCHEE, C. F. (1994), Glycosidase activities of the 293 and NSO cell lines, and of an antibody-producing hybridoma cell line, *Biotechnol. Bioeng.* **43**, 423–428.

GU, M. B., TODD, P., KOMPALA, D. S. (1996), Metabolic burden in recombinant CHO cells: effect of *dhfr* gene amplification and *lacZ* expression, *Cytotechnology* **18**, 159–166.

HARRIS, R. J., MURNANE, A. A., UTTER, S. L., WAGNER, K. L., COX, E. T., POLASTRI, G. D., HELDER, J.C., SLIWKOWSKI, M. B. (1993), Assessing genetic heterogeneity in production cell lines: detection by peptide mapping of a low level Tyr to Gln sequence variant in a recombinant antibody, *Bio/Technology* **11**, 1293–1297.

HASSELL, T. E., BRAND, H. N., RENNER, G. L., WESTLAKE, A. J., FIELD, R. P. (1992), Stability of production of recombinant antibodies from glutamine synthetase amplified CHO and NSO cell lines. in: *Animal Cell Technology: Developments, Processes and Products* (SPIER, R. E., GRIFFITHS, J. B., MacDONALD, C., Eds.), pp. 42–47. Oxford: Butterworth-Heinemann.

HOLLIS, G. F., MARK, G. E. (1995), Homologous recombination antibody expression system for murine cells, *Int. Patent* WO95/17516.

ICH (1996), *Topic Q5B: Quality of Biotechnological Products: Analysis of the Expression Con-*

struct in Cell Lines Used for Production of r-DNA Derived Protein Products.

JAKOBVITS, A. (1995), Production of fully human antibodies by transgenic mice, *Curr. Opin. Biotechnol.* **6**, 561–566.

JENKINS, N., PAREKH, R. B., JAMES, D. C. (1996), Getting the glycosylation right: implications for the biotechnology industry, *Curr. Opin. Biotechnol.* **4**, 975–981.

KALUZA, B., LENZ, H., RUSSMAN, E., HOCK, H., RENTROP, O., MAJDIC, O., KNAPP, W., WEIDLE, U. H. (1991), Synthesis and functional characterization of a recombinant monoclonal antibody directed against the α-chain of the human interleukin-2 receptor, *Gene* **107**, 297–305.

KANG, K. A., RYU, D. D. Y. (1991), Studies on scale-up parameters of an immunoglobulin separation system using protein A affinity chromatography, *Biotechnol. Progr.* **7**, 205–212.

KARREMAN, S., KARREMAN, C., HAUSER, H. (1997), Construction of recombinant cell lines with defined properties using FLP recombinase driven gene replacement, in: *Animal Cell Technology: From Vaccines to Genetic Medicine* (CARRONDO, M. I. T., GRIFFITHS, J. B., MOREIRA, J. L. P., Eds.), pp. 511–517. Dordrecht: Kluwer Academic Publishers.

KEEN, M. J., HALE, C. (1996), The use of a serum-free medium for the production of functionally active humanized monoclonal antibody from NSO mouse myeloma cells engineered using glutamine synthetase as a selectable marker, *Cytotechnology* **18**, 207–217.

KEEN, M. J., RAPSON, N. T. (1995), Development of a serum-free culture medium for the large-scale production of recombinant protein from a Chinese hamster ovary cell line, *Cytotechnology* **17**, 153–163.

KENNEY, A. C. (1989), Large-scale purification of monoclonal antibodies, in: *Monoclonal Antibodies: Production and Application*, pp. 143–160. New York: Alan Liss.

KETTLEBOROUGH, C., SALDANHA, J., HEATH, V. J., MORRISON, C. J., BENDIG, M. M. (1991), Humanization of a mouse monoclonal antibody by CDR-grafting: the importance of framework residues on loop formation, *Protein Eng.* **4**, 773–783.

KIRKPATRICK, R. B., GANGULY, S., ANGELICHO, M., GRIEGO, S., SHATZMAN, A. R., SILVERMAN, C., ROSENBERG, M. (1995), Heavy chain dimers as well as complete antibodies are efficiently formed and secreted from *Drosophila* via a BiP-mediated pathway, *J. Biol. Chem.* **270**, 19800–19805.

KOPP, K., SCHLÜTER, M., NOÉ, W., WERNER, R. (1997), Glycosylation patterns of recombinant

therapeutic proteins produced in two mammalian cell lines, in: *Animal Cell Technology: From Vaccines to Genetic Medicine* (CARRONDO, M. J. T., GRIFFITHS, J. B., MOREIRA, J. L. P., Eds.), pp. 503–509. Dordrecht: Kluwer Academic Publishers.

LADYMAN, H. M., RITTER, M. A. (1995), *Production of Monoclonal Antibodies: Production, Engineering and Clinical Applications.* Cambridge: Cambridge University Press.

LEES, G., DARLING, A. (1995), Biosafety considerations, in: *Monoclonal Antibodies: Principles and Applications* (BIRCH, J. R., LENNOX, E. S., Eds.), pp. 267–298. New York: Wiley-Liss.

LEVINE, H. L., TARNOWSKI, S. J., DOSMAR, M., FENTON, D. M., GARDNER, J. N., HAGEMAN, T. C., LU, P., STEININGER, B. (1992), Industry perspective on the validation of column-based separation processes for the purification of proteins, *J. Parent Sci. Technol.* **46**, 87–97.

LIFELY, M. R., HALE, C., BOYCE, S., KEEN, M. J., PHILLIPS, J. (1995), Glycosylation and biological activity of CAMPATH-1H expressed in different cell lines and grown under different culture conditions, *Glycobiology* **5**, 813–822.

LUBINIECKI, A. S., WIEBE, M. E., BUILDER, S. E. (1990), Process validation for cell culture-derived pharmaceutical proteins, in: *Large-Scale Mammalian Cell Culture Technology* (LUBINIECKI, A. S., Ed.), pp. 515–541. New York: Marcel Dekker.

MA, J. K., HIATT, A., HEIN, M., VINE, H. D., WANG, F., STABILA, P., van DOLLEWEERD, C., MOSTOV, K., LEHNER, T. (1995), Generation and assembly of secretory antibodies in plants, *Science* **268**, 716–719.

MATHER, J. P. (1990), Optimizing cell and culture environment for production of recombinant proteins, *Methods Enzymol.* **185**, 567–577.

MCA. Medicines Control Agency (1993), *The Rules and Guidance for Pharmaceutical Manufactures,* UK: HMSO.

MCA. Medicines Control Agency (1994), Production and Quality Control of Monoclonal Antibodies, *MCA EuroDirect Publication* No. 5271/94.

MORRIS, C. B., WARBURTON, S. (1994), Serum – Screening and Selection, in: *Cell and Tissue Culture: Laboratory Proceedures* (DOYLE, A., GRIFFITHS, J. B., NEWELL, D. G., Eds.), 2B: 1.1–1.5. Chichester: John Wiley & Sons.

MOUNTAIN, A., ADAIR, J. R. (1992), Engineering antibodies for therapy, *Biotechnol. Gen. Eng. Rev.* **10**, 1.

NG, P., MITRA, G. (1994), Removal of DNA contaminants from therapeutic protein preparations, *J. Chromatogr.* **658**, 459–463.

NYGREN, P.-A., STAHL, S., UHLEN, M. (1994), Engineering proteins to facilitate bioprocessing, *TIBTECH* **12**, 184–188.

NYYSSÖNEN, E., PENTTILÄ, M., HARKKI, A., SALOHEIMO, A., KNOWLES, J. K. C., KERÄNEN, S. (1993), Efficient production of antibody fragments by the filamentous fungus *Trichoderma reesi, Bio/Technology* **11**, 591–596.

PAGE, M. J., SYDENHAIN, M. A. (1991), High level expression of the humanized monoclonal antibody Campath-1H in Chinese hamster ovary cells, *Bio/Technology* **9**, 64–68.

PALLAVICINI, M. G., DETERESA, P. S., ROSETTE, C., GRAY, J. W., WURM, F. M. (1990), Effects of methotrexate on transfected DNA stability in mammalian cells, *Mol. Cell Biol.* **10**(1), 401–404.

PARK, S. Y., LEE, G. M. (1995), Enhancement of monoclonal antibody production by immobilized hybridoma cell culture with hyperosmolar medium, *Biotechnol. Bioeng.* **48**, 699–705.

PEAKMAN, T. C., WORDEN, J., HARRIS, R. H., COOPER, H., TITE, J., PAGE, M. J., GEWERT, D. R., BARTHOLEMEW, M., CROWE, J. S., BRETT, S. (1994), Comparison of expression of a humanized monoclonal antibody in mouse NSO myeloma cells and Chinese hamster ovary cells, *Hum. Antibodies Hybridomas* **5**, 65–74.

PENDSE, G. J., KARKARE, S., BAILEY, J. E. (1992), Effect of cloned gene dosage on cell growth and hepatitis B surface antigen synthesis and secretion in recombinant CHO cells, *Biotechnol. Bioeng.* **40**, 119–129.

Pharmacia (1993), *Monoclonal Antibody Purification Handbook*. Uppsala, Sweden.

POILEY, J. A., BIERLEY, S. T., HILLESUND, T., NELSON, R. E., MONTICELLO, T. M., RAINERI, R. (1994), Methods for estimating retroviral burden, *BioPharm.* **5**, 32–35.

PROUDFOOT, K. A., TORRANCE, C., LAWSON, A. D. G., KING, D. J. (1992), Purification of recombinant chimeric B72.3 Fab' and F(ab')₂ using streptococcal Protein G, *Protein Expr. Purif.* **3**, 368–373.

RACHER, A. J., MOREIRA, J. L., ALVES, P. M., WIRTH, M., WEIDLE, U. H., HAUSER, H., CARRONDO, M. J. T., GRIFFITHS, J. B. (1994), Expression of recombinant antibody and secreted alkaline phosphatase in mammalian cells. Influence of cell line and culture system upon production kinetics, *Appl. Microbiol. Biotechnol.* **40**, 851–856.

RAO, P., KROON, D. J. (1993), Orthoclone OKT3: Chemical mechanisms and functional effects of degradation of a therapeutic monoclonal antibody, in: *Stability and Characterization of Protein and Peptide Drugs – Case Histories* (WANG, J. Y., PERALMAN, R., Eds.), pp. 135–156. New York: Plenum Press.

RAPER, J., DOUGLAS, Y., GORDON-WALKER, N., CAULCOTT, C. A. (1992), Long-term stability of expression of humanized monoclonal antibody CAMPATH1-H in Chinese hamster ovary cells, in: *Animal Cell Technology: Developments, Processes and Products* (SPIER, R. E., GRIFFITHS, J. B., MACDONALD, C., Eds.), pp. 51–53. Oxford: Butterworth-Heinemann.

RAY, N., RIVERA, R., GUPTA, R., MUELLER, D. (1997), Large-scale production of humanized monoclonal antibody expressed in a GS-NSO cell line, in: *Animal Cell Technology: From Vaccines to Genetic Medicine* (CARRONDO, M. J. T., GRIFFITHS, J. B., MOREIRA, J. L. P., Eds.), pp. 235–241. Dordrecht: Kluwer Academic Publishers.

REFF, M. E. (1994), Fully impaired consensus Kozak sequences for mammalian expression, *Int. Patent* WO 94/11523.

RHODES, P. M., BIRCH, J. R. (1988), Large-scale production of proteins from mammalian cells, *Bio/Technology* **6**, 518–523.

ROBINSON, D. K., SEAMANS, T. C., GOULD, S. L., DISTEFANO, D. J., CHAN, C. P., LEE, D. K., BIBILA, T. A., GLAZOMITSKY, K., MUNSHI, S., DAUGHERTY, B., O'NEIL PALLADINO, L., STAFFORD-HOLLIS, J., HOLLIS, G. F., SILBERKLANG, M. (1994), Optimization of a fed-batch process for production of a recombinant antibody, *Ann. NY Acad. Sci.* **745**, 285–296.

ROE, S. D. (1994), Protein A leakage from affinity absorbents, in: *Separations for Biotechnology* 3 (PYLE, D. L., Ed.). London: Royal Society of Chemistry.

SEELY, R. J., WIGHT, H. D., FRY, H. H., RUDGE, S. R., SLAFF, G. F. (1994), Biotechnology product validation, part 7: Validation of chromatography resin useful life, *BioPharm* **7**, 41–48.

SHADLE, P. J., MCALLISTER, P. R., SMITH, T. M., LUBINIECKI, A. S. (1995), Viral validation strategy for recombinant products drived from established animal cell lines, in: *Animal Cell Technology – Developments towards the 21st Century* (BEUVERY, E. C., GRIFFITHS, J. B., ZEIJLEMAKER, W. P., Eds.), pp. 631–635. Dordrecht: Kluwer Academic Publishers.

SINACORE, M. S., CHARLEBOIS, T. S., HARRISON, S., BRENNAN, S., RICHARDS, T., HAMILTON, M., SCOTT, S., BRODEUR, S., OAKES, P., LEONARD, M., SWITZER, M., ANAGNOSTOPOULOS, A., FOSTER, B., HARRIS, A., JANKOWSK, M., BOND, M., MARTIN, S., ADAMSON, S. R. (1996), CHO DUKX cell lineages preadapted to growth in serum-free suspension culture enable rapid development of cell culture processes for the manufacture of recombinant proteins, *Biotechnol. Bioeng.* **52**, 518–528.

SUZUKI, E., OLLIS, D. F. (1990), Enhanced antibody production at slowed growth rates: experimental demonstration and a simple structural model, *Biotechnol. Progr.* **6**, 231–236.

THÖMMES, J., HALFAR, M., LENZ, S., KULA, M.-R. (1995), Purification of monoclonal antibodies from whole hybridoma fermentation broth by fluidized bed adsorption, *Biotechnol. Bioeng.* **45**, 205–211.

TRILL, J. J., SHATZMAN, A. R., GANGULY, S. (1995), Production of monoclonal antibodies in COS and CHO cells, *Curr. Opin. Biotechnol.* **6**, 553–560.

UNDERWOOD, P. A., BEAN, P. A. (1988), Hazards of the limiting dilution method of cloning hybridomas, *J. Immunol. Methods* **107**, 119–128.

WITTINGTON, P. N. (1990), Fermentation broth clarification techniques, *Appl. Biochem. Biotechnol.* **23**, 91–121.

WHITTLE, N., ADAIR, J., LLOYD, C., JENKINS, L., DEVINE, J., SCHLOM, J., RAUBITSCHEK, A., COLCHER, D., BODMER, M. (1987), Expression in COS cells of a mouse–human chimeric B72.3 antibody, *Protein Eng.* **1**, 499–505.

WIEBE, M. E., MAY, L. H. (1990), Cell banking, *Bioproc. Technol.* **10**, 147–160 (review).

WOOD, C. R., BOSS, M. A., KENTEN, J. H., CALVERT, J. E., ROBERTS, N. A., EMTAGE, J. S. (1985), The synthesis and *in vivo* assembly of functional antibodies in yeast, *Nature* **314**, 446–449.

XIE, L., WANG, D. I. C. (1994a), Stoichiometric analysis of animal cell growth and its application in medium design, *Biotechnol. Bioeng.* **43**, 1164–1174.

XIE, L., WANG, D. I. C. (1994b), Fed-batch cultivation of animal cells using different medium design concepts and feeding strategies, *Biotechnol. Bioeng.* **43**, 1175–1189.

XIE, L., WANG, D. I. C. (1996), High cell density and high monoclonal antibody production through medium design and rational control in a bioreactor, *Biotechnol. Bioeng.* **51**, 725–729.

9 Use of Antibodies for Immunopurification

DAVID J. KING

Slough, UK

1 Introduction 276
2 Immunopurification 276
3 Choice of Antibody for Immunopurification 276
4 Immobilization of Antibody for Immunopurification 277
  4.1 Immobilization through Chemical Coupling to Amine Groups 278
  4.2 Site-Specific Immobilization 279
5 Design of Immunoaffinity Chromatography Procedures 280
6 Applications of Immunopurification 282
7 Summary 285
8 References 285

1 Introduction

Antibodies are key reagents for almost all biological research laboratories, are widely used in diagnostic applications and are increasingly finding a role in a wide range of therapeutic applications. The specificity of monoclonal antibodies for their respective antigens has led to their application in the laboratory in the detection, measurement, purification and analysis of a wide range of substances. The detection and measurement of antigens by immunoassay techniques have been described (HARLOW and LANE, 1988; LEFKOVITS, 1997). Immunoassay and methods are described in chapter II. This review focuses on the application of antibodies for immunopurification and related techniques.

2 Immunopurification

The use of antibodies for the immunopurification of a wide range of substances is a valuable technique in biotechnological research. This is usually operated as a column chromatographic method known as immunoaffinity chromatography. The application of immunoaffinity chromatography for the purification of proteins is often the method of choice for rapid, simple isolation of proteins for research purposes from either natural or recombinant sources. The use of this technique avoids the need to develop multi-step conventional purification protocols for every protein of interest, which can be a slow and difficult process. In some cases immunoaffinity chromatography is also used commercially for the purification of high value products. It is particularly useful for purification of proteins present in small amounts with a high level of contaminating protein, as the highly specific nature of the antibody–antigen interaction allows isolation of pure protein, essentially in a single step.

In immunoaffinity chromatography, antibody with specificity for the protein of interest is coupled to a solid support and packed into a column. The solution containing the protein of interest is then passed through the column under suitable conditions for antibody-binding to take place. Contaminant proteins can then be washed through the column and the bound protein eluted by changing the composition of the buffer to conditions which promote dissociation of the antibody–antigen interaction (Fig. 1).

3 Choice of Antibody for Immunopurification

Although it is possible to use polyclonal antisera for immunopurification, the technique has become far more widely used since the advent of monoclonal antibody technology. Monoclonal antibodies offer several advantages compared to polyclonal reagents for immunopurification purposes. Monoclonal antibodies can be produced with impure antigen, or even antigen expressed on the surface of whole cells, whereas polyclonal reagents require purified antigen, if a specific reagent is to be obtained, which may defeat the object of the immunopurification! Also, in contrast to polyclonal reagents, monoclonal antibodies can be produced in potentially unlimited quantities and are identical from batch to batch. In addition, a panel of antibodies can easily be produced from which individual antibodies can be selected which have optimal characteristics for immunopurification.

Selection of the antibody from a panel of monoclonal antibodies is an attractive option as not all antibodies are suitable for immunopurification purposes. For example, very high affinity antibodies are not ideal. Although binding of the antigen from dilute solutions is efficient, extremely harsh conditions are usually required for elution which may result in irreversible damage to the antibody and/or the protein of interest. Therefore, low to moderate affinity antibodies with affinity constants in the range of $10^{-4}-10^{-8}$ M^{-1} have been suggested to be optimal for efficient binding and elution from the column (PHILLIPS, 1989; JACK et al., 1987). Antibodies from a panel can be chosen on the basis of

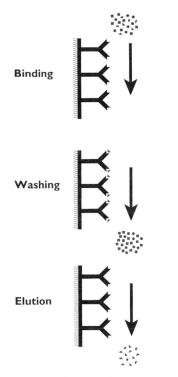

Fig. 1. Immunopurification is achieved by application of a complex mixture to immobilized antibody. After binding, contaminants are washed away followed by elution of the purified material.

their specificity and affinity alone, or tested in small scale experiments for binding and elution characteristics. This is easily accomplished in small scale purification experiments which require approximately 1–10 mg of purified antibody. These experiments can be performed in batch format, in which a small amount of antibody is coupled to a solid matrix, added to the solution containing antigen, and recovered by centrifugation in a bench top centrifuge. Several eluents can then be screened to help identify the optimal reagents for use in the column format. Alternatively, an ELISA format assay can be devised to test the antibodies ability to bind and elute antigen even before small amounts of purified antibody are available (BONDE, 1991). This can also be achieved using BIAcore format binding assays (MALMBORG and BORREBAECK, 1995). Although such ELISA

and BIAcore experiments are useful as a preliminary screen, it is still important to test immobilization to the solid support at a small scale, as unexpected losses in antigen-binding activity of the antibody on immobilization are common.

4 Immobilization of Antibody for Immunopurification

There exists a large number of commercially available matrices for immobilization of antibodies for use in immunoaffinity chromatography, which can be obtained pre-activated to allow rapid, simple preparation of the immunoaffinity matrix. Desirable characteristics of a matrix for use in immunoaffinity chromatography are shown in Tab. 1. The most popular matrices are based on cross-linked beaded agaroses such as Sepharose, which have good chemical and mechanical stability over conditions likely to be used during immunoaffinity chromatography. However, many other types of material are available including a range of synthetic matrices and coated silicas for use in high-performance immunoaffinity chromatography (GROMAN and WILCHEK, 1987; PHILLIPS, 1989).

Alternative techniques to column chromatography are also available. Antibodies can be immobilized on membranes which are robust and have the advantage of being able to be used at high flow rates allowing faster purifications using less antibody (PAK et al., 1984; NACHMAN et al., 1992). Achieving high capacity membranes is difficult, but this can be compensated for by fast recycling using an automated system. Another alternative is to couple antibody to a polymeric material such as polyethylene glycol which can then be used in 2-phase liquid–liquid extraction protocols (STOCKS and BROOKS, 1988). A further refinement of such 2-phase extraction methods is to attach antibody to a polymer which undergoes a phase transition at an increased temperature. This allows efficient mixing in a

Tab. 1. Properties of Desirable Matrices for Immobilization of Antibodies for Immunoaffinity Chromatography

Rigidity	For use at high flow rates in column chromatography
Hydrophilicity	To minimize non-specific binding to the matrix via hydrophobic interactions
Porosity and permeability	For efficient chromatography a high surface area is required and thus porous beads with pores large enough to allow diffusion of protein molecules are desirable
Ease of derivatization	To allow simple attachment of antibody. Many matrices are available in pre- derivatized form
Stability	Able to withstand extremes of pH and chaotropic agents used as eluents

single phase until the temperature is raised to separate the phases, which may allow affinity extractions to be achieved with high efficiencies (TAKEI et al., 1994).

4.1 Immobilization through Chemical Coupling to Amine Groups

Immobilization of antibodies to the solid matrix can be achieved through a variety of different chemical approaches. Most commonly, antibodies are attached through amine groups, primarily of lysine residues. Several criteria need to be taken into account when selecting the best activation chemistry (Tab. 2). Coupling via cyanogen bromide activated supports such as CNBr-Sepharose™ is a commonly used method which is carried out under mild conditions and results in high antibody coupling efficiencies, routinely >90% in the author's laboratory. The major drawbacks of this technique are a tendency for a low level of antibody leakage from the matrix due to the instability of the isourea bond. Also the linkage results in a positively charged matrix at neutral pH. This is usually insignificant compared to the antibody charge but in some cases may affect the level of non-specific binding to the matrix through action as an anion exchanger. Alternative chemical linkages can be more stable than cyanogen bromide activated materials and thus result in less leakage from the column (see Sect. 5). For example, N-hydroxysuccinimide ester activated materials such as Sepharose™ and Affigel™

have been reported to have excellent stability (BEER et al., 1995).

The attachment of antibodies to matrices using chemical attachment through lysines results in the random orientation of antibodies attached to the matrix. Thus although the chemical coupling has been carried out under mild conditions, minimizing any chemical damage, only a proportion of the antibody binding sites is available for antigen-binding. In most cases these are approximately 10–15% of the total binding sites (FOWELL and CHASE, 1986). This inefficiency is thought to be due to incorrect orientation of a proportion of antibody molecules, multi-site attachment to the matrix preventing antigen-binding, and steric restrictions on antigen-binding by closely packed antibody molecules. This latter steric restriction is particularly prevalent at high substitution ratios and can be eased at lower coupled antibody density (FOWELL et al., 1986; LIAPIS et al., 1989; PFEIFFER et al., 1987; HEARN and DAVIES, 1990).

Tab. 2. Issues in the Selection of Chemical Coupling Techniques for Immobilizing Antibodies for Use in Chromatography

Ease of use
Antibody coupling efficiency
Activity of coupled antibody
Leakage of coupled antibody
Non-specific binding characteristics
Alterations to physical properties of the matrix, e.g., extra cross-linking

4.2 Site-Specific Immobilization

In attempts to improve the proportion of immobilized antibody available for antigen-binding, several methods for site-specific immobilization have been developed. These include immobilization via bacterial immunoglobulin binding proteins, attachment to carbohydrate and thiol directed coupling of antibody fragments (Fig. 2). Immobilized staphylococcal protein A is widely used for antibody purification and binds many IgG molecules via the Fc region at a site between the CH_2 and CH_3 domains (DEISENHOFER, 1981). The antigen-binding sites of the antibody are consequently free to bind antigen. Thus antibodies which are capable of binding protein A can be immobilized through binding to immobilized protein A followed by non-specific cross-linking with glutaraldehyde (GYKA et al., 1983) or preferably with dimethylpimelimidate (SISSON and CASTOR, 1990). Preparation of oriented matrices in this way has resulted in considerable improvement in the efficiency of antibody utilization with 20-fold increases reported (GYKA et al., 1983). However, care must be taken that the cross-linking of antibody to protein A is efficient, as otherwise problems with leakage of the antibody from the matrix during elution result. However, cross-linking conditions need to be mild to avoid antibody damage. Another limitation of this technique is that the sample from which the antigen is to be purified must be free of protein-A-binding molecules, such as endogenous IgG. Otherwise contamination of the eluted protein will result. If an antibody is to be used which does not bind protein A, other bacterial immunoglobulin-binding proteins can be used, for example, streptococcal protein G has a wide range of immunoglobulin-binding specificities (BJORK and AKERSTROM, 1990).

An alternative site-specific method is to attach the antibody via carbohydrate. Immunoglobulins are glycoproteins with the carbohydrate attached usually at a single site in the CH_2 domain of the antibody within the Fc region. Therefore, attachment at this site also assists orientation of the antibody allowing more active antigen binding sites to be retained. Coupling via carbohydrate is achieved

a) Random attachment   **b) Attachment through carbohydrate**

c) Attachment via Protein A or G

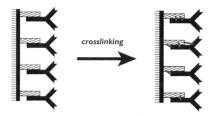

crosslinking

d) Attachment of Fab' via hinge thiol

Fig. 2. Antibody immobilized through amine groups is oriented randomly on the solid support (a), resulting in only a minor proportion of the antibody retaining binding activity. Site-specific attachment can be achieved through coupling to Fc carbohydrate (b), through binding of the Fc region to protein A or G, followed by chemical cross-linking (c) or by attaching Fab' through a hinge thiol group (d).

through firstly oxidizing vicinal hydroxyl groups on the carbohydrate with periodate to form aldehydes. This is followed by coupling to a matrix derivatized with a hydrazide functional group (HOFFMAN and O'SHANESSY, 1988; O'SHANESSY, 1990). Although coupling of the aldehyde to amine containing supports could also be achieved using this methodology, hydrazides are used as these are reactive at low pH where the antibody amino groups are protonated thus preventing oxidized antibody reacting with another antibody molecule in solution to produce aggregates. The resulting hydrazone can be further stabilized by reduction with cyanoborohydride to form a hydrazine linkage. However, this reduction may result in damage to the antibody and in some cases has been shown to be unnecessary. Using this strategy, the theoretical maximum binding of two antigen molecules per immobilized antibody molecule has been claimed (DOMEN et al., 1990). One problem with this approach is the stability of the antibody to periodate oxidation. Some antibodies lose antigen-binding properties due to oxidation of essential amino acid residues.

Site-specific coupling can also be achieved through hinge thiol groups. Many monoclonal antibodies can be readily digested with enzymes such as pepsin or bromelain to form F(ab′)$_2$ fragments which retain all of the antigen-binding characteristics of the IgG (PARHAM, 1986). Such F(ab′)$_2$ fragments can be selectively reduced to form Fab′ fragments which contain free thiol groups at the hinge region (PRISYAZHNOY et al., 1988). The Fab′ fragments generated can then be used to specifically attach the Fab′ to a thiol reactive matrix such as a maleimide activated or iodoacetyl activated material, resulting in a stable thioether linkage (DOMEN et al., 1990; PRISYAZHNOY et al., 1988). Intact IgG can also be reduced at the hinge region to generate free thiol groups for coupling to a solid support. However, this strategy is generally less successful than the use of Fab′ fragments as the two halves of the antibody molecule formed by reduction of the hinge region remain associated through strong non-covalent interactions between the two CH$_3$ domains in the Fc region. Reaction of the hinge thiol groups with solid supports is thus more difficult both due to steric restrictions and to the tendency for hinge disulphide bonds to reform.

Although increased antigen-binding ability per molecule is achieved with site-specifically immobilized immunoaffinity matrices, there is usually a reduction in antibody coupling efficiency compared to random coupling methods (HOFFMAN and O'SHANESSY, 1988; ORTHNER et al., 1991). Also, the benefits are most pronounced at low coupling ratios due to the avoidance of steric restrictions on antigen-binding by closely packed antibody

molecules mentioned above. Therefore, the benefits promised by site-specific immobilization are not always realized. An alternative approach which is feasible in a few cases is to protect the antibody-binding site during the coupling reaction by preforming an antibody–antigen complex which is then immobilized. This is feasible only when the antigen is not a protein itself. For example, an antigen-binding peptide (without relevant functional groups) can be used (VELANDER et al., 1992).

5 Design of Immunoaffinity Chromatography Procedures

Immunoaffinity chromatography is essentially a simple procedure as illustrated in Fig. 1. However, care in the design of the experimental conditions used can greatly improve the efficiency of the purification and the quality of the purified protein. Leakage of antibody from the immunoaffinity matrix takes place to a greater or lesser extent in all cases. The degree of leakage is dependent on the chemistry of the immobilization method used and the nature of the eluent. The leakage on exposure to eluent in the first cycle of use is always more pronounced than on subsequent uses. It is, therefore, recommended to prewash an immunoaffinity column with the intended eluent before equilibration in a neutral buffer prior to sample loading. The sample loading step is normally carried out at a low flow rate to maximize binding efficiency, at neutral pH or slightly above, and in the presence of sodium chloride. Antibodies carry a number of charged groups and consequently immobilized antibody can operate as an ion exchange material. Therefore, salt is often added to reduce ionic interactions. However, the concentration of salt should not be too high as non-specific interactions through hydrophobic forces may then be promoted. Concentrations up to 0.5 M sodium chloride are routinely used, although in some cases salt may inhibit effective antigen-binding. In addition, it is sometimes desirable to reduce non-specific binding interactions by the addition of detergent or polyethylene glycol in the buffer (FLEMINGER et al., 1990), provided that any contaminating detergent can be easily removed from the protein at a later stage. After loading, the column is washed to remove contaminants, initially with the same buffer used for sample loading. It may be beneficial to include an extra washing stage to remove any non-specifically or loosely bound material from the column, using conditions which will not elute the bound antigen. The conditions for this step depend on the strength of the antibody–antigen interaction and the conditions necessary for elution, and may be, e.g., washing at a low concentration of a weak chaotropic agent or with an intermediate pH between that for binding and elution.

Elution from the column is the most difficult part of the procedure and usually requires the testing of several conditions before an optimal procedure is developed. These can be tested as part of the small scale antibody selection procedure described in Sect. 3. Conditions should be sufficient to break the antibody–antigen interaction without irreversibly damaging antigen or antibody. The forces involved in the antibody–antigen interaction are a combination of ionic, hydrogen bonds, hydrophobic and van der Waals interactions. Therefore, elution conditions should be able to disrupt these interactions. However, the mix of types of interactions and the overall strength of binding varies from one antibody–antigen interaction to another, and thus the severity of elution conditions required also

varies widely. These interactions are the same as those involved in maintaining protein structure, and hence all elution buffers are denaturing to some degree. Examples of agents used for elution are shown in Tab. 3 and may be used on their own or in some cases in combination. It is sometimes possible to reduce the concentration, and hence the harshness of an eluant by using combinations of agents.

The simplest and most commonly used elution method employs a low pH buffer such as glycine, acetate or citrate at pH values down to 2. Propionic acid is a particularly useful reagent as it also reduces polarity which helps to disrupt hydrophobic interactions. Chaotropic agents such as thiocyanate are also widely used. Their effectiveness has been compared on the scale below (PHILLIPS, 1989)

$$CCl_3COO^- > SCN^- > CF_3COO^- > ClO_4^- > I^- > Cl^-$$

Less widely used are harsh denaturants such as urea and guanidine hydrochloride as these rarely allow the elution of active material. Organic solvents have been used in some cases as they are effective at disrupting hydrophobic interactions, although irreversible denaturation is also a problem for most solvents (ANDERSSON et al., 1978). Other techniques such as high pressure (OLSON et al., 1989) or electrophoretic elution (MORGAN et al., 1978) have not been widely used due to the relatively complex equipment required.

One of the most gentle elution techniques is the selective elution of antigen from conformation-specific antibodies. For example, factor IX has been purified using an antibody that recognizes the protein only when com-

Tab. 3. Commonly Used Eluents for Immunoaffinity Chromatography

Extremes of pH	Citric acid (0.1 M) pH 2, glycine-HCl pH 2–3 (0.1 M) acetic acid (0.1–1 M) pH 1–3, propionic acid (0.1–1 M) triethylamine (0.1 M) pH 11, ethanolamine (0.1 M)
Chaotropes	Potassium thiocyanate (2–3 M), magnesium chloride (4.5 M), sodium iodide (2–3 M)
Denaturants	Urea (6–8 M), guanidine hydrochloride (3–4 M)
Organic solvents	Ethylene glycol (up to 50%), dioxane (5–10%)
Specific eluents	EDTA (2–10 mM) for metal dependent antibodies, competitive peptides

plexed with calcium. Factor IX could be eluted by the addition of the chelating agent EDTA to remove calcium ions resulting in a conformational change (LIEBMAN et al., 1985). A similar approach has been used to purify protein C (NAKAMURA and SAKATA, 1987), and is the basis of the FLAG epitope system (HOPP et al., 1988). The FLAG epitope consists of the sequence Asp-Tyr-Lys-(Asp)$_4$-Lys and can be fused to recombinant proteins and used as a "tag" for simple immunopurification. The expressed tagged protein can then be purified by use of a monoclonal antibody which recognizes a calcium complex of the FLAG epitope, allowing mild elution by the addition of EDTA or by a competitive peptide (HOPP et al., 1996).

Leakage of antibody from the immunoaffinity matrix can be a significant problem, particularly for commercial applications. Antibody which leaks from the matrix may be difficult to remove from the purified protein at a later stage. As mentioned above leakage takes place to a greater or lesser extent in all cases and the degree of leakage is dependent on the matrix material, the chemistry of the immobilization method used and the nature of the eluent (PENG et al., 1986). Leakage can occur via breakdown of a component of the matrix material, breakage of the antibody–matrix linkage or by degradation of the antibody itself, and is best minimized by selection of a matrix material and coupling chemistry which is stable under the intended eluent buffer conditions.

Several limited comparisons of matrix materials and coupling chemistries with respect to leakage have been reported, however, there is no comprehensive study comparing currently available materials. In a study of goat anti-apolipoprotein antibody immobilized on Sepharose, divinyl sulfone coupling was found to be more stable than tresyl which was more stable than cyanogen bromide at pH 7.3 (UBRICH et al., 1992). At pH 2.8 however, the stability of tresyl and divinyl sulfone coupling was reversed. In another study comparing different matrices and immobilization chemistry for mouse antibodies to human factor IX, it was demonstrated that the antibody itself was a major determinant of the stability of the coupled material (BESSOS et al., 1991).

In one of the most comprehensive studies examining leakage, an immobilized rat monoclonal antibody recognizing human thyrotropin was attached to cross-linked agarose activated with cyanogen bromide, hydrazido, N-hydroxy-succinimide ester and 1,1′-carbonyl-diimidazole (BEER et al., 1995). These methods all resulted in affinity columns which efficiently purified thyrotropin. Other activation chemistries attempted in this study [*p*-benzoquinone, divinylsulfone, 1,4-bis(2:3-epoxypropane)butane, *p*-nitrophenylchloroformate and trichloro-*S*-triazine] resulted in very poor immunoaffinity columns, which were unable to purify thyrotropin. Stability of the immunoaffinity columns was in the order N-hydroxy-succinimide ester > cyanogen bromide > 1,1′-carbonyldiimidazole > hydrazido. Leakage was also shown to be dependent on the eluent buffer used. With the better materials the leakage of antibody was very low. For example, the cyanogen bromide activated matrix resulted in approximately 0.1% of the antibody leaking per run. Such stability implies the column could be used 500 times before a 5% loss of capacity was seen (BEER et al., 1995). In general, lower immobilized densities of antibody also lead to lower leakage into the product (SATO et al., 1987). It has also been suggested that the stability of the antibody itself may be increased by chemically cross-linking the protein chains of the molecule with glutaraldehyde (KOWAL and PARSONS, 1980) or a bismaleimide (GOLDBERG et al., 1991). Although the use of such reagents may reduce the activity of the immobilized antibody and hence the capacity of the immunoaffinity matrix.

6 Applications of Immunopurification

Immunopurification is now very widely used for the purification of proteins for research purposes and is probably the method of choice for the purification of proteins from sources of low abundance or where alternative affinity purification methods are unavailable. For example, proteins present in low abundance from natural sources purified in this way include growth hormones (BERGHMAN et al., 1996), fibrin fragments (SOBEL and GAWINOWICZ, 1996), muscle-specific proteins (TURNACIOGLU et al., 1996) and lipoproteins (AGUIE et al., 1995). Immunopurification is also widely used for the purification of proteins from recombinant sources. In this case it is not always necessary to have, or to raise, a specific antibody for the protein of interest. Purification "tags" can be added to the gene sequence before expression such that a fusion protein is prepared which allows generic immunoaffinity columns to be used for the rapid isolation of the recombinant product. For example, hemagglutinin epitopes have been used in the purification of methionine aminopeptidase (LI and CHANG, 1995). Commercially available tag systems, such as the FLAG epitope mentioned in Sect. 5, can be used and have been designed such that elution from the column can be achieved under very mild conditions. This system has found wide application, e.g., in the purification of tumor necrosis factor (SU et al., 1992), transcription factors (CHIANG and ROEDER, 1993), β_2-adrenergic receptors (GUAN et al., 1992), interleukin-3 (PARK et al., 1989), interleukin-5 receptors (BROWN et al., 1995) and antibody fragments (KNAPPIK and PLUCKTHUN, 1994). Antibody fragments are also often purified using an alternative commercially available peptide tag system known as E-tag (MARKS et al., 1992).

For the unique specificity and high degree of purification achieved, immunopurification would be the method of choice for manufacturing therapeutic proteins. However, the high cost of manufacturing large amounts of antibody for purification purposes is restrictive. Nevertheless, immunopurification has been used, and continues to be used, for the preparation of some high value protein therapeutics which are required in small quantities and are difficult to purify by other means, such as interferons, factor VIII and factor IX (BAILON and ROY, 1990; GRIFFITH, 1991; KIM et al., 1992). Indeed recombinant factor VII produced by Novo Nordisk and approved for human use as recently as 1996, uses immunopurification as the key step in its purifica-

tion. Key problems for the use of immunopurification of antibodies for production of therapeutics include the leakage issue and manufacturing. Leakage of antibody from the column must be carefully monitored and any contaminating antibody removed from the product. Also, regulatory authorities demand that the antibody to be used for immunopurification must be produced under GMP conditions to a standard as stringent as that of the therapeutic itself. This is obviously a major cost for any therapeutic product.

One strategy to reduce the cost of immunopurification reagents is the use of antibody fragments. The region of the monoclonal antibody which binds antigen is a minor proportion of the entire antibody molecule and is restricted to the paired variable domains (Vh and Vl) or Fv fragment of the antibody. Production of Fv fragments of antibodies is very difficult by proteolysis of IgG, but can be readily achieved by recombinant DNA technology (SKERRA and PLUCKTHUN, 1988). Bacterial systems such as *E. coli* can be used for the expression of active Fv molecules in very high yields which can lead to an inexpensive production system (KING et al., 1993). In addition, Fv molecules have more binding sites per milligram of protein than intact IgG and have other potential advantages such as their small size which allows easier immobilization within the pores of some support materials. Successful immobilization and immunoaffinity chromatography have been demonstrated for several Fv fragments (BERRY et al., 1991; SPITZNAGEL and CLARK, 1993). In a comparison of IgG Fab' and Fv fragments it was found that immobilization of Fv resulted in the highest binding capacity, presumably due to the small size of the fragment (SPITZNAGEL and CLARK, 1993). Attempts to reduce the size of the antibody even further have not proved successful. In some cases a single antibody variable domain, such as Vh, can retain antigen-binding activity, and such domains have been termed single domain antibodies or DABs (WARD et al., 1989). However, immobilization of a DAB recognizing hen lysozyme did not result in a successful immunopurification reagent as a high level of non-specific binding to the material was observed (BERRY and DAVIES, 1992). This was probably the result of exposure of hydrophobic patches on the surface of the Vh domain which normally associate with the Vl domain. One problem with the use of Fv fragments for immunopurification is that although associated by non-covalent forces, the Vh and Vl domains are not covalently linked, and thus leakage of non-coupled domains is likely under elution conditions. Several approaches have been developed to stabilize Fv fragments including the introduction of a disulphide bond between the Vh and Vl domains (GLOCKSHUBER et al., 1990) and the use of a peptide linker connecting the polypeptides of the two domains into a single polypeptide, known as a single chain Fv (BIRD et al., 1988; HUSTON, et al., 1988). The use of scFv fragments in place of Fvs has been shown to reduce the amount of antibody fragment leaking from the column and allows improved capacity on reuse of immunoaffinity columns (BERRY and PIERCE, 1993). scFv fragments can also be produced cheaply as soluble, active molecules in *E. coli* (DESPLANCQ et al., 1994).

In some cases it is not even necessary to purify the antibody fragment first. For example, an interesting development is the construction of Fv fragments fused to a peptide biotin mimic, known as Strep-Tag (SCHMIDT and SKERRA, 1993). Expression of the Fv fragment fused to the Strep-Tag sequence allows simple affinity purification of the Fv fragment using a streptavidin affinity column. Fv fragments against target proteins can be expressed as Strep-Tag fusions and used to purify the target protein directly. The membrane protein complexes cytochrome c oxidase and ubiquinol:cytochrome c oxidoreductase from *Paracoccus denitrificans* have been purified in this way (KLEYMANN et al., 1995). In this case *E. coli* periplasmic extract containing the tagged Fv fragment was simply mixed with a membrane preparation from *Paracoccus denitrificans* and the resulting complexes were purified in one step by binding to an immobilized streptavidin column.

The use of such low cost antibody fragments may permit an entire range of new immunopurification applications to become commercially viable. In addition, the development of phage display technology for the rapid isolation and selection of panels of new antibodies directly as antibody fragments (reviewed by WINTER et al., 1994; BURTON and BARBAS, 1994) will greatly increase the speed with which new immunopurification reagents can be produced. Antibody fragments with affinities into the subnanomolar range can be isolated from libraries of variable region genes without immunization using such techniques (VAUGHAN et al., 1996). A variety of techniques exist to isolate antibodies with desired binding affinities (HOOGENBOOM, 1997). Therefore, suitable antibodies for immunopurification should be readily achievable.

Applications of immunopurification are not restricted to proteinaceous substances, and the technique is increasingly used for the extraction and measurement of drugs, toxins and other low molecular weight materials. For example, immunopurification has recently been used to detect aflatoxin contamination in cheeses (DRAGACCI et al., 1995), for the determination of salbutamol and clenbuterol in tissue samples (POU et al., 1994), for the detection of chloramphenicol (GUDE et al., 1995), and in the determination of levels of cannabis metabolites in saliva (KIRCHER and PARLAR, 1996). For such measurements immunoaffinity chromatography is often coupled with other techniques such as high-performance liquid chromatography (HPLC), gas chromatography (GC) or mass spectrometry (MS).

The immunopurification of whole cells is also a useful procedure in which marker antibodies can be used to separate populations of cells for both research and clinical applications. This is usually achieved using fluorescently labeled antibody which binds to the cell, and the population of labeled cells can be isolated using a fluorescent activated cell sorter. Antibody to the stem cell marker CD34 has recently been used in this way to isolate a CD34+ bone marrow cell population for use in transplantation studies (RONDELLI et al., 1996).

Immunopurification can also be used for the specific removal of substances such as the removal of traces of specific contaminant proteins, e.g., the removal of traces of the highly toxic contaminant ricin B chain from ricin A prepared for *in vivo* use as an immunotoxin (FULTON et al., 1986) or the removal of bovine IgG from mouse monoclonal preparations. Such negative immunopurification can be extended to the removal of bacteria. The removal of bacteria from foodstuffs using immunoaffinity chromatography with immobilized antibody fragments has been suggested as a potentially economic process, and the removal of *Pseudomonas aeruginosa* from milk has been demonstrated to be effective (MOLLOY et al., 1995).

Several clinical applications for the removal of substances from blood using immunoaffinity chromatography have also been suggested. In these cases blood is pumped from the body through an immobilized antibody column and then back into the patient. Potential applications for such technology are the removal of IgE from the plasma of patients with allergic or other hyper-IgE syndromes (SATO et al., 1989), and the removal of radiolabeled antibodies from blood following tumor radioimmunotherapy (JOHNSON et al., 1991; NORRGREN et al., 1991). In the latter application radiolabeled anti-tumor antibodies are injected into the patient and allowed to localize to the tumor. After localization has taken place the majority of radiolabeled antibody is removed from the blood, thus reducing non-specific toxicity to the bone marrow and other organs and increasing the therapeutic index of the treatment.

7 Summary

The ability to generate specific monoclonal antibodies has allowed improvement in immunopurification techniques and an increase in its application from laboratory use to commercial and clinical uses.

8 References

AGUIE, G. A., RADER, D. J., CLAVEY, V., TRABER, M. G., TORPIER, G., KAYDEN, H. J., FRU-
CHART, J. C., BREWER, H. B., CASTRO, G. (1995), Lipoproteins containing apolipoprotein B isolated from patients with abetalipoproteinemia and homozygous hypobetalipoproteinemia: identification and characterization, *Atherosclerosis* **118**, 183–191.

ANDERSSON, K., BENYAMIN, Y., DOUZOU, P., BALNY, C. (1978), Organic solvents and temperature effects on desorption from immunosorbents, *J. Immunol. Methods* **23**, 17–21.

BAILON, P., ROY, S. K. (1990), Recovery of recombinant proteins by immunoaffinity chromatography, in: *Protein Purification* (LADISCH, M. R., WILLSON, R. C., PAINTON, C. C., BUILDER, S. E., Eds.), pp. 150–167. Washington, DC: ACS.

BEER, D. J., YATES, A. M., RANDLES, S. C., JACK, G. W. (1995), A comparison of the leakage of a monoclonal antibody from various immunoaffinity chromatography matrices, *Bioseparation* **5**, 241–247.

BERGHMAN, L. R., LESCROART, O., ROELANTS, I., OLLEVEIR, F., KUNH, E. R., VERHAERT, P. D., DeLOOF, A., VAN LEUVEN, F., VANDESANDE, F. (1996), One-step immunoaffinity purification and partial characterization of hypophyseal growth hormone from the African catfish, *Clarias gariepinus* (Burchell), *Comp. Biochem. Physiol.* **113**, 773–780.

BERRY, M. J., DAVIES, J. (1992), Use of antibody fragments in immunoaffinity chromatography: comparison of Fv fragments, VH fragments and paralog peptides, *J. Chromatogr.* **597**, 239–245.

BERRY, M. J., PIERCE, J. J. (1993), Stability of immunoadsorbents comprising antibody fragments: comparison of Fv fragments and single-chain Fv fragments, *J. Chromatogr.* **629**, 161–168.

BERRY, M. J., DAVIES, J., SMITH, C. G., SMITH, I. (1991), Immobilization of Fv antibody fragments on porous silica and their utility in affinity chromatography, *J. Chromatogr.* **587**, 161–169.

BESSOS, H., APPLEYARD, C., MICKLEM, L. R., PEPPER, D. S. (1991), Monoclonal antibody leakage from gels: effect of support, activation and eluent composition, *Prep. Chromatogr.* **1**, 207–220.

BIRD, R. E., HARDMAN, K. D., JACOBSON, J. W., JOHNSON, S., KAUFMAN, B. M., LEE, S. M., LEE, T., POPE, S. H., RIORDAN, G. S., WHITLOW, M. (1988), Single chain antigen binding proteins, *Science* **242**, 423–426.

BJORK, L., AKERSTROM, B. (1990), Streptococcal protein G, in: *Bacterial Immunoglobulin Binding Proteins* (BOYLE, M. D. P., Ed.), pp. 113–126. San Diego, CA: Academic Press.

BONDE, M., FROKIER, H., PEPPER, D. S. (1991), Selection of monoclonal antibodies for immunoaffinity chromatography: model studies with antibodies against soy bean trypsin inhibitor, *J. Biochem. Biophys. Methods* **23**, 73–82.

BROWN, P. M., TAGARI, P., ROWAN, K. R., YU, V. L., O'NEILL, G. P., MIDDAUGH, C. R., SANYAL, G., FORD-HUTCHINSON, A. W., NICHOLSON, D. W. (1995), Epitope-labeled soluble human interleukin-5 receptors. Affinity cross-link labeling, IL-5 binding and biological activity, *J. Biol. Chem.* **270**, 29236–29243.

BURTON, D. R., BARBAS, C. F. (1994), Human antibodies from combinatorial libraries, *Adv. Immunol.* **57**, 191–280.

CHIANG, C. M., ROEDER, R. G. (1993), Expression and purification of general transcription factors by FLAG epitope tagging and peptide elution, *Peptide Res.* **6**, 62–64.

DEISENHOFER, J. (1981), Crystallographic refinement and atomic models of a human Fc fragment and its complex with fragment B of protein A from *Staphylococcus aureus* at 2.9 and 2.8 Å resolution, *Biochemistry* **20**, 2361–2370.

DESPLANCQ, D., KING, D. J., LAWSON, A. D. G., MOUNTAIN, A. (1994), Multimerization behavior of single chain Fv variants for the tumor binding antibody B72.3, *Protein Eng.* **7**, 1027–1033.

DOMEN, P. L., NEVENS, J. R., MALLIA, A. K., HERMANSON, G. T., KLENK, D. C. (1990), Site-directed immobilization of proteins, *J. Chromatogr.* **510**, 293–302.

DRAGACCI, S., GLEIZES, E., FREMY, J. M., CANDLISH, A. A. (1995), Use of immunoaffinity chromatography as a purification step for the determination of aflatoxin M1 in cheeses, *Food Addit. Contam.* **12**, 59–65.

FLEMINGER, G., SOLOMON, B., WOLF, R., HADAS, E. (1990), Effect of polyethylene glycol on the non-specific adsorption of proteins to Eupergit C and agarose, *J. Chromatogr.* **510**, 271–279.

FOWELL, S. A., CHASE, H. A. (1986), A comparison of some activated matrices for preparation of immunosorbents, *J. Biotechnol.* **4**, 355–368.

FULTON, R. J., BLAKELEY, D. C., KNOWLES, P. P., UHR, J. W., THORPE, P. E., VITETTA, E. S. (1986), Isolation of pure ricin A1, A2 and B chains and characterization of their toxicity, *J. Biol. Chem.* **261**, 5314–5319.

GLOCKSHUBER, R., MALIA, M., PFITZINGER, I., PLÜCKTHUN, A. (1990), A comparison of strategies to stabilize immunoglobulin Fv fragments, *Biochemistry* **29**, 1362–1367.

GOLDBERG, M., KNUDSEN, K. L., PLATT, D., KOHEN, F., BAYER, E. A., WILCHEK, M. (1991), Specific interchain cross-linking of antibodies using bismaleimides: repression of ligand leakage in immunoaffinity chromatography, *Bioconjug. Chem.* **2**, 275–280.

GRIFFITH, M. (1991), Ultrapure plasma factor VIII produced by anti-factor VIIIc immunoaffinity chromatography and solvent/detergent viral inactivation, *Ann. Hematol.* **63**, 131–137.

GROMAN, E. V., WILCHEK, M. (1987), Recent developments in affinity chromatography supports, *Trends Biotechnol.* **5**, 220–224.

GUAN, X. M., KOBILKA, T. S., KOBILKA, B. K. (1992), Enhancement of membrane insertion and function in a type IIIb membrane protein following introduction of a cleavable signal peptide, *J. Biol. Chem.* **267**, 21995–21998.

GUDE, T., PREISS, A., RUBACH, K. (1995), Determination of chloramphenicol in muscle, liver, kidney and urine of pigs by means of immunoaffinity chromatography and gas chromatography with electron capture detection, *J. Chromatogr.* **673**, 197–204.

GYKA, G., GHETIE, V., SJOQUIST, J. (1983), Cross-linkage of antibodies to staphylococcal protein A matrices, *J. Immunol. Methods* **57**, 227–233.

HARLOW, E., LANE, D. (1988), *Antibodies. A Laboratory Manual.* New York: Cold Spring Harbor Laboratory Press.

HEARN, M. T. W., DAVIES, J. R. (1990), Evaluation of factors which affect column performance with immobilized monoclonal antibodies, *J. Chromatogr.* **512**, 23–39.

HOFFMAN, W. L., O'SHANESSY, D. J. (1988), Site-specific immobilization of antibodies by their oligosaccharide moieties to new hydrazide derivatized solid supports, *J. Immunol. Methods* **112**, 113–120.

HOOGENBOOM, H. R. (1997), Designing and optimizing library selection strategies for generating high-affinity antibodies, *Trends Biotechnol.* **15**, 62–70.

HOPP, T. P., PRICKETT, K. S., PRICE, V., LIBBY, R. T., MARCH, C. J., CERRETTI, P., URDAL, D. L., CONLON, P. J. (1988), A short polypeptide marker sequence useful for recombinant protein identification and purification, *Bio/Technology* **6**, 1204–1210.

HOPP, T. P., GALLIS, B., PRICKETT, K. S. (1996), Metal-binding properties of a calcium dependent monoclonal antibody, *Mol. Immunol.* **33**, 601–608.

HUSTON, J. S., LEVINSON, D., MUDGETT-HUNTER, M., TAI, M. S., NOVOTNY, J., MARGOLIES, M. N., RIDGE, R. J., BRUCCOLERI, R. E., HABER, E., CREA, R., OPPERMANN, H. (1988), Protein engineering of antibody binding sites: recovery of specific activity in an anti-digoxin single-chain Fv analogue produced in *Escherichia coli*, *Proc. Natl. Acad. Sci. USA* **85**, 5879–5883.

JACK, G. W., BLAZEK, R., JAMES, K., BOYD, J. E., MICKLEM, L. R. (1987), The automated production by immunoaffinity chromatography of the human pituitary glycoprotein hormones thyrotropin, follitropin and lutropin, *J. Chem. Technol. Biotechnol.* **39**, 45–58.

JOHNSON, T. K., MADDOCK, S., KASLIWAL, R., BLOEDOW, D., HARTMANN, C., FEYERBAND, A., DIENHART, D. G., THICKMAN, D., GLENN, S., GONZALEZ, R., LEAR, J., BUNN, P. (1991), Radioimmunoadsorption of KC-4G3 antibody in peripheral blood: implications for radioimmunotherapy, *Antibody Immunoconj. Radiopharm.* **4**, 885–893.

KIM, H. C., MCMILLAN, C. W., WHITE, G. C., BERGMAN, G. E., HORTON, M. W., SAIDI, P. (1992), Purified factor IX using monoclonal immunoaffinity technique, *Blood* **79**, 568–575.

KING, D. J., BYRON, O. D., MOUNTAIN, A., WEIR, N., HARVEY, A., LAWSON, A. D. G., PROUDFOOT, K. A., BALDOCK, D., HARDING, S. E., YARRANTON, G. T., OWENS, R. J. (1993), Expression, purification and characterization of B72.3 Fv fragments, *Biochem. J.* **290**, 723–729.

KIRCHER, V., PARLAR, H. (1996), Determination of delta 9-tetrahydrocannabinol from human saliva by tandem immunoaffinity chromatography high-performance liquid chromatography, *J. Chromatogr.* **677**, 245–255.

KLEYMANN, G., OSTERMEIER, C., LUDWIG, B., SKERRA, A., MICHEL, H. (1995), Engineered Fv fragments as a tool for the one-step purification of integral multisubunit membrane protein complexes, *Bio/Technology* **13**, 155–160.

KNAPPIK, A., PLUCKTHUN, A. (1994), An improved affinity tag based on the FLAG epitope for detection and purification of recombinant antibody fragments, *Biotechniques* **17**, 754–761.

KOWAL, R., PARSONS, R. G. (1980), Stabilization of proteins immobilized on Sepharose from leakage by glutaraldehyde cross-linking, *Anal. Biochem.* **102**, 72–76.

LEFKOVITS, I., Ed. (1997), *Immunology Methods Manual. The Comprehensive Sourcebook of Techniques.* San Diego, CA: Academic Press.

LI, X., CHANG, Y. H. (1995), Amino-terminal protein processing in *Saccharomyces cerevisiae* is an essential function that requires two distinct methionine aminopeptidases, *Proc. Natl. Acad. Sci. USA* **92**, 12357–12361.

LIAPIS, A. I., ANSPACH, B., FINDLEY, M. E., DAVIES, J., HEARN, M. T. W., UNGER, K. K. (1989), Biospecific adsorption of lysozyme onto monoclonal antibody ligand immobilized on nonporous silica particles, *Biotechnol. Bioeng.* **34**, 467–477.

LIEBMAN, H. A., LIMENTANI, S. A., FURIE, B. C., FURIE, B. (1985), Immunoaffinity purification of factor IX (Christmas factor) by using conformation specific antibodies directed against the factor IX-metal complex, *Proc. Natl. Acad. Sci. USA* **82**, 3879–3883.

MALMBORG, A. C., BORREBAECK, C. A. K. (1995), BIAcore as a tool in antibody engineering, *J. Immunol. Methods* **183**, 7–13.

MARKS, J. D., GRIFFITHS, A. D., MALMQVIST, M., CLACKSON, T. P., BYE, J. M., WINTER, G. (1992), By-passing immunization: building high-affinity antibodies by chain shuffling, *Bio/Technology* **10**, 779–783.

MOLLOY, P., BRYDON, L., PORTER, A. J., HARRIS, W. J. (1995), Separation and concentration of bacteria with immobilized antibody fragments, *J. Appl. Bacteriol.* **78**, 359–365.

MORGAN, M. R. A., KERR, E. J., DEAN, P. D. G. (1978), Electrophoretic desorption: preparative elution of steroid specific antibodies from immunoadsorbents, *J. Steroid Biochem.* **9**, 767–770.

NACHMAN, M., AZAD, A. R. M., BAILON, P. (1992), Efficient recovery of recombinant proteins using membrane based immunoaffinity chromatography, *Biotechnol. Bioeng.* **40**, 564–571.

NAKAMURA, S., SAKATA, Y. (1987), Immunoaffinity purification of protein C by using conformation-specific monoclonal antibodies to protein C-calcium ion complex, *Biochim. Biophys. Acta* **925**, 85–93.

NORRGREN, K., STRAND, S. E., NILSSON, R., LINDGREN, L., LILLIEHORN, P. (1991), Evaluation of extracorporeal immunoadsorption for reduction of the blood background in diagnostic and therapeutic applications of radiolabelled monoclonal antibodies, *Antibody Immunoconj. Radiopharm.* **4**, 907–914.

OLSON, W. C., LEUNG, S. K., YARMUSH, M. L. (1989), Recovery of antigens from immunoadsorbents using high pressure, *Bio/Technology* **7**, 369–373.

ORTHNER, C. L., HIGHSMITH, F. A., THARAKAN, J., MADURAWE, R. D., MORCOL, T., VELANDER, W. H. (1991), Comparison of the performance of immunosorbents prepared by site-directed or random coupling of monoclonal antibody, *J. Chromatogr.* **558**, 55–70.

O'SHANESSY, D. J. (1990), Hydrazido-derivatized supports in affinity chromatography, *J. Chromatogr.* **510**, 13–21.

PAK, K. Y., RANDERSON, D. H., BLASZCZYK, M., SEARS, H. F., STEPLEWSKI, Z., KOPROWSKI, H. (1984), Extraction of circulating gastrointestinal cancer antigen using solid phase immunoadsorption system of monoclonal antibody coupled membrane, *J. Immunol. Methods* **66**, 51–58.

PARHAM, P. (1986), Preparation and purification of active fragments from mouse monoclonal antibodies, in: *Handbook of Experimental Immunology,* Vol. 1, *Immunochemistry* (WEIR, D. M.,

Ed.), 4th Edn., pp. 14.1–14.23. Oxford: Blackwell Scientific Publications.

PARK, L. S., FRIEND, D., PRICE, V., ANDERSON, D., SINGER, J., PRICKETT, K. S., URDAL, D. L. (1989), Heterogeneity in human interleukin 3 receptors: a subclass that binds human granulocyte/macrophage colony stimulating factor, *J. Biol. Chem.* **264**, 5420–5427.

PENG, L., CALTON, G. J., BURNETT, J. W. (1986), Stability of antibody attachment in immunosorbent chromatography, *Enzyme Microb. Technol.* **8**, 681–685.

PFEIFFER, N. E., WYLIE, D. E., SCHUSTER, S. M. (1987), Immunoaffinity chromatography utilizing monoclonal antibodies. Factors which influence antigen-binding capacity, *J. Immunol. Methods* **97**, 1–9.

PHILLIPS, T. M. (1989), High-performance immunoaffinity chromatography, *Adv. Chromatogr.* **29**, 133–173.

POU, K., ONG, H., ADAM, A., LAMOTHE, P., DELAHAUT, P. (1994), Combined immunoextraction approach coupled to a chemiluminescence enzyme immunoassay for the determination of trace levels of salbutamol and clenbuterol in tissue samples, *Analyst* **119**, 2659–2662.

PRISYAZHNOY, V. S., FUSEK, M., ALAKHOV, Y. (1988), Synthesis of high-capacity immunoaffinity sorbents with oriented immobilized immunoglobulins or their Fab' fragments for isolation of proteins, *J. Chromatogr.* **424**, 243–253.

RONDELLI, D., ANDREWS, R. G., HANSEN, J. A., RYNCARZ, R., FAERBER, M. A., ANASETTI, C. (1996), Alloantigen presenting function of normal human CD34+ hematopoietic cells, *Blood* **88**, 2619–2675.

SATO, H., KIDAKA, T., HORI, M. (1987), Leakage of immobilized IgG from therapeutic immunosorbents, *Appl. Biochem. Biotechnol.* **15**, 145–158.

SATO, H., WATANABE, K., AZUMA, J., KIDAKA, T., HORI, M. (1989), Specific removal of IgE by therapeutic immunoadsorption system, *J. Immunol. Methods* **118**, 161–168.

SCHMIDT, T. G. M., SKERRA, A. (1993), The random peptide library assisted engineering of a carboxyl terminal affinity peptide, useful for the detection and purification of a functional Ig Fv fragment, *Protein Eng.* **6**, 109–122.

SISSON, T. H., CASTOR, C. W. (1990), An improved method for immobilizing IgG antibodies to protein A agarose, *J. Immunol. Methods* **127**, 215–220.

SKERRA, A., PLÜCKTHUN, A. (1988), Assembly of a functional immunoglobulin Fv fragment in *Escherichia coli, Science* **240**, 1038–1041.

SOBEL, J. H., GAWINOWICZ, M. A. (1996), Identification of the alpha chain lysine donor sites involved in factor XIIIa fibrin cross-linking, *J. Biol. Chem.* **271**, 19288–19297.

SPITZNAGEL, T. M., CLARK, D. S. (1993), Surface-density and orientation effects on immobilized antibodies and antibody fragments, *Bio/Technology* **11**, 825–829.

STOCKS, S. J., BROOKS, D. E. (1988), Development of a general ligand for immunoaffinity partitioning in two phase aqueous polymer systems, *Anal. Biochem.* **173**, 86–92.

SU, X., PRESTWOOD, A. K., MCGRAW, R. A. (1992), Production of recombinant porcine tumor necrosis factor alpha in a novel *E. coli* expression system, *Biotechniques* **13**, 756–762.

TAKEI, Y. G., MATSUKA, M., AOKI, T., SANUI, K., OGATA, N., KIKUCHI, A., SAKURAI, Y., OKANO, T. (1994), Temperature-responsive bioconjugates. 3. Antibody-poly(N-isopropylacrylamide) conjugates for temperature-modulated precipitations and affinity bioseparations, *Bioconj. Chem.* **5**, 577–582.

TURNACIOGLU, K. K., MITTAL, B., SANGER, J. M., SANGER, J. W. (1996), Partial characterization of zeugmatin indicates that it is part of the Z-band region of titin, *Cell Motil. Cytoskeleton* **34**, 108–121.

UBRICH, N., HUBERT, P., REGNAULT, V., DELLACHERIE, E., RIVAT, C. (1992), Compared stability of Sepharose based immunosorbents prepared by various methods, *J. Chromatogr.* **584**, 17–22.

VAUGHAN, T. J., WILLIAMS, A. J., PRITCHARD, K., OSBOURN, J. K., POPE, A. R., EARNSHAW, J. C., MCCAFFERTY, J., HODITS, R. A., WILTON, J., JOHNSON, K. S. (1996), Human antibodies with sub-nanomolar affinities isolated from a large non-immunized phage display library, *Nature Biotechnol.* **14**, 309–314.

VELANDER, W. H., SUBRAMANIAN, A., MADURAWE, R. D., ORTHNER, C. L. (1992), The use of Fab masking antigens to enhance the activity of immobilized antibodies, *Biotechnol. Bioeng.* **39**, 1013–1023.

WARD, E. S., GUSSOW, D., GRIFFITHS, A. D., JONES, P. T., WINTER, G. (1989), Binding activities of a repertoire of single immunoglobulin variable domains secreted from *Escherichia coli, Nature* **341**, 544–546.

WINTER, G., GRIFFITHS, A. D., HAWKINS, R. E., HOOGENBOOM, H. R. (1994), Making antibodies by phage display technology, *Annu. Rev. Immunol.* **12**, 433–455.

10 Preclinical Testing of Antibodies: Pharmacology, Kinetics and Immunogenicity

ROLY FOULKES
SUE STEPHENS

Slough, UK

1 Introduction 290
2 Development of Assays 291
2.1 Immunoassays 291
2.2 Cell-Based Assays 291
2.3 Validation of Assays 291
3 Product Selection 292
3.1 Choice of Starting Antibody 292
3.2 Selection of Isotype 293
3.3 Whole Antibody vs. Fragments 294
4 Preclinical Determination of Activity 295
4.1 Species Cross-Reactivity 295
4.2 Use of Parallel Reagents 295
4.3 Use of Xenograft and Transgenic Models 297
5 Pharmacokinetics 297
6 Immunogenicity 299
7 Summary 300
8 References 300

List of Abbreviations

ADCC	antibody-dependent cell-mediated cytotoxicity
AUC	area under the plasma concentration time curve
CDR	complementarity determining regions
DFM	di-Fab-maleimide
EAE	experimental allergic encephalomyelitis
ELISA	enzyme-linked immunosorbant assay
FACS	fluorescence activated cell scanner
HAMA	human anti-mouse antibody
HMVECS	human dermal microvascular endothelial cell
HUVEC	human umbilical vein endothelial cell
IBD	inflammatory bowel disease
IL	interleukin
LPS	lipopolysaccharide
MAB	monoclonal antibody
MHC II	major histocompatibility complex class II
OKT3	anti-CD3 MAB
PEM	polymorphic epithelial mucin
RIA	radioimmunoassay
SCID	severe combined immunodeficiency
SPLAT-1	human anti-E-selectin MAB
TFM	tri-Fab-maleimide
TNF	tumor necrosis factor

1 Introduction

Monoclonal antibodies (MABs) have potential as therapeutic agents because of their high affinity and selectivity for the target antigen. Approaches to making MABs have been discussed earlier and a range of products from murine monoclonal antibodies to recombinant human antibodies can be produced. Indeed murine MABs are used clinically in acute diseases such as the use of OKT3 (an anti-CD3 MAB) for the treatment of acute rejection episodes following trans-

plantation (Ortho Multicenter Transplant Study Group, 1985). However, the major drawback for these murine MABs is the high prevalence of human anti-mouse antibody responses in the recipient (the so called HAMA response). This has two clinical consequences, it results in rapid clearance of the MAB after the first infusion and has the potential to cause adverse responses upon subsequent administration of any murine MAB. Such MABs are therefore unsuitable for the treatment of chronic diseases where repeated therapy is required. Immunogenicity can be reduced using human MABs containing reduced numbers of murine residues, the vast majority of which are contained in the CDR regions. Such engineered human antibodies have a far greater potential for use in the treatment of chronic diseases.

In order to develop an engineered human antibody for clinical development, from the starting murine MAB, there exists a number of key issues in the preclinical program which must be addressed. These include development of appropriate assays, choice of the best starting MAB and appropriate isotype, maintenance of potency during the engineering process, demonstration of potential effectiveness in disease models, and confirmation of *in vivo* efficacy of candidate reagents, demonstration of adequate pharmacokinetic profile and assessment of immunogenicity.

The reactivity profile of the antibody with human tissues and wider safety testing is dealt with in Sect. 5.

To address all of these satisfactorily, a range of *in vitro* and *in vivo* assays needs to be set up. More time consuming may be the need to make parallel reagents i.e., MABs with specificity for the same or similar antigen in other species to allow studies in a range of animal species. This is particularly important when there is limited species cross-reactivity of the selected MAB. This chapter discusses these issues with examples from a range of projects where engineered human MABs have been entered into preclinical development.

2 Development of Assays

Assays are required at all stages of therapeutic antibody development. This includes the initial screening of candidate antibodies, the engineering of chosen antibodies, for supporting *in vitro* and *in vivo* potency studies, pharmacokinetics, immunogenicity, cross-reactivity, toxicity, and efficacy studies, and finally to support clinical trials. The type of assay selected will depend on the target antigen, e.g., soluble vs. cell associated, and whether this is available in a purified form. It will also depend on the type of activity required of the antibody, e.g., whether a blocking or neutralizing activity alone is required or whether complement activation or cell internalization is needed.

2.1 Immunoassays

For soluble antigens, radioimmunoassays (RIA) or, more conveniently, enzyme-linked immunosorbant assays (ELISA) are commonly used. A number of potential formats are possible. Conventional sandwich assays are useful for hybridoma screening, pharmacokinetic assays, assessment of antigen levels etc. (STEPHENS et al., 1995). Competition assays using antigen coated plates and labeled competing antibodies are useful for monitoring binding efficiency of grafted or modified antibodies relative to the parent molecule. They can also be used to investigate the neutralizing effect of anti-idiotypic immune responses (HAKIMI et al., 1991; ISAACS et al., 1992). Competitive formats are also used to investigate blocking activity of antibodies for receptor–ligand interactions (e.g., anti-E-selectin antibody-blocking, binding of E-selectin to sLex) and to determine the effects of the presence of other ligands on antibody-binding, for example, the effect of TNF receptors on the binding of an anti-TNF antibody to TNF. A third format, the double-antigen sandwich format has proved a highly specific means of measuring immune responses to antibodies (and other therapeutic recombinant proteins), for example, anti-cytokine antibodies (STEPHENS et al., 1995) and antibodies to

cancer antigens (BUIST et al., 1995; LOBU-GLIO et al., 1989). This method has the advantage that affinity purified antisera from other species such as rabbit can be used as standards and responses quantified relative to these reagents (COBBOLD et al., 1990). This assay format, however, is unsuitable in circumstances where the antigen has multiple epitopes and high levels of circulating antigen are likely to be present, as is the case when using an anti-PEM (polymorphic epithelial mucin) antibody in ovarian cancer where high levels of circulating PEM may be present. In these instances, conventional sandwich assays (using antigen coated plates and anti-IgM or -IgG) should be used on serial dilutions of samples and results expressed as increases in titers over pretreatment levels.

2.2 Cell-Based Assays

Where the target antigen is cell associated, and cannot be extracted or produced in a suitable recombinant form, immunoassay formats may still be suitable, using fixed cells or cell lysates. Where this proves unsatisfactory fluorescence activated cell scanners (FACS) provide a useful tool for examining antibody-binding. Cell lines are frequently available which express the antigens but care should be taken to confirm activity using freshly isolated cells.

Binding activity using an immunoassay is rarely sufficient on its own for preclinical evaluation of an antibody. A number of *in vitro* cell-based assays may also be required, to assess secondary functions of the antibody such as cell killing or neutralization, complement activation, and Fc receptor binding. Such assays are important not only in the selection of the starting MAB (see Sect. 3.1) but also for monitoring of MAB performance as it is developed through the engineering procedure.

2.3 Validation of Assays

Once an appropriate assay format has been determined and each stage of the assay optimized, the assay must be validated to determine performance characteristics and robustness. This is essential if the assay is to be used for preclinical toxicology, product stability studies and eventually clinical studies. The levels of validation will depend on the stage of development for the antibody. For early screening and grafting studies it is sufficient to perform minimal testing to determine within assay precision (generally <10%), recovery values of analyte in the buffers used (typically 85–115%), and the working range and limit of detection for the assay. The working range for an immunoassay is typically considered to be the concentration range where the coefficient of variation for measured values is less than 10% (minimum of 10 replicates, Fig. 1). The limit of detection is generally the mean +2.5 times the standard deviation of the zero standard; in practice, however, it is better to avoid ascribing values to samples which are below the lowest standard.

For assays being used to support regulatory submissions, more detailed performance studies will be required on precision (repeatability and reproducibility), specificity (cross-reactivity and interference), accuracy (recovery and parallelism), and sensitivity. For later stages of clinical development, investigation of robustness criteria should include variations in reagent batches and storage conditions, operators and equipment, sample collection and storage, incubation time and temperature, and reagent volumes. Of particular relevance in preclinical pharmacology studies is the investigation of matrix effects of serum or plasma as these can interfere in the recovery of many analytes including antibodies and cytokines. A further point to bear in mind is the potential effect of *in vivo* responses to the antibody which may affect binding activities. For example, anti-idiotypic responses to an antibody may block binding to the antigen in a pharmacokinetic assay based on antigen coated plates. Conversely, the presence of circulating therapeutic antibody may mask detection of an anti-idiotypic response.

3 Product Selection

3.1 Choice of Starting Antibody

The most important decision for the development of an antibody product candidate is the choice of the starting MAB. The selection of the target antigen will depend not only on the disease indication and desired therapeutic effects but also upon patent considerations. Selection of MABs may be determined by affinity, potency or other activities such as the epitope-binding, neutralizing ability, signaling capacity or the likelihood of internalization. Since affinity can be a major contributor to

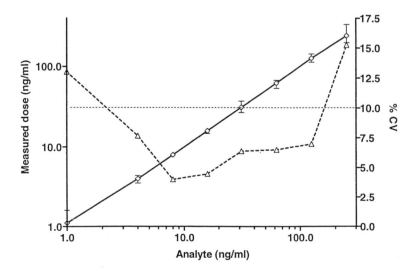

Fig. 1. Working range of an ELISA is described as the concentration range where the coefficient of variation (%CV, ...Δ...) for the measured dose (mean ± range for 10 replicates, −◊−) is less than 10%.

potency and, therefore, therapeutic dose, generally the highest affinity MABs available at the time are selected as starting reagents. This can be determined by Scatchard analysis or, more recently, by using surface plasmon resonance analysis (e.g., using BIAcore machines). The latter is dependent on being able to immobilize purified antigen in the correct conformation on the chip. However, affinity measurements do not necessarily indicate which are the best neutralizing MABs and thus the selection process must take place in concert with potency comparisons in *in vitro* biological or cell-based assays. For example, from a panel of murine MABs raised against human TNFα, a range of different inhibitory profiles in cell-based assays was obtained. The MABs were initially assayed in an L929 cytotoxicity assay, a murine fibroblast cell line expressing TNF receptors which is killed by TNF of various species. MABs with potent neutralizing activity for human TNF were then also studied in further bioassays, such as their ability to prevent TNF-mediated E-selectin expression in either HUVECs (human umbilical vein endothelial cells) or HMVECS (human dermal microvascular endothelial cells, Tab. 1). For MABs of murine origin, this is also important baseline data from

which to monitor progress through any humanization procedure and to ensure the potency is maintained in the final molecule.

3.2 Selection of Isotype

Engineering of MABs also offers the opportunity to select the MAB isotype and therefore MAB effector function. Fragments, such as F(ab')$_2$, often have the same *in vitro* potency as the parent IgG but are restricted in their *in vivo* potential in terms of long-term neutralization of antigen because of their much reduced circulating half-life. However, when the MAB needs to be cleared from the circulation rapidly as in the case of some radioimmunotherapeutic MABs, an antibody fragment may be more appropriate (see Sect. 3.3). For whole antibodies, a human IgG$_1$ is more efficient than an IgG$_4$ in complement mediated cytotoxicity and antibody-dependent cell-mediated cytotoxicity (ADCC). For example, with a MAB to E-selectin, an adhesion molecule expressed on vascular endothelial cells, lysis of target cells by the MAB is undesirable. Therefore, γ4 is the isotype of choice. For other MABs, the reverse may be true, especially where binding to cell surface

Tab. 1. Relative Potencies of a Panel of Anti-Human TNF Antibodies on Affinity and in a Range of Cellular Assays (see text for details)

Antibody	L929 IC$_{90}$ [ng mL^{-1}]	Affinity KD [$\cdot10^{-10}$ M]	HUVEC IC$_{50}$ [ng mL^{-1}]	HMVEC IC$_{50}$ [ng mL]$^{-1}$
HTNF26	9.45	3.5	1,400	1,800
THNF39	12.4	7.8	2,000	130
HTNF33	14.3	9.1	2,000	300
HTNF12	40	1.8	>10,000	5,000
HTNF31	41	32	160	60
HTNF1	60	3.8	ND	ND
CDP571	80	3.2	1,200	1,000
HTNF27	200	39.9	ND	2,200
HTNF28	350	24.2	ND	ND
HTNF38	365	50	ND	1,200
HTNF24	500	9.2	ND	ND
HTNF23	1,000	6.6	ND	250
HTNF22	2,000	6.6	ND	ND
HTNF25	5,000	16.0	6,000	2,500

ND: not determined.

antigens can enhance cellular clearance. However, this may not be so clear with information from cell-based assays alone, and an early indication *in vivo* of the relative potency of candidate MABs and the effect of isotype can be useful. Although this can be limited by poor species cross-reactivity of MABs, *in vivo* bioassays can be used. For anti-human TNF this was done by monitoring the ability of a panel of MABs to ameliorate the transient pyrexia induced by the i.v. administration of recombinant human TNF to conscious rabbits (SHAW et al., 1991). MABs could be ranked in terms of their relative potency in reducing the pyrexia and in their ability to neutralize TNF in the plasma (Fig. 2). This approach also allowed an early insight into the effects of isotype variation. Mouse–human chimeric versions of anti-human TNF MABs were tested in the rabbit pyrexia model. Both an IgG$_1$ and an IgG$_4$ were equally efficient in neutralizing TNF in the plasma, but only the γ4 was able to reduce the pyrexia; the γ1 actually exacerbated the temperature rise. Indeed the pyrexia could be reproduced following administration of preformed immune complexes of TNF/γ1 but not TNF/γ4 (SUITTERS et al., 1994). Although the explanation for this is unclear, it seems dependent on the type of immune complex formed and serves to highlight the benefit of the early use of *in vivo* assays. A further insight into the effect of isotype choice on the therapeutic potential of a MAB can be gained in animal models of disease using analogous reagents (see Sect. 4.2).

3.3 Whole Antibody vs. Fragments

For some MABs, neutralization of the target antigen is not the only consideration. MABs directed against tumor antigens can be used to deliver cytotoxic agents or radioisotopes to the tumor site. For radioimmunotherapy a prime safety consideration is the balance between adequate tumor targeting and the avoidance of significant amounts of radiolabel accumulating in the bone marrow. For this reason MAB fragments are often used which are more rapidly cleared from the circulation than whole antibody. In this case

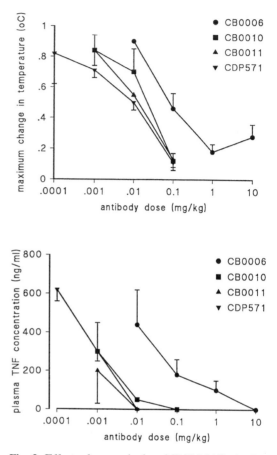

Fig. 2. Effect of a panel of anti-TNF MABs in the rabbit pyrexia model. Conscious rabbits were given either CB0006, CB0010, CB0011 or CDP571 i.v. 15 min prior to rec hu TNF (1 μg kg^{-1}). The upper panel shows the maximum change in temperature for each antibody and the lower panel the amount of rec hu TNF remaining in the plasma at 60 min. These data highlight the difference in *in vivo* potency within a panel of MABs. Values are means $\pm$ SEM (upper) or 95% confidence intervals (C.I.) (lower), ($n=4$–6 per point).

the early determination of *in vivo* biodistribution is essential. MAB targeted radioimmunotherapeutics are in clinical development and some of these are based on di-Fab or tri Fab-maleimide (DFM or TFM) constructs which can be used to attach a macrocycle to contain the isotope (^{125}I or ^{90}Y). Studies with the MAB A33 (which recognizes a tumor associated antigen on human colorectal carcinoma) have shown good tumor targeting with limited bone marrow and kidney accumulation of the TFM compared to the IgG (ANTONIW et al., 1996).

4 Preclinical Determination of Activity

The rationale for the potential use of neutralizing MABs in the clinic is often initially based on information relating to the known biological effects of target antigens. For soluble antigens such as cytokines, enhanced mRNA expression or protein production seen in disease states may be powerful evidence. In contrast, for cell surface antigens, a knowledge of the physiological role of the antigen may make it an attractive target e.g., CD4, CD3, HLA-DR or its increased expression on certain tumors, e.g., CD33 in acute myeloid leukemia or PEM in ovarian, breast and lung cancer. Nevertheless, proof of principle in animal models of disease is desirable and *in vivo* testing of monoclonal antibodies to demonstrate clinical utility and efficacy is required.

4.1 Species Cross-Reactivity

It is critical, at an early stage, to determine the cross-reactivity profile of the MAB in a variety of animal species. Ideally, both proof of principle and efficacy testing should be carried out in preclinical models with the MAB proposed for clinical use. There are examples of MABs with sufficient cross-species reactivity to allow testing of the proposed therapeutic MAB in animal models, although these are often restricted to primates (WATKINS et al., 1997; HINSHAW et al., 1990). There are even some MABs that have sufficient cross-reactivity with the antigen in lower species for useful experiments to be done. BAYX1351 has been extensively studied in pigs and yielded important information about the role of TNF in a variety of models of shock (JESMOK et al., 1992).

The ability to recognize the antigen in other species is not the only consideration. The relative potency against that antigen compared to that against the human antigen is important if models are being used to confirm the efficacy of the MAB. If, e.g., the MAB was some 100-fold inferior in neutralizing activity against marmoset TNF compared to human TNF then the doses to be used would be far higher and the relevance of that choice of species would be questionable for predicting efficacious doses in man.

4.2 Use of Parallel Reagents

However, there is often limited species cross-reactivity of the proposed therapeutic reagent. In these instances, there is a need to investigate the pharmacology of the therapeutic principle using parallel or analogous reagents, i.e., to use MABs that see the same or similar antigen in laboratory animals. Often these reagents are already available from other sources but if not then it is a critical part of the preclinical program to generate these antibodies. Indeed some of the pioneering work to elucidate the physiological and pathophysiological role of TNF in sepsis was done with anti-mouse TNF MABs (TRACEY et al., 1987; WILLIAMS et al., 1992). Since then, anti-TNF MABs have been used in a variety of species and a variety of disease models (WILLIAMS et al., 1992; LEWTHWAITE et al., 1995; WARD et al., 1993; SEU et al., 1991; OPAL et al., 1990; PIQUET et al., 1987; BAKER et al., 1994). Much early work was done with a hamster anti-mouse TNF MAB known as TN3 19.12 (SHEEHAN et al., 1989), and in order to address the potential for chronic treatment and the effect of isotype choice in animal models, the genes for this MAB have been cloned and chimeric versions of either the mouse $\gamma1$ or $\gamma2a$ isotype have been made. The importance of this approach is exemplified by the observation that in a mouse LPS-induced shock model the IgG_1 isotype (non-complement fixing, Fc receptor-binding) was far better at preventing death than was the IgG_{2a} isotype (Fig. 3). This was further investigated in a neutropenic rat sepsis model (induced with *Pseudomonas aerugi-*

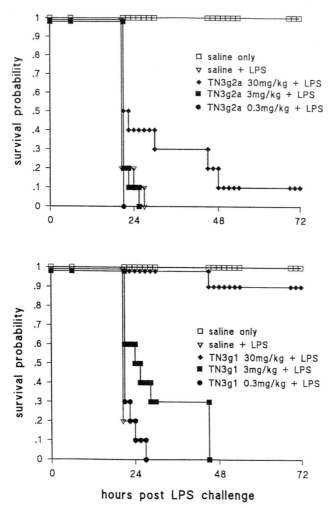

Fig. 3. Effect of different isotypes of chimeric TN3 on survival after LPS-induced shock in mice. Animals were given LPS (50 mg kg^{-1} i.v.) 60 min after the IgG$_{2a}$ isotype (upper panel) or the IgG$_1$ isotype (lower panel). Values are means ± SEM ($n = 10$ per group).

nosa) in which cTN3γ1 or γ2a were given after the symptoms of sepsis were manifest. Despite similar levels of bacteremia in all groups of rats, the cTN3γ2a offered marginal survival benefit (27% survival) over the control group (0% survival), whereas the cTN3γ1 treatment resulted in significant (70%) survival of the rats (SUITTERS et al., 1994).

These chimeric MABs have also been used in a variety of chronic situations, to test primarily the concept of a role for TNF in disease and to determine whether MAB isotype is important. Efficacy has been shown in mouse collagen induced arthritis (WARD et al., 1995), mouse experimental allergic encephalomyelitis (EAE) (BAKER er al., 1994)

and rat acetic acid (WARD et al., 1993) or degraded carrageenan induced colonic inflammation with chronic administration. Only in the shock type models did the choice of isotype make a difference to the outcome.

Similarly, MABs specific for rabbit TNF have been developed to test further the concept for a role for TNF in arthritis since one of the better models of arthritis is the antigen-induced arthritis model in rabbits. The ability of anti-TNF to inhibit the early inflammatory responses associated with this model has been demonstrated, changes that are not readily reversed by most clinically used anti-inflammatory agents (LEWTHWAITE et al., 1995). The antibody to rabbit TNF was also used to study the effects of anti-TNF treatment in a conscious rabbit model of hemorrhagic shock, a model that was far superior in terms of sensitivity and practicality than any available in rodents. Use of these parallel reagents allowed generation of valuable information without the immediate need to do studies in similar primate models (FOULKES et al., 1994).

A different approach to that above is needed when no suitable animal model of the disease in question exists. In this situation, it is important to study the effects of the MAB on certain biological features known to be associated with the disease. For example, there are no animal models of psoriasis. This indication is characterized by an influx of inflammatory and immune cells into the skin resulting in a typical raised psoriatic plaque. The adhesion molecule, E-selectin, is thought to be important but in order to study the rationale for the use of an anti-E-selectin MAB in psoriasis, studies were carried out in pigs using parallel reagents. The ability of the MAB YT11.1, an antibody to pig E-selectin, to prevent lymphocyte and neutrophil infiltration into the skin of pigs treated with topical dinitrofluorobenzene (i.e., a delayed type hypersensitivity reaction) was used to assess the potential of this therapeutic approach (OWENS et al., 1997).

Occasionally there exist examples of the proposed clinical disease in animals where the candidate therapeutic reagent can be studied. When this occurs then the dual objectives of concept testing and demonstration of efficacy of the chosen molecule can be tested in a single study. For anti-TNF, e.g., much work on its potential as an agent for the treatment of inflammatory bowel disease (IBD) was done in rodents. However, a spontaneous form of ulcerative colitis has been described in cotton top tamarins, a new world primate related to the marmoset. Advantage of this was taken to demonstrate the efficacy and potential of repeated dosing of the engineered anti-human TNF MAB CDP571 which cross-reacts only with primate TNF (WATKINS et al., 1997). These studies showed the ability of anti-TNF to reduce markedly the severity of this disease in a relevant species.

4.3 Use of Xenograft and Transgenic Models

Where the degree of cross-reactivity of an antibody with other species is so limited that it makes efficacy testing impossible, the use of xenograft models or transgenic mice expressing the human antigen are becoming increasingly common. For example, the engineered human anti-E-selectin MAB, SPLAT-1 is being developed for use in psoriasis and cross-reacts only with some primate E-selectin. In SCID mice transplanted with human skin grafts in which the human vasculature is retained, i.e., endothelial cells capable of expressing human E-selectin, the anti-E-selectin MAB SPLAT-1 was shown to inhibit leucocyte infiltration into the skin following intradermal injections of recombinant human TNF and IL-1 (OWENS et al., 1997).

Such xenograft models offer an opportunity for preclinical *in vivo* efficacy testing in human based systems. For MAB targeted anticancer reagents, human tumor xenograft models often offer the only avenue for *in vivo* efficacy testing. The ^{90}Y labeled TFM molecule, A33, was studied in nude mice with established tumors (SW1222 human colon carcinoma cells). Following a single i.v. injection, this antibody markedly attenuated the growth of the xenograft (ANTONIW et al., 1996).

Examples of transgenic mice with potential as preclinical models include the transgenic mouse expressing human TNFα developed by KOLLIAS and colleagues (KEFFER et al., 1991) which exhibits the symptoms of progressive arthritis. This animal model allows the assessment of a MAB not only on its ability to inhibit TNF but also in the appropriate disease model.

5 Pharmacokinetics

Pharmacokinetics of engineered human antibodies have shown considerable improvements in half-life over murine antibodies with a concomitant reduction in immunogenicity (STEPHENS et al., 1995; HAKIMI et al., 1991; LOBUGLIO et al., 1989). However, at present there is no accurate and reliable small animal model which is predictive of the pharmacokinetics in man. In general, pharmacokinetics have been established in the toxicology species of choice, e.g. a rodent species such as rat and a non-human primate such as the cynomolgus monkey. Radiolabeling of the antibody enables the pharmacokinetic profile to be followed more easily but this is not always possible. In rodents, antigen-based immunoassays using anti-human immunoglobulin (Ig) second layers are the method of choice, and these can be used to confirm the binding activity of the injected antibody. Where antigen is not available for such assays, anti-Ig capture and detection can be used (e.g., anti-human Fc followed by anti-κ).

In non-human primates and man, antigen capture followed by an isotype specific second antibody frequently gives sufficient specificity for pharmacokinetic testing, although care should be taken to test background levels as autoantibodies, e.g., to cytokines such as TNF, have been reported (FOMSGAARD et al., 1989). Where antigen is not available, polyclonal or monoclonal anti-idiotypic reagents can be used for capture and/or detection. Anti-Ig formats are not possible in these instances, as the anti-human Ig reagents will cross-react with circulating primate Ig.

When injected i.v., antibodies are typically eliminated with a biexponential decline, comprising an α- or tissue distribution phase and a β- or elimination phase. Half-lives should be calculated using model independent pharmacokinetic parameters and area under the plasma concentration time curve (AUC) calculated using the trapezoidal rule (STEPHENS et al., 1995; HAKIMI et al., 1991). Modeled systems are also frequently used (STRAND et al., 1993) but these are prone to error when limited data points are available. Repeated doses have indicated that α- and β-phases occur after each dose (Fig. 4) (OWENS et al., 1997). A number of factors will influence the length of the half-life, including size of the molecule (IgG vs. Fab' for example), Fc receptor binding (particularly neonatal Fc receptor, FcRn), immunogenicity and, perhaps most dramatically, presence of the antigen. For example in a cynomolgus monkey, a human IgG_4 antibody to human MHC II has a half-life of <4 h due to the high level of expression of antigen on the surface of cells, whereas a similarly grafted antibody to E-selectin has an elimination half-life of approximately 2 weeks (Fig. 5). Although the number of engineered antibodies which have been injected into man is small, the half-life generally seems to be 50% longer than that seen in primates for the same molecule (STEPHENS et al., 1995; CAVACINI et al., 1994; WOLFE et al., 1996).

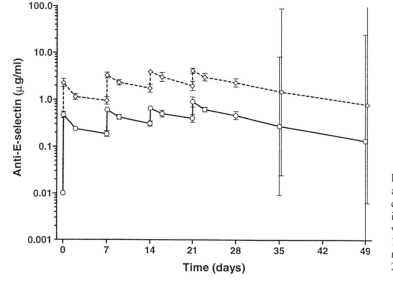

Fig. 4. Pharmacokinetics of anti-E-selectin antibody in cynomolgus monkeys given 4 i.v. infusions at weekly intervals. –o– 20 mg kg^{-1}; ...$\Diamond$... 100 mg kg^{-1}. (Geometric means ±95% C.I.; days 0–28, $n=8$; days 35–49, $n=2$).

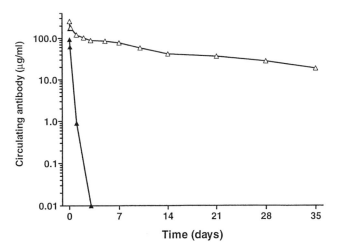

Fig. 5. Pharmacokinetics in healthy cynomolgus monkeys following a single i.v. infusion of 10 mg kg^{-1}: – anti-E-selectin, –Δ–; anti-class II MHC, –♦–.

I.v. infusion may not necessarily be the most suitable route, particularly where frequent injections are required. Pharmacokinetics following other routes of administration have been investigated (WARD et al., 1992) and although the peak circulating MAB levels may be lower following intraperitoneal, intramuscular or subcutaneous injection, elimination half-lives are unaffected and AUC values may be maintained by increasing the dose.

6 Immunogenicity

The use of murine monoclonal antibodies in man has been severely limited by HAMA responses directed against the constant (and variable) regions of the molecule, which generally result in accelerated elimination of antibody from 7–10 d onwards (FOON et al., 1984; CHEUNG et al., 1987). The HAMA response may also preclude repeated dosing, although several doses of murine antibodies have been administered over a short period without untoward effects (KHAZAELI et al., 1988). Development of chimeric and other engineered or human antibodies has considerably reduced but not always eliminated immunogenicity (LoBUGLIO et al., 1989; HAKIMI et al., 1991). The majority of the described responses to these products are directed towards the idiotope and tend to be of

the IgM class rather than IgG (STEPHENS et al., 1995). These responses may diminish with increasing dose of the therapeutic antibody (STEPHENS et al., 1995; SEARS et al., 1987).

As with pharmacokinetics, no ideal model to predict immunogenicity in man exists. So far, non-human primates have been the best indicator. Factors known potentially to affect immunogenicity are the type, location, and quantity of the antigen recognized (ISAACS and WALDMANN, 1994), antibody Fc functions such as receptor-binding and complement activation and glycosylation (REITAN and HANNESTAD, 1995; HEYMAN, 1990; ROUTLEDGE et al., 1995), and the number of "murine" residues remaining in the variable region of an engineered human antibody. Where labels or other compounds are attached to antibodies, there exists the possibility that these may be immunogenic (e.g., linkers for radioisotopes, prodrugs or cytotoxic agents) (HIRD et al., 1991). Assays should be developed which will distinguish between those immune responses to the antibody and those to the attached molecule. This can be done by comparing the molecule similarly attached to an unrelated protein as a screen, in addition to the unconjugated antibody.

When immune responses to engineered antibodies do occur, it is important for the development of the product to determine against which part of the molecule they are directed. This can readily be done using the assay used to detect the immune response, into which excess of molecules containing defined homologous regions are competed. It is also relevant to know the isotype of the immune response, since when a class switch to IgG is seen after the 1st dose, this is more likely to result in increased clearance following subsequent doses than when the response is confined to the IgM class.

7 Summary

Many of the activities required to develop antibodies for clinical use have been outlined in this chapter. *In vitro* testing will provide an indication of binding affinity and selectivity for the target antigen but may not always indicate the consequences of secondary Fc-mediated effector functions. These may be determined *in vivo* using models as relevant as possible to the proposed therapeutic indication. However, the highly specific nature of many antibodies for their target antigen can result in limited species cross-reactivity. Where only human antigens are recognized, antibodies to similar antigens in other species may need to be developed in order to test the proof of principle or xenograft and transgenic models used. Although ultimate proof of activity will be derived from clinical studies, a rigorous but focused preclinical development program allows the most appropriate reagent to be selected.

8 References

ANTONIW, P., FARNSWORTH, A. P. H., TURNER, A., HAINES, A. M. R., MOUNTAIN, A. et al. (1996), Radioimmunotherapy of colorectal carcinoma xenografts in nude mice with yttrium-90 A33 IgG and Tri-Fab (TFM), *Br. J. Cancer* **74**, 513–524.

BAKER, D., BUTLER, D., SCALLON, B. J., O'NEILL, J. K., TURK, J. L., FELDMANN, M. (1994), Control of established experimental allergic encephalomyelitis by inhibition of tumor necrosis factor (TNF) activity within the central nervous system using monoclonal antibodies and TNF

receptor–immunoglobulin fusion proteins, *Eur. J. Immunol.* **24**, 2040–2048.

BUIST, M. R., KENEMANS, P., VAN KAMP, G. J., HAISMA, H. J. (1995), Minor human antibody response to a mouse and chimeric monoclonal antibody after a single i.v. infusion in ovarian carcinoma patients: a comparison of five assays, *Cancer Immunol. Imunother.* **40**, 24–30.

CAVACINI, L. A., POWER, J., EMES, C. L., MACE, K., TREACY, G., POSNER, M. R. (1994), Plasma pharmacokinetics and biological activity of a human immunodeficiency virus type 1 neutralizing human monoclonal antibody, F105, in cynomolgus monkeys, *J. Immunother. Emph. Tumor Immunol.* **15**, 251–256.

CHEUNG, N.-K. V., LAZARUS, H., MIRALDI, F. D., ABRAMOWSKI, C. R., KALLICK, S. et al. (1987), Ganglioside GD2 specific monoclonal antibody 3F8: a phase I study in patients with neuroblastoma and malignant melanoma, *J. Clin. Oncol.* **5**, 1430–1440.

COBBOLD, S. P., REBELLO, P. R., DAVIES, H. F., FRIEND, P. J., CLARK, M. R. (1990), A simple method for measuring patient anti-globulin responses against isotypic or idiotypic determinants, *J. Immunol. Methods* **127**, 19–24.

FOMSGAARD, A., SVENSON, M., BENDTZEN, K. (1989), Auto-antibodies to tumor necrosis factor α in healthy humans and patients with inflammatory diseases and gram-negative bacterial infection, *Scand. J. Immunol.* **30**, 219–223.

FOON, K. A., SCHROFF, R. W., BUNN, P. A., MAYER, D., ABRAMS, P. G. et al. (1984), Effects of monoclonal antibody therapy in patients with chronic lymphocytic leukemia, *Blood* **64**, 1085–1093.

FOULKES, R., HUGHES, B., KINGABY, R., WOODGER, R., VETTERLEIN, O. (1994), Anti-TNF treatment reduce severity of organ damage in a conscious rabbit model of haemorrhagic/traumatic shock, *Eur. Cytokine Net.* **5**, 120.

HAKIMI, J., CHIZZONITE, R., LUKE, D. R., FAMILLETTI, P. C., BAILON, P. et al. (1991), Reduced immunogenicity and improved pharmacokinetics of humanized anti-Tac in cynomolgus monkeys, *J. Immunol.* **147**, 1352–1359.

HEYMAN, B. (1990), Fc-dependent IgG-mediated suppression of the antibody response: fact or artefact? *Scand. J. Immunol.* **31**, 601–607.

HINSHAW, L. B., TEKAMP-OLSON, P., CHANG, A. C. K., LEE, P. A., TAYLOR, F. B. JR. et al. (1990), Survival of primates in LD_{100} septic shock following therapy with antibody to tumor necrosis factor (TNFα), *Circ. Shock* **30**, 279–292.

HIRD, V., VERHOEYEN, M., BADLEY, R. A., PRICE, D., SNOOK, D. et al. (1991), Tumor localization with a radioactively labeled reshaped human monoclonal antibody, *Br. J. Cancer* **64**, 911–914.

ISAACS, J. D., WALDMANN, H. (1994), Helplessness as a strategy for avoiding antiglobulin responses to therapeutic monoclonal antibodies, *Ther. Immunol.* **1**, 303–312.

ISAACS, J. D., WATTS, R. A., HAZLEMAN, B. L., HALE, G., KEOGAN, M. T. et al. (1992), Humanized monoclonal antibody therapy for rheumatoid arthritis, *Lancet* **340**, 748–752.

JESMOK, G., LINDSEY, C., DUERR, M., FOURNEL, M., EMERSON, T. (1992), Efficacy of monoclonal antibody against human tumor necrosis factor in *E. coli* challenged swine, *Am. J. Pathol.* **141**, 1197–1207.

KEFFER, J., PROBERT, L., CAZLARIS, H., GEORGOPOULOS, S., KASLARIS, E. et al. (1991), Transgenic mice expressing human tumor necrosis factor: A predictive genetic model of arthritis, *EMBO J.* **10**, 4025–4031.

KHAZAELI, M. B., SALEH, M. N., WHEELER, R. H., HUSTER, W. J., HOLDEN, H. et al. (1988), Phase I trial of multiple large doses of murine monoclonal antibody CO17-1A. II Pharmacokinetics and immune response, *J. Natl. Cancer Inst.* **80**, 937–942.

LEWTHWAITE, J., BLAKE, S., HRDINGHAM, T., FOULKES, R., STEPHENS, S. et al. (1995), Role of TNFα in the induction of antigen induced arthritis in the rabbit and the anti-arthritic effect of species specific TNFα neutralizing monoclonal antibody, *Ann. Rheum. Dis.* **54**, 366–374.

LoBUGLIO, A. F., WHEELER, R. H., TRANG, J., HAYNES, A., ROGERS, K. et al. (1989), Mouse/human chimeric monoclonal antibody in man: kinetics and immune response, *Proc. Natl. Acad. Sci. USA* **86**, 4220–4224.

OPAL, S. M., CROSS, A. S., KELLY, N. M., SADOFF, J. C., BODMER, M. W. et al. (1990), Efficacy of a monoclonal antibody directed against tumor necrosis factor in protecting neutropenic rats from lethal infection with *Pseudomonas aeruginosa*, *J. Inf. Dis.* **161**, 1148–1152.

Ortho Multicenter Transplant Study Group (1985), A randomized clinical trial of OKT3 monoclonal antibody for acute rejection of cadaveric renal transplants, *N. Eng. J. Med.* **313**, 337–342.

OWENS, R., BALL, E., GANESH, R., NESBITT, A., BROWN, D. et al. (1997), The *in vivo* and *in vitro* characterization of an engineered human antibody to E-selectin, *Immunotechnolgy* **3**, 107–116.

PIGUET, P.-F., GRAU, G. E., ALLET, B., VASSALLI, P. (1987), Tumor necrosis factor/cachectin is an effector of skin and gut lesions of the acute phase of graft-versus-host disease, *J. Exp. Med.* **166**, 1280–1289.

REITAN, S. K., HANNESTAD, K. (1995), A syngeneic idiotype is immunogenic when borne by IgM but tolerogenic when joined to IgG, *Eur. J. Immunol.* **25**, 1601–1608.

ROUTLEDGE, E. G., FALCONER, M. E., POPE, H., LLOYD, I. S., WALDMANN, H. (1995), The effect of aglycosylation on the immunogenicity of a humanized therapeutic CD3 monoclonal antibody, *Transplantation* **60**, 847–853.

SEARS, H. F., BAGLI, D. J., HERLYN, D., DEFREITAS, E., SUZUKI, H. et al. (1987), Human immune response to monoclonal antibody administration is dose-dependent, *Arch. Surg.* **122**, 1384–1388.

SEU, P., IMAGAWA, D. K., WASEF, E., OLTHOFF, K. M., HART, J. et al. (1991), Monoclonal antitumor necrosis factor-α antibody treatment of rat cardiac allografts: synergism with low-dose cyclosporine and immunohistological studies, *J. Surg. Res.* **50**, 520–528.

SHAW, N., FOULKES, R., MORGAN, A., STEPHENS, S., HIGGS, G., BODMER, M. (1991), Investigation of the efficacy of a panel of anti-human TNF antibodies in the conscious rabbit model, *Br. J. Pharmacol.* **104**, 89.

SHEEHAN, K. C. F., RUDDLE, N. H., SCHREIBER, R. D. (1989), Generation and characterization of hamster monoclonal antibodies that neutralize murine tumor necrosis factors, *J. Immunol.* **142**, 3884–3893.

STEPHENS, S., EMTAGE, S., VETTERLEIN, O., CHAPLIN, L., BEBBINGTON, C. et al. (1995), Comprehensive pharmacokinetics of a humanized antibody and analysis of residual anti-idiotypic responses, *Immunology* **85**, 668–674.

STRAND, S. E., ZANZONICO, P., JOHNSON, T. K. (1993), Pharmacokinetic modelling, *Med. Phys.* **20**, 515–527.

SUITTERS, A. J., FOULKES, R., OPAL, S. M., PALARDY, J. E., EMTAGE, J. S. et al. (1994), Differential effect of isotype on efficacy of anti-tumor necrosis factor α chimeric antibodies in experimental septic shock, *J. Exp. Med.* **179**, 849–856.

TRACEY, K. J., FONG, Y., HESSE, D. G., MANOGUE, K. R., LEE, A. T. et al. (1987), Anti-cachectin/TNF monoclonal antibodies prevent septic shock during lethal bacteremia, *Nature* **330**, 662–664.

WARD, P. S., VETTERLEIN, O., BODMER, M., FOULKES, R. (1992), Is the intravenous route the only therapeutic route for monoclonal antibodies? *Br. J. Pharmacol.* **107**, 43.

WARD, P. S., WOODGER, S. R., BODMER, M., FOULKES, R. (1993), Anti-tumor necrosis factor α monoclonal antibodies (anti-TNF MAB) are therapeutically effective in a model of colonic inflammation, *Br. J. Pharmacol.* **110**, 77P.

WARD, P. S., BODEN, T., WOODGER, R., FOULKES, R. (1995), Isotype variation of TNF monoclonal antibodies. Is there a therapeutic difference in rheumatoid arthritis? *Br. J. Rheumatol.* **34** (Suppl.), 294.

WATKINS, P. E., WARREN, B. F., STEPHENS, S., WARD, P., FOULKES, R. (1997), Treatment of ulcerative colitis in the cottontop tamarin using antibody to tumor necrosis factor alpha, *Gut* **40**, 628–633.

WILLIAMS, R. O., FELDMANN, M., MAINI, R. N. (1992), Anti-tumor necrosis factor ameliorates joint disease in murine collagen-induced arthritis, *Proc. Natl. Acad. Sci. USA* **89**, 9784–9788.

WOLFE, E. J., CAVACINI, L. A., SAMORE, M. H., POSNER, M. R., KOZAIL, C. et al. (1996), Pharmacokinetics of F105, a human monoclonal antibody, in persons infected with human immunodeficiency virus type 1, *Clin. Pharmacol. Therapeut.* **59**, 662–667.

11 Preclinical Testing of Antibodies: Safety Aspects

RONALD W. JAMES

Huntingdon, UK

1 Introduction   304
2 Regulatory Testing Requirements   305
3 Timing of Studies in Relation to Clinical Trials   306
4 Tissue Specificity and Cross-Reactivity   306
5 Pharmaco-Toxicological Testing   307
  5.1 Number and Choice of Species   308
  5.2 Dosage Selection, Frequency of Administration and Study Duration   308
  5.3 Reproduction Toxicity Studies   308
  5.4 Genotoxicity Studies   309
  5.5 Oncogenicity Carcinogenicity Studies   309
  5.6 Local Tolerance   309
  5.7 Special Studies   309
6 References   309

1 Introduction

During the 20th century there has been enormous progress in the understanding of the molecular basis of both the pathogenesis and the pharmacological modulation of disease processes. At the same time the technological advances which have occurred in the field of biotechnology have been successfully applied to the commercial production of naturally occurring molecules having utility in the diagnosis and/or treatment of disease (TOMLINSON, 1992). Monoclonal antibodies (MABs) represent a significant proportion of the entities which have been produced by this new technology (WALDMAN, 1991). Products that consist of or contain MABs can or might prove useful in the diagnosis and treatment of a wide array of human diseases such as septic shock, AIDs, multiple sclerosis, and cancer. They have a substantial theoretical advantage, particularly in cancer treatment, due to the selectivity and specificity for tumor associated antigens, and in the general lack of affinity that specific MABs have for normal cells. While it is more than 20 years since KÖHLER and MILSTEIN (1975) described the production of immortalized cell lines capable of secreting MABs, the number of *in vivo* diagnostic or therapeutic MABs that have been approved for marketing is relatively few. The reasons for this include factors such as lack of familiarity, both on the part of developers and regulators, with MABs as therapeutics, the unique aspects of developing MABs, and safety and clinical trial design issues (GORE et al., 1996).

The toxicological principles on which preclinical safety testing guidelines are based have evolved over the last 50 years or more. It has become apparent that many of the assumptions made in the early days of the testing of chemically synthesized xenobiotics were not valid for monoclonal antibodies. As a result the stereotyped approaches followed by several generations of industrial and regulatory toxicologists are in the process of being discarded. There is increasing resistance to the development of rigid inflexible guidelines, and a science-based "case-by-case" approach to the preclinical safety evaluation of biotech-

nology-derived products is being promoted (ROMMEBERGER, 1990).

As with any other therapeutic entity the manifestations of toxicity of a MAb can be attributed to:

- The presence of product related impurities or variants and process related residual or extraneous materials auch as host cell protein/DNA, viral particles, etc.
- Extension of the intended pharmacodynamic action(s).
- Idiosyncratic effects (e.g., hypotension) unrelated to the primary pharmacological/immunological activity.

The potential safety issues which arise due to the presence of product and/or process related contaminants are normally addressed by the application of appropriate quality control tests, which may include animal tests when considered appropriate, on each batch of MAB destined for clinical usage (SULLIVAN, 1990). Preclinical safety tests to characterize the intrinsic toxicity, whether it be pharmacologically mediated or idiosyncratic, are normally only required to be conducted once using a representative batch of clinical grade material. These tests, sometimes referred to as the "pivotal" safety tests are not usually repeated unless there is a substantial change in the nature of a MAB as a result of modifications made to the production methodology. Whenever there are changes to the production process rigorous *in vitro* and biochemical functional observations and animal pharmacokinetic studies are used whenever possible to demonstrate that the entities made by the old and new methods are equivalent.

It can be very difficult, because of the often highly species-specific properties of the antibody and the antigen against which it is targeted, to specify animal safety tests for MABs which are relevant to the prediction of safety for humans. Often, only small quantities of a MAB can be produced and this imposes practical constraints on toxicological assessment, especially during the early stages of development. A pragmatic approach is often necessary which distinguishes between "toxicity" tests employing a range of doses encompassing both the NOAEL (no observable adverse effect level) and overtly toxic doses and "safety" tests conducted at one or two doses which represent arbitrary multiples of the anticipated therapeutic dose to be given to humans. These doses should not elicit overt manifestations of toxicity in the test animals.

2 Regulatory Testing Requirements

The US Food and Drug Administration (FDA), Commission of the European Communities and the Japanese Ministry of Health and Welfare (MOHW) have all published guidance documents which discuss the regulatory requirements for preclinical safety tests to support the therapeutic use of biotechnology-derived therapeutics including MABs:

1) **FDA.** *"Points to Consider in the Manufacture and Testing of Monoclonal Antibody Products for Human Use"* (FDA, 1994).
2) **Commission of the European Communities.** *"Preclinical Biological Safety Testing of Medicinal Products Derived from Biotechnology (and Comparable Products Derived from Chemical Synthesis)"* (Commission of the European Communities, 1989).
3) **Japan Ministry of Health and Welfare.** *"Draft Guidelines for Preparation of Data Required on Application for Approval of Drugs Manufactured by the Application of Cell Culture Techniques"* (MOHW, 1988).

The safety testing of biotechnological products is one of the topics currently under consideration by the International Conference on Harmonization of Technical Requirements for Registration of Pharmaceuticals for Human Use (ICH). It is hoped that a tripartite guidance document will be produced in the near future which will be adopted by the European, Japanese and US drug regulatory authorities. While these guidance documents will provide the basis for a regulatory framework, it is not anticipated that they will contain specific or detailed study designs since this would not be consistent with the flexible case-by-case approach. The nature and extent of the animal testing required for regulatory approval can only be determined after careful consideration of physico-chemical properties, the mechanism by which the intended therapeutic effect is mediated, and the nature of the patient population affected by the disease target. Unless this approach is followed it is only too easy to generate data which are both inappropriate and misleading.

The primary goals of any preclinical safety testing program must be to

- identify an initial safe starting dose and subsequent dose escalation scheme in humans,
- identify potential target organs of toxicity and reversibility of toxicity,
- identify parameters for clinical safety monitoring.

Monoclonal antibodies may be used either as aids to diagnosis or as therapeutic modalities. The antibody may be administered "naked" or as an immuno-conjugate formed by combination with a radionuclide, toxin, enzyme, recombinant protein or drug. Naked antibodies are used to target specific proteins or receptors. The role of the antibody when administered as an immuno-conjugate is to act as a delivery device ensuring that the conjugated molecule reaches its cellular target. When MABs are conjugated to radionuclides, toxins, enzymes, recombinant proteins, or drugs with established diagnostic or therapeutic uses, then it should only be necessary to investigate the toxicity potential of the immuno-conjugate and/or the naked antibody. As a general principle it is advisable to carry out the pivotal preclinical safety tests using the same or an essentially similar formulation to that intended for administration to humans. One does, however, have to be alert to the possibility that animals may react differently to humans when formulations containing molecules such as human serum albumin (HSA) are used. MATSUZAWA et al. (1992) have published details of the preclinical safety testing scheme which they applied to a human monoclonal antibody against cytomegalovirus (MAB 23) which serves as an example of the extent and scope of the toxicological testing that may be requested by regulatory authorities for this category of product. Consideration must be given to the need for appropriate safety pharmacology and pharmacokinetic studies, single and repeated dose toxicity studies in one or more species, reproduction toxicity studies, genotoxicity studies, carcinogenicity studies, local tolerance studies and special investigations including assessment of the immunogenicity potential. Careful consideration of the MAB itself, the nature of the disease target and the patient population, the frequency of administration etc. will help to clarify the need for, or the justification to exclude, specific studies. A unique and major safety concern which applies to the safety evaluation of MABs is the potential for cross-reactivity with receptors in non-target tissues. Thus a scheme for investigating tissue cross-reactivity is a paramount feature of the preclinical safety testing of these agents.

3 Timing of Studies in Relation to Clinical Trials

As with any other entity undergoing evaluation as a potential therapy it is important to integrate the preclinical safety testing program into the overall development plan. The toxicologist has, therefore, to have an appreciation of the chronology, objectives and design of the studies to be conducted in humans before advising on the scope, and extent of the preclinical safety tests needed to support the development plan. If this is done then it is possible to specify not only the tests which are required but also the sequence in which they should be undertaken.

Initially only those studies which are absolutely necessary to support first in man studies should be conducted. Tissue cross-reactivity, safety pharmacology, and single dose toxicity studies will be a high priority while other study types are likely to be less important at this stage. Priority attention should be given to the need to characterize both the pharmacokinetics and the immunogenicity of MABs

during the early phases of the preclinical safety testing. An understanding of the plasma–time concentration profile and the potential to form serum antibodies is an integral aspect of the evaluation of the animal studies and of the extrapolation of the reaulte to humans.

The need for more extended toxicity testing will not only depend upon the nature and proposed use of the MAB, be it diagnostic or therapeutic, but also upon the findings generated from the tissue cross-reactivity, safety pharmacology, and single dose toxicity studies, as well as the effects seen in the first in man studies.

4 Tissue Specificity and Cross-Reactivity

A major safety concern for monoclonal antibodies is that the antigenic determinant (or a structurally similar epitope) may be present on human cells or tissues other than the intended target tissue, resulting in a potential for undesirable cross-reactions. Accordingly, laboratory tests should be conducted to assess this possibility and, when cross-reactions are encountered, the resultant potential hazard and/or risk to potential recipients should be evaluated. The FDA (1994) specifically mentioned that in the special case of bispecific antibodies, each parent antibody should be evaluated individually, in addition to testing of the bispecific product. Quick-frozen adult human cells or tissues are surveyed *in vitro* using immunocytochemical or immunohistochemical techniques. Demonstration of cross-reactivity triggers the need for a comprehensive *in vivo* investigation, first in animals and subsequently in humans. It has to be emphasized that the techniques which are applied to characterizing the potential interaction of MABs with non-target (normal) tissues are influenced by numerous technical factors which can result in both false positive and false negative results. The need to standardize both the techniques which are employed and the conduct of these investigations in order to overcome these difficulties is a matter of priority to industry and regulators.

In general, the technique used should be optimized using a positive control tissue. Where expression of the antigen is widespread, this can be straightforward (e.g., class II MHC) but where the antigen is not expressed on normal tissue, a sample of diseased tissue could be used or cell lines or antigen spiked samples could be generated. Positive tissues should be fixed and treated in the same way as the treated tissues. Second antibody reagents used in assays need careful evaluation when testing engineered human antibodies. Anti-isotype reagents may be used for less common isotypes such as IgG_4, but can show cross-reactivity with human Ig in the tissues. Alternatively, anti-idiotypic reagents can be developed, or the test antibody can be labeled, for example with biotin.

The list of tissues which are specified by FDA (1994) for the study of cross-reactivity are listed in Tab. 1.

Surgical samples are preferred but *post mortem* samples are acceptable provided tissue preservation is adequate. Tissues from at least three unrelated human donors should be evaluated in order to screen for polymorphism and several concentrations of the product should be tested. Appropriate positive and negative controls should be used to confirm the acceptable condition of tissues and adequacy of the assay. Antibody affinities as well as expected achievable peak plasma concentrations should be considered when choosing the concentrations used for tissue binding studies. Attempts should also be made to compare the ratio of target to cross-reactive tissue binding. Non-specific binding may be observed and the specificity of potential cross-reactions should be assessed using inhibition assays with purified antigen if available.

The substitution of antibodies of similar specificity for cross-reactivity testing is not encouraged, and if a derivatized or fragmented antibody is to be used clinically then it should be tested in that form whenever feasible. When cross-reactivity is encountered the possibility of polymorphism to the target antigen should be investigated using a larger donor panel to ascertain their frequency. Comparison of *in vitro* cross-reactivity using animal tissues becomes necessary in order to select appropriate species for the extended toxicity testing which becomes necessary for MABs which exhibit cross-reactivity with non-target human tissues.

5 Pharmaco-Toxicological Testing

The development of appropriate models for pharmaco-toxicological assessment should be focused on:

1. the selection of the test species,
2. the manner of delivery, including the dose, the route of administration and the treatment regimen.

Pivotal toxicology studies, i.e., those upon which safety judgements are based, are generally expected to be performed in compliance with good laboratory practice (GLP) regulations. However, it is recognized that this may be difficult or even impossible when special-

Tab. 1. List of Tissues Specified by FDA for the Study of Cross-Reactivity

Adrenal	Endothelium	Lymph node	Spinal cord
Bladder	Eye	Ovary	Spleen
Blood cells	Fallopian tube	Pancreas	Striated muscle
Bone marrow	Gastrointestinal tract	Parathyroid	Testis
Breast	Heart	Placenta	Thymus
Cerebellum	Kidney	Pituitary	Thyroid
Cerebral cortex	Liver	Prostate	Ureter
Colon	Lung	Skin	Urerus

ized test systems which are not routinely available have to be employed. In such circumstances areas of non-compliance need to be identified and commented upon by a recognized preclinical expert.

A detailed assessment of the *in vitro* and *in vivo* pharmacological actions which are relevant to the proposed indication should be provided together with a rationale for predicting the anticipated effective clinical dose. A pharmacological basis for the selection of one or more test species which recognize the MAB should also be provided.

Experience with polyclonal antibodies has shown that acute cardio-respiratory reactions may be precipitated in both animals and humans. *In vivo* safety pharmacology tests should therefore always be performed prior to administering a novel MAB product to either volunteers or patients.

The study of pharmacokinetics and biodistribution poses special challenges due to the nature of MABs and the potential to obtain bimodal or paradoxical kinetics. The methodologies employed may yield assay dependent kinetics, and the formation of antibodies may facilitate accelerated immune-mediated clearance (TOON 1996). A number of these issues has been discussed in more detail in Chapter 10, this Volume.

5.1 Number and Choice of Species

The number and choice of species used for toxicity testing must be based on pharmacological relevance, i.e., it is pointless to conduct studies in species which are non-responsive or which exhibit irrelevant tissue cross-reactivity. A single species can be justified when there is sound evidence of pharmacological relevance. The view is sometimes expressed that two species should be used if it is difficult to determine species relevance. Unless such studies reveal idiosyncratic mechanisms of toxicity which could not be predicted from an understanding of the primary mechanism of therapeutic effect they are probably meaningless and should not be encouraged. On the other hand, it can be argued that primates should be used if it is known that lower species do not respond to human specific antibodies. A decision to use a primate species should not be made before it is clearly apparent that less sentient species are unsuitable because they are unresponsive to the candidate MAB. As the science of transgenics advances it is possible that the problems of species specificity may eventually be overcome.

5.2 Dosage Selection, Frequency of Administration, and Study Duration

The classical approaches used for traditional xenobiotics are totally inappropriate in the case of MABs. As already mentioned, it may prove difficult or impossible to elicit overt manifestations of toxicity even when high doses of a MAB are administered to animals. Given the frequent limitations of test material supply and the lack of demonstrable toxicity end points upon which to base dose selection, it is reasonable to choose arbitrary multiples of the dose which is anticipated to be clinically relevant.

Daily dosing is also inappropriate and the frequency of administration should reflect the duration of biological activity following a single administration. A MAB may persist and exert its intended biological actions for long periods after a single administration. The duration of the biological activity can be established on the basis of pharmacokinetic measurements and, if appropriate, relevant bioassay procedures. The aim should be to ensure that biological activity is sustained in the test animals over a period of time which is commensurate with the anticipated human exposure. Thus it is likely that repeated dose studies will employ intermittent dosing schedules over the period required to support the anticipated duration of human exposure.

5.3 Reproduction Toxicity Studies

Reproduction studies should not be performed unless adverse effects are seen on gonadal histology in the general toxicity studies and/or the MAB will be administered to women who may become pregnant and in whom it would be difficult or inappropriate to contraindicate treatment during pregnancy.

If it is felt that reproduction studies are necessary or appropriate, then the same considerations already discussed in the context of the choice of species and the selection of doses for general toxicity studies will apply. The experimental design must be adjusted to minimize immunological complications and it may be necessary to subdivide the treatment groups in order to administer one or more doses at specific time periods etc. Reproduction studies should not be conducted using primates unless there is no alternative choice. The reproductive toxicity testing of therapeutic biotechnology agents, including MABs, has recently been discussed by HENCK et al. (1996).

5.4 Genotoxicity Studies

The genotoxicity studies which are routinely conducted for traditional pharmaceuticals are generally not applicable to monoclonal antibodies. Limited studies may be indicated where there is cause for concern based upon inability to remove impurities, the presence of specific contaminants, or the presence of an organic molecule conjugated to the MAB.

5.5 Oncogenicity/Carcinogenicity Studies

Oncogenicity/carcinogenicity studies are almost certainly unwarranted in the case of MABs. At the present time the target indications, patient populations, and the duration of exposure do not meet the minimum criteria triggering the need for such studies. It is feasible that in the future MABs will be developed for administration to patients for long periods of time etc. One would hope, however, that as understanding and knowledge of the mechanisms of carcinogenesis become clearer that the anachronistic rodent carcinogenicity testing protocol will be abandoned or at least modified before these products arrive.

5.6 Local Tolerance

Monoclonal antibody preparations are given by parenteral routes, and the local tolerance of the actual formulation to be administered to humans should be tested in appropriate animal models prior to being administered to humans. When the intended clinical formulation has been used for the single and/or repeated dose toxicity studies and no problems are identified at the injection sites, then additional local tolerance studies are not necessary. If, however, the actual clinical formulation was not used or reactions are seen at injection sites it may be necessary to undertake specific local tolerance studies using the standard rabbit models.

5.7 Special Studies

It is likely that administration of a MAB to animals will elicit an immunogenic response. It is, therefore, essential to measure the production and time course of serum antibodies during the conduct of repeated dose toxicity studies. Antibody responses should be characterized as neutralizing or non-neutralizing, and their appearance correlated with any observed pharmacological or toxicological changes. The predictive relevance for humans of serum antibodies to a MAB in animals may be questionable especially if the antibodies are non-neutralizing, and there is no dose–response relationship, tolerance is induced, pharmacokinetics and biodistribution are not altered, and safety issues are not apparent. Tests for anaphylaxis employing guinea pigs are not recommended and are considered to be of little value in predicting human safety to MABs.

6 References

Commission of the European Communities (1989), Preclinical biological safety testing of medicinal products derived from biotechnology (and comparable products derived from chemical synthesis), *Notes to Applicants for Marketing Authorizations* Draft 5, April 1989.

FDA (1994), *Points to Consider in the Manufacture and Testing of Monoclonal Antibody Products for Human Use,* US Food and Drug Administration, Office of Biologics Research and Review Centre for Drugs and Biologics, August 1994.

GORE, M. R., HANSEN, S. K., MILLER, L. L. (1996), Developing a regulatory strategy for monoclonal antibodies, *Drug Information J.* **30**, 257–267.

HENCK, J. W., HILBISH, K. G., SERABIAN, M. A., et al. (1996), Reproductive toxicity testing of therapeutic biotechnology agents, *Teratology* **53**, 185–195.

KÖHLER, G., MILSTEIN, C. (1975), Continuous culture of fused cells secreting antibody of predefined specificity, *Nature* **256**, 495 ff.

MATSUZAWA, K., KOYAMA, T., SUGAWARA, S., et al. (1992), The preclinical safety evaluation of human monoclonal antibody against cytomegalovirus, *Fundam. Appl. Toxicol.* **19**, 26–32.

MOHW (1988), Draft guidelines for preparation of data required on application for approval of drugs manufactured by the application of cell culture techniques, *MOHW Notification No. 10* June 6, 1988.

ROMMEBERGER, H. (1990), Toxicological testing of monoclonal antibodies, *Dev. Biol. Stand.* **71**, 185.

SULLIVAN, S. F. (1990), Regulatory issues with respect to monoclonal antibodies – an industry perspective, *Dev. Biol. Stand.* **71**, 207–211.

TOMLINSON, E. (1992), Impact of the new biologies on the medical and pharmaceutical sciences, *J. Pharm. Pharmacol.* **44**, 147–149.

TOON, S. (1996), The relevance of pharmacokinetics in the development of biotechnology products, *Eur. J. Drug Metab. Pharmacokinet.* **21**, 93–103.

WALDMAN, T. A. (1991), Monoclonal antibodies in diagnosis and therapy, *Science* **252**, 1657–1662.

12 Therapeutic Applications of Monoclonal Antibodies: A Clinical Overview

MARK SOPWITH

Slough, UK

1 Introduction   312
2 Targets for Intervention   314
3 Clinical Indications   315
   3.1 Cardiovascular and Pulmonary Systems   315
   3.2 Sepsis Syndrome and Infection   316
   3.3 Immunosuppression and Transplantation   317
   3.4 Rheumatoid Arthritis and Autoimmune Disorders   318
   3.5 Cancer   320
4 Conclusions   322
5 References   322

1 Introduction

The achievements of pharmaceutical research over the past 100 years have been remarkable. This success is exemplified by the flow of novel, small chemical entities derived by organic synthesis. Many of these products are highly effective in inhibiting the activity of target enzymes and in blocking the interaction between low molecular weight signaling molecules and their respective receptors.

By contrast, the development of small synthetic inhibitors of protein-protein and protein-carbohydrate interactions has proved more challenging. Typically, in these interactions substantial molecular surfaces are approximated. Thus a putative inhibitor of small size, however great its affinity for the target, may be capable of only limited activity.

Recently, important advances have been made in elucidating the molecular interactions that underly cell–cell and cell–matrix adhesion and signaling. Antibody reagents have been vital to this work. In turn, antibodies with the same or similar specificity to those effective in animal models can be tested in patients. The use of anti-ICAM-1 antibodies (discussed below) to interfere with white cell adhesion and T lymphocyte activation illustrates the potential that exists for progressing rapidly from the laboratory into studies in the clinic.

In similar fashion, antibodies can block signaling by cytokines and chemokines. For example, antibodies reactive with tumor necrosis factor (TNF) have shown clinical benefit in inflammatory indications such as rheumatoid arthritis and Crohn's disease. Alternatively, the receptor rather than the ligand can be targeted, as in the use of anti-IL-2 receptor antibodies to induce immune suppression.

Thirdly, antibodies can be selected that, far from blocking the function of their target, are able to activate it. Applications include anti-p185HER2 and anti-EGF receptor antibodies for the therapy of tumors that express these molecules on their cell surface (BASELGA et al., 1996; DIVGI et al., 1991); and, potentially, anti-insulin receptor activating antibodies for the treatment of rare forms of diabetes caused by failure of the receptor to recognize its ligand or to signal (KROOK et al., 1996).

Fourthly, antibodies that recognize a cell surface target can be used to deliver biological activity. Experimental therapies for cancer have used this approach. The toxicity delivered may reside in the antibody itself, if an isotype is selected that is able to mediate complement activation or cellular cytotoxicity. Or the toxicity may reside in a toxin or radionuclide attached to the antibody. Examples include the choice of human IgG1 isotype for an anti-CD33 treatment of acute myeloid leukemia (CARON et al., 1994), and the attachment of the toxic antibiotic calicheamicin to another anti-CD33 antibody for the same purpose (APPELBAUM et al., 1992).

For many therapeutic applications advantage is taken of the extraordinary specificity of binding that characterizes antibodies. This specificity is a feature that small pharmaceuticals struggle to match. An individual cytokine or an isoform of a cell surface molecule can be targeted with great precision by an antibody. Important biological understanding may result from analyzing the effects of such a defined intervention. Moreover, therapeutic benefits may be accompanied by fewer unwanted effects than is possible using less specific pharmacological agents.

Antibody therapy is not without problems, however.

The great specificity of antibodies may sometimes be less than ideal. Most disease is the expression of an extremely complex network of dysfunction. Many mediators and cell types are involved. The neutralization of an individual signal may be too limited to impact clinically. Can the individual elimination of TNF or IL-1 activity much modify the course of a complicated multisystem disorder like septic shock? Likewise, whether blocking a single adhesion molecule such as E-selectin or $\alpha4\beta1$ can so alter leukocyte trafficking that disorders, such as psoriasis and multiple sclerosis, can be ameliorated has yet to be shown. Fine specificity may be a drawback, too, versus microbial targets. The neutralizing effect of the antibody may be overcome by a single mutation in the microorganism. The evolution of hepatitis B virus surface antigen in the face of monoclonal anti-HBs illustrates this point (MCMAHON et al., 1992).

Fine specificity attending, perhaps, the selection of an antibody of the greatest available affinity may present problems of species specificity. An antibody chosen for its binding properties against a human target may possess little or no reactivity with animal homologs. For example, antibodies to TNF have been selected for clinical trial which react only with the TNF of highest primates. It is obviously difficult in convenient animal models to define the activity of these antibodies. Animal models present a further difficulty. If the sequence of the antibody is similar to human it will be immunogenic in laboratory species. This will limit extended animal dosing. These problems may be sidestepped by studying, say, in the mouse a mouse antibody that possesses equivalent activity in that species to the clinical product in man. An alternative to using parallel reagents is to use transgenic animals in which the human antigen is expressed with an intensity and distribution similar to that found in man. Nevertheless, the best efforts to predict the likely effects of the product in man may be confounded.

However, even highly specific antibodies may possess broad biological effects because of the distribution of their target. Antibodies specific for CD18, a subunit common to 4 separate integrin heterodimers (VAN DER VIEREN et al., 1995), have wide-ranging effects on leukocyte trafficking and T-lymphocyte activity. This profile of activity is potentially valuable. If use is prolonged, however, the therapeutic benefits of anti-CD18 activity must be set against the potential for substantial immunosuppression.

The 1st generation of antibody products was mostly of mouse origin. These, not surprisingly, induced an immune response in patients. The immune response blocked the binding of the antibody to its target and, secondly, caused the rapid clearance of any antibody remaining in the circulation. Generation of an immune response also posed the risk of a clinically important allergic reaction either immediately or should the same or another mouse antibody be given in the future. The introduction of antibodies that are virtually indistinguishable from those of human origin has diminished these concerns. Such antibodies still tend to induce low-level antiidiotypic immune responses, however. It may be difficult to avoid these responses completely, for theoretical reasons (JERNE, 1984), and because recombinant human products like insulin, growth hormone and inferon β show residual immunogenicity. Whether low-level antiidiotypic responses are clinically meaningful, however, will depend on the antibody in question and its application.

Variations in the antibody molecule have been introduced with the objective of achieving more favorable plasma pharmacokinetics and a more favorable cost of goods. Antibody fragments that can be manufactured microbially, rather than by mammalian cells, are less expensive to make and can be produced on a larger scale. Antibody fragments may be preferred to whole antibody when the clinical application is as well or better served by rapid clearance from the blood. This is the case, e.g., for tumor imaging *in vivo*. If longer plasma half life is desired, either whole antibody may be used or the clearance of antibody fragments may be extended by changing their size and structure. $F(ab')_2$ may be chosen as opposed to a smaller fragment, or antibody fragments may be modified by the attachment of polyethylene glycol (SUZUKI et al., 1984) or inexpensive protein.

A further challenge to antibody therapy is the route of administration. Oral or topical treatment may occasionally be appropriate (WHITTUM-HUDSON et al., 1996). Most uses will involve parenteral administration, however. The intravenous route may be optimal in acute care, but intramuscular or subcutaneous administrations may be preferred if feasible. Duration of treatment, convenience and cost will all inform the decision.

The benefits of ReoPro (abciximab; anti-GPIIb/IIIa) as an adjunct to coronary artery angioplasty, the value of Zenapax (dacliximab; anti-IL-2 receptor) and other antibodies in transplantation, and the positive results from antibody-based cancer treatments are clinical and commercial successes which underscore the coming of age of antibody therapeutics.

2 Targets for Intervention

Several mechanistic themes underlie the numerous and varied therapeutic applications of antibodies.

Monoclonal antibodies have been key to our increasing understanding of the interactions that underly immune function. Panels of antibodies have been raised to molecules on the surface of immunologically active cells and to the mediators that they secrete. It has been a relatively simple step to examine the effect of such antibodies in patients in whom modification of the immune response is sought (HELDERMAN, 1995). Initially, polyclonal antilymphocyte sera were used in transplantation to induce immune suppression and to reverse graft rejection. Subsequently, monoclonal antibodies against the lymphocyte marker CD3 and the leukocyte marker CD52 were shown to have useful activity in indications both within and outside transplantation (GOLDSTEIN et al., 1985; ISAACS et al., 1992). Somewhat more selective in their outcome, anti-CD4 antibodies that react with helper T-lymphocytes have been studied. Many of these anti-CD4 antibodies cause marked and extended depletion of CD4-expressing cells, and the risk of opportunistic infection or of lymphoproliferative disorders is a concern. The development of non-depleting anti-CD4 antibodies may alleviate these worries (ETTENGER and YADIN, 1995).

In the search for more selective immune modulation several avenues are being explored. Firstly, interest has focussed on cell surface markers such as the interleukin-2 receptor (KUPIEC-WEGLINSKI et al., 1988) and CD134 (WEINBERG et al., 1996) whose expression is upregulated on activated lymphocytes. Secondly, the discovery in the mouse that CD4-expressing cells can be divided into Th1 and Th2 subsets based on patterns of cytokine expression has suggested an alternative set of targets (ABBAS et al., 1996). Th1 cells are thought to be responsible for marshalling defences against microbial infection. However, Th1-dominant immune responses are associated with inflammation and tissue injury; and they may contribute to the pathogenesis of chronic inflammatory disorders such as arthritis and inflammatory bowel disease. Interleukin-4 (IL-4) and interleukin-5 (IL-5) are the signature cytokines of Th2 cells. IL-4 is the major inducer of B-cell switching to IgE production, whilst IL-5 is the principal eosinophil activating cytokine. Th2 cells are, therefore, considered important in the body's defence against pathogenic helminths. These cells and their cytokines are likely to drive clinical allergic disorders, too, and anti-IL-5 antibodies will shortly be tested in patients with asthma. Because Th1 and Th2 cells are derived from uncommited lineages, their differentiation is open to manipulation. Interleukin-12 and IL-4, which appear pivotal to Th1 and Th2 differentiation respectively, are the evident initial targets.

Thirdly, recently recognized co-stimulatory pathways involved in T-cell signaling offer novel avenues for immune modulation. These interactions between lymphocytes and antigen presenting cells occur in parallel with the binding of the T-cell receptor to peptide presented by a major histocompatibility complex (MHC) molecule. Co-stimulatory pathways include that between CD28 and CTLA-4 on the lymphocyte with B7-1 and B7-2 on the antigen presenting cell (REISSER and STADECKER, 1996). Antagonists – including antibodies – are effective in animal models of diabetes, systemic lupus erythematodes and transplantation. Another costimulatory pathway recently identified is the interaction between CD40 and gp39. Blockade of gp39- and B7-mediated pathways simultaneously in animal models of transplantation is highly effective in preventing graft rejection (LARSEN et al., 1996). Lastly, a co-stimulatory pathway already subject to clinical trials is the interaction between the lymphocyte surface molecule LFA-1 (CD11 a/CD18) and ICAM-1 (ETTENGER and YADIN, 1995).

Many of the molecules responsible for the initial adhesive contact between a circulating white cell and a capillary endothelial cell, and for the subsequent signaling events, tight binding and movement of the white cell into the extravascular space, have been identified using antibodies. A paradigm of sequential contact and signaling steps that directs the emigration of appropriate cells into individual tissues has been popularized by, among others, SPRINGER (1994). The trafficking of a skin-homing subset of T-cells, for example, appears to depend at least in part on the expression by activated capillary endothelial cells of the adhesion molecule E-selectin (PICKER et al., 1991; SHIMIZU et al., 1991). Similarly, trafficking of lymphocytes through the gut is directed by endothelial cell expression of mucosal addressin cell adhesion molecule 1 (MAdCAM-1) (STREETER et al., 1988). Antibodies to these adhesion molecules, for the treatment of chronic inflammatory disorders of the skin and bowel, may suborn the immune system with a previously unattainable selectivity.

The therapeutic use of an antibody that has been raised against an individual patient's lymphoma B-cell idiotype illustrates lymphocyte-directed therapy at its most specific. Individualizing treatment in this way is at present unappealing. Nevertheless, if, say, the T-cell receptor was identified that defined the lymphocyte clone upon which an autoimmune disease depended, antibodies to that receptor would provide highly specific immune therapy. Similarly, antibodies to a pathogenetic peptide, or to the MHC molecule that presents that peptide, might be used. However, unlike in certain animal models of disease, it has proved very difficult in man to identify key lymphocyte clones responsible for driving autoimmune disease – or, indeed, to identify pivotal antigenic peptide sequences (ANTEL et al., 1996).

The expression of many of the cell surface targets mentioned above is subject to control by cytokine molecules. Interleukin-1 and TNF, e.g., upregulate expression of the adhesion molecules E-selectin, ICAM-1 and VCAM-1, in addition to possessing other broadly pro-inflammatory activities (OPPEN-HEIMER-MARKS and LIPSKY, 1996). Anticytokine therapeutics have been tested with early success in rheumatoid arthritis and Crohn's disease. Antibodies against another pro-inflammatory molecule, interleukin-8 (IL-8; a neutrophil attractant), are being tested in pulmonary inflammatory disorders (DONNELLY et al., 1993).

Cancer therapies seek to exploit all these principles. Antibodies that bind to tumor cell surface antigens are being tested for clinical activity. This activity may be either intrinsic to the antibody or conferred by attaching cytotoxic molecules or radioactive isotopes (FRANKEL et al., 1996). Antibodies against adhesion molecules involved in the new blood vessel formation that is necessary for tumor growth and for clinically significant metastatic spread are being tested also: anti-$\alpha v \beta 3$ is an example (BROOKS et al., 1994). Antibodies that interfere with cell matrix interactions, and which thus interfere with cell migration and other functions, are also likely targets for antibody treatment (HART and SAINI, 1992). And antibodies are in clinical trial that react either with tumor growth factors or with their receptors. Interleukin-6 (BATAILLE et al., 1995), epidermal growth factor receptor (DIVGI et al., 1991), vascular derived growth factor (VEGF) and its receptor-expressed by proliferating blood vessels as well as by some tumor cells, and p185HER2 (BASELGA et al., 1996), are all promising examples of such targets.

These themes of cell targeting and of interference with adhesion and cell–cell signaling are further exemplified in the following section, in which the therapeutic use of antibodies across a range of differing clinical indications is outlined.

3 Clinical Indications

3.1 Cardiovascular and Pulmonary Systems

Disorders of the cardiovascular system make an appropriate start to this section on specific clinical indications. Abciximab (Reo-Pro) is an antibody product directed against integrin GPIIb/IIIa, an adhesion molecule involved in the final common pathway for platelet aggregation. This antibody Fab fragment is the most important antibody therapeutic to have received market authorization to date. A description of the product appears in Chapter 16 of in this volume. Clinical trials of abciximab given at the time of percutaneous transluminal coronary angioplasty

(PTCA) have shown it to improve near and medium term outcome in patients at high risk of cardiac complications (TOPOL et al., 1994) and, in subsequent trials, in less selected patient groups and in patients with unstable refractory angina. Other agents that limit platelet aggregation or coagulation, and which might prevent arterial reocclusion or venous thrombosis, include antibodies against von Willebrand factor and tissue factor.

A complication of PTCA is progressive restenosis of the previously dilated artery. Part of the problem is neointimal hyperproliferation which may be so gross as to encroach profoundly into the arterial lumen over a period of several months. The mechanisms that underly this exuberant response to arterial wall trauma remain opaque. In animal models of restenosis, however, anti-platelet derived growth factor (anti-PDGF) antibodies have shown efficacy (FERNS et al., 1991).

That restoration of blood flow, following a period of local or systemic circulatory arrest, may of itself cause tissue damage has been contentious. However, by using antibodies that interfere with the endothelial margination and activation of neutrophils, the existence of reperfusion injury has been both defined and a component of its cause identified. Reperfusion injury may occur on reopening previously closed coronary, cerebral or peripheral arteries in association with hemorrhagic shock, following head injury, and during major vascular surgery and transplantation. A number of anti-adhesion molecule antibodies have been proposed for clinical use. Perhaps most advanced are anti-CD8 antibodies, which have shown strikingly beneficial effects in animal models (VEDDER et al., 1990; MILESKI et al., 1990). Anti-CD11a and anti-CD11b antibodies and antibodies to L-selectin, P-selectin and ICAM-1, 2 and 3 have also all been proposed for use in situations of ischemia-reperfusion (FRENETTE and WAGNER, 1996). Antibodies that neutralize the neutrophil chemotactin IL-8 or the C'5 component of complement may prevent the lung damage associated with cardiopulmonary bypass (SEKIDO et al., 1993; KAWAHITO et al., 1995; FITCH et al., 1996).

Asthma and other allergic disorders involving the respiratory system are conditions in which the immune system is biased towards the production of IgE and the mobilization and local activation of eosinophils. Anti-IgE antibodies are currently in clinical trial for seasonal rhinitis (SHIELDS et al., 1995); and, in animal models at least, the ability of anti-IgE-ricin to eliminate IgE-secreting B-lymphocytes is being examined (LUSTGARTEN et al., 1996). In a primate model the reduction by anti-IL-5 antibodies of the airways hyperresponsiveness that functionally characterizes asthma, in addition to the predicted effects of the antibody on eosinophil numbers and activation, suggests that this approach might translate into physiological and clinical effectiveness (MAUSER et al., 1995). The entry of inflammatory cells into the airways depends, naturally, on intercellular adhesion systems. For the eosinophil, the interaction between VLA4 ($\alpha 4\beta 1$ integrin) and VCAM-1 is important and relatively specific. Antibodies that bind to either target may reduce eosinophil recruitment into the airways and may also interfere with the recruitment of T-lymphocytes (LOBB and HEMLER, 1994).

3.2 Sepsis Syndrome and Infection

Sepsis syndrome, or systemic inflammatory response syndrome (SIRS), is a constellation of clinical features that occurs with serious systemic infection (BONE et al., 1992). Because a similar stereotyped picture may follow not only infection but trauma, hemorrhage, pancreatitis and immune-mediated tissue injury, a common pathway of endothelial activation, cytokine release and systemic inflammation has been proposed (CASEY et al., 1993). It is accompanied by progressive microcirculatory disruption, hypotension, metabolic decompensation, multiple organ dysfunction and death. That certain cytokines, such as TNF, can mimic many of the features of sepsis supports the concept of a common inflammatory pathway activated by diverse insults.

In many clinical trials either proposed or undertaken in SIRS an individual cytokine or cellular mediator has been targeted. As well as biological products able to neutralize interleukin-1, a range of different antibody and antibody-based products has been tested that neutralize TNF and prevent mortality in animal models of sepsis (TRACEY et al., 1987; HINSHAW et al., 1990). TNF-neutralizing biological products comprising soluble TNF-receptor fused to IgG constant region are also in clinical trial. After mixed clinical results, the deciding test of this approach may be the large phase III trial in North America in progress as this chapter is in preparation, in which selected patients with septic shock are being treated with the mouse anti-human TNF antibody BAY×1351. Other products of interest include anti-IL-8, particularly in pulmonary inflammation; anti-complement antibodies; and anti-ICAM-1 and anti-E-selectin antibodies that have the potential to reduce neutrophil-mediated damage.

The causes of sepsis may also be blocked by antibodies. In particular, clinical trials in SIRS of anti-endotoxin antibodies have been undertaken. These antibodies are evidently a potential treatment for gram-negative sepsis. Further, SIRS might respond, whatever its trigger, if endotoxin translocation across the bowel wall contributes to the progression of multiple organ dysfunction. Antibodies of this kind continue in clinical trial in sepsis and, in more focussed studies, in victims of trauma and in systemic meningococcal infection. Nevertheless, the failure of anti-endotoxin antibodies to affect mortality in sepsis syndrome has been one of the disappointments of antibody therapy to date (WARREN et al., 1992). Some commentators consider that neutralizing a single effector cannot change the outcome of so complex a pathogenesis. Others have emphasized the heterogeneity of the patient groups that have been studied; the late investigation of therapy; and the unrealistic assumptions of likely treatment benefit that have underlain the modest size of some studies. The failure to date of antibody therapy in the treatment of sepsis is discussed in Chapter 13 of this volume.

Using antibodies to neutralize infective organisms or their toxins has a distinguished history. Polyclonal sera or high-titer immunoglobulin preparations are used still for the prophylaxis of viral A hepatitis and to prevent *in utero* infection with herpes varicella-zoster (HVZ) in exposed pregnant women.

More recently, anti-cytomegalovirus (CMV) immunoglobulin has been licensed for use in at-risk transplant recipients. Monoclonal antibodies have been developed against a variety of infections, including HVZ and CMV, herpes simplex, papilloma virus, hepatitis B and human immunodeficiency virus. Antibodies in this field are generally IgG, but anti-respiratory syncytial virus IgA antibodies are in development.

Finally, more subtle interventions may prove valuable. Recent delineation of Th1/Th2 helper subsets has made possible notions of modifying the body's responses to persistent infection with *Leishmania* and *Schistosoma* species, mycobacteria, and helminths (ABBAS et al., 1996).

3.3 Immunosuppression and Transplantation

Solid organ transplantation offers many examples of the direct translation of animal experimentation to the clinical situation.

Polyclonal antilymphocytic sera are still used clinically although standardized and engineered antibody products are progressively taking their place. For the most part, antibodies are used as a component of therapeutic regimens designed to induce immunosuppression at the time of grafting. Not only may they improve graft survival, but they allow lower doses of toxic agents like cyclosporin to be used. The anti-CD3 antibody muromonab-CD3 (Orthoclone OKT3) is one such antibody, introduced initially to treat graft rejection episodes refractory to high dose steroid therapy (GOLDSTEIN et al., 1985). This antibody is very effective, but suffers from two disadvantages. Firstly, it is a mouse antibody and the immune response generated towards it is clinically significant. Secondly, use of OKT3 is associated with a first-dose effect during which a wave of cytokine release from activated T-cells is associated with mild or more severe cardiorespiratory effects. Finally, OKT3 is so effective a global immunosuppressive agent that the dangers of unwanted infection, secondary lymphoproliferation and lymphoma must be taken due account of.

As discussed in Sect. 2, in transplantation the present goal is immunosuppression of a more selective kind or the induction of specific immune tolerance. Impressive progress has been made in the laboratory, especially in mice. For example, combinations of anti-CD4 and anti-CD8 antibodies, and interference with co-stimulatory pathways, have been shown to induce long-term specific immune suppression.

A second marketed monoclonal antibody product appears likely to join OKT3. The humanized anti-IL-2 receptor antibody daclizimab (Zenapax) (KIRKMAN et al., 1989) which is directed toward activated T-cells, has been reported to reduce the incidence of subsequent acute rejection episodes in patients when added to a combination of cyclosporin and steroids.

Previously, dacliximab had proved disappointing in preventing the occurrence of donor lymphocyte-mediated graft-versus-host disease (GVHD) in patients undergoing bone marrow transplantation. However, some anti-cytokine antibodies may be more useful in GVHD. Anti-TNF antibodies in particular have shown encouraging activity (HERVÉ et al., 1992).

Solid organ transplantation also provides a platform for testing interventions designed to limit ischemia-reperfusion injury, as discusseding in Sect. 3.1.

Thus far, immunosuppression has been discussed in the context of allografting. However, in the near future transplantation across species is likely, e.g., to man from transgenic pigs that have been engineered to prevent complement-mediated hyperacute rejection. Immunologically this is a situation very different from allotransplantation and new immunosuppressive combinations – perhaps including antibody therapies – will be required.

3.4 Rheumatoid Arthritis and Autoimmune Disorders

Autoimmune disorders offer further instances of the emerging clinical benefits of antibody therapy directed towards the regulatory and effector cells of the immune system, and their cytokines.

Rheumatoid arthritis in particular has been the subject of clinical trials of antibody therapy. Current views of the pathogenesis of rheumatoid arthritis hold that lymphocytes are necessary for initiation of the disease and, probably, for its maintenance. In addition to activated T-cells, the involved joint contains macrophages, modified fibroblasts, other inflammatory cells and a complex mixture of potent inflammatory mediators, cytokines and proteases that are believed to contribute to the acute clinical picture and to progressive joint destruction. Experimental interventions directed towards T- and B-lymphocytes include antibodies, variously mouse, chimeric or human-like, towards CD3, CD4, CD5, CD7, CD19, CD21, CD25 (IL-2 receptor), and CD52 (KINGSLEY et al., 1991; ISAACS et al., 1992). Antibodies against lymphocyte co-stimulatory signals, discussed in Sect. 2, or to manipulate Th1–Th2 ratios provide further options.

The most striking effects of antibody treatment in rheumatoid arthritis have been observed, however, from the use of anti-TNF antibodies. Rapid clinical improvement accompanied by reductions in non-specific laboratory markers of inflammation have been reported consistently across several studies (EL-LIOTT et al., 1994; RANKIN et al., 1995). Results from the clinical trial of one of these antibodies are described more fully in Chapter 14 of this volume. The clinical responses seen with TNF neutralization may not only result in practical antibody therapy for rheumatoid arthritis. They have confirmed the thesis that TNF plays a pivotal role in the disease process and have opened up an exciting opportunity for pharmaceutical exploitation. Other antibody specificities being explored include antibodies toward the cytokines IL-1 and IL-6 (WENDLING et al., 1993); and, returning to interventions directed towards the cellular component of the disease, antibodies towards ICAM-1 and VCAM-1 that interfere with leukocyte–endothelial adhesion and, potentially, the recruitment of cells into the inflamed joint.

In severe systemic vasculitis refractory to other treatment, important success has been reported from combining anti-CD52 and anti-CD4 antibody therapy (MATHIESON et al.,

1990). Systemic lupus erythematodes has been treated with anti-CD4 (HIEPE et al., 1991). The threatening nature of this disease may be judged from the use of anti-CD5-ricin conjugate experimentally to target both T- and B-lymphocytes (STAFFORD et al., 1994).

Similar treatment principles may be applied to other disorders known or believed to be driven by the immune system. In psoriasis, studies have been undertaken, or are being undertaken, with anti-CD4 (MOREL et al., 1992) and anti-CD11a antibodies, as well as with anti-E-selectin antibodies which may be expected not only to inhibit neutrophil recruitment into the skin but also, specifically, to limit the localization of the skin-homing subset of lymphocytes. Anti-CD4 antibodies have been tried in uveitis, a condition in which inflammation of varied etiologies threatens vision (THURAU et al., 1994).

Clinical trials suggest that anti-TNF antibodies may be effective in Crohn's disease and ulcerative colitis (VAN DULLEMEN et al., 1995; STACK et al., in press; EVANS et al., 1996). An alternative approach, with effects on lymphocyte recruitment that may be bowel subset-selective, is the use of anti-MAd-CAM-1 antibodies or antibodies to the relevant integrin ligand, $\alpha 4\beta 7$ (SPRINGER, 1994).

Insulin-dependent diabetes mellitus is the consequence of autoimmune pancreatic islet B-cell destruction. In the non-obese diabetic mouse the disease can be prevented by variously interfering with immune function, including treatment with anti-CD11b and anti-CD18 antibodies. Unfortunately, in patients, insulin-dependent diabetes only becomes overt as the destructive process approaches completion. However, prediction of frank diabetes in at-risk individuals is becoming increasingly confident and will allow earlier intervention. Patients with established diabetes are subject to serious long-term complications. Diabetic retinopathy is characterized by proliferation of small vessels, bleeding from which may cause permanent visual impairment. VEGF has been implicated as a mediator of angiogenesis in diabetic proliferative eye disease (AIELLO et al., 1994), so that the therapeutic use of anti-VEGF and anti-VEGF-receptor antibodies has been suggested. Blockade of αv integrins may also in-

hibit small vessel proliferation (HAMMES et al., 1996). Another feared complication of diabetes is kidney disease. Antibodies, raised against a glycosylated intermediate of albumin that is increased in diabetes, can reduce microalbuminuria in animal models. More provocatively, these antibodies given early prevent the development of morphological renal changes (COHEN et al., 1995). Whether such antibodies can benefit established nephropathy is not clear yet – nor is the utility of these antibodies in patients.

The treatment of neurological diseases completes this section.

Myasthenia gravis is a classical system-specific autoimmune disorder, in which antibodies are generated that block the muscarinic cholinergic receptor. Experimental therapies such as anti-CD4 antibodies have been evaluated in myasthenia gravis refractory to standard treatment with anticholine esterases and immune suppressive agents (ÅHLBERG et al., 1994).

The most important inflammatory disorder of the nervous system is multiple sclerosis. This chronic disease is characterized by acute foci of T-cell and macrophage infiltration with subsequent demyelination and axonal degeneration. In experimental allergic encephalomyelitis (EAE) in animals a number of antibodies have shown efficacy (LIU et al., 1996). These include anti-lymphocyte antibodies like anti-CD4, B7 and CD134 (WEINBERG et al., 1996); and in the clinic anti-CD52 antibodies have shown promise (MOREAU et al., 1994). Alternatively, anti-$\alpha 4\beta 1$ integrin antibodies may allow some selectivity against inflammatory infiltration of the central nervous system (YEDNOCK et al., 1992). In animals, anti-TNF antibodies can prevent EAE and can intervene in ongoing inflammation (RAINE, 1995). Clinical studies of anti-TNF therapy are in progress. In EAE induced by immunizing animals with myelin preparations or with myelin basic protein, the immune response may be directed towards highly limited immunological determinants. Reflecting this, the encephalitogenic T-cell response may be highly skewed towards use of one or only a small number of T-cell receptor β-chains. In principle, therefore, selective antibody therapy directed towards the β-chain(s) utilized most prominently in the T-cell response might be feasible. Using in-bred strains of mice this has been achieved. Unfortunately, in man, in multiple sclerosis as with other putative autoimmune diseases, T-cell receptor usage by the time of clinical presentation appears less restricted than in animal models. Anti-MHC antibodies have been found effective in animal models of EAE (STEINMAN et al., 1981; JONKER et al., 1988) and their clinical use is awaited with interest.

3.5 Cancer

With the availability of monoclonal antibodies exciting advances in the treatment of cancer were promised. As the difficulties entailed in progressing from promise to reality became better appreciated, however, some critics went so far as to dismiss the approach. Now, antibody-based therapy is in resurgence. Clinical trials of antibody and antibody targeted therapies have established proof of principle, and marketing authorization was recently granted in Europe for the first time to an antibody therapeutic product, antibody 17-1A (Panorex).

What are the barriers that have delayed the emergence of practicable cancer therapy using antibodies? Typically, an antibody-based treatment for patients with cancer comprises an antibody linked to a cytotoxic modality. An antibody is chosen that reacts with an antigen expressed on the surface of tumor cells. A cytotoxic drug, or a radioisotope, is chosen that is capable of killing cells when delivered in this way. Already a number of issues are evident. The impressive selectivity of monoclonal antibodies reinforced the hope that antibodies might be raised that were specific for tumor antigens and unreactive with normal cells. However, tumor cells rarely express on their surface an antigen that is specific to their transformed status. In principle, tumor specificity might reside in a mutated proto-oncogene sequence presented on the cell surface in the context of MHC class 1. This presentation might suffice to allow T-cell attack, but has not yet proved exploitable by antibody directed therapy. Too idealized an expectation of tumor targeting is similarly under-

mined by the fact that an administered antibody moiety will be captured by receptors in the liver and kidney that normally clear immunoglobulin from the circulation. Many tumor cell surface antigens are shed; as well as blocking the binding of conjugate to tumor, the binding of conjugate to circulating antigen will result in clearance of the conjugate into normal tissues. The toxin, too, may localize the conjugate into normal tissues. For example, non-specific cellular uptake mediated by non-deglycosylated ricin causes not only end organ damage, especially of the liver, but vascular leak syndrome (HERTLER and FRANKEL, 1989). Lastly, the target antigen must be expressed in a consistent way by tumor cells and at a minimum density, and, if a toxin or an isotope with a short radiation pathway is to kill the cell, the target antigen must mediate the internalization of the conjugate.

If vital tissues express the antigen, serious problems may arise. The immunoconjugate OVB3-*Pseudomonas* exotoxin, whose toxicity could in retrospect be ascribed to reactivity with normal brain, is just one example of this (PAI et al., 1991). By contrast, the choice of CD33 for acute myeloid leukemia may prove to be more apt. Its normal tissue expression is virtually limited to myeloid bone marrow precursors yet, crucially, it is not expressed by pluripotent stem cells. Several antigens suggested as targets for the treatment of B-cell lymphoma, such as CD20, similarly possess highly restricted tissue distribution and clinical trials have shown tumor responses at dosage levels that are acceptably tolerated by patients (MALONEY et al., 1994). The choice of p185HER2 and of Lewis Y (PAI et al., 1996) has been demonstrably successful, too.

Successful tumor imaging using labeled antibodies suggests that favorable ratios of antibody localization to tumor versus key vital organs can be attained. Nevertheless, killing still requires that sufficient toxicity is loaded into tumor tissue. For small amounts of accessible tumor, the choice of an active immunoglobulin isotype such as human IgG1 may prove sufficient. This choice has been made for an anti-CD33 antibody now in clinical trial for the treatment of minimal residual acute myeloid leukemia, for the antibody 17-1A for the postoperative adjuvant treatment of colorectal carcinoma (REITHMÜLLER et al., 1994) and for an anti-p185HER2 antibody for therapy of breast and other cancers. More commonly, a radioisotope or toxin is attached to the antibody to confer cytotoxicity. Radioisotopes with long path length radiation such as ^{131}iodine and ^{90}yttrium are not only able to kill cells to which antibody binds, but adjacent cells also. For the treatment of lymphoma or for bone marrow ablation isotopes such as these can be effective (PRESS et al., 1995). For most solid tumors, however, it is not clear that adequate tumor loading of these isotopes can be obtained. Conventional cytotoxic drugs, such as doxorubicin, mitomycin and vinca alkaloids, have been attached to antibodies with some success. However, toxins that are orders of magnitude more toxic than these familiar agents have greater potential. Calicheamicin, CC-1065-related compounds, diphtheria toxin, gelonin, genestein, maytansinoids, momordin, *Pseudomonas* exotoxin, ricin, saporin and *Staphylococcal* enterotoxin A are examples of such highly toxic entities (FRANKEL et al., 1996). These toxins are generally attached to an antibody by individualized linking chemistry. This chemistry is key to the conjugate being adequately stable in solution, yet cleavable with release of the toxin in the intracellular environment. Alternatively, protein toxins may be genetically engineered directly to the sequence of an antibody fragment, as illustrated by the fusion of a *Pseudomonal* exotoxin fragment to an anti-Lewis Y Fv of the antibody B3 with retention of cytotoxic activity (PAI and PASTAN, 1993). Others have used bispecific antibodies: one arm of the heterodimer reacts with a chosen target antigen and the other arm is specific for a toxin, such as saporin, which it delivers. Bispecific antibodies have similarly been designed, one arm of which binds to host effector cells such as cytotoxic T-cells or macrophages via, say, anti-CD3 or anti-CD64 specificity (FEATHERSONE, 1996). The host's own effector systems may thus be brought to bear against the tumor.

Immunoglobulins are large molecules. This is likely to reduce their penetration of poorly vascularized solid tumor. To improve tumor penetration, antibody fragments have been recommended. However, fragments are cleared more rapidly from the circulation; especially through the kidneys, an organ in which antibody fragments may be actively retained. These features may enhance tumor imaging by reducing background activity. For therapy, on the other hand, not only may net tumor loading be diminished by the more rapid clearance of antibody fragments from the plasma, but renal toxicity may also offset any advantage.

A different way of using antibodies to target toxic drug is antibody directed enzyme prodrug therapy (ADEPT) (BAGSHAWE et al., 1994). In this strategy, an antibody–enzyme conjugate is injected. Having allowed the conjugate to locate to tumor, and to clear from the circulation, a cytotoxic prodrug is injected. The prodrug is activated by the enzyme component of the conjugate previously given. The effect of the 2 steps is to localize active cytotoxic drug to the tumor. In animal models this procedure can be made to work nicely, but successful clinical application of ADEPT is awaited.

Because therapy of most cancers requires repeated cycles of treatment, the need for therapeutic molecules of low immunogenicity is self-evident. This principle must be qualified, however. The development of an anti-anti-idiotypic response, a response in effect against the antigen target of the antibody originally administered, has the potential to mediate anti-cancer activity. This mechanism may contribute to the beneficial effects of the mouse antibody 17-1A. Anti-idiotype antibodies themselves are now being administered therapeutically.

To liberate antibody therapy from the constraints of tumor penetration and loading, a quite different tactic is being evaluated. Antigen expressed, not by tumor cells, but by tumor vasculature is being targeted (THORPE and BURROWS, 1995). Antigens overexpressed by this vasculature are likely to be shared by other activated or proliferating blood vessels. Nevertheless, antibodies that react with VEGF receptor (YUAN et al., 1996), with the integrins $\alpha v \beta 3$ and $\alpha v \beta 5$ (FRIEDLANDER et al., 1995), with endosialin and fibronectin (RETTIG et al., 1992), and with cytokine upregulated MHC or adhesion molecules, offer promise. An antibody against a vascular target may be active against many different kinds of solid tumor. Likewise, antibodies, such as F19 (GARIN-CHESA et al., 1990), to tumor stromal targets may find application across many solid tumor types.

Antibodies that neutralize growth factors may also possess anti-tumor effects. Examples include the use of anti-IL-6 in myeloma. Or a growth factor receptor may be targeted, as with the use of anti-EDGF receptor, anti-VEGF receptor or anti-p185HER2 antibodies.

Finally, antibodies can be used to purify CD34-positive stem cells for marrow repopulation; to purge marrow prior to transplantation; and to block multiple drug resistance due to P-glycoprotein overexpression.

4 Conclusions

Antibody and antibody-based therapies work. Their combination of potency with extraordinary specificity, the development of less immunogenic products, the potential design of constant region-mediated activity and of pharmacokinetic characteristics optimal for the use envisaged, the progressive improvements in manufacturing and cost of goods, and the continuing identification of novel targets, all these factors indicate an exciting future for antibody products.

This chapter took as its starting point the remarkable success of applying organic chemistry to the synthesis of novel molecules for pharmaceutical use. Now new combinatorial chemistry techniques are being introduced to generate novel structural leads in vastly greater number and to refine those leads. In using antibodies we are exploiting a natural combinatorial chemistry whose power and scope has been developed over time to meet the most demanding requirements for diversity and activity.

5 References

ABBAS, A. K., MURPHY, K. M., SHER, A. (1996), Functional diversity of helper T lymphocytes, *Nature* **383**, 787–793.

ÅHLBERG, R., YI Q., PIRSKANEN, R., MATELL, G., SWERUP, C., RIEBER, E. P., RIETHMÜLLER, G., HOLM, G., LEFVERT, A. K. (1994), Treatment of myasthenia gravis with anti-CD4 antibody: improvement correlates to decreased T-cell autoreactivity, *Neurology* **44**, 1732–1737.

AIELLO, L. P., AVERY, R. L., ARRIGG, P. G., KEYT, B. A., JAMPEL, H. D., SHAH, S. T., PASQUALE, L. R., THIEME, H., IWAMOTO, M. A., PARK, J. E., NGUYEN, H. V., AIELLO, L. M., FERRARA, N., KING, G. L. (1994), Vascular endothelial growth factor in ocular fluid of patients with diabetic retinopathy and other retinal disorders, *N. Engl. J. Med.* **331**, 1480–1487.

ANTEL, J. P., BENER, B., OWENS, T. (1996), Immunotherapy for multiple sclerosis: from theory to practice, *Nature Med.* **2**, 1074–1075.

APPELBAUM, F. R., MATTHEWS, D. C., EARY, J. F., BADGER, C. C., KELLOGG, M., PRESS, O. W., MARTIN, P. J., FISHER, D. R., NELP, W. B., THOMAS, E. D., BERNSTEIN, I. D. (1992), The use of radiolabeled anti-CD33 antibody to augment marrow irradiation prior to marrow transplantation for acute myelogenous leukemia, *Transplantation* **54**, 829–833.

BAGSHAWE, K. D., SHARMA, S. K., SPRINGER, C. J. (1994), Antibody directed enzyme prodrug therapy (ADEPT). A review of some theoretical, experimental and clinical aspects, *Ann. Oncol.* **5**, 879–891.

BASELGA, J., TRIPATHY, D., MENDELSOHN, J., BAUGHMAN, S., BENZ, C. C., DANTIS, L., SKLARIN, N. T., SEIDMAN, A. D., HUDIS, C. A., MOORE, J., ROSEN, P. P., TWADDELL, T., HENDERSON, I. C., NORTON, L. (1996), Phase II study of weekly intravenous recombinant humanized anti-p185HER2 monoclonal antibody in patients with HER2/neu-overexpressing metastatic breast cancer, *J. Clin. Oncol.* **14**, 737–744.

BATAILLE, R., BARLOGIE, B., ZU, Z. Y., ROSSI, J. F., LAVABRE-BERTRAND, T., BECK, T., WIJDENES, J., BROCHIER, J., KLEIN, B. (1995), Biologic effects of anti-interleukin-6 murine monoclonal antibody in advanced multiple myeloma, *Blood* **86**, 685–691.

BONE, R. C. and members of the American College of Chest Physicians/Society of Critical Care Medicine Consensus Conference Committee (1992), American College of Chest Physicians/ Society of Critical Care Medicine Conference:

definitions for sepsis and organ failure and guidelines for the use of innovative therapies in sepsis, *Crit. Care Med.* **20**, 864–873.

BROOKS, P. C., CLARK, R. A. F., CHERESH, D. A. (1994), Requirement of vascular integrin $\alpha v \beta 3$ for angiogenesis, *Science* **264**, 569–571.

CARON, P. C., SCHWARTZ, M. A., CO, M. S., QUEEN, C., FINN, R. D., GRAHAM, M. C., DIVGI, C. R., LARSON, S. M., SCHEINBERG, D. A. (1994), Murine and humanized constructs of monoclonal antibody M195 (anti-CD33) for the therapy of acute myelogenous leukemia, *Cancer* **73** (Suppl.), 1049–1056.

CASEY, L. C., BALK, R. A., BONE, R. C. (1993), Plasma cytokine and endotoxin levels correlate with survival in patients with the sepsis syndrome, *Ann. Intern. Med.* **119**, 771–778.

COHEN, M. P., SHARMA K., JIN, Y., HUD, E., WU, V.-Y., TOMASZEWSKI, J., ZLYADEH, Z. N. (1995), Prevention of diabetic nephropathy in db/db mice with glycated albumin antagonists, *J. Clin. Invest.* **95**, 2338–2345.

DIVGI, C. R., WELT, S., KRIS, M., REAL, F. X., YEH, S. D. J., GRALLA, R., MERCHANT, B., SCHWEIGHART, S., UNGER, M., LARSON, S. M., MENDELSOHN, J. (1991), Phase I and imaging trial of indium 111-labeled anti-epidermal growth factor receptor monoclonal antibody 225 in patients with squamous cell lung cancer, *J. Natl. Cancer Inst.* **83**, 97–104.

DONELLY, S. C., STRIETER, R. M., KUNKEL, S. L., WALZ, A., ROBERTSON, C. R., CARTER, D. C., GRANT, I. S., POLLOK, A. J., HASLETT, C. (1993), Interleukin-8 and development of adult respiratory distress syndrome in at-risk patient groups, *Lancet* **341**, 643–647.

ELLIOTT, M. J., MAINI, R. N., FELDMANN, M., KALDEN, J. R., ANTONI, C., SMOLEN, J. S., LEEB, B., BREEDVELD, F. C., MACFALANE, J. D., BIJL, H., WOODY, J. N. (1994), Randomized double-blind comparison of chimeric monoclonal antibody to tumour necrosis factor α (cA2) versus placebo in rheumatoid arthritis, *Lancet* **344**, 1105–1110.

ETTENGER, R. B., YADIN, O. (1995), The potential role of therapeutic antibodies in the regulation of rejection, *Transplant. Proc.* **27** (Suppl. 1), 13–17.

EVANS, R. C., CLARK, L., HEATH, P., RHODES, J. M. (1996), Treatment of ulcerative colitis with an engineered human anti-TNFα antibody, *CDP571, Gastroenterology* **110**, A-15.

FEATHERSTONE, C. (1996), Bispecific antibodies: the new magic bullets, *Lancet* **348**, 536.

FERNS, G. A. A., RAINES, E. W., SPRUGEL, K. H., MOTANI, A. S., REIDY, M. A., ROSS, R. (1991), Inhibition of neointimal smooth muscle accumulation after angioplasty by an antibody to PDGF, *Science* **253**, 1129–1132.

FITCH, J. C. K., ELEFTERIADES, J. A., RINDER, H. M., MATIS, L. L., EVANS, M. J., ROLLINS, S. A., ALFORD, B. L., HINES, R. L. (1996), Safety, pharmacokinetics, and immunogenicity of intravenous administration of H5G1. 1-SCFV in humans, *Blood* **88** (Suppl. 1), 654a.

FRANKEL, A. E., FITZGERALD, D., SIEGALL, C., PRESS, O. W. (1996), Advances in immunotoxin biology and therapy: a summary of the Fourth International Symposium on Immunotoxins, *Cancer Res.* **1996**, 926–932.

FRENETTE, P. S., WAGNER, D. D. (1996), Adhesion molecules – part II: blood vessels and blood cells, *N. Engl. J. Med.* **335**, 43–45.

FRIEDLANDER, M., BROOKS, P. C., SHAFFER, R. W., KINCAID, C. M., VARNER, J. A., CHERESH, D. A. (1995), Definition of two angiogenic pathways by distinct αv integrins, *Science* **270**, 1500–1502.

GARIN-CHESA, P., OLD, L. J., RETTIG, W. J. (1990), Cell surface glycoprotein of reactive stromal fibroblasts as a potential antibody target in human epithelial cancers, *Proc. Natl. Acad. Sci. USA.* **87**, 7235–7239.

GOLDSTEIN, G. and the Ortho Multicenter Transplant Study Group (1985), A randomized clinical trial of OKT3 monoclonal antibody for acute rejection of cadaveric renal transplants, *N. Engl. J. Med.* **313**, 337–342.

HAMMES, H.-P., BROWNLEE, M., JONCZYK, A., SUTTER, A., PREISSNER, K. T. (1996), Subcutaneous injection of a cyclic peptide antagonist of vitronectin receptor-type integrins inhibits retinal neovascularization, *Nature Med.* **2**, 529–533.

HART, I. R., SAINI, A. (1992), Biology of tumour metastasis, *Lancet* **339**, 1453–1457.

HELDERMAN, J. H. (1995), Review and preview of anti-T-cell antibodies, *Transplant. Proc.* **27** (Suppl. 1), 8–9.

HERTLER, A. A., FRANKEL, A. E. (1989), Immunotoxins: a clinical review of their use in the treatment of malignancies, *J. Clin. Oncol.* **7**, 1932–1942.

HERVÉ, P., FLESCH, M., TIBERGHIEN, P., WIJDENES, J., RACADOT, E., BORDIGONI, P., PLOUVIER, E., STEPHAN, J. L., BOURDEAU, H., HOLLER, E., LIOURE, B., ROCHE, C., VILMER, E., DEMEOCQ, F., KUENTZ, M., CAHN, J. Y. (1992), Phase I–II trial of a monoclonal anti-tumor necrosis factor α antibody for the treatment of refractory severe acute graft-versus-host disease, *Blood* **79**, 3362–3368.

HIEPE, F., VOLK, H.-D., APOSTOLOFF, E., BAEHR VON, R., EMMRICH, F. (1991), Treatment of severe systemic lupus erythematosus with anti-CD4 monoclonal antibody, *Lancet* **338**, 1529–1530.

HINSHAW, L. B., TEKAMP-OLSON, P., CHANG, A. C. K., LEE, P. A., TAYLOR, F. B. JR., MURRAY, C. K., PEER, G. T., EMERSON, T. E. JR., PASSEY, R. B., KUO, G. C. (1990), Survival of primates in LD100 septic shock following therapy with antibody to tumor necrosis factor (TNFα), *Circ. Shock* **30**, 279–292.

ISAACS, J. D., WATTS, R. A., HAZLEMAN, B. L., HALE, G., KEOGAN, M. T., COBBOLD, S. P., WALDMANN, H. (1992), Humanized monoclonal antibody therapy for rheumatoid arthritis, *Lancet* **340**, 748–752.

JERNE, N. K. (1984), Idiotypic networks and other preconceived ideas, *Immunol. Rev.* **79**, 5–24.

JONKER, M., LANBALGEN VAN, R., MITCHELL, D. J., DURHAM, S. K., STEINMAN, L. (1988), Successful treatment of EAE in rhesus monkeys with MHC class II specific monoclonal antibodies, *J. Autoimmun.* **1**, 399–414.

KAWAHITO, K., KAWAKAMI, M., FUJIWARA, T., ADACHI, H., INO, T. (1995), Interleukin-8 and monocyte chemotactic activating factor responses to cardiopulmonary bypass, *J. Thorac. Cardiovasc. Surg.* **110**, 99–102.

KINGSLEY, G., PANAYI, G., LANCHBURY, J. (1991), Immunotherapy of rheumatic diseases – practice and prospects, *Immunol. Today* **12**, 177–179.

KIRKMAN, R. L., SHAPIRO, M. E., CARPENTER, C. B., MILFORD, E. L., RAMOS, E. L., TILNEY, N. L., WALDMAN, T. A., ZIMMERMAN, C. E., STROM, T. B. (1989), Early experience with anti-Tac in clinical renal transplantation, *Transplant. Proc.* **21**, 1766–1768.

KROOK, A., SOOS, M. A., KUMAR, S., SIDDLE, K., O'REILLY, S. (1996), Functional activation of mutant human insulin receptor by monoclonal antibody, *Lancet* **347**, 1586–1590.

KUPIEC-WEGLINSKI, J. W., DIAMANTSTEIN, T., TILNEY, N. L. (1988), Interleukin 2 receptor-targeted therapy – rationale and applications in organ transplantation, *Transplantation* **46**, 785–792.

LARSEN, C. P., ELWOOD, E. T., ALEXANDER, D. Z., RICHIE, S. C., HENDRIX, R., TUCKER-BURDEN, C., CHO, H. R., ARUFFO, A., HOLLENBAUGH, D., LINSLEY, P. S., WINN, K. J., PEARSON, T. C. (1996), Long-term acceptance of skin and cardiac allografts after blocking CD40 and CD28 pathways, *Nature* **381**, 434–438.

LIU, X., MASHOUR, G. A., KURTZ, A. (1996), Recent developments in the treatment of encephalomyelitis, *Exp. Opin. Ther. Patents* **6**, 457–470.

LOBB, R. R., HEMLER, M. E. (1994), The pathophysiologic role of α4 integrins *in vivo*, *J. Clin. Invest.* **94**, 1722–1728.

LUSTGARTEN, J., WAKS, T., ESHHAR, Z. (1996), Prolonged inhibition of IgE production in mice following treatment with an IgE-specific immunotoxin, *Mol. Immunol.* **33**, 245–251.

MALONEY, D. G., LILES, T. M., CZERWINSKI, D. K., WALDICHUK, C., ROSENBERG, J., GRILLO-LOPEZ, A., LEVY, R. (1994), Phase I clinical trial using escalating single-dose infusion of chimeric anti-CD20 monoclonal antibody (IDEC-C2B8) in patients with recurrent B-cell lymphoma, *Blood* **84**, 2457–2466.

MATHIESON, P. W., COBBOLD, S. P., HALE, G., CLARK, M. R., OLIVEIRA, D. B. G., LOCKWOOD, C. M., WALDMANN, H. (1990), Monoclonal antibody therapy in systemic vasculitis, *N. Engl. J. Med.* **323**, 250–254.

MAUSER, P. J., PITMAN, A. M., FERNANDEZ, X., FORAN, S. K., ADAMS, G. K., KREUTNER, W., EGAN, R. W., CHAPMAN, R. W. (1995), Effects of an antibody to interleukin-5 in a monkey model of asthma, *Am. J. Resp. Crit. Care Med.* **152**, 467–472.

MCMAHON, G., EHRLICH, P. H., MOUSTAFA, Z. A., MCCARTHY, L. A., DOTTAVIO, D., TOLPIN, M. D., NADLER, P. I., OSTBERG, L. (1992), Genetic alterations in the gene encoding the major HBsAg: DNA and immunological analysis of recurrent HBsAg derived from monoclonal antibody-treated liver transplant patients, *Hepatology* **15**, 757–766.

MILESKI, W. J., WINN, R. K., VEDDER, N. B., POHLMAN, T. H., HARLAN, J. M., RICE, C. L. (1990), Inhibition of CD18-dependent neutrophil adherence reduces organ injury after hemorrhagic shock in primates, *Surgery* **108**, 206–212.

MOREAU, T., THORPE, J, MILLER, D., MOSELEY, I., HALE, G., WALDMAN, H., CLAYTON, D., WING, M., SCOLDING, N., COMPSTON, A. (1994), Preliminary evidence from magnetic resonance imaging for reduction in disease activity after lymphocyte depletion in multiple sclerosis, *Lancet* **344**, 298–301.

MOREL, P., REVILLARD, J. P., NICOLAS, J. F., WIJDNES, J., ROZOVA, H., THIVOLET, J. (1992), Anti-CD4 monoclonal antibody therapy in severe psoriasis, *J. Autoimmun.* **5**, 465–477.

OPPENHEIMER-MARKS, N., LIPSKY, P. E. (1996), Adhesion molecules as targets for the treatment of autoimmune diseases, *Clin. Immunol. Immunopathol.* **79**, 203–210.

PAI, L. H., PASTAN, I. (1993), Immunotoxin therapy for cancer, *JAMA* **269**, 78–81.

PAI, L. H., BOOKMAN, M. A., OZOLS, R. F., YOUNG, R. C., SMITH, J. W., LONGO, D. L., GOULD, B., FRANKEL, A., MCLAY, E. F., HOWELL, S., REED, E., WILLINGHAM, M. C., FITZGERALD, D. J., PASTAN, I. (1991), Clinical evaluation of intraperitoneal *Pseudomonas* exotoxin immunoconjugate OVB3-PE in patients with ovarian cancer, *J. Clin. Oncol.* **9**, 2095–2103.

PAI, L. H., WITTES, R., SETSER, A., WILLINGHAM, M. C., PASTAN, I. (1996), Treatment of advanced solid tumors with immunotoxin LMB-1: an antibody linked to *Pseudomonas* exotoxin, *Nature Med.* **2**, 350–353.

PICKER, L. J., KISHIMOTO, T. K., SMITH, C. W., WARNOCK, R. A., BUTCHER, E. C. (1991), ELAM-1 is an adhesion molecule for skin-homing T cells, *Nature* **349**, 796–799.

PRESS, O. W., EARY, J. F., APPELBAUM, F. R., MARTIN, P. J., NELP, W. B., GLENN, S., FISHER, D. R., PORTER B., MATTHEWS, D. C., GOOLEY, T., BERNSTEIN, I. D. (1995), Phase II trial of ^{131}I-B1 (anti-CD20) antibody therapy with autologous stem cell transplantation for relapsed B cell lymphomas, *Lancet* **346**, 336–340.

RAINE, C. S. (1995), Multiple sclerosis: TNF revisited, with promise, *Nature Med.* **1**, 211–214.

RANKIN, E. C. C., CHOY, E. H. S., KASSIMOS, D., KINGSLEY, G. H., SOPWITH, A. M., ISENBERG, D. A., PANAYI, G. S. (1995), The therapeutic effects of an engineered human anti-tumour necrosis factor alpha antibody (CDP571) in rheumatoid arthritis, *Br. J. Rheumatol.* **34**, 334–342.

REISER, H., STADECKER, M. J. (1996), Costimulatory B7 molecules in the pathogenesis of infectious and autoimmune diseases, *N. Engl. J. Med.* **335**, 1369–1377.

REITMÜLLER, G., SCHNEIDER-GÄDICKE, E., SCHLIMOK, G., SCHMIEGEL, W., RAAB, R., HÖFFKEN, K., GRUBER, R., PICHLMAIER, H., HIRCHE, H., PICHLMAYR, R., BUGGISCH, P., WITTE, J. and the German Cancer Aid 17-1A Study Group (1992), Randomised trial of monoclonal antibody for adjuvant therapy of resected Dukes' C colorectal carcinoma, *Lancet* **343**, 1177–1183.

RETTIG, W. J., GARIN-CHESA, P., HEALEY, J. H., SU, S. L., JAFFE, E. A., OLD, L. J. (1992), Identification of endosialin, a cell surface glycoprotein of vascular endothelial cells in human cancer, *Proc. Natl. Acad. Sci. USA* **89**, 10832–10836.

SEKIDO, N., MAKAIDA, N., HARADA, A., NAKANI SHI, I., WATANABE, Y., MATSUSHIMA (1993), Prevention of lung reperfusion injury in rabbits by a monoclonal antibody against interleukin-8, *Nature* **365**, 654–657.

SHIELDS, R. L., WHETHER, W. R., ZIONCHECK, K., O'CONNELL, L., FENDLY, B., PRESTA, L. G., THOMAS, D., SABAN, R., JARDIEU, P. (1995), Inhibition of allergic reactions with antibodies to IgE, *Int. Arch. Allergy Immunol.* **107**, 308–312.

SHIMIZU, Y., SHAW, S., GRABER, N., GOPAL, T. V., HORGAN, K. J., SEVENTER VAN, G. A., NEWMAN, W. (1991), Activation-independent binding of human memory T cells to adhesion molecule ELAM-1, *Nature* **349**, 799–802.

SPRINGER, T. A. (1994), Traffic signals for lymphocyte recirculation and leukocyte emigration: the multistep paradigm, *Cell* **76**, 301–314.

STACK, W., MANN, S. D., ROY, A. J., HEATH, P., SOPWITH, M., FREEMAN, J., HOLMES, G., LONG, R., FORBES, A., KAMM, M. A., HAWKEY, C. I. (in press), Randomized controlled trial of CDP571 antibody to tumour necrosis factor in Crohn's disease, *Lancet*.

STAFFORD, F. J., FLEISHER, T. A., LÉE, G., BROWN, M., STRAND, V., AUSTIN, H. A., BALOW, J. E., KLIPPEL, J. H. (1994), A pilot study of anti-CD5 ricin A immunoconjugate in systemic lupus erythematosus, *J. Rheumatol.* **21**, 2068–2170.

STEINMAN, L., ROSENBAUM, J. T., SRIRAM, S., MCDEVITT, H. O. (1981), *In vivo* effects of antibodies to immune response gene products: prevention of experimental allergic encephalitis, *Proc. Natl. Acad. Sci. USA* **78**, 7111–7114.

STREETER, P. R., LAKEY-BERG, E., ROUSE, B. T. N., BARGATZE, R. F., BUTCHER, E. C. (1988), A tissue-specific endothelial cell molecule involved in lymphocyte homing, *Nature* **331**, 41–46.

SUZUKI, T., KANBARA, N., TOMONO, T., HAYASHI, N., SHINOHARA, I. (1984), Physicochemical and biological properties of poly(ethylene glycol)-coupled immunoglobulin G, *Biochim. Biophys. Acta* **788**, 248–255.

THORPE, P. E., BURROWS, F. J. (1995), Antibody-directed targeting of the vasculature of solid tumors, *Breast Cancer Res. Treat.* **36**, 237–251.

THURAU, S. R., WILDNER, G., REITER, C., RIETHMÜLLER, G., LUND, O. E. (1994), Treatment of endogenous uveitis with anti-CD4 monoclonal antibody: first report, *Ger. J. Ophthalmol.* **3**, 409–413.

TOPOL, E. J., CALIFF, R. M., WEISMAN, H. F., ELLIS, S. G., TCHENG, J. E., WORLEY, S., IVANHOE, R., GEORGE, B. S., FINTEL, D., WESTON, M., SIGMON, K., ANDERSON, K. M., LEE, K. L., WILLERSON, J. T. and the EPIC investigators (1994), Randomized trial of coronary intervention with antibody against platelet IIb/IIIa integrin for reduction of clinical restenosis: results at six months, *Lancet* **343**, 881–886.

TRACEY, K. J., FONG, Y., HESSE, D. G., MANOGUE, K. R., LEE, A. T., KUO, G. C., LOWRY, S. F., CERAMI, A. (1987), Anti-cachectin/TNF monoclonal antibodies prevent septic shock during lethal bacteremia, *Nature* **330**, 662–664.

VAN DER VIEREN, M., TRONG, H. L., WOOD, C. L., MOORE, P. F., ST. JOHN, T., STAUNTON, D. E., GALLATIN, W. M. (1995), A novel leukointegrin, αdβ2, binds preferentially to ICAM-3, *Immunity* **3**, 683–690.

VAN DULLEMEN, H. M., VAN DEVENTER, S. J. H., HOMMES, D. W., BIJL, H. A., JANSEN, J., TYTGAT, G. N. J., WOODY, J. (1995), Treatment of Crohn's disease with anti-tumor necrosis factor chimeric monoclonal antibody (cA2), *Gastroenterology* **109**, 129–135.

VEDDER, N. B., WINN, R. K., RICE, C. L., CHI, E. Y., ARFORS, K.-E., HARLAN, J. M. (1990), Inhibition of leukocyte adherence by anti-CD18 monoclonal antibody attenuates reperfusion injury in the rabbit ear, *Proc. Natl. Acad. Sci. USA* **87**, 2643–2646.

WARREN, H. S., DANNER, R. L., MUNFORD, R. S. (1992), Anti-endotoxin monoclonal antibodies, *N. Engl. J. Med.* **326**, 1151–1157.

WEINBERG, A. D., BOURDETTE, D. N., SULLIVAN, T. J., LEMON, M., WALLIN, J. J., MAZIARZ, R., DAVEY, M., PALIDA, F., GODFREY, W., ENGLEMAN, E., FULTON, R. J., OFFNER, H., VANDENBARK, A. A. (1996), Selective depletion of myelin-reactive T cells with anti-OX-40 antibody ameliorates autoimmune encephalomyelitis, *Nature Med.* **2**, 183–189.

WENDLING, D., RACADOT, E., WIJDNES, J. (1993), Treatment of severe rheumatoid arthritis by anti-interleukin 6 monoclonal antibody, *J. Rheumatol.* **20**, 259–262.

WHITTUM-HUDSON, J. A., AN, L.-L., SALTZMAN, W. M., PRENDERGAST, R. A., MACDONALD, B. (1996), Oral immunization with an anti-idiotypic antibody to the exoglycolipid antigen protects against experimental *Chlamydia trachomatis* infection, *Nature Med.* **2**, 1116–1121.

YEDNOCK, T. A., CANNON, C., FRITZ, L. C., SANCHEZ-MADRID, F., STEINMAN, L., KARIN, N. (1992), Prevention of experimental autoimmune encephalomyelitis by antibodies against α4β1 integrin, *Nature* **356**, 63–66.

YUAN, F., CHEN, Y., DELLIAN, M., SAFABAKHSH, N., FERRARA, N., JAIN, R. K. (1996), Time-dependent vascular regression and permeability changes in established human tumor xenografts induced by an anti-vascular endothelial growth factor/vascular permeability factor antibody, *Proc. Natl. Acad. Sci. USA* **93**, 14765–14770.

Case Studies

13 Antibodies for Sepsis: Some Lessons Learnt

STEVEN M. OPAL

Pawtucket, RI, USA

1 Introduction   330
2 Rationale for Monoclonal Antibody Therapy in Sepsis   330
    2.1 Anti-Endotoxin Monoclonal Antibodies   331
    2.2 Anti-TNF Monoclonal Antibodies   332
3 Lessons Learnt from Preclinical Studies and Clinical Trial Design   335
    3.1 Lessons Learnt from Animal Models of Sepsis   335
    3.2 Problems with Definitions of Sepsis   336
    3.3 Lessons Learnt about Clinical Trial Design   337
4 Summary and Recommendations   338
5 References   339

1 Introduction

The impetus for the development of monoclonal antibodies (MABs) for the treatment of sepsis began with a report by ZIEGLER et al. (1982) with *E. coli* J5 antisera. In this seminal report, a polyvalent human antiserum raised against a rough LPS mutant strain of *E. coli* 0111:B4 was shown to be protective in a double-blind, placebo controlled, clinical trial. A statistically significant reduction in mortality was found in patients with gram-negative bacteremia and septic shock. The results of this clinical trial, and the proposed mechanism of action of the polyclonal antisera sparked a 15-year debate which continues to the present time (CROSS, 1994).

The prevailing theory for the mechanism of action of the immune-based therapy was that human antibodies could be raised to the highly conserved, inner core glycolipid structures of bacterial endotoxin. It was reasoned that such antibodies would protect septic patients from the lethal effects of endotoxin (ZIEGLER et al., 1991). It was abundantly clear that pooled antisera from human volunteers would not be a practical long-term solution for the treatment of human sepsis. Lot to lot variations of antibody levels from different volunteers, production and purification problems, and the omnipresent risk of transmission of blood-borne pathogens make human serum-based therapy an unattractive therapeutic option for septic shock.

Assuming that the mechanism of action of the *E. coli* J5 antisera was correct, a solution to the antibody production problems appeared to be readily at hand. The availability of monoclonal antibody technology provided an opportunity to generate high concentrations of specific antibodies which would bind with high affinity to the common core structures of bacterial endotoxin. This approach culminated in the generation of the first two monoclonal antibodies for human experimentation for sepsis, the human HA-1A MAB (TENG et al., 1985) and the murine E5 monoclonal antibody (YOUNG et al., 1989).

With the discovery of the central role of the proinflammatory cytokines in the pathogenesis of human sepsis (BEUTLER and CERAMI, 1987; TRACEY et al., 1987), it was speculated that monoclonal antibodies targeted against endogenous inflammatory mediators such as TNF might benefit the septic patient. Many research groups in the biotechnology and pharmaceutical industry have pursued this therapeutic approach for sepsis and extensive experience now exists with a variety of monoclonal antibody strategies against human tumor necrosis factor. The clinical experience with the two antiendotoxin antibodies and the many anti-TNF MABs which followed will be the focus of the remainder of this chapter.

The net results of the numerous clinical trials for sepsis with monoclonal antibodies against bacterial endotoxin and tumor necrosis factor have generally been disappointing (NATANSON et al., 1994; CROSS and OPAL, 1994). Careful review of these clinical trials reveals tantalizing evidence that some septic patients may benefit from the monoclonal antibody therapy (OPAL and FISHER, 1996; ABRAHAM et al., 1995). However, pitfalls in the conduct and interpretation of human clinical trials in septic shock and an incomplete understanding of the basic mechanisms of septic shock have contributed to failed clinical trials thus far. It is appropriate to review the lessons learnt from these clinical trials. It is hoped that future trials in septic shock will focus upon a more clearly defined patient population with improved therapeutic agents for the management of sepsis.

2 Rationale for Monoclonal Antibody Therapy in Sepsis

The therapeutic rationale for monoclonal antibody therapy in human sepsis is deceptively simple. If bacterial endotoxin (LPS) is the principal mediator of gram-negative sepsis and septic shock, then a monoclonal antibody which specifically binds and neutralizes bacterial endotoxin should benefit patients with gram-negative sepsis. Likewise, if tumor necrosis factor is the principal host-derived inflammatory mediator of sepsis, then a monoclonal antibody which specifically neutralizes TNF activity should also benefit patients in septic shock. Monoclonal antibodies as a treatment strategy for septic shock appear rather straight forward and amenable to specific testing in controlled clinical trials. However, a number of mechanistic and study design issues has clouded the interpretation of the clinical experience with currently available monoclonal antibodies. The potential problems with passive immunotherapy directed against bacterial endotoxin and tumor necrosis factor are reviewed in the following sections.

2.1 Anti-Endotoxin Monoclonal Antibodies

The importance of bacterial endotoxin in the pathogenesis of septic shock in humans has been appreciated since the beginning of the 20th century. Extensive preclinical and clinical experience attests to the pathophysiologic consequences of LPS exposure (MORRISON and RYAN, 1987).

LPS is a complex structure (RIETSCHEL and BRADE, 1992) composed of three distinct elements:

- a lipid A component;
- a core oligosaccharide component with specific sequences of heptose and hexose sugars;
- a repeating polysaccharide side chain covalently linked to the core glycolipid structure.

The lipid A molecule consists of specific fatty acids bound by ester or amide linkages to a β-(1-6) linked di-N-acetylglucosamine backbone. The diphosphorylated structure is principally responsible for the overall toxicity of LPS. The lipid A structure is covalently linked by 2 or 3 KDO moieties (2-keto 3-deoxy-octonate) to a series of specific heptose and hexose sugars. There exists a limited number (approximately 8) of chemotypes of this core oligosaccharide which together with lipid A make up the core glycolipid structure

of LPS. Some bacteria express only the core glycolipid form of LPS and are known as rough mutants with incomplete LPS. They are categorized as Ra to Re mutants depending upon the chain length of the core sugars.

A series of unique, repeating oligosaccharide side chains make up the polysaccharide part of the LPS molecule. These are the immunodominant epitopes on the LPS molecule and give serotype specificity to different bacterial strains. These outer membrane structures are surface exposed and are the major antigenic determinants of the LPS molecule.

Developing a monoclonal antibody therapeutic strategy against LPS is complicated by the fact that antibodies are most easily directed against serotype-specific polysaccharide side chains. There are hundreds of serotypes within gram-negative bacterial species; and serotype-specific MABs offer no cross-protection against other serotypes of bacterial LPS. Passive protection using monoclonal antibodies would require a cocktail of many different monoclonal antibodies directed against multiple serotypes of gram-negative bacteria.

Alternatively, the highly conserved, core glycolipid structure of LPS could be used as the immunogen to develop monoclonal antibodies against a broad range of gram-negative bacterial species. This has proven difficult as lipid structures are generally not good immunogens. Antibodies that have been raised against the lipid A component of bacterial endotoxin often express low affinity and non-specific binding to other phospholipid molecules (CROSS and OPAL, 1994). Despite these difficulties, a number of monoclonal antibodies has been successfully developed, HA-1A (TENG et al., 1985) and E5 (YOUNG et al., 1989; DEPADOVA et al., 1993), which bind to the core structure of bacterial endotoxin. Another similar monoclonal antibody, known as T 88, has been developed against the enterobacterial common antigen (DAIFUKU et al., 1992). This antibody has also undergone extensive clinical evaluation.

A variety of immunochemical and production problems has plagued the clinical development of anti-endotoxin monoclonal antibodies in sepsis. Some of the antibodies have rather low binding avidities and do not effectively neutralize the biological effects of LPS in standard endotoxin detection assays (the *Limulus* reaction). This makes it difficult to determine the specific activity of each lot of monoclonal antibody for activity and specificity. The monoclonal antibodies tested thus far are IgM class immunoglobulins. These immunoglobulins have a volume of distribution which is restricted to the intravascular space. Their large size prevents extensive permeability into the interstitial spaces of the extracellular fluid where LPS can interact with immune cells and, consequently, this may limit the potential therapeutic efficacy of these agents. In contrast to serotype-specific MABs, core glycolipid monoclonal antibodies do not promote the opsonophagocytosis of intact gram-negative bacteria. The binding to the core glycolipid structure of LPS in viable bacteria is limited as this portion of LPS is buried within the outer membrane of gram-negative bacteria. The core regions of LPS are not surface exposed and are, therefore, not readily accessible for binding by large IgM monoclonal antibodies. Even if the primary mechanism of action was limited to binding to free circulating endotoxin, the therapeutic rationale for the anti-endotoxin MABs would remain problematic since it has been difficult to demonstrate high levels of circulating endotoxin, even in patients with documented gram-negative bacteremia (DANNER et al., 1991).

One of the monoclonal antibodies that has been extensively investigated in spectic shock (HA-1A) demonstrates significant binding to the cell membrane of human B cells (WARREN et al., 1992; CROSS, 1994) and in at least one animal model of sepsis in dogs, HA-1A appeared to worsen the outcome (QUEZADO et al., 1993).

The rationale for the use of the anti-endotoxin monoclonal antibody therapies is further complicated by the fact that gram-negative bacteria account for only 40–50% of the cases of septic shock in recent clinical trials (OPAL and FISHER, 1996). This restricts the potential therapeutic efficacy of these antibodies to only half the entire septic population. An additional drawback with the anti-endotoxin monoclonal antibody strategy is the rapidity (within minutes) with which endotoxin leads to pathologic injury in experi-

mental studies. It is difficult for the clinician to recognize that a patient is septic or to confirm the early phases of gram-negative sepsis in patients in order to initiate the monoclonal antibody therapy within the first few hours of the onset of sepsis. It is possible that the "therapeutic window" in which an anti-endotoxin strategy would be effective is passed before sepsis is identified at the bedside, eliminating the opportunity to provide the monoclonal antibody at the optimal time for the greatest benefit.

Despite all the potential pitfalls of anti-endotoxin monoclonal antibody therapy in clinical medicine, several trials have been undertaken in the last decade to determine if this therapeutic strategy will be useful for septic shock. These studies are reviewed in Tab. 1.

2.2 Anti-TNF Monoclonal Antibodies

Following its initial cloning and characterization in the mid-1980s (BEUTLER and CERAMI, 1987), it became immediately apparent that TNF was an important endogenous mediator of septic shock. A large number of animal studies and clinical investigations has revealed that TNF is a highly toxic molecule when released in excess quantities in the systemic circulation. There is now compelling evidence that tumor necrosis factor is one of the principal mediators in the pathophysiology of sepsis. It has become a logical target for new innovative therapies in the treatment of human sepsis.

A variety of monoclonal antibodies have been developed that specifically bind and neutralize the bioactivity of human tumor necrosis factor (FISHER et al., 1993; ABRAHAM et al., 1995; REINHART et al., 1996). Anti-TNF strategies for sepsis have a major theoretical advantage over anti-endotoxin strategies. While anti-endotoxin monoclonal antibodies can only be expected to be effective in gram-negative sepsis, TNF inhibition would be expected to be effective regardless of the causative microorganism. TNF inhibition has been shown to be an effective treatment in experimental studies where gram-positive bacterial

Tab. 1. Clinical Trials in Sepsis with Anti-Endotoxin Monoclonal Antibodies

Author/Year	Antibody	Study	No. of Patients	Outcome
Ziegler et al., 1991	HA-1A – human anti-lipid A MAB	Phase III	534	Overall – no benefit; the shock and bacteremia subgroup appeared to improve ($p = 0.01$)
McCloskey et al., 1994	CHESS – HA-1A	Phase III	2,199	Overall – no benefit
Greenman et al., 1991	E5 – murine anti-lipid A MAB	1st Phase III	468	Overall – no benefit, patients with gram-negative infections without shock appeared to benefit
Bone et al., 1995	E5	2nd Phase III	830	Overall – no benefit, patients with organ failure appeared to improve
	E5	3rd Phase III	~2,000	Ongoing study – should complete in 1997
Press release 1994	T88 – anti-enterobacterial common antigen MAB	Phase III	826	Overall – no benefit

organisms were used as the septic challenge (Hinshaw et al., 1992). Gram-positive organisms express a number of superantigens which activate a large number of CD4 cells and precipitate the generation of proinflammatory cytokine networks. The ability of anti-TNF monoclonal antibodies to be of potential benefit in both gram-negative or gram-positive bacterial sepsis is a major advantage of this antibody approach. It is often difficult to distinguish by clinical and/or hemodynamic parameters between gram-positive and gram-negative bacterial sepsis (Natanson et al., 1994). Therefore, anti-TNF strategies would have broader applications in sepsis and could be employed successfully prior to knowing the identity of the causative organism. This could save precious time and obviate the need to know the infecting microorganism prior to initiation of anti-sepsis treatments.

A theoretical disadvantage of anti-TNF strategies for sepsis is the finding that TNF serves as an important proinflammatory cytokine in the host response to microbial invasion (Beutler and Cerami, 1987). In certain experimental animal systems, anti-TNF strategies have resulted in an exacerbation of

systemic infection. This is particularly true with invasive intracellular pathogens such as *Listeria* (Nakane et al., 1988) or fungi (Allendoerfer et al., 1993) and in experimental settings, combinations of inhibitors for both IL-1 and TNF will exacerbate infection and increase lethality in some gram-negative models (Opal et al., 1996).

Further, C3H/HeJ mice which have a transcriptional and translational block in TNF synthesis, and TNF receptor type 1 knockout mice are both resistant to the lethal effects of endotoxemia, yet highly susceptible to lethality from invasive bacterial infections (Cross et al., 1989; Rothe et al., 1993). First degree relatives of patients with fatal meningococcal disease have been shown to have low TNF synthetic capacity when compared to relatives of patients with non-lethal meningococcal infections (Westendorp et al., 1997). This points to a potential therapeutic dilemma with the use of anticytokine therapy in sepsis.

Anticytokine therapy should prevent the injurious effects of excess cytokine activation by the host immune system; yet, cytokines also play an important protective role in the primary immune response to invasive bacterial and fungal pathogens. Viewed from this prospective, anticytokine therapy can function as a "double-edged" sword in the treatment of septic shock from invasive microbial pathogens (DINARELLO et al., 1993). The situation is somewhat akin to the use of corticosteroids as anti-inflammatory agents. Corticosteroids may benefit patients with excess systemic inflammation yet be deleterious for patients who require a rigorous host immune response to microbial pathogens. Disruption of the host response to bacterial infection, particularly among gram-positive pathogens, has been proposed as an explanation for the adverse consequences of the type II soluble TNF receptor: Fc fusion peptide in septic shock (FISHER et al., 1996). If the experimental observations with anti-TNF antibodies are ultimately shown to be applicable to human sepsis, it will substantially confound patient selection for anti-TNF monoclonal antibody therapy in sepsis.

To further complicate matters following initial TNF activation in the early phases of sepsis, TNF is suppressed in the later phases of septic shock. The production of a variety of anti-inflammatory cytokines and specific cytokine inhibitors (i.e., soluble TNF receptors, IL-1 receptor antagonist) serve to down-regulate TNF production as septic shock develops (VAN DEUREN et al., 1994; DINARELLO et al., 1993).

It has also been demonstrated that individuals differ in their intrinsic ability to generate a tumor necrosis factor response to a variety of inflammatory stimuli (STÜBER et al., 1996; WESTENDORP et al., 1997). These findings would indicate that anti-TNF therapy might be efficacious in some patients (high TNF responders) while anti-TNF MABs might be ineffective or detrimental in other patient populations (lower TNF responders). Similarly, a TNF inhibitor strategy would appear to be better suited in clinical situations where TNF levels are elevated than where TNF synthesis is suppressed by endogenous regulatory cytokines. TNF synthetic capacity is controlled by a number of parameters including DNA polymorphisms at the regulatory elements of the human TNF gene (STÜBER et al., 1996).

There is an emerging concensus that moderate amounts of TNF production promote a coordinated and appropriate host immune response to infecting microorganisms. However, excess systemic quantities of TNF induce widespread immune activation and endothelial cell injury with deleterious clinical consequences. This creates a therapeutic dilemma when initiating an anti-TNF strategy in the treatment of septic shock. It may be advantageous to attenuate TNF production yet not block all TNF synthesis in the presence of an ongoing systemic infectious process. Thus, the timing, dosage, and patient selection become critical elements to the optimal use of anti-TNF monoclonal antibodies in sepsis.

Unfortunately, clinical and hemodynamic assessments do not accurately distinguish between septic patients with high or low TNF production. Furthermore, measurement of circulating TNF levels in septic patients generally reveals disappointingly low levels of TNF (FISHER et al., 1993; CALANDRA et al., 1991). TNF has a short half-life in the blood and much of the toxicity of TNF is mediated by local TNF production in the extravascular space. For these reasons, many groups have attempted to find surrogate markers for TNF excess in the design of clinical trials with anti-TNF MABs.

Early septic shock with systemic hypotension was indicated in the Bayer Anti-TNF MAB (BAYX1351) study (Norasept I) as a subgroup of septic patients most likely to respond to the anti-TNF MAB therapy (ABRAHAM et al., 1995). However, recent results from a study of 1900 patients with septic shock failed to confirm a mortality benefit in the group treated with BAYX1351. Other groups have focused upon IL-6 levels as a marker of TNF activation (REINHART et al., 1996). Whether IL-6, physiologic parameters, or other surrogate markers (e.g., soluble cytokine receptors, other cytokines, procalcitonin, inflammatory cell activation markers, etc.) will ultimately prove to be useful in the definition of the target population for anti-TNF MABs remains to be demonstrated. A review of recent clinical trials with anti-TNF monoclonal antibody strategies is found in Tab. 2.

Tab. 2. Clinical Trials in Sepsis with Anti-TNF Monoclonal Antibodies

Author/Year	Antibody	Study	No. of Patients	Outcome
FISHER et al., 1993	CB0006 – murine MAB	Phase II	80	Overall – no benefit; subgroup with elevated TNF levels may have benefited
ABRAHAM et al., 1995	Bayer X1351 – murine MAB	Norasept I Phase III	994	Overall – no benefit; improved outcome over the first 72 h; some benefit in patients with septic shock
COHEN et al., 1996	Bayer X1351	Intersept Phase III	564	Overall – no benefit; reversal of shock in some patients
	Bayer X1351	Norasept II Phase III	~1,900	Overall – no benefit in patients with septic shock
REINHART et al., 1996	MAK195F – Fab 2'MAB	Phase II	122	Overall – no benefit; patients with elevated IL-6 levels may have benefited
DHAINAUT et al., 1995	CPD571 – "humanized" murine MAB	Phase II	42	Overall – no benefit
FISHER et al., 1996	P75 TNFR:Fc – type II TNF receptor: IgG	Phase II	141	Overall – worse outcome in treatment group ($p = 0.016$)
GLAUSER et al., 1996	P55 TNFR:FC – type I TNF receptor: IgG	Phase II	500?	Overall – no benefit, patients without refractory shock appeared to benefit independent of IL-6 levels

3 Lessons Learnt from Preclinical Studies and Clinical Trial Design

It has been difficult to replicate the favorable effects of monoclonal antibody therapy seen in experimental settings into successful clinical trials for human septic shock. This can be partly explained by the hazards of extrapolation of favorable animal experiments into clinical effectiveness in sepsis trials. Other explanations can be found in the vagaries and practical realities of clinical trial design of prospective, double-blind, placebo controlled multicenter studies.

3.1 Lessons Learnt from Animal Models of Sepsis

Why do experimental agents often work convincingly in animal models yet fail to provide benefits in human clinical trials? It is lamentable that animal models have misled clinical investigators in the design and implementation of clinical trials in human sepsis.

It should be recognized that animal models are purposely weighted in favor of detecting beneficial effects of the experimental agent and are carefully designed to limit the number of confounding variables. This isolates possible variables and allows the experimental agent to be tested in a clear and unambiguous manner. The microbial challenge, genotype, and experimental conditions of the animals are tightly controlled in animal models of sepsis. Most importantly, the timing of administration of the experimental agent can be carefully controlled to maximize the treatment effect (NATANSON et al., 1994). Many models are designed to test preventative treatments, with the experimental agent given before the septic challenge. This is the optimal time to prevent the development of the pathophysiology of sepsis. In clinical medicine, however, the number of potential confounding co-morbidities in a septic population is enormous (OPAL and FISHER, 1996). Sepsis is a heterogenous syndrome caused by many different microorganisms from multiple, different septic foci, in patients with a multitude of complicating underlying illnesses. The cause of morbidity and mortality in human sepsis is often related to the severity of the patient's underlying disease and not to the septic physiology itself.

In the clinical setting, experimental monoclonal antibody studies are conducted after established septic injury has occurred in study patients. Sepsis may have been present for hours or even days before the experimental agent is given. This unavoidable delay in treatment intervention may preclude any chance for the experimental MAB to reverse the injurious effects of early septic shock. Some of the more important differences between animal models of sepsis and actual human septic shock are given in Tab. 3. These factors need to be considered in designing protocols and determining the analytical plans and power calculations for efficacy trials in phase III human studies of sepsis.

3.2 Problems with Definitions of Sepsis

The American College of Chest Physicians/Society for Critical Care Medicine consensus definitions for sepsis provided a conceptual framework for definition of clinical syndromes in septic patients (BONE et al., 1992). While the intentions of these concensus definitions were laudable, the practical consequences of sepsis definitions were often confusing and occasionally misleading.

The concept of SIRS (systemic inflammatory response syndrome) is useful and clinically appropriate. Systemic inflammation manifest by fever, chills, leukocytosis, rapid respiratory rate and heart rate is related to the systemic activation of inflammatory mediators from a variety of infectious and non-infectious stimuli. Simultaneously, a compensatory anti-inflammatory state, now known as CARS (compensatory antagonistic response syndrome) also occurs. It has proven difficult to distinguish SIRS from CARS by clinical criteria alone. Furthermore, the definition of SIRS is so broad and non-specific that it loses its discriminatory function. Every patient with the "flu" may meet the criteria for SIRS. Marathon runners often have exercise-induced hyperthermia, increased heart rate and respiratory rate, diffuse myalgias, weakness, and confusion and would meet the definition for SIRS. Clearly, such patients would not meet

Tab. 3. Differences between Animal Models and Actual Human Sepsis

Parameter	Animal Models	Clinical Sepsis
Age, diet, genetic background	Tightly controlled	Highly variable
Co-morbidities	None or single pathologic lesion	Common and highly variable
Infecting organism	Usually a single defined pathogen	Highly variable
Time of onset of sepsis	Usually know precisely	Highly variable
Level of supportive care	Minimal after acute phase	Extensive and sophisticated
Cause of mortality	Septic insult	Underlying disease and sepsis

the intended meaning of SIRS by intensivists. The heterogeneity of the SIRS population in hospitalized patients is very broad indeed.

The clinical definition of shock continues to be problematic. How much of a fluid challenge is needed to state that a patient has fluid non-responsive hypotension? What dose of vasopressor substance is considered necessary to support the patient's blood pressure? These parameters vary from patient to patient, and it is difficult to establish a uniformly accepted threshold for a definition of shock in clinical medicine. Perhaps other hemodynamic parameters such as cardiac index or systemic vascular resistance, or other laboratory measures (i.e., base deficit, plasma lactate level, tissue oxygenation, etc.) should be used to support the clinical diagnosis of septic shock. The clinical definition of septic shock remains imprecise and subject to differing interpretations by intensivists and physiologists alike.

To further complicate the definitions of sepsis, the etiologic agent primarily responsible for sepsis may be difficult to clearly define. A typical patient in a sepsis trial would include a patient with a major intraabdominal abscess complicated by severe sepsis. Such a patient may have multiple pathogens isolated from the intraabdominal abscess. The surgical wound cultures may show other potential pathogens. Such a patient might have different organisms found in the endotracheal tube suction specimen in association with diffuse pulmonary infiltrates of uncertain etiology. A urinary catheter is often present in such patients, and cultures drawn via a catheter may have other potential uropathogens. Which organism(s) is (are) primarily responsible for sepsis in such a patient?

The causative organism for sepsis may not be readily apparent even in those patients with positive blood cultures. Contaminants are frequently found in blood cultures drawn through indwelling vascular catheters in septic patients. The frequent isolation of coagulase-negative staphylococci from the blood of such patients leads to diagnostic confusion. Is the presence of a small number of coagulase-negative staphylococci in blood cultures clinically significant? This remains open to interpretation and debate. For these reasons, it is

difficult to confidently assign a patient to a specific category of sepsis and even accurately describe the causative agent of sepsis.

Despite continued problems with the imprecise definitions of sepsis, other substantial trial design issues are principally responsible for misinterpretations of clinical sepsis studies. A bewildering array of co-morbidities, a multitude of sites and types of infecting organisms, and a variety of different time intervals between the onset of sepsis and the initiation of the experimental agent; these variables markedly affect the potential protective efficacy of experimental agents in sepsis.

3.3 Lessons Learnt about Clinical Trial Design

The marked heterogeneity of septic populations creates a real challenge in the design and analysis of the large phase III sepsis trials required to register a therapy. A number of confounding variables must be taken into account in the organization of the analytical plan. The primary end point for studies is mortality but the predicted risk of mortality in patients who meet sepsis definitions may vary from less than 5% to greater than 95% (BONE et al., 1992).

Consider a young, previously healthy woman with postpartum bacteremia and sepsis from a catheter-related *E. coli* urinary tract infection. Such a patient would be at an extremely low risk of death from sepsis yet might meet entry criteria for many sepsis trials. Since the power calculations of sepsis trials are based upon 28 day all-cause mortality in the placebo group, large numbers of patients at low risk of mortality severely limit the likelihood of a positive clinical trial. Even if the experimental agent did improve outcome in such patients, it would take tens of thousands of patients to convincingly demonstrate efficacy in a controlled clinical trial.

Alternatively, consider an elderly patient with diabetes, chronic renal insufficiency, congestive heart failure and refractory acute leukemia with bacteremic pneumonia due to *Pseudomonas aeruginosa*. The predicted risk of mortality in such a patient would be greater than 90% regardless of the antisepsis treatment administered. Such patients might meet entry criteria for a clinical sepsis trial but would provide little discriminatory information about the efficacy of the study drug. The patient's risk of mortality is exceedingly high, and the underlying disease is severe and not reversible.

Since landmark mortality data at 28 days have become the standard upon which the phase III clinical trial for sepsis is based, clinical investigators must limit the study population to severely ill patients. This is necessary to ensure that a sufficiently high event rate (mortality) will occur in the placebo arm of the study. Therefore, only severely ill patients with advanced sepsis are eligible for clinical trials. Limitation of sepsis studies to patients with advanced sepsis eliminates the opportunity to intervene in the early phases of sepsis.

There are two basic strategies to handle the marked heterogeneity of septic populations in clinical trial design. In order to control for the multitude of potential randomization imbalances, one strategy is to perform a large trial with many thousands of patients in the treatment and placebo groups. The large sample size should eliminate substantial imbalances in the treatment groups of the study.

The other trial design strategy is to perform a study in a tightly defined and highly controlled study population. An example of such a strategy would be a study of a single pathogen (e.g., meningococcal sepsis or *E. coli* sepsis) or a study focusing upon a specific disease (i.e., intraabdominal abscess, meningitis, pneumonia, etc.).

Both strategies have strengths and weaknesses as outlined in Tab. 4, and both types of clinical trials are in progress by different study groups at the present time. The more successful strategy for trial design will be determined as the results of clinical trials become available in the next few years.

4 Summary and Recommendations

Clinical trials for new sepsis therapies are intrinsically difficult to organize and analyze. It is quite possible that truly effective antisepsis agents have already been tested and discarded because of problems in clinical trial design and analysis. The following suggestions are proposed for future trials with experimental agents in the treatment of human sepsis.

(1) Do not overinterpret preclinical data on the efficacy of experimental sepsis drugs.

Tab. 4. Clinical Trial Design Problems

Large, Simple Trials ("Mega Trials")		Small, Tightly Defined Study Population	
Advantages	Disadvantages	Advantages	Disadvantages
Limits randomization imbalances	Logistics difficult and expensive	Easier to organize and conduct	Smaller numbers increase the play of chance within study
Power to detect small differences	Difficult to control quality at multiple study sites	Easier to control quality of study	May select the wrong target population and miss optimal one
Approaches actual clinical practice	May miss important subgroup effects	Few motivated and selected study sites	May not be generalizable to all septic patients
Straightforward design and statistical analysis	Limited mechanistic information available	Homogenous study group allows detailed data collection	Greater likelihood that bias and random events will affect the results

Animal models generally overestimate efficacy as they are weighted in favor of the experimental agent.

(2) Have a clear therapeutic rationale and reasonable understanding of the mechanisms, action and potential hazards of the experimental treatment. The mechanism of action of the anti-sepsis drug should be clear before embarking on a clinical trial. Potential toxicity and dosing requirements should be estimated with reasonable accuracy from preclinical studies.

(3) Study the experimental agent in a variety of animal models with differing infectious challenges. This will limit the risk of species-specific effects of the anti-sepsis drug and will diminish the possibility that untoward reactions will be missed in the preclinical evaluation of experimental agents. Since over 50% of patients in recent clinical trials of sepsis have had infectious diseases other than gram-negative bacteria, preclinical investigations which examine polymicrobial and gram-positive causes of sepsis should be undertaken.

(4) Do not rely upon preclinical studies which administer the experimental agent before the septic challenge is administered. Experimental studies that attempt to salvage animals with established sepsis are more relevant to the clinical situation in human sepsis and provide a more realistic impression of the expected clinical efficacy of the experimental agent.

(5) Be conservative in the power calculations of sample size of the clinical trials.

(6) Expect differential effects in different patient subpopulations. The heterogeneity of septic populations makes it likely that subgroups will vary in their responsiveness to the experimental agent. Be prepared to perform a large trial which will balance variable effects within the study population or organize a targeted study in a highly selected subgroup of patients. If a relatively homogenous group of septic patients can be identified, it would greatly facilitate the clinical evaluation for novel treatments for sepsis.

(7) Plan an interim analysis by an independent safety and efficacy monitoring committee. In sepsis trials, the potential for unexpected toxicities is a real possibility. For safety reasons, an interim analysis is suggested to limit the potential adverse consequences of an unrecognized immunotoxicity of the experimental agent.

Despite all the hazards and set backs of clinical sepsis trials, a new improved therapy for sepsis is a critical unmet need. Carefully designed clinical trials with thoroughly tested experimental agents for sepsis will be necessary to improve the outcome of septic patients in the future.

5 References

ABRAHAM, E., WUNDERINK, R., SILVERMAN, H., et al. (1995), Efficacy and safety of monoclonal antibody to human tumor necrosis factor-alpha in patients with sepsis syndrome, a randomized, controlled, double-blind multicenter clinical trial, *JAMA* **273**, 934–941.

ALLENDOERFER, R., MAGEE, D. M., SMITH, J. G., et al. (1993), Induction of tumor necrosis factor-alpha in murine *Candida albicans* infection, *J. Infect. Dis.* **167**, 1168–1172.

BEUTLER, B., CERAMI, A. (1987), Cachectin: more than a tumor necrosis factor, *N. Engl. J. Med.* **316**, 379–385.

BONE, R. C., BALK, R. A., CERRA, F. B., et al. (1992), Definitions for sepsis and organ failure and guidelines for the use of innovative therapies for sepsis, *Chest* **101**, 1644–1655.

BONE, R. C., BALK, R. A., FEIN, A. M., et al. (1995), A second large controlled clinical study of E5, a monoclonal antibody to endotoxin: results of a prospective, multicenter, randomized, controlled trial, *Crit. Care Med.* **23**, 994–1006.

CALANDRA, T., GERAIN, J., HEUMANN, D., et al. (1991), High circulating levels of interleukin-6 in patients with septic shock: evolution during sepsis, prognostic value, interplay with other cytokines, *Am. J. Med.* **91**, 23–29.

COHEN, J., CARLET, J., and the INTERSEPT Study Group (1996), An international, multi-center, placebo-controlled, trial of monoclonal antibody to human tumor necrosis factor-alpha in patients with sepsis, *Crit. Care Med.* **24**, 1431–1440.

CROSS, A. S. (1994), Antiendotoxin antibodies: a dead end? *Ann. Intern. Med.* **121**, 58–60.

CROSS, A. S., OPAL, S. M. (1994), Therapeutic intervention in sepsis with antibody to endotoxin: is there a future? *J. Endotoxin. Res.* **1**, 57–69.

CROSS, A. S., SADOFF, J. C., KELLY, N. M., et al. (1989), Pretreatment with recombinant murine tumor necrosis factor alpha/cachectin and murine interleukin-1 alpha protects mice from lethal bacterial infection, *J. Exp. Med.* **169**, 2021–2027.

DAIFUKU, R., HAENTFLING, K., YOUNG, J., et al. (1992), Phase I study of antilipopolysaccharide human monoclonal antibody MAB-T88, *Antimicrob. Agents Chemother.* **36**, 2349–2351.

DANNER, R. L., ELIN, R. J., ROSSEINI, J. M., et al. (1991), Endotoxemia in human septic shock, *Chest* **99**, 169–175.

DEPADOVA, F. E., BARCLAY, R., BRAUDE, H., et al. (1993), SD2219-800: a chimeric broadly cross-reactive and cross-neutralizing anti-core LPS antibody, *Circ. Shock* **1** (Suppl.), 47 (abstract 12.3).

DHAINAUT, J.-F. A., VINCENT, J.-L., RICHARD, C., et al. (1995), CDP571, a humanized antibody to human tumor necrosis factor-alpha: safety, pharmacokinetics, immune response, and influence of the antibody on cytokine concentrations in patients with septic shock, *Crit. Care Med.* **22**, 1461–1469.

DINARELLO, C. A., GELFAND, J. A., WOLFF, S. M. (1993), Anticytokine strategies in the treatment of systemic inflammatory response syndrome, *JAMA* **269**, 1825–1835.

FISHER, C. J., JR., OPAL, S. M., DHAINAUT, J. F., et al. (1993), Influence of an anti-tumor necrosis factor monoclonal antibody on cytokine levels in patients with sepsis, *Crit. Care Med.* **21**, 318–327.

FISHER, C. J., AGOSTI, J. M., OPAL, S. M., et al. (1996), Treatment of sepsis with the tumor necrosis factor receptor: Fc fusion protein, *N. Engl. J. Med.* **334**, 1697–1702.

GLAUSER, M., ABRAHAM, E., LATERRO, P. F., et al. (1996), Ro 45-2081 in the treatment of severe sepsis and septic shock: correlates of treatment, baseline cytokine levels and outcome, *Interscience Conference on Antimicrobial Agents and Chemotherapy,* p. 144, New Orleans, LA (abstract #G10).

GREENMAN, R. L., SCHEIN, R. N. H., MARTIN, M. A., et al. (1991), A controlled clinical trial of E5 murine monoclonal IgM antibody to endotoxin in the treatment of gram-negative sepsis, *JAMA* **266**, 1097–1102.

HINSHAW, L. B., EMERSON, T. E., TAYLOR, F. B. JR., et al. (1992), Lethal *Staphylococcus aureus*-induced shock in primates: prevention of death with anti-TNF antibody, *J. Trauma* **33**, 568–573.

MCCLOSKEY, R. V., STRAUBE, R. C., SANDERS, C., SMITH, C. R., and the CHESS Trial Study Group (1994), Treatment of septic shock with human monoclonal antibody HA1A: randomized, double-blind, placebo-controlled trial, *Ann. Intern. Med.* **120**, 1–5.

MORRISON, D. C., RYAN, J. L. (1987), Endotoxin and disease mechanisms, *Annu. Rev. Med.* **38**, 417–32.

NAKANE, A., MINAGUWA, T., KATO, K. (1981), Endogenous tumor necrosis factor (cachectin) is essential to host resistance against *Listeria monocytogenes* infection, *Infect. Immun.* **56**, 2563–2569.

NATANSON, C., HOFFMAN, W. D., SUFFREDINI, A. F. et al. (1994), Selected treatment strategies for septic shock based on proposed mechanisms of pathogenesis, *Ann. Intern. Med.* **120**, 771–783.

OPAL, S. M., FISHER, C. J., JR. (1996), Clinical trials with novel therapeutic agents for sepsis. Why did they fail; *Intensivmed.* **33**, 160–166.

OPAL, S. M., CROSS, A. S., JHUNG, J. W., et al. (1996), Potential hazards of combination immunotherapy in the treatment of experimental septic shock, *J. Infect. Dis.* **173**, 1415–1421.

QUEZADO, Z. M., NATANSON, C., ALLING, D. W., et al. (1991), A controlled trial of HA-1A in a canine model of gram-negative shock, *JAMA* **269**, 2221–2227.

REINHART, K., WIEGAND-LÖHNERT, C., GRIMMINGER, F., et al. (1996), Assessment of the safety and efficacy of the monoclonal anti-tumor necrosis factor antibody-fragment, MAK195F, in patients with sepsis and septic shock: a multicenter, randomized, placebo-controlled, dose-ranging study, *Crit. Care Med.* **24**, 733–742.

RIETSCHEL, E. T. H., BRADE, H. (1992), Bacterial endotoxins, *Sci. Am.* **267**, 26–30.

ROTHE, J., LESSLAUER, W., LÖTSCHER, H., et al. (1993), Mice lacking the tumor necrosis factor receptor 1 are resistant to TNF-mediated toxicity but highly susceptible to infection by *Listeria monocytogenes, Nature* **364**, 798–802.

STÜBER, F., PETERSEN, M., BOKALMANN, F., SCHADE, U. (1996), A genomic polymorphism within the tumor necrosis factor locus influences plasma tumor necrosis factor-alpha concentrations and outcome of patients with severe sepsis, *Crit. Care Med.* **24**, 281–384.

TENG, N. N. H., KAPLAN, H. S., HEBERT, J. M., et al. (1985), Protection against gram-negative bacteremia and endotoxemia with human monoclonal IgM antibodies, *Proc. Natl. Acad. Sci. USA* **82**, 1790–1794.

TRACEY, K. J., FONG, Y., HESSE, D. G., et al. (1987), Anti-cachectin/TNF monoclonal antibodies prevent septic shock during lethal bacteremia, *Nature* **330**, 662–664.

VAN DEUREN, M., VAN DER VEN-JONGEKRIJG, J., DEMACKER, P. N. M., et al. (1994), Differential expression of proinflammatory cytokines and their inhibitors during the course of meningococcal infections, *J. Infect. Dis.* **169**, 157–161.

WARREN, H. S., DANNER, R. L., MUNFORD, R. S. (1992), Anti-endotoxin antibodies, *N. Engl. J. Med.* **326**, 1153–1157.

WESTENDORP, R. G. J., LANGERMANS, J. A. M., HUIZINGA, T. W. J., ELOUALI, A. H., VERWEIJ, C. L., BOOMSMA, D. I., VANDENBROUKE, J. P. (1997), Genetic influence of cytokine production and fatal meningococcal disease, *Lancet* **349**, 170–173.

YOUNG, L. S., GASCON, R., ALAM S., BERMUDEZ, L. E. M. (1989), Monoclonal antibodies for the treatment of gram-negative infections, *Rev. Infect. Dis.* **11** (Suppl. 2), S1564–1571.

YOUNG, L. S., MARTIN, W. J., MEYER, R. D., et al. (1997), Gram-negative rod bacteremia: microbiologic, immunologic, and therapeutic considerations, *Ann. Intern. Med.* **86**, 456–569.

ZIEGLER, E. J., MCCUTCHAN, J. A., FIERER, J., et al. (1982), Treatment of gram-negative bacteremia and shock with human anti-serum to a mutant, *Escherichia coli, N. Engl. J. Med.* **307**, 1225–1230.

ZIEGLER, E. J., FISHER, C. J., JR., SPRUNG, C. L., et al. (1991), Treatment of gram-negative bacteremia and septic shock with HA1A human monoclonal antibody against endotoxin, *N. Engl. J. Med.* **324**, 429–436.

14 An Engineered Human Antibody for Chronic Therapy: CDP571

MARK SOPWITH
SUE STEPHENS

Slough, UK

1 Introduction 344
2 Generation of Engineered Human Antibody, CDP571 345
3 Preclinical Studies with CDP571 in Non-Human Primates 345
4 Studies with CDP571 in Humans 346
    4.1 Safety, Pharmacokinetics and Immunogenicity in Human Volunteers 346
    4.2 Single Dose Studies in Patients 346
    4.3 Repeated Dose Study in Patients with Rheumatoid Arthritis 347
    4.4 Pharmacokinetics and Immunogenicity 348
5. Conclusions 352
6 References 352

1 Introduction

Most episodes of disease are brief and the patient can look forward to complete recovery. Other disorders run a chronic course. Persistent and unremitting poor health for months or years on end may be complicated by the accumulation of dysfunction, frank disability or early death. With aging populations, diseases of this kind pose increasing problems to individuals, to carers, and to health care systems worldwide.

Rheumatoid arthritis (RA) is a chronic systemic disorder that is characterized by inflammation and progressive destruction of joints. The genesis of the disease involves an interplay between inherited predispositions and the environment. Additional factors, also poorly understood, are believed important in maintaining the disease process. No curative treatment is available. Medication is predominantly symptomatic and is likely to be prescribed for many months or years. Likewise, additional medication for acute exacerbations of disease will be required not just once but many times.

In consequence, any treatment for RA that has a useful effect will be prescribed repeatedly. Long duration of action, safety, and sustained efficacy with repeated dosing are greatly to be desired of any new remedy. Convenience in use and benefits sufficient to persuade health care providers to purchase are further attributes of any therapy that is to be widely accessible.

The pivotal role that tumor necrosis factor (TNF) may play in the pathogenesis and clinical features of RA has aroused considerable excitement (MAINI et al., 1993). TNF is a highly conserved and important signaling molecule. The significance of this cytokine is illustrated by, e.g., the sensitivity of transgenic mice lacking TNF receptor 1 (p55) to certain infections (PFEFFER et al., 1993). Paradoxically, if secreted in large amounts, TNF may be damaging. The RA joint contains many biological mediators that might contribute to inflammation and tissue destruction but, of these, recent research has emphasized the importance of TNF (BRENNAN et al., 1989). TNF is expressed in the RA joint by macrophages and lymphocytes, by cells with fibroblast characteristics, and by vascular endothelial cells. The expression of TNF is increased especially at the pannus–cartilage junction, a site at which destruction of the joint is particularly aggressive (CHU et al., 1991). Activities of TNF that suggest that it may contribute to inflammation and disease progression include the upregulation of cellular adhesion molecules (such as E-selectin, VCAM, ICAMs, CD18), stimulation of chemokine expression (including RANTES, IL-8, platelet activating factor), upregulation of MHC class I and class II, and the direct activation of leukocytes. Mediators whose synthesis and release are increased by TNF include prostaglandins, nitric oxide and tissue matrix metalloproteinases whose activities in turn may contribute to the resorption of cartilage and bone. That TNF can stimulate angiogenesis and synoviocyte proliferation is further evidence, albeit still circumstantial, of the possible contribution of TNF to the long-term destruction of the RA joint (SOPWITH, 1995). The most persuasive observations, however, have been replicated in several different animal models of arthritis, in which inflammation was diminished and joint destruction prevented by anti-TNF neutralizing antibodies (HENDERSON et al., 1992; PIGUET et al., 1992; WILLIAMS et al., 1992).

Studies of anti-TNF antibodies have been undertaken in patients with RA. The positive outcome of these studies has been striking (ELLIOTT et al., 1994; RANKIN et al., 1995).

Encouraged by the unambiguous involvement of TNF in pathogenesis, the role of TNF in other human diseases has also excited interest. Clinical situations in which excessive TNF may have clinically deleterious effects include septic shock (systemic inflammatory response syndrome), inflammatory bowel disease, multiple sclerosis, transplant rejection and graft-versus-host disease. In all these situations, even in acute septic shock – in which plasma TNF concentrations may spike over periods of several days (CALANDRA et al., 1990), extended or repeated neutralization of TNF is likely to be necessary therapeutically.

An engineered human antibody may allow long-term repeated dose treatment. When patients receive antibodies that are plainly foreign, e.g., monoclonal antibodies of mouse origin, a marked immune response is induced. This response is sufficient to limit the administration of the antibody either to a single dose or to no more than short-term use. The immune response clears the therapeutic antibody from the circulation. If the antibody should be redosed its plasma concentration is greatly reduced and immune complex formation may cause adverse events. Further, antigen binding and the biological activity of the antibody are blocked. To diminish their foreignness to the human immune system, antibodies have been subjected to increasing degrees of "humanization". Murine antibodies have been superceded by chimeric mouse/human antibodies and, more recently, by antibodies whose sequence is virtually indistinguishable from that of a human antibody. Antibodies of this kind possess plasma half-lives that approach those of normal human immunoglobulins. Even so, when given to man, experience suggests that a humanized antibody may still be registered by the recipient; the hypervariable region of the administered antibody is liable to give rise to at least a low-level antiidiotypic response. Only clinical testing can confirm whether a meaningful immune response will be induced in practice.

The remainder of this chapter describes the development of CDP571, an engineered human antibody that neutralizes human TNF, and the use of this antibody in man in single and repeated doses.

2 Generation of Engineered Human Antibody, CDP571

The mouse parent monoclonal antibody (CB0010) was raised against recombinant human TNF (rhTNF) and selected for potency in a TNF cytotoxicity assay using L929 mouse fibroblast cells. Complementarity determining regions, together with framework residues defined by a computer graphic model as able to affect antigen binding activity, were trans-

ferred into the framework of the human antibody Eu with human κ light and $\gamma4$ heavy chain constant regions (STEPHENS et al., 1995). Selection of constant region isotype was based on comparisons of effector functions in mouse disease models using mouse–hamster chimeric versions of the anti-mouse TNF antibody TN3 19.12. In models of sepsis or shock, the inactive (in terms of complement fixation and Fc receptor binding) mouse isotype $\gamma1$ was more effective in reducing mortality than the active mouse isotype $\gamma2a$ (SUITTERS et al., 1994). In models of bowel inflammation and collagen-induced arthritis, both isotypes were equally protective (WARD et al., 1993; 1995). Comparisons of human $\gamma4$ and $\gamma1$ isotypes in a rhTNF induced rabbit pyrexia model confirmed the advantage of the inactive isotype (SUITTERS et al., 1994).

An expression vector was constructed with a glutamine synthetase selectable marker as described previously (STEPHENS et al., 1995) and the plasmid was transferred into NS0 myeloma cells. The engineered antibody retained full binding and biological activity for human TNF and neutralized non-human primate TNF.

3 Preclinical Studies with CDP571 in Non-Human Primates

The pharmacokinetics and immunogenicity of CDP571 were compared with the mouse parent antibody (CB0010) in cynomolgus monkeys. After a single dose of 0.1 mg kg^{-1} of CDP571, the half-life was 66 h vs. 27 h for the murine CB0010. Immunogenicity of the idiotype was reduced compared to CB0010. Responses to the constant regions were absent. The anti-idiotype response was predominantly IgM 14 d after dosing, but had switched to an IgG response by day 35 (STEPHENS et al., 1995). In a second study, repeated doses of 4 and 20 mg kg^{-1} CDP571 were administered at 72 h intervals for 30 d. At 20 mg kg^{-1}, peak and trough antibody levels in 7/8 animals remained constant throughout the dosing period and antibodies to CDP571 were low or undetectable; one animal only in this group mounted an anti-CDP571 response. At 4 mg kg^{-1}, peak and trough antibody levels declined from the 4th dose onwards and this coincided with the development of antibodies to the idiotype and, in some animals, to constant and/or framework regions. These results suggest that the non-responsiveness at 20 mg kg^{-1} may be due to high-dose tolerance (STEPHENS et al., 1997).

4 Studies with CDP571 in Humans

4.1 Safety, Pharmacokinetics and Immunogenicity in Human Volunteers

CDP571 was infused intravenously into 24 healthy male human subjects in an ascending dosage, placebo controlled study to examine the toleration and safety, pharmacokinetics and immunogenicity of single doses of the antibody. All infusions were well tolerated and no unwanted effects were detected.

The antibody was eliminated with a half-life ranging from 5 d at the lowest dose (0.1 mg kg^{-1}) to 14 d at higher doses (up to 10 mg kg^{-1}) (STEPHENS et al., 1995). Antibodies to CDP571 were detectable in subjects receiving lower doses but levels decreased with increasing dosage of CDP571. In contrast to the cynomolgus monkey, in which there was a switch to IgG these antibodies were predominantly IgM and were directed entirely against the idiotype. Circulating CDP571 not only remained detectable throughout the rise and fall in titer of these antibodies, but was still able to bind TNF.

4.2 Single Dose Studies in Patients

CDP571 has been administered as a single intravenous infusion to four kinds of patients.

The first clinical trial of CDP571 was conducted in patients with septic shock. Subsequently, trials were undertaken in patients with Crohn's disease, with ulcerative colitis, and with non-insulin dependent diabetes mellitus (NIDDM). In addition, a study in patients with RA was undertaken in which up to 4 infusions of CDP571 were administered to individual patients over a period of 8–10 months. This study is described separately.

Substantial data implicate TNF as a key mediator in the systemic inflammatory response syndrome. In particular, anti-TNF antibodies have been shown to prevent death in animal models of systemic infection (HINSHAW et al., 1990; TRACEY et al., 1987). A preliminary study of CDP571 was undertaken in patients with sepsis syndrome and shock. A total of 42 patients was entered. Each patient received either 0.1, 0.3, 1.0 or 3 mg kg^{-1} of CDP571 or placebo. CDP571 caused a rapid reduction in circulating TNF concentrations with concomitant decreases in IL-1β and IL-6. With such small groups of patients no effect on mortality could be discerned. No ill effects of CDP571 were observed (DHAINAUT et al., 1995).

TNF has been implicated in the pathogenesis of inflammatory bowel disease (IBD), specifically Crohn's disease and ulcerative colitis. TNF neutralization in several different animal models of bowel inflammation has been shown by ourselves and others to moderate both the clinical picture and the morphological severity of bowel inflammation (WARD et al., 1993; WARREN et al., 1994). Preliminary studies of CDP571 were carried out separately in patients with Crohn's disease (STACK et al., in press) or with ulcerative colitis (EVANS et al., 1996). In each study CDP571 was given as a single infusion at a dosage level of 5 mg kg^{-1}. In the study in Crohn's disease, 20 patients with active disease received CDP571 and a reference group of 10 patients was randomized to receive human albumin solution as placebo. In the CDP571-treated group, the Crohn's Disease Activity Index (CDAI) was reduced from a median of 263 to 167 two weeks after infusion ($p \leq 0.001$), the time point specified in advance as of principal interest. By comparison, in the placebo treated group the CDAI changed only from 253 to 247. After CDP571, 6 patients fulfilled the criterion for disease remission of CDAI <150; and a further 3 patients had a CDAI of ≤156 at 2 weeks. The corresponding numbers of patients in the placebo group were 0 and 1. While the improvements in the treated group were maximal at week 2, some patients showed improvement throughout the 8 weeks of the follow-up period.

In ulcerative colitis, 15 patients with mild/moderate disease, some of whom had previously failed to respond adequately to oral steroid therapy, received CDP571 in open fashion. By week 1 the Powell-Tuck disease assessment score had fallen from a mean of 6.4 to 4.7, accompanied by a reduction in number of fecal movements per day. On sigmoidoscopy, scheduled at week 2, the appearance of the rectal mucosa had improved from a score of 2.3 pretreatment to 1.2. These improvements persisted for several weeks. In each of these studies in patients with IBD, circulating C-reactive protein diminished after CDP571 – a finding consistent with the beneficial clinical effects of the antibody.

Fourthly, the effects of CDP571 were tested in patients with NIDDM (OFEI et al., 1996). The work of HOTAMISLIGIL et al. (1993), in rodent models of obesity/diabetes has indicated that excessive TNF expression in adipose tissue contributes to the animals' resistance to insulin action. In the most telling experiment, peripheral insulin resistance was rapidly reversed *in vivo* by TNF neutralization. The finding that human adipose tissue similarly overexpresses TNF in proportion to degree of adiposity suggested that TNF might play a similar role in the pathogenesis of NIDDM (HOTAMISLIGIL et al., 1995). In a placebo controlled trial, 10 patients with NIDDM were randomized to receive CDP571 5 mg kg^{-1} and 11 patients to receive saline placebo. Circulating CDP571 concentrations capable of neutralizing TNF in a cytotoxicity bioassay were present to at least day 8 after infusion. However, no change in insulin sensitivity, measured using a short insulin sensitivity test at baseline, 1 and 4 weeks, nor any change in glycemic control, could be detected with CDP571. Evidently the antibody, and, by implication, neutralization of TNF by whatever means over a period of 1–2 weeks, does not moderate peripheral insulin resistance in patients with NIDDM.

4.3 Repeated Dose Study in Patients with Rheumatoid Arthritis

The clinical studies described above were all of relatively short duration and involved a single dosing of antibody. Because of the relatively slow clearance from plasma of engineered human antibodies such as CDP571 a relatively long duration of action can be obtained following a single infusion. Nevertheless, to suppress TNF activity long term and so maintain clinical benefit for months or years, repeated dosing will be necessary.

The ability of CDP571 to deliver longer-term benefits was explored by undertaking a repeated dose study in patients with RA (RANKIN et al., 1995). For the first dose of the investigational drug, 36 patients were divided into 3 dosage groups (0.1, 1 and 10 mg kg^{-1}) and studied sequentially. Each group of 12 patients was randomized to receive in double-blinded fashion either human albumin placebo (4 patients) or CDP571 (8 patients). Patients were assessed 1–3 weeks prior to treatment and at 1, 2, 4 and 8 weeks post infusion. Disease activity was measured using European League Against Rheumatism core criteria, which included assessor scores of the number of tender and swollen joints (maximum 28 joints) and the patients' own scores of pain and disease activity. Laboratory measures included ESR and C-reactive protein (CRP) and the supportive or exploratory end points of plasma IL-6, circulating levels of the metalloproteinase stromelysin, and urinary excretion of type 1 and type 2 mature collagen fragments as markers of bone and articular cartilage degradation.

The remainder of the study was conducted in open fashion. 30 of the original 36 patients received a second treatment, either 1 or 10 mg kg^{-1} CDP571. They were assessed as above, except that collagen fragments were not measured. Lastly, in an extension of the study, 16 patients received a third infusion and 14 patients received a fourth infusion of antibody at the same dosage as before. Safety monitoring and limited laboratory data only were collected.

After the first infusion, patients who received placebo did not improve. As described by RANKIN et al. (1995), however, patients who received CDP571 obtained dose-dependent relief of the symptoms and signs of their arthritis, accompanied within several days of dosing by a markedly improved sense of wellbeing. For example, in patients who received 10 mg kg^{-1} CDP571, the reduction in the number of tender joints recorded by the assessor and the reduction in patients' own scoring of pain, using a visual analogue scale, were both statistically significant by comparison with the placebo treated group by 2 weeks ($p = 0.048$ and 0.024 respectively, both adjusted for multiple comparisons). These and other clinical benefits were sustained for between 1 and 2 months (Tab. 1). Symptoms during CDP571 infusion were no more frequent or severe than with placebo. During follow-up fewer adverse events were reported in patients who received CDP571, an observation accounted for largely by the fewer episodes of RA exacerbation recorded after antibody treatment. Improvements in laboratory variables supported these clinical findings. Median plasma CRP levels were reduced by week 1 virtually to within the normal range after 1 or 10 mg kg^{-1} and remained diminished for up to 8 weeks. Reductions in ESR were statistically significant 2 weeks after 10 mg kg^{-1} ($p = 0.015$, adjusted). Plasma IL-6 concentrations fell in parallel. Further, after CDP571 treatment circulating stromelysin levels were lowered and the urinary excretion of fragments of type 1 and type 2 collagen was diminished for up to 8 weeks after dosing.

These last observations encourage the notion that TNF neutralization may modify progressive RA joint destruction, as well as providing immediate and welcome clinical relief. These benefits would only be really worthwhile, however, if they could be maintained with repeated CDP571 dosing. The effects of a second dose of antibody were therefore examined. Patients' arthritis again improved, and CRP and ESR were again reduced (Tab. 2). Inspection of individual patient data suggested that patients who had received 10

mg kg^{-1} on both first and second occasions were especially benefited. The duration of clinical and laboratory improvements appeared similar after the second infusion to those obtained after the first. Following either 1 or 10 mg kg^{-1} CDP571, stromelysin levels were reduced. After 10 mg kg^{-1} the plasma stromelysin concentration was still less at week 8 than the value prior to the second treatment (SOPWITH, 1995). Because of disease progression requiring increased therapy, a total of 8 patients withdrew from the study during follow-up: 6 patients had received 1 mg kg^{-1} and 2 had received 10 mg kg^{-1} CDP571. This trend to dose dependency supports the view that redosing with the antibody was clinically effective.

Further experience of repeated CDP571 dosing was obtained by acceding to requests for an open extension phase of the study. In this final phase a self-selected group of patients received a third and fourth dose of CDP571. Patients were followed after each treatment for 8 weeks to monitor safety. The CDP571 infusions were well tolerated. CRP and ESR both fell with antibody treatment. The repeated administration of CDP571 at 10 mg kg^{-1} was especially effective (Tab. 3). IL-6, too, was measured and its plasma concentration with time mirrored the sustained reductions measured in acute phase proteins.

4.4 Pharmacokinetics and Immunogenicity

The single dose studies described had indicated that the level of antibodies detectable was inversely proportional to the dose of CDP571, and this observation was confirmed in the repeated dosing RA study. Most patients produced detectable antibodies to CDP571 at some point during the course of therapy, but there was variation in the levels and the effect that these antibodies had on the clearance of CDP571. The elimination profiles for all patients receiving each dose in the relevant cycle, regardless of which dose of CDP571 they received in the first cycle, are illustrated in Fig. 1. The overall profile does not appear to change after repeated doses,

Tab. 1. Results in Rheumatoid Arthritis Patients Following First Infusion of CDP571 at 1 mg kg^{-1} or 10 mg kg^{-1} or Placebo

Visit	Placebo					1 mg kg^{-1} Group					10 mg kg^{-1} Group				
	n	Tender Joints	Swollen Joints	Pain [cm]	CRP [mg L^{-1}]	n	Tender Joints	Swollen Joints	Pain [cm]	CRP [mg L^{-1}]	n	Tender Joints	Swollen Joints	Pain [cm]	CRP [mg L^{-1}]
Pre-Infusion	12	12.5 (7–28)	17.5 (7–24)	6.2 (3.2–8.0)	80.0 (6–142)	8	16.0 (3–28)	15.0 (6–22)	5.5 (4.1–8.3)	37.0 (13–116)	8	16.5 (10–28)	16.5 (8–25)	8.4 (2.9–9.9)	50.5 (3–118)
Week 1	12	13.0 (7–28)	17.0 (5–25)	5.7 (1.7–7.8)	68.5 (4–135)	8	17.0 (0–24)	16.5 (7–20)	4.2 (1.6–6.3)	10.5 (2–61)	8	11.0 (3–28)	11.5 (7–25)	4.3 (0.6–9.7)	11.0 (2–22)
Week 2	12	14.5 (9–28)	16.0 (7–25)	7.9 (0.8–9.5)	59.0 (2–104)	8	14.0 (0–27)	16.0 (10–23)	5.0 (2.0–8.3)	12.5 (3–62)	8	11.0 (3–28)	11.5 (10–25)	3.6 (0.9–9.9)	12.5 (1–30)
Week 4	11	16.0 (4–28)	20.0 (0–26)	5.5 (1.4–9.1)	55.0 (3–118)	7	15.0 (1–21)	17.0 (2–24)	7.8 (1.4–8.6)	41.0 (4–128)	8	10.0 (2–28)	14.0 (6–25)	5.7 (0.2–9.9)	19.0 (1–30)
Week 8	11	20.0 (2–28)	19.0 (2–25)	8.5 (0–9.9)	31.0 (2–115)	7	13.0 (4–26)	17.0 (6–23)	8.3 (6.6–9.7)	37.0 (11–152)	8	12.0 (2–28)	14.5 (8–24)	7.2 (0.6–9.5)	27.5 (1–47)

Tab. 2. Results in Rheumatoid Arthritis Patiens Following Second Infusion of CDP571 at 1 mg kg^{-1} or 10 mg kg^{-1}

Visit	1 mg kg^{-1} Group					10 mg kg^{-1} Group				
	n	Tender Joints	Swollen Joints	Pain [cm]	CRP [mg L^{-1}]	n	Tender Joints	Swollen Joints	Pain [cm]	CRP [mg L^{-1}]
Pre-infusion	17	13.0 (2–28)	18.0 (6–25)	8.3 (0.8–9.8)	48.0 (3–151)	12	20.0 (4–28)	19.5 (3–24)	7.6 (0.5–9.7)	40.0 (2–120)
Week 1	16	9.5 (0–28)	17.5 (2–25)	4.6 (0.3–9.7)	24.0 (2–169)	12	7.0 (1–28)	14.0 (2–22)	3.2 (0.0–7.5)	5.0 (1–125)
Week 2	15	7.0 (1–28)	16.0 (2–27)	4.1 (0.3–9.8)	20.0 (5–160)	12	6.5 (2–28)	13.5 (5–24)	2.2 (0.2–9.1)	11.5 (1–107)
Week 4	13	11.0 (0–28)	18.0 (4–25)	4.0 (0.3–9.6)	21.0 (1–125)	12	4.0 (0–28)	13.0 (3–23)	1.7 (0.4–9.8)	13.5 (1–192)
Week 8	11	8.0 (1–28)	17.0 (4–26)	5.6 (0–9.9)	34.5 (3–135)	12	11.0 (0–28)	19.5 (0–25)	7.2 (0.2–9.9)	44.5 (2–111)

Tab. 3. C-Reactive Protein Results in Rheumatoid Arthritis Patients Following Third and Fourth Infusions of CDP571 at 1 mg kg^{-1} and 10 mg kg^{-1}

Visit	1 mg kg^{-1}		10 mg kg^{-1}	
	n	CRP [mg L^{-1}]	n	CRP [mg L^{-1}]
Pre-3rd Infusion	7	77.0 (9–116)	9	51.0 (1–120)
Week 2	7	38.0 (3–132)	8	21.0 (3–74)
Week 4	7	43.0 (3–109)	9	7.0 (1–134)
Week 8	7	40.0 (3–120)	8	21.5 (3–104)
Pre-4th Infusion	7	45.0 (3–109)	7	12.0 (1–80)
Week 2	7	25.0 (3–184)	7	3.0 (1–29)
Week 4	6	49.5 (16–86)	7	7.0 (1–32)
Week 8	7	47.0 (3–104)	7	10.0 (3–45)

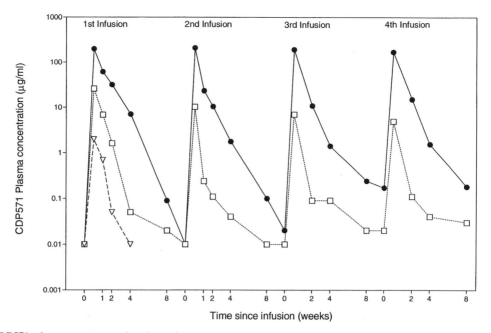

Fig. 1. CDP571 plasma concentration in patients with rheumatoid arthritis following up to 4 infusions at the following doses: $--\nabla--$ 0.1 mg kg^{-1}, $\cdots\square\cdots$ 1.0 mg kg^{-1}, $-\bullet-$ 10.0 mg kg^{-1}.

with half lives calculated as approximately 6 d. However, by examining individual patient data additional informative observations can be made. For patients completing 3 or 4 cycles at 10 mg kg^{-1}, the majority showed consistent or prolonged elimination profiles: CDP571 was still detectable 8 weeks after the 4th infusion. 1 patient received 10 mg kg^{-1} in the first cycle followed by 3 repeated doses at 1.0 mg kg^{-1} and again showed constant elimination rates. The class of anti-CDP571 produced in the 10 mg kg^{-1} patients was IgM and little or no switch to IgG was detected. In contrast, patients receiving the lowest dose of 0.1 mg kg^{-1} in the first cycle generally showed accelerated CDP571 clearance in subsequent cycles. These patients tended to class-switch to IgG anti-CDP571 production by 8 weeks following their first dose. Subsequent doses of CDP571 boosted specific IgG production, and the clearance of CDP571 was accelerated. Patients receiving 1 mg kg^{-1} in the first cycle were variable in their pharmacokinetic profiles, with similar numbers showing constant elimination rates or accelerated clea-

rance. This suggests that at the higher doses a state of tolerance can be achieved at which anti-CDP571 antibodies become undetectable and the half-life is prolonged.

In some of the RA patients, cross-reacting antibodies to CDP571 were detected at entry into the study, presumably due to the presence of rheumatoid factor. This had not been the case in volunteer subjects or in patients with IBD or sepsis. Epitope mapping of these cross-reacting antibodies indicated that they were directed both against γ4 constant regions and against the CDP571 idiotype. However, their presence was not predictive of subsequent development of a specific immune response. In those patients who did make a specific response, the antibodies were directed against the idiotype of CDP571, and there was no increase in antibodies to the γ4 constant regions. Finally, there was no evidence of antibodies directed against the carbohydrate moiety.

5 Conclusions

These studies with the engineered human antibody CDP571 have shown that neutralizing TNF can be beneficial in patients with several different forms of inflammatory disease. That the antibody can be given repeatedly with useful effect and with safety, and with the induction of only limited immune responses, shows that antibodies of this kind may prove useful and practicable for treating patients with chronic disorders for which current medication is of only modest value.

Acknowledgement

The authors wish to thank all the investigators who took part in the clinical studies, and OLIVIA VETTERLEIN for her contribution to the assays of antibody and immune response.

6 References

BRENNAN, F. M., CHANTRY, D., JACKSON, A., MAINI, R., FELDMANN, M. (1989), Inhibitory effect of TNFα antibodies on synovial cell interleukin-1 production in rheumatoid arthritis, Lancet II, 244–247.

CALANDRA, T., BAUMGARTNER, J.-D., GRAU, G. E., WU, M.-M., LAMBERT, P.-H., SCHELLEKENS, J., VERHOEF, J., GLAUSER, M. P., and the Swiss-Dutch J5 Immunoglobulin Study Group (1990), Prognostic values of tumor necrosis factor/cachectin, interleukin-1, interferon-α, and interferon-γ in the serum of patients with septic shock, J. Infect. Dis. 161, 982–987.

CHU, C. Q., FIELD, M., FELDMANN, M., MAINI, R. N. (1991), Localization of tumor necrosis factor α in synovial tissues and the cartilage pannus junction in patients with rheumatoid arthritis, Arthritis Rheum. 34, 1125–1132.

DHAINAUT, J.-F. A., VINCENT, J.-L., RICHARD, C., LEJEUNE, P., MARTIN, C., FIEROBE, L., STEPHENS, S., NEY, U. M., SOPWITH, M., and the CDP571 Sepsis Study Group (1995), CDP571, a humanized antibody to human tumor necrosis factor-α: safety, pharmacokinetics, immune response, and influence of the antibody on cytokine concentrations in patients with septic shock, Crit. Care Med. 23, 1461–1469.

ELLIOTT, M. J., MAINI, R. N., FELDMANN, M., KALDEN, J. R., ANTONI, C., SMOLEN, J. S., LEEB, B., BREEDVELD, F. C., MACFARLANE, J. D., BIJL, H., WOODY, J. N. (1994), Randomizeed double-blind comparison of chimeric monoclonal antibody to tumor necrosis factor α (cA2) versus placebo in rheumatoid arthritis, Lancet 344, 1105–1110.

EVANS, R. C., CLARK, L., HEATH, P., RHODES, J. M. (1996), Treatment of ulcerative colitis with an engineered human anti-TNFα antibody, CDP571, Gastroenterology 110, A-15.

HENDERSON, B., FOULKES, R., BLAKE, S., LEWTHWAITE, J., BROWN, D., ANDREW, D., STEPHENS, S. (1992), TNF and related cytokines in lapine antigen-induced arthritis, Eur. Cytokine Netw. 261, 1992.

HINSHAW, L. B., TEKAMP-OLSON, P., CHANG, A. C. K., LEE, P. A., TAYLOR, F. B. JR., MURRAY, C. K., PEER, G. T., EMERSON, T. E. JR., PASSEY, R. B., KUO, G. C. (1990), Survival of primates in LD100 septic shock following therapy with antibody to tumor necrosis factor (TNFα), Circ. Shock 30, 279–292.

HOTAMISLIGIL, G.S., SHARGILL, N. S., SPIEGELMAN, B. M. (1993), Adipose expression of tumor necrosis factor-α: direct role in obesity-linked insulin resistance, Science 259, 87–91.

HOTAMISLIGIL, G. S., ARNER, P., CARO, J. F., ATKINSON, R. L., SPEIGELMAN, B. M. (1995), Increased adipose tissue expression of tumor necrosis factor-α in human obesity and insulin resistance, J. Clin. Invest. 95, 2409–2415.

MAINI, R. N., BRENNAN, F. M., WILLIAMS, R., CHU, C. Q., COPE, A. P., GIBBONS, D., ELLIOTT, M., FELDMANN, M. (1993), TNF-α in rheumatoid arthritis and prospects of anti-TNF therapy, Clin. Exp. Rheumatol. 11 (Suppl. 8), S173–175.

OFEI, F., HUREL, S., NEWKIRK, J., SOPWITH, M., TAYLOR, R. (1996), Effects of an engineered human anti-TNFα antibody (CDP571) on insulin sensitivity and glycemic control in patients with NIDDM, Diabetes 45, 881–885.

PFEFFER, K., MATSUYAMA, T., KUNDIG, T. M., WAKEHAM, A., KISHIHARA, K., AHAHINIAN, A., WIEGMANN, K., OHASHI, P. S., KRONKE, M., MAK, T. W. (1993), Mice deficient for the 55kD tumor necrosis factor receptor are resistant to endotoxic shock, yet succumb to L. monocytogenes infection, Cell 73, 457–467.

PIQUET, P. F., GRAU, G. E., VESIN, C., LOETSCHER, H., GENTZ, R., LESSLAUER, W. (1992), Evolution of collagen arthritis in mice is arrested by treatment with anti-tumor necrosis factor (TNF) antibody or a recombinant soluble TNF receptor, *Immunology* **77**, 510–514.

RANKIN, E. C. C., CHOY, E. H. S., KASSIMOS, D., KINGSLEY, G. H., SOPWITH, A. M., ISENBERG, D. A., PANAYI, G. S. (1995), The therapeutic effects of an engineered human anti-tumor necrosis factor alpha antibody (CDP571) in rheumatoid arthritis, *Br. J. Rheumatol.* **34**, 334–342.

SOPWITH, A. M. (1995), Engineered human anti-tumor necrosis factor α antibody in rheumatoid arthritis: prospects for chronic therapy, in: *Proc. Early Decisions in DMARD Development IV* (STRAND, V., Ed.), pp. 99–109. Atlanta, GA: Arthritis Foundation.

STACK, W., MANN, S. D., ROY, A. J., HEATH, P., SOPWITH, M., FREEMAN, J., HOLMES, G., LONG, R., FORBES, A., KAMM, M. A., HAWKEY, C. J. (in press), Randomized controlled trial of CDP571 antibody to tumor necrosis factor in Crohn's disease, *Lancet*.

STEPHENS, S., EMTAGE, S., VETTERLEIN, O., CHAPLIN, L., BEBBINGTON, C., NESBITT, A., SOPWITH, M., ATHWAL, D., NOVAK, C., BODMER, M. (1995), Comprehensive pharmacokinetics of a humanized antibody and analysis of residual anti-idiotypic response, *Immunology* **85**, 668–674.

STEPHENS, S., VETTERLEIN, O., SOPWITH, M. (1997), CDP571, an engineered antibody to human tumor necrosis factor, in: *Antibody Therapeutics* (HARRIS, W. J., ADAIR, J. R., Eds), pp. 317–340. Boca Raton, FL: CRC Press.

SUITTERS, A. J., FOULKES, R., OPAL, S. M., PALARDY, J. E., EMTAGE, J. S., ROLFE, M., STEPHENS, S., MORGAN, A., HOLT, A. R., CHAPLIN, L. C., SHAW, N. E., NESBITT, A. M., BODMER, M. W. (1994), Differential effect of isotype on efficacy of anti-tumor necrosis factor α chimeric antibodies in experimental septic shock, *J. Exp. Med.* **179**, 849–856.

TRACEY, K. J., FONG, Y., HESSE. D. G., MANOGUE, K. R., LEE, A. T., KUO, G. C., LOWRY, S. F., CERAMI, A. (1987), Anti-cachectin/TNF monoclonal antibodies prevent septic shock during lethal bacteremia, *Nature* **330**, 662–664.

WARD, P. S., WOODGER, S. R., BODMER, M., FOULKES, R. (1993), Anti-tumor necrosis factor α monoclonal antibodies (anti-TNF MAb) are therapeutically effective in animal model of colonic inflammation, *Br. J. Pharmacol.* **110**, 77P.

WARD, P. S., BODEN, T., WOODGER, R., FOULKES, R. (1995), Isotype variation of TNF monoclonal antibodies. Is there a therapeutic difference in rheumatoid arthritis?, *Br. J. Rheumatol.* **34** (Suppl.), 294.

WARREN, B. F., WATKINS, P. E., FOULKES, R., WARD, P., STEPHENS, S., BODMER, M. (1994), Anti-TNF alpha treatment of a model of human ulcerative colitis, *Gut* **35** (Suppl. 2), 1994.

WILLIAMS, R. O., FELDMANN, M., MAINI, R. N. (1992), Anti-tumor necrosis factor ameliorates joint disease in murine collagen-induced arthritis, *Proc. Natl. Acad. Sci. USA* **89**, 9784–9788.

15 Antibody Targeted Chemotherapy

MARK S. BERGER

Radnor, PA, USA

PHILIP R. HAMANN

Pearl River, NY, USA

MARK SOPWITH

Slough, UK

1 Introduction   356
2 Calicheamicin   356
3 CMA-676   358
  3.1 Rationale for CMA-676 Development   358
  3.2 Patient Enrollment   359
  3.3 CMA-676 Safety Data   359
  3.4 CMA-676 Efficacy Data   361
4 CMB-401   361
  4.1 Imaging Studies   361
5 Conclusions   363
6 References   363

List of Abbreviations

AML   acute myeloid leukemia, acute myelogenous leukemia

BMT   bone marrow transplantation

CMA-676   immunoconjugate consisting of hP67.6, linker with hydrolytic release site, and calicheamicin

CMB-401   immunoconjugate consisting of hCTMO1, linker without hydrolytic release site, and calicheamicin

CNS   central nervous system

CR   complete remission

CTM01   anti-polyepithelial mucin antibody

hCTM01   humanized anti-polyepithelial mucin antibody

hP67.6   humanized anti-CD33 antibody

PEM   polymorphic epithelial mucin

1 Introduction

Wyeth-Ayerst Research and Celltech Therapeutics are collaborators in the development of two immunoconjugates directed against proteins expressed on hematologic and solid tumors. The rationale for development is based on the selective targeting of a potent cytotoxic drug to tumor cells in order to increase the specificity of therapy and reduce the level of unwanted side effects. The immunoconjugates consist of an antibody component directed against a tumor marker, linked to the cytotoxic agent calicheamicin. The immunoconjugates in development are CMA-676 for the treatment of acute myeloid leukemia (AML) and CMB-401 for epithelial tumors including ovarian and non-small cell lung carcinoma.

Both antibody components are engineered human IgG$_4$ antibodies, one (hP67.6) directed against the CD33 antigen on leukemic cells and the second (hCTMO1) against polymorphic epithelial mucin (PEM) which is overexpressed on many epithelial cell tumors.

This case study presents information on calicheamicin and then discusses the present status of preclinical and clinical studies with the antibody targeted chemotherapy agents CMA-676 and CMB-401.

2 Calicheamicin

Calicheamin γ_1^I (Fig. 1) is a member of a family of very potent "enediyne" antitumor antibiotics (LEE et al., 1987a, b). This unusual molecule contains 4 sugar residues, 3 of which were seen for the first time in the calicheamicins, a hexasubstituted aromatic ring containing an iodine, a methyl trisulfide, and a structural unit that was truly unique at the time of its elucidation, a bicyclic enediyne. This assemblage binds in a sequence-specific manner to the minor groove of DNA (ZEIN et al., 1989) by virtue of the carbohydrate portion which is predisposed to a helix complementary to that of DNA (WALKER et al., 1994).

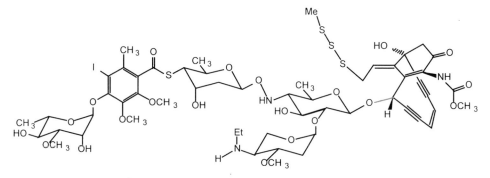

Fig. 1. Calicheamicin γ_1^I.

When the sulfur–sulfur bonds, referred to as the "trigger", are reduced, the "warhead" undergoes rearrangement, ultimately leading to a very reactive, short-lived diradical species (**4**, Fig. 2). The intermediate thiol is undetectable at low temperature, and compound **3** has a calculated half-life under physiological conditions of about 4 s (DE VOSS et al., 1990). Deuterium labeling experiments have shown that the observed double-stranded DNA breaks are caused by the abstraction of specific hydrogen atoms from the deoxyribose rings of DNA by diradical calicheamicin (Fig. 2, **4**) (HANGELAND et al., 1992). This damage to the DNA ultimately leads to cell death, proposed to be through apoptosis (NICOLAOU et al., 1994) or through the exhaustion of NAD$^+$ levels as a result of poly-ADP polymerase activity (ZHAO et al., 1990).

The methyl trisulfide of calicheamicin γ_1^I or any of the related derivatives can be reacted with thiols to release CH_3SSH and form a disulfide as shown in structures **5** and **6** (Fig. 3) (ELLESTAD et al., 1989). Disulfide formation can be used to introduce a new functional group as part of "R" in structure **6** that allows for the attachment of the calicheamicins to antibodies. This method of functionalization leaves the warhead and the calicheamicin sugars intact yet retains an acceptable "trigger" in the form of a disulfide.

The bioreduction/triggering of the calicheamicins is also presumed to be due to a reaction with free thiols. The most prevalent free thiol in higher organisms is the reduced form of glutathione (L-γ-glutamyl-L-cysteinyl-glycine, GSH). Lower amounts of L-γ-glutamyl-L-cysteine and cysteine are also present in biological fluids as free thiols. The total concentration of these thiols is about 15–20 μM in blood and in a low mM range intracellularly (MAMSOOR et al., 1992; MISTRY and HARRAP, 1991). The predominant reaction of the calicheamicins with these biological thiols is also disulfide formation as shown in Fig. 3 (MYERS et al., 1994). The triggering shown in Fig. 2 also occurs as a result of reaction with thiols, but at a significantly slower rate.

Glutathione disulfides can also be made when GSH reacts with a disulfide calicheamicin derivative leading to disulfide exchange. The glutathione disulfide of γ_1^I calicheamicin has been made and compared biologically to γ_1^I calicheamicin itself. The two are equally effective at cutting DNA, while the glutathione disulfide is somewhat less toxic to various cancer cell lines in culture and slightly more toxic to mice when administered by intraperitoneal injection (G. ELLESTAD, S. CARVAJA, C. BEYER, personal communication). Although no feasible method has been developed for the detection of the glutathione

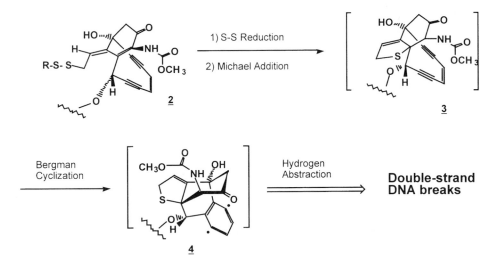

Fig. 2. Triggering of the calicheamicin warhead.

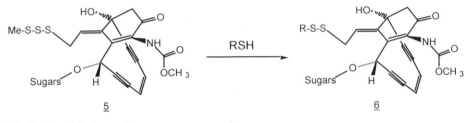

Fig. 3. Disulfide formation.

disulfides of the calicheamicin derivatives in biological systems due to the extremely low concentrations, the activity seen for the glutathione disulfide of γ_1^I calicheamicin is consistent with the intracellular involvement of this class of compounds.

Conjugates of calicheamicin and monoclonal antibodies have been produced by attaching hydrazide derivatives to the oxidized carbohydrate units (RODWELL and MCKEARN, 1987) or by acylating the lysines of the antibodies. Therefore, conjugates have been studied which either do or do not have a site of hydrolytic release. Two other variables that have been studied are the stability of the disulfide linkage and the specific calicheamicin derivative used to make the conjugate. The following generalities can be made based on many years of work with different antibodies:

- Conjugates of N-acetyl calicheamicin are slightly less potent but have better therapeutic indices in experimental models (HINMAN et al., 1993).
- Conjugates with stabilized disulfides also have better therapeutic indices (HINMAN et al., 1995).
- Whether a site of hydrolytic release is an advantage or not depends on the antibody in use.

For an anti-polyepithelial mucin antibody, CTM01, hydrolytic release was found not to be an advantage (HAMANN et al., unpublished results). However, for the anti-CD33 antibody P67.6, hydrolytic release was found to be essential for good biological activity (HAMANN et al., unpublished results).

The conjugate that was taken forward to clinical trials in leukemia, therefore, contains

calicheamicin N-acetyl γ_1^I, a disulfide stabilized by two adjacent methyl substituents, and a site of hydrolytic release. This conjugate, which is curative of HL-60 xenograft tumors in nude mice, is designated CMA-676 (CMA-676) while the conjugate that has been taken forward to clinical trials in solid tumors, CMB-401 (CMB-401) does not contain a site of hydrolytic release.

3 CMA-676

3.1 Rationale for CMA-676 Development

The rationale for the development of CMA-676 and CMB-401 involves selective targeting of potent cytotoxic drugs to tumor cells. In the case of CMA-676, an engineered human anti-CD33 antibody is utilized to expose leukemic cells to active calicheamicin, sparing cells that do not express CD33. CD33 is only produced by hematopoietic cells, so toxicity is largely limited to blood forming cells. Most patients with AML are potential candidates for CMA-676 treatment, as approximately 80% of patients with AML have leukemia cells that express CD33. CD33 is internalized on binding to anti-CD33 antibody, and thus CMA-676 is brought into the cell where active calicheamicin is released. Initial biodistribution studies utilized a murine anti-CD33 antibody labeled with ^{131}I to demonstrate that the antibody targeted the radioactive label to the bone marrow (APPELBAUM et al., 1992). Subsequently an engineered human anti-CD33 antibody was produced and linked to calicheamicin to produce CMA-676, which can be used in studies that involve repeated administration of this immunoconjugate.

3.2 Patient Enrollment

CMA-676 has been administered in one open label, ascending dose, safety and tolerance study in the USA (SIEVERS et al., 1997a, b). This study enrolled patients with CD33-positive acute myelogenous leukemia (AML) who were in relapse after standard therapy, refractory to standard therapy, or in relapse following bone marrow transplantation (BMT). Each patient received a maximum of 3 doses at least 14 d apart during a course of CMA-676, and there were at least 3 patients at each dosage level. The data presented here are from the 40 patients enrolled between April 1995 and May 1997 who received CMA-

676 at doses of 0.25, 0.5, 1, 2, 4, 5, 6 and 9 mg protein per m^2 (one patient received a second course of CMA-676 and thus, although 41 courses of CMA-676 have been given, 40 individual patients have received this drug). A maximum tolerated dose of CMA-676 was not reached. Tab. 1 summarizes the characteristics of patients enrolling in this study.

3.3 CMA-676 Safety Data

The most common non-hematologic adverse event seen in this study has been a post-infusion syndrome of fever and chills, occasionally accompanied by rigors, seen in 32 of the 40 patients. This syndrome began during the 2 h CMA-676 infusion and, except in 4 cases, ended within 6 h of the start of the CMA-676 infusion. The most frequently reported non-hematologic study drug-related adverse events are shown in Tab. 2.

Tab. 1. Enrollment in CMA-676 Ascending Dose Study

	CMA-676 Dose [mg m^{-2}]							
Dosage group	0.25	0.50	1	2	4	5	6	9
No. patients	4	3	4	3	6	6	8	7
Age range, years	24–48	42–65	26–70	31–70	28–61	38–73	23–57	27–69
Patients receiving 3 cycles (no.)	1	1	1	2	1	3[a]	1	2

[a] Dose adjustment to 4 mg m^{-2} on subsequent doses for one patient.

Tab. 2. Most Frequently Reported Drug-Related Adverse Events (No. Patients)

	CMA-676 Dose [mg m^{-2}]							
	0.25	0.5	1	2	4	5	6	9
Event	($n=4$)	($n=3$)	($n=4$)	($n=3$)	($n=6$)	($n=6$)	($n=7$)	
Grade I[a] fever/chills	2	3	1	1	5	–	–	–
Grade II fever/chills	–	–	2	2	1	4	3	6
Grade III[b] fever/chills	–	–	–	–	–	–	–	1
Grade IV fever/chills	–	–	–	–	–	–	1	–
Grade III liver enzyme elevation (SGOT, SGPT)	–	–	–	–	–	1	2	1
Grade II hyperbilirubinemia	–	–	–	–	1	1	–	–
Grade II nausea	–	–	–	–	–	–	1	1

[a] Grades are defined according to modified WHO toxicity grading scale.
[b] Modified to include fever $<40°C$ with asymptomatic hypotension.

In the group which received 6 mg m^{-2}, one patient already being treated with low doses of dopamine for asymptomatic hypotension developed fever to 40.5 °C, chills, and symptomatic hypotension within 3 h after the infusion. This patient responded to fluids and pressor administration. The investigator judged this episode of worsening hypotension/fever to be probably related to CMA-676 infusion, but whether the patient may have been predisposed to this reaction is unclear. The patient was removed from the study as a result of the adverse event. In the 9 mg m^{-2} dosage group one patient developed fever to a maximum of 39.4 °C with asymptomatic hypotension occurring approximately 5 h after starting the first cycle of CMA-676. The second cycle of CMA-676 administration in this patient was complicated by apparent sepsis starting just prior to CMA-676 administration. Symptomatic hypotension occurred approximately 10 h after the start of CMA-676 administration, but the contribution of CMA-676 to the hypotensive episode is difficult to determine.

Hematologic adverse events have been difficult to discern in the patients who did not achieve remissions because of the hematologic effects of acute leukemia itself. However, there appeared to be a relationship between CMA-676 administration and the development of thrombocytopenia in a patient who received CMA-676 at 4 mg m^{-2} and then experienced a significant loss of blood from a nose bleed. This patient also developed thrombocytopenia without nose bleed approximately 1 week after receiving a second cycle of CMA-676. The patients who have achieved clinical remissions (as defined in Tab. 3) have shown clear evidence of granulocytopenia and thrombocytopenia related to CMA-676 administration. In general, responding patients have recovered from grade IV granulocytopenia by approximately 14–21 d after the last dose of CMA-676. However, one patient who received 3 cycles of CMA-676 at 9 mg m^{-2} experienced grade IV granulocytopenia for 6 weeks after the third cycle was administered. An additional patient who received 2 cycles of CMA-676 at 9 mg m^{-2} developed blood culture proven sepsis while granulocytopenia after the second cycle, and experienced grade IV granulocytopenia and thrombocytopenia after that cycle for a total of 7 weeks prior to expiring from infection. In the later patient, the episode of sepsis may have contributed to the length of the bone marrow suppression.

One patient who had a complete remission for 4 months after receiving 3 cycles of CMA-676 (1 mg m^{-2}) was eligible for a second course of CMA-676 after bone marrow relapse occurred. This patient received a second course of CMA-676 in the 6 mg m^{-2} dosage group. He was known to have a cardiomyopathy from previous anthracycline. Approximately 5 min after the second cycle of CMA-676 was administered this patient experienced mild shortness of breath and chest tightness lasting no more than 10 min. Treatment consisted of the administration of oxygen for a short period. Immune response studies docu-

Tab. 3. Preliminary Results of Phase I Study of CMA-676 in AML

	CMA-676 Dose [mg m^{-2}]							
Dosage group	0.25	0.50	1	2	4	5	6	9
No. patients	4	3	4	3	6	6	8	7
Complete remissions[a] (no.)	0	0	1	0	1	0	0	0
Morphologic remissions[b] (no.)	0	0	0	0	0	1	2	4

[a] Complete remission is defined as (1) leukemic blasts absent from the peripheral blood; (2) percentage of blasts in the bone marrow $\leq 5\%$; (3) normocellular bone marrow; (4) peripheral counts of ≥ 9 g dL^{-1} hemoglobin, $\geq 100,000$ platelets μL^{-1}, ANC $\geq 1,500$ μL^{-1}; and (5) red cell and platelet transfusion independence.

[b] Morphologic remission is defined the same way as complete remission except that peripheral blood counts have not fully recovered.

mented a significant rise in titers of antibody to the calicheamicin–linker complex. Thus this patient may have had a respiratory syndrome associated with immune reaction to a portion of CMA-676.

3.4 CMA-676 Efficacy Data

Preliminary data from 40 patients enrolled between April 1995 and May 1997 are presented in Tab. 3.

In the phase I study two of the 33 patients receiving CMA-676 (at 1 and 4 mg m^{-2}) experienced objective complete remissions (CRs) (SIEVERS et al., 1997a, b). Of the two CRs, one (at 1 mg m^{-2}) was maintained for 140 d, until chloroma of the hip occurred. When this patient developed bone marrow relapse, he received a second course of CMA-676 (6 mg m^{-2}), and after one cycle had no leukemic blasts detectable in the peripheral blood or bone marrow. After the second cycle a significant titer of antibodies to calicheamicin was detected, and no further CMA-676 was administered. The other patient who experienced a CR (at 4 mg m^{-2}) remained in remission for 214 d. The patient subsequently relapsed with a testicular chloroma and CNS leukemic involvement and received local therapy.

7 patients (1 at 5 mg m^{-2}, 2 at 6 mg m^{-2}, and 4 at 9 mg m^{-2}) achieved morphologic remission in the bone marrow. Morphologic remissions were up to approximately 10 weeks in duration after the last dose of CMA-676. These data include the one patient with CR for 4 months who achieved morphologic remission when CMA-676 was given again at relapse (see above). (The definitions for CR and morphologic remission are shown in Tab. 3.)

4 CMB-401

4.1 Imaging Studies

By comparison with anti-CD33 antibody directed therapy of acute myeloid leukemia, the use of antibodies to target calicheamicin to solid tumor is a more demanding concept. Access of antibody to such tumors is likely to be limited by poor vascular supply, and by factors intrinsic to the solid nature and structure of these tumors that will curtail penetration by circulating antibody.

Further, many of the potentially attractive antigen targets expressed by solid tumors are not only present on their cell surfaces but are shed. The elevated local and plasma concentrations of free antigen that result may reduce loading of antibody onto tumor cells. Polymorphic epithelial mucin, or PEM (the product of the *muc-1* gene), the internalizable glycoprotein target of CTM01, is shed in this way. Finally, PEM, although overexpressed by many tumors of secretory epithelial origin, is not a tumor-specific antigen. It is also expressed by many normal secretory cells; e.g., by the breast, gut, pancreas and kidney, although on these normal cells the expression of PEM is confined to their apical surface and thus may be segregated from circulating anti-PEM antibody.

Because of uncertainties as to the differential access of CTM01 to tumor and to normal tissues, it was considered important before initiating clinical studies of the hCTM01–calicheamicin conjugate, CMB-401, to examine the ability of hCTM01 to target to tumor in patients.

Two biodistribution studies of hCTM01 have been undertaken, one in patients with epithelial ovarian cancer (VAN HOF et al., 1996) and one in patients with lung cancer.

For both clinical biodistribution studies hCTM01 was labeled with ^{111}In, a γ emitter, via the covalently linked macrocycle 9N3. This radiolabel was selected as a tracer likely to model realistically the accumulation within cells of internalized CMB-401 in a future study. The distribution of ^{111}In was assessed by external γ camera counting. The resulting body images were inspected to obtain an indication of hCTM01 distribution and, separately, the external counts were quantitated over regions of interest such as vital organs. In patients with ovarian cancer in whom planned surgery was performed 6 d after the i.v. infusion of labeled antibody, samples of tumor and of selected normal tissues were taken and were directly counted for radioactivity to obtain more exact information on antibody uptake.

In the ovarian cancer biodistribution study, dose levels of 0.1 and 1 mg kg^{-1} labeled hCTM01 were initially examined. Subsequently a schedule in which the infusion of 0.1 mg kg^{-1} ^{111}In labeled hCTM01 was preceded by the infusion of 1 mg kg^{-1} unlabeled antibody was studied. External γ counting confirmed preferential uptake of hCTM01 by tumor, which was imaged in all but 3 out of the 28 patients in whom epithelial ovarian cancer was confirmed at surgery. Increased uptake versus background was also noted in the liver, spleen, bone marrow and breasts. Liver uptake occurred rapidly, being prominent by 1 h after antibody infusion. Several mechanisms may have contributed to liver uptake of antibody. These mechanisms include normal IgG clearance pathways and the clearance by the liver of complexes formed between infused antibody and precirculating PEM antigen; such complexes indeed were observed. With the intention of reducing liver uptake of labeled antibody, and with a view to trials of CMB-401, the effect of a preinfusion of 1 mg kg^{-1} unlabeled antibody on the biodistribution of ^{111}In labeled antibody was investigated.

Direct tissue counting revealed the uptakes of ^{111}In shown in Tab. 4.

For all 3 dosage groups the differences between tumor uptake and uptake by normal tissues, and between tumor uptake and blood content, were statistically significant. Thus

these direct quantitations confirmed preferential uptake of hCTM01 by epithelial ovarian tumor. The apparent reductions in liver uptake of labeled antibody with increasing protein dose level and with unlabeled antibody predosing were supported by similar trends in liver uptake data in patients in whom frank malignancy was not confirmed at surgery or in whom a liver biopsy had not been performed, but quantitation of the external γ counts in the region of interest of the liver was undertaken.

The majority of patients taking part in the lung cancer biodistribution study had non-small cell lung cancer. ^{111}In labeled hCTM01 was given to an initial group of patients at a dose level of 1 mg kg^{-1}; following this, the biodistribution of labeled hCTM01 given in a dose of 0.1 mg kg^{-1} after unlabeled antibody predosing was examined. Overall, tumor previously identified by computerized X-ray tomography were imaged unambiguously in 12 patients out of 17 with non-small cell lung cancer. Uptake by normal organs, as regions of interest, such as the liver, were comparable with the outcome of the study in women with ovarian cancer.

As a result of these studies of the biodistribution in patients of hCTM01, clinical studies of the safety and tolerance of CMB-401 have now been initiated in women with epithelial ovarian cancer and in patients with non-small cell lung cancer. The dosage schedules being used in the studies of CMB-401 are based on the hCTM01 biodistribution results. Thus, in the study ovarian cancer women are receiving

Tab. 4. Tissue Uptakes of ^{111}In Labeled hCTM01 (Mean % Injected Dose per kg) in Patients with Ovarian Cancer

Tissue	^{111}In hCTM01 Dosage Groups		
	0.1 mg kg^{-1}	1 mg kg^{-1}	0.1 mg kg^{-1} and Antibody Predose
Tumor	4.66 ($n=10$)	8.11 ($n=10$)	8.89 ($n=7$)
Normal tissues[a]	0.66 ($n=10$)	0.66 ($n=10$)	1.03 ($n=7$)
Blood	2.20 ($n=10$)	1.12 ($n=10$)	1.73 ($n=7$)
Liver	21.60 ($n=7$)	15.55 ($n=4$)	10.86 ($n=3$)

[a] Mean of uptakes by skin, adipose tissue, skeletal muscle, normal peritoneum.

CMB-401 with hCTM01 preinfusion to limit liver uptake of the active conjugate.

5 Conclusions

Clinical studies with CMA-676 have been scientifically significant as they have demonstrated proof-of-principle for the use of calicheamicin-based immunoconjugates in cancer therapy. Further studies with CMA-676 will determine what role this drug may play when added to the available treatments for acute myeloid leukemia. Clinical studies with CMB-401 are underway to determine whether the same approach can be effective when used for solid tumors.

Acknowledgements

We would like to acknowledge the work of the staffs at the Fred Hutchinson Cancer Research Center (particularly Dr. ERIC SIEVERS, Dr. IRV BERNSTEIN, and Dr. FRED APPELBAUM) and the City of Hope Medical Center (particularly Dr. RICARDO SPIELBERGER and Dr. STEVE FORMAN) in the performance of the phase I CMA-676 clinical trial.

6 References

APPELBAUM, F. R., MATTHEW, D. C., EARY, J. F. et al. (1992), The use of radiolabeled anti-CD33 antibody to augment marrow irradiation prior to marrow transplantation for acute myelogenous leukemia, *Transplantation* **54**, 829–833.

CMA-676 *Conjugates of Methyltrithio Antitumor Agents and Intermediates for their Synthesis*, USA 5773001.

CMB-401 *Targeted Forms of Methyltrithio Antitumor Agents*, USA 5053394.

DE VOSS, J. J., HANGELAND, J. J., TOWNSEND, C. A. (1990), Characterization of the *in vitro* cyclization chemistry of calicheamicin and its relation to DNA cleavage, *J. Am. Chem. Soc.* **112**, 4554–4556.

ELLESTAD, G. A., HAMANN, P. R., ZEIN, N. et al. (1989), Reactions of the trisulfide moiety in calicheamicin, *Tetrahedron Lett.* **30**, 3033–3036.

HANGELAND, J. J., DE VOSS, J. J., HEATH, J. A. et al. (1992), Specific abstraction of the 5'S- and 4'-deoxyribosyl hydrogen atoms from DNA by calicheamicin γ_1^I, *J. Am. Chem. Soc.* **114**, 9200–9202.

HINMAN, L. M., HAMANN, P. R., MENENDEZ, A. T. et al. (1993), Preparation and characterization of monoclonal antibody conjugates of the calicheamicins: A novel and potent family of antitumor antibiotics, *Cancer Res.* **53**, 3336–3342.

HINMAN, L. M., HAMANN, P. R., UPESLACIS, J. (1995), *Preparation of Conjugates to Monoclonal Antibodies in Enediyne Antibiotics as Antitumor Agents* (BORDERS, D., DOYLE, T., Eds.), pp. 87–105. New York: Marcel Dekker.

LEE, M. D., DUNNE, T. M., SIEGEL, M. M. et al. (1987a), Calicheamicins, a novel family of antitumor antibiotics, 1: Chemistry and partial characterization of γ_1^I, *J. Am. Chem. Soc.* **109**, 3464–3466.

LEE, M. D., DUNNE, T. M., CHANG, C. C. et al. (1987b), Calicheamicins, a novel family of antitumor antibiotics, 2: Chemistry and structure of γ_1^I, *J. Am. Chem. Soc.* **109**, 3466–3468.

MAMSOOR, M. A., SVARDAL, A. M., UELAND, P. M. (1992), Determination of the *in vivo* redox status of cysteine, cysteinylglycine, homocysteine, and glutathione in human plasma, *Anal. Biochem.* **200**, 218–229.

MISTRY, P., HARRAP, K. R. (1991), Historical aspects of glutathione and cancer chemotherapy, *Pharm. Ther.* **49**, 125–132.

MYERS, A. G., COHEN, S. B., KNOWN, B. M. (1994), A study of the reaction of calicheamicin γ_1 with glutathione in the presence of double-stranded DNA, *J. Am. Chem. Soc.* **116**, 1255–1271.

NICOLAOU, K. C., LI, T., NAKADA, M. et al. (1994), Calicheamicin γ_1^I through molecular design to a compound that cleaves DNA selectively and efficiently and initiates cell death, *Angew. Chem. (Int. Edn. Engl.)* **106**, 183–186.

RODWELL, J. D., MCKEARN, T. J. (1987), *US Patent* No. 4671958.

SIEVERS, E., BERNSTEIN, I., SPIELBERGER, R. et al. (1997a), Dose escalation Phase I study of recombinant engineered human anti-CD33 antibody-calicheamicin drug conjugate (CMA-676) in patients with relapsed or refractory acute myeloid leukemia (AML), *Proceedings ASCO* **16**, 3a.

SIEVERS, E., APELBAUM, F., SPIELBERGER, R. et al. (1997b), Selective ablation of acute myeloid leukemia using an anti-CD33 calicheamicin immunoconjugate, *Blood* **10**, 5041.

VAN HOF, A. C., MOLTHOFF, C. F. M., DAVIES, Q. et al. (1996), Biodistribution of [111]indium-labeled engineered human antibody CTMO1 in ovarian cancer patients: Influence of protein dose, *Cancer Res.* **56**, 5179–5185.

WALKER, S., GANGE, D., GUPTA, V., KAHNE, D. (1994), Analysis of hydroxylamine glycosidic linkages: structural concequences of the NO bond in calicheamicin, *J. Am. Chem. Soc.* **116**, 31997–32006.

ZEIN, N., PONCIN, M., NILAKANTAN, R., ELLESTAD, G. (1989), Calicheamicin γ_1^I and DNA: Molecular recognition process responsible for site-specificity, *Science* **244**, 697–699.

ZHAO, B., KONNO, S., WU, J. M., ORONSKY, A. L. (1990), Modulation of nicotinamide adenine dinucleotide and poly(adenosine diphosphoribose) metabolism by calicheamicin γ_1 in human HL-60 cells, *Cancer Lett.* **50**, 141–147.

16 ReoPro Clinical Development: A Case Study

HARLAN F. WEISMAN

Malvern, PA, USA

1 Introduction   366
2 Trials   367
   2.1 The EPIC Trial   367
   2.2 The EPILOG Trial   374
   2.3 The CAPTURE Trial   376
3 Summary and Conclusions   378
4 References   379

1 Introduction

In the latter decades of this century, the understanding of the pathobiology of a number of diseases has markedly increased and therapeutic focus has shifted to specific, targeted pharmacologic interventions for preventing and ameliorating these diseases. The challenge of the pharmaceutical industry, particularly the biopharmaceutical industry, as we approach the next millennium is to develop innovative new drugs in the context of high development costs, limited human resources, intense competition, a constantly evolving regulatory landscape and rapidly changing commercial market dynamics. To remain profitable, the pharmaceutical industry must not only develop safe and effective new therapies, but therapies that represent a meaningful improvement over current treatment and must do so in a compressed development cycle with every effort to minimize development costs. Failure to meet these challenges will result in expensive drugs offering only marginal improvement over existent therapies that are unlikely to be adopted by the medical community or reimbursed by third party payers.

Formerly, 7–10 year development cycles for new drugs were considered acceptable and many drugs took even longer to develop. The ReoPro™ (abciximab, c7E3 Fab) development program may provide a model of how innovative new drugs can be brought to market rapidly within the ever changing dynamics of managed care, health economics, and outcomes management. Tab. 1 shows a historical overview of the ReoPro development and approval processes for the first indication for this product, adjunctive therapy for patients undergoing percutaneous coronary intervention at high risk of thrombotic complications. The Investigational New Drug (IND) application for the Fab fragment of the human/mouse chimeric monoclonal 7E3 IgG antibody (c7E3 Fab), the molecule that was to become ReoPro, was filed by Centocor with the FDA in February 1990. The first phase I dose escalation trial began at Northwick Park Hospital in England and the Durham, North Carolina Veterans Administration Hospital in

Tab. 1. ReoPro Development and Approval Process; Historical Overview

Milestones	Cumulative Development Time [a]
Investigational New Drug Application	0.0
Phase I completed	1.0
Phase II completed	1.75
Phase III (EPIC) enrollment completed	2.75
Phase III (EPIC) analysis	3.25
Product License Application	4.0
FDA Advisory Committee	4.5
US Approval and Product Launch	5.0

April 1990. Phase II trials in high-risk patients undergoing percutaneous coronary angioplasty and in patients with acute myocardial infarction commenced in February 1991 and were completed by August 1991. The phase III trial that led to the first approval in high-risk coronary intervention, known as the EPIC (*E*valuation of c7E3 Fab to *P*revent *I*schemic *C*omplications) trial, began enrollment in November 1991 and completed enrollment 1 year later in November 1992. The analysis of the 30-day primary endpoint was completed and the database unblinded in March 1993 with the first presentation of results at the American College of Cardiology Scientific Sessions later that same month. The Product License Application for the high-risk coronary intervention indication was filed with the United States Food and Drug Administration (FDA) in December 1993. Dossiers were submitted to the 12 European member states of the Committee of Proprietary Medicinal Products (CPMP) by February 1994 along with submissions to 9 other countries. The application for approval was reviewed by the FDA Cardio-Renal Advisory Committee in June 1994 and by the European CPMP in November 1994. The first approval of ReoPro came by November 1994 in Sweden with FDA approval and CPMP approval occurring simultaneously in December 1994. The product was launched by Centocor's marketing partner, Eli Lilly, in the United States and several European countries, in February 1995. Thus, from IND filing to marketing, ReoPro development occurred within a 5 year time frame. No other new cardiovascular drug has proceeded through worldwide development and the approval process more quickly.

The marketing approval of ReoPro was the culmination of the combined efforts of basic research scientists, clinicians and health economists in the academic community and industry. The development of ReoPro presented a rare opportunity for these scientists to open a new area of biomedical research, the chance to take receptor pharmacology from the laboratory to the bedside with the first of a new class of agents in the most pressing area of acute cardiovascular treatment, namely the prevention of arterial thrombosis. One fortunate aspect of ReoPro development was having a well-defined receptor target, i.e., the platelet GPIIb/IIIa receptor. The number of GPIIb/IIIa receptors that were blocked following ReoPro dosing, as well as the biological effect of this blockade, namely, the inhibition of platelet aggregation, were quantifiable not only in animal studies, but also in clinical trials in humans (JORDAN et al., 1997). Thus, the receptor effects (receptor blockade) and biological activity (inhibition of platelet aggregation) could be precisely correlated with clinical efficacy (prevention of cardiac ischemic complications). In addition to these unique characteristics of ReoPro, excellence in design and execution of clinical trials were the important guiding principles that lead to the success of the ReoPro clinical development program.

From the beginning of ReoPro clinical development, effort was placed on identifying and meeting the needs of present and future customers. These customers include the regulatory agencies that review the marketing license applications, the physicians who use the product, the hospitals in which the product is used, the patients who receive the therapy, and increasingly the third party payers who ultimately pay for the product. These third party payers, include private insurance companies, for-profit and not-for-profit managed health care organizations, and government agencies. As greater emphasis is placed on medical cost containment, who pays varies and continuously changes within national regions as well as between countries, so identifying these customers and their requirements is a constant challenge. Furthermore, what is desired by one constituent of this diverse group of regulatory, medical, and payer customers may be at odds with the others. Nevertheless, a common element for all customers is the desire for credibility and quality of the product, and Centocor and the investigators conducting ReoPro clinical trials believed that following the highest standards of clinical trial design was the best means of establishing this confidence in the product. However, with novel agents, standard templates for clinical trials are not always available and this was true for ReoPro.

2 Trials

2.1 The EPIC Trial

The EPIC trial at the time of its inception had unique features which have since become standard in subsequent interventional cardiology trials studying antithrombotic agents. EPIC was a three arm, placebo controlled, dose comparison, efficacy trial and not many of these had been conducted in the past and none of this size had ever been conducted in interventional cardiology. Thus, considerable effort was put into the trial's design.

In the EPIC trial, patients received one of three dose regimens: placebo bolus plus a 12 h placebo infusion, a ReoPro bolus of 0.25 mg kg^{-1} plus a 12 h placebo infusion, or a ReoPro bolus of 0.25 mg kg^{-1} plus a 12 h ReoPro infusion of 10 $\mu g\ min^{-1}$; therapy was initiated 10–60 min prior to first device activation of a balloon angioplasty or atherectomy catheter. High-dose, non-weight-adjusted heparin was given concomitantly with ReoPro, as was standard practice at the time. A significant reduction in the composite primary endpoint of the trial comprising death, myocardial infarction (MI), or urgent repeat coronary intervention (including percutaneous coronary intervention or coronary artery bypass) was demonstrated at 30 d (The EPIC investigators, 1994). The 30 d endpoint event rates for each treatment group by intention-to-treat analysis are shown in Tab. 2. The primary event rate was 12.8% in the placebo treatment group, 11.5% in the ReoPro bolus treatment group, and 8.3% in the ReoPro bolus plus infusion treatment group. The 34.8% reduction in the primary endpoint rate in the ReoPro bolus plus infusion treatment group was statistically significant ($p = 0.008$ vs placebo), but there was no statistically significant difference in the primary endpoint event rate between the ReoPro bolus and placebo treatment groups ($p = 0.428$).

In terms of the components of the primary endpoint, death was relatively rare and occurred with similar frequency in each group. The greatest dose-response effects were seen in the MI and urgent intervention event rates. Patients who received the ReoPro bolus plus infusion treatment had a 39.4% reduction in the incidence of MI ($p = 0.014$ vs placebo) and a 49.1% reduction in the incidence of urgent intervention ($p = 0.003$ vs placebo). The reduction in MI was observed in both Q-wave and non-Q-wave MI. Consistent reductions in the occurrence of endpoints were seen in all prospectively defined subgroups: elderly as well as younger patients, women as well as men, and patients with stable coronary disease as well as those with acute, unstable coronary syndromes. Tab. 2 also shows that these beneficial effects of the ReoPro bolus plus infusion regimen were maintained during 6-month and 3-year follow up (TOPOL et al., 1994).

Several important lessons can be drawn from the EPIC trial experience. In beginning the design of a trial, an unequivocal hypothesis is important to ensure that unequivocal results are obtained. This means that clinically important, unambiguous endpoints are essential. In the case of EPIC, a composite endpoint was chosen comprising death, myocardial infarction, and severe myocardial ischemia requiring urgent repeat coronary intervention (coronary artery bypass surgery or coronary angioplasty). The components of this composite represent the serious, irreversible complications of angioplasty, and, therefore, this composite met the criteria of being clinically important and unambiguous. It is also important to choose a well-defined patient population. In EPIC, high-risk patients were selected to increase the probability of success in the clinical trial. Although choosing high-risk patients somewhat narrowed the initial indication for ReoPro, selecting this well-defined patient group ensured that those reviewing the results at the end of the trial would be clear in whom ReoPro worked.

Many clinical trials of new therapeutic agents have failed because of inadequate

Tab. 2. 30-Day, 6-Month and 3-Year Outcomes by Treatment Group in the EPIC Trial

	Placebo	Bolus	Bolus + Infusion
Patients randomized	696	695	708
Patients with death, MI or urgent intervention within 30 days	89 (12.8%)	79 (11.5%)	59 (8.3%)
% change vs placebo		−10.4%	−34.8%
p-value vs placebo		0.428	0.008
Patients with death, MI or repeat revascularization within 6 months	241 (35.1%)	224 (32.6%)	189 (27.0%)
% change vs placebo		− 7.1%	−22.9%
p-value vs placebo		0.276	0.001
Patients with death, MI or repeat revascularization within 3 years	319 (47.2%)	321 (47.4%)	283 (41.4%)
% change vs placebo		0.3%	−13.0%
p-value vs placebo		0.779	0.009

powering. That is, sample size in these trials was not adequate to detect a clinically meaningful efficacy benefit. As already mentioned, at the time, EPIC was the largest, prospective coronary intervention trial ever conducted. Avoiding bias is another obvious guiding principle that is necessary to ensure the success of large clinical trials. In EPIC, the intention-to-treat method of analysis was employed. That is, all patients were included in the analysis whether or not they actually received treatment with the assigned study agent or whether they actually underwent a coronary intervention. The rationale behind intention-to-treat analysis is often confusing to the layperson and to some experienced clinicians. However, it safeguards against bias when deciding who should or should not be included in the analysis, since all patients are included. This emphasizes the principle of ensuring adequate powering of the trial. A sufficient number of patients must be included to protect the trial from randomized patients who do not actually follow all aspects of the study protocol, but are included in the analysis. In the case of EPIC, significant reductions in both 30 d and 6 month events were obtained by the intention-to-treat analysis. Not surprisingly, even more impressive results were obtained when the analysis was confined to those patients who actually received treatment. This is reassuring because it means that benefit was seen in randomized patients that actually received ReoPro, but that no benefit was seen in the patients who were randomized to receive ReoPro, but in fact did not.

A believable control group and making sure that investigators, patients and the sponsor are blinded are also important factors for credible comparisons between the groups of patients receiving and not receiving active treatment. In the case of EPIC, the placebo group received conventional treatment comprising aspirin and high dose heparin. This presented a dilemma in designing the EPIC trial because the need for double-blinding demanded that the patients treated with ReoPro also receive conventional treatment. An obvious concern was that the use of high dose heparin and aspirin combined with ReoPro might lead to an increase in bleeding. Fortunately, although an increase in bleeding complications was observed in the EPIC trial among those patients who received ReoPro, the bleeding events were largely confined to the vascular access site in the groin and were easily manageable.

Because the trial sponsor (Centocor) wanted the EPIC trial results to be accepted by regulators, physicians, hospitals and third party payers, the sponsor kept as much distance from the conduct of the trial as possible to avoid any appearance of potential bias being introduced. The use of a double-blind also meant that the investigators and the clinical coordinating centers were kept distant from knowledge of the trial results while it was ongoing. Therefore, independent oversight committees to review endpoints and to monitor safety and efficacy during the enrollment phase of the trial were employed. Not until the database was unlocked was it unblinded and made available to the investigators and to Centocor for final analysis.

A number of steps were taken prospectively to help meet and ensure the credibility and quality of the conduct and analysis of the EPIC trial. A two volume analytic plan delineated the trial design and conduct. This analytical plan was reviewed by leading statisticians experienced in cardiovascular disease and was also reviewed by the FDA to make sure that they were in agreement with the plans for the conduct and analysis of the trial. As already mentioned, there were layers of safeguards protecting the integrity of the trial. An independent blinded Clinical Endpoint Committee reviewed and confirmed all efficacy and major safety events. An independent Safety and Efficacy Monitoring Committee was responsible for ongoing safety monitoring and the review of interim efficacy and safety analyses. This ongoing review of data was important because ReoPro was a potent new agent that affected hemostasis and was being combined with high dose heparin and aspirin. Therefore, the Safety and Efficacy Monitoring Committee was empowered to make judgements about the ongoing safety of the trial and weigh this against any evidence of ReoPro's efficacy. The use of double-blinding has already been mentioned, and numerous measures were taken to ensure that this blinding was maintained. The trial was also audited several times before, during, and after its completion. The quality assurance group at Centocor, which is independent of the clinical research unit, audited the investigative sites and they reviewed the procedures of the trial before and during the course of the trial, and once again after the trial was completed. This audit was repeated by the independent quality assurance group of Eli Lilly, Centocor's marketing partner for ReoPro. Finally, to make sure that no stones were left unturned, an independent quality assurance consultant, repeated the audit. After all of these audits were completed, the FDA conducted their own audit of the trial and confirmed the high standards and quality of the study.

The final step in the process leading to the approval of ReoPro was the preparation and submission of the reports of the ReoPro manufacturing process, preclinical studies and clinical trial results for regulatory agency review. The ReoPro Product License Application submitted to the FDA comprised 157 volumes of written text that was backed by a computerized submission on CD-ROM. The written submission and the CD-ROM disks contained all of the analysis data sets used to analyze the EPIC trial, as well as all the case report forms from the EPIC trial, and additional information that permitted the FDA medical reviewers to reproduce the analyses that were presented in the written document, as well as to perform their own analyses and review individual case records.

The rapidity of the ReoPro approval process, which was probably faster than any previous application for a new cardiovascular drug, was facilitated by these efforts on the quality, conduct, analysis and presentation of the results of the ReoPro clinical trials. As stated earlier, the culmination of this effort was the approval of ReoPro in the United States and in European countries within 1 year of the application. This should have been the end of the story and several years ago it may have been. But again, the reality of drug development in the 1990's is a new emphasis on health economics and outcomes management. Thus, since the EPIC trial was completed, analyzed, and submitted for regulatory approval, a continued effort was fo-

cused on understanding the health economics of ReoPro as shown in the EPIC trial.

The EPIC trial demonstrated both short and long-term benefits, initially over 6 months, and more recently this follow up has been extended to a median of more than 3 years (see Tab. 2). The long term follow up included not only clinical outcome data, but also economic outcome data. Fig. 1 shows the accrual of follow up cost savings comparing ReoPro vs placebo over 6 months. As shown in Fig. 1, cost savings related to ReoPro progressively increase over time. By 6 months, almost the entire cost of the drug has been recovered because ReoPro reduces health care costs by reducing rehospitalizations for not only the clinical endpoints of death, myocardial infarction, and revascularization procedures, but also for other cardiovascular morbidities including heart failure and unstable angina. This reduction in hospitalizations is shown in Fig. 2. The data shown in Figs. 1 and 2 emphasize one of the vexing problems of the 1990s. Hospital pharmacies pay for the drug, however, as shown in Fig. 1, the EPIC data suggest little accrual of cost savings during the initial hospitalization. Patients and third party payers clearly benefit from the improved clinical outcome and decreased health care costs associated with ReoPro treatment, but this benefit, which accrues over time, is underwritten by the hospitals providing the initial care. In a completely capitated health

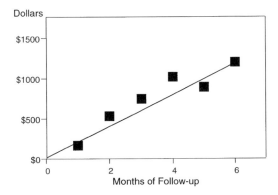

Fig. 1. Accumulation of cost savings observed in the EPIC trial with bolus plus infusion therapy relative to placebo therapy over 6 months exclusive of the cost of ReoPro, ■ savings with ReoPro.

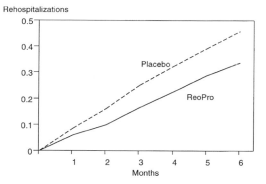

Fig. 2. Average number of rehospitalizations per patient for the placebo and ReoPro bolus plus infusion treatment groups in the EPIC trial.

care environment, ReoPro can save money, but few patients in the United States or other countries are under a fully capitated medical care plan.

To further illustrate the cost-effectiveness of ReoPro, a cost-effectiveness analysis was undertaken (HILLEGASS, 1996; VAN HOUT and SIMOONS, 1995; MARK et al., 1996). In a cost-effectiveness approach, a cost figure is obtained that describes the cost of therapy in terms of its clinical effectiveness, i.e., a cost per unit of efficacy. Tab. 3 shows the average cost-effectiveness of ReoPro compared to placebo. The costs that are shown in the table represent the average total cost derived from all patients within the placebo and ReoPro bolus plus infusion groups of the EPIC trial. Over 6 months, placebo treatment costs

$17,988. The cost of ReoPro therapy, including the price of the ReoPro bolus plus infusion regimen was $18,143. The difference between placebo and the ReoPro bolus plus infusion arm was $145, so the net cost of ReoPro was $145 spread over 6 months for the average patient. Tab. 3 also shows 6 month efficacy defined as survival without death or repeated coronary intervention over 6 months. In the placebo group, 64.9% of patients were event-free for the 6 month follow up compared to 73.8% of the ReoPro treated patients. To compute the average cost-effectiveness, this effectiveness measure was divided into the total cost. This is shown in the last column of Tab. 3 and was $27,732 for placebo or conventional treatment and was $24,593 for the ReoPro treated patients. The bottom line is that over 6 months, ReoPro saves $3,148 per successful outcome including the cost of the drug. These results are applicable to all patients enrolled in the EPIC trial averaged over 6 months. However, these results may not apply to individual hospitals and individual patients and that is one of the challenges under current investigation.

ReoPro is cost-effective in reducing the ischemic complications of coronary angioplasty, but this cost-effectiveness is spread over the long-term. Hospitals and hospital pharmacies are currently in an environment in which they must consider only the initial hospitalization in managing their costs and reimbursement. Long-term clinical outcome for the patient as well as the long-term costs are less meaningful to hospitals as they consider

Tab. 3. EPIC Trial Health Economic Analysis

	6 Month Average Total Cost[a]	6 Month Efficacy[b]	6 Month Cost per Desired Outcome[c]
Placebo	$17,998	64.9%	$27,732
ReoPro	$18,143	73.8%	$24,583

[a] Total costs of initial and subsequent hospitalzation including estimated physician fees; for ReoPro group the average cost of a course of ReoPro treatment is included.

[b] Percentage of surviving patients without myocardial infarction or repeated coronary revascularization.

[c] Average cost-effectiveness calculated by dividing 6 month average total cost by 6 month efficacy for each treatment group.

their dwindling financial resources and tightening budgets. Assessing the effectiveness of ReoPro in this environment is difficult in a randomized, double-blind clinical trial setting. The reason for this is that double-blinding and the importance of having all patients adhere to the same protocol, make it difficult to ascertain whether a treatment can alter current management practices for specific patient types. For the hospital pharmacist, hospital administrator, catheterization laboratory director, and interventional cardiologist, the EPIC results demonstrate that ReoPro improves clinically meaningful outcomes and has economic benefit in a clinical trial setting. However, these results do not provide them with specific guidance of how ReoPro will affect their current practice in the invasive cardiology laboratory and how ReoPro's introduction will affect their hospital budget. ReoPro is just one of a number of competing new therapeutic modalities that can provide therapeutic benefit to their patients, but also may severely impact their financial picture.

As we approach the year 2000, health care institutions need to focus on explicit clinical presentations and develop specific management guidelines and clinical pathways. Thus, recent efforts have been undertaken to work with hospital pharmacies and interventional cardiologists to help them in analyzing the impact of the use of ReoPro in their hospitals. In the EPIC trial, ReoPro treatment was highly effective in all prespecified subgroups, but may offer distinct advantages over conventional therapy in a few of these (AGUIRRE et al., 1996). Two of these subgroups are of particular interest. The first is the treatment of patients with acute coronary syndromes where ReoPro was shown to have a highly significant mortality benefit, as well as a significant reduction in post-procedure myocardial infarction (LINCOFF et al., 1997c). Among the almost 500 patients in EPIC with acute coronary syndromes followed for 3 years, mortality was reduced by more than 75% from 11.6% to 2.2% ($p = 0.002$). The second area of focus has been in patients with visible intracoronary thrombus identified during the diagnostic angiographic procedure, including patients with degenerated vein grafts. In both of these situations, ReoPro may elim-

inate the need for intracoronary thrombolytics or prolonged heparin therapy before proceeding with the coronary intervention. Therefore, ReoPro offers the potential of improving clinical outcome and decreasing initial hospital costs.

ReoPro may reduce costs associated with percutaneous coronary interventions in patients with unstable angina by allowing earlier intervention and lowering the risk of thrombotic complications, thereby permitting shorter time to hospital discharge following the procedure. As an illustration of this potential, intensive care unit costs and length of hospital stay before and after angioplasty have been assessed comparing conventional treatment to ReoPro treatment. This is illustrated in Tab. 4. The dollar figures used in Tab. 4 have been derived from several large institutions that are currently using ReoPro and were provided by members of the cardiology departments from these hospitals. Because conventional treatment requires a period of stabilization to cool symptoms before proceeding to cardiac catheterization, several days in the CCU (cardiological care unit) with heparin monitoring are often required. However, because ReoPro potently prevents platelet aggregation and thrombosis (which are the causative factors leading to the unstable symptoms), there is the potential to reduce CCU days, heparin treatment, and heparin monitoring. The total cost savings shown in Tab. 4 associated with ReoPro treatment exceed the cost of a typical course of ReoPro treatment ($1,350). Similarly because of the avoidance of ischemic complications, CCU days and post-procedure hospital stay can be reduced with ReoPro treatment, again reducing total costs associated with the hospitalization. Although systematic study of this approach definitively demonstrating that ReoPro can lead to improved economics in all hospitals has not been completed, this preliminary analysis demonstrates the type of cost savings benefit assessment that can be employed by hospitals and may provide a basis for evaluating the cost effectiveness of ReoPro in individual institutions.

A similar analysis has been undertaken in selected institutions in which patients with degenerated saphenous vein grafts are treated

Tab. 4. Potential for Reduced Intensive Care Costs and Hospital Days in Unstable Angina Patients Treated with ReoPro

	Estimated Length of Study		
	Conventional Treatment [d]	ReoPro [d]	Potential Savings[a] [$]
Pre PTCA			
CCU days	2–3	0–1	1,348–4,044
Heparin and monitoring	2–3	0	448– 575
Total costs pre PTCA			1,796–4,619
Post PTCA			
CCU days	1	0	1,348
Hospital room	2	1	762
Total costs post PTCA			2,110

[a] Potential savings exclude the cost of ReoPro ($1,350 for an average course of treatment). The estimated length of stay and cost savings are based on data provided by 5 hospitals comparing conventional management to management with ReoPro (see text for explanation).

with ReoPro to improve clinical outcome and reduce costs associated with percutaneous coronary intervention. In this situation, ReoPro can improve clinical outcome and decrease total hospital costs by allowing earlier intervention and lowering the risk of thrombotic complications, again reducing the time from intervention to discharge. This is illustrated in Tab. 5. Patients with degenerated saphenous vein grafts filled with thrombus often are treated with intracoronary urokinase for up to 24 h in a monitored bed. As shown in Tab. 5, by permitting a patient to proceed from diagnostic catheterization to coronary intervention without preceding intracoronary thrombolytics, ReoPro can reduce total costs by an amount that exceeds the price of a typical course of ReoPro treatment. Again, this approach has not been studied prospectively, but this type of analysis might be applicable in assessing the cost-effectiveness of ReoPro in individual institutions.

In the future, attention will be directed not only to the types of health economic studies that were conducted in EPIC and the individual patient management approach just reviewed, but also to examining the types of issues of interest to other decision-makers in the health care delivery system. Up to now, the assessment of direct health care costs has

Tab. 5. Potential for Reduced Intensive Care Costs and Hospital Days in Patients with Degenerated Saphenous Vein Grafts Treated with ReoPro

	Estimated Length of Stay		
	Conventional Treatment [d]	ReoPro [d]	Potential Savings[a] [$]
Monitored bed	½–1	0	381– 762
Intracoronary UK	½–1	0	1,430
Total costs			1,811–2,192

[a] Potential savings exclude the cost of ReoPro ($1,350 for an average course of treatment). The estimated length of stay and cost savings are based on data provided by 5 hospitals comparing conventional management to management with ReoPro (see text for explanation).

been emphasized through the measurement of costs of hospitalization and physician fees. But increasingly, government agencies, private managed care organizations and large employers are concerned with the indirect costs of health care including the productivity of employees. They want to know how a new treatment impacts whether employees are go-ing to work everyday and how productive they are when they are at work. These types of measurements are not as clear as those already reviewed, but they are increasingly important to address and will be an area of focus in future ReoPro clinical trials.

2.2 The EPILOG Trial

The EPIC trial results demonstrated not only that ReoPro improved acute outcome following percutaneous coronary intervention, but also had maintained benefit for as long as 3 or more years following the initial angioplasty procedure. The major down-side to ReoPro treatment in EPIC was an increased rate of bleeding events primarily at the local vascular access site (AGUIRRE et al., 1995). In addition, the patient population in EPIC was confined to individuals presumed to be at high risk of ischemic complications. On the basis of a small pilot trial known as PROLOG (LINCOFF et al., 1997a), which suggested that the increased rate of groin bleeding episodes that were observed in the EPIC trial can be attenuated by weight adjustment of heparin and by improved patient management guidelines, without impacting the efficacy of ReoPro, a large clinical trial known as EPILOG (**E**valuation of **P**TCA to **I**mprove **L**ong-term **O**utcome by c7E3 **G**PIIb/IIIa receptor blockade) was undertaken (The EPILOG Investigators, 1997).

The EPILOG trial was originally designed to enroll 4,800 patients undergoing percutaneous coronary intervention. Unlike the EPIC trial, EPILOG enrolled a wide spectrum of patients including high-risk and low-risk patients. The only patients excluded were patients with acute myocardial infarction or refractory unstable angina. The reason for the exclusion of patients with acute coronary syndromes is that these patients had statistically significant reductions in death and myocardial infarction in the EPIC trial and it was felt to be unethical to randomize these patients to placebo treatment. In EPILOG, patients were randomized into one of three treatment groups. The first group received placebo plus aspirin and standard dose weight-adjusted he-

parin. The other two groups each received a ReoPro regimen that was similar to the bolus plus infusion regimen used in the EPIC trial (0.25 mg kg^{-1} bolus followed by a 0.125 μg kg^{-1} min^{-1} infusion to a maximum of 10 μg min^{-1} for 12 h). Patients in one of these ReoPro groups received the same weight-adjusted heparin regimen as the placebo group, while patients in the other ReoPro group received a lower dose of weight-adjusted heparin. In those patients receiving the standard weight-adjusted dose the initial heparin bolus was 100 U kg^{-1}, and the heparin was then adjusted to maintain the activated clotting time (ACT) at approximately 300–350 s. In the ReoPro treatment group that received the low weight-adjusted heparin dose, the initial heparin bolus was 70 U kg^{-1} with a target ACT set at 200–250 s.

Although EPILOG was designed to enroll 4,800 patients, it was stopped after an interim analysis of the first 1,500 enrolled patients, upon the recommendation of the trial's independent Safety and Efficacy Monitoring Committee, because of a highly significant reduction associated with ReoPro treatment in the rate of the composite endpoint of death or MI. At the time the trial was stopped, 2,792 patients had been randomized. Tab. 6 shows the results of the final efficacy analysis of the primary endpoints of death or MI at 30 d, death, MI or urgent intervention at 30 d and death, MI or any intervention at 6 months. The event rate for the composite of death or MI at 30 d was reduced by 58.8% in the ReoPro plus low-dose heparin group ($p < 0.0001$) and by 54.4% in the ReoPro plus standard-dose heparin group ($p < 0.0001$). The event rate for the composite endpoint of death, MI or urgent revascularization at 30 d was reduced by 55.8% in the ReoPro plus low-dose heparin group ($p < 0.0001$) and by 54.1% in the ReoPro plus standard-dose heparin group ($p < 0.0001$). Tab. 6 also shows the number of patients with death, MI, or repeated revascularization (urgent and non-urgent) at 6 months post randomization. The event rate for death, MI, or repeated revascularization (urgent and non-urgent) at 6 months post randomization was 25.8% in the placebo group, 22.8% in the ReoPro plus low-dose heparin group ($p = 0.034$), and

Tab. 6. 30-Day and 6-Month Outcomes by TreatmentGroup in the EPILOG Trial

	Total	Placebo	ReoPro + Low-Dose Heparin	ReoPro- Standard- Dose Heparin	Combined ReoPro
Patients randomized	2,792	939	935	918	1,853
Patients with death or MI within 30 d	158 (5.7%)	85 (9.1%)	35 (3.8%)	38 (4.2%)	73 (4.0%)
% change vs placebo			−58.8%	−54.4%	−56.6%
p-value vs placebo			<0.001	<0.001	<0.001
Patients with death, MI, or urgent revascularization within 30 d	206 (7.4%)	109 (11.7%)	48 (5.2%)	49 (5.4%)	97 (5.3%)
% change vs placebo			−55.8 %	−54.1%	−54.9%
p-value vs placebo			<0.001	<0.001	<0.001
Patients with death, MI, or re-peated revasculari-zation within 6 months					
% change vs placebo	656 (23.6%)	241 (25.8%)	212 (22.8%)	203 (22.3%)	415 (22.5%)
p-value vs placebo			−11.7%	−13.7%	−12.6%
			0.034	0.020	0.011

22.3% in the ReoPro plus standard-dose heparin group ($p = 0.020$).

One of the major aims of EPILOG was to test whether the benefit of ReoPro in reducing serious complications in high-risk patients could be extended to low-risk patients. Thus, an important subgroup analysis in EPILOG, which unlike EPIC included low as well as high-risk patients, included examination of the primary endpoints of the trial by risk classification. This analysis showed that the composite endpoint of death or MI endpoint was reduced at 30 d in ReoPro patients in both the low-risk group (64.1% reduction, $p < 0.001$) and high-risk group (53.8% reduction, $p < 0.001$). Similarly, for the 30 d endpoint of death, MI or urgent revascularization, ReoPro once again demonstrated efficacy in both risk groups. In the high-risk pa-

tients, the composite endpoint event rate was reduced by 52.9% ($p < 0.001$) and in low-risk patients, the composite endpoint event rate was reduced 60.3% ($p < 0.001$). The reductions in these events were maintained at 6 months for both high- and low-risk patients.

In EPIC, which was conducted from 1991 through 1992, the primary coronary intervention was balloon angioplasty. EPILOG, which was conducted in 1995, contained newer interventions including coronary stenting, which afforded the opportunity to study ReoPro with newer interventional devices. With respect to the type of intervention performed in EPILOG, the composite of death or MI at 30 d was reduced to a similar degree among patients undergoing balloon angioplasty (6.9% in the placebo group vs 2.8% in the combined ReoPro groups, $p < 0.001$); stent implantation (19.5% in the placebo group vs 7.1% in the combined ReoPro groups, $p < 0.001$); or directional coronary atherectomy (DCA), transluminal extraction catheter (TEC) atherectomy, or laser atherectomy (19.2% in the placebo group vs 8.2% in the combined ReoPro groups, $p = 0.027$). Similar reductions for these devices were seen in the other composite endpoints and these benefits were maintained at 6 months.

In addition to establishing the ReoPro treatment benefit in high- and low-risk patients, the EPILOG trial was designed to test strategies to lower the incidence of major bleeding. The results in EPIC suggested that high doses of heparin may have been a major contributing factor to the bleeding rate seen in the EPIC trial, consequently in EPILOG, patients were randomized to receive either low-dose weight-adjusted heparin or standard-dose weight-adjusted heparin concomitantly with ReoPro bolus plus infusion. Early sheath removal and access care guidelines were recommended for all patients. Major bleeding events (including those associated with coronary artery bypass grafts [CABG] and non-CABG-related) occurred in 2.9% of patients, 3.1% in the placebo group, 2.0% in the ReoPro plus low-dose heparin group, and 3.5% in the ReoPro plus standard-dose heparin group. This is in comparison to a rate of 14.0% major bleeding in ReoPro bolus plus infusion treated patients in the EPIC trial.

The results of the EPILOG trial extend the efficacy findings of the EPIC trial to low-risk patients, as well as high-risk patients, and to all types of percutaneous coronary intervention. The results also demonstrate that weight-adjustment of heparin, in combination with other strategies such as a strong recommendation for early sheath removal, specific vascular access site and patient management guidelines, and weight-adjustment of the ReoPro infusion, reduce the rate of major bleeding to that of patients receiving placebo and standard-dose weight-adjusted heparin, while preserving efficacy. As with the EPIC trial, a prospectively defined health economic analysis was performed and the preliminary results have recently been presented (LIN-COFF et al., 1997b). As expected, the enhanced efficacy and reduced bleeding compli-

cations observed with ReoPro in EPILOG compared to EPIC resulted in cost saving during the index hospitalization in which the coronary intervention was performed.

2.3 The CAPTURE Trial

Since the completion of EPIC, ReoPro has also been studied in the setting of unstable angina, testing the hypothesis that the potent platelet inhibition provided by ReoPro, in addition to standard intensive medical treatment, could reduce ischemic complications in patients with refractory unstable angina by stabilizing them and preventing post-coronary intervention thrombotic complications. The **C**himeric **A**nti-**P**latelet **T**herapy in **U**nstable Angina **R**efractory to Standard Medical Therapy (CAPTURE) trial was a phase III, multicenter, double-blind, placebo controlled randomized study in patients with refractory unstable angina who were eligible for angioplasty conducted in collaboration with the European Cooperative Study Group (The CAPTURE Investigators, 1997).

In CAPTURE, diagnostic coronary angiography was performed within 48 h of an episode of myocardial ischemia in suitable, hospitalized patients with unstable angina which was refractory to standard treatment with aspirin, intravenous heparin and nitrates. If the culprit coronary lesion was suitable for PTCA (percutaneous transluminal coronary angioplasty) and the PTCA could be performed within 24 h after the start of the study medication, the patient was subsequently randomized to one of two treatment groups: ReoPro bolus (0.25 mg kg^{-1}) followed by ReoPro infusion (10 µg min^{-1}) or placebo bolus followed by a placebo infusion. Treatment began within 24 h of the enrolling angiography and continued for 18–26 h; PTCA was to be performed between 18 and 24 h after the start of treatment with study medication, although study agent administration could be continued if the intervention was delayed. Treatment was then continued until 1 h following the end of the PTCA. Guidelines for weight-adjusted heparin administration in the catheterization laboratory recommended 100 U kg^{-1}

or 10,000 U, whichever was less, with additional doses administered based on the results of activated clotting time or activated partial thromboplastin time (APTT). Unless there was a known aspirin allergy, 250 mg of aspirin were to be administered 2 h prior to the procedure and the administration was to be continued (50–500 mg) on a daily basis through 30 d post randomization. Sheath site care and patient management guidelines were implemented to reduce the risk of bleeding.

The occurrence of myocardial infarction, urgent intervention for recurrent ischemia or death during the 30 d after randomization constituted the primary composite efficacy endpoint of the trial. Total planned enrollment for the trial was 1,400 patients. However, the trial was stopped on the recommendation of the independent Safety and Efficacy Monitoring Committee after an interim analysis of 1,050 patients because of a highly significant reduction in the primary endpoint of the trial in the ReoPro group. At the time the trial was stopped, 1,267 patients had been enrolled.

Tab. 7 shows the intention-to-treat analysis of the CAPTURE primary endpoint of death, MI or urgent intervention within 30 d of randomization. ReoPro treatment resulted in a 28.9% reduction in primary endpoint events from 15.9% in the placebo treatment group to 11.3% in the ReoPro treatment group ($p = 0.012$). Tab. 7 also shows that ReoPro treatment resulted in lower event rates for all components of the primary endpoint with the greatest effect seen in the occurrence of MI. There was a 49.6% reduction in the number of patients with MI from 8.2% in the placebo

treatment group to 4.1% in the ReoPro treatment group ($p = 0.002$).

As in the EPILOG trial, ReoPro treatment reduced the primary endpoint event rate regardless of the type of intervention – balloons alone, stents or atherectomy. The ReoPro treatment benefit in reduction of the primary endpoint event rate was also consistent across prespecified patient subgroups including time between start of study treatment and the most recent prior angina attack, single vs multiple vessel disease, gender, and age.

Not only did ReoPro reduce the rate of ischemic complication through 30 d, ReoPro treatment prior to percutaneous coronary intervention also stabilized patients and produced the following benefits: (1) reduced the number of patients experiencing MI before, during and after percutaneous coronary intervention (Fig. 3); (2) reduced the number of patients having PTCA performed urgently before their planned PTCA (from 14–9%) and the number of patients with endpoint events among the patients who went urgently to PTCA by 80.6% ($p = 0.037$); (3) reduced the rate of PTCA failure by 46.5% ($p = 0.001$) and the number of patients with endpoint events among patients with failed PTCA by 75%; and (4) improved coronary flow and reduced intracoronary thrombus between the qualifying and index angiograms ($p = 0.017$).

The percentage of patients with major bleeding events in CAPTURE was slightly higher in the ReoPro treatment group (placebo, 2.5%; ReoPro, 4.5%, $p = 0.119$), but much less (3.3%) than that reported for the bolus plus infusion treatment group in the EPIC trial (14.0%). Nevertheless, major bleeding in

Tab. 7. CAPTURE Trial Primary Endpoint Event Rates for All Randomized Patients

	Total ($n = 1,265$)	Placebo ($n = 635$)	ReoPro ($n = 630$)	% Change vs Placebo	p-Value
Patients who died, had MI or urgent intervention[a]	172 (13.6%)	101 (15.9%)	71 (11.3%)	−28.9%	0.012
Death	14 (1.1%)	8 (1.3%)	6 (1.0%)	−25.2%	0.603
MI	78 (6.2%)	52 (8.2%)	26 (4.1%)	−49.6%	0.002
Urgent intervention	118 (9.4%)	69 (10.9%)	49 (7.8%)	−28.0%	0.054

[a] Patients were counted once within a component, but could have been counted in more than one component.

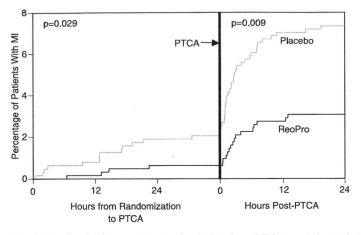

Fig. 3. Kaplan-Meier event rates for MI before PTCA and through the 24 h after PTCA in randomized patients in the CAPTURE trial. ReoPro reduced the percentage of patients with MIs by 69% in the period from randomization to the start of the procedure ($p = 0.029$). During and after percutaneous intervention, the number of MIs greatly increased from 17 to 51. However, ReoPro reduced the percentage of MIs by 50% during and after the PTCA, as well ($p = 0.009$). These results indicate that ReoPro stabilizes patients prior to PTCA and reduces the number of MIs during and after percutaneous intervention.

the ReoPro treatment group occurred much less frequently in the CAPTURE trial than in the EPIC trial. The excess bleeding occurred most often at the femoral artery access site and was generally managed with application of pressure and change of dressing. There was no increase in surgery in response to bleeding.

The results of the CAPTURE trial demonstrate that a 24 h infusion regimen of ReoPro can be used to stabilize unstable angina patients who are unresponsive to standard medical therapy and reduces the incidence of acute ischemic events prior to, during and after PTCA. The ReoPro regimen used in CAPTURE was associated with substantially less bleeding than was observed in EPIC, but more than in EPILOG. Given that the EPILOG trial has demonstrated that sheath site care, patient management and lower weight-adjusted heparin regimens can reduce the rate of major bleeding in ReoPro treated patients to equal that of placebo treated patients, the positive margin between benefit and risk observed in the CAPTURE trial can be viewed as a minimum and has the potential to become even larger.

3 Summary and Conclusions

ReoPro is the first of a new class of cardiovascular drugs targeted at the platelet GPIIb/IIIa receptor. The ReoPro clinical development program was built on the foundation of recent advances in molecular biology, receptor biochemistry and thrombosis research which permitted extensive characterization of the ReoPro molecule and its target receptor. The clinical trials program capitalized on the unique pharmacologic aspects of the ReoPro molecule and translated the knowledge of the dose response characteristics in preventing thrombosis learned in animal models to a human setting in which there was clear need for improved therapy. The underpinning of the clinical development program was an understanding of the needs of various customers who would be ultimately affected by the product including patients, physicians, other health care providers, regulatory agencies and third party payers. Every aspect of the EPIC trial design was aimed at this appreciation of customer needs, including the choice of sample size, careful selection of clinically relevant primary and secondary endpoints, use of independent committees to adjudicate endpoints and oversee safety, a statistical and health economic analysis plan subjected to review by leading clinical trial design authorities, and redundant quality assurance procedures to ensure that the highest standards were met in reporting the results to the scientific, medical and regulatory communities. Because of this planning and effort, ReoPro received marketing approval on the basis of the EPIC trial results more rapidly than any previous novel cardiovascular therapeutic agent. Despite this success, many unanticipated issues were left unresolved and needed to be addressed before the complete potential for commercial success could be realized. These factors included the rapidly changing health care systems in the United States and to a lesser extent in Europe, in which the cost and quality of health care are increasingly being scrutinized by government and private agencies responsible for financially underwriting medical services. In this environment, only those therapies which offer clear advantages over existing treatments will be adopted. Thus, subsequent clinical development of ReoPro including the EPILOG and CAPTURE trials not only focused on further delineating the medical applications of this therapy in widened indications, but also emphasized the demonstration of significant improvement over current therapy in both clinical and economic outcomes. In addition to percutaneous coronary intervention and unstable angina, clinical development of ReoPro is now aimed at new therapeutic applications including a number of other conditions in which arterial thrombosis plays an important pathophysiologic role. These conditions include acute myocardial infarction, peripheral vascular disease and cerebral thrombosis, including patients with stroke and transient ischemic attacks. The potency of ReoPro for completely inhibiting platelet aggregation, its immediate onset of action, as well as its prolonged duration of action make ReoPro uniquely suited for use in these indications.

4 References

AGUIRRE, F. V., TOPOL, E. J., FERGUSON, J. J., ANDERSON, K., BLANKENSHIP, J. C. et al. Investigators (1995), Bleeding complications with the chimeric antibody to platelet glycoprotein IIb/IIIa in patients undergoing percutaneous coronary intervention, *Circulation* **91**, 2882–2890.

AGUIRRE, F. V., TOPOL, E. J., ANDERSON, K. M., KLEIMAN, N. S., WEISMAN, H. F. et al. (1996), Clinical benefit within patient subgroups receiving c7E3 *fab* (abciximab) during percutaneous coronary revascularization: subgroup analysis from the EPIC trial, *J. Invas. Cardiol.* **8** (Suppl. B), 21B–29B.

HILLEGASS, W. B. (1996), The economics of IIb/IIIa therapy, *J. Invas. Cardiol.* **8** (Suppl. B), 30B–33B.

HOUT, B. A., VAN SIMOONS, M. L. (1995), Costs and effects of c7E3 in high-risk PTCA patients. An indirect analysis for the Netherland, *Eur. Heart J.* **16** (Suppl. I), 81–85.

JORDAN, R. E., MASCELLI, M. A., NAKADA, M., WEISMAN, H. F. (1997), The pharmacology and clinical development of abciximab (c7E3 Fab): A monoclonal antibody inhibitor of GPIIb/IIIa and $\alpha_v \beta_3$, in: *New Therapeutic Agents in Thrombosis and Thrombolysis.* New York: Marcel Dekker.

LINCOFF, A. M., TCHENG, J. E., CALIFF, R. M., BASS, T., POPMA, J. et al. for the the PROLOG Investigators (1997a), Standard versus low-dose weight-adjusted heparin in patients treated with the platelet GPIIb/IIIa receptor antibody fragment abciximab (c7E3 Fab) during percutaneous coronary revascularization, *Am. J. Cardiol* **79**, 286–291.

LINCOFF, A. M., MARK, D. B., CALIFF, R. M., TCHENG, J. E., ELLIS, S. G. et al. (1997b), Economic assessment of platelet glycoprotein IIb/IIIa receptor blockade during coronary intervention in the EPILOG trial, *J. Am. Coll. Cardiol.* **29**(A), 240A.

LINCOFF, A. M., CALIFF, R. M., ANDERSON, K. M., WEISMAN, H. F., AGUIRRE, F. V. et al. for the EPIC Investigators (1997c), Evidence for prevention death and myocardial infarction with platelet membrane glycoprotein IIb/IIIa receptor blockade by abciximab (c7E3 Fab) among patients with unstable angina undergoing percutaneous coronary revascularization, *J. Am. Coll. Cardiol.* **30**, 149–156.

MARK, D. B., TALLEY, J. D., TOPOL, E. J., BOWMAN, L., LAM, L. C. et al. for the EPIC investigators (1996), Economic assessment of platelet glycoprotein IIb/IIIa inhibition for prevention of ischemic complications of high-risk coronary angioplasty, *Circulation* **94**, 629–635.

The CAPTURE Investigators (1997), Randomized placebo-controlled trial of abciximab before and during coronary intervention in refractory unstable angina: The CAPTURE study, *Lancet* **349**, 1429–1435.

The EPIC Investigators (1994), Use of a monoclonal antibody directed against the platelet glycoprotein IIb/IIIa receptor in high-risk coronary angioplasty, *N. Engl. J. Med.* **330,** 956–961.

The EPILOG Investigators (1997), Platelet glycoprotein IIb/IIIa receptor blockade and low-dose heparin during precutaneous coronary revascularization, *N. Engl. J. Med.* **336**, 1689–1696.

TOPOL, E. J., CALIFF, R. M., WEISMAN, H. F., ELLIS, S. G., TCHENG, J. E et al. (1994), Randomised trial of coronary intervention with antibody against platelet IIb/IIa integrin for reduction of clinical restenosis: results at six months, *Lancet* **343**, 881–886.

Gene Therapy

17 Overview of Gene Therapy

ANDREW MOUNTAIN

Keele, UK

1 Introduction 384
  1.1 What is Gene Therapy? 384
  1.2 Why is Gene Therapy Important? 384
2 Candidate Diseases 385
3 Approaches to Gene Delivery 385
4 History and Clinical Status of Gene Therapy 386
5 Technical Issues for Gene Therapy Development 387
  5.1 Poor Access to Target Cells 387
  5.2 Low Transfection Efficiency 388
  5.3 Limited Transgene Size 389
  5.4 Inadequate Transgene Expression 389
  5.5 Poor Targeting 390
  5.6 Immune Responses 391
  5.7 Manufacturing and Safety Issues 391
6 The Future of Gene Therapy 392
7 General References 393

List of Abbreviations

AAV	adeno-associated virus
Ad	adenovirus
AFP	α fetoprotein
CEA	carcinoembryonic antigen
CNS	central nervous system
CTL	cytotoxic T lymphocyte
EBV	Epstein Barr virus
GDEPT	gene directed enzyme prodrug therapy
HSVTK	thymidine kinase of herpes simplex virus
kb	kilobase
PSA	prostate specific antigen
Rv	retrovirus
SV40	Simian virus 40
VEGF	vascular endothelial growth factor

1 Introduction

Since the first clinical studies began in 1990 gene therapy has attained a high profile with public and investors, and has become the focus of a whole new industry. Opinions on its potential range from the view that it represents a revolution in medicine, through fears of eugenic applications to the view that it is merely playing expensively and ineffectually with technology.

This overview and Chapters 18–21 of this volume will deal only with gene therapy of somatic cells (i.e., non-reproductive cells). For ethical and technical reasons there are no approved clinical protocols for germ line gene therapies at the present time.

In this overview the main approaches to gene therapy, the lessons learned in the first 8 years of clinical experience, and how it may develop in the future will be summarized. At first though an introduction to the other gene therapy chapters in this volume will be provided. I will begin though by attempting answers to two key questions which are not usually covered in the many reviews of this area.

1.1 What is Gene Therapy?

There are many definitions. I define gene therapy as the introduction of DNA or RNA into cells with the intention of clinical benefit (which can be either therapeutic or prophylactic). Introduction can take place *in vivo* or *ex vivo*, the latter being a type of cell therapy. I include DNA vaccines but exclude oligonucleotide therapy. The genes transferred in gene therapy are referred to as "transgenes". The use of reporter transgenes for the purpose of determining the fate of transfected cells and pharmacokinetics or biodistribution of cells and transgenes should more properly be termed "gene marking". Nevertheless, it is highly relevant to gene therapy and will not be excluded from this overview.

In earlier days the term gene therapy was often assumed to mean replacement of a defective gene with a functional one. This should really be termed "gene replacement therapy" and is one of the two main types of gene therapy, the other being "gene addition therapy". The latter is technically much easier and all the ongoing clinical protocols for gene therapies involve gene addition rather than gene replacement.

1.2 Why is Gene Therapy Important?

Gene therapy is regarded by many as a revolution in medicine because it promises to allow a unique degree of subtlety in therapeutic intervention. This is partly because it will eventually allow much longer duration of action than conventional drugs through sustained expression of the transgene and ultimately permanent correction of defects causing disease. It is partly also because gene therapy will result in biological effects which are more potent and better localized to the most appropriate cells and tissues for clinical benefit than any other form of therapy. Ultimately with gene therapy it should be possible to arrange for expression of the transgene at the optimal time for achieving benefit. Gene therapy thus promises to allow benefits for diseases which are presently untreatable, and substantial improvements in therapeutic ratio and proportion of cures for diseases which are currently poorly managed.

2 Candidate Diseases

Tab. 1 summarizes the more obvious disease targets for gene therapy. It does not provide an exhaustive list, but it does indicate that a wide range of diseases and cell types will be involved. Gene therapy vectors are complex entities and will probably remain so. Because manufacturing costs for these vectors are high, it is likely that for the next decade gene therapies will be aimed largely at life threatening or disabling diseases. Gene therapy will only be applied to other diseases when vectors giving much more efficient gene delivery and sustained expression have been developed, and when much more clinical experience has demonstrated a good safety profile.

3 Approaches to Gene Delivery

Gene therapies involve gene delivery either *ex vivo* or *in vivo*. In the *ex vivo* approach cells are removed from patients and trans-fected. Therapy involves administration of the engineered cells. The *ex vivo* approach has the advantages of giving more efficient gene delivery, requiring smaller quantities of vector and allowing propagation of the cells to generate higher cell doses. It suffers from the disadvantages of being largely patient-specific due to cell immunogenicity and more costly, because cell manipulation adds manufacturing and quality control difficulties.

Most *ex vivo* applications are patient-specific, i.e., they use autologous cells, but allogeneic cells are being evaluated as cell vaccines in several clinical studies. In general large pharmaceutical companies are not pursuing patient-specific *ex vivo* therapies (or cell therapies at all) because this is an unfamiliar business arena with limited profit margins due to heavy manufacturing costs and a high service component. Most *ex vivo* approaches involve either blood or tumor cells because these are the easiest to obtain from the patient, to propagate and to return. *Ex vivo* concepts are also under development with liver hepatocytes, muscle myoblasts and skin fibroblasts. *Ex vivo* gene therapy will become more broadly useful when it becomes possible to identify, purify and manipulate stem cells for more tissues and to achieve sustained lineage-specific expression.

The *in vivo* approach involves direct administration of the gene delivery vector to patients. Because work on targeting of gene delivery vectors and on cell type specific transgene expression is at an early stage all proto-

Tab. 1. Candidate Diseases, Target Cell Types and Approved Clinical Protocols

Disease	Target Cells	Number of Protocols Approved in USA by February 1998
Cancer	tumor cells antigen presenting cells	138
Monogenic inherited disorders	lung epithelial cells, macrophages T cells, blood progenitor cells, hepatocytes, muscle cells	33
HIV infection	T cells, blood progenitor cells	23
Cardiovascular	endothelial cells, vascular smooth muscle cells	4
Rheumatoid arthritis	synovial lining cells	1
Cubital tunnel syndrome	nerve cells	1

cols approved so far involve local injection for localization to appropriate cells, rather than systemic administration. The *in vivo* approach is not patient-specific and is attracting more interest from large pharmaceutical companies. Its main disadvantages concern a need for larger quantities of vector and relatively inefficient gene delivery at this stage.

There, are three main types of gene delivery vehicle – viral, non-viral and physical. At the present stage of development viral vectors tend to give much more efficient gene delivery, so most of the ongoing clinical studies involve viral vectors. Many different viruses are being adapted as delivery vehicles, but the most advanced and commonly used ones are undoubtedly retrovirus (Rv), adenovirus (Ad) and adeno-associated virus (AAV). Chapter 18 in this volume covers in detail the pros and cons of these three vector systems, together with vector design, manufacturing, preclinical and clinical experience with them.

Many different approaches are also being taken to develop non-viral vectors. These also fall into three main categories, involving naked DNA, DNA condensed by cationic lipids and particles comprising DNA condensed by cationic polymers (and in some cases contained in anionic liposomes). The most important physical methods involve electroporation and biolistic (particle bombardment) approaches. The latter is usually referred to as the gene gun. Chapter 19 of this volume covers non-viral and physical methods in depth.

4 History and Clinical Status of Gene Therapy

At the time of writing approximately 300 clinical protocols have been approved world wide since the first trial began in 1990. 232 of these have been approved in the USA, of which 32 have no therapeutic intcnt. As indicated in Tab. 1 about two thirds of the 200 genuine therapy protocols are directed at cancer, and all but 6 of the others at either inher-

ited monogenic disorders or HIV infection. Chapter 21 in this volume covers HIV in depth. Half of the protocols directed to inherited disorders concern cystic fibrosis. This distribution partly reflects the lack of effective alternative therapies for these diseases.

For approximately 90% of all approved protocols the therapeutic concept involves the killing of disease related cells (by direct or immunological means) rather than the long term restoration of missing or defective proteins. This is because the present generation of vectors gives only inefficient delivery and transient expression. About 80% of all approved protocols involve viral vectors, with retroviral vectors accounting for 60%.

For quite a large proportion of the approved protocols no patients have been treated in the 6–9 months following approval. This illustrates the difficulties encountered in manufacturing vectors, especially viral vectors, in the required quality and quantity. This reflects the complexity of these vectors and demonstrates that manufacturing is a major issue for gene therapy. To address the problem the US government has established three expert centers for development and manufacture of gene delivery vectors.

More than 3,000 patients have been administered with gene therapies at the time of writing, and although there are several anecdotal reports of clinical responses and many reports of encouraging biological changes based on surrogate markers there has been no conclusive evidence of clinical benefit so far. Only a handful of gene therapies has progressed into phase II clinical studies. These include: an *in vivo* cancer vaccine approach in which cationic lipids are used to deliver a protein for increased immunogenicity of tumor cells (sponsored by Vical); an *in vivo* HIV vaccine in which Rv is used to deliver genes encoding HIV proteins (Viagene/Chiron); an *in vivo* approach to cancer which involves intratumoral injection of Ad to deliver the gene for the p53 tumor suppressor to induce cytostatic and apoptotic effects (Schering Plough, RPR Gencell/Introgene); an *ex vivo* approach to HIV which involves injection of cytotoxic T cells engineered for specific recognition of HIV infected cells (Cell Genesys); an *in vivo* approach to ischemic heart disease which involves injection of Ad or naked DNA encoding angiogenic factors into the coronary artery; and an *in vivo* approach to cancer which employs a virus only able to replicate in and lyse tumor cells defective for the tumor suppressor p53 (Onyx). It is notable that this short list includes a wide spectrum of different approaches. Only one phase III study is ongoing at present (Novartis/Genetic Therapy Inc). Interestingly this involves *in vivo* gene delivery but with a cellular inoculum: mouse cells producing a recombinant Rv for delivery of the HSVTK gene (thymidine kinase gene of herpes simplex virus) are inoculated into the brain of patients with terminal glioblastoma, and subsequent administration of the prodrug gancyclovir leads to killing of the HSVTK expressing cells. This is one example of a popular approach termed GDEPT (gene directed enzyme prodrug therapy).

The great majority of ongoing or completed clinical studies are small phase I/II studies in which the main objective is to demonstrate safety and obtain information to guide dose selection for subsequent efficacy trials. In general these studies have led to positive conclusions regarding safety aspects of gene therapies. Very few adverse events have been observed, and these have mostly been inflammatory reactions to Ad vectors given in large doses.

There are several reasons for the failure to obtain conclusive evidence for clinical benefit of gene therapies thus far. One of them is the intractable nature of the diseases being tackled and the terminal stage of many of the patients treated. It is unreasonable to expect complete clinical responses in phase I studies for AIDS and cancer, diseases which have resisted all other therapeutic modalities for so long. Another reason is that animal models tend to be less relevant for complex biological agents and preclinical data, therefore, less predictive of clinical results. However, the most important reason concerns the inadequacy of the first generation of gene delivery vectors. Lack of positive data in clinical studies supported by the USA's National Institute of Health prompted that body to form a committee of experts to scrutinize the available data in late 1995. The committee recommended that resources be diverted away from premature clinical studies with inadequate gene delivery vectors and into the development of improved vectors. This adversely affected public and investor confidence in gene therapy at the time, but it has probably had a positive effect overall, in that it has led to a more productive use of resources for the progression of gene therapy.

5 Technical Issues for Gene Therapy Development

These early preclinical and clinical experiences, though disappointing, have helped to define the main technical hurdles which need to be addressed for gene therapies to be meaningfully assessed in the clinic. These hurdles are:

1. poor access to target cells,
2. low transfection efficiency,
3. limited transgene size,
4. inadequate transgene expression,
5. poor targeting,
6. immune responses,
7. manufacturing and safety issues.

5.1 Poor Access to Target Cells

This is a major hurdle only for *in vivo* gene delivery – good access to target cells is a major advantage of the *ex vivo* approach. In general part of the problem concerns inactivation of the delivery system by components of body fluids prior to reaching the target cells. Mouse Rv's, e.g., are rapidly inactivated by the human complement system, although recent preclinical data suggest this susceptibility can be overcome. Human Ad-based vectors appear to be prevented from accessing many target cell types by neutralizing antibodies. This applies even to the first administration, since most humans have prior exposure to Ad. A notable exception to this limitation concerns intratumoral injection – data from clinical trials in which Ad is used to deliver a wild-type p53 gene suggests the efficiency of gene delivery on repeat dosing is not substantially reduced by strong anti-Ad antibody responses which are boosted as a consequence of the first administration.

For non-viral vectors the access problem often has two components. The first is the tendency of many formulations to aggregate, and the second is their tendency to be rapidly inactivated by components of serum. The latter is not well characterized, but is presumed to involve binding of anionic serum proteins.

The large size of gene delivery vehicles is likely to lead to poor access to many target cell types for purely physical reasons. The penetration of solid tumors following intravenous injection is likely to be very limited for all vector types, as is distribution through solid tissues after a single localized injection. In the short-term, therefore, the most attractive target cells from this viewpoint are those in the peripheral circulation, endothelial cells and hepatocytes (accessible through the sinusoidal liver vessels).

Electroporation and gene gun are both capable of fairly efficient gene delivery to several cell types *ex vivo,* but their *in vivo* use is limited to target cells in the immediate vicinity of exposed epithelial cells. Langerhans cells in the skin is the most important *in vivo* application of these physical methods for vaccines.

5.2 Low Transfection Efficiency

Assuming reasonable access to the surface of the target cells the second major hurdle for many vectors concerns achieving efficient and productive gene delivery. For viral vectors cell entry requires the cell to possess the appropriate receptor on its surface. A wide variety of human cell types has the natural receptors for the commonly used Rv, Ad and AAV vectors, for which cell entry appears to be very efficient. For untargeted non-viral vectors the mechanism of cell penetration is often unclear, but it is widely assumed to involve electrostatic interaction between the anionic cell surface and the cationic vector surface. For some cell type and non-viral vector combinations cell entry has been shown to

be efficient, >30% of all cells being penetrated.

The second main factor affecting transfection efficiency concerns trafficking to the site allowing expression. This is generally in the nucleus, although vectors allowing expression in the cytoplasm without nuclear access are being developed. The commonly used viral vectors traffic to the nucleus very efficiently, although by different routes. Once there Ad does not generally integrate (the efficiency of integration is <1 in a million), but efficiently gives expression of the transgene, i.e., nuclear penetration is generally productive. Rv requires integration to effect transduction (the term used for transfection by viruses) and this only occurs during mitosis. Rv, therefore, will only transduce proliferating cells, which is a major limitation of this vector. This has led to a major effort to develop alternative integrating viral vectors. Good progress has been made in developing lentiviral vectors, which appear capable of integrative transduction of both dividing and non-dividing cells. AAV is capable of fairly efficient transduction of a variety of human cell types. AAV transduction is less well characterized than that by Ad or Rv, but it appears likely that AAV is capable of both episomal and integrative transgene expression. Integration appears usually to be a very slow and inefficient process occurring after weeks of persistence as an episome, perhaps as a response to certain forms of insult to the cell.

In general productive intracellular trafficking to sites allowing transgene expression is much less efficient for non-viral vectors. Although characterization of trafficking has been rudimentary to date, it is clear that most non-viral vectors penetrate the cytoplasm via the endosome, and escape from the latter is undoubtedly a major hurdle for them. It is likely that penetrating the nuclear membrane is also a substantial hurdle for non-viral vectors. This probably means that non-viral vectors transfect proliferating cells more efficiently than quiescent ones, although it has proven very difficult to demonstrate this unequivocally. Major efforts are under way to improve trafficking from endosome to nucleus for non-viral vectors, often by incorporating components from viruses.

5.3 Limited Transgene Size

This is a major limitation for the first generation of viral vectors, for which insert capacities are around 4.5 kb for AAV, 6.5 kb for Rv and 8.5 kb for Ad. It is likely to prove difficult or impossible to increase these capacities for AAV and Rv, because even the first generation vectors contain only minimal viral sequences. Many candidate genes are much larger than this – the muscle dystrophin gene for muscular dystophy, e.g., is several hundred kb in length. The gene for the cystic fibrosis transmembrane receptor is 4.5 kb, and can only be delivered by AAV along with very short gene expression signals. There is much more scope for increasing the insert capacity for Ad vectors, because the virus has a much bigger genome. Vectors accomodating 13 kb have already been made, as have the first attempts at vectors in which most of the viral genome is removed. These "gutless" or "delta" vectors will eventually be able to accommodate up to 38 kb.

Non-viral vectors have much larger capacity for transgenes and gene expression signals, and the main limitation is likely to arise from difficulties in achieving high manufacturing yields for plasmids in excess of 50–100 kb.

5.4 Inadequate Transgene Expression

Transgene expression in transfected cells is often inadequate in terms of either the level or the duration of expression. Non-integrating vectors (Ad, AAV sometimes, non-viral) are capable of giving a transient burst of high level expression for the transgene over several days when strong constitutive promoters (typically derived from viruses) are used. Much more prolonged expression, up to several months, has been observed following Ad or AAV delivery to certain slowly dividing tissues, notable CNS and muscle.

Expression over a few hours or days is probably adequate for strategies which involve cell killing (such as GDEPT for cancer) or vaccination, but not for therapeutic strategies which involve longer-term restoration of missing proteins (such as correction of inherited monogenic disorders).

Integrating vectors (Rv, lentivirus) tend to give more sustained expression, but typically at a lower level. The expression level is determined partly by the number of integrated copies, partly by the choice of promoter element, but also partly by endogenous transcriptional activity at the chromosomal insertion site. Transgene expression usually ceases within a few days to weeks due to either methylation of the inserted DNA or to conversion of the region of integration from euchromatin to transcriptionally inactive heterochromatin. Recent data suggest these two epigenetic effects are interrelated. Rather slow progress is being made in the incorporation into these vectors of insulator sequences to protect transgene expression from local chromatin remodeling, in order to allow more sustained and predictable expression. An alternative approach involves building into the vector locus control regions which are capable of maintaining chromatin in a transcriptionally active conformation indefinitely. At their present level of characterization, however, these locus control regions are too large for delivery by Rv or lentivirus vectors. Progress has recently been reported in the development of non-viral vectors which integrate into the preferred insertion site of AAV (on human chromosome 19) by virtue of expression of the replication/recombination gene of the latter.

An alternative approach to achieving sustained, high level expression involves the use of replicating vectors. Progress has been made in the last 2 years in the development of such "episomal" vectors derived from oncogenic viruses such as Epstein Barr virus (EBV) and Simian virus 40 (SV40) and also in the development of minichromosome vectors. The former are relatively small but pose safety issues, while the latter are very large and present serious difficulties in manufacturing and delivery. Very recently a sophisticated vector has been developed which involves Ad delivery of a "proepisome" released from the Ad genome by site-specific recombination. This vector has been shown to allow replication and stable maintenance under drug selection over many generations *in vitro,* and also following delivery by direct injection into tumors. This is a highly promising approach which combines the high efficiency of Ad delivery with sustained expression in daughter cells which, following cell division, do not express the Ad genes responsible for provoking unwanted immune destruction of the cells.

5.5 Poor Targeting

One of the major conceptual advantages of gene therapy approaches to medicine is the possibility of achieving transgene expression restricted to the most appropriate target cells and to the most appropriate times for conferring clinical benefit. Ultimately this will be achieved by targeting at three levels. One involves local injection of the vector, the second involves targeted uptake into cells via specific cell surface receptors, and the third involves cell type specific transcription. At the present time the second and third of these are at a fairly early stage of development.

Incorporation of particular ligands into Rv envelopes has given targeted binding to cells expressing the cognate receptors, but targeting of producive infection is proving much more difficult to achieve, and the viral titers are reduced to levels well below useful levels. In contrast it has recently become clear that Ad is capable of infecting cells efficiently following retargeting through several receptors. There are published examples of retargeting through the use of bifunctional reagents which block interaction with the usual Ad receptor and promote binding to other cell surface molecules, and also of retargeting through engineering-specific peptide ligands into one of the Ad cell recognition proteins. This retargeting is likely to allow significant extension of clinical utility for Ad vectors because it should reduce systemic toxicity, especially in the liver, which is the organ of limiting toxicity for Ad. Very little information is available concerning the possibility of retargeting AAV.

In principle targeted uptake should prove easier to accomplish for non-viral vectors because these allow attachment of ligands by chemical conjugation. Although there are several reports of targeted transfection for cells *in vitro,* there are very few *in vivo* because this requires rcsolution of several other problems which lead to low transfection efficiency for the present generation of non-viral delivery vehicles.

Three types of approach are being taken to achieve targeted transcription. One involves the use of naturally occuring promoter elements which show activity restricted to particular cell types, or more commonly to particular tumor types. Examples are the AFP promoter (liver), PSA promoter (prostate), VEGF receptor (prolifering endothelial cells), CEA promoter (several tumor types expressing the CEA antigen) and her2 promoter (breast and ovarian tumors). Unfortunately most of these promoters give relatively weak expression in the permissive cell types and may require further engineering to be genuinely useful.

The second approach involves the use of locus control regions, which in addition to giving sustained expression resistant to heterochromatization give tissue specific expression. Elements of this type have been reported which are capable of restricting expression to T cells, erythroid cells, antigen presenting cells, macrophages, brain and muscle. These elements promise to be extremely useful once integrating non-viral vectors have been developed. In the meantime their utility on episomal vectors is being explored.

The third approach involves promoters allowing induction to expression in response to exogenous substances. Many such systems are being developed, and the most advanced ones are induced by erythropoietin, tetracycline, rapamycin or steroids. The last two have been shown to allow repeated induction over a period of months with oral administration of the inducer following delivery with AAV vectors in mice. These studies are the closest approximation yet to proof of principle for the holy grail of gene therapy – long-term transgene persistence with expression turned on and off at will by orally administered compounds.

While these targeting methodologies are being developed in a clinically acceptable format many clinical protocols are ongoing in which targeting is achieved by local injection into the target tissues (e.g., into the heart for angiogenic approaches to atherosclerosis, into superficial tumors for cancer) or into local blood vessels supplying them, or by natural tropisms (e.g., exploiting the restriction of Rv transduction to dividing cells in brain tumor protocols).

5.6 Immune Responses

Immune responses to both the delivery vehicle and the transgene product are major hurdles for gene therapy. Responses to the delivery vehicle are generally believed to be a much greater problem for viral vectors than non-viral ones and it is clear that this is a major factor limiting the utility of Ad vectors. Strong immune and inflammatory responses have been observed in animals and in essentially all patients exposed to Ad vectors. Indeed the only serious adverse events observed in all the clinical studies on gene therapy thus far have been inflammatory responses to Ad administered in large doses into the respiratory tract of patients with cystic fibrosis. Although these responses are still rather poorly characterized the speed at which they develop suggest they arise through production of inflammatory cytokines in response to Ad coat proteins or to receptor engagement. Nevertheless it is very clear from animal studies that Ad administration leads to strong and neutralizing antibody responses to coat proteins, and to CTL responses directed to cells expressing Ad genes and transgenes with heterologous products. The antibody and CTL responses reduce the effectiveness of repeated dosing and the CTL responses are probably responsible for the termination of transgene expression in some cases.

Rv and AAV appear to be much less immunogenic than Ad, partly due to the nature of their protective coats and partly because the vectors do not give expression of viral genes. However, there is much less clinical experience with these vectors than Ad for *in vivo* protocols. CTL responses to heterologous transgene encoded proteins have been observed for T cells engineered *ex vivo* in patients after a single dose, even in seriously immunocompromised AIDS patients. These responses drastically reduced the circulating half-life of the engineered cells on repeated dosing. These observations strongly suggest that CTL responses directed to cells expressing proteins recognized as foreign will be a major problem for gene therapy, interfering with sustained expression and repeated dosing.

Approaches being taken to circumvent the neutralizing antibody responses include increasing doses to achieve an excess of vector over neutralizing capacity, covalent attachment of polyethyleneglycol, and in the case of Ad switching human serotypes or using Ad derived from other species. Several approaches are being taken to avoid CTL reponses. Most of these involve exploiting the genes and sequences which human viruses employ to prevent presentation of viral proteins on class I HLA molecules. Alternative approaches involve using antibodies to effect a transient block of T cell costimulation or non-specific immunosuppression during repeated dosing. These approaches are all at a fairly early stage of development and evaluation at present.

5.7 Manufacturing and Safety Issues

All gene therapy vectors are complex entities for which manufacturing costs are presently very high and product yields relatively low. This is particularly true for viral vectors because these are biological entities for which achieving robust, high yielding processes for production, purification and quality control of product is a formidable challenge. Trials for many approved clinical protocols have not begun within 6 months of approval because of this difficulty in producing the vectors in adequate quality and quantity. Rv in particular has proven very difficult to produce with suitably high titers. As indicated in Chapter 18 for almost all applications viral vectors and producer cell lines have to be engineered such that the vectors are replication defective for safety reasons. Avoiding contamination of product with replication competent virus is thus a significant hurdle from the safety and regulatory standpoint. At the present time exploration of the clinical utility of Ad vectors at high doses is being delayed by this problem.

No doubt these problems will be resolved in the next few years. Manufacturing costs, however, will always be much higher for gene therapy vectors than for small molecule therapies. The cheapest gene therapies at the moment involve gene gun delivery of small doses of naked DNA for vaccination. Under the managed health care system which prevails in the USA it is likely the more expensive gene therapies will need to show dramatically superior benefit as sole therapies in order to become widely used. For example, the system will probably not support the use of an expensive gene therapy in addition to that of an expensive bone marrow transplant and multiple chemotherapies in combination for cancer. Fortunately, there is every prospect that gene therapies will give dramatically improved outcomes.

6 The Future of Gene Therapy

Thus far the clinical results with gene therapy have disappointed many people, but this is partly because their expectations were unrealistic. Since the period of hype in the early 1990s, and negativity in 1995 and 1996 the field has settled into a phase of real and steady progress.

Gene therapies are complex in nature, and relevant animal models are not available in most cases. It is likely that extensive series of clinical studies will be required in order to evaluate the many changes with potential for improving gene delivery and expression characteristics. This may require an unconventional regulatory framework.

The most promising therapeutic concepts in the short term concern direct killing of tumor cells with locoregional administration of Ad vectors for certain localized cancer types, immunotherapies such as DNA vaccines and stimulation of angiogenesis for ischemic heart

disease because these require only relatively inefficient delivery and short lived expression. Several *ex vivo* approaches using sustained expression in blood progenitor cells following Rv delivery may also become established in the fairly near term. The *ex vivo* approach will be applicable to many more indications when the isolation and proliferation of stem cells for more tissue types can be achieved, and if the problem of immune responses to allogenic cells is overcome so that cell therapies are no longer patient-specific.

At the present time Rv is the vector of choice for *ex vivo* therapies because it gives the most stable expression in expanded cell populations. Lentivirus vectors hold more promise for these applications, but in the short term at least these present regulatory issues because they are derived from immunodeficiency viruses such as HIV. In the medium term episomal vectors will be developed to the point at which an adequate duration of expression can also be achieved for *ex vivo* applications following non-viral delivery.

For *in vivo* applications Ad is presently the vector of choice because of its capability for efficient gene delivery, and real progress is being made towards overcoming its major disadvantages of short lived transgene expression and immunogenicity. In the long term most gene therapy products of substantial commercial value are likely to involve non-viral vectors because of their advantages in manufacturing and safety.

No doubt most of the new clinical protocols will involve gene addition therapy for the next few years. Although not covered in this review exciting developments have recently been reported which involve heritable correction of single base mutations in cells in tissue culture using triplex forming oligonucleotides. This is the first step on the long road to gene replacement therapy, which is no longer in the realms of science fiction.

From the medical viewpoint the ideal gene therapy for many diseases would involve permanent inheritance of the transgene in the most appropriate cell in a safe manner, with expression induced by a cheap, orally administered agent at the most appropriate time(s) for clinical benefit or by an endogenous disease episode-related inducer. Although it is clear that many hurdles remain to be overcome this scenario appears to be achievable in the longer term. I anticipate that the medical (and commercial) potential of gene therapy, together with the remarkable advances being made in molecular and cell biology, will ensure that the achievable becomes reality.

Meanwhile the genomics revolution will ensure that an ever increasing range of candidate diseases and transgenes are available and characterized sufficiently to permit gene therapy approaches. There is no doubt in my mind that genomics and gene therapy together represent a genuine revolution in medicine.

7 General References

FELGNER, P. L. (1997), Novel strategies for gene therapy, *Sci. Am.* June, 86–95

FRIEDMANN, T. (1997), Overcoming the obstacles to gene therapy, *Sci. Am.* June, 80–85.

BOULIKAS, T. (Ed.) (1998), *Gene Therapy and Molecular Biology – From Basic Mechanisms to Clinical Applications,* Palo Alto, CA: Gene Therapy Press.

ANDERSON, W. F. (1998), Human gene therapy, *Nature* **392** (Supplement on Therapeutic Horizons), 25–30.

GAGE, F. H. (1998), Cell therapy, *Nature* **392** (Supplement on Therapeutic Horizons), 18–24.

LIN, M. A. (1998), Vaccine developments, *Nature Medicine* **4** (Vaccine Supplement), 515–519.

BLAU, H., KHAVARI, P. (1997), Gene therapy: progress, problems, prospects, *Nature Medicine* **3**, 612–613.

MELCHER, A., GARCIA-RIBAS, I., VILE, R. (1997), Gene therapy for cancer – managing expectations, *BMJ* **315**, 1604–1607.

MARTIN, P., THOMAS, S. (1998), The commercial development of gene therapy in Europe and the USA, *Human Gene Ther.* **9**, 87–114.

VERMA, I., SOMIA, N. (1997), Gene therapy – promises, problems and prospects, *Nature* **389**, 239–242.

18 Viral Vectors for Gene Therapy

BARRIE J. CARTER

Seattle, WA, USA

1 Overview of Viral Vectors 397
2 Retrovirus Vectors 398
  2.1 Retrovirus Structure and Biology 399
    2.1.1 Biology of Retrovirus Life Cycle 399
    2.1.2 Structure of Virus and its Genome 399
    2.1.3 Mode of Cell Entry and Host Tropism 400
    2.1.4 Integration and Persistence 400
  2.2 Design of Retrovirus Vectors 400
    2.2.1 Elements of Virus Genome Required 400
    2.2.2 Elements of Particle Structure that Influence Vector Properties 401
    2.2.3 Design of Transgene Expression Cassettes 402
  2.3 Retrovirus Vector Manufacturing 402
    2.3.1 Complementation Systems 402
    2.3.2 Packaging Cells 403
    2.3.3 RCR Contamination 403
    2.3.4 Purification 404
  2.4 Preclinical Studies with Retrovirus Vectors 404
    2.4.1 *In vitro* Studies 404
    2.4.2 *In vivo* Studies 404
      2.4.2.1 Target Cell Specificity and Expression 404
      2.4.2.2 Host Responses to Vectors 405
  2.5 Clinical Studies with Retrovirus Vectors 405
3 Adenovirus Vectors 406
  3.1 Adenovirus Structure and Biology 406
    3.1.1 Biology of Adenovirus 406
    3.1.2 Structure of Adenovirus and its Genome 406
    3.1.3 Adenovirus Life Cycle 407
    3.1.4 Latency and Persistence 408
  3.2 Design of Adenovirus Vectors 408
    3.2.1 Elements of Virus Genome Required 408
    3.2.2 Elements of Particle Structure that Influence Vector Properties 408
    3.2.3 Design of Transgene Expression Cassettes 409

3.3 Adenovirus Vector Manufacturing 409
    3.3.1 Complementation Systems 409
    3.3.2 Packaging Cells 409
    3.3.3 RCA Contamination 410
    3.3.4 Purification of Adenovirus Vectors 411
3.4 Preclinical Studies with Adenovirus Vectors 411
    3.4.1 *In vitro* Studies 411
    3.4.2 *In vivo* Studies 411
        3.4.2.1 Organ and Cell Targeting 411
        3.4.2.2 Expression and Persistence 412
        3.4.2.3 Host Immune Responses, Inflammation and Toxicity 412
3.5 Clinical Studies with Adenovirus Vectors 413
4 Adeno-Associated Virus Vectors 413
4.1 Adeno-Associated Virus Structure and Biology 413
    4.1.1 Biology of Parental Virus and its Life Cycle 413
    4.1.2 Structure of AAV and its Genome 414
    4.1.3 Mode of Cell Entry and Host Tropism 415
    4.1.4 Replication or Persistence and Integration 415
4.2 Design of Adeno-Associated Virus Vectors 416
    4.2.1 Elements of Virus Genome Required 416
    4.2.2 Elements of Particle Structure that Influence Vector Properties 417
    4.2.3 Design Rules or Limitations for Gene Cassettes and Regulatory Elements 417
4.3 Adeno-Associated Virus Vector Manufacturing 417
    4.3.1 Complementation Systems 417
    4.3.2 Packaging Cells 418
    4.3.3 Replication Competent or Wild-Type AAV 418
    4.3.4 Purification 419
4.4 Preclinical Studies with Adeno-Associated Virus Vectors 419
    4.4.1 *In vitro* Studies 419
    4.4.2 *In vivo* Studies 420
        4.4.2.1 Cell Targeting, Expression and Persistence 420
        4.4.2.2 Host Responses and Toxicity 420
4.5 Clinical Studies with Adeno-Associated Virus Vectors 421
5 References 421

List of Abbreviations

AAV	adeno-associated virus
Ad	adenovirus
bp	base pair
CFTR	cystic fibrosis transmembrane receptor
CMV	cytomegalovirus
CNS	central nervous system
CTL	cytotoxic lymphocyte
DC	double copy
EBV	Epstein Barr virus
EMC	encephalomyocarditis virus
EPO	erythropoietin
GaLV	gibbon ape leukemia virus
HIV	human immunodeficiency virus
HIV1	human immunodeficiency virus type 1
HPLC	high pressure liquid chromatography
HSV	herpes simplex virus
HyTK	fusion product of the genes *hygro* and *tk*
IRES	internal ribosome entry site
ITR	inverted terminal repeat
kb	kilobase
LTR	long terminal repeat
MHC	major histocompatibility complex
MLP	major late promoter
Mo-MLV	Moloney murine leukemia virus
Mo-MSV	Moloney murine sarcoma virus
ORF	open reading frame
OTC	ornithine transcarbamylase
pfu	plaque forming unit
RCA	replication competent adenovirus
RCAAV	pseudo wild-type AAV
RCR	replication competent retrovirus
RF	replicating form
SIN	self-inactivating
SS	single-strand
tk	thymidine kinase gene
TNF	tumor necrosis factor
VSV	vesicular stomatitis virus

1 Overview of Viral Vectors

This chapter will be limited to the discussion of retrovirus vectors, adenovirus (Ad) vectors and adeno-associated virus (AAV) vectors since these are the only viral vectors that have advanced to clinical trials aimed at developing gene therapy products. For all three vectors, there is beginning to be a base of clinical data that will help to determine the level of confidence that preclinical testing provides for their safety and biological function. This will lead to a more rapid realization of the advantages and limitations of the vectors and promote optimization of the applications for which each may be suited best. Herpes viruses may soon be introduced into clinical trials but there is so far little formal preclinical biology or toxicology information. Vaccinia vectors are being used in clinical trials but they are generally considered as vaccines, and in the United States have not been reviewed by the NIH Recombinant Advisory Committee.

Rather than provide an exhaustive review of the literature, citations are provided to illustrate major themes and to guide access to more extensive information. Several recent reviews have discussed viral vectors generally (JOLLY, 1994; SMITH, 1995) or with emphasis on clinical trials (BRENNER, 1995; CRYSTAL, 1995). In addition there are various recent reviews on retrovectors (MULLIGAN, 1991; BUNNELL and MORGAN, 1996; HAVENGA et al., 1997), adenovirus vectors (BERKNER, 1988; CHENGVALA et al., 1991; WILSON, 1996; HITT et al., 1997) and AAV vectors (CARTER, 1992; MUZYCZKA, 1992; FLOTTE and CARTER, 1995) which are useful guides to much of the earlier literature.

Viruses have complex structures and life cycles and many are pathogens, but several viruses are highly efficient gene delivery vehicles. Non-viral vectors generally cannot approach the efficiency of viral vectors when considered on the basis of gene copies that must be presented to the target cell in order that one or several copies can be expressed in the cell nucleus. For viral vectors, the usual approach is to remove the unneeded or pathogenic features while retaining the efficiency of gene delivery, expression and persistence where appropriate. The goal of non-viral vector design is usually to add features to DNA to mimic those functions which viruses already perform well.

The design of viral vectors addresses many common considerations for both safety and biological activity including removal of virulence genes, absence of replication competent parental virus and host responses to components of the vector. Also, particular considerations for individual classes of vectors reflect the properties of the parental virus, the target organ or disease, and the gene to be delivered. Retrovectors integrate efficiently, so insertional mutagenesis was considered an important issue. Adenovirus vectors generate significant host cellular immune responses and require strategies to remove or prevent expression of adenovirus genes or possibly to develop approaches to transient immune modulation. For AAV vectors an important issue has been to develop efficient production systems.

Development of viral vectors requires analysis of the biology and life cycle of the parental virus and the structure of the virus and its genome. Genetic analysis is also required to determine which viral elements are required in *cis* and which genes can be supplied in *trans*. This allows the design of complementation systems to produce viral vectors which use generic packaging cells to supply the *trans* complementing functions. Packaging cell lines can then be used as producer cells for the generation of particular vectors. The main issues to be confronted are increasing the titer of vector and decreasing or eliminating the production of replication competent or wild-type virus. Purification of vectors is important for safety and product identity and because impurities may severely impact the biology of the vector. Once viral vectors can be produced, then the design rules or limitations for gene cassettes and regulatory elements can be determined.

Vectors can be assessed with *in vitro* experiments but these may not be predictive of their behavior *in vivo*. Many cells used in *in vitro* culture are transformed, and even primary cells do not necessarily reflect the function of the same cells *in vivo*. Thus, it is essential to perform extensive *in vivo* experiments in animals to assess organ and cell targeting, the efficiency, specificity and persistence of gene expression and the likelihood of toxicity and host immune and inflammatory responses. Experiments in animals are also important because the properties of vectors may vary significantly from those that are exhibited by the parental virus. However, there is a paucity of defined, disease-specific animal models. Also, there is no well-established track record for determining which animal species or models are most useful for determining the potential biological efficiency of gene delivery and expression or the likelihood of dose-limiting toxicities.

Gene therapy in patients began with *ex vivo* approaches by taking advantage of the stable integration of retrovectors to modify rapidly dividing cells (notably lymphocytes) that were then returned to the patient to treat either rare monogenic diseases, e.g., adenosine deaminase deficiency, or a common acquired disease such as cancer. However, *in vivo* delivery systems have rapidly been adopted as is reflected in the use of adenovirus and AAV vectors.

2 Retrovirus Vectors

Retrovectors integrate with high efficiency and contain no viral genes so they can mediate long-term expression and avoid host cellular immune responses. Retrovectors are limited by an inability to transduce non-dividing cells, difficulty in obtaining high-titer production, and instability of these lipid enveloped viruses which renders purification and concentration difficult. The viral LTR (long terminal repeat) sequence may interfere with gene expression of various gene cassettes.

2.1 Retrovirus Structure and Biology

2.1.1 Biology of Retrovirus Life Cycle

Retroviruses are found in a wide variety of animal species and some are causally associated with particular diseases. They include three subfamilies, the oncovirinae such as the avian Rous sarcoma virus and Moloney murine leukemia (Mo-MLV) virus, the lentivirinae such as HIV and the spumavirinae such as simian foamy virus. All current retrovectors are based upon the Mo-MLV retrovirus and like all members of this subfamily do not infect non-dividing cells. Viruses of the other two subfamilies can infect non-dividing cells and initial attempts to develop vectors based on these viruses are beginning (ZUFFEREY et al., 1997) but will not be described here.

Oncoretroviruses enter cells via an interaction between the viral envelope protein and a cellular receptor which controls the host tropism and target cell specificity of retroviral vectors. After entry into the cytoplasm, the viral RNA is converted to a double-strand DNA which integrates into the host cell genome. The integrated provirus generates new RNA genomes which are condensed into core particles and further encased in the lipoprotein envelope by budding from the cell surface. Oncoretroviruses may cause diseases by carrying a homolog of a cellular oncogene or by activating the transcriptional promoter of a cellular oncogene.

2.1.2 Structure of Virus and its Genome

Retroviruses are complex particles with two copies of a positive strand RNA genome condensed within a core containing several viral proteins and are enclosed in a lipid envelope containing both viral and host cell proteins (COFFIN, 1984). The viral RNA genome (Fig. 1) is about 9 kb long and contains 3 genes, in order *gag, pol* and *env*, bounded by terminal *cis*-acting control regions. These *cis*-acting regions include, at either end of the genome, the same long terminal repeat (LTR) region which contains a subregion that has transcriptional enhancer and promoter activities followed by a polyadenylation site. There is a splice donor site between the 5′ LTR and the *gag* gene and a splice acceptor site in the 3′ end of the *pol* gene upstream of the *env* gene. The splice donor site is followed by a region called Ψ+, that overlaps the 5′ end of *gag* and is required in *cis* for encapsidation of the viral RNA.

Transcription of retroviruses occurs from the integrated DNA provirus initiating at the 5′ LTR as the transcription promoter and terminating at the polyadenylation site at the 3′ LTR. The *gag* and *pol* genes are transcribed as a single unit. The *env* gene is expressed from a transcript that is spliced to eliminate both *gag* and *pol*. All three retroviral genes are multifunctional. The Gag protein is processed by proteolytic cleavage to generate p15 (matrix protein), p12, p30 (capsid protein) and p10 (nucleocapsid protein). The Pol protein encodes protease, reverse

Fig. 1. Genome structure of Moloney murine leukemia virus. The negative strand of the viral RNA genome is shown schematically. The genes are *gag, pol,* and *env*. The LTR regions contain the transcription enhancer and promoter (EP), the tRNA primer binding site (−P) or the positive strand initiation site (+P). The splice donor (SD), splice acceptor (SA), and packaging signal (Ψ) are indicated. The Gag protein is processed to generate 4 proteins; matrix (MA), p12, capsid (CA) and nucleocapsid (NC). The *pol* gene contains individual domains for protease function (PR), reverse transcriptase (RT), and integrase (IN). The Env polyprotein yields the gp70 surface protein (SU) and the p15E transmembrane protein (TM) (BUNNELL and MORGAN; 1996).

transcriptase and integrase activities. The Env protein encompasses a gp70 surface protein and a p15 transmembrane protein.

2.1.3 Mode of Cell Entry and Host Tropism

The initial binding of retroviruses to host cells depends upon interaction between the viral gp70 Env protein and a cellular receptor, and this is a key issue for retroviral vector design (MILLER, 1996). The host range of retrovectors can be altered by changing the Env protein in a process referred to as pseudotyping. The normal ecotropic Env protein of Mo-MLV does not bind to human cells but retrovectors can be packaged using an *env* gene from an amphotropic virus that infects human cells. This has been done in all retroviruses used clinically, except in one case where an *env* gene from gibbon ape leukemia virus (GaLV) has been used for clinical application. The GaLV Env receptor occurs with high frequency on human lymphocytes, stem cells, and producer cells using this *env* gene appear to yield improved vector titers (MILLER et al., 1991).

2.1.4 Integration and Persistence

After entry into the cytoplasm the retroviral RNA is used as a template by the reverse transcriptase contained in the viral core to generate a duplex viral DNA genome. This DNA duplex is then integrated into the host cell genome in a recombination process mediated by the viral integrase that is also contained within the viral core particle. For oncoretroviruses, this DNA cannot cross the nuclear membrane and these viruses only integrate into host cell DNA of dividing cells as they go through mitosis to disrupt the nuclear membrane. This limits the use of retrovirus vectors to dividing cells.

2.2 Design of Retrovirus Vectors

Retroviral vectors and packaging systems have been investigated by a large body of investigators over the last 15 years, and vectors produced in at least 5 different systems have now been used in clinical trials. However, there has been no consistent effort on comparative analysis of design rules for retrovector, and retrovector development retains, a large element of empiricism. Nevertheless, general considerations can be identified and clinical applications have three main challenges; vector titers must be maximized to enhance specific productivity and decrease biomass requirements in manufacturing; transgene expression must be optimized to enhance potency; and production of replication competent retrovirus (RCR) must be avoided.

2.2.1 Elements of Virus Genome Required

The viral 5' and 3' LTR regions are required in *cis* for viral replication functions. Another required *cis* function is the packaging signal, which generally is the extended $\Psi+$ region, because it usually leads to higher titers. Many vectors have both a LTR and an $\Psi+$ based on the Mo-MLV virus as in the prototype vector N2. In the N2 vector the LTR is used as the transcription promoter, but several modifications have been made to enhance expression, to increase vector titer and to decrease recombinational generation of RCR.

In one series of vectors, LNL6 (HOCH et al., 1989), the AUG initiation codon of the *gag* gene was mutated to a TAG termination codon. This ensures that recombinants that might arise would be crippled in initiating *gag* expression and also enhance translation from the inserted gene located downstream. Also, a hybrid LTR was constructed using elements of the Mo-MLV and Mo-MSV viruses to improve the packaging titer and decrease homology with sequences in the packaging cells. In subsequent vectors, such as LN derived from LNL6 (MILLER and ROSMAN, 1989) and MFG (JAFFEE et al., 1993), the remaining 3' regions of the *env* gene were deleted to decrease homologous recombination with sequences in the packaging line. The MFG vector also retains the splice donor site in the extended $\Psi+$ region followed downstream by the splice acceptor sites and the AUG initiation codon of the *env* gene. In this vector the gene of interest is inserted such that its translation is initiated from the *env* AUG codon. Both the LNL6 and MFG type vectors yield good titer preparations that are free of RCR if used with the appropriate packaging cell line, and both can yield robust gene expression.

The above vectors use the LTR as the transcription promoter. Other vectors have been designed to use inserted internal promoters but this may lead to interference from the LTR (see Sect. 2.2.3). This has in turn led to vectors in which the LTR is inactivated after integration by taking advantage of a peculiarity of the mechanism of integration of retroviruses, namely that the 5′ LTR of the provirus is derived from the 3′ LTR of the viral genome. Self-inactivating (SIN) vectors are deleted in the proximal U5 transcription enhancer region of the 3′ LTR to generate an inserted provirus in which the 5′ LTR is deleted for transcription promoter functions. Alternately, in double copy (DC) vectors the gene cassette and promoter are inserted into the 3′ LTR region to generate an integrated provirus which has both 5′ and 3′ LTRs containing the inserted gene cassette to generate two transcriptionally active copies of the desired gene. Neither SIN nor DC vectors have been used clinically because the titers of such vector preparations are often low and there is some concern about possible rearrangements of the inserted gene cassettes.

2.2.2 Elements of Particle Structure that Influence Vector Properties

The particle structure of retrovectors primarily influences the packaging limit and the host range. The Mo-MLV genome is about 10 kb, but the LTRs and the extended $\Psi+$ region are required. So the limit for the payload sequence is about 8 kb, and internal transcription promoters or other tissue-specific regulatory sequences reduce the packaging limit accordingly. There is no absolute requirement for a full-size vector genome, but some DNA sequences have proved difficult to incorporate into stable retroviruses. This may reflect instability of particular sequences in the retrovirus reverse transcription cycle.

The *env* gene controls the host range and specificity of retrovirus. The Env protein of the Mo-MLV retroviruses exhibits an ecotropic host range and infects only rodent cells. Most retrovectors used in human clinical trials use the *env* gene from an amphotropic retrovirus, strain 4070, which permits infection of human cells via the Ram receptor protein. One exception is the vector derived from a packaging line PG13 in which the *env* gene is from GaLV. The GaLV Env protein binds to a receptor Glv which is also present on human cells, but the GaLV pseudotyped vector appears to be more efficient in transducing lymphocytes, and producer lines using the GaLV *env* gene tend to generate higher titer vector supernatants (MILLER et al., 1991). Both Ram and Glv are phosphate transporters, and starving the cells of phosphate may enhance vector entry. Retrovectors can be pseudotyped using the G protein from vesicular stomatitis virus (VSV) to take advantage of the very broad host range of the VSV protein and the increased stability of VSV. However, the yield of such pseudotyped vectors has been poor because the VSV G protein is toxic to cells, and generation of packaging cell lines has been difficult (ORY et al., 1996).

Retrovectors are unstable in human serum and are rapidly inactivated by human complement which has severely limited their application to *in vivo* systemic delivery. Inactivation by human complement of viruses produced in rodent or non-primate cell lines may be due to Gal(α1–3)galactosyl sugar on the Env proteins which is recognized by antibodies present in human serum. One way to avoid this problem is to produce the vectors in a packaging cell based on a human cell line using the envelope gene from the cat virus RD114. Thus resistance to human complement may be achieved by appropriate choice of both the cell line and the *env* gene.

2.2.3 Design of Transgene Expression Cassettes

In general, the optimum design of a retrovirus vector must be determined on a case-by-case basis. Some sequences inserted into retrovector backbones may be unstable and rearrange during replication or integration or may yield low titer vector. It is difficult to define clear rules to avoid such instability and very discrete changes in the vector backbone may be important.

The LTR is a relatively strong promoter and is often used. Transcription from the LTR generally cannot be regulated by physiologic events, and particularly for *in vivo* applications, expression from the LTR may be subject to transcriptional silencing, and this may prevent long-term expression.

Internal transcription promoters are employed in many retrovectors because they can be designed for tissue-specificity, for appropriate regulation by physiologic responses, and they may be less susceptible to transcriptional silencing. However, the LTR is generally still present in the vector, and this may interfere with a downstream internal promoter by promoter occlusion or by transcriptional silencing via global effects on the chromatin structure surrounding the integrated vector genome. Generally, insertion of the gene expression cassette in the same transcriptional orientation as the retroviral vector LTR provides better titer and vector stability, and the inclusion of intron sequences usually improves expression.

Internal transcription promoters may be used in retrovectors which express two genes, especially when one of the genes is a selective marker gene such as *neo, tk* or *hygro.* The selective marker gene may be beneficial in selecting for the transduced cell population and also provides a convenient way to titrate vector stocks. An alternative approach to design of 2-gene vectors is to use internal ribosome entry site (IRES) sequences derived from polio and EMC virus which allow ribosomal entry and translation from an internal region of the mRNA. Thus, two genes can be transcribed in tandem from a single transcription unit, but each can then be separately translated. Two genes may also be expressed as a single fusion protein. For example, a protein called HyTK, formed as a fusion of the positive selective marker, *hygro,* and the negative selective marker HSV thymidine kinase (*tk*), ensures that cells positively selected with hygromycin to contain the vector can be ablated by negative selection with ganciclovir. This was designed as a safety feature for early clinical trials (see Sect. 2.5).

Other ways to maximize expression include ensuring a favorable context for translation initiation as noted above in the MFG vector. In the HyTK vector, the region surrounding the AUG codon of the HyTK fusion gene was modified to conform to an optimal Kozak sequence often associated with efficient initiation of protein synthesis in mammalian cells (RIDDELL et al., 1996).

2.3 Retrovirus Vector Manufacturing

2.3.1 Complementation Systems

Retrovectors are produced by complementation with a stable packaging cell line that expresses the *gag* and *pol* genes to permit vector genome replication and formation of the viral particle core and an *env* gene which pseudotypes the vector and defines its receptor specificity and host cell tropism (MILLER, 1996). The vector genome is introduced stably into the complementing packaging cells to derive a producer cell line. Producer cells continuously generate vector particles which are released into the cell growth medium to provide vector supernatants. Producer cells can be generated by directly transfecting the vector plasmid into packaging cells and selecting for a producer cell clone. Alternatively, the packaging cells are transiently transfected with the vector plasmid to generate a supernatant containing vector particles which are then used to infect the packaging cells and to derive stable producer cell clones. This latter approach has the advantage that the vector is stably integrated into the producer cell via the normal retroviral integration process, and this generally leads to producer cell lines with higher titers.

In the packaging cell lines based on murine or canine cell lines, the retroviral supernatants generated by transient transfection of the vector plasmid often have relatively low titers. Recently, transient transfection systems based on human 293 packaging cells have been shown to readily generate supernatants with good titers. These systems may be useful for rapidly screening many vectors.

In any genetic complementation system recombination may occur. For retrovector production recombination might lead to production of RCR which could have significant potential for pathogenicity (see Sect. 2.3.3). Prevention of recombination events leading to generation of RCR has been addressed in several ways, through modification to both packaging cells and vectors, and the combination of both must be considered carefully. It is important to reduce sequence homology between the vector and any viral components of the packaging cells to decrease the likelihood of homologous recombination. Alteration of components of the complementation systems and separation of the *gag-pol* gene from the *env* gene can reduce homology with the vector and increase the number or recombination events required to generate RCR. The most recent packaging cells (split-gene packaging) contain the *gag-pol* and *env* genes cassettes introduced separately to ensure, that they will be located at individual chromosome sites. In the subsequently derived producer cells, the two gene cassettes and the vector represent three separate entities requiring multiple recombination events to generate RCR.

2.3.2 Packaging Cells

Packaging cell lines representing several generations of improvements have been described extensively (JOLLY, 1994; BUNNELL and MORGAN, 1996). However, only a very limited number of these packaging cells has been used to generate vectors for clinical trials. The earliest packaging cells contained the entire Mo-MLV genome having only a deletion of the ψ sequence. Such cells can readily produce RCR by a single recombination event and cannot be used for production for clinical trials. The first and most widely used cell line for clinical trials is the PA317 cell line (MILLER and BUTTIMORE, 1986). In addition to the ψ deletion, this cell line has a modified 5' LTR that is inactivated for integration, and all of the viral genome downstream of the *env* gene (including the 3' LTR) is deleted and replaced with an SV40 polyA signal. When used with an appropriately designed vector, this cell line does not yield RCR which would require a minimum of two recombination events. However, scale up of vector production may increase the probability of recombination and additional packaging cells have been designed which would require three or more recombination events to generate RCR. Two packaging cell lines, ψCRIP (DANOS and MULLIGAN, 1988) and gp+AM12 (MARKOWITZ et al., 1988), have a split-gene design in which the *gal-pol* and *env* genes are present in separate cassettes each having the SV40 polyA signal at the 3' end and a LTR at the 5' end. Both cell lines have been used for vector production for clinical trials.

All of the above described cells are NIH 3T3 murine cells and have *gag-pol* genes derived from Mo-MLV and the *env* gene from the amphotropic 4070 virus. The PG13 cell line, although 3T3-based, is a split-gene cell line containing the GaLV *env* gene and has been used in clinical trials to produce vectors for transduction of peripheral blood stem cells (MILLER et al., 1991). Murine cell lines carry a variety of endogenous retrovirus sequences, which potentially might be problematic. Thus, two additional cell lines used for the production of clinical vectors are DA and CFA which are split-gene lines containing no LTRs and constructed in the canine cell lines, D17 and CF2, respectively (JOLLY, 1994).

2.3.3 RCR Contamination

The Mo-MLV causes T cell neoplasia in rodents, and it is important to avoid generation of RCR in vector production. Whether RCR is pathogenic for humans is uncertain but studies in non-human primates clearly highlighted the possibility (ANDERSON, 1993). Three of 10 rhesus monkeys inoculated with supernatants containing both a retroviral vector and a high titer of amphotropic RCR developed T cell lymphoma. In this study the virus supernatant was used to transduce bone marrow progenitor cells that were returned to the severely immunosuppressed monkeys, whose bone marrow had been ablated by total body irradiation.

The current design of retroviral producer lines has greatly decreased the likelihood of recombination events that can generate RCR as discussed above. Also, *ex vivo* transduction of cells for clinical trials is only performed by exposure to carefully tested retroviral supernatants. Sensitive tests to detect RCR in vector supernatants have been developed, and extensive testing has been mandated by the US Food and Drug Administration (WILSON et al., 1997). This testing has been largely successful, such that on several occasions generation (or "breakout") of RCR during production has been readily detected. Additionally, transduced cells can be tested for RCR before infusion into the patient, and there is no evidence that any patient has received a retrovector contaminated with RCR.

2.3.4 Purification

Retroviral vectors are released directly into the cell culture medium at relatively low concentration, typically at titers of 10^6 or perhaps 10^7 transducing units per mL. Thus, the specific productivity may be as low as several transducing units per cell. Further, retroviral particles are relatively fragile and cannot readily be concentrated or purified to a high degree. This presents significant challenges for manufacturing scale up and also imposes limitations on their use for direct *in vivo* delivery (JOLLY, 1994). Therefore, it is clearly important to obtain the highest possible yields from the producer cells and to carefully consider the components of the medium, such as serum, into which the vector is secreted.

2.4 Preclinical Studies with Retrovirus Vectors

2.4.1 *In vitro* Studies

Retrovectors are capable of transducing a wide array of cells (from many mammalian species) grown *in vitro*. The primary limitation is that the cells must be dividing, but otherwise transduction is generally efficient. Transduction is usually quite stable and expression is maintained following cell passage. Retroviruses are often used for studies in which it is desired to introduce genes into cells to obtain stable expression at low copy number. *In vitro* transduction assays are useful to test basic elements of vector design and as a rapid test system for optimizing expression levels. However, *in vitro* experiments have significant limitations and often are not predictive of the behavior of the vector in *in vivo* animal experiments.

2.4.2 *In vivo* Studies

2.4.2.1 Target Cell Specificity and Expression

There are several significant limitations to the *in vivo* use of retrovectors. Retrovectors are produced at low titers and prohibitively large volumes of supernatant would be required for many direct *in vivo* delivery applications. Inhibition by complement, inactivation by other components in blood, or very rapid viral clearance also limit *in vivo* delivery. The inability of retroviruses to transduce non-dividing cells limits the potential targets for gene therapy. Thus, retrovectors cannot readily be used for direct delivery to targets such as liver, lung epithelium or brain. However, dividing tumor cells in these organs may preferentially be targeted. The amphotropic receptors are widely distributed on most human cells so there is little intrinsic targeting specificity. Attempts to modify targeting specificity by binding of antibodies or incorporation of ligands as Env fusion proteins have typically resulted in very low titer production or low particle infectivity (SALMONS and GUNZBURG, 1993; KASAHARA et al., 1994; HANENBERG et al., 1996; HALL et al., 1997).

In many *in vivo* applications, expression from retrovectors is relatively short-lived. In some cases, this probably reflects transcriptional silencing either due to the chromosomal location of the integrated vector or to interfering effects of the LTR. In addition, host immune responses to the transgene may play a role in elimination of gene expression, but often this has not carefully been evaluated.

The limitations on *in vivo* delivery of retrovectors lead to significant focus on *ex vivo* transduction of cells followed by reimplantation or reinfusion of the modified cells (BRENNER, 1995; CRYSTAL, 1995; HAVENGA et al., 1997; BLAESE et al., 1995). In animal models, *ex vivo* transduction has been used to transduce and reimplant cells such as hematopoietic progenitors, lymphocytes, or *ex vivo* expanded hepatocytes. In addition, many tumor "vaccine" models have been developed, in which retroviruses are used to modify tumors with antigens, cytokines or costimulatory molecules to enhance immune rejection of the tumor.

2.4.2.2 Host Responses to Vectors

When retrovectors are used for *ex vivo* modification, there appears to be relatively little host response directly to components of the vector. The absence of retroviral coding sequences avoids expression of antigens that would induce CTL (cytotoxic lymphocyte) responses. Because transduced cells are grown through several passages *ex vivo,* the retroviral protein components are presumably significantly depleted, thus decreasing the possibility of humoral immune responses. For direct *in vivo* delivery, retroviral protein components are likely to generate antibody responses. Retroviral vectors delivered intramuscularly can induce antibody responses against the vector Env protein, but this did not prevent subsequent delivery of the vector (MCCORMACK et al., 1997). The sensitivity of

retrovectors to human complement has been mentioned above.

An important consideration is host immune responses to the transgene. Retrovectors expressing the HIV Env protein can induce both humoral and cellular responses to HIV Env following intramuscular injection in rodents and non-human primates, and this is the basis of a current clinical trial aimed at generating a vaccine for HIV (MCCORMACK et al., 1997). However, marker genes such as *neo, hygro* or *tk* have been widely used in clinical trials both to mark cells or to provide positive selective pressure *ex vivo* or permit ablation of the cells *in vivo*. These generally are foreign antigens and lead to immune responses (RIDDELL et al., 1996).

2.5 Clinical Studies with Retrovirus Vectors

There are two main examples of *in vivo* delivery. In one case, a retrovector expressing the HIV Env protein is being administered by direct intramuscular injection to AIDS patients to induce humoral or cellular immune responses, and this has reached phase II clinical trials. In another trial, currently in phase III, murine retroviral cells producing a vector containing the HSV *tk* gene are directly injected into the brain to treat inoperable gliomas by subsequent administration of ganciclovir (RAM et al., 1997).

Most clinical applications of retroviral vectors employ *ex vivo* modification and reinfusion or reimplantation of the modified cells as in the first two gene therapy experiments which used modified lymphocytes (BLAESE et al., 1995; BRENNER, 1995; HAVENGA et al., 1997). In one case, the goal was to correct a monogenic deficiency of adenosine deaminase and thus allow normal T cell maturation. This trial has resulted in long-term persistence of the modified CD4 and CD8 cells in 2 patients in whom treatment originally began about 6 years ago (BLAESE et al., 1995). In the other clinical trial, tumor infiltrating lymphocytes were modified with retrovirus vectors expressing a cytokine gene with the goal of enhancing an immune effector response against acquired malignant disease. Another clinical trial introduced the HyTK fusion gene into clones of autologous, antigen-specific (HIV-*gag*) CTLs for reinfusion into AIDS patients. This trial used the HyTK gene as a selective marker and for safety as a suicide gene if needed, but produced an interesting result in the treated patients. A specific CTL response was elicited against epitopes in the HyTK fusion protein which emphasizes the importance of considering immune responses against the transgene (RIDDELL et al., 1996). Clinical studies using retrovectors to introduce genes into hematopoietic stem cells by *ex vivo* transduction for reimplantation have not been successful in human trials (ORLIC et al., 1996). This reflects our lack of knowledge in identifying and manipulating true human hematopoietic stem cells. Retrovectors may also be applied in the context of tumor vaccines to transduce either autologous tumor cells or allogenic cells with various combinations of antigens, cytokines or costimulatory molecules. The modified cells are then implanted into the patient with the goal of inducing tumor-specific immune responses.

3 Adenovirus Vectors

Adenovirus (Ad) vectors (HITT et al., 1997) can efficiently infect cells and mediate high levels of gene expression and are readily produced and purified at high titer. The main limitations to Ad vectors concern their lack of persistence and sustained transgene expression which is further compounded by strong host cell inflammatory and cellular and humoral immune responses. These host responses result from the large number of viral genes present in most Ad vectors and the relatively large number of protein components of the viral particles.

3.1 Adenovirus Structure and Biology

3.1.1 Biology of Adenovirus

Adenoviruses were originally isolated from human adenoid tissue and generally cause a range of mild respiratory diseases, conjunctivitis and some enteric diseases. There are over 40 serotypes isolated from humans. Group B adenoviruses, Ad4 and Ad7, have been delivered as enteric-coated capsules to vaccinate military recruits against respiratory disease. Group A adenoviruses, Ad12, 18, and 31, are tumorigenic in animals. Most Ad vectors are based on Ad2 and Ad5 which have very similar DNA but are not tumorigenic in animals. Adenoviruses infect and grow in both dividing or non-dividing cells. In natural infections, adenovirus may generally infect quiescent epithelial cells and may persist for prolonged periods in adenoidal and tonsillar tissue. The biology of adenoviruses has been extensively reviewed by HORWITZ (1990a, b).

3.1.2 Structure of Adenovirus and its Genome (Fig. 2)

Adenoviruses are non-enveloped icosahedral particles about 80 nm in diameter. The particles have a complex structure with a 35 kb linear duplex DNA genome and at least 12 different structural proteins. One group of proteins forms an internal core containing the condensed DNA and the remaining proteins form the outer capsid, of which the major proteins are the hexon, penton base and fiber. The hexon is the primary structural protein of each of the 20 sides of the capsid. The fiber protein projects outward from each of the 12 vertices and is anchored to the capsid by the penton base protein. The number and complexity of the protein components of adenovirus has important implications both for generation of vectors in complementation systems and as antigens to elicit humoral immune responses.

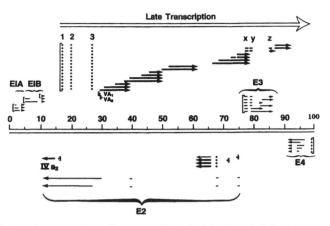

Fig. 2. Structure and transcription map of the adenovirus type 5 genome. The double-stranded Ad5 DNA genome is shown by the duplex line on a scale of 100 map units. The genome is 35 kb in length and one map unit is equal to 350 bp. The early genes are transcribed into the mRNAs shown as E1A, E1B, E2, E3, and E4. The late region is transcribed primarily from the major late promoter at map unit 16 and the late messages contain the tripartite leader shown as 1, 2, 3, and various exons derived by alternate splicing. The arrows indicate the direction of transcription and the arrow heads show the 3' end of the mRNA.

The Ad genome is a linear DNA duplex of approximately 35 kb and contains a short inverted terminal repeat (ITR) at each end which functions as the replication origin. The remainder of the genome contains coding sequences for about 40–50 proteins. These sequences comprise about 12 mRNA families expressed from 6 main transcription units all of which give rise to multiple mRNAs via alternate splicing programs. The E1A, E1B, E3 and late regions are transcribed in the rightward direction. The E2, E4 and IVa2 regions are transcribed in the leftward orientation on the other strand. There is a complex set of events that regulate adenovirus gene expression after the adenovirus core penetrates to the nucleus. E1A is the first gene transcribed and the E1A protein then transactivates transcription of the intermediate early genes E1B, E2, E3, and E4. These genes code for proteins that provide regulatory functions to modify the physiologic state of the cell and catalytic or stoichiometric functions required for Ad DNA replication. After DNA replication begins, the major late promoter (MLP) is activated to transcribe about 75% of the genome length in the rightward direction. This late mRNA transcript is processed by alternate splicing and polyadenylation to yield 5

families (L1, 2, 3, 4, 5) that are translated to yield all of them are structural proteins.

Nearly all of them are required either for genome replication or generation of Ad particles which presents a complex complementation problem. Thus, most Ad vectors contain a large number of viral genes which may be expressed and elicit undesirable cellular immune responses *in vivo*. The E3 genes are not required for replication and virus generation *in vivo* but E3 may be important in down-regulating cellular immune responses. The utility of retaining the E3 region in Ad vectors has been debated because its removal increases the payload of the vector.

3.1.3 Adenovirus Life Cycle

Adenovirus binds to cells via an interaction (at least for Ad2 and Ad5) between the fiber protein and a recently identified receptor, the 46 Kd CAR protein, which is also the cellular receptor for Coxsackie virus B3 (BERGELSON et al., 1997). A second interaction between the Ad penton base protein and the vitronectin-binding integrins is then required to mediate internalization by receptor-mediated endocytosis. The DNA core is rapidly transported to the nucleus. The receptor specificity of adenoviruses may limit cell tropism but also provides an opportunity to redesign the vector via modification of fiber or penton base to modify infection efficiency or tropism.

After entry to the nucleus, early gene transcription activates DNA replication within about 6–8 h. This requires the E2B CBP protein bound at the ITR as a primer and the other E2B protein, the Ad DNA polymerase. A third protein, the E2A 72 Kd DNA-binding protein is also required, probably stoichiometrically for bulk replication. After DNA replication begins, the late transcription unit generates all the structural proteins, and assembly and production of progeny particles occurs from about 24–48 h. Adenovirus replication leads to fragility of nuclear membranes, and several adenovirus gene products ultimately mediate cell death, probably through TNF apoptosis pathways.

3.1.4 Latency and Persistence

Adenoviruses persist for prolonged periods in adenoids and tonsils but the mechanism of this persistence is poorly understood (HORWITZ, 1990a). Similarly, adenovirus genomes can be shown to integrate, but at very low frequency, especially in non-permissive cells. However, adenovirus vectors are not generally considered to be persistent and unlike retroviruses or AAV, do not appear to have developed a specific episomal or integration mechanism for efficient persistence. Most Ad vectors have failed to provide persistent gene expression, which may reflect both the lack of a specific Ad mechanism for persistence and significant host inflammatory and cellular immune responses. Currently, Ad vectors may be more suited for applications where short-term, high level expression is required in either dividing or non-dividing cells.

3.2 Design of Adenovirus Vectors

3.2.1 Elements of Virus Genome Required

Virtually all Ad vectors have been based on serotype 2 and 5 and all the discussion here refers to these serotypes. The potential packaging capacity of Ad vectors is almost 35 kb, because the only *cis*-acting regions absolutely required are the 102 base pair (bp) ITRs at each end and the packaging signal (bp 194–358) near the left hand ITR. In practice, the regions which can be deleted from the Ad genome have been severely limited by the lack of packaging cell lines to provide these sequences by complementation. Ad particles can package genomes up to 105% of genome length and 2 kb of additional sequence can be inserted without any deletion of Ad sequence. Deletion of the E1A and E1B regions provides room to package about 5 kb of sequence. E3 can also be deleted to give a total packaging capacity of about 7.5 kb, but this has engendered some debate. One of the E3 proteins can down-regulate presentation of MHC class I molecules at the cell surface and help to suppress the likelihood of cellular immune responses to viral proteins. To retain E3 but increase the packaging capacity one group deleted all but one essential open reading frame (*orf*6) from the E4 region (ARMENTANO et al., 1997). Interestingly, this has had an unexpected effect on persistence of transgene expression (see Sect. 3.4.2.2). E1 deleted vectors are replication-deficient, and all vectors for gene therapy have this deletion.

3.2.2 Elements of Particle Structure that Influence Vector Properties

All of the 12 or more structural proteins of adenovirus are required in vector particles, and none appears dispensable. Protein IX, coded from the same region as E1B, was deleted in some early vectors but is required for particle stability and is now generally retained in the vector. The fiber and penton base control both receptor specificity and internalization. These can be modified by altering the head regions of the fiber protein with regions from other Ad serotypes and the penton base specificity can be modified by replacing its RGD domain with other receptor ligands. Alternatively, the fiber protein can be altered to include a new ligand such as a heparin-binding domain to target widely expressed heparin containing cellular receptors. Ad particles contain no lipid and are relatively robust and can be readily concentrated and purified by CsCl gradient centrifugation or more conventional biochemical techniques.

3.2.3 Design of Transgene Expression Cassettes

Most Ad vectors have the foreign gene inserted at the left end of the genome in place of E1A and B genes. A wide variety of promoters and enhancers has been used, including the Ad E1A or MLP promoter, other viral promoters such as the commonly used CMV immediate early promoter or various mammalian cell transcription promoters such as those from β-actin or the phosphoglycerokinase genes. Promoters in Ad vectors have generally been chosen based on the particular application and the target tissue, and vector induced promoter shutdown is usually not seen. However, recent studies suggest that some tissue or cell-specific promoters may have their specificity altered in the context of an adenovirus (ARMENTANO et al., 1997).

The adenovirus ITRs do not appear to interfere with transcription. However, the protein IX gene is usually retained downstream of the inserted gene and transcription from the inserted gene cassette may interfere with transcription of protein IX and lead to low-titer vector preparations.

3.3 Adenovirus Vector Manufacturing

3.3.1 Complementation Systems

The production systems for Ad vectors have nearly all been based on the use of a human cell line, 293 cells, originally derived by transfection of a left terminal fragment of Ad5 into primary human embryonic kidney cells (LOUIS et al., 1997). The 293 cell line has inserted on chromosome 19 the Ad5 sequences from nucleotides 1–4,344 which include the left hand ITR, the Ad packaging signal and the E1 region. The cells express E1A and E1B proteins but not protein IX. This cell line can complement E1 deleted Ad and is the primary cell line in which Ad vectors are produced.

To generate the vector construct, one of several strategies can be employed (HITT et al., 1997). The transgene cassette can be inserted into a shuttle plasmid containing Ad flanking sequence. Thus, for an E1 deletion–insertion vector, the transgene cassette would be flanked by the left hand ITR and the packaging signal, and by downstream pIX Ad sequences at the right hand end. This can, in some cases, be ligated *in vitro* to a fragment containing the rest of the Ad genome, and recombinant vector is recovered by transfection and propagation in 293 cells. More generally, the shuttle plasmid can be cotransfected with a plasmid containing the Ad genome which is deleted for E1 and perhaps E3. The vectors are rescued after *in vivo* recombination in 293 cells. To aid in identification of recombinant vectors, marker genes inserted into the E1 region of the viral DNA have been used. Following recombination with the vector shuttle plasmid, the marker gene is deleted.

3.3.2 Packaging Cells

All of the so-called first generation Ad vectors for gene therapy are E1 deleted vectors generated in the 293 cell line and can readily be produced at titers of 10^{10}–10^{11} mL^{-1}. Thus, a seed stock of vector can be readily reinfected into 293 cells so that the system is potentially scalable for manufacturing. However, three problems have beset this production system. First, replication competent adenovirus (RCA) can be generated by recombination between the vector and Ad sequences in the 293 cell line. Second, the vectors still retain most of the adenovirus early genes and all the late genes, and though the E1 deleted vectors are relatively crippled for replication, they may still express significant amounts of Ad early and late genes that induce strong cellular immune responses. Third, the packaging capacity of E1 deleted vectors is limited to about 5 kb. To address these issues, several different approaches have been attempted by altering the vector design and by developing additional packaging cell lines.

The most direct way to decrease RCA is to design the vector to avoid direct overlapping homology with the E1A and surrounding sequences in 293 cells. For instance, the protein IX gene can be deleted from the vector backbone or relocated to the E4 region (HEHIR et al., 1996). This significantly reduced but did not eliminate generation of RCA using conventional 293 cells for large scale production.

Effort has been spent in designing complementation systems to increase packaging size and decrease immune responses and other deleterious effects on the cell by supplying additional complementary Ad genes from cell lines. The E2A and E4 genes are difficult to supply from complementing cell lines, because constitutive expression cannot be tolerated, and they must be supplied by inducible systems. However, 293 cells can be modified to express E4 *orf*6 from regulated promoters (BROUGH et al., 1996).

293 cells contain both an Ad ITR and an Ad packaging signal and thus always have some remaining sequence homology with the vector. Other E1 complementing cell lines have been derived from transfecting human A549 cells with Ad5 E1 sequence but avoiding the ITR and packaging signals. However, these cell lines complemented 10-fold less well than 293 cells. GARZIGLIA et al. (1996) combined an Ad vector deleted for E1, E2A and E3 with an A549-based cell line containing the E1 and E2A regions under inducible

promoter which should produce less RCA. Also, the absence of E2A reduces host cell responses to this protein and eliminates its effect in enhancing late gene expression. Alternately, E2A expression can be decreased by using a vector with a temperature sensitive mutation in E2A, but this mutation has extensive biochemical leakiness at 37 °C.

Several groups have attempted to generate Ad vectors deleted for most viral genes, so-called "gutless" vectors. The general approach is to delete most or all of the Ad coding regions and insert the transgene cassette flanked by the Ad packaging sequence and the ITRs. A helper Ad virus is then used which may be deleted for E1 and packaging signals. All of these helper-dependent systems still generate significant levels of RCA or helper virus. The systems rely on the helper virus, and any RCA having a different buoyant density, which permits final purification of the vector by CsCl density gradient centrifugation. A further approach to elimination of Ad genes from vectors takes advantage of Cre recombinase to eliminate any sequence flanked by lox (Cre recombinase recognition) sites by recombination. One application of the Cre recombinase system to helper-independent vector production was described by LIEBER et al. (1996). In this system, the vector contained a transgene in place of the E1 region, an E3 deletion and lox sites upstream of the MLP and downstream of the E2A gene. Growth of this vector on 293 cells stably expressing the results in elimination of most of the Ad genes in about half of the particles produced to leave vector particles containing only the transgene and the E4 gene. These particles could be enriched by CsCl gradient purification and the helper virus particles were reduced to about 0.5 %. Surprisingly, however, the vector dit not give persistent expression *in vivo*.

3.3.3 RCA Contamination

RCA is disadvantageous because it may outgrow the vector, it may modify or increase vector expression, it may engender host immune response or may cause mobilization or dissemination of the vector. Finally, RCA is a human pathogen. Elimination of RCA is important because currently the US FDA required for clinical use lots of Ad vectors to contain less than one RCA per patient dose. This has proved a significant limitation and in practice generally has prevented dosing at 10^{10} or above. Several strategies described above have been employed to reduce RCA production (HEHIR et al., 1996). In summary, careful choice of vector design compared to sequences in the packaging cells with respect to the E1 region may be able to significantly reduce generation of RCA to levels below 1 in 10^{10}. However, production of vectors deleted for most Ad genes in the presence of similarly low levels of helper adenovirus has not been achieved.

3.3.4 Purification of Adenovirus Vectors

The classical method for purification of adenovirus particles has generally involved several cycles of freeze-thaw to lyse the cells followed by ultracentrifugation to equilibrium in CsCl gradients. More recently, benzonase has been used to degrade DNA and reduce the viscosity of the lysate before the CsCl banding. However, CsCl may not be useful for commercial scale-up production of vectors and this, incidentally, places a severe limitation on production of current helper-dependent gene deleted Ad vectors. A more conventional pharmaceutical approach to purification of vectors replaces CsCl purification with column chromatography as described by HUYGHE et al. (1995). Column chromatography can be much more readily scaled up in a reproducible fashion.

Scale-up of production requires high throughput assays especially at the process development stage. A rapid HPLC assay for quantitative and qualitative assessment of adenovirus particles has been described recently (SHABRAM et al., 1997). Assays of vector infectivity or bioactivity are also important in order to determine the particle to infectivity ratio which has often been reported to be about 100:1. However, conventional infectivity assays may be misleading. A detailed investigation of quantitative assays of Ad vectors shows that the true ratio is as low as 10:1 and may approach unity (MITTEREDER et al., 1996).

3.4 Preclinical Studies with Adenovirus Vectors

3.4.1 *In vitro* Studies

Ad vectors can express genes in a wide variety of dividing or quiescent cells in culture. Expression is generally efficient and a wide array of genes related to many diseases has been studied as recently tabulated by HITT et al. (1997). The propensity to express in many cell types *in vitro* has in turn encouraged testing of Ad vectors for many *in vivo* applications.

Although *in vitro* cell culture is not very predictive for *in vivo* function and cannot measure host immune responses, these studies are useful for testing various aspects of vector design and function. For instance, Ad vectors expressing the cystic fibrosis transmembrane receptor (CFTR) cDNA can correct the chloride channel defect in airway epithelial cells taken from cystic fibrosis patients and grown in monolayer *in vitro*.

3.4.2 *In vivo* Studies

3.4.2.1 Organ and Cell Targeting

Ad vectors have been tested in a large range of organs in many animal models as tabulated in detail by HITT et al. (1997). The animal models used include various strains of both immune-competent and immune-deficient mice, rats, cotton rats, rabbits, and non-human primates including rhesus macaques and baboons. Most of these animal species, including the non-human primates, are non-permissive for replication of the human adenovirus on which most Ad vectors are based. This may limit their usefulness for some toxicity studies. For instance, expression of Ad late genes such as hexon or fiber requires

DNA replication. Thus using *in vivo* models not permissive for Ad replication might underestimate the extent of cellular immune responses to expression of viral genes. One exception is the cotton rat which is permissive for Ad replication and in which Ad2 or Ad5 can induce pneumonia (ROSENFELD et al., 1993). However, even in this animal model, the 50% lethal dose is at least 10^9 pfu (plaque forming units) which is probably many orders of magnitude above the infectious dose for humans.

Ad vectors have been used successfully to achieve transgene expression in a remarkable array of tissues including lung, liver and hepatobiliary tree, heart, arteries, brain, retina, and muscle. A variety of routes of administration has been tested including intramuscular, intracranial, intravenous and administration by both bronchoscopy or aerosol to airway epithelium.

3.4.2.2 Expression and Persistence

In vivo delivery of Ad vectors is generally capable of yielding good, and often very robust, levels of gene expression. However, the expression level generally drops, often by several orders of magnitude, within a few days or 2 weeks. This has mostly been thought to be due to host immune responses leading to elimination of transduced cells and this concept is supported by more prolonged persistence in some immune-deficient animal models (CRYSTAL, 1995; WILSON, 1996). However, other non-immune mechanisms probably also play an important role in lack of persistence of Ad vectors.

It was noted in Sect. 3.1.4 that adenovirus may persist for prolonged periods of time in its natural human host in tissues such as adenoids. The mechanism of this persistence is unknown but one possibility is through a low level of Ad genome replication. Interestingly, Ad vectors which were generated in a *cre-lox* system and were deleted for all adenovirus genes except E4 showed greatly decreased persistence compared to first generation Ad vectors deleted only for E1 (LIEBER et al., 1996). It is suggested that the first generation vectors may persist longer due to low level re-

plication which is not completely eliminated by the E1 deletion. In contrast, an Ad vector which was modified in the E4 region to contain only the E4 *orf6* gene showed greatly decreased persistence of gene expression from a CMV promoter in mouse lung, although the vector genome was present (ARMENTANO et al., 1997). Thus, Ad proteins other than E4 may be important for activation of some promoters. Clearly, non-immunologic mechanisms may have a significant impact but these remain to be well defined.

3.4.2.3 Host Immune Responses, Inflammation and Toxicity

Host responses to Ad vectors have been studied in many animal species after delivery of the vector to lung, muscle, liver or intravenously. These studies showed that there are a variety of host responses, but the precise interpretation of some of this work has been clouded by immune responses to the transgene, species and strain differences and possibly different degrees of permissiveness of the host for replication of the vector (WILSON, 1996; YANG et al., 1995; SONG et al., 1997).

In addition to responses to the transgene, there are responses both to components of the vector particle and due to expression of viral genes. Binding of the vector particle to cells appears to directly induce expression of the inflammatory cytokines IL-6 and IL-8. Humoral antibody responses to components of the particle are readily seen, and these include neutralizing IgG in the serum and neutralizing IgA antibody in the lung. Neutralizing antibody responses may eventually be limiting for repeated delivery of vectors. Immune responses to expression of viral genes have included both $Cd8^+$ CTLs and $CD4^+$ cells of both Th, and Th2 types. Similarly, both types of cellular immune responses may also occur to the transgene. The cellular responses may lead to undesirable lymphocyte infiltration and also to elimination of the transduced cell. The induction of both inflammatory cytokines and T lymphocyte response may also lead to significant toxicity.

Several strategies are being developed to attempt to avoid these host responses and to overcome the dose limiting toxicity observed for Ad vectors. Strategies based upon redesign of the vectors to decrease expression of viral genes as described above would aim to reduce cellular immune responses. A variety of approaches is being tested to induce transient immune suppression at the time of vector delivery. Attempts to decrease CTL responses include cyclosporin or a soluble chimeric CTLA4-Ig that binds CTLA4 and blocks costimulatory signals required for T cell activation (KAY et al., 1995). Deoxysper-gualin can reduce both CD4 T cell responses and humoral antibody (KAPLAN and SMITH, 1997). Antibody to CD40 ligand has been observed to reduce IgG antibody, as well as IL-12 to decrease IgA antibody (YANG et al., 1996).

3.5 Clinical Studies with Adenovirus Vectors

Ad vectors have been used in phase I clinical trials for several target diseases including two monogenic diseases, cystic fibrosis and OTC deficiency (CRYSTAL, 1995). Ad vectors are also being tested for cancer applications including delivery of a normal p53 tumor suppressor gene or a cytosine deaminase gene to activate a prodrug. These initial trials have helped to delineate some of the potential limitations and host responses to adenovirus vectors. As a result of the varied observations regarding immune responses in some of these trials, a clinical trial of an adenovirus vector in normal human subjects has begun (CRYSTAL, personal communication).

One of the two most extensively studied applications so far is cystic fibrosis (ROSENFELD and COLLINS, 1996). Ad vectors have been used to deliver the CFTR cDNA to nasal epithelium and to the lung by either localized bronchoscopy or by aerosol. There have been some indications of a transient partial correction of the transmembrane potential difference in nasal epithelium in some trials but dose limiting toxicity in others. In the lung there have been significant IL-6-based

inflammatory responses. However, expression of the CFTR gene has been difficult to detect and has persisted for only several days.

Other clinical application of Ad vectors concerns the delivery of a normal p53 gene by direct injection into tumors (NGUYEN et al., 1997). Several phase I studies in different cancer types have been completed, with encouraging safety and efficacy results. Phase II studies are ongoing (WILSON et al., 1997).

4 Adeno-Associated Virus Vectors

The advantages of AAV vectors are that the parental virus does not cause disease, the vectors can readily transduce non-dividing cells and persist long-term, and because they contain no viral genes, they do not elicit host cellular immune responses. Thus, AAV vectors can mediate impressive long-term gene expression *in vivo* and can easily be purified and concentrated. AAV vectors are the smallest particulate gene delivery system. Disadvantages of AAV vectors are the limited payload capacity of about 4.5 kb and until recently the lack of good producer systems that could generate high titer virus. A series of extensive discussions of the background and some applications of AAV vectors can be found in BERNS and GIRAUD (1996).

4.1 Adeno-Associated Virus Structure and Biology

4.1.1 Biology of Parental Virus and its Life Cycle

Adeno-associated viruses are small, DNA containing viruses which belong to the family parvoviridae within the genus dependovirus. General reviews of AAV may be found in CARTER (1989a), CARTER et al. (1989), and BERNS (1990). AAV is a defective parvovirus that grows only in cells in which certain functions are provided by a coinfecting helper virus which is generally an adenovirus or some herpes virus. AAV has a broad host range and replicates in many cell lines of human, simian or rodent origin, provided an appropriate helper virus is present. The nature of the helper function is not known but appears to be some indirect effect of the helper virus which renders the cell permissive for AAV replication. This concept is supported by the observation that in certain cases AAV replication may occur at a low efficiency in the absence of helper virus coinfection, if the cells are treated with agents that are either genotoxic or that disrupt the cell cycle.

Infection of cells by AAV in the absence of helper functions results in persistence of AAV as a latent provirus integrated into the host cell genome. In such cell lines the integrated AAV genome may be rescued and replicated to yield a burst of infectious progeny AAV particles, if the cells are superinfected with a helper virus such as adenovirus. AAV exhibits a high preference for integration at a specific region, the AAVS1 site on human chromosome 19. This is mediated by the AAV *rep* gene (LINDEN et al., 1996), but no integrated AAV provirus has yet been demonstrated in humans. Most AAV vectors do not retain this degree of specificity for integration into the chromosome 19 region.

AAV has not been associated with the cause of any disease, but has been isolated from humans, generally in association with an infection by adenovirus (BLACKLOW, 1988). There are at least 5 defined serotypes of AAV. Most of the U.S. population over age 10 is seropositive for AAV2 and AAV3. Seroconversion to AAV was observed in young children during the course of an adenovirus infection. AAV appears to be transmitted primarily in nursery populations in conjunction with the helper adenovirus, and appears to be defective in its natural human host.

4.1.2 Structure of AAV and its Genome (Fig. 3)

AAV is a non-enveloped particle about 20 nm in diameter with icosahedral symmetry which is stable to heat, mild proteolytic digestion and non-ionic detergents. The AAV par-

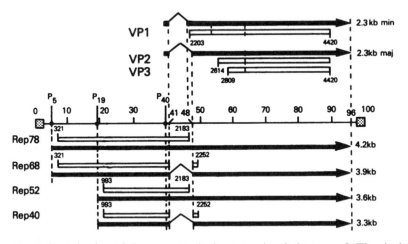

Fig. 3. Organization of the genome of adeno-associated virus type 2. The single-stranded DNA genome is 4,681 nucleotides in length and is shown as the single bar on a scale of 100 map units. One map unit equals approximately 47 nucleotides. The stipled boxes indicate the ITRs. The transcription promoters p_5, p_{19} and p_{40} are shown by solid circles. mRNAs are shown by solid bars with the intron indicated by the caret and the polyadenylation site at the 3′ end shown by the arrow head. The coding regions for the Rep78, 68, 52, 40 and VP1, 2, 3 are shown by open boxes with the initiation and termination codons indicated by the nucleotide number. Reprinted from SMUDA and CARTER (1991).

ticle comprises a protein coat containing the 3 capsid proteins VP1, VP2, and VP3 which encloses a linear single-stranded DNA genome. The AAV DNA genome has a molecular weight of about $1.5 \cdot 10^6$ Da, and for AAV2 it is 4,681 nucleotides long. Strands of either complementary sense are packaged into individual particles, but either type of particle is equally infectious.

The AAV2 genome has one copy of the 145 nucleotides long ITR at each end and a unique sequence region of 4,381 nucleotides that contains 2 main open reading frames for the *rep* and *cap* genes. The unique region contains 3 transcription promoters p_5, p_{19}, and p_{40} that are used to express the *rep* and *cap* genes. The ITR sequences are required in *cis* to provide functional origins of replication (*ori*) as well as signals for encapsidation, integration into the cell genome and rescue from either host cell chromosomes or recombinant plasmids.

The *rep* and *cap* genes are required in *trans* to provide functions for replication and encapsidation of viral genomes, respectively (FLOTTE and CARTER, 1995; MUZYCSKA, 1992). The *rep* gene is expressed as a family of 4 proteins, Rep78, Rep68, Rep52 and Rep40, that comprise a common internal region sequence but differ in their N- and C-terminal regions. Rep78 and Rep68 are required for AAV duplex DNA replication, bind specifically to the AAV ITR, and possess several enzymatic activities required for replication at the AAV termini. Rep52 and Rep40 have none of these properties, but are needed for progeny, single-strand DNA accumulation.

The Rep proteins exhibit several pleiotropic regulatory activities including positive and negative regulation of AAV genes and expression from some heterologous promoters, as well as inhibitory effects on cell growth. The AAV p_5 promoter is negatively autoregulated by Rep78 or Rep68 (PERRIERA and MUZYCZKA, 1997). Because of the inhibitory effects of expression of *rep* on cell growth, expression of *rep* in stable cell lines was difficult to achieve and delayed development of AAV packaging cell lines.

The proteins VP1, VP2, and VP3 coded from the *cap* gene share a common overlapping sequence, but VP1 and VP2 contain additional N-terminal sequences. All three are required for capsid production. Mutations which eliminate all three proteins prevent accumulation of single-strand progeny AAV DNA, whereas mutations in the VP1 N-terminus permit single-strand DNA accumulation but prevent assembly of stable, infectious particles.

4.1.3 Mode of Cell Entry and Host Tropism

AAV appears to have a broad host range and grows *in vitro* in many human cells, and also in a variety of simian and rodent cell lines if a helper virus with the appropriate host range is present. The cellular receptor for AAV is not clearly defined. However, different AAV serotypes grow *in vitro* in a wide array of cell lines so that, even if the different serotypes have different receptors, these receptors may be very promiscuous. AAV also infects various animal species. For instance, human isolates of AAV will grow in mice or monkeys if the appropriate mouse or monkey adenovirus is also present. It is not yet known if AAV exhibits any cell or tissue tropism during *in vivo* growth, but this may be better understood as more data accumulate from *in vivo* studies with AAV vectors.

4.1.4 Replication or Persistence and Integration

There are two distinguishable phases of the AAV life cycle that reflect the two modes of survival for an AAV genome in either "productive" or "non-productive" conditions. In productive cells, in the presence of helper virus, an infecting AAV genome is greatly amplified to generate a large burst of infectious progeny. In non-productive cells, in the absence of helper, AAV persists in a latent state from which it can be activated into a productive replication cycle. The events which may occur following infection of cells with AAV are complex, and the two alternatives, essentially represent the two opposite ends of a more complex spectrum of events.

In a productive infection, the infecting parental AAV single-strand genome is converted to a parental duplex replicating form (RF) by a self-priming mechanism, which takes advantage of the ability of the ITR to form a hairpin structure. This process can occur in the absence of helper virus, because it is apparently performed by a cellular DNA polymerase, but the process may be enhanced by helper virus. The parental RF molecule is then amplified to form a large pool of progeny RF molecules in a process which requires both the helper functions and the AAV *rep* gene products, Rep78 and Rep68. AAV RF genomes are a mixture of head-to-head or tail-to-tail multimers or concatemers and are precursors to progeny single-strand (SS) DNA genomes, that are packaged into preformed empty AAV capsids composed of VP1, VP2 and VP3 proteins.

In the absence of a helper virus the AAV genomes reach the cell nucleus with the same efficiency, but bulk replication generally does not occur. The infecting genomes may persist as single-strands or be converted to double-stranded DNA which allows persistence as free unintegrated genomes that may be detected for a considerable number of passages of the cells. Ultimately, the AAV genome integrates into the host cell chromosome with a high frequency at the AAVS1 region on the q arm of chromosome 19. In general, the integrated AAV genome is present at low copy number and exists as tandem head-to-tail repeats. Current evidence suggests that AAV vectors that contain no AAV *rep* coding sequences have reduced efficiency and specificity for integration at the chromosome 19 site (LINDEN et al., 1996; KEARNS et al., 1996).

The AAV integration event appears to require for both specificity and efficiency the AAV Rep protein which binds to the ITR and to a site in the AAVS1 chromosomal site (WEITZMAN et al., 1994). Whether any replication of the AAV genome is required for integration is unclear, but recent evidence suggests that there is generation of head-to-tail concatem that are precursors of integrated genomes (T. QUINTON, C. LYNCH, S. LUPTON, B. CARTER, personal communication).

How the proliferative or physiologic state of the host cell effects the integration of AAV or conversion from SS to duplex, and thus expression from AAV vectors, remains unresolved. However, AAV vectors can provide long-term expression *in vivo* in non-dividing or slowly proliferating cells.

4.2 Design of Adeno-Associated Virus Vectors

The generation of AAV vectors is based upon molecular cloning of double-strand AAV DNA in bacteria (reviewed in CARTER, 1989b). Transfection of such AAV recombinant plasmids into mammalian cells, that are also infected with adenovirus, results in rescue of the AAV genome free of any plasmid sequence to yield a burst of infectious AAV particles. This rescue may occur by a mechanism analogous to that used in rescue of a latent provirus after superinfection of cells with adenovirus.

4.2.1 Elements of Virus Genome Required

The general principles of AAV vector construction have been reviewed recently (CARTER, 1992; MUZYCZKA, 1992; FLOTTE and CARTER, 1995). AAV vectors are constructed in AAV recombinant plasmids by substituting the AAV coding sequence with foreign DNA to generate a vector plasmid. In the vector, the *cis*-acting ITR sequences must be retained intact. The vector plasmid is introduced into producer cells which are also infected with an appropriate helper virus such as adenovirus. In order to achieve replication and encapsidation of the vector genome into AAV particles, the vector plasmid must be complemented for the AAV functions required in *trans, rep* and *cap*, that were deleted in construction of the vector plasmid. AAV vector particles can be purified and concentrated from lysates of such producer cells.

4.2.2 Elements of Particle Structure that Influence Vector Properties

The most significant effect of the particle structure is the limit of about 5 kb of DNA that can be packaged in an AAV vector particle. This places some design constraints and is likely to prevent very large cDNAs from being expressed in AAV vectors. As noted above, the AAV receptor is undefined, and it remains unclear if different serotypes have different receptors and thus different specificities. The capsid is likely to induce humoral immune responses, and it is possible that alternate serotypes might be useful in avoiding this. It remains unknown if humoral immune responses might be limiting, but if so, strategies discussed above for Ad vectors could be employed.

4.2.3 Design Rules or Limitations for Gene Cassettes and Regulatory Elements

Except for the limitation on packaging and the requirements for ITRs, there are no obvious limitations on design of gene cassettes in AAV vectors. The ITRs can function as weak transcription promoters (FLOTTE et al., 1993b) but do not interfere with other promoters, and tissue-specific promoters appear to retain specificity. Introns function and may enhance expression. More than one promoter and gene cassette can be inserted in the same vector. Transcription from AAV does not seem to be susceptible to *in vivo* silencing.

4.3 Adeno-Associated Virus Vector Manufacturing

4.3.1 Complementation Systems

AAV vectors for use in human gene therapy must be generated at high titer since many *in vivo* applications of AAV may require doses in excess of 10^{10}. Also, the vector preparations must be free of wild-type AAV virus, because the presence of wild-type AAV may alter the properties or efficiency of AAV vectors (FLOTTE and CARTER, 1995).

Development of AAV vector production systems was slow because it proved difficult to generate packaging cell lines for AAV. Consequently, AAV vector production was initially based on transfection of a vector plasmid and complementing plasmids into adenovirus-infected cells, usually 293 cells. These transfection systems are unwieldy, often yield only low titer and can generate wild-type AAV or recombinant chimeric, pseudo wild-type AAV (RCAAV) through recombination. There have been significant improvements, and some transfection-based systems now produce reasonable titers of vector, but all transfection systems may be likely to generate RCAAV at some level. Transfection systems most likely will be less desirable for scale-up for commercial production.

The first AAV vectors (CARTER, 1992; MUZYCZKA, 1992) were packaged by cotransfection with plasmids that expressed the AAV *rep* and *cap* genes and had significant overlapping homology with the vector. These approaches generated vector particles contaminated with wild-type AAV due to homologous recombination, and the vectors exhibited low transduction efficiencies. Packaging plasmids with no overlapping AAV sequence homology with the vector reduced but did not eliminate generation of wild-type AAV. For example, SAMULSKI et al. (1989) constructed AAV vectors based on the plasmid pSub201 by deleting all of the AAV sequence between AAV nucleotide 190–4,490. *Rep* and *cap* functions were supplied using a packaging plasmid, pAAV/Ad, which contained the AAV nucleotide sequence from nucleotide 190–4,490 enclosed at each end with one copy of the adenovirus ITR. In pAAV/Ad, *rep* and *cap* are expressed from the normal AAV promoters and the adenovirus ITR apparently enhances the expression of AAV capsid proteins. This pAAV/Ad packaging system has been widely used, but although there is no overlapping homology between the vector plasmid and pAAV/Ad, the entire AAV sequence is present. This abutting homology leads to the generation of wild-type AAV at significant levels (FLOTTE et al., 1995; ALLEN et al., 1997).

FLOTTE et al. (1995) described a packaging plasmid, pRS5, which expresses the AAV Rep68 and 78 proteins from a non-AAV promoter, the human immunodeficiency virus type 1 (HIV1) LTR promoter. This combination of vector plasmid and packaging plasmid, which does not contain an intact AAV sequence since the region containing the P5 promoter (AAV nucleotides 144–263) was not present in either plasmid, prevents the generation of wild-type AAV. However, even in this system, some pseudo wild-type AAV (RCAAV) may be generated at very low frequency by non-homologous recombination (ALLEN et al., 1997). This non-homologous recombination could be decreased to undetectable levels by using a packaging plasmid carrying *rep* and *cap* genes in separate cassettes (split-gene packaging) and having divergent transcriptional orientations (ALLEN et al., 1997). This split-gene system would require 3 or 4 recombination events to generate RCAAV.

Other approaches to modifying AAV vector production in human 293 cells have included replacing the adenovirus helper with a plasmid containing the Ad E2A, E4 and VA RNA genes. This is useful in avoiding use of adenovirus but makes the transfection even more unwieldy (COLOSI et al., 1995). An alternate approach to avoid DNA transfection would be to insert the *rep* and *cap* genes and the AAV vector into the adenovirus genome. However, Ad vectors containing the AAV *rep* gene may be unstable. An alternate approach would be to generate stable cell lines containing both *rep* and *cap* genes and the vector.

4.3.2 Packaging Cells

Cell lines containing AAV vectors in an EBV-based episomal plasmid were constructed, and vectors could be produced by transfection with a packaging plasmid and infection with adenovirus (LEBKOWSKI et al., 1992). However, it is difficult to ensure the stability of vectors contained in an episome. Alternatively, cell lines with the vector plasmid stably integrated can be used for vector production by transfecting the pRS5 plasmid and infecting with adenovirus (FLOTTE et al., 1995). This demonstrated the concept of rescue of stably integrated vector from a producer line.

Generation of AAV packaging cell lines expressing AAV *rep* and *cap* genes was due to the inhibitory cytostatic function of *rep*, which prevents generation of stable cell lines having constitutive *rep* expression (YANG et al., 1994). The first stable cell lines containing a *rep* gene capable of generating functional *rep* protein were constructed by YANG et al. (1994) who replaced the p_5 promoter with a regulated metallothion gene promoter. CLARK et al. (1995) generated Hela cell lines containing the *rep* and *cap* genes having the normal AAV transcription promoters but deleted for AAV ITRs. When these packaging cells are infected with adenovirus, *rep* and *cap* gene expression are activated and can complement a transfected vector plasmid to generate vector. Alternatively, the vector plasmid can be stably incorporated into the packaging cells to yield AAV vector producer cell lines. These producer cells need only be infected with adenovirus to generate vector. They provide a scalable AAV vector production system that does not require manufacture of DNA and may reduce generation of RCAAV.

4.3.3 Replication Competent or Wild-Type AAV

Generation of wild-type or RCAAV needs to be avoided for several reasons. Wild-type AAV is not a human pathogen, but the presence of this in vector preparations may increase the likelihood of vector mobilization following a helper virus infection in the patient. Also, the wild-type AAV could increase the likelihood of AAV gene expression and thus increased immune responses to capsid protein. Finally, wild-type AAV can express the *rep* gene, and this significantly alters the biology of the vector because of the positive or negative regulatory effects of *rep* proteins.

4.3.4 Purification

AAV has generally been purified by proteolytic digestion of cell lysates in the presence of detergents followed by banding in CsCl gradients to concentrate and purify the particles and separate adenovirus particles. This remains the basis of most vector purification schemes but for scale-up CsCl gradients will not be useful. A more pharmaceutically acceptable procedure will probably involve column chromatography. TAMAYOSE et al. (1996), e.g., have described the use of sulfonated cellulose column chromatography to purify AAV vectors.

4.4 Preclinical Studies with Adeno-Associated Virus Vectors

4.4.1 *In vitro* Studies

AAV vectors have been extensively studied in transduction of cells *in vitro*. As with other vectors, these *in vitro* assays have been useful to study vector design and transduction mechanisms but may not be predictable for behavior *in vivo*. AAV vectors have been used to express a variety of reporter genes or human genes in stable human cell lines and primary human cells. Many of these studies have been summarized in KOTIN (1994) and FLOTTE and CARTER (1995). In general, AAV vectors can provide expression of a transduced gene in a large proportion of cells at relatively low multiplicities of about 1,000 particles per cell. However, the efficiency of this transduction may be altered depending on whether the cells are stable, transformed cell lines or primary cultures and whether the cells are stationary or dividing, but there are no clear general rules. One primary factor determining efficiency is conversion of the infecting single-stranded genome to a duplex template suitable for transcription, and this may occur at different rates in different cells *in vitro*. A second factor determining efficiency of transduction is whether the AAV genome can persist. As noted above, vectors deleted for *rep* may integrate non-specifically and with low efficiency. Thus, in cells which are rapidly dividing in culture, AAV vectors

may give very efficient transient transduction but be rapidly diluted out as cell division proceeds. In contrast, in cells that are quiescent, transduction may be initially inefficient but increases over time as formation of duplexes can occur without being diluted out by rapid cell division. This latter situation is important because many *in vivo* gene therapy targets comprise non-dividing or slowly turning over differentiated cells and, as *in vivo* studies show (see Sect. 4.4.2), these are very good targets for AAV transduction and long-term expression.

In quiescent cells, it appears that the inefficient transduction is partly related to the longer time required for conversion of single-strands to duplexes. This process presumably requires cellular DNA polymerase and can be enhanced by procedures that stimulate DNA synthesis including DNA damaging agents such as UV, γ-irradiation, genotoxic agents such as hydroxyurea, topoisomerase inhibitor drugs or the adenovirus E4 *orf*6 gene. While it is likely that some of the stimulatory effects of these agents reflect enhancement of single-strand conversion, it might also reflect amplification of double-strand templates in the presence of *rep* expressed by contaminating wild-type AAV. For many *in vivo* cases, these treatments are not required because in non-dividing cells there is time for generation of double-strand templates to occur (see Sect. 4.4.2.1).

In vitro studies have been useful for studying vector design and for demonstrating biological efficacy of the vector. For instance, human airway epithelial cells from cystic fibrosis patients grown *in vitro* and transduced with an AAV vector expressing the CFTR cDNA were functionally corrected for their electrophysiological defect in chloride channel function (FLOTTE et al., 1993b).

A wide variety of human cells and cell lines can be transduced *in vitro* by AAV vectors, but in some cases cells may be quite resistant. In particular, primary human hematopoietic cells and especially CD34 $^+$ cells may be a difficult target, because some groups report successful AAV-mediated transduction but others have been completely unsuccessful. Most studies with AAV vectors have focused on direct *in vivo* delivery.

4.4.2 *In vivo* Studies

4.4.2.1 Cell Targeting, Expression and Persistence

AAV vectors have proven remarkably useful in providing efficient, long-term gene expression *in vivo*. Most of these *in vivo* studies have addressed differentiated target cells which are either slowly proliferating or post-mitotic and non-dividing, and these types of cells may be the best targets for AAV gene therapy. AAV vectors expressing the human CFTR cDNA were delivered by bronchoscopy to the airway epithelium in lungs of rabbits and rhesus macaques and mediated expression of CFTR (FLOTTE et al., 1993a, b; CONRAD et al., 1996) for at least 6 months. Direct stereotactic injection to the rodent or non-human primate brain of AAV vectors expressing reporter genes or tyrosine hydroxylase can mediate gene expression for at least 3–6 months (KAPLITT et al., 1994). Subcutaneous injection into the eye can mediate robust, persistent expression in the retinal pigment cells (FLANNERY et al., 1997). Delivery of AAV vectors expressing a clotting factor IX cDNA to the liver by tail vein or partial vein injection in mice mediated expression of factor IX protein and therapeutic serum levels were maintained in the serum for over 6 months (HERZOG et al., 1997). Robust and prolonged expression of reporter genes for up to 18 months was achieved in immunocompetent mice by direct intramuscular injection of AAV vectors (XIAO et al., 1996). Prolonged persistence of therapeutic levels of human erythropoietin (EPO) was achieved by intramuscular injection of an AAV huEpo vector (KESSLER et al., 1996).

These *in vivo* studies show that robust and therapeutic levels of gene expression can be obtained, and the conversion of SS DNA to duplex transcription clearly occurs. Interestingly in some of these studies, particularly in muscle, expression rises over the first several weeks and then is maintained at a constant level. This reflects both a slow process of conversion to duplex genomes and then to concatemeric structures. Interestingly, in rhesus macaques, an AAV CFTR vector was present 3 months after instillation in the lung and appeared to be a dimeric concatemer (AFIONE et al., 1996). In muscle after several weeks high molecular weight head-to-tail concatemers form (XIAO et al., 1996). These reflect the mechanism of the AAV persistence pathway discussed in Sect. 4.1.4. In the *in vivo* studies, it has been difficult to rigorously demonstrate if any of the persistent form of the AAV vector is actually integrated rather than episomal.

Although some *in vitro* experiments suggested that conversion of SS to duplex could be enhanced by a variety of treatments that damage DNA, this has not proven necessary in most *in vivo* studies. Indeed when this was tested, for instance in delivery to mouse liver or brain, expression was not greatly improved and the phenomenon was dependent on the presence of wild-type AAV (XIAO et al., 1997). However, it is important to avoid wild-type AAV in vector preparations because it may drastically impact transduction and provide grossly misleading conclusions. In one study, a contaminating wild-type AAV prevented transduction by AAV vectors in rabbit lungs (HALBERT et al., 1997).

4.4.2.2 Host Responses and Toxicity

In all of the *in vivo* studies thus far with AAV vectors in rodents, rabbits or rhesus macaques, there has been no evidence of inflammatory or cytokine-mediated responses and no evidence of cellular immune responses such as CTL or CD4 TH responses. Surprisingly, in some cases such as mice, there has not been an immune response to an expressed foreign reporter gene. Whether AAV is a poor adjuvant or does not readily infect professional antigen presenting cells remains to be determined. There is very little information on the likelihood for generation of humoral antibody responses to the AAV capsid protein and more extensive studies will be required to assess, whether neutralizing antibody induction will provide some limitations to AAV vectors.

The toxicity of AAV vectors has been most extensively tested by delivery of AAV CFTR vector particles directly to one lobe of the lung. In rabbits (FLOTTE et al., 1993a) the vector persisted and expressed CFTR for at least 6 months, but no short- or long-term toxicity was observed and there was no indication of T cell infiltration or inflammatory responses. In rhesus macaques (CONRAD et al., 1996), AAV CFTR vector particles were delivered directly to one lobe of a lung and also expressed for 6 months. No toxicities were observed by pulmonary function testing, by radiological examination, analysis of blood gases and cell counts and differential counts in bronchoalveolar lavage or by gross morphological examination or histopathological examination of organ tissues. There was very minimal spread of the vector to organs outside of the lung, no vector was detected in gonads and no toxicity noted in any organ. Studies in rhesus macaques were also performed to determine if the AAV CFTR vector could possibly be shed or rescued from a treated individual (AFIONE et al., 1996). AAV CFTR particles were delivered to the lower respiratory tract and a high dose of adenovirus and wild-type AAV particles were administered to the nose of the animals. Subsequent analysis suggested that the probability of vector shedding and transmission to others is likely to be low.

4.5 Clinical Studies with Adeno-Associated Virus Vectors

AAV vectors have now been introduced into clinical trials in which an AAV CFTR vector is delivered to the maxillary sinus, nasal epithelium or lung of cystic fibrosis patients. These are the first and so far only clinical trials with an AAV vector (CARTER, 1996). Based on the *in vivo* biology which is now being performed, it is likely that future clinical trials with AAV vectors may address hemophilia B (Factor IX deficiency), diabetic or inherited retinopathy, or possibly CNS defects such as Parkinson's disease. Whether AAV vectors should be used for cancer targets such as direct injection into brain gliomas

to ablate tumors using a suicide gene such as *tk* needs careful thought because the propensity of AAV to transduce and persist in nondividing cells might lead to serious toxicity.

5 References

AFIONE, S. A., CONRAD, C. K., KEARNS, W. G., CHUNDURU, S., ADAMS, R. et al. (1996), *In vivo* model of adeno-associated virus vector persistence and rescue, *J. Virol.* **70**, 3235–3241.

ALLEN, J. A., DEBELAK, D. J., REYNOLDS, T. C., MILLER, A. D. (1997), Identification and elimination of replication-competent adeno-associated virus (AAV) that can arise by non-homologous recombination during AAV vector production, *J. Virol.* **71**, 6816–6822.

ANDERSON, F. W. (1993), What about those monkeys that got T-cell lymphoma? *Hum. Gene Ther.* **4**, 1–2.

ARMENTANO, D., ZABNER, J., SACKS, C., SOOKDEO, C. C., SMITH, M. P. et al. (1997), Effect of the E4 regions on the persistence of transgene expression from adenovirus vectors, *J. Virol.* **71**, 2408–2416.

BERGELSON, J., CUNNINGHAM, J. A., DROGUETT G., KURT-JONES, E. A., KRITHIVAS A. et al. (1997), Isolation of a common receptor for coxsackie B viruses and adenoviruses 2 and 5, *Science* **275**, 1320–1323.

BERKNER, K. L. 1988), Development of adenovirus vectors for the expression of heterologous genes, *Biotechniques* **6**, 616–626.

BERNS, K. I. (1990), Parvoviridae and their replication, in: *Virology* (FIELDS, B. N., KNIPE, D. M., et al., Eds.), pp. 1743–1764. New York: Raven Press.

BERNS, K. I., GIRAUD, C. (Eds.) (1996), *Adeno-Associated Virus (AAV) Vectors in Gene Therapy, Current Topics in Microbiology in Microbiology and Immunology.* Berlin: Springer-Verlag.

BLACKLOW, N. R. (1988), Adeno-associated viruses of humans, in *Parvoviruses and Disease* (PATTISON, J. R., Ed.), pp. 165–174. Boca Raton, FL: CRC Press.

BLAESE, M., BLANKENSTEIN, T., BRENNER, M., COHEN-HOGUENAUER, O., GANSBACHER, B. et al. (1995), Vectors in cancer therapy: how will they deliver? *Cancer Gene Ther.* **2**, 291–297.

BRENNER, M. K. (1995), Human somatic gene therapy: progress and problems, *J. Intern. Med.* **237**, 229–239.

BROUGH, D., LIZANOVA, A., HSU, C., KULESA, V. A., KOVESDI, I. (1996), A gene transfer vector–cell line system for complete functional complementation of adenovirus early regions E1 and E4, *J. Virol.* **70**, 6497–6501.

BUNNELL, B. A., MORGAN, R. A. (1996), Retrovirus-mediated gene transfer, in: *Viral Genome Methods* (ADOLPH, K. W., Ed.), pp. 3–23. Boca Raton, FL: CRC Press.

CARTER, B. J. (1989a), The growth cycle of adeno-associated virus, in *Handbook of Parvoviruses*, Vol. I. (TJISSEN, P., Ed.), pp. 155–168. Boca Raton, FL: CRC Press.

CARTER, B. J. (1989b), Parvoviruses as vectors, in *Handbook of Parvoviruses*, Vol. II (TJISSEN, P., Ed.), pp. 155–168. Boca Raton, FL: CRC Press.

CARTER, B. J. (1992), Adeno-associated virus vectors, *Curr. Opin. Biotechnol.* **3**, 533–539.

CARTER, B. J. (1996), The promise of adeno-associated virus vectors, *Nature Biotechnology* **14**, 1725–1726.

CARTER, B. J., MENDELSON, E., TREMPE, J. P. (1989), AAV DNA replication, integration and genetics, in: *Handbook of Parvoviruses*, Vol. I (TJISSEN, P., Ed.), pp. 169–226. Boca Raton, FL: CRC Press.

CHENGVALA, M. V. R., LUBECK, M. D., SELLING, B. J., NATUK, R. J., HSU, K. L. et al. (1991), Adenovirus vectors for gene expression, *Curr. Opin. Biotechnol.* **2**, 718–722.

CLARK, K. R., VOULGAROPOULOU, F., FRALEY, D. M., JOHNSON, P. R. (1995), Cell lines for the production of recombinant adeno-associated virus, *Hum. Gene Ther.* **6**, 1329–1341.

COFFIN, J. (1984), Structure of the retroviral genome, in: *RNA Tumor Viruses,* 2nd Edn. (WEISS, R., TEICH, N., VARMUS, H., COFFIN, J., Eds.), pp. 261–368. New York: Cold Spring Harbor Laboratory Press.

COLOSI, P., ELLIGER, S., ELLIGER, C., KURTZMAN, G. (1995), AAV vectors can be efficiently produced without helper virus, *Blood* **86**, 627a.

CONRAD, C. K., ALLEN, S. S., AFIONE, S. A., REYNOLDS, T. C., BECK, S. E. et al. (1996), Safety of single-dose administration of an adeno-associated virus (AAV-CFTR) vector in the primate lung, *Gene Ther.* **3**, 658–668.

CRYSTAL, R. G. (1995), Transfer of genes to humans: early lessons and obstacles to succes, *Science* **270**, 404–410.

DANOS, O., MULLIGAN, R. C. (1988), Safe and efficient generation of recombinant retrovirus with amphotropic and ecotropic host ranges, *Proc. Natl. Acad. Sci. USA* **85**, 6460–6464.

FLANNERY, J.G., ZOLOTUKHIN, S., VAQUERO, M.I., LaVAIL, M. M., MUZYCZKA, N., HAUS-

WIRTH, W. W. (1997), Efficient photoreceptor-targeted gene expression *in vivo* by recombinant adeno-associated virus, *Proc. Natl. Acad. Sci. USA* **94**, 6916–6921.

FLOTTE, T. R., CARTER, B. J. (1995), Adeno-associated virus vectors for gene therapy, *Gene Ther.* **2**, 357–362.

FLOTTE, T. R., AFIONE, S. A., SOLOW, R., MCGRATH, S. A., CONRAD, C. et al. (1993a), *In vivo* delivery of adeno-associated vectors expressing the cystic fibrosis transmembrane conductance regulator to the airway epithelium, *Proc. Natl. Acad. Sci. USA* **93**, 10163–10617.

FLOTTE, T. R., ZEITLIN, P. L., SOLOW, AFIONE, S., OWENS, R. A. et al. (1993b), Expression of the cystic fibrosis transmembrane conductance regular from a novel adeno-associated virus promoter, *J. Biol. Chem.* **268**, 3781–3790.

FLOTTE, T. R., BARRAZZA-ORTIZ, X., SOLOW, R., AFIONE, S. A., CARTER, B. J., GUGGINO, W. B. (1995), An improved system for packaging recombinant adeno-associated virus vectors capable of *in vivo* transduction, *Gene Ther.* **2**, 39–47.

GARZIGLIA, M. K., KODAN, M. J., YEI, S., LIM, J., LEE, G. M. et al. (1996), Elimination of both E1 and E2A from adenovirus vectors further improves prospects for *in vivo* human gene therapy, *J. Virol.* **70**, 4173–4178.

HALL, F. L. (1997), Targeting retroviral vectors to vascular lesions by genetic engineering of the Mo-MLV qp70 envelope protein, *Hum. Gene Ther.* **8**, 2183–2192.

HANENBERG, H., XIAO, X. L., DILLOO, D., HASHINO, K., KATO, I., WILLIAMS, D. A. (1996), Colocalization of retrovirus and target cells on specific fibronectin fragments increases genetic transduction of mammalian cells, *Nature Medicine* **2**, 876–882.

HAVENGA, M., HOOGERBRUGGE, P., VALERIO, D., VAN ES, H. H. G. (1997), Retroviral stem cell gene therapy, *Stem Cells* **15**, 162–179.

HEHIR, K., ARMENTANO, D. M., CARDOZA, L. M., CHOQUETTE, T. L., BERTHELETTE, P. B. et al. (1996), Molecular characterization of replication-competent variants of adenovirus vector and genome modifications to prevent their occurrence, *J. Virol.* **70**, 8459–8467.

HALBERT, C. L., STONDAERT, T. A., AITKEN, M. L., MERANDER, I. E., RUSSELL, R. W., MILLER, A. D. (1997), Transduction by adeno-associated virus vectors in the rabbit airway: efficiency, persistence and administration, *J. Virol.* **71**, 5932–5941.

HERZOG, R. W., HAGSTROM, J. N., KUNG, S. H., TAI, S. J., WILSON, J. M. et al. (1997), Stable gene transfer and expression of human blood coagulation factor IX after intramuscular injection of recombinant adeno-associated virus, *Proc. Natl. Acad. Sci. USA* **94**, 5804–5805.

HITT, M. M., ADDISON, C. L., GRAHMAM, F. L. (1997), Human adenovirus vectors for gene transfer into mammalian cells, in: *Advances in Pharmacology* Vol. 40 (AUGUST, T., Ed.), pp. 137–206. San Diego, CA: Academic Press.

HOCH, R. A., MILLER, A. D., OSBORNE, W. R. A. (1989), Expression of human adenovirus deaminase from various strong promoters after gene transfer into human hematopoietic cell lines, *Blood* **74**, 876–881.

HORWITZ, M. S. (1990a), Adenoviruses, in: *Virology,* 2nd Edn. (FIELDS, B. N., KNIPE, D. M., Eds.), pp. 1723–1740. New York: Raven Press.

HORWITZ, M. S. (1990b), Adenoviruses and their replication, in: *Virology,* 2nd Edn. (FIELDS, B. N., KNIPE, D. M., Eds.), pp. 1679–1721. New York: Raven Press.

HUYGHE, B. G., LIU, X., SUTJIPTO, S., SUGARMAN, B. J., HORN, M. T. et al. (1995), Purification of type 5 recombinant adenovirus encoding human p53 by column chromatography, *Hum. Gene Ther.* **6**, 1403–1416.

JAFFEE, E. M., DRANOFF, G., COHEN, L. K., HAUDER, K. M., MULLIGAN, R. C., PARDOL, D. (1993), High-efficiency gene transfer into primary human tumor explants without cell selection, *Cancer Res.* **53**, 2221–2226.

JOLLY, D. (1994), Viral vector systems for gene therapy, *Cancer Gene Ther.* **1**, 51–64.

JOLLY, D., WERNER, J. (1997), Anti-vector immunoglobulin induced by retroviral vectors, *Hum. Gene Ther.* **8**, 1263–1273.

KAPLAN, J. M., SMITH, A. E. (1997), Transient immunosuppression with deotyspergualin improves longevity of transgene expression and ability to readminister adenoviral vector to the mouse lung, *Hum. Gene Ther.* **8**, 1095–1104.

KAPLITT, M. G., LEONE, P., SAMULSKI, R. J., XIAO, X., PFAFF, D. et al. (1994), Long-term gene expression and phenotypic correction using adeno-associated virus vectors in the mammalian brain, *Nature Genetics* **8**, 148–154.

KASAHARA, N. A., DOZY, A. M., KAN, Y. W. (1994), Tissue-specific targeting of retroviral vectors through ligand receptor interactions, *Science* **266**, 1373–1376.

KAY, M. A., HOLTEMAN, A.-X., MEUSE, L., GOWEN, A., OCHS, H. D.et al. (1995), Long-term hepatic adenovirus-mediated gene expression in mice following CTLA4 Ig administration, *Nature Genetics* **11**, 191–195.

KEARNS, W. G., AFIONE, S. A., FULMER, S. B., PANG, M. G., ERIKSON, D. et al. (1996), Recombinant adeno-associated virus (AAV-CFTR) vectors do not integrate in a site-specific fashion in an immortalized epithelial cell line, *Gene Ther.* **3**, 748–755.

KESSLER, P. D., PODSAKOFF, G. M., CHEN, X., McQUISTON, S. A., COLOSI, P. C. (1996), Gene delivery to skeletal muscle results in sustained expression and systemic delivery of a therapeutic protein, *Proc. Natl. Acad. Sci. USA* **93**, 14082–14087.

KOTIN, R. M. (1994), Prospects for the use of adeno-associated virus as a vector for human gene therapy, *Hum. Gene Ther.* **5**, 793–801.

KOZAK, M. (1987), At least six nucleotides preceding the AUG initiator codon enhance translation in mammalian cells, *J. Mol. Biol.* **196**, 94–950.

LEBKOWSKI, M., NALLY, M. A., OKARMES, T. (1992), Production of recombinant adeno-associated virus vectors, *U.S. Patent* 5173414.

LIEBER, A., HE, C. Y., KIRILLOVA, I., KAY, M. A. (1996), Recombinant adenovirus with large deletions generated by *cre*-mediated excision exhibit different biological properties compared with first-generation vectors *in vitro* and *in vivo*, *J. Virol.* **70**, 8944–8960.

LINDEN, R. M., WARD, P., GIRAUD, C., WINOCOUR, E., BERNS, K. I. (1996), Site-specific integration by adeno-associated virus, *Proc. Natl. Acad. Sci. USA* **93**, 11288–11294.

LOUIS, N., EVELEGH, C., GRAHAM, F. L. (1997), Cloning and sequencing of the cellular-viral junctions from the human adenovirus type 5 transformed 293 cell line, *Virology* **233**, 423–429.

MARKOWITZ, D., GOFF, S., BANK, A. C. (1988), Construction and use of a safe and efficient amphotropic packaging cell line, *Virology* **167**, 400–406.

McCORMACK, J. E., MARTINEAU, D., DePOLO, N., MAIFERT, S., AKABARION, L. et al. (1997), Anti-vector immunoglobin induced by retroviral vectors, *Hum. Gene Ther.* **8**, 1263–1273.

MILLER, A. D. (1996), Cell-surface receptors for retroviruses and implications for gene transfer, *Proc. Natl. Acad. Sci. USA* **93**, 11407–11413.

MILLER, A. D., BUTTIMORE, C. (1986), Redesign of retrovirus packaging cell lines to avoid recombination leading to helper virus production, *Mol. Cell. Biol.* **6**, 2895–2902.

MILLER, A. D., ROSMAN, G. J. (1989), Improved vectors for gene transfer and expression, *Biotechniques* **7**, 980–985.

MILLER, A. D., GARCIA, J. V., VON SUHR, N., LYNCH, C., WILSON, C., EDEN, M. V. (1991), Construction and properties of retrovirus packaging cells based on gibbon ape leukemia virus, *J. Virol.* **65**, 2220–2224.

MITTEREDER, N., MARCH, K. L., TRAPNELL, B. C. (1996), Evaluation of the concentration and bioactivity of adenovirus vectors for gene therapy, *J. Virol.* **70**, 7498–7509.

MULLIGAN, R. C. (1991), *Gene Transfer and Gene Therapy: Principles, Prospects and Perspective, Etiology of Human Disease at the Human Level* (LINDSTEIN, J., PETTERSSON, U., Eds.), pp. 143–189. New York: Raven Press.

MUZYCZKA, N. (1992), Use of adeno-associated virus as a generalized transduction vector in mammalian cells, *Curr. Top. Microbiol. Immunol.* **158**, 97–129.

NGUGEN, D., WIEHLE, S., KOCH, P., BRANCH, C., YEN, N. et al. (1997), Delivery of the p53 tumor suppressor gene into lung cancer cells by an adenovirus DNA complex, *Cancer Gene Ther.* **4**, 191–198.

ORLIC, D., GIRARD, L. J., JORDAN, C. T., ANDERSON, S. M., CLINE, A. P., BOCLINE, D. M. (1996), The level of mRNA encoding the amphotropic retrovirus receptor in mouse and human hematopoietic stem cells is low and correlates with the efficiency of retrovirus transduction, *Proc. Natl. Acad. Sci. USA* **93**, 11097–11102.

ORY, D. S., NEUGEBORN, B. A., MULLIGAN, R. C. (1996), A stable human-derived packaging cell line for production of high-titer retrovirus/vesicular stomatitis virus pseudotypes, *Proc. Natl. Acad. Sci. USA* **93**, 11400–11406.

PEREIRA, D. J., MUZYCZKA, N. (1997), The adeno-associated virus type 2 p40 promoter requires a proximal Sp1 interaction and a p19 CArG-like element to facilitate Rep transactivation, *J. Virol.* **71**, 4300–4309.

RAM, Z., CULVER, K. W., OSHIRO, E. M., VIOLA, J. J., DE VROOM, H. L. et al. (1997), Therapy of malignant Grain tumors by intratumoral implantation of retroviral vector producing cells, *Nature Medicine* **3**, 1354–1361.

RIDDELL, S. R., ELLIOT, M., LEWINSOHN, D., GILBERT, M. J., WILSON, L. et al. (1996), T-cell mediated rejection of gene-modified HIV-specific cytotoxic T lymphocytes in HIV-infected patients, *Nature Medicine* **2**, 216–222.

ROSENFELD, M. A., COLLINS, F. S. (1996), Gene therapy for cystic fibrosis, *Chest* **109**, 241–252.

ROSENFELD, M. A., SIEGFRIED, W., YOSHIMURA, K., YONEYAMA, K., FUKAYAMA, M. et al. (1991), Adenovirus-mediated transfer of recombinant alpha1-antitrypsin gene to lung epithelium *in vivo, Science* **252**, 431–434.

SALMONS, B., GUNZBURG, W. (1993), Targeting of retroviral vectors for gene therapy, *Hum. Gene Ther.* **4**, 129–141.

SAMULSKI, R. J., CHANG, L. S., SHENK, T. E. (1989), Helper-free stocks of recombinant adeno-associated viruses: normal integration does not require viral gene expression, *J. Virol.* **63**, 3822–3828.

SHABRAM, P. W., GIROUX, D., GOUDREAU, A. M., GREGORY, R. J., HORN, M. T. et al. (1997), Analytical anion-exchange HPLC of recombinant type-5 adenoviral particles, *Hum. Gene Ther.* **8**, 453–465.

SMITH, A. E. (1995), Viral vectors in gene therapy, *Annu. Rev. Microbiol.* **49**, 807–838.

SMUDA, J. W., CARTER, B. J. (1991), Adeno-associated viruses having nonsense mutations in the capsid gene: growth in mammalian cells containing an inducible amber suppressor, *Virology* **184**, 310–318.

SONG, W., KONG, H. L., TRAKTMAN, P., CRYSTAL, R. (1997), Cytotoxic T lymphocyte responses to proteins coded by heterologous transgenes transferred *in vivo* by adenoviral vectors, *Hum. Gene Ther.* **8**, 1207–1217.

TAMAYOSE, K., HIRAI, Y., SHIMADA, T. (1996), A new strategy for large-scale preparation of high-titer recombinant adeno-associated virus vector by using packaging cell lines and sulfonated cellulose column chromatography, *Hum. Gene Ther.* **7**, 507–513.

WEITZMAN, M. D., KYOSTIO, S. R. M., KOTIN, R. M., OWENS, R. A. (1994), Adeno-associated virus (AAV) Rep proteins mediate complex formation between AAV DNA and its integration site in human DNA, *Proc. Natl. Acad. Sci. USA* **91**, 5808–5812.

WILSON, J. (1996), Adenoviruses as gene-delivery vehicles, *N. Engl. J. Med.* **334**, 1185–1187.

WILSON, C. A., NG, T. H., MILLER, A. E. (1997), Evaluation of recommendations for replication competent retrovirus testing associated with use of retroviral vectors, *Hum. Gene Ther.* **8**, 869–874.

WILSON, D., MERRITT, J., CLAYMAN, G., SWISHER, S., NEMUNAITIS, J. et al. (1998), Clinical gene therapy strategies: phase I/II results with adenoviral p53 (INGN 201) gene transfer in advanced head and neck and non-small cell lung cancer. *Presentation at 1st American Society of Gene Therapy, Seattle, Washington,* May 1998.

XIAO, X., LI, J., SAMULSKI, R. J. (1996), Efficient long-term gene transfer into muscle tissue of immunocompetent mice by adeno-associated virus vector, *J. Virol.* **70**, 8098–8108.

XIAO, X., LI, J., MCCOWN, T. J., SAMULSKI, R. J. (1997), Gene transfer by adeno-associated virus vectors into the central nervous system, *Exp. Neurol.* **144**, 113–124.

YANG, Q., CHEN, F., TREMPE, J. P. (1994), Characterization of cell lines that inducibly express the adeno-associated virus Rep proteins, *J. Virol.* **68**, 4847–4856.

YANG, Y., LI, Q., ERTL, H. C. J., WILSON, J. (1995), Cellular and humoral immune responses to viral antigens create barriers to lung-directed gene therapy with recombinant adenovirues, *J. Virol.* **69**, 2004–2015.

YANG, YM, SU, Q., GEROD, I. S., SCHILZ, R., FLAVELL, R. A., WILSON, J. M. (1996), Transient subversion of CD40 ligand function diminishes immune responses to adenovirus vectors in mouse liver and lung tissues, *J. Virol.* **70**, 6370–6377.

ZUFFEREY, R., NAGY, D., MANDEL, R. J., NALDINI, L., TRONO, D. (1997), Multiply attenuated lentiviral vector achieves efficient gene delivery *in vivo, Nature Biotechnology* **15**, 871–875.

19 Non-Viral Vectors for Gene Therapy

NEIL WEIR

Slough, UK

1 Introduction 428
2 Components for Non-Viral Gene Delivery and Expression 429
3 Transfection Systems 429
   3.1 Direct Gene Transfer (Naked DNA) 429
   3.2 Microinjection of DNA 430
   3.3 Calcium Phosphate Precipitation 430
   3.4 Non-Targeted Cationic Polymers 431
   3.5 Electroporation 431
   3.6 Particle Bombardment 432
   3.7 Cationic Liposomes 432
   3.8 Anionic and Neutral Liposomes 434
   3.9 Liposome–HVJ Complexes 434
   3.10 Receptor-Mediated Endocytosis (RME) 434
      3.10.1 Nucleic Acid 435
      3.10.2 DNA-Binding and Compaction 435
      3.10.3 Targeting Moiety 435
      3.10.4 Endosome Release 436
4 Components of the DNA to be Delivered 437
5 Conclusions 438
6 References 438

List of Abbreviations

AAT	α-1-antitrypsin
ARDS	adult respiratory distress syndrome
ASGPr	asialoglycoprotein
ASOR	asialoorosmucoid
CF	cystic fibrosis
CFTR	cystic fibrosis transmembrane conductance regulator
CS	circumsporite protein
DC-Chol	3 β-[N-(N′,N′- dimethylaminoethane)carbamoyl]cholesterol
DDAB	dimethyldioctadecylammonium-bromide
DEAE	diethylaminoethyl
DMRIE	1,2-dimyristyloxypropyl-3-dimethylhydroxyethylammonium bromide
DMSO	dimethylsulfoxide
DOPE	dioleoylphosphatidylethanol-amine
DORI	N-[1-(2,3-dioleoyloxy)propyl-N-1-(2-hydroxy)ethyl]-N,N-dimethyl ammonium iodide
DOSPA	2,3-dioleyloxy-N-[2-spermine-(carboxamido)ethyl]-N,N-dimethyl-1-propanaminium trifluoro-acetate
DOTAP	N-[1-(2,3-dioleoyloxy)propyl]-N,N,N-trimethylammonium chloride
DOTMA	N-(2,3-dioleoyloxy)propyl-N,N,N-trimethylammonium chloride
EBV	Epstein-Barr virus
EGF	epidermal growth factor
EGFr	epidermal growth factor receptor
EM	electron microscopy
HMG-1	high mobility group 1 DNA-binding protein
HSV	herpes simplex virus
HVJ	hemagglutinating virus of Japan
MHC	major histocompatibility complex
pLL	poly L-lysine
RME	receptor-mediated endocytosis
TK	thymidine kinase
Tn	Tn antigen (N-acetyl-D-galactosamine α 1-O-Ser/Thr)

1 Introduction

The eventual success or failure of any gene therapy approach may depend largely on matching the gene delivery vector from the rapidly expanding list of viral and non-viral systems described with the proposed therapeutic gene and target cell population. At present there is certainly no one optimal vector for all applications. Some strategies demand the pragmatic approach of testing more than one vector while others, probably the majority, conclude that the existing technology is not up to the task and improvements have to be made. The choice of viral or non-viral vectors provides a good starting point in the decision making process. Recombinant viruses, particularly retroviruses, have been the vectors of choice for most early gene therapy studies on the basis of transfection efficiencies and duration of gene expression in primary human cells. Indeed viruses provide the inspiration for many of the improvements in non-viral systems, improvements which, if continued, will shift the balance of opinion away from viruses to non-viral methods. Protocols using plasmid DNA now account for 23% of recorded clinical protocols, 21% of individuals treated and 38% of protocols using *in vivo* gene delivery. Vector formulations and cell therapy protocols are ultimately aimed at a market where detailed characterization and controlled manufacture of components are critical, as is the demonstration of the exclusion of extraneous material and any potentially adventitious agents. These requirements can be most clearly met using synthetic vectors. Optimized non-viral vectors may end up resembling man-made viruses and their great advantage from the viewpoint of clinicians and the public may be just that they will be man-made.

2 Components for Non-Viral Gene Delivery and Expression

Before analyzing the various non-viral systems it is worthwhile to consider the common barriers which have to be crossed, or in some cases bypassed, by all synthetic systems designed to deliver DNA in a transcription competent state to the nucleus of the intended host cell. These can be summarized:

1. Barriers to gene delivery
   - bioavailability
   - avoidance of recipients' immune system response to gene delivery vehicle
   - extravasation
   - access to intended target cell
   - interaction/binding to target cell
   - cell entry (transfer across cell membrane or internalization into phagosome/endosome)
   - escape from endosome
   - transfer to cell nucleus
   - dissociation of DNA from carrier and/or conversion to transcription active form
2. Barriers to suitable gene expression
   - promoter function in cell type of interest
   - appropriate transcription level
   - appropriate translation
   - duration of expression/silencing, replication, integration
   - host cell immune response to foreign gene expression and avoidance thereof

Very few if any non-viral approaches presently address all of the above potential requirements. Some ensure potential barriers are avoided by simply bypassing them, e.g., *ex vivo* transfection could be considered to start at gene delivery barrier "access to intended target cell", direct injection of DNA into the cell nucleus is equivalent to starting the process at gene delivery barrier "dissociation of DNA from carrier and/or conversion to transcriptional active form". The down side to this approach is that it generally results in reduced applicability.

Rather than specifically bypassing barriers it is possible to saturate them with material; e.g., a gene delivery system which makes no particular attempt to enhance the transport of DNA to the nucleus or to protect it from nuclease on the way, may still be successful in transfecting the target cell, provided sufficient DNA is delivered in the first place. One of the great advantages of gene therapy is that one copy of a gene successfully delivered to the nucleus of the target cell and actively transcribed may be sufficient to bring about the desired therapeutic effect. There are few if any cases where the same can be said for conventional drugs.

The identification of potential barriers to cell transfection should not imply that all gene delivery vehicle development has been done by rational design of mechanisms to promote endosome disruption, nuclear trafficking, etc. Basic transfection procedures have generally been optimized using gene expression as the readout. However, it is becoming clear that most delivery systems have reached the stage where the most significant improvements in efficiency will be made by identifying and addressing specifically the principal barriers to transfection.

3 Transfection Systems

In attempting to review comprehensively non-viral gene delivery methods it is difficult to avoid creating simply a list of known procedures to transfect mammalian cells *in vitro* and *in vivo*. While not avoiding this obligatory list, the following examination of the state of the art concentrates on the contribution each method has made to the evolution, understanding and eventual application of gene delivery.

3.1 Direct Gene Transfer (Naked DNA)

Purified plasmid DNA formulated in 5% sucrose, injected directly into skeletal muscle has been shown to transfect muscle cells without the addition of any transfection enhancing agent (WOLFF et al., 1990). Gene expression following direct injection of DNA *in vivo* has most commonly been demonstrated in, but is not restricted to skeletal and cardiac muscle and other tissues including dermis (HENGGE et al., 1995) and lung. The duration of expression achieved in muscle is quite remarkable. Luciferase activity has been observed up to 19 months post gene delivery (WOLFF et al., 1992) which is in marked contrast to most tissues where, irrespective of the delivery mechanism, periods of expression are normally measured in days. The level of reporter gene expression obtained with naked DNA *in vivo* was reported to be similar to that achieved with an equivalent amount of DNA delivered using the commercially available cationic lipid formulation Lipofectin (BRL) to a readily transfectable cell line *in vitro* (WOLFF et al., 1990; FELGNER et al., 1987). Direct yield comparisons between the two systems may not entirely be appropriate. However, when it is considered that DNA alone will give little or no transfection *in vitro* this is certainly one instance where *in vivo* delivery may be advantageous. Why muscle cells should be particularly suited to uptake and continued expression of DNA is not known. WOLFF et al. (1992) have postulated two explanations; first the structure of the muscle, with multinucleated cells, sarcoplasmic reticulum and transverse tubule system giving extracellular fluid access deep into the cell, may be advantageous for DNA entry. Alternatively, muscle cells may incur damage at the time of administration giving DNA direct access to the cytosol followed by membrane recovery. Whatever the reason, expression in muscle cells is not silenced as in other tissues and this is not due to replication of the plasmid DNA which retains its prokaryotic methylation pattern within the muscle.

One important application of this form of DNA delivery is DNA vaccination. Expression of a gene within a host cell permits proteosome degradation of the gene product and MHC restricted presentation of constituent peptides. ULMER et al. (1993) used direct injection of a plasmid encoding an influenza A virus antigen to generate a protective cellular immune response to the virus. Several clinical studies involving the use of naked DNA infection to elicit protective or therapeutic immune responses to viral infection are ongoing.

3.2 Microinjection of DNA

Injection of plasmid DNA directly into the cell nucleus, as described by CAPECCHI (1980) results in very high-efficiency transient expression, in this case of HSV TK, the thymidine kinase gene of herpes simplex virus. Microinjection into the cytoplasm of 1,000 cells resulted in no expression of TK suggesting that the transfer of DNA from the cytosol to the nucleus is at best rather inefficient. Stable expression of HSV TK following direct intranuclear injection was again a rather infrequent event occurring in one of 500–1,000 cells injected. This work has been useful in identifying rate limiting steps in the process of transfection and clearly identifies the transport of DNA from the cytosol to the nucleus and for protection of the DNA during this process as critical.

3.3 Calcium Phosphate Precipitation

Transfection of cells by precipitating DNA with $CaPO_4$ at neutral pH and applying this particulate material to cell monolayers is one of the earliest methods of transferring DNA into cells (GRAHAM and VAN DER EB, 1973). This general method is still commonly used, yet the precise details of the mechanism of action are still unknown. However, it is known that at the appropriate pH and DNA concentration $CaPO_4$ performs a function common to many DNA delivery systems in that it precipitates DNA into small particles which may be phagocytosed or endocytosed by the target cell. LOYTER et al. (1982) used fluorescence microscopy to follow endocytosis of precipitated DNA and were able to demonstrate that the pH at which complexes were formed and the concentration of DNA in the complex influenced DNA uptake, in line with the enhancement of gene transfer previously determined for these variables. Additionally and again in common with other processes, $CaPO_4$ facilitates adsorption of DNA onto the surface of the cell (LOYTER et al., 1982). $CaPO_4$ complexes can also be formed slowly in the culture medium, a procedure which is reported to be significantly more efficient at generating stable transformants of certain cell lines (CHEN and OKAYAMA, 1987).

Further alterations to the original method have included post-transfection treatment of the cells with glycerol, DMSO, tubulin inhibitors and lysosomal inhibitors. The common feature for all of these is their influence on the trafficking of DNA from the cell surface or endosome to the cytosol.

3.4 Non-Targeted Cationic Polymers

Cationic polymers are used in several gene delivery systems, most notably receptor-mediated transfection, and are considered here as a separate group because they illustrate two of the key features of non-viral transfection systems. Polycations such as polybrene have the property of condensing DNA at appropriate molar ratios (MANNING, 1978). DNA is condensed to a size that can be endocytosed, and an excess of positive charge brings about an electrostatic interaction with the cell. The compacting agent may also protect the DNA from nucleases. Transfection enhancement by DEAE dextran (ISHIKAWA and HOMCY, 1992) and polybrene (KAWAI and NISHIZAWA, 1984) is analogous to that mediated by $CaPO_4$.

More recently, novel dendrimeric polycations such as polyamidoamine cascade polymers (HAENSLER and SZOKA, 1993) have been added to the list of transfection enhancing polycations. Following internalization these species encounter a common barrier to efficient transfection, namely endosome escape. It is, therefore, not surprising that elaborations of the basic method include a means to escape the endosome, such as endosomolytic peptides (HAENSLER and SZOKA, 1993) or, as described by KAWAI and NISHIZAWA (1984), the inclusion of DMSO which is thought to increase membrane permeability thereby potentially evading or disrupting the endosome.

3.5 Electroporation

The chemical methods for transfection described in Sects. 3.1–3.4 depend largely on active phagocytosis or endocytosis. Electroporation (NEUMANN et al., 1982) differs in that it is thought to generate pores or local areas of cell membrane breakdown (KINOSITA and TSONG, 1977) through which DNA enters the cell making the method more applicable to less active phagocytotic cells. Several parameters may be varied to optimize conditions for electroporation of a given cell line including electric field strength, pulse duration, temperature regime, DNA state and medium composition. Preferred conditions vary significantly from one cell type to another, but numerous procedures have been published making it generally possible to find a useful starting point. Linearization of DNA prior to electroporation results in a higher frequency of stable transfectants.

Electroporation efficiencies obtained where DNA is transferred from the bulk medium in which the cells are suspended, may in certain cases be improved by complexing the DNA on the cell surface or in endocytotic vesicles prior to application of the electric field (CRAIG et al., 1995).

Perhaps surprising electroporation is not restricted in its application to *in vitro/ex vivo* protocols and recent developments have opened up the prospect of *in vivo* transfection. ZHANG et al. (1996) described the application of a pulsed electric field to the skin of an animal combined with topical application of naked DNA resulting in transfection of defined depths of the dermis according to the nature of the applied field. Without the applied field only transfection of the hair follicles was achieved. Application of the electric field to the skin combined with i.v. infusion of DNA to bring about *in vivo* transfection of erythrocytes and lymphocytes has also been proposed (HOFFMAN, 1996).

3.6 Particle Bombardment

Physical disruption of the target cell plasma membrane to deliver DNA directly to the cytosol may also be achieved using particle bombardment. YANG et al. (1990) described a procedure in which 1–3 μm gold particles coated with DNA are fired into tissue or cell cultures using a high voltage electric discharge device. After optimization the method is reported to be no more deleterious to cell viability than methods such as lipofection or $CaPO_4$ precipitation. *In vivo,* bombardment has been shown to achieve transfection of up to 20% of epidermal cells in the skin. Lower levels of expression in muscle post surgical exposure were postulated to be due to poor penetration of microparticles through the muscle fascia, a view supported by the fact that larger particles and higher voltage discharge were reported to increase expression. The combination of access without surgery, the levels of expression achieved and the abundance of antigen presenting cells make the skin an attractive target for DNA immunization using particle bombardment. Such immunizations have been used to generate humoral as well as cell-mediated immune responses. The latter again exemplify the benefit of host cell expression leading to MHC class I restricted antigen presentation (HAYNES et al., 1996).

From a more theoretical point of view particle bombardment offers a useful tool for dissecting the reasons for resistance to transfection observed with certain cell types. Those which have proved difficult to transfect by non-viral means, including many primary cells in culture, may provide particular barriers to the entry of DNA into the cytosol. Because of low/no endocytosis, the generation of extracellular matrix or particularly active or early nuclease activity in the maturing endosome. The ability to bypass such challenges in large numbers of cells will at least help to determine whether the problem arises pre or post delivery of DNA to the cytosol.

3.7 Cationic Liposomes

The introduction of cationic lipids resolved many of the problems encountered with neutral and anionic liposomes (see Sect. 3.8) which had been used previously to transfer DNA into cells. Cationic liposomes spontaneously interact and capture DNA with very high efficiencies driven by the charge attraction. This process avoids the need for active encapsulation during liposome formation or reformation so the cationic vesicles can be formed independently and then mixed with the DNA solution. High-efficiency entrapment obviates the need for removal of empty vesicles after DNA association. At the optimized ratio of DNA to lipid, the complexes formed have the appropriate charge to associate with cell surfaces. Resulting complexes are, however, no longer liposomes, but undefined complexes of DNA and lipid, the structure of which varies according to the lipids involved, ratios used and indeed conditions in which the complexes are made. STERNBERG et al. (1994) described the structures generated as spaghetti and meatballs while GERSHON et al. (1993) showed electron microscopic (EM) evidence of more homogenous lipid coating and compaction at appropriate DNA–plasmid ratios. Lipid–DNA complexes were originally thought to fuse directly with the cell membrane, but more recently EM evidence has indicated that cationic complexes are endocytosed (ZHOU and HUANG, 1994). Inclusion of DOPE as a neutral lipid or chloroquine in the transfection medium, both capable of disrupting endosomes, can enhance the transfection efficiency of cationic liposomes (FELGNER et al., 1994) further indicating an endosomal route of cell entry (FARHOOD et al., 1995). The first compounds described in this general class of reagents, sometimes called cytofectins, contained a single quaternary amine. Transfection levels achieved using monovalent cationic lipids such as DOTMA, DORI, DMRIE, DDAB, and DOTAP were improved by the introduction of lipids containing polycationic polar head groups such as polylysine (ZHOU et al., 1991; ZHOU and HUANG, 1994; DUZGUNES et al., 1989) or sperimine, the latter being the cationic component of DOSPA (HAWLEY-

NELSON et al., 1993). Both these polycations are efficient DNA condensing agents in their own right. Transfection efficiencies with DOSPA are reported to be approximately 30-fold higher than those achieved with DOT-MA.

Cationic liposome–DNA complexes have been used to deliver genes to a variety of tissues *in vivo* including pulmonary and cardiac endothelial cells and tumor cells. In particular, delivery to the lung has widely been studied via both intratracheal and intravenous administration. BRIGHAM et al. (1989) initially demonstrated reporter gene function in the lung following i.t. and i.v. administration. Successful delivery and transcription of the cystic fibrosis transmembrane regulator (CFTR) gene (YOSHIMURA et al., 1992) and expression of CFTR leading to partial correction of deficient cAMP related chloride responses in a mouse model (ALTON et al., 1993) have also been demonstrated. These significant advances in the development of *in vivo* non-viral gene therapy for cystic fibrosis (CF) have led to a number of clinical trials being approved to test the system in man. Early results (CAPLAN et al., 1995) indicated that it is possible to obtain some restoration in the electrophysiological deficit between CF and non-CF patients at least in the nasal epithelium. However, the authors caution that even in this comparatively accessible region transfection efficiency and duration of expression will need to be improved to see any clinical benefit. Perhaps more important, given the nature of this initial trial, was the fact that no histological or immunohistological changes were observed in epithelial biopsies. This is consistent with findings from animal experiments which show no significant toxicity of cationic liposome–DNA complexes in the lung (CANONICO et al., 1994). These latter authors have shown expression of human α-1 antitrypsin (AAT) in the lung following both i.v. and aerosolized administration and cationic lipid delivery of this gene to the lung is now being tested in clinical trials involving patients with adult respiratory distress syndrome (ARDS).

Absence of toxicity in animals and lack of adverse effects in humans give great encouragement to the proponents of these systems given the early concerns arising from the *in vitro* toxicity of cationic lipids. In studies to assess the impact of such apparent toxicities SENIOR et al. (1991) demonstrated erythrocyte hemolysis after contact with cationic liposomes containing DOTMA or stearylamine. Liposomes complexed with DNA at optimum ratios for high transfection efficiency and low toxicity may of course present less of a problem. In addition to showing greater transfection activity than lipofectin on a variety of cell lines, the novel cationic lipid DC Chol (GAO and HUANG, 1991) is reported to be less toxic *in vitro*. DC Chol formulated with the neutral phospholipid DOPE has now been used in clinical trials for cancer AAT deficiency and CF.

Cationic lipids at present lead the field in non-viral *in vivo* gene delivery systems and as a result of their safety profile to date, the feasibility of their large-scale manufacture and the efficiency of DNA encapsulation may continue to do so for some time, at least where non-targeted topical or i.v. pulmonary delivery are concerned.

The principal outstanding limitations of cationic liposomes are their poor stability (complexes are prepared immediately prior to use), heterogeneity of size, transfection efficiency and stability in serum. However, compaction of the DNA using a cationic polymer such as polylysine (pLL) prior to association with the liposomes (GAO and HUNG, 1996; VITIELLO et al., 1996) has been shown to improve all of the above limitations. Transfection efficiencies are reported to be up to 50-fold greater than those achieved with uncompacted DNA (VITELLO et al., 1996). Contributing to this improvement may be the generation of smaller more homogenous complexes (<100 nm) leading to greater endocytosis, increased resistance to DNAse activity in serum and the intracellular environment (GOA and HUNG, 1996; VITIELLO et al., 1996), and trafficking to the nucleus due to pLL resembling to nuclear localization signals.

The cationic nature of the DNA–liposome complex is thought to result in non-specific ionic interactions with biological surfaces, possibly mediated by anionic membrane components such as heparansulfate proteoglycans (LABAT-MOLEUR et al., 1996). KAO et al.

(1996) have shown that it is possible to achieve some degree of specificity of transfection using both antibodies and a carbohydrate ligand to direct cell-specific interaction and internalization.

3.8 Anionic and Neutral Liposomes

Transfection of mammalian cells with neutral and anionic liposomes preceded that with their more commonly known cationic equivalents (FRALEY et al., 1980). The exponential growth in the use of cationic lipids both for research and proposed therapeutic purposes contrasts with a comparatively low but steady level of interest in anionic and neutral liposomes. There are several reasons for this, but predominant among these is the difficulty of encapsulating DNA in the vesicle. Unlike the cationic system DNA does not spontaneously associate with preformed anionic and neutral liposomes, therefore, the lipid vesicle has to be formed or reformed around the DNA (NICOLAU and CUDD, 1989). Such processes are generally inefficient and/or result in very large liposomes and often cause DNA damage. However, it has been shown that another apparent disadvantage, namely the failure to associate with negatively charged cell surfaces, can be turned to an advantage if targeting is desired. Targeted transfection using antibody conjugated (immuno) liposomes, particularly pH sensitive vesicles containing the hexagonal phase forming lipid DOPE, has been demonstrated by a number of groups (WANG and HUANG, 1987; HOLMBERG et al., 1994). Targetable transfection using small molecular weight ligands, e.g., the glycolipid lactosylceramide has also been demonstrated (SORIANO et al., 1983). The prospects for targeted anionic liposomes were enhanced significantly by the finding by LEE and HUANG (1996) that precondensed DNA could be made to associate spontaneously with preformed liposomes including a targeting agent. Such agents promise to combine the DNA compaction and encapsulation efficiency of optimized cationic systems with the non-synthetic lipid formulations, targetability and stability of anionic formulations.

3.9 Liposome–HVJ Complexes

The hemagglutinating virus of Japan (HVJ) offers a direct means of entry into the cytoplasm avoiding the endosome. Neutral liposomes, formed in the presence of DNA condensed with a non-histone DNA binding protein (HMG-1) and complexed with UV inactivated HVJ, form very efficient gene transfer agents *in vitro* (KANEDA et al., 1989). These liposome-based complexes have the advantage of being able to co-deliver protein thus extending the possibility of active enhancement of DNA trafficking to the nucleus.

3.10 Receptor-Mediated Endocytosis (RME)

The work of WU and WU (1987) describing gene expression following receptor-mediated endocytosis of DNA began an exponentially expanding effort in this field. The great attraction offered by this technology is the ability to transfect specifically a defined cell population on the basis of a chosen targeting ligand. The collection of technologies falling under this category generally contains 3–4 common components, each of which mimics a genetic or physico-chemical function seen in viruses. As such, this technology approaches the concept of building the man-made virus. The common components are:

1. the genetic information,
2. a DNA binding component generally also responsible for condensing the DNA into discrete particles,
3. a targeting moiety to drive interaction with, and internalization into, the target cell which is generally attached to the DNA compacting agent, and
4. an agent to mediate disruption of, or specific escape from the endosome.

3.10.1 Nucleic Acid

The great majority of work in this field describes the delivery of plasmid DNA. However, at the lower end of the range, an antisense oligonucleotide has been delivered to and shown to inhibit viral replication in hepatitis infected cells (WU and WU, 1992). One of the great attractions of this form of gene delivery is that in contrast to commonly used viral systems there is no defined limit on the size of DNA which may be inserted. COTTEN et al. (1992) have shown that it is possible to deliver a 48 kb circular DNA cosmid using this route, although, as these authors point out, expression of a small gene from a large vector is not the same as using the cosmid to deliver and express a large gene.

3.10.2 DNA-Binding and Compaction

Poly L-lysine (PLL) is the most commonly used DNA binding component since the ε amino groups on the PLL provide a convenient means of coupling the polymer to the intended ligand (WAGNER et al., 1990). Synthetic peptides (GOTTSHALK et al., 1996) and protamine are examples of other polycationic species which have been used in the role of DNA binding and compaction (WAGNER et al., 1990; CHEN et al., 1995). In the latter case a Fab targeting component was expressed as a fusion to human protamine. The nature of the complex formed between the nucleic acid and the binding component is critical and generally small uniform particles, which may adopt structures such as toroids, rods or spheres, appear to transfect most efficiently. Several methods for creating the desired compacted species have been described (PERALES et al., 1994; WAGNER et al., 1991) which demonstrate that the important factors are the relative concentrations of DNA and compacting agent, their order of addition, the salt concentration in the compacting buffer and the nature and quality of the ligand–DNA binding element. Confirmation of complex formation can be achieved by agarose gel retardation assays, EM, circular dichroism and laser light scattering studies. The potential immune response to PLL-condensed DNA has received considerable attention. Studies carried out by FERKOL et al. (1996), however, indicate that following repeated administration of Fab–PLL–DNA complexes the humoral immune response was restricted to the targeting moiety and no complement activation could be detected.

3.10.3 Targeting Moiety

In addition to directing association with the target cell, the ligand–receptor interaction should result in internalization into the cell. The internalization event can be the result of a constitutive turnover of the receptor or may be induced by multivalent binding of the complex on the cell surface. Successful transfection has been accomplished via transfer into both endosomes and calveolae (GOTTSHALK et al., 1994). The work of WU and WU (1988) which established this technology, targeted the hepatocyte-specific asialoglycoprotein receptor (ASGPr) using asialoorosmucoid (ASOR) conjugated to PLL. This receptor represents an attractive target for *in vivo* targeted delivery because of its high level of cell surface expression restricted to the liver parenchyma (WU and WU, 1988). Following this initial proof of principle numerous cell surface receptors have been successfully targeted for RME. Tab. 1 lists some of these receptors and target cell types.

Included in this rapidly expanding list of competent targeting agents are two categories which merit particular interest: firstly, the small molecule (non-protein) ligands such as lactose, galactose and folate which offer the prospect of developing synthetic, comparatively low cost and highly characterized targeted transfection agents; secondly, monoclonal antibodies, or fragments thereof, which offer the prospect of targeting, using generic technology, to a wide range of cell types. Thus a high affinity interaction specific for the cell type of interest with known internalization properties rather than relying on a fortuitous, tissue restricted ligand-binding event can often be selected.

Tabl. 1. Targeting Ligands for Receptor-Mediated Endocytosis

Ligand	Receptor	Cell Type	Reference
ASOR	ASGPr	hepatocyte/hepatoma	WU and WU, 1987
Tetra antennary galactose	ASGPr	hepatocyte	PLANK et al., 1992
Lactose	galactose-binding lectin	hepatocyte	MIDOUX et al., 1993
Mannose	macrophage mannose receptor	macrophage	FERKOL et al., 1996
Transferrin	transferrin receptors	various	WAGNER et al., 1990
Folate	folate receptor (overexpressed)	certain tumor lines	GOTTSHALK et al., 1993
Insulin	insulin receptor	hepatoma	ROSENKRANZ et al., 1992
EGF	EGFr	lung cancer cell lines	CRISTIANO and ROTH, 1996
Malarial circumsporite protein (CS)	CS receptor	hepatoma HeLa, NIH 3T3 K562	DING et al., 1995
Lung surfactant associated proteins A and B		airway epithelial cells	ROSS et al., 1995 BAATZ et al., 1994
Antibodies	cell surface thrombomodulin	airway epithelial cells	TRUBETSKOY, 1992a TRUBETSKOY, 1992b
	EGF receptor	squamous cell carcinoma	CHEN et al., 1994; SHIMIZU et al., 1996
	CD5	Jurkat	MERWIN et al., 1995
	Tn antigen	Jurkat	THURNHER et al., 1994
	CD 3	CD3+ve PBMCs	BUSCHLE et al., 1995
Antibody fragments	polymeric immunoglobulin receptor	primary human epithelial cells	FERKOL et al., 1993
Fab	polymeric immunoglobulin receptor	rat airway epithelial cells	FERKOL et al., 1995
	gp120	HIV infected Jurkat	CHEN et al., 1995

3.10.4 Endosome Release

While the first three components are, on their own, capable of mediating gene delivery, dramatic improvements in the efficiency of gene expression have been achieved by adding this fourth function. Common to other transfection systems involving cell entry via endocytosis release of DNA from the endosome prior to fusion with the lysosome and prior to the resultant exposure to nucleases and proteases is essential. Two general approaches to this problem have been employed. Compounds known to influence the acidification and maturation of the endosome, in particular chloroquine, have been shown to enhance markedly transfection levels (COTTEN et al., 1990). However, this enhancement is not common to all cells and the exact mechanism of action is likely to be more than simply buffering the endosome. Moreover, one of the attractive features of receptor-mediated endocytosis is, that it may ultimately be applicable to *in vivo* therapies, therefore, the endosome escape mechanism should be contained within the complex itself. Once again reference to the systems used by viruses to negotiate this barrier proved invaluable. Inclusion of replication incompetent adenovirus particles in a transferrin-mediated transfection system (transferrinfection) was shown by CURIEL et al. (1991) to enhance gene expression 2,000-fold. This property is not a function of viral infection since inactivated particles are also active and the specificity of transfection remained directed by transferrin binding. High viral titers were, however, required presumably to ensure cooccupation of an endosome by complex and virus particle. Direct attachment of the inactivated virus to the PLL–transferrin complex (WAGNER et al., 1992a) resulted in similarly enhanced efficiencies thereby generating a self-contained transfection complex.

The concept of building a non-viral vector is clearly compromised if it is necessary to tag on an intact, albeit inactivated, virus. Therefore, the desire to define functional components mediating viral endosome escape and graft these onto the man-made complex continues to attract much interest and significant success. Work has focused on peptides based on the amino terminus of hemagglutinin, the protein responsible for influenza viral envelope fusion with the endosome. These peptides have been conjugated to PLL and shown to significantly enhance transferinfection (WAGNER et al., 1992). It seems likely that such peptides will render the complexes immunogenic unless they can be shielded until entry into the endosome.

4 Components of the DNA to be Delivered

So far considered how to get the DNA into the cell of interest; this section will briefly touch on the many interesting choices that go into deciding what these DNA molecules should comprise. Principal factors to be considered are:

1. the size of each component, including coding regions, additional (e.g., suicide) genes, transcription regulation elements and replication function(s),
2. level and duration of expression required,
3. the possible need for regulation of expression (temporal or cell specific).

Each of these factors is interlinked; e.g., increased duration of expression may require a significant increase in construct size and high level expression may compromise duration and regulation. The level of transgene expression limits most gene therapy applications. Therefore, strong constitutive viral promoters are commonly used. Such promoter elements are subject to silencing and promoter preferences and ultimately their effects have to be determined *in vivo* (MCLACHLAN et al., 1995). Regulatable promoters, e.g., those restricting expression primarily to tumor cells (RICHARDS et al., 1995), promise to become increasingly useful in offering a further level of control of gene expression contributing to both safety and the therapeutic window. The transient nature of gene expression is another common problem associated with non-viral vectors. Options available to extend the duration of expression, other than choice of promoter, are chromosomal integration or controlled extrachromosomal replication of the DNA, both options drawing on precedents set by viruses. Terminal repeats from the genome of the adeno-associated virus which integrates into human chromosome have been engineered into a plasmid vector which shows significantly prolonged transgene expression (PHILIP et al., 1994). Efficient integration of vectors carrying transgenes introduces the possibility of using chromosomal functions, in particular locus control regions which offer position independent controlled expression of the transgenes (DILLON and GROSVELD, 1993). Extrachromosomal replicating vectors offer the advantages of sustained expression without the potentially mutagenic events associated with random chromosomal insertion. Sustained gene expression can be obtained by the construction of an artificial chromosome containing the transgene (HUXLEY, 1994). Such constructs are generally thought to integrate into the host genome although there are reports of stable extrachromosomal maintenance (FEATHERSONE and HUXLEY, 1993). These constructs are, however, inevitably very large, making the DNA technically difficult to prepare. More promising, in terms of technical feasibility and ease of characterization are extrachromosomal replicating vectors comprising the nuclear retention and replicating functions of similarly maintained viral genomes such as EBV (CALOS, 1996).

5 Conclusions

Despite the exponential increase in interest and effort in the field of non-viral vector development, the basic objectives remain similar to those set out by ANDERSON (1992). Ultimately, for gene therapy to be widely used and to have a significant medical impact vectors need to be developed which can safety be administered *in vivo*. Implicit in this requirement is that manufacturing methods should be reproducible, generating well characterized vectors which mediate efficient but controlled transgene expression. No clear preferred vector system has emerged. Liposomes and ligand targeted complexes show great promise in terms of delivery of DNA to particular cell populations. It appears that there will still be a wide range of technologies adapted to their own particular niche, principally because no two cell types are alike in terms of accessibility, surface receptors and internalization properties. If a more generic technology is to emerge it will have to offer the capacity to tune functions such as size, charge, surface ligands and endosomolytic properties to suit the particular application.

While technology to deliver DNA to the cell has significantly avanced this has yet to be combined satisfactorily with methods to control level and duration of expression. In summary, all the pieces of the jigsaw puzzle appear to have been identified, but putting them all together remains a great challenge.

6 References

ALTON, E. W., MIDDLETON, P. G., CAPLAN, N. J., SMITH, S. N., STEEL, D. M. et al. (1993), Non-invasive liposome-mediated gene delivery can correct the ion trasport defect in cystic fibrosis mutant mice, *Nature Genetics* **5**, 135–142.

ANDERSON, W. F. (1992), Human gene therapy, *Science* **256**, 808–813.

BAATZ, J. E., BRUNO, M. D., CIRAOLO, P. J., GLASSER, S. W., STRIPP, B. R. et al. (1994), Utilization of modified surfactant-associated protein B for delivery of DNA to airway cells in culture, *Proc. Natl. Acad. Sci. USA* **91**, 2547–2551.

BRIGHAM, K. L., MEYRICK, B., CHRISTMAN, B., MAGNUSON, M., KING, G., BERRY, L. C. Jr. (1989), *In vivo* transfection of murine lungs with a functioning prokaryotic gene using a liposome vehicle, *Am. J. Med. Sci.* **298**, 278–281.

BUSCHLE, M., COTTEN, M., KIRLAPPOS, H., MECHTLER, K., SCHAFFNER, G. et al. (1995), Receptor-mediated gene transfer into human T-lymphocytes via binding of DNA/CD3 antibody particles to the CD3 T cell receptor complex, *Hum. Gene Ther.* **6**, 753–761.

CALOS, M. P. (1996), The potential of extrachromosomal replicating vectors for gene therapy, *Trends Genet.* **12**, 463–466.

CANONICO, A. E., PLITMAN, J. D., CONARY, J. T., MEYRICK, B. O., BRIGHAM, K. L. (1994), No lung toxicity after repeated aerosol or intravenous delivery of plasmid–cationic liposome complexes, *J. Appl. Physiol.* **77**, 415–419.

CAPLAN, N. J., ALTON, E. W., MIDDLETON, P. G., DORIN, J. R., STEVENSON, B. J. et al. (1995), Liposome-mediated CFTR gene transfer to the nasal epithelium of patients with cystic fibrosis, *Nature Medicine* **1**, 39–46.

CAPPECCHI, M. R. (1980), High efficiency transformation by direct microinjection of DNA into cultured mammalian cells, *Cell* **22**, 479–488.

CHEN, C., OKAYAMA, H. (1987), High-efficiency transformation of mammalian cells by plasmid DNA, *Mol. Cell. Biol.* **7**, 2745–2752.

CHEN, J., GAMOU, S., TAKAYANAGI, A., SHIMIZU, N. (1994), A novel gene delivery system using EGF receptor-mediated endocytosis, *FEBS Lett.* **338**, 167–169.

CHEN, S. Y., ZANI, C., KHOURI, Y., MARASCO, W. A. (1995), Design of a genetic immunotoxin to eliminate toxin immunogenicity, *Gene Ther.* **2**, 116–123.

COTTEN, M., LANGLE-ROUAULT, F., KIRLAPPOS, H., WAGNER, E., MECHTLER, K. et al. (1990), Transferrin-polycation-mediated introduction of DNA into human leukemic cells: stimulation by agents that affect the survival of transfected DNA or modulate transferrin receptor levels, *Proc. Natl. Acad. Sci. USA* **87**, 4033–4037.

COTTEN, M., WAGNER, E., ZATLOUKAL, K., PHILLIPS, S., CURIEL, D. T., BIRNSTIEL, M. L. (1992), High-efficiency receptor-mediated delivery of small and large (48 kilobase gene) constructs using the endosome-disruption activity of defective or chemically inactivated adenovirus particles, *Proc. Natl. Acad. Sci. USA* **89**, 6094–6098.

CRAIG, R. K., ANTONIOU, M., DJEHA, H. (1995), *International Patent Application* WO 9506129.

CRISTIANO, R. J., ROTH, J. A. (1996), Epidermal growth factor mediated DNA delivery into lung cancer cells via the epidermal growth factor receptor, *Cancer Gene Ther.* **3**, 4–10.

CURIEL, D. T., AGARWAL, S., WAGNER, E., COTTEN, M. (1991), Adenovirus enhancement of transferrin-polylysine-mediated gene delivery, *Proc. Natl. Acad. Sci. USA* **88**, 8850–8854.

DILLON, N., GROSVELD, F. (1993), Transcriptional regulation of multigene loci: multilevel control, *Trends Genet.* **9**, 134–137.

DING, Z. M., CRISTIANO, R. J., ROTH, J. A., TAKACS, B., KUO, M. T. (1995), Malarial circumsporozoite protein is a novel gene delivery vehicle to primary hepatocyte cultures and cultured cells, *J. Biol. Chem.* **270**, 3667–3676.

DUZGUNES, N., GOLDSTEIN, J. A., FRIEND, D. S., FELGNER, P. L. (1989), Fusion of liposomes containing a novel cationic lipid, N-[2,3-(dioleyloxy)propyl]-N,N,N trimethylammonium: induction by multivalent anions and asymmetric fusion with acidic phospholipid vesicles, *Biochemistry* **28**, 9179–9184.

FARHOOD, H., SERBINA, N., HUANG, L. (1995), The role of dioleoyl phosphatidyl ethanolamine in cationic liposome-mediated gene transfer, *Biochim. Biophys. Acta* **1235**, 289–295.

FEATHERSTONE, T., HUXLEY, C. (1993), Extrachromosomal maintenance and amplification of yeast artificial chromosome DNA in mouse cells, *Genomics* **17**, 267–278.

FELGNER, P. L., GADEK, T. R., HOLM, M., ROMAN, R., CHAN, H. W. et al. (1987), Lipofection: a highly efficient, lipid-mediated DNA-transfection procedure, *Proc. Natl. Acad. Sci. USA* **84**, 7413–7417.

FELGNER, J. H., KUMAR, R., SRIDHAR, C. N., WHEELER, C. J., TSAI, Y. J. et al. (1994), Enhanced gene delivery and mechanism studies with a novel series of cationic lipid formulations, *J. Biol. Chem.* **269**, 2550–2561.

FERKOL, T., KAETZEL, C. S., DAVIS, P. B. (1993), Gene transfer into respiratory epithelial cells by targeting the polymeric immunoglobulin receptor, *J. Clin. Invest.* **92**, 2394–2400.

FERKOL, T., PERALES, J. C., ECKMAN, E., KAETZEL, C. S., HANSON, R. W., DAVIS, P. B. (1995), Gene transfer into the airway epithelium of animals by targeting the polymeric immunoglobulin receptor, *J. Clin. Invest.* **95**, 493–502.

FERKOL, T., PELLICENA-PALLE, A., ECKMAN, E., PERALES, J. C., TRZASKA, T. et al. (1996), Immunologic responses to gene transfer into mice via the polymeric immunoglobulin receptor, *Gene Ther.* **3**, 669–678.

FRALEY, R., SUBRAMANI, S., BERG, P., PAPAHADJOPOULOS, D. (1980), Introduction of liposome-encapsulated SV40 DNA into cells, *J. Biol. Chem.* **255**, 10431–10435.

GAO, X., HUANG, L. (1991), A novel cationic liposome reagent for efficient transfection of mammalian cells, *Biochem. Biophys. Res. Commun.* **179**, 280–285.

GAO, X., HUANG, L. (1996), Potentiation of cationic liposome-mediated gene delivery by polycations, *Biochemistry* **35**, 1027–1036.

GERSHON, H., GHIRLANDO, R., GUTTMAN, S. B., MINSKY, A. (1993), Mode of formation and structural features of DNA–cationic liposome complexes used for transfection, *Biochemistry* **32**, 7143–7151.

GOTTSHALK, S., CRISTIANO, R. J., SMITH, L. C., WOO, S. L. C. (1994), Folate receptor-mediated DNA delivery into tumour cells: potosomal disruption results in enhanced gene expression, *Gene Ther.* **1**, 185-191.

GOTTSHALK, S., SPARROW, J. T., HAUER, J., MIMMS, M. P., LELAND, F. E. et al. (1996), A novel DNA–peptide complex for efficient gene transfer and expression in mammalian cells, *Gene Ther.* **3**, 448–457.

GRAHAM, F. L., VAN DER EB, A. J. (1973), A new technique for the assay of infectivity of human adenovirus 5 DNA, *Virology* **52**, 456–460.

HAENSLER, J., SZOKA, F. C. Jr. (1993), Polyamidoamine cascade polymers mediate efficient transfection of cells in culture, *Bioconj. Chem.* **4**, 372–379.

HAWLEY-NELSON, P., CICCARONE, V., GEBEYEHU, G., JESSE, J., FELGNER, P. L. (1993), Lipofectamine TM reagent: a new, higher efficiency polycationic liposome transfection reagent, *Focus* **15**, 73–79.

HAYNES, J. R., McCABE, D. E., SWAIN, W. F., WIDERA, G., FULLER, J. T. (1996), Particle-mediated nucleic acid immunization, *J. Biotechnol.* **44**, 37–42.

HENGGE, U. R., CHAN, E. F., FOSTER, R. A., WALKER, P. S., VOGEL, J. C. (1995), Cytokine gene expression in epidermis with biological effects following injection of naked DNA, *Nature Genetics* **10**, 161–166.

HOFFMAN, G. A. (1996), *International Patent Application*, WO 9612520.

HOLMBERG, E. G., REUER, Q. R., GEISERT, E. E., OWENS, J. L. (1994), Delivery of plasmid DNA to glial cells using pH-sensitive immunoliposomes, *Biochem. Biophys. Res. Commun.* **201**, 888–893.

HUXLEY, C. (1994), Mammalian artificial chromosomes: a new tool for gene therapy, *Gene Ther.* **1**, 7–12.

ISHIKAWA, Y., HOMCY, C. J. (1992), High efficiency gene transfer into mammalian cells by a double transfection protocol, *Nucleic Acids Res.* **20**, 4367.

KANEDA, Y., IWAI, K., UCHIDA, T. (1989), Increased expression of DNA cointroduced with nuclear protein in adult rat liver, *Science* **243**, 375–378.

KAO, G. Y., LUNG -JI, C., ALLEN, T. M. (1996), Use of targeted cationic liposomes in enhanced DNA delivery to cancer cells, *Cancer Gene Ther.* **3**, 250–256.

KAWAI, S., NISHIZAWA, M. (1984), New procedure for DNA transfection with polycation and dimethyl sulfoxide, *Mol. Cell. Biol.* **4**, 1172–1174.

KINOSITA, K. Jr., TSONG, T. T. (1977), Hemolysis of human erythrocytes by transient electric field, *Proc. Natl. Acad. Sci. USA* **74**, 1923–1927.

LABAT-MOLEUR, F., STEFFAN, A.-M., BRISSON, C., PERRON, H., FEUGAS, O. et al. (1996), An electron microscopy study into the mechanism of gene transfer with lipopolyamines, *Gene Ther.* **3**, 1010–1017.

LEE, R. J., HUANG, L. (1996), Folate-targeted, anionic liposome-entrapped polylysine condensed DNA for tumor cell-specific gene transfer, *J. Biol. Chem.* **271**, 8481–8487.

LOYTER, A., SCANGOS, G. A., RUDDLE, F. H. (1982), Mechanisms of DNA uptake by mammalian cells: fate of exogenously added DNA monitored by the use of fluorescent dyes, *Proc. Natl. Acad. Sci. USA* **79**, 422–426.

MANNING, G. S. (1978), The molecular theory of polyelectrolyte solutions with applications to the electrostatic properties of polynucleotides, *Q. Rev. Biophys.* **11**, 179–246.

McLACHLAN, G., DAVIDSON, D. J., STEVENSON, B. J., DICKINSON, P., DAVIDSON-SMITH, H. et al. (1995), Evaluation *in vitro* and *in vivo* of cationic liposome expression construct complexes for cystic fibrosis gene therapy, *Gene Ther.* **2**, 614–622.

MERWIN, J. R., CARMICHAEL, E. P., NOELL, G. S., DeROME, M. E., THOMAS, W. L. et al. (1995), CD5-mediated specific delivery of DNA to T lymphocytes: compartmentalization augmented by adenovirus, *J. Immunol. Methods* **186**, 257–266.

MIDOUX, P., MENDES, C., LEGRAND, A., RAIMOND, J., MAYER, R. et al. (1993), Specific gene transfer mediated by lactosylated poly-L-lysine into hepatoma cells, *Nucleic Acids Res.* **21**, 871–878.

NEUMANN, E., SCHAEFER-RIDDER, M., WANG, Y., HOFSCHNEIDER, P. H. (1982), Gene transfer into mouse myeloma cells by electroporation in high electric fields, *EMBO J.* **1**, 841–845.

NICOLAU, C., CUDD, A. (1989), Liposomes as carriers of DNA, *Crit. Rev. Ther. Drug Carrier Syst.* **6**, 239–271.

PERALES, J. C., FERKOL, T., MOLAS, M. HANSON, R. W. (1994), An evaluation of receptor-mediated gene transfer using synthetic DNA–ligand complexes, *Eur. J. Biochem.* **226**, 255–266.

PHILIP, R., BRUNETTE, E., KILINSKI, L., MURUGESH, D., McNALLY, M. A. et al. (1994), Efficient and sustained gene expression in primary T lymphocytes and primary and cultured tumor cells mediated by adeno-associated virus plasmid DNA complexed to cationic liposomes, *Mol. Cell. Biol.* **14**, 2411–2418.

PLANK, C., ZATLOUKAL, K., COTTEN, M., MECHTLER, K., WAGNER, E. (1992), Gene transfer into hepatocytes using asialoglycoprotein receptor-mediated endocytosis of DNA complexed with an artificial tetra-antennary galactose ligand, *Bioconj. Chem.* **3**, 533–539.

RICHARDS, C. A., AUSTIN, E. A., HUBER, B. E. (1995), Transcriptional regulatory sequences of carcinoembryonic antigen: identification and use with cytosine deaminase for tumour-specific gene therapy; chimeric gene construction for use in tumour-specific suicide therapy, *Hum. Gene Ther.* **6**, 881–893.

ROSENKRANZ, A. A., YACHMENEV, S. V., JANS, D. A., SEREBRYAKOVA, N. V., MURAV'EV, V. I. et al. (1992), Receptor-mediated endocytosis and nuclear transport of a transfecting DNA construct, *Exp. Cell Res.* **199**, 323–329.

ROSS, G. F., MORRIS, R. E., CIRAOLO, G., HUELSMAN, K., BRUNO, M. et al. (1995), Surfactant protein A–polylysine conjugates for delivery of DNA to airway cells in culture, *Hum. Gene Ther.* **6**, 31–40.

SENIOR, J. H., TRIMBLE, K. R., MASKIEWICZ. R. (1991), Interaction of positively charged liposomes with blood: implications for their application *in vivo*, *Biochim. Biophys. Acta* **1070**, 173–179.

SHIMIZU, N., CHEN, J., GAMOU, S., TAKAYANAGI, A. (1996), Immunogene approach toward cancer therapy using crythrocyte growth factor receptor-mediated gene delivery, *Cancer Gene Ther.* **3**, 113–120.

SORIANO, P., DIJKSTRA, J., LEGRAND, A., SPANJER, H., LONDOS-GAGLIARDI, D. et al. (1983),

Targeted and nontargeted liposomes for *in vivo* transfer to rat liver cells of a plasmid containing the preproinsulin I gene, *Proc. Natl. Acad. Sci. USA* **80**, 7128–7131.

STERNBERG, B., SORGI, F. L., HUANG, L. (1994), New structures in complex formation between DNA and cationic liposomes visualized by freeze-fracture electron microscopy, *FEBS Lett.* **356**, 361–366.

THURNHER, M., WAGNER, E., CLAUSEN, H., MECHTLER, K., RUSCONI, S. et al. (1994), Carbohydrate receptor-mediated gene transfer to human T leukemic cells, *Glycobiology* **4**, 429–435.

TRUBETSKOY, V. S., TORCHILIN, V. P., KENNEL,S. J., HUANG, L. (1992a), Use of N-terminal modified poly(L-lysine)–antibody conjugate as a carrier for targeted gene delivery in mouse lung endothelial cells, *Bioconj. Chem.* **3**, 323–327.

TRUBETSKOY, V. S., TORCHILIN, V. P., KENNEL, S., HUANG,L. (1992b), Cationic liposomes enhance targeted delivery and expression of exogenous DNA mediated by N-terminal modified poly(L-lysine)–antibody conjugate in mouse lung endothelial cells, *Biochim. Biophys. Acta* **1131**, 311–313.

ULMER, J. B., DONNELLY, J. J., PARKER, S. E., RHODES, G. H., FELGNER, P. L. et al. (1993), Heterologous protection against influenza by injection of DNA encoding a viral protein, *Science* **259**, 1745–1749.

VITIELLO, L., CHONN, A., WASSERMAN, J. D., DUFF, C., WORTON, R. G. (1996), Condensation of plasmid DNA with polylysine improves liposome-mediated gene transfer into established and primary muscle cells, *Gene Ther.* **3**, 396–404.

WAGNER, E., ZENKE, M., COTTEN, M., BEUG, H., BIRNSTIEL, M. L. (1990), Transferrin–polycation conjugates as carriers for DNA uptake into cells, *Proc. Natl. Acad. Sci. USA* **87**, 3410–3414.

WAGNER, E., COTTEN, M., FOISNER, R., BIRNSTIEL, M. L. (1991), Transferrin–polycation-DNA complexes: the effect of polycations on the structure of the complex and DNA delivery to cells, *Proc. Natl. Acad. Sci. USA* **88**, 4255–4259.

WAGNER, E., ZATLOUKAL, K., COTTEN, M., KIRLAPPOS, H., MECHTLER, K. (1992a), Coupling of adenovirus to transferrin–polylysine/DNA complexes greatly enhances receptor-mediated gene delivery and expression of transfected genes, *Proc. Natl. Acad. Sci. USA* **89**, 6099–6103.

WAGNER, E., PLANK, C., ZATLOUKAL, K., COTTEN, M., BIRNSTIEL, M. L. (1992b), Influenza virus hemagglutinin HA-2 N-terminal fusogenic peptides augment gene transfer by transferrin–polylysine-DNA complexes: toward a synthetic virus-like gene transfer vehicle, *Proc. Natl. Acad. Sci. USA* **89**, 7934–7938.

WANG, C. Y., HUANG, L. (1987), pH-Sensitive immunoliposomes mediate target-cell specific delivery and controlled expression of a foreign gene in mouse, *Proc. Natl. Acad. Sci. USA* **84**, 7851–7855.

WOLFF, J. A., MALONE, R. W., WILLIAMS, P., CHONG, W., ACSADI, G. et al. (1990), Direct gene transfer into mouse muscle *in vivo*, *Science* **247**, 1465–1468.

WOLFF, J. A., LUDTKE, J. J., ACSADI, G., WILLIAMS, P., JANI, A. (1992), Long-term persistence of plasmid DNA and foreign gene expression in mouse muscle, *Hum. Mol. Genet.* **1**, 363–369.

WU, G. Y., WU, C. H. (1987), Receptor-mediated *in vitro* gene transformation by a soluble DNA carrier system, *J. Biol. Chem.* **262**, 4429–4432.

WU, G. Y., WU, C. H. (1988), Receptor-mediated gene delivery and expression *in vivo*, *J. Biol. Chem.* **263**, 14621–14624.

WU, G. Y., WU, C. H. (1992), Specific inhibition of hepatitis B viral gene expression *in vitro* by targeted antisense oligonucleotides, *J. Biol. Chem.* **267**, 12436–12439.

YANG, N. S., BURKHOLDER, J., ROBERTS, B., MARTINELL, B., MCCABE, D. (1990), *In vivo* and *in vitro* gene transfer to mammalian somatic cells by particle bombardment, *Proc. Natl. Acad. Sci. USA* **87**, 9568–9572.

YOSHIMURA, K., ROSENFELD, M. A., NAKAMURA, H., SCHERER, E. M., PAVIRANI, A. et al. (1992), Expression of the human cystic fibrosis transmembrane conductance regulator gene in the mouse lung after *in vivo* intratracheal plasmid-mediated gene transfer, *Nucleic Acids Res.* **20**, 3233–3240.

ZHANG, L., LI, L., HOFFMANN, G. A., HOFFMAN, R. M. (1996), Depth-targeted efficient gene delivery and expression in the skin by pulsed electric fields: an approach to gene therapy of skin aging and other diseases, *Biochem. Biophys. Res. Commun.* **220**, 633–636.

ZHOU, X., HUANG, L. (1994), DNA transfection mediated by cationic liposomes containing lipopolylysine: characterization and mechanism of action, *Biochim. Biophys. Acta* **1189**, 195–203.

ZHOU, X. H., KLIBANOV, A. L., HUANG, L. (1991), Lipophilic polylysines mediate efficient DNA transfection in mammalian cells, *Biochim. Biophys. Acta* **1065**, 8–14.

20 Issues of Large-Scale Plasmid DNA Manufacturing

MARTIN SCHLEEF

Hilden, Germany

1 Introduction 444
2 Safety and Regulatory Aspects of Plasmid DNA Therapeutics 445
3 Basic Concepts for Plasmid Fermentation and Purification 446
  3.1 Plasmids 446
  3.2 *E. coli* Host Cells 447
  3.3 Plasmid Isolation at the Laboratory Scale 447
  3.4 Fermentation of Plasmid Containing Cells 448
  3.5 How to Design a Plasmid Manufacturing Process 449
  3.6 Plasmid Stability 451
4 The GMP Plasmid DNA Manufacturing Process 452
  4.1 Vector Selection and Preclinical Work 452
  4.2 Selection of Producer Strains and Master Cell Banks 452
  4.3 Pilot Production 453
  4.4 Fermentation and Harvesting 453
  4.5 Downstream Processing 454
  4.6 Upscaling of Specific Process Steps 456
  4.7 Types of Manufacturing Processes 458
  4.8 Quality Control of Plasmid DNA 458
    4.8.1 Restriction Analysis of Plasmid DNA Manufacturing Batch 459
    4.8.2 Sequencing of Plasmid DNA Manufacturing Batch 459
    4.8.3 Sterility of Plasmid DNA Manufacturing Batch 459
    4.8.4 DNA Quantity of Plasmid DNA Manufacturing Batch 459
    4.8.5 Other Quality Control of Plasmid DNA Manufacturing Batch 459
5 Pharmaceutical Requirements: GMP 460
  5.1 What Does GMP Mean? 460
  5.2 Manufacturing Facility Requirements 461
  5.3 Process Validation 462
  5.4 In-Process Control 463
  5.5 Quality Assurance and Quality Control 463
  5.6 Documentation 465
6 The Future of Therapeutic Plasmid DNA 465
7 References 466

List of Abbreviations

CBER	Center for Biologics Evaluation and Research
CCC	covalently closed circular
CDER	Center for Drug Evaluation and Research
CFR-21	Code of Federal Regulations
cGMP	current good manufacturing practice
cv	column volumes
DEAE	N,N-diethylaminoethyl
DMF	drug master file
DQ	design quality
DSP	downstream processing
EC	European Community
EMEA	European Medicines Evaluation Agency
EPC	end of production cell sample
FDA	Food and Drug Administration
GMP	good manufacturing practice
HPLC	high-performance liquid chromatography
IND	investigational new drug
IPC	in-process control
IQ	installation qualification
ISS	immunostimulatory sequences
kbp	kilo base pairs
LAL	*Limulus* amebocyte lysate
LPS	lipopolysaccharide
MCB	master cell bank
MW	molecular weight
NIH	National Institutes of Health
OC	open circular
OD	optical density
OQ	operation qualification
PQ	performance qualification
QA	quality assurance
QC	quality control
SDS	sodium dodecyl sulfate
SOP	standard operating procedures
WCB	working cell bank
WHO	World Health Organization

1 Introduction

By virtue of the developments made recently in biotechnology and molecular medicine, the speculation of 20 years ago that gene technology would become a powerful tool to cure disease directly, has become reality (TATUM, 1966). The potential to cure disease at the level of the specific genetic defect rather than at the conventional phenotypic level, induced a widespread search for suitable administration routes to target affected cells at the genetic level. The first gene therapy approach was used on September 14, 1990, at the NIH to treat a patient suffering from ADA (adenosine deaminase deficiency). Viral vector systems were used to administer the recombinant DNA. In gene therapy and genetic vaccination approaches clinicians and researchers use gene technology as the means to either cure inherited genetic diseases such as cystic fibrosis and cancer (CAPLEN et al., 1994), or to preventively or curatively treat against infectious diseases such as hepatitis B and C (MICHEL et al., 1995; MAJOR et al., 1995; LE BORGNE et al., 1998; GREGORIADIS, 1998; see Chapter 21, this volume). A considerable amount of progress has been made by using gene transfer technology in which viruses are modified and their well-developed system for transferring genetic information into eukaryotic cells is used to introduce transgenes into cells for either stable or transient expression (GÜNZBURG and SALMONS, 1997; Chapters 18 and 19, this volume). Various transfection techniques involving non-viral plasmid vectors have also been developed to transfer genetic information into eukaryotic cells (GRAHAM and VAN DER EB, 1973; NEUMANN et al., 1982; FELGNER et al., 1987; TANG et al., 1996; for review of different delivery systems see PRASRAMPURIA and HUNT, 1998). The importance of plasmid DNA as a pharmaceutical substance has increased considerably since it was shown that naked DNA injected into muscle tissue is expressed *in vivo,* and that the introduction of immunogenic sequences can result in animal vaccination against the encoded peptide (WOLFF et al., 1991; VOGEL and SARVER, 1995; DONNELLY et al., 1997). The administration of "the blueprint" of an immunogenic epitope was not just a revolutionary finding for the field of immunology, but also was the key for further expansion of the large vaccine market for nucleic acid technologies. This finding has in turn made the demand for purificd plasmid DNA and technology transfer for plasmid DNA purification at the industrial scale even greater. The following points are imperative:

(1) the product quality must exceed that found in research laboratories,
(2) the technology must have the capacity for scale-up in the long term, and
(3) the product and technology should satisfy the requirements of manufacturers, regulatory authorities, and the rapidly growing gene therapy market.

Proper manufacturing conditions are essential for the production of any substance to be used in the treatment of humans or animals. This applies to every material introduced into the manufacturing process, starting from any raw material to the pharmaceutically active ingredient (the plasmid DNA itself). Therefore, the manufacturing procedure must include stringent quality assurance (QA) and quality control (QC) (SCHORR et al., 1995) to guarantee safety and functionality of the product. In this chapter the recent developments in large scale pharmaceutical plasmid DNA production under current good manufacturing practice (cGMP) conditions are described. Also described are the manufacturing quality assurance and quality criteria for the production of clinical-grade plasmid DNA. An overview of points that should be taken into account before and during manufacturing of plasmid DNA for such intended applications is also presented.

2 Safety and Regulatory Aspects of Plasmid DNA Therapeutics

The development of a plasmid intended for any use as a pharmaceutical component, as well as a vector for the expression of proteins in producer cells (prokaryotic or eukaryotic) to obtain biomolecules (e.g., proteins, RNA) active in a medicinal capacity, requires long-term planning of vector design and construction, and evaluation of the results obtained.

At this stage it is not essential to perform research under GMP working conditions. Indeed, GMP conditions should not be employed during the research phase. However, parts of the developing system can be extracted and utilized at a later stage. The most important factor is full documentation of the system development process. This feature is also essential for the patenting department of the research organization, so that the patent awarding body will be able to protect intellectual property rights.

The very first step should be accurate vector design using the information available in guidelines such as *Points to Consider on Plasmid DNA Vaccines for Preventive Infective Disease Indications* (CBER, 1996), *Guidance for Human Somatic Cell Therapy and Gene Therapy* (CBER, 1998) from the FDA, *Guidelines for Assuring the Quality of DNA Vaccines* (WHO, 1997) from the WHO, and contacts with experts in the field (for review see MEAGER and ROBERTSON, 1998 and CICHUTEK and KRÄMER, in press). Any sequence segment introduced into (or not removed from) the plasmid construct should be as small as possible and fully characterized (at least by sequencing), to be able to assess its functionality in context with the other co-assembled sequences.

The design of a therapeutic plasmid directly and significantly influences the manufacturing process. It will be shown in Sect. 3.2 that the combination of a certain plasmid with a producer strain causes sizable quality and quantity differences arising from each individual plasmid. In essence, the sequence and size of a plasmid can lead to quality differences and/or the need for a larger manufacturing scale in order to obtain the appropriate amount of plasmid DNA. Depending on the desired use for the plasmid DNA (gene therapy or vaccination), certain safety issues must be carefully considered in parallel. Those that influence the large scale manufacturing process will be discussed in Sect. 3.5, while the safety and regulatory aspects of plasmid design are discussed by ROBERTSON (1994), MEAGER and ROBERTSON (1998), and in Chapter 23, this volume.

In order to design a safe manufacturing process for pharmaceutical-grade plasmid DNA, the manufacturer has to ensure that the process conforms to every applicable law or regulation, and that it can be performed reproducibly with respect to at least quality thresholds and specifications. Products from the manufacture of biopharmaceuticals with the exception of small molecule pharmaceutical substances, are difficult to qualify at the finished product level by existing quality control. This is because additional testing techniques have to be developed first. Hence, the process itself has to be safe enough to compensate for this, thereby making in-process control and validation essential prerequisites for the manufacturing process.

3 Basic Concepts for Plasmid Fermentation and Purification

3.1 Plasmids

Plasmids are circular duplex DNA molecules that are stably maintained as episomal genetic information within bacteria (HELINSKI, 1979; SUMMERS, 1996). Their size ranges between approx. 1.5 and approx. 120 kbp, and their copy number per bacterial cell is at least one (DAVIS et al., 1980). The copy number can, in the case of small plasmids, reach >1000 copies per cell. Replication does not depend on any plasmid-encoded protein and is not synchronous with replication of the bacterial host chromosome (DAVIS et al., 1980). The plasmid size depends on the conformity of the molecule. If linear, a 3 kbp plasmid has a molecular weight of $2 \cdot 10^6$ Da and a length of 1 μm (DAVIS et al., 1980). The exact conformity of a plasmid prepared from *Escherichia coli* (*E. coli*) cells depends on their integrity [covalenty closed circular (CCC), or nicked, and open circular (OC)] and the result of the replication process: monomeric or multimeric (see also Sect. 4.8). Multimeric chains of two or more molecules

(catenates) appear less frequently in plasmid preparations and depend on the *E. coli* host strain and the growth phase during which the cells were lysed. Multimeric large molecules in which two plasmids are present as a head-to-tail fusion within one circle (concatemers; likely generated by replication mistakes), are observed more frequently and typically migrate in agarose gel electrophoresis close to or with the OC band. The topological configuration of CCC plasmids is defined by their linking number, which is a measure of times the two DNA strands of the double helix are interwound. If plasmid is underwound (negatively supercoiled) this results in a more compact molecule (TSE-DINH et al., 1997). Plasmids can be genetically engineered to contain foreign sequence information (SCHWAB, 1993), and can be maintained and amplified in microbial cells, typically *E. coli*. The encoded genes may, depending on their regulatory elements, be expressed within these cells. In manufacturing processes in which recombinant proteins are produced, such gene expression is desired and induced. Although plasmid manufacturing processes typically employ the plasmid constructs usually used for expression in an eukaryotic system, gene expression is not wanted. Only those sequences that control the maintenance and yield of the plasmid should be expressed if necessary. Typically, this represents at least an antibiotic resistance gene (selective factor) within a portion of the plasmid. This gene is required (forms part of the backbone) to carry the therapeutic sequence unit. With respect to its intended use, the FDA and WHO recommend to totally avoid the use of antibiotic resistance genes. If the use of such genes cannot be circumvented, a gene encoding any resistance to antibiotics of the β-lactam type, such as ampicillin, should not be selected. Preferably the kanamycin resistance gene should be used instead. The selective factor is required during the plasmid construct development phase, when transforming into *E. coli* host cells, and when qualifying the stability of the plasmid throughout the entire manufacturing process (Sect. 3.6).

3.2 *E. coli* Host Cells

A host strain appropriate for all research work and/or industrial scale pharmaceutical manufacturing does not exist *per se*. In general, it is necessary to work with a clone derived from a host strain stock that has been completely characterized, is free of any contamination, and is not harmful to the environment, the isolated product, the exposed patients, and the manufacturing employees and health care personnel. Over the years, a significant amount of experience has been generated with *E. coli* K12 (NEIDHARDT, 1996), especially in the field of molecular cloning and DNA techniques. *E. coli* K12 fulfills the role of a safe, well-characterized host strain for DNA production.

Systematic analysis of various *E. coli* substrains demonstrated quality and quantity differences between all tested strains (Fig 1). Specifically, these differences relate to the amount of plasmid DNA per biomass and the isoform distribution of the plasmid.

The observed differences in isoforms depend on the particular plasmid as well as on the host strain. This means that not only the genetic background of the host causes the differences of the plasmid isoform distribution, but also the plasmid itself to a certain extent.

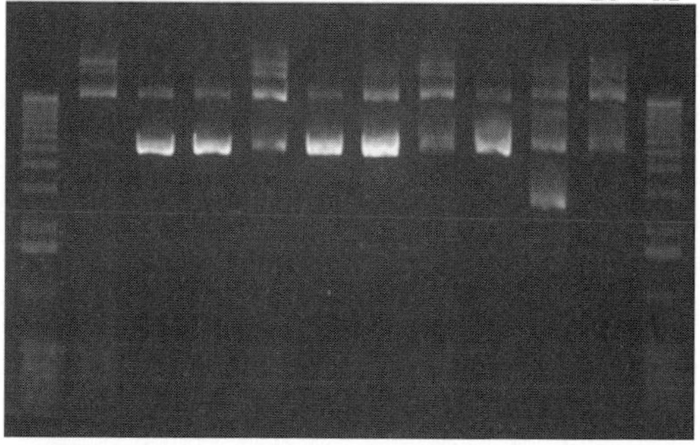

Fig. 1. Differences in plasmid DNA quality and quantity, derived from different *E. coli* host strains. M: 1 kb ladder; 1: NM522; 2: DH5; 3: XL1-blue; 4: DH1; 5: DH5-α; 6: JM108; 7: JM110; 8: HB101; 9: JM83; 10: JM101. Equal amounts of plasmid DNA solution from parallel plasmid preparations were applied to each lane of a 1% agarose gel. DNA was stained with ethidium bromide.

3.3 Plasmid Isolation at the Laboratory Scale

Since the discovery that plasmids play an important role in gene technology, laboratory scale methods to purify plasmid DNA for applications such as transformation, sequencing, and transfection studies, have improved dramatically. Consequently, plasmid purification techniques have become easier and less time consuming, particularly since the development of specially optimized kits.

The technique most commonly used to purify plasmid DNA from *E. coli* is the alkaline lysis procedure (Fig. 2). In this approach, a cleared lysate is generated which contains all components of a disrupted cell including plasmid DNA, but with minimal amounts of cell wall components and genomic DNA (BIRNBOIM and DOLY, 1979; ISH-HOROWICS and BURKE, 1981). The second classic approach is a boiling procedure in which cells are lysed by heating in the presence of a detergent (typically SDS, as used in the alkaline lysis procedure). In both cases (SAMBROOK et al., 1989), subsequent CsCl density gradient centrifugation is performed. Since this technique is expensive, time consuming, and uses substances toxic for both the environment and the product (e.g., CsCl, ethidium bromide, and phenol), it is only of historic significance. For laboratory scale purification, diverse cleanup procedures based on DNA-binding matrices (anion exchange), silica matrices such as membranes or slurry, glass fiber fleece, adsorbing membranes, magnetic beads, or selective precipitation approaches, are applicable (for a summary see *<http://www.the scientist. library.open.edu/ yr1997/august/profile2-970818.html>*.) Cell lysis approaches based on mechanical disruption or ultrasonic techniques cause shearing of at least the chromosomal DNA and are not recommended. In some applications, enzymes such as lysozyme are used to support cell lysis.

The first stage of alkaline bacterial lysis for plasmid preparation involves complete resuspension of cells in a defined buffer solution. The cells must be completely separated from each other to avoid large cell aggregates which will produce incomplete and inefficient

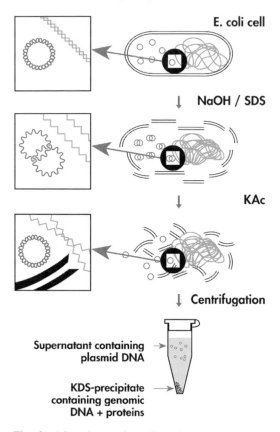

E. coli cell

↓ **NaOH / SDS**

↓ **KAc**

↓ **Centrifugation**

Supernatant containing
plasmid DNA →

KDS-precipitate →
containing genomic
DNA + proteins

Fig. 2. After harvesting, *E. coli* cells are resuspended in buffer P1 (50 mM TrisHCl, pH 8.0; 10 mM EDTA; 100 μg mL^{-1} RNase A). Subsequent alkaline treatment with buffer P2 (200 mM NaOH; 1% SDS) results in cell lysis and the release and denaturation of cellular components, e.g., proteins, genomic DNA, and plasmid DNA. Addition of buffer P3 (3.0 M potassium acetate, pH 5.5) neutralizes the lysate and causes plasmid DNA (but not genomic DNA) to reanneal and remain in solution. The high salt concentration causes SDS to precipitate, and the denatured proteins, genomic DNA, and cell debris become trapped in salt detergent complexes (KDS). The KDS precipitate is removed (see text), and the cleared lysate contains the purified plasmid DNA.

lysis. Complete lysis is absolutely essential. The lysis solution consisting of SDS and sodium hydroxide disrupts the cells, solubilizes the cell membrane components, denatures the majority of biomolecules including plasmids, and results in a viscous suspension. Sub-

sequently, this alkaline suspension is neutralized by the addition of acidic potassium acetate (pH 5.5). This high potassium acetate concentration facilitates renaturation of the plasmid, and aggregate formation occurs between the genomic DNA, *E. coli* proteins, membrane components, and SDS. These aggregates are up to 10 mm in size, appear like flakes, and are typically flocculant. The resulting liquid phase contains the plasmid DNA plus additional molecules that are not aggregated.

3.4 Fermentation of Plasmid Containing Cells

For laboratory scale plasmid isolation, bacterial cultivation is typically performed in culture bottles with volumes of up to 2 L. Growth media and conditions for optimal bacterial growth and plasmid yields, resulting in optical density (OD_{600}) values of approximately 3–6 OD units in complex bacterial growth media have been developed. For purposes such as research-grade plasmid preparation, this procedure is sufficient and the final analysis of the prepared DNA is usually performed by agarose gel electrophoresis, DNA quantification, and an identity test (restriction digestion).

The culture bottle system is a batch culture with no type of on-line monitoring or regulation. The growth conditions are adjusted before inoculating the medium, and the medium is unattended for usually 16–20 h. Moreover, no pH monitoring or adjustment is performed, and oxygen and carbon source levels are neither measured nor regulated. Due to the loss of essential substrates, the culture ultimately dies as self-toxification takes place. Overgrown cultures even contain degraded cellular components (including plasmids).

To overcome these culture problems for isolated recombinant proteins, high-performance fermentation technology was developed over some years. Fermentation processes require different growth media from small scale bottle cultures. The possibility of monitoring the growth conditions allows replenishment of essential media components before they run out (feeding) and the opportunity to maintain constant pH control and oxygen supply (KORZ et al., 1995; LAHIGANI et al., 1996). Besides the effects of such regulation, the culturing process became more defined and the pharmaceutical requirements on documentation could be fulfilled. A further feature of fermentation technology for large scale plasmid production is the performance of high-density fermentation (CHEN et al., 1997) to obtain large amounts of biomass. Experimental work on the composition of bacterial growth media for bottle cultures and fermentation has demonstrated that the choice of fermentation conditions and growth media strongly influences the yields and quality of plasmids from *E. coli* cells (SCHMIDT and SCHLEEF, unpublished results). The main focus was placed on the amount of plasmid per cell (copy number) that can be monitored on-line by capillary gel electrophoresis (SCHMIDT et al., 1996).

3.5 How to Design a Plasmid Manufacturing Process

Every industrial manufacturing process must start at the laboratory and pilot scales. It is very important at these stages that all components and working conditions are as close as possible to those that will ultimately be used for manufacturing. The transfer from research scale technology to manufacturing scale requires management of the upscaling process. Such upscaling is not just a simple multiplication of the relevant factors, but instead requires highest competence, time investment, and incurs costs. For example, the performance of a step such as neutralization of alkaline lysates on ice (research scale) would require expensive cooling instrumentation at a 100 L or 1000 L scale. Related aspects will be discussed in Sect. 4.6.

In order to design a manufacturing process, a general protocol must exist. Absolutely no planning should be started before an appropriate upscaling has been done and approved. The logistics of such manufacturing has to conform with the capacity and capability of the manufacturing site. The site needs to be able to monitor the process at every step [in-process control (IPC), quality assurance, and quality control], to be permitted to do so (legal and regulatory aspects). Furthermore, costs must be of a reasonable level (see also Sect. 6) and the instrumentation should be adequate.

The following process steps are required for plasmid manufacture. Each step must be compatible with previous and subsequent steps.

In cases where a plasmid was constructed and selected for manufacturing, and a complete characterization was performed, the first step is to evaluate the characterization data. A critical question to ask in plasmid manufacturing is the risk assessment with respect to biosafety in manufacturing. The containment level of plasmid clones is a consequence of the host cell type and the sequence of the plasmid. The cell type in all published cases is *E. coli* K12, which is considered to be a biosafety strain without any risk. The plasmid has to fulfill the appropriate prerequisite to be of a biosafety level that can be handled in the desired manufacturing facility. In cases requiring higher biosafety levels, an enormous increase in manufacturing costs will be generated. Therefore, exact classification is essential and critical, particularly because plasmid–vaccine constructs naturally contain pathogen-derived sequences. For the plasmid manufacturing process, safety aspects influence the process design for safety levels S1 and S2 (BL1 and BL2). For the plasmid as a molecule, no regulation relating to genetically modified microorganisms is applicable, although the author recommends protection of personnel from uptake (air, skin) of plasmid DNA.

For other parameters such as replication and growth behavior of cells containing the plasmid, quality, quantity and stability, small scale pilot cultivation experiments can generate enough data for preliminary process design. To ensure the reliability of product yield calculations, the pilot scale data should be generated using identical raw materials and the biological materials (cell stabs) as for later stages. This means that cells from the working cell bank (WCB) should, e.g., be used to determine inoculation rates. The scale of the fermenter, together with the expected biomass yield and product content, are key parameters in designing a manufacturing process, if the amount of product finally required is already known. Since individual plasmid clones show different productivity, only rough estimations can be given here. Typically, 0.5–1 g plasmid DNA per 1 kg of wet weight biomass is obtained from cultivation media using high copy number plasmids. A lower yield of plasmid DNA per biomass should be a signal for further research to increase the yield (e.g., changing the construct design). Immense costs and unrealistic upscaling requirements will in the long term cause termination of a manufacturing project. This demonstrates that time, volume streams, upscaling considerations, and requirements for process monitoring are additional essential factors for successful manufacture planning. Currently, with a 200 L working volume fermentation unit, up to 50 kg of wet weight biomass yielding approximately 25 g of plasmid can be produced per run.

At the end of fermentation, the cells must be isolated from the cultivation medium. The bacteria should be separated from the residues of the growth medium and metabolic products, and the volume containing the cells has to be reduced to give a more easily managed cell mass for the next steps (storage and lysis).

Research scale separation within centrifuge bottles (available up to 1 L volume) are too time consuming for batch volumes >20 L. Six times 0.8 L culture in 1 L bottles each taking 30 min per centrifugation run would require at least 2.5 h concentration time (with approximately 0.5 h hands-on time). In such cases, and below 200 L scales, the cell culture may be concentrated and undesired residues of the fermentation washed out by cross-flow filtration. Volumes greater than 200 L require a flow-through or continuous centrifuge for time efficient harvesting. In the case of manufacturing different plasmids (which are considered to be different products), the flow-through centrifuge has to be cleaned and sanitized completely. In batch centrifuge or cross-flow applications, single use of the disposable device is recommended. Finally the cell suspension should be, if possible, at least 10 times concentrated in comparison with the initial fermentation product.

To obtain cell lysis and a solution containing plasmid DNA with minimized contamination from other biomolecules or cell and medium components, the alkaline lysis procedure described in Sect. 3.3 can be used. In any case, the solid floating or sedimenting particles have to be removed from the lysate. This can be achieved by either centrifugation at high g forces (not applicable at the manufacturing scale) or by filtration techniques. Depth filters, absolute filters (e.g., sterile filters), filter fleeces or filter slurry are the materials of choice. Each material used should be of a quality suitable for use in pharmaceutical manufacturing. The use of some materials and devices may be restricted by intellectual property regulations. At this stage, the volume of the cleared lysate may be within the 100 L range. To reduce the required mass transfer, and to limit the manufacturing costs, either a capturing step is required, or if possible, selective adsorption of plasmids should be performed. In this regard, the existing technology for pilot scale manufacturing (≤ 10 mg) is diverse. Some techniques use anion exchange chromatography for capturing, followed by one or more polishing steps (CHANDRA et al., 1992). Others use gel filtration approaches that exclude upscaling at a later manufacturing scale, and are less relevant (MARQUET et al., 1995b). Further variants of pilot scale technology use reversed-phase chromatography but employ hazardous or toxic substances (GREEN et al., 1997). A combination of capturing and polishing within one single chromatographic step with integrated efficient removal of lipopolysaccharide (LPS) molecules (COLPAN et al., 1995) is the methodology of choice. This approach also saves time (one step instead of at least two) and there is no limit on the scale of plasmid purification achievable.

Chromatography steps within bioprocessing have to be chosen carefully. Typically, chromatographic materials that can perform and fulfill all other requirements have to be selected and tested in a set of pilot experiments (DASARTHY et al., 1996). Important factors are the dynamic capacity, selectivity, cleanability, and lifetime. For the isolation of plasmid DNA, the major contaminants that are either positively charged or do not bind specifically to the chromatography matrix, will flow through the anion exchange chromatography column and are, therefore, removed. Those molecules that are negatively charged (as are plasmids), but bind to the column with different ionic strength, will either flow through in the loading step, or will be subsequently removed by increased selectivity in washing steps or gradient elution.

Classic anion exchange matrices as summarized in DASARTHY et al. (1996) are used in multistep plasmid isolation approaches, and have the capability to select between plasmid DNA and RNA. The elution points for plasmid DNA are positioned at ranges where host proteins are anticipated to elute. The surface charge and the specific, e.g., N,N-diethylaminoethyl (DEAE) group of specially designed plasmid purification matrices (COLPAN et al., 1995) allow plasmid DNA elution at high salt conditions (1.6 M NaCl, pH 7.0). This, in turn, means that at lower ionic strength, the host proteins and RNA will be washed away. Additionally, the binding–elution properties allow the selection between not only RNA and DNA, but also between double-stranded plasmid DNA and any single-stranded DNA.

A final step for plasmid recovery following chromatography is the transfer of the DNA into the required solution for storage until needed. The removal of a high salt load is essential. This can be achieved by dialysis or precipitation of the DNA.

3.6 Plasmid Stability

Contrary to recombinant protein manufacturing in which the integrity and presence of the encoding gene(s) are monitored, in plasmid manufacturing the product does not need to be induced. Moreover, if replicated to higher amounts per cell, the product does not have to be stored or removed from the cytosol. The amount of plasmid DNA per cell is limited, and depends on certain parameters (SUMMERS, 1996).

The pathway from a defined, tested, and released research cell stab to kg quantities of biomass for plasmid purification is long. At no stage the structure of the enclosed plasmid should be changed. Consistency is monitored by quality assurance (see Sect. 5.5). The accumulation of cells not containing the desired plasmid at an adequate copy number severely affects the productivity of the cultivation process. It is inferred from recombinant protein manufacturing data that the absence of burdening plasmids within host cells enables them to potentially grow faster, and consequently to outgrow the plasmid-containing host cells (KITTLE and PIMENTEL, 1997). It is worthwhile starting a cultivation only with a cell bank that has been demonstrated to have at least 70% plasmid stability (preferably what should be done) if no antibiotic or other selective factor such as in repressor titration systems (WILLIAMS et al., 1998) is used in cell propagation. Two points have to be distinguished: first, the functional presence of the selective factor that is tested by replica plating of at least 100 colonies from any stage of the cultivation, and second, the constancy of the plasmid copy number which is determined as a ratio of plasmid DNA per cell mass. Since DNA quantification is affected by certain parameters (WILFINGER et al., 1997; SAUER et al., 1998) and the determination of the biomass is relative (OD measuring, cell counting or weight determination) early in process planning, at least one of the tests should be performed to work with an internal reference value.

As a part of process validation of cell banking and cultivation (NAGLAK et al., 1994), plasmid stability is tested in the master cell bank (MCB), the working cell bank (WCB), and in the end of production cell sample (EPC) where cells are collected from the fermented biomass and are cultured beyond the typical number of passages required for the manufacturing. This overcultivation generates the data necessary for further upscaling of cultivation and confidence in the reproducibility of the current process with respect to plasmid stability.

4 The GMP Plasmid DNA Manufacturing Process

4.1 Vector Selection and Preclinical Work

Plasmid vectors intended for use in gene therapy and genetic vaccination are designed and constructed in research laboratories. Here a first type of quality control is performed by the researcher: restriction digestion and sequencing of the construct. Plasmid design should consider the FDA and WHO guidelines (CBER, 1996; WHO, 1997) for a safe application in clinical trials and manufacturing considerations discussed in this chapter. The desired plasmid is subsequently amplified in *E. coli* cells and, typically, transfection studies are performed to ensure plasmid functionality. This work constitutes basic research and ends when enough research data has been collected to be confident of having developed a vector that will perform well *in vivo*. In order to prove the concept, and to guarantee the absence of unexpected side effects, the quality of DNA used for any type of preclinical animal treatment must be the same as that subsequently used in human applications. Additionally, formulation and storage conditions must be defined and should be validated prior to the phase I and II trials. To ensure that the plasmid DNA quality remains consistent from this step onwards, it is first subjected to an identity test.

An important parameter that frequently influences process planning at a very late stage and, therefore, usually leads to unexpected delays and costs, is the size of the required plasmid. The influence of plasmid size is discussed here. Large plasmids (>10 kbp) significantly influence the total copy number per cell. This means that even plasmids with the backbone of a typical high-copy number plasmid (containing the origin of replication) replicate at low copy number levels. In cases were this cannot be avoided because the insert is initially derived from a large gene and cannot be used at reduced sizes, the process design requires between 10 and 100-fold upscaling.

4.2 Selection of Producer Strains and Master Cell Banks

Once the plasmid construct has been designed and characterized, it is transformed into a range of different *E. coli* K12 strains to determine the strain that performs best with the individual plasmid. The *E. coli* K12 strains should be characterized and derived from a qualified source. Comparison and selection of a suitable host strain is based on the production capacity and quality of the plasmid DNA, which in turn varies considerably with the *E. coli* strain and the individual plasmid construct. The initial choice of vector backbone is also an important issue at this stage, since low copy number plasmids can cause tremendous additional work (see Sect. 4.6) and costs (see Sect. 6) for the manufacturing process.

Selected clones are amplified and are used to generate a master cell bank. This and all following steps are performed under full cGMP working conditions (STEUR and OSTROVE, 1996). The MCB consists of approximately 200 vials, is extensively characterized (Tab. 1), and forms the source for all further production. As such, it must be treated very carefully and storage has to be documented completely. One vial from the MCB is used to produce a second cell bank called the working cell bank. The WCB is subject to the same quality control as the MCB. Usually, the WCB also consists of approximately 200 vials, and any further inoculation for production of the same plasmid DNA is performed using one of these vials. This means that from this stage onwards, it is possible to inoculate reproducibly as long as vials are left over. When all 200 WCB vials have been used, a new WCB is established using one MCB vial. In this way, the manufacturing process can be repeated reproducibly up to 40,000 times (200·200). This ensures long-term consistency of seed material. Ideally, such cell banks must be stored in cryotubes in the gas phase of liquid nitrogen at more than one location under fully documented and validated (preferably full GMP) conditions.

Tab. 1. Quality Control of MCB and WCB. Each Cell Bank is Fully Characterized by the Tests Indicated

Test	Specification	Method
Microbiological characterization	*E. coli* *E. coli*	API 20E API ZYM
Absence of contamination	typical colony form typical no spores	microscopy (visual inspection) plating on agar disks spore assay
Plasmid identity	fragment size conforms sequence identity	DNA isolation, restriction digests sequencing
Plasmid stability	70% stability	replica plating
Absence of bacteriophages	no bacteriophages	plating (plaque assay)

4.3 Pilot Production

A clinical grade plasmid DNA purification system using anion exchange chromatography is flexible, and batch sizes of milligram to gram amounts are possible with today's technology. The use of fermentation devices is necessary due to the high biomass required. Large scale *E. coli* fermentation systems for plasmid production have been developed which operate and perform differently from classic protein purification methods. Critical factors to consider are the maintenance of high copy numbers (SCHMIDT et al., 1996) together with the homogeneity of the plasmid isoform distribution. In all cases, a pilot DNA production is required if a new MCB or WCB was produced. This ensures constant growth characteristics of the cells, especially with respect to plasmid stability and yield. It is also vital to be confident that the subsequent manufacturing process will start with the same quality of material as that expected from the finished product.

The data of the pilot run are used for planning of fermentations, the required batch sizes, and any further step of processing until quality control for the plasmid. One main issue is the testing of its potency in animal or tissue culture to ensure functionality before process scale manufacturing. Pilot batches should be as close as possible to the subsequent GMP conforming process but (if possible) smaller in scale.

4.4 Fermentation and Harvesting

Fermentation for pharmaceutical manufacturing processes requires defined, high quality raw materials (see Sect. 5.5), instrumentation that is validated and personnel capable of using the instrumentation, and has to follow appropriate guidelines, regulations, and laws.

After preparing the growth medium, typically a mock-fermentation is performed. This involves all activities excluding cell inoculation, to prove that the culture medium and instrument can be maintained sterile in the specific environment being used. The cultivation is then typically started with a characterized and released WCB stock vial on a preculture that, depending on the scale finally intended, may be the only preculture or one of many. Over the whole propagation of growth, at least the following parameters are monitored and documented: temperature, pH, pO_2, speed of stirring propeller, use of antifoam, use of other solutions such as acid, base, in some cases feeding solution, optical density (OD_{600}), wet cell weight, dry cell weight, and plasmid content.

For batch fermentations, overnight runs are typically performed and, depending on cell growth and plasmid type, amounts of 20–25 g of wet cell weight per liter are obtained. Cultures like this typically reach OD_{600} values of between 10 and 30.

Fed-batch cultivations lead to significantly higher biomass yields (approx. factor 10 higher) avoiding the loss of product. The amount of product per cell (or per cell weight) in particular strongly influences the downstream processing (DSP) scale, and, therefore, defines an other important part of the production costs. The upscaling of the fermentation process is discussed in Sect 4.6.

The history plot of the fermentation run, the IPC results, and environmental data mentioned above form part of a manufacturing batch record that documents the fermentation process. Harvesting of the cells in GMP manufacturing is performed with one-time-use cross-flow cassettes or continuous centrifuges. The concentrated cells are subject to extensive QA work to determine product yield, integrity, and the absence of contamination.

The appropriate storage of the biomass is essential to ensure unchanged quality of the cells containing the plasmid DNA. Aliquots of concentrated cells (e.g., 1 L) should be filled into sterile polypropylene containers and frozen immediately after harvesting. Storage time of at least 6 months does not significantly change product quality. It is important to continuously monitor the temperature at the storage place and to include this information into the respective documentation.

4.5 Downstream Processing

The following process description is summarized in Fig. 3. The biomass used for processing must be thawn and the cells resuspended in an appropriate buffer. Large amounts of biomass are processed in suitable vessels such as 5 L glass bottles for scales up to 100 g of wet cell weight, or styring vessels for kg scales.

Large lysis–suspension volumes are cleared by a combined sedimentation, floating and filtration procedure. In this approach, the plasmid-containing clear phase is filtered using a

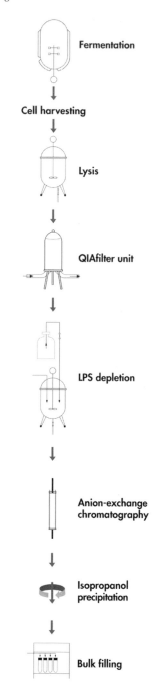

Fig. 3. Flow diagram of the plasmid manufacturing process described in detail in Sects. 4.4 and 4.5.

filtration unit (QIAfilter™) that removes the solid aggregates to obtain a sterile and homogenous cleared lysate. The unit consists of a prefilter (to prevent blocking), a filter–slurry bed, and a sterile filter (0.2 µg, e.g., cellulose acetate). The use of such a device is essential, and can be applied at any scale. The resulting solution will be used for the chromatography step for plasmid isolation. At this point, the processed plasmids within the cleared lysate are available for the first time within the process as a sterile solution. The process flow requires an air class change from the class D environment into the class C area, where chromatography should be performed (for the definition of air classes see AUTERHOFF, 1997). Since by definition the cleared lysate does not contain any living microorganisms or spores at this stage, this is a suitable point for transfer to the class C area. To remove residual contaminants such as nucleotides, proteins, RNA, single-stranded DNA, and *E. coli* lipopolysaccharides, a one-step anion exchange chromatography procedure (COLPAN et al., 1995) with integrated endotoxin removal has been developed. The integration of this removal step combines the convenience and cost effectiveness of performing only one step with the advantage of not loosing plasmid DNA, as occurs with classic types of LPS removal. Any filtration or adsorption of LPS molecules to such classic devices results in considerable product loss, since both DNA and LPS molecules are negatively charged, rendering differentiation difficult. LPS removal is essential because endotoxins such as *E. coli* LPS can have cytotoxic effects on mammalian cells *in vitro* and *in vivo* (COTTEN et al., 1994; WICKS et al., 1995). If present in sufficiently large amounts *in vivo,* they can cause symptoms of toxic shock syndrome (MORRISON and RYAN, 1987) and activation of the complement cascade (VUKAJLOVICH et al., 1987). It has been demonstrated that LPS contamination of DNA directly influences transfection efficiencies in sensitive cultured cells, and that different cells show variable sensitivity to LPS-contaminated DNA (WEBER et al., 1995).

Within the chromatography process, plasmid DNA binds to the positively charged groups at the surface of the anion exchange resin, and contaminants are removed by washing with a specific buffer solution. The chromatography procedure is a charge elution step gradient procedure. Fig. 4 shows the relevant chromatographic steps. After washing, the plasmid DNA is eluted by a high-salt buffer and precipitated with isopropanol. The precipitated DNA is then washed, dried, and resuspended in an appropriate buffer system. If required, the DNA solution is aliquoted into glass vials or higher volume storage containers. The purification steps are described in detail within the following example protocol. In each step, extensive quality assurance is performed by taking samples for IPC analysis.

The following protocol is an example for the isolation of approximately 0.5–1 g plasmid DNA from 1 kg of wet weight biomass.

(1) 1 kg of wet weight biomass is resuspended completely in 15 L buffer P1 (50 mM Tris Cl, pH 8.0; 10 mM EDTA; 100 µg mL^{-1} RNase A).

(2) 15 L buffer P2 (200 mM NaOH, 1% SDS) are added, mixed carefully, and subsequently incubated for at least 5 min.

(3) 15 L buffer P3 (3.0 M potassium acetate, pH 5.5) are added, mixed carefully, and the lysate is incubated at room temperature for at least 30 min.

(4) The lysate is cleared by filtration through a QIAfilter™ unit, and the proprietary buffer ER (QIAGEN) for integrated LPS removal is added.

(5) The cleared lysate is incubated at 4 °C for at least 1 h plus the loading time for the following chromatographic step.

(6) A process chromatography column is packed with 1.4 kg QIAGEN® anion exchange resin for plasmid manufacture, and equilibrated with 1.5 column volumes (cv) buffer QBT (750 mM NaCl; 50 mM MOPS, pH 7.0; 15% isopropanol; 0.15% Triton® X-100) at a linear flow rate of 0.5–0.6 cm min^{-1}.

(7) The column is loaded with 40 L of cleared lysate at a flow rate of 0.1–0.2 cm min^{-1}.

(8) Washing of the loaded column is performed with 8 cv of buffer QC (1.0 M

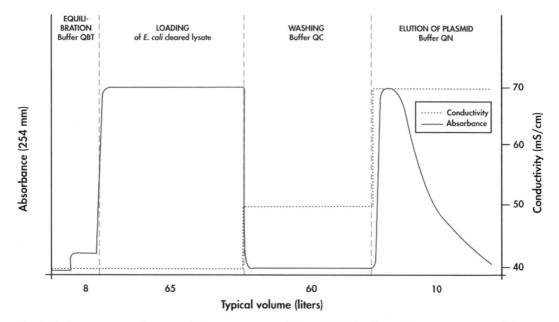

Fig. 4. A chromatogram from a typical manufacture of plasmid DNA. Cleared lysate was prepared from a 50 L culture (~1.6 kg wet weight biomass) of *E. coli* harboring a high copy number plasmid. The lysate was loaded onto a 7 L process column filled with 2.8 kg of QIAGEN resin. After washing with buffer QC (1 M NaCl, pH 7.0), the DNA was eluted with buffer QN (1.6 M NaCl, pH 7.0).

NaCl, 50 mM MOPS, pH 7.0; 15% isopropanol) at a flow rate of 0.8–1.0 cm min^{-1}.

(9) A step elution of plasmid DNA is performed with 1.5 cv of buffer QN (1.6 M NaCl; 50 mM MOPS, pH 7.0; 15% isopropanol) at a flow rate of 0.2–0.3 cm min^{-1}.

(10) Plasmid DNA is precipitated by adding 0.7 volumes of isopropanol, and batch centrifugation at 11,000 g for 30 min.

(11) The DNA precipitate is washed with 70% ethanol, rinsed, dried, and resuspended in a suitable buffering solution. The DNA is bulk filled into appropriate storage vials (glass or polypropylene). All buffers and solutions as well as all instrumentation and devices have to be sterile and pyrogen-free, so as to avoid contamination of the purified plasmid DNA.

4.6 Upscaling of Specific Process Steps

Upscaling of a process is required to overcome the limitations of process steps that cannot be expanded linearly due to in parallel or exponentially expanding costs, or simply because the parameters influencing productivity or product quality do not follow this linear increase. Cell cultivation is an example that demonstrates upscaling at a specific process step: a research scale cultivation of 2 L has to be scaled up, as the biomass of 10–40 g cells at, e.g., 5 OD units represents only 0.1% of the required amount. This means that a factor 1,000 upscaling is required. The use of a 2,000 L glass culture bottle is simply not practical or sensible (and it is also difficult and expensive to manufacture). Additionally, the use of 1,000 times a 2 L culture bottle raises logistical and cost problems. Rather, a system suitable for such a scale is fermentation technology. This system not only allows an in-

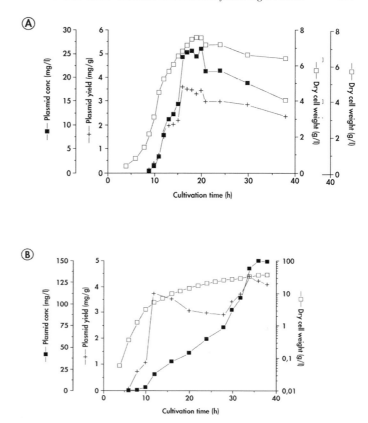

Fig. 5. Growth panels of a 7 L batch fermentation (**A**) and a fed-batch fermentation (**B**) for the isolation of plasmid pUK21-CMVβ (M. SCHLEEF, QIAGEN, unpublished results) from *E. coli* DH5α. In both diagrams, the plasmid concentration (black squares), the plasmid weight per biomass (+), and the biomass yield per culture volume (open squares) are presented over the cultivation time. Batch cultivation was performed at 37 °C with 5 L min^{-1} air, pH 7.0, $2 \cdot 10^4$ Pa overpressure, and pO$_2$ was set to 90%. Fed-batch cultivation started with comparable conditions and feeding was initialized at the end of the batch phase (data: SCHMIDT et al., unpublished results, University of Bielefeld, QIAGEN).

crease in volume within one vessel, but also enables exact regulation of cell growth to give higher cell densities. In the calculation model mentioned above, the use of a 100 L fermenter with high cell density fermentation (e.g., 100 OD units) increases the biomass by a factor of 1,000. As long as the plasmid yield per biomass remains at least constant, this is the preferable approach. Fig. 5 shows the upscaling of a batch fermentation process by developing a fed-batch strategy for plasmid manufacturing. Each upscaling requires validation of the new process step, and the entire process has to be re-evaluated on account of the

change(s). If the process that has been modified was covered by a drug master file at the FDA, then the file must also be updated accordingly.

Other process steps that are worthwhile to upscale, are those in which batch handling is performed. This refers to, e.g., any centrifugation step with precipitation bottles. The industrial scale technology for cell harvesting or debris removal from the lysis surpasses (THEODOSSIOU et al., 1997, in press) batch centrifugation by using instrumentation that allows fast flow rates (continuous flow centrifuge of filtration). Whereas anion exchange chromatography is readily scaleable, other techniques such as gel filtration are difficult to scale up beyond a certain level.

4.7 Types of Manufacturing Processes

Over the last years, an increased demand for therapeutic-grade plasmid DNA has emerged. Certain research groups mainly from industry have modified, scaled up, redesigned, or developed a new processing technology at a 10 to 100-fold scale compared with research scale techniques. Although this was termed large scale purification, plasmid DNA quantities <0.5 g were being produced. Nearly all processes use two or more chromatographic steps to obtain quality levels acceptable to the regulatory authorities. In all except one technology, the alkaline lysis procedure is employed. In one particular technique (THATCHER et al., 1997), plasmid-containing cells are harvested at a specific growth phase, the pH is adjusted exactly for alkaline lysis so as to prevent local overdenaturation of plasmids. The plasmid DNA is captured by fluidized bed anion exchange chromatography, and the product is polished by a second anion exchange chromatographic or gel filtration step. Another multichromatographic technology uses an alkaline lysis procedure and bacterial debris removal which is achieved by batch centrifugation. As outlined previously, the scaleability of batch centrifugation is limited. The subsequent steps are either anion exchange chromatography followed by gel filtration, or two anion exchange steps (CHANDRA et al., 1992). Bovine RNase A (DNase free) is used optionally. In both cases, isocratic elution of the captured DNA followed by a chromatographic step with linear gradients are performed to elute the plasmid from the chromatography column. Finally, the DNA is recovered by ethanol batch precipitation.

A non-alkaline lysis process uses a flow-through heat exchanger followed by batch centrifugation to release the plasmids and remove debris (LEE and SAGAR, 1996). The lysate is microfiltered and loaded onto an anion exchange column after diafiltration. Reversed phase chromatography is applied for polishing. Some technologies that perform the alkaline lysis procedure were modified at that level in replacing SDS by, e.g., Tween® 80 and not using enzymes at that step (MARQUET et al., 1995a, b). A major problem associated with these processes is the broad plasmid elution peak in the linear gradient elution of the anion exchange or the gel filtration chromatography. Adding polyethylene glycol to the mixture improves the technology (HORN et al., 1998). The fractions subsequently collected containing plasmid DNA are then further purified by gel filtration. This last step limits the scale of manufacturing, as discussed in Sect. 4.6. Certain individual steps are designed differently from one technology to another. Some publications cite the use of hydroxyapatite chromatography matrices as a means to plasmid capturing (KUHNE, 1997), while others present the combination of alkaline lysis, diafiltration, and triple helix affinity chromatography. In the latter, a sequence-specific interaction between an oligonucleotide and the specific plasmid leads to the formation of a pH-dependent triple helix (WILS et al., 1997). A polishing step follows.

The process described in Sect. 4.5 is used for various clinical phase I/II manufacturing runs, is covered by a drug master file (DMF) (No BB-MF6224) at the FDA, and uses a silica-based anion exchange chromatography step to purify plasmids from a filtered lysate (COLPAN et al., 1995; SCHLEEF et al., 1996; ORR, 1998). With this single chromatographic step procedure, the specific surface charge permits selective isolation of double-stranded plasmid DNA, simultaneous removal of LPS, and allows for high flow rates because of the low compression of the chromatography matrix.

4.8 Quality Control of Plasmid DNA

DNA therapeutics are chemically highly defined and, therefore, can be analyzed by chemical and biochemical as well as by physical assays.

At the time that DNA manufacturing projects first started, only research criteria for plasmid DNA quality were available. Usually, gel electrophoresis tests and sequencing were performed to estimate the quality of research-grade material, by molecular weight (MW) comparison, and restriction digestion.

More recently, a set of quality criteria (SCHORR et al., 1995) was established which are to date well-accepted in the scientific community. Regulatory aspects for designing, manufacturing, quality assurance, and quality control of vaccination vectors are summarized in the references WHO (1997) and CBER (1998).

Before lot release, the following quality control is performed on each plasmid DNA manufacturing batch as described in Sect. 4.5.

4.8.1 Restriction Analysis of Plasmid DNA Manufacturing Batch

The plasmid DNA is digested to completion by using different restriction enzymes following the instructions of the manufacturer. The size of the restriction fragments and the total plasmid size are determined by agarose gel electrophoresis.

4.8.2 Sequencing of Plasmid DNA Manufacturing Batch

The complete nucleotide sequence of both DNA strands is determined by DNA sequencing. All steps are performed following standard operating procedures and the data are documented in a sequencing report. The amount of sequencing work performed depends on the specific vector construct. The Center for Biologics Evaluation and Research (CBER, 1996) recommend complete sequencing of the plasmid directly isolated from the MCB, and sequencing of the finished product at least once within each manufactured lot. Further productions require this at least for the relevant encoding sequences of the construct including flanking regions.

With any changes in sequence, the resulting plasmid is considered to be a new product and, therefore, must be fully recharacterized.

If point mutations have changed the relevant sequence portions without negative or with even positive effects, this may be evaluated as a modification. In such cases, complete documentation demonstrating evidence for an improvement is required. For these reasons, the FDA evaluates single events such as these on a case-by-case basis.

4.8.3 Sterility of Plasmid DNA Manufacturing Batch

The absence of any contaminating microorganisms or spores is tested by USP23 (1995) or Anonymous (1998) conforming sterility tests, or the load with microorganisms is determined by bioburden assays.

4.8.4 DNA Quantity of Plasmid DNA Manufacturing Batch

The DNA concentration is determined by spectrophotometric analysis within 10 mM Tris Cl pH 8.5, and calculated from its absorbance at 260 nm (see also SAUER et al., 1998).

4.8.5 Other Quality Control of Plasmid DNA Manufacturing Batch

Besides analysis of identity (restriction and sequencing), spectrophotometric scans between 220 and 320 nm are used to detect salt and organic contamination (WILFINGER et al., 1997). A main issue is the appearance of the DNA following agarose gel electrophoresis. The isoform distribution (by agarose gel electrophoresis) and the DNA concentration are determined. The contents of RNA, protein, genomic DNA, and LPS, are determined by HPLC, BCA protein assay (Pierce, Rockford, IL), Southern blot analysis, and the ki-

Tab. 2. Quality Control Tests on Bulk Plasmid DNA

Test	Specification	Method
Appearance	clear, colorless solution	visual inspection
Identity of the plasmid	restriction fragment size conforms	restriction enzyme digest and agarose gel electrophoresis (visual inspection)
	sequence identity	sequencing
DNA homogeneity	>90% CCC form	agarose gel electrophoresis (visual inspection)
DNA concentration	depends on application	A_{260} in 1 mM Tris Cl, pH 8.5
A_{260}/A_{280}	1.80–1.95	A_{260}/A_{280} ratio in TE buffer
Scan 220–320 nm	conforms (peak at 260 nm)	absorbance scan
Presence of bacterial host DNA	<5%	agarose gel electrophoresis and blot hybridization
Presence of protein	<5%	BCA colorimetric assay
Presence of RNA	not detectable by	HPLC
Presence of endotoxin (LAL)	<0.1 EU/μg	USP23 (1995)
Absence of microorganisms	<1 microorganism	Bioburden assay (USP23, 1995; Anonymous, 1998)

netic QCL test kit (Biowhittaker, Walkersville, MD) for endotoxin (LPS) quantification.

A summary of the lot release criteria for clinical phase I and II trials is provided in Tab. 2. These quality issues were discussed at the FDA/WHO *Workshop on the Control and Standardization of Nucleic Acid Vaccines,* at the US National Institute of Allergy and Infectious Diseases in Bethesda, MD, on February 8, 1996 (SMITH et al., 1997). Future quality control will contain advanced analysis techniques. The plasmid isoform distribution can be monitored and distinct isoforms can be quantified by capillary gel electrophoresis (SCHMIDT et al., unpublished results) as illustrated in Fig. 6.

5 Pharmaceutical Requirements: GMP

5.1 What Does GMP Mean?

A prerequisite for manufacture of pharmaceutical substances is to at least work under clean, reproducible, and documented conditions. In order to ensure uniformity of manufacturing standards, legal and regulatory aspects must be considered. Good manufacturing practices (GMP) are in their current form (cGMP) defined by regulatory authorities such as the Food and Drug Administration (FDA) for the USA, or by the European

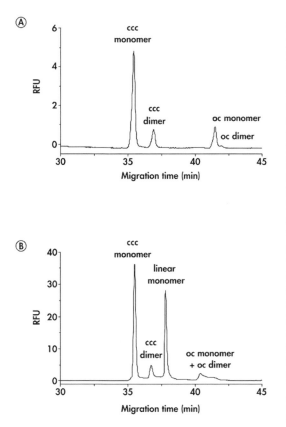

Fig. 6. Capillary gel electrophoresis of plasmid pCMV-S2S DNA (MICHEL et al., 1995) (**A**), and the same sample spiked with a linearized probe of the same plasmid (**B**). Different plasmid isoforms can be detected, and the peak area is correlated with the isoform quantity. RFU: relative fluorescence units (data: SCHMIDT et al., unpublished results, University of Bielefeld, QIAGEN).

Agency for the Evaluation of Medicinal Products (EMEA) for the European Community (EC). The manufacturing of biopharmaceuticals is basically governed by the regulations already known from classic drugs, but with specific modifications of the existing guidelines. For the EC, the principles and guidelines for GMP manufacture of medical products are outlined within the directive 91/356/EEC (for human use materials) and directive 912/412/EEC (for veterinary use materials). Detailed guidelines in accordance with principles described in the directives are published in the *EC Guide to Good Manufacturing*

Practice for Medical Products (AUTERHOFF, 1997). GMP with respect to the FDA must follow the Code of Federal Regulations, Title 21, Part 210 (CFR-21, 1991).

In contrast to conventional medicinal products that are produced using highly consistent chemical and physical techniques, biopharmaceutical manufacturing involves biological processes and materials showing more variability, such as the cultivation of microorganisms for the isolation of biomolecules. This may lead to inherent variability, and the different range and nature of by-products is later on observed as contamination within the desired product. To at least monitor these parameters, a functional and precise in-process control system (see Sect. 5.4), as required if working under GMP conditions, has to be established. In general, manufacturing processes for pharmaceutical products to be tested or used, should be covered by a detailed manufacturing description. This includes all available data on the process, its components, quality assurance, the manufacturing facility, and validation studies of the process as described within Sects. 5.2–5.6.

Ideally, such data are summarized in a DMF that is filed with the appropriate body such as the FDA in the USA. Such a DMF combines the advantages of fully disclosing proprietary process data to the authorities with, at the same time, protection of the data from competitor access.

5.2 Manufacturing Facility Requirements

Pharmaceutical production processes require specific manufacturing conditions. These conditions are defined by laws and regulations described in the GMP guidelines (AUTERHOFF, 1997). One main point is the design and organization of the facility (TOLBERT et al., 1996; METZNER, 1997) where such operations are performed for plasmid DNA manufacturing.

In any case, a manufacturer needs a manufacturing authorization which simultaneously fulfills GMP guideline requirements. The facility considerations and requirements for

GMP are described in the Code of Federal Regulations (CFR-21, 1991). General requirements for these facilities are constant environmental conditions with measurable parameters such as temperature, humidity, airborne particle counts, and microbial burden (LEE, 1989; ROSCIOLI et al., 1996). The design of such a facility (floor plan, air venting system) or of general processes (e.g., cleaning and changeover procedures, quality assurance, material and personnel flow) needs to be focused on the intended use to manufacture a specific product. This is common for fully developed drugs that need to be produced in large lot sizes. However, clinical samples of plasmid DNA for phase I or II studies, which are essentially developed pilot scale products, may be needed for a small number of patients only.

Small or young pharmaceutical companies or promising biotech companies in the field of biomedicine may lack experience and sufficient financing and personnel. Although they also must ensure the feasibility of their intended drug application, they may not wish to invest time and money into designing, building, and maintaining their own manufacturing facility. In these instances, companies need to subcontract plasmid production to an experienced contract manufacturer (MESSING, 1998; BADER et al., 1992).

A clear spatial distinction should be made between zones in which living organisms are handled (e.g., cultivation) and zones for subsequent process steps such as DSP, polishing, and filling operations. The different areas should have separate air locks for material and personnel to avoid any risk of cross-contamination events. These areas should be large enough to have space for work and instrumentation, and should be free of any equipment that will not be used for the current project. Of course, the manufacture of only one plasmid is possible at any given time.

Clear material flow in manufacturing is another prerequisite for proper work. One issue that requires extensive planning is the requirement of air pressure differentials between the different areas. This protects the product from environmental influences, namely microbial contamination, viruses, and endotoxins, during more downstream processes. This is usually achieved by slight overpressure within those containments that need more stringent protection from their surrounding neighborhood. One classic example is a fermenter that is run at, e.g., $2 \cdot 10^4$–$5 \cdot 10^4$ Pa overpressure, to ensure product protection.

In addition, the cultivation of recombinant microorganisms such as plasmid containing *E. coli* strains requires protection of the environment under certain conditions (European Council Directive, 1990) prohibiting the exit of any unclassified cell. The use of pressure cascades within air locks or house-in-house solutions efficiently overcomes this problem.

More general essential features for GMP conforming facility operations are:

(1) The personnel are appropriately qualified to do the work.
(2) Training is performed regularly and additionally as necessary.
(3) Equipment and area design are state of the art, or at least validated to perform as intended.
(4) Raw materials are documented, tested, under quarantine or released after testing, and are stored properly.
(5) Containers and labels are released for use and the quality assurance and quality control departments have all services and tests available to do their work.
(6) Complete documentation exists.
(7) Even minor process steps follow standard operating procedures (SOPs).

After manufacturing, the plasmid DNA has to be stored and transported under suitable and documented conditions.

5.3 Process Validation

Validation of a manufacturing process (CHAPMAN, 1991), steps of the process, e.g., fermentation (NAGLAK et al., 1994) or chromatography (SEELY et al., 1994; BARRY and CHOJNACKI, 1995), instruments, or any work performed to monitor the process (e.g., QA/QC methods) are defined as a systematic control of basic steps and instrumentation in production and control. The aim is to ensure that the manufactured plasmid is safe and can be produced reproducibly under the defined manufacturing and QA/QC conditions. Validation starts early: critical paramctcrs and their limits are determined and evaluated within product development. At production scales, the critical steps are analyzed and additional (or revised) parameters are defined. In particular upscaling, which is not necessarily linear, requires additional validation work. One very helpful tool is the historic data, even from the research and developmental phase of the plasmid construct, for comparison and interpretation of any discrepancies (see also Sect. 5.6).

The design qualification (DQ) of an instrument describes what an instrument should be specifically qualified to do. If it is applied to a manufacturing facility, the installation qualification (IQ) ensures that it is positioned correctly and fulfills all necessary requirements (see DQ). After IQ, the instrument has to be able to operate as expected (operation qualification; OQ), and subsequent performance qualification (PQ) ensures the intended quality within the manufacturing process.

The validation of a plasmid manufacturing process should be performed at a mg scale whenever possible, and is preferably performed prospectively. At least enough material for quality determination should be generated. For DNA quantification and QC (see Sect. 4.8), only μg amounts are necessary. If potency studies have to be performed, the quantity of test substance depends on the application. Large amounts of DNA solution are used for reliable sterility or bioburden assays.

Especially the depletion of living microorganisms (mainly the *E. coli* host cells) throughout the process is important. With 0.2 μm sterile filtration, the size of plasmid molecules typically used in formulation solutions [e.g., water for injection (not recommended), or 1 mM Tris Cl pH 8.0, 0.1 mM EDTA] leads to tremendous product loss in the majority of cases. Other techniques such as thermal sterilization or γ-irradiation are not applicable. Consequently, it is not enough to only test the finished bulk plasmid solution for sterility ["Quality cannot be inspected or tested into the finished product", (CDER,

1987), but "Quality, safety, and effectiveness must be designed and built into the product" (CDER, 1987)]. This in fact means that from the last applicable sterile filtration step onwards, any activity is undertaken to keep the bioburden status constant.

5.4 In-Process Control

The validation of certain process steps described above (Sect. 5.3) requires sampling of raw material, intermediate product, and the final product. The comparative analysis generates information regarding differences. Unlike the QC of raw materials or final products (see Sects. 4.8 and 5.5), the IPC investigates intermediate products to ensure quality and to generate data for process monitoring and steering. This means that the QC rules over "release" or "no release" of materials and the final product. By comparison, IPC results steer or influence the process at variable steps. Such steps are preferably avoided to ensure consistency, but in case of plasmid manufacturing, the use of one process scheme for the manufacture of certain different plasmids causes differences from one plasmid construct to another. One example is plasmid copy number, which not only differs with individual constructs, but also can increase or decrease slightly with the fermentation phase and harvesting time.

The data of the IPCs form part of the manufacturing documentation (Sect. 5.6). Testing procedures follow SOPs and are described in Sect. 4.8; see Tabs. 3 and 4 for an overview of the different IPCs of a plasmid manufacturing process.

5.5 Quality Assurance and Quality Control

QA is a concept which covers all aspects that influence the quality of a manufactured product. This means the GMP conforming organization for manufacturing, which is at least responsible for overseeing this independently. The QA manager is a registered qualified person responsible for the final release of the

Tab. 3. Overview of In-Process Controls in Plasmid Production – Upstream Controls for Plasmid Quality and Microbial Integrity

Characteristic	Strain Selection and RCB[a]	MCB[b]	WCB[c]	Pilot Fermentation	Fermentation	Assay
Growth control	✓	✓	✓	✓	✓	photometry
Plasmid quality	✓	✓	✓	✓	✓	agarose gel, spectrometry
Plasmid quantity	✓			✓	✓	agarose gel
Plasmid stability	✓	✓	✓			plating assay
Microbial purity		✓	✓	✓	✓	microscopy
Phage contamination		✓	✓	✓	✓	plating assay
Microbial contamination		✓	✓	✓	✓	colorimetric assay
Vital cell number		✓	✓	✓	✓	plating assay
Plasmid identity	✓					sequencing

[a] RCB: research cell bank,   [b] MCB: master cell bank,   [c] WCB: manufacturers' working cell bank

Tab. 4. Overview of In-Process Controls in Plasmid Production – Downstream Controls for Plasmid Quality and Microbial Integrity

Source of Test Sample	DNA Quality	Microbes Absent
Lysate pool	✓	
Cleared lysate after QIAfilter™	✓	
Cleared lysate after sterile filtration	✓	✓
Cleared lysate after incubation with buffer ER	✓	
QBT equilibration buffer		✓
QBT buffer eluate after column equilibration	✓	
Last eluate fraction from cleared lysate	✓	
Eluate pool from cleared lysate	✓	
Last eluate fraction from buffer QC wash	✓	
Pool of buffer QC wash	✓	
Buffer QN elution	✓	
Resuspension buffer (variable)		✓
Resuspended DNA	✓	
DNA solution at final concentration	✓	✓

product and any raw material being incorporated into the manufacturing process. Additionally this person supervises the document control procedures, and organizes self-inspections and auditing procedures. Also, environmental monitoring of the facility as well as QC of raw materials and disposables, and IPCs are performed.

For example: A 70% ethanol solution for washing DNA precipitation pellets is required for the manufacturing process. It is ordered at and delivered by a selected and preferably certified (ISO 9000 system) supplier. Upon arrival of the shipment at the specified place, documentation and entrance control should be performed. A sample of ethanol is taken for identity testing at least. The idea behind this is that it should be emphatically excluded that the clear colorless solution within the bottle might also be methanol or isopropanol caused by a mix-up or by incorrect labeling by the supplier. Furthermore, the quality concerning given parameters has to be rechecked and compared with the shipped documents and certificates of analysis. As long as this sample is under investigation, it is "not released", meaning that it remains "under quarantine", and can, therefore, not be used. A color-coded label on the bottle indicates this status, and only authorized persons are allowed to use such labels.

More than 100 different raw materials, disposables, and instruments are used for the manufacture of plasmid DNA, all of which have to pass such entrance control. The QC of the IPC samples is performed as described for the finished product in Sect 4.8. Samples are taken at each critical step.

5.6 Documentation

GMP conforming manufacture requires complete documentation. The basis for this is to be able to follow each step of the manufacturing process even years afterwards, and to assure that any deviations from the intended process are recognized in time to guarantee product quality as wholly as possible.

The following are general issues:

(1) The personnel are qualified to do the documentation, meaning that they should understand how and why.
(2) Any deviation or anomalous results have to be fully documented and should cause immediate action of the responsible persons or institutions.
(3) The full documentation has to be stored under controlled conditions, and copies will be required in case of subsequent clinical evaluation of plasmid DNA produced. In case of an investigational new drug (IND) application, the full documentation of each and every production step (manufacturing batch record) and even historic data on research and development will be required for authorization (DESAIN, 1991). The DMF (see Sect. 5.1) of the manufacturing process and the facility constitutes the backbone of this documentation.

6 The Future of Therapeutic Plasmid DNA

Large scale pharmaceutical manufacturing technology for nucleic acids under full cGMP working conditions is currently underway. The process described here in detail (Sects. 4.2–4.8) does not employ any toxic substances such as ethidium bromide, CsCl, or phenol. One major consideration in the development of this technology was to avoid time consuming and costly centrifugation, as well as multiple chromatographic column runs. Extensive in-process control ensures constant monitoring of quality and reproducibility within each manufacturing step. The stringent quality criteria are the result of extensive developmental work. The removal of endotoxins (*E. coli* LPS) from DNA is a key step in the process and is essential for use of the DNA in clinical phase I and II studies. This clinical-grade plasmid DNA manufacturing procedure has been approved for human clinical phase I studies in the UK (CALPEN et al., 1994) and other European countries, as well as in the USA by the FDA (ISNER et al., 1995).

Future developments in the use of plasmid DNA as a medical product indicate that the prospective field of application will be vaccination in animals and curative vaccination in humans at the very least. Modified vector and delivery systems are currently underway in which plasmids are formulated with other substances, e.g., liposome or dendrimer technology (BEHR, 1994; TANG et al., 1996), or modified by, e.g., immunostimulatory sequences (ISSs) (SATO et al., 1996) or CpG motifs (KRIEG et al., 1995; KRIEG, 1996) where functional short sequence stretches strongly influence the success of application.

Multicomponent vaccines, provided no antigenic competition occurs, are a further advancement towards the development of a general process for the manufacturing of a therapeutic substance using a single standard process. Together with an improved application technique, the dose per patient may be further reduced. By the way of illustration, initial experimental analysis by direct injection of naked plasmid DNA into muscle (LE

BORGNE et al., 1998; GREGORIADIS, 1998) required milligram amounts of plasmid per dose, whereas the use of biolistic approaches ("gene gun") has lowered the amount required to microgram quantities per dose.

The unique and intriguing feature of plasmid DNA is that it can be manufactured following a single identical processing scheme in spite of vastly differing sequence content. By comparison, the manufacturing of protein therapeutics in prokaryotes requires a specific process for each individual substance. For each DNA product, the costs for process development have to be calculated by pharmaceutical companies, and influence the interest in developing a particular product. The amount of desired product is the critical parameter for process scale, and economy of scale is of great importance.

If, e.g., 300,000 patients are affected by a disease that should be cured using 10 mg DNA per patient and year, this implies a total requirement of 3 kg plasmid DNA. For a proposed time range of treating the patients over 10 years (in this example), this would require 300 g per year. In case of 2 week manufacturing projects (20 per annum may be possible), each has to have a scale of 15 g. Estimating a yield of 1 g plasmid DNA per 1 kg of wet cell weight (biomass) for a high-copy number plasmid, this means that 300 kg biomass is required. 300 kg can be obtained from fermentation volumes of at least 1,200 L. To obtain 15 g DNA, 15 kg biomass would have to be generated and processed in a 2-week range. This corresponds to a fermentation volume of 60 L. This volume and its associated costs plus those of the downstream process (which are typically more expensive) form a convincing incentive to search for and develop application techniques in which less material is required. Fundamentally, if only 10 µg instead of 10 mg could be used per patient, then this scale of production would be possible using today's technology. As an alternative solution, the plasmid–biomass ratio could be enhanced in order to make the project more economically viable. Whatever the approach, there is a need for further development of manufacturing processes capable of producing plasmids suitable for clinical phase III and IV trials and for marketing authorization.

Acknowledgements

I wish to thank G. M. BOWDEN for critically reviewing the article, M. MERKEL, J. HUCKLENBROICH, P. VOIGTLÄNDER, L. BREITKOPF and N. SCHOLLE for technical support, M. H. MÜLLER, R. BAIER, P. MORITZ, T. KLÜTZ, B. ANGST, J. SCHORR as well as T. SCHMIDT, K. FRIEHS, E. FLASCHEL (University of Bielefeld), H. ZIEHR (GBF, Braunschweig), H. SIEDE and W. MÜHLHARDT (Strathmann Biotech, Hannover) for critical discussions, and R. KERKHOFF for help with the manuscript.

7 References

Anonymous (1998), Europäisches Arzneibuch, Nachtrag 1998. Stuttgart: Deutscher Apotheker Verlag; Eschborn: Govi-Verlag – Pharmazeutischer Verlag GmbH.

AUTERHOFF, G. (1997), *EC Guide to Good Manufacturing Practice for Medical Products,* 2nd Edn. Aulendorf: ECV.

BADER, F. G., BLUM, A., GARFINKE, B. D., MACFARLANE, D., MASSA, T., COPMANN, T. L. (1992), Multiuse manufacturing facilities for biologicals, *Biopharm* 5, 32–40.

BARRY, A. R., CHOJNACKI, R. (1995), Biotechnology product validation, Part 8: Chromatography media and column qualification, *Pharm. Technol. Eur.* 6, 32–38.

BEHR, J. P. (1994), Gene transfer with synthetic cationic amphiphiles: prospects for gene therapy, *Bioconj. Chem.* 5, 382–389.

BIRNBOIM, H. C., DOLY, J. (1979), A rapid alkaline extraction procedure for screening recombinant plasmid DNA, *Nucleic Acids Res.* 7, 1513–1523.

CAPLEN, N. J., GAO, X., HAYES, P., ELASWARAPU, R., FISHER, G. et al. (1994), Gene therapy for cystic fibrosis in humans by liposome-mediated DNA transfer: U.K. regulatory process and production of resources, *Gene Ther.* 1, 139–147.

CBER (1996), *Points to Consider on Plasmid DNA Vaccines for Preventive Infectious Disease Indications,* Rockville, MD: Center for Biologics Evaluation and Research (HFM-630), FDA.

CBER (1998), *Guidance for Industry: Guidance for Human Somatic Cell Therapy and Gene Therapy,* Rockville, MD: Center for Biologics Evaluation and Research, FDA.

CDER (1987), *Guidelines on General Principles of Process Validation,* Rockville, MD: Center for Drug Evaluation and Research, FDA, Division of Manufacturing and Product Quality (HFD-320), <*http://www.fda.gov/cder/pv.htm*>.

CFR-21 (1991), Current Good Manufacturing Practice in Manufacturing, Processing, Packaging, or Holding of Drugs; General, *Code of Federal Regulations,* Title 21, Part 210.

CHANDRA, G., PATEL, P., KOST, T. A., GRAY, J. G. (1992), Large-scale purification of plasmid DNA by fast protein liquid chromatography using a Hi-load Q sepharose column, *Anal. Biochem.* **203**, 169–172.

CHAPMAN, K. G. (1991), A history of validation in the United States: Part I, *Pharm. Technol.* **15**, 82–96.

CHEN, W., GRAHAM, C., CICCARELLI, R. B. (1997), Automated fed-batch fermentation with feed-back controls based on dissolved oxygen (DO) and pH for production of DNA vaccines, *J. Ind. Microbiol. Biotechnol.* **18**, 43–48.

CICHUTEK, K., KRÄMER, I. (in press), Gene therapy in Germany and in Europe: regulatory issues, *Qual. Assur. J.*

COLPAN, M., SCHORR, J., MORITZ, P. (1995), Process for producing endotoxin-free or endotoxin-poor nucleic acids and/or oligonucleotides for gene therapy, *WO 95/21177.*

COTTEN, M., BAKER, A., SALTIK, M., WAGNER, E., BUSCHLE, M. (1994), Lipopolysaccharide is a frequent contamination of plasmid DNA preparations and can be toxic to primary cells in the presence of adenovirus, *Gene Ther.* **1**, 239–246.

DASARTHY, Y., RAMBERG, M., ANDERSSON, M. (1996), A systematic approach to screening ion-exchange chromatography media for process development, *Biopharm* **9**, 42–45.

DAVIS, B. D., DULBECCO, R., EISEN, H. H., GINSBERG, H. S. (1980), *Microbiology,* 3rd Edn. Philadelphia, PA: Harper and Row.

DESAIN, C. (1991), Master production batch records, *Biopharm* **4**, 20–24.

DONNELLY, J. J., ULMER, J. B., LIU, M. A. (1997), *DNA Vaccines Life Sci.* **60**, 163–172.

European Council Directive (1990), Council Directive of April 23, 1990 on the contained use of genetically modified microorganisms (90/219/EEC) (Amtsblatt der EG, Nr. L117 vom 8. Mai, 1990, page 1 ff., and the adaption thereof (94/51/EG) from November 07, 1994).

FELGNER, P. L., GADEK, T. R., HOLM, M., ROMAN, R., CAHN, H. W. et al. (1987), Lipofectin: A highly efficient, lipid-mediated DNA transfection procedure, *Proc. Natl. Acad. Sci. USA* **84**, 7413–7417.

GRAHAM, F. L., VAN DER EB, A. J. (1973), A new technique for the assay of infectivity of human adenovirus 5 DNA, *Virology* **52**, 456–467.

GREEN, A. P., PRIOR, G. M., HELVESTON, N. M., TAITTINGER, B. E., LIN, X., THOMPSON, J. A. (1997), Preparative purification of supercoiled plasmid DNA for therapeutic applications, *Biopharm* **10**, 52–62.

GREGORIADIS, G. (1998), Genetic vaccines: Strategies for optimization, *Pharm. Res.* **15**, 661-670.

GÜNZBURG, W. H., SALMONS, B. (Eds.) (1997), *Gentransfer in Säugetierzellen.* Heidelberg: Spektrum Akademischer Verlag.

HELINSKI, D. (1979), Bacterial plasmids: Autonomous replication and vehicles for gene cloning, *CRC Crit. Rev. Biochem.* **7**, 83–101.

HORN, N., BUDAHAZI, G., MARQUET, M. (1998), *U.S. Patent 5 707 812.*

ISH-HOROWICS, D., BURKE, J. F. (1981), Rapid and efficient cosmid cloning, *Nucleic Acids Res.* **9**, 2989–2998.

ISNER, J. M., WALSH, J., SYMES, A., PIECZEK, A., TAKESHITA, S. et al. (1995), Arterial gene therapy for therapeutic angiogenesis in patients with peripheral artery disease, *Circulation* **91**, 2687–2692.

KITTLE, J. D., PIMENTEL, B. J. (1997), Testing the genetic stability of recombinant DNA cell banks, *Biopharm* **10**, 30–35.

KORZ, D. J., RINAS, U., HELLMUTH, K., SANDERS, E. A., DECKWER, W.-D. (1995), Simple fed-batch technique for high cell density cultivation of *Escherichia coli, J. Biotechnol.* **39**, 59–65.

KRIEG, A. M. (1996), Lymphocyte activation by CpG dinucleotide motifs in prokaryotic DNA, *Trends Microbiol.* **4**, 73–77.

KRIEG, A. M., YI, A.-K., MATSON, S., WALDSCHMIDT, T. J., BISHOP, G. A. et al. (1995), CpG motifs in bacterial DNA trigger direct B-cell activation, *Nature* **374**, 546–549.

KUHNE, W. (1997), Process for preparing purified nucleic acid and the use thereof, *WO 97/29113.*

LAHIJANI, R., HULLEY, G., SORIANO, G., HORM, N. A., MARQUET, M. (1996), High-yield production of pBR322-derived plasmids intended for human gene therapy by employing a temperature-controllable point mutation, *Hum. Gene Ther.* **7**, 1971–1980.

LE BORGNE, S., MANCINI, M., LE GRAND, R., SCHLEEF, M., DORMONT, D. et al. (1998), *In vivo* induction of specific cytotoxic T lymphocytes in mice and rhesus macaques immunized with DNA vector encoding an HIV epitope fused with hepatitis B surface antigen, *Virology* **240**, 304–315.

LEE, J. Y. (1989), Environmental requirements for clean rooms, *Biopharm* **2**, 42–45.

LEE, A. L., SAGAR, S. (1996), A method for large-scale plasmid purification, *WO96/36706*.

MAJOR, M. E., VITVITSKI, L., MINK, M. A., SCHLEEF, M., WHALEN, R. G. et al. (1995), DNA-based immunization with chimeric vectors for the induction of immune responses against the hepatitis C virus nucleocapsid, *J. Virol.* **69**, 5798–5805.

MARQUET, M., HORN, N., MEEK, J., BUDAHAZI, G. (1995a), Production of pharmaceutical-grade plasmid DNA, *WO 95/21250*.

MARQUET, M., HORN, N., MEEK, J. (1995b), Process development for the manufacture of plasmid DNA vectors for use in gene therapy, *Biopharm* **8**, 26–37.

MEAGER, A., ROBERTSON, J. S. (1998), Regulatory and standardization issues for DNA and vectored vaccines, *Curr. Res. Mol. Ther.* **1**, 262–265.

MESSING, S. H. (1998), Outsourcing biopharmaceutical manufacturing operations, *Biopharm* **11**, 28–30.

METZNER, B. (1997), Criteria for the design of cleanrooms in biotechnology, *Swiss Pharma* **19**, 5–8.

MICHEL, M.-L., DAVIS, H. L., SCHLEEF, M., MANCINI, M., TIOLLAIS, P., WHALEN, R. G. (1995), DNA-mediated immunization to the hepatitis B surface antigen in mice: Aspects of the humoral response mimic hepatitis B viral infection in humans, *Proc. Natl. Acad. Sci. USA* **92**, 5307–5311.

MORRISON, D. C., RYAN, J. L. (1987), Endotoxins and disease mechanism, *Annu. Rev. Med.* **38**, 417–432.

NAGLAK, T. J., KEITH, M. G., OMSTEAD, D. R. (1994), Validation of fermentation processes, *Biopharm* **7**, 28–36.

NEIDHARDT, F. C. (Ed.) (1996), *Escherichia coli and Salmonella typhimurium,* 2nd Edn., Vols. 1 and 2. Washington, D.C.: ASM Press.

NEUMANN, E., SCHAEFER-RIDDER, M., WANG, Y., HOFSCHNEIDER, P. H. (1982), Gene transfer into mouse lyoma cells by electroporation in high electric fields, *EMBO J.* **7**, 841–845.

ORR, T. (1998), Product recovery and purification, *Gen. Eng. News* **18**, No. 5.

PARASRAMPURIA, D. A., HUNT, C. A. (1998), Therapeutic delivery issues in gene therapy, Part 1: Vectors, *Biopharm* **11**, 38–45.

ROBERTSON, J. S. (1994), Safety considerations for nucleic acid vaccines, *Vaccine* **12**, 1526–1528.

ROSCIOLI, N. A., RENSHAW, C. A., GILBERT, A. A., KERRY, C. F., PROBST, P. G. (1996), Environmental monitoring considerations for biological manufacturing, *Biopharm* **9**, 32–40.

SAMBROOK, J., FRITSCH, E. F., MANIATIS, T. (1989), *Molecular Cloning,* 2nd Edn. Cold Spring Harbor, FL: CSH Laboratory Press.

SATO, Y., ROMAN, M., TIGHE, H., LEE, D., CORR, M. et al. (1996), Immunstimulatory DNA sequences necessary for effective intradermal gene immunization, *Science* **273**, 352–354.

SAUER, P., MÜLLER, M., KANG, J. (1998), Quantification of DNA, *QIAGEN News* **2**, 23–26.

SCHLEEF, M., MORTIZ, P., SCHORR, J. (1996), How to produce nucleic acids for human gene therapy and genetic vaccination under full cGMP pharmaceutical manufacturing conditions, *Eur. J. Pharm. Sci.* **4**, S25.

SCHMIDT, T., FRIEHS, K., FLASCHEL, E. (1996), Rapid determination of plasmid copy number, *J. Biotechnol.* **49**, 219–229.

SCHORR, J., MORITZ, P., SEDDON, T., SCHLEEF, M. (1995), Plasmid DNA for human gene therapy and DNA vaccines, *N. Y. Acad. Sci.* **772**, 271–273.

SCHWAB, H. (1993), Principles of genetic engineering for *Escherichia coli,* in: *Genetic Engineering of Microorganisms* (PÜHLER, A., Ed.), pp. 1–54. Weinheim: VCH.

SEELY, R. J., WIGHT, H. D., FRY, H. H., RUDGE, S. R., SLAFF, G. F. (1994), Biotechnology product validation, Part 7: Validation of chromatography resin useful life, *Pharm. Technol. Eur.* **6**, 32–38.

SMITH, H. A., GOLDENTHAL, K. L., VOGEL, F. R., RABINOVICH, R., AGUADO, T. (1997), Workshop on the control and standardization of nucleic acid vaccines, *Vaccine* **15**, 931–933.

STEUR, A., OSTROVE, J. M. (1996), Establishing cell banks under current good manufacturing practices, *Biopharm* **9**, 40–44.

SUMMERS, D. K. (1996), *The Biology of Plasmids.* Oxford: Blackwell Science.

TANG, M. X., REDEMANN, C. T., SZOKA, F. C. (1996), *In vitro* gene delivery by degraded polyamidoamine dendrimers, *Bioconj. Chem.* **7**, 703–714.

TATUM, E. L. (1966), Molecular biology, nucleic acids, and the future of medicine, *Perspect. Biol. Med.* **10**, 19–32.

THATCHER, D. R., HITCHCOCK, A. G., HANAK, J. A., VARLEY, D. L. (1997), Method of plasmid DNA production and purification, *WO 97/29190*.

THEODOSSIOU, I., COLLINS, I. J., WARD, J. M., THOMAS, O. R. T., DUNNILL, P. (1997), The processing of a plasmid-based gene from *E. coli.* Primary recovery by filtration, *Bioprocess Eng.* **16**, 175–183.

THEODOSSIOU, I., THOMAS, O. R. T., DUNNILL, P. (in press), Methods of enhancing the recovery of plasmid genes from neutralized cell lysate, *Bioprocess Eng.*

TOLBERT, W. R., MERCHANT, B., TAYLOR, J. A., PERGOLIZZI, R. G. (1996), Designing an initial gene therapy manufacturing facility, *Biopharm* **9**, 32–40.

TSE-DINH, Y.-C., QI, H., MENZEL, R. (1997), DNA supercoiling and bacterial adaption: thermotolerance and thermoresistance, *Trends Microbiol.* **5**, 323–326.

USP23 (1995), *The United States Pharmacopoeia,* 23. Rockville, MD: United States Pharmacopoeia Convention, Inc.

VOGEL, F. R., SARVER, H. (1995), Nucleic acid vaccines, *Clin. Microbiol. Rev.* **8**, 406–410.

VUKAJLOVICH, S. W., HOFFMAN, J., MORRISON, D. (1987), Activation of human serum complement by bacterial lipopolysaccharides: Structural requirements for antibody independent activation of the classic and alternative pathways, *Mol. Immunol.* **24**, 319–331.

WEBER, M., MÖLLER, K., WELZECK, M., SCHORR, J. (1995), Effects of lipopolysaccharide on transfection efficiency in eukaryotic cells, *Biotechniques* **19**, 930–940.

WHO (1997), *Guidelines for Assuring the Quality of DNA Vaccines,* Geneva; WHO Technical Report *<c:\appli\wp\docs\guidna. Fin>.*

WICKS, I. P., HOWELL, M. L., HANCOCK, T., KOHSAKA, H., OLEE, T., CARSON, D. A. (1995), Bacterial lipopolysaccharide copurifies with plasmid DNA: implications for animal models and human gene therapy, *Hum. Gene Ther.* **6**, 317–323.

WILFINGER, W. W., MACKEY, K., CHOMCZYNSKI, P. (1997), Effect of pH and ionic strength on the spectrophotometric assessment of nucleic acid purification, *Biotechniques* **22**, 474–481.

WILLIAMS, S. G., CRANENBURGH, R. M., WEISS, A. M. E., WRIGHTON, C. J., SHERRATT, D. J., HANAK, J. A. J. (1998), Repressor titration: a novel system for selection and stable maintenance of recombinant plasmids, *Nucleic Acids Res.* **26**, 2120–2124.

WILS, P., ESCRIOU, V., WARNERY, A., LACROIX, F., LAGNEAUX, D. et al. (1997), Efficient purification of plasmid DNA for gene transfer using triple-helix affinity chromatography, *Gene Ther.* **4**, 323–330.

WOLFF, J. A., WILLIAMS, P., ACSADI, G., JIAO, S., JANI, A., CHONG, W. (1991), Conditions affecting direct gene transfer into rodent muscle *in vivo, Biotechniques* **11**, 474–485.

21 Gene Therapy for HIV Infection

Mark C. Poznansky

Boston, MA, USA

Myra McClure
Gregor B. Adams

London, UK

1 Introduction: HIV-1 Infection as an Acquired Genetic Defect   472
2 Strategies Interfering with Viral Replication   473
  2.1 Design of Optimal Anti-HIV Constructs   473
      2.1.1 RNA Inhibitors: Antisense RNA, Ribozymes and Decoy RNA   474
      2.1.2 Protein Inhibitors: Dominant-Negative Proteins and Intracellular
            Antibodies   476
  2.2 Design of the Optimal Vector for the Delivery of Anti-Viral Constructs   479
      2.2.1 Amphotropic Murine Leukemia Virus (Mo-MLV) Based Vectors   479
      2.2.2 Novel Retroviral and Adeno-Associated Virus (AAV) Based Vectors   480
3 Strategies Augmenting the Immune Response to HIV   481
4 Ongoing Phase I Clinical Trials of Gene Therapies for HIV Infection   482
  4.1 Biosafety and Ethical Issues Involved in the First Generation of Phase I Trials   483
5 Refining Gene Therapy Protocols   484
      5.1 Optimized Vector Design and Tissue Specific Expression of the Transduced
          Gene   484
      5.2 Improved Transduction Protocols for Primary Human T Cells   485
      5.3 Gene Therapy for Progenitor Cells and Stem Cells from HIV Infected
          Individuals   485
5.4 The Next Generation of Gene Therapy Trials for HIV Infection   486
6 References   487

List of Abbreviations

AAV	adeno-associated virus
ADA	adenosine deaminase
AZT	azido thymidine
BMT	bone marrow transplantation
CMV	cytomegalovirus
CTL	cytotoxic lymphocyte
ER	endoplasmic reticulum
HFV	human foamy retrovirus
HIV	human immunodeficiency virus
HIV-IT TAF	HIV-1 IIIB *env*-transduced autologous fibroblasts
HSV	herpes simplex virus
IL	interleukin
IRE	internal ribosomal entry site
LCR	locus control region
LDLR	low-density lipoprotein recepter
LTR	long terminal repeat
Mo-MLV	amphotropic murine leukemia virus
OKT3	CD4-binding monoclonal antibody
PBMC	peripheral blood mononuclear cells
RCR	replication competent retrovirus
RRE	*rev* response element
TAR	transactivation region
TCR	T cell receptor
TK	thymidine kinase

1 Introduction: HIV-1 Infection as an Acquired Genetic Defect

In the face of the continuing human immunodeficiency virus type 1 (HIV-1) pandemic and the limited success of current interventions in controlling either the spread or the effects of HIV infection, novel strategies are being developed to combat the retrovirus, such as gene therapy (FIELDS, 1994; DROPULIC and JEANG, 1994; ANDERSON, 1994; YU et al., 1994).

Gene therapy exploits the fact that biological and physical vectors can be used to transfer genetic material into mammalian somatic cells (WOLFF and LEDERBERG, 1994; ANDERSON, 1984; BALTIMORE, 1988). A variety of disorders resulting from congenital and acquired genetic diseases has been considered for treatment with gene therapy, including adenosine deaminase (ADA) deficiency, cystic fibrosis, cancer and more recently, cytomegalovirus (CMV) and HIV-1 infection (BLAESE, 1993; KANTOFF et al., 1986; BORDIGNON et al., 1993; WEATHERALL, 1992; WEATHERALL, 1994; VILE and RUSSELL, 1994; DRUMM et al., 1990; ROSENFELD et al., 1992).

HIV-1 represents an appropriate target for gene therapy because it integrates its genome into the human host DNA and can, therefore, be considered to cause an acquired genetic defect (HASELTINE, 1991; FAUCI, 1993; JOHNSON and HOTH, 1993). The genome of HIV-1 and its encoded proteins have been defined in detail and so there are several clear targets for a gene therapy (HASELTINE, 1990; HASELTINE, 1991; WONG-STAAL, 1991; WONG-STAAL and GALLO, 1985; COHEN and SUBRAMANIAN, 1994). It has been demonstrated that the CD4$^+$ T lymphocyte and other immune cells bearing the CD4 receptor, such as monocytes and dendritic cells, are infectable by HIV-1 *in vitro* and have been found to be infected *in vivo* (DALGLEISH et al., 1984; LANGHOFF et al., 1991; PATTERSON and KNIGHT, 1987; MERRILL and CHEN, 1991; MANN et al., 1990; SHNITTMANN et al., 1989; POZNANSKY et al., 1991). Therefore, any proposed gene therapy could be appropriately aimed at inhibiting replication of the virus in CD4 bearing T cells *in vivo*.

Two principal strategies can be used to treat HIV infection by gene therapy. One involves the delivery of anti-viral constructs which directly interfere with the retroviral replicative cycle, and the other involves gene-based immunotherapy, i.e., the transfer of genes which augments the immune response to the virus (DROPULIC and JEANG, 1994; ANDERSON, 1994; YU et al., 1994; BALTIMORE, 1988) (Fig. 1).

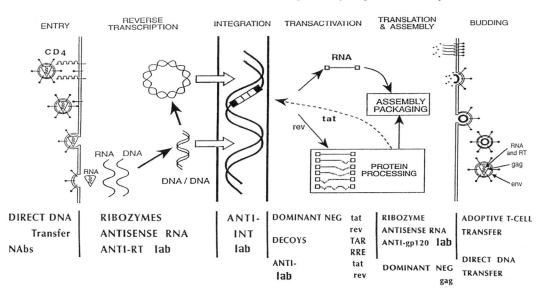

Fig. 1. Antiviral constructs for HIV gene therapy: steps in the replicative cycle of HIV, including binding and entry to the CD4$^+$ T cell, reverse transcription of HIV RNA, integration of the proviral DNA into the host genome, transactivation of the viral genome and expression of viral proteins, assembly and packaging of new virions and budding of virus from the surface of the cell. Antiviral constructs are also listed which interrupt various stages in the cycle, including neutralizing antibodies (Nab), ribozymes, antisense RNA, decoy RNA and intracellular antibodies (Iab) directed against HIV proteins including reverse transcriptase (RT), integrase or p17. Gene therapy can also be used to augment immune responses to HIV by direct injection of expression plasmids for immunogenic HIV proteins or by adoptive transfer of HIV specific CTLs.

2 Strategies Interfering with Viral Replication

2.1 Design of Optimal Anti-HIV Constructs

The life cycle of HIV (Fig. 1) is a multi-step process involving several HIV proteins including the retroviral structural proteins encoded in *gag, pol,* and *env* and accessory proteins involved in viral RNA transcription and regulation of viral infectivity including *tat* and *rev* (HASELTINE, 1990; COHEN and SUBRAMANIAN, 1994; WONG-STAAL and HASELTINE, 1992). HIV-1 consists of a bilipid envelope with the inserted HIV proteins gp120 and gp40 which surround a core particle formed from capsid proteins and containing the HIV RNA genome. Capsid proteins include p24 and the reverse transcriptase enzyme bound onto the double stranded RNA genome. HIV infects T cells through the binding of its envelope proteins to the CD4 receptor molecule on the helper T cell. The viral particle then fuses with the membrane of the cell and the core particle of HIV-1 is released into the cytoplasm of the cell. The RNA genome of the virus is then reverse transcribed into DNA, the proviral DNA enters the nucleus of the cell and the viral enzyme integrase assists the integration of the proviral DNA into the host human genome. Expression of the HIV proteins from the proviral DNA is controlled by a number of cellular and viral factors which interact with promoter and enhancer sequences encoded within the long terminal repeat (LTR) of the proviral genome. Of greatest significance in this respect are the HIV proteins Tat and Rev which activate transcription from the proviral DNA and assist the formation of full length viral RNA, respectively. Tat and Rev exert their effects by binding to specific sequences or transcriptional control elements which are present in proviral DNA or RNA called the *tat* transactivation region (TAR) and the *rev* response element (RRE). Following the activation of the proviral genome, HIV proteins are synthesized and core particles assemble in the cytoplasm of the cell. Newly formed HIV RNA is incorporated into the core particle through the binding of core proteins to a specific nucleotide sequence or packaging signal on the nascent viral RNA. Infectious HIV particles ultimately bud off from the infected cell membrane with an intact bilipid envelope containing inserted HIV envelope proteins.

Each of the steps in the viral replicative cycle described above has extensively been studied and, therefore, presents itself as target for inhibition by anti-viral constructs (DROPULIC and JEANG, 1994; ANDERSON, 1994; RATNER et al., 1985) (Fig. 1). Any anti-viral construct optimized to inhibit the activity of HIV RNA or protein should also ideally fulfill the following conditions:

1. The construct should be expressed in the same subcellular compartment as the HIV molecule whose function it will inhibit.
2. The correct conformation of the anti-viral molecule should be maintained in the subcellular compartment in which the antiviral effect of the molecule is exerted.
3. The gene therapy should be designed so that HIV cannot readily escape by mutation from the inhibitory action of the anti-viral construct or constructs.
4. The anti-viral construct should be capable of inhibiting primary isolates of HIV found in humans *in vivo,* not merely laboratory isolates.
5. The anti-viral construct should inhibit HIV replication in a cell which is already infected by HIV and/or inhibits infection and replication of HIV in uninfected cells following challenge by exogenous virus.
6. The anti-viral construct should be non-toxic, non-immunogenic and its expression in the cell should have no effect on normal cell function.

The various approaches interfering with viral replication and assembly, and the constructs concerned are described below and are shown diagramatically in Fig. 2. Gene therapy approaches aimed at interfering with the HIV replicative cycle use two types of inhibitors, RNA or protein.

2.1.1 RNA Inhibitors: Antisense RNA, Ribozymes and Decoy RNA

Antisense RNA inhibits gene expression by hybridizing to its homologous sequence in mRNA (GREEN et al., 1986; KINCHINGTON et al., 1992; BORDIER et al., 1992) (Fig. 2). Depending on its sequence, antisense RNA may hybridize with "incoming" HIV RNA in a newly infected cell and inhibit the process of infection and/or hybridize with "outgoing" HIV RNA from an integrated provirus and thereby inhibit the production of HIV virions in an infected cell, although the latter would require higher concentrations of the antisense RNA molecule (KINCHINGTON et al., 1992; BORDIER et al., 1992; HATTA et al., 1993; WANG and DOLNICK, 1993). Antisense RNA constructs have been designed which target both RNA encoding the transcriptional control elements of the HIV genome (VICKERS et al., 1991; CHATERGEE et al., 1992; KRUGER et al., 1982) and HIV RNA sequences encoding the structural (HATTA et al., 1993; RITNER and SZACKIEL, 1991) and regulatory proteins (HATTA et al., 1993; RHODES and JAMES, 1991). For example, CHATERGEE et al. (1992) demonstrated that the expression of antisense RNA to TAR caused a 70–90% reduction in HIV production in cells transfected with the HIV genome. A specific drawback of the antisense RNA approach is the requirement for a large excess of antisense molecules to inhibit HIV replication (WANG and DOLNICK, 1993).

A second form of RNA inhibitory molecule is the ribozyme or catalytic RNA which cleaves HIV-1 RNA in a sequence-specific manner and, therefore, may be designed to interfere at different points in the viral replication cycle (YU et al., 1994) (Fig. 2). This inherent versatility of ribozymes allows multi-

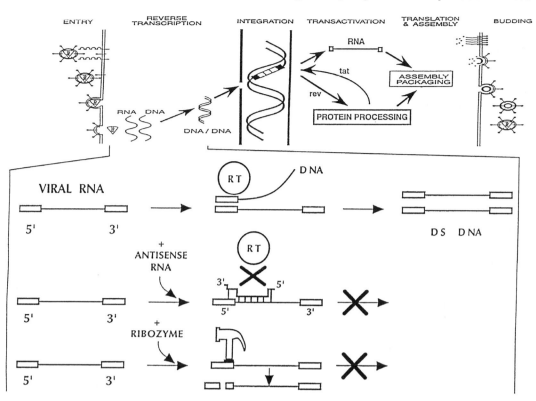

Fig. 2. The anti-viral actions of ribozymes and antisense RNA: demonstration how antisense RNA or ribozyme homologous to the 5′ CAP sequence of the HIV RNA can inhibit the early events involved in HIV replication, including reverse transcription. Equally, antisense RNA and ribozyme can inhibit the translation of postinfection events such as translation of HIV RNA generated from the expression of genes encoded in proviral DNA.

targeting within the HIV genome including the transcriptional control and packaging sequences of HIV RNA (YU et al., 1993; WEERASINGHE et al., 1991; DROPULIC et al., 1992; SULLENGER and CECH, 1993) and the structural genes such as *gag* (SARVER et al., 1990; TAYLOR and ROSSI, 1991) and *env* (CHEN et al., 1992). Ribozymes which target both non-coding and coding HIV RNA sequences are equally effective in inhibiting HIV replication *in vitro*. WONG-STAAL et al. (1994) developed an expression vector encoding a ribozyme which cleaves at the site on HIV RNA which stabilizes the RNA molecule in the cytoplasm (i.e., the CAP site of HIV RNA) and disrupts both "incoming" viral RNA as well as the production of nascent HIV RNA which both contain that sequence

(YU et al., 1994). Several groups have tested ribozymes which cleave *env*, *gag* and LTR-containing RNAs and demonstrated up to a 20 d delay in HIV replication in cells stably expressing the particular ribozyme following challenge with wild-type HIV-1 (DROPULIC et al., 1992; SULLENGER and CECH, 1993; SARVER et al., 1990; TAYLOR and ROSSI, 1991; CHEN et al., 1992; SULLENGER et al., 1993; LEE et al., 1992). There is the theoretical disadvantage that ribozymes may cleave non-viral or cellular RNAs with RNA sequences which are homologous to HIV RN (YU et al., 1994). In addition, ribozymes require stringent conditions, including pH, calcium and magnesium ion concentrations, to efficiently cleave RNA and these may not always be met in the cell (WANG and DOL-

NICK, 1993). Little is currently known about the stability and secondary structure of the anti-viral moiety in a subcellular compartment, be it ribozyme, antisense RNA or polymeric RNA decoy (DROPULIC and JEANG, 1994). Finally, the activity of ribozymes as anti-viral constructs *in vivo* remains to be fully assessed.

HIV replication can be inhibited by the sequestration of retroviral transactivating molecules such as Tat and Rev (Fig. 1) which play a central part in the retroviral life cycle (SULLENGER et al., 1993; LEE et al., 1992). This concept forms the basis for the use of RNA decoys to TAR and the RRE of the HIV-1 genome. TAR and RRE RNA decoys mimic the RNA structures of the HIV genome which bind Tat and Rev (Fig. 3). The RNA decoys consist of multiple copies of the transcription factor binding site and are designed to sequester *tat* and *rev*, thereby inhibiting

HIV replication. It has been demonstrated that when a TAR decoy RNA is stably expressed by cells, they become resistant to HIV replication for over 20 weeks (LISKIEWICZ et al., 1993). RNA decoys have the potential drawback of sequestering cellular transcriptional factors which could, in turn, interfere with normal cellular function. However, until now stable expression of RNA decoys in human cells has not been shown to effect the morphology of the cells or growth kinetics (SMITH et al., 1993; JEANG et al., 1991).

2.1.2 Protein Inhibitors: Dominant-Negative Proteins and Intracellular Antibodies

HIV-1 replication can be inhibited by the expression of mutant viral proteins lacking an

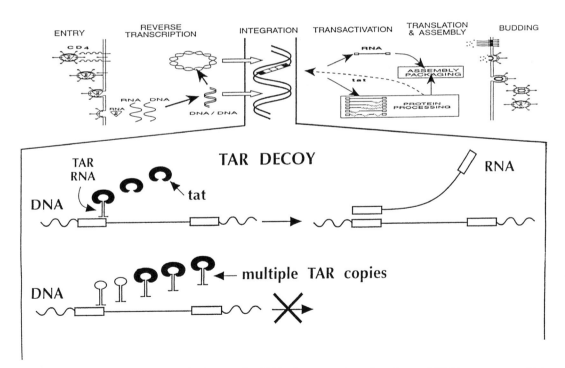

Fig. 3. The anti-viral action of TAR decoys. The TAR decoy consists of engineered DNA sequences which when transcribed generate multiple copies of the TAR sequence of the HIV genome. The viral transactivator protein, Tat, therefore, binds to and is sequestered by the TAR sequences and the decoy RNA, therefore, competes against the transactivation of the viral genome by Tat produced in the HIV infected cell.

active domain essential for function but dominant with regards to the expression of mutant proteins (FRIEDMANN et al., 1988) (Fig. 4). So-called dominant-negative mutants have been constructed for the Gag, Rev, Tat and Env proteins of HIV-1. BUCHSCHACHER et al. (1992) utilized a dominant-negative mutant of *env* which was incapable of binding CD4 and inhibited the production of infectious viral particles. Cell lines expressing the *rev, gag* and *env* dominant-negative mutants have been shown to be resistant to HIV infection (MALIM et al., 1989; MALIM et al., 1992). Dominant-negative mutants have several potential disadvantages including the fact that they may be processed as antigenic proteins and result ultimately in a cytotoxic lymphocyte (CTL) response causing destruction of those cells expressing the mutant protein (WOFFENDIN, 1995; GREENBERG, 1995). In addition, the dominant-negative mutant pro-

tein must be processed in the same compartment as the wild-type protein and not must exert toxic effects on the cell at that location (WINTER and MILSTEIN, 1991).

Intracellular antibodies are engineered from human monoclonal antibodies which selectively bind and inactivate HIV proteins (Fig. 5). An intracellular antibody consists of the antigen-binding domains of the heavy and light chains of the IgG molecule joined by an interchain linker (WINTER and MILSTEIN, 1991). MARASCO et al. (1993) derived an intracellular antibody from the group-specific human monoclonal antibody F105 which competes for the CD4-binding site on gp120 (WINTER and MILSTEIN, 1991; CHEN et al., 1994; MARASCO et al., 1993). The monoclonal antibody, F105, can bind the envelope glycoprotein of many HIV-1 strains including several primary isolates (POSNER et al., 1991). The intracellular antibody sFv105 consists of

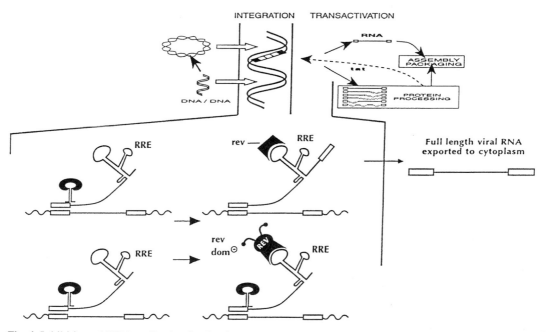

Fig. 4. Inhibition of HIV replication by dominant-negative HIV proteins. The dominant-negative Rev protein (Rev dom⁻) can interfere with the correct splicing and transport of HIV RNA from the nucleus by interfering with the function of wild-type rev. Rev dom⁻ binds to the *rev* response element (RRE) of the HIV RNA which is transcribed from the proviral RNA and thereby blocks the correct export of full length HIV RNA to the cytoplasm. Dominant-negative Tat and Rev principally inhibit postintegration events such as HIV RNA transport and HIV genome transactivation.

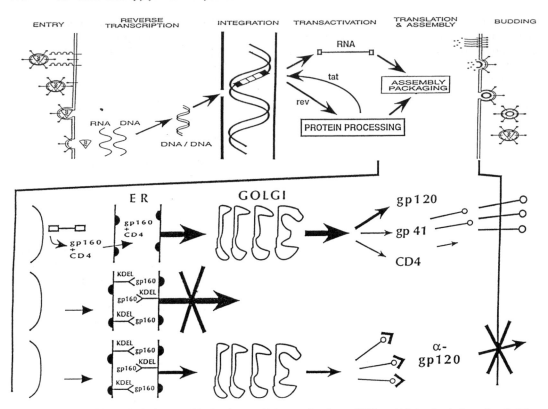

Fig. 5. The inhibition of viral assembly by intracellular antibody to HIV gp120 in the infected cell. The figure demonstrates the normal assembly of the HIV envelope glycoproteins gp140 and gp120 from gp160 in the ER and Golgi body and then insertion of the envelope proteins of HIV into the host cell membrane. An intracellular antibody to gp120 with an attached KDEL sequence binds the gp160 in the ER and thereby inhibits the posttranslational processing and cleavage of viral envelope proteins in the Golgi and the ultimate transport of gp1 and gp120 to the host cell membrane. The intracellular antibody to gp120 is, therefore, active in inhibiting the postintegration steps of protein processing and viral assembly and packaging. Intracellular antibodies have also been produced which inhibit the actions of the HIV proteins p17 and Tat and which, therefore, inhibit both pre- and postintegration events in the replicative cycle of HIV.

both the heavy and light chain variable regions of F105, joined by an interchain linker and containing a KDEL sequence which consists of a specific sequence of amino acids and which directs the sFv105 protein to the lumen of the endoplasmic reticulum (ER) (DUAN et al., 1995). The intracellular antibody has been shown to sequester gp120 in the ER and interferes with the production of viral particles. HIV replication was inhibited for up to 4 weeks in cells in which this intracellular antibody was stably expressed (CHEN et al., 1994). More recently, the G1Na Mo-MLV based vector (see Sect. 2.2.1) has been used to deliver sFv105 to both CD4$^+$ human T cell lines and primary human CD4$^+$ T cells. Inhibition of HIV replication in both cell types up to 30 d post challenge with wild-type HIV was demonstrated (MARASCO, 1995, personal communication). MARASCO et al. (1993), DUAN et al. (1995), and POZANSKY et al. (1998) have shown that HIV replication is inhibited in CD4$^+$ T cell lines stably expressing intracellular antibodies to *rev, tat* or *gag* (MHASHLIKAR et al., 1995; DUAN et al., 1995). The potential drawbacks of intracellular antibodies are that an immune response may be generated against them, although they are human-derived proteins which can be engineered to be exclusively expressed in the intracellular compartment. In addition, escape mutants may occur that render the intracellular antibody ineffective, even if it has been engineered to bind highly conserved epitopes on the target HIV protein.

2.2 Design of the Optimal Vector for the Delivery of Anti-Viral Constructs

If any of the aforementioned anti-viral constructs are to be effective as therapeutic agents, then the genes encoding them must be transduced into cells which are susceptible to HIV infection *in vivo* (BALTIMORE, 1988; DROPULIC and JEANG; 1994; ANDERSON, 1994; YU et al., 1994). Moreover, since HIV induces a chronic infection in humans, the vector must facilitate stable expression of the construct. An optimized vector should fulfill the following criteria:

1. The vector must efficiently and stably transduce target cells for HIV infection. These are the primary CD4$^+$ T cell, dendritic cells and cells of the monocyte/macrophage lineage.
2. The vector must result in the stable expression of the encoded antiviral construct at levels in the transduced cell which are high enough to inhibit HIV infection and/or replication.
3. The vector must be safe for human usage and should not itself be capable of inducing deleterious effects in transduced cell populations, or in the host in general.
4. In the case of viral-based vector systems it must be clearly demonstrated that the recombinant replication-competent virus is not generated in the packaging system and that there is no evidence of insertional mutagenesis induced by retroviral-based vectors.

2.2.1 Amphotropic Murine Leukemia Virus (Mo-MLV) Based Vectors

Wild-type Mo-MLV (see Chapter 18, this volume) is capable of delivering genes encoding Env, Gag and Pol proteins to a wide variety of eukaryotic cells. Engineered defects in the genome of the Mo-MLV uncluding deletion of genes coding the viral structural proteins allow it to deliver only the "therapeutic" gene of interest to eukaryotic cells. Defective amphotrophic Mo-MLV has been used as a vector for the transduction of a variety of human cell lines and primary human cell lineages (MULLIGAN, 1993; MILLER and ROSMAN, 1989; MILLER, 1990; MILLER, 1992; CORNETTA et al., 1993; JAFFEE et al., 1993; KOTANI et al., 1994; LEAVITT et al., 1994). The Mo-MLV vector has provided the basis for gene therapies for many human diseases other than HIV infection (BLAESE, 1993; WEATHERALL, 1992; VILE and RUSSELL, 1994; LOZIER and BRINKHOUS, 1994; STRAUSS, 1994; GROSSMAN and WILSON, 1993). Mo-MLV-based retroviral vectors have been used to deliver a variety of anti-HIV constructs such as ribozymes, RNA decoys, and intracellular antibodies to human CD4$^+$ T cell lines and primary human CD4$^+$ T cells derived from peripheral blood, *in vitro* and more recently *in vivo* (WEERASINGHE et al., 1991; SULLENGER et al., 1993; LISKIEWICZ et al., 1993; LEAVITT et al., 1994; BEVEC et al., 1994).

In general, the use of Mo-MLV-based vectors to deliver anti-viral constructs to human T cells *in vitro* has yielded promising results. WEERASINGHE et al. (1991) utilized a Mo-MLV-based vector encoding a ribozyme which cleaved the 5′ leader sequence of HIV-1 RNA. Ribozyme production was driven by the thymidine kinase promoter of herpes simplex virus (HSV) and contained a promoter which was activated by the HIV protein Tat. When human CD4$^+$ T cells derived from a cell line were transduced with the vector, the cells expressing the ribozyme were shown to be resistant to productive HIV infection for 22 d following challenge with wild-type HIV-

1. In experiments by WONG-STAAL et al. (1994) primary human CD4$^+$ T cells and human T cell lines were transduced with a LNL-6-based Mo-MLV vector encoding an anti HIV ribozyme whose production was driven by an internal tRNA promoter (LEAVITT et al., 1994; YAMADA et al., 1994). The transduced T cell lines and primary T cells were resistant to HIV infection up to 35 and 10 d respectively, following a single challenge with both laboratory strains and primary isolates of HIV-1. Mo-MLV based vectors have also been used to deliver TAR- and RRE-based RNA decoys to CD4$^+$ T-cells. SULLENGER et al. (1993) demonstrated that CEM cells transduced with the TAR decoy RNA were resistant to HIV infection for up to 25 d following challenge with wild-type HIV-1. Similarly, LISKIEWICZ et al. (1993) and BEVEC et al. (1994) demonstrated that T cells transduced using a Mo-MLV-based vector encoding a combined polymeric TAR RNA and ribozyme which cleaved the RRE of HIV-1 or a polymeric RRE decoy RNA showed a greater than 99% inhibition in HIV production up to 14 months following challenge with HIV.

For each system described it is encouraging that Mo-MLV delivered constructs have remained stably expressed in cells for between 3 and 6 months. Furthermore, no effects on T cell proliferation, as measured by ^{3}H thymidine uptake, CD4 expression or cellular morphology following the transduction with Mo-MLV-based vectors have been reported (WEERASINGHE et al., 1991; LISKIEWICZ et al., 1993; LEAVITT et al., 1994; YAMADA et al., 1994; BEVEC et al., 1994).

Mo-MLV-based vectors, not surprisingly, have drawbacks which have been overcome to a limited extent by technical modifications in the transduction process. It has proven difficult to generate titers of Mo-MLV-based vectors of greater than 10^6 per mL (giving multiplicities of infection in the range of 1:10). This is disappointing since high titer vector appears to be required for the transduction of primary T cells to achieve full protection for T cells susceptible to HIV infection. For this reason efforts have been made to concentrate vector and optimize transduction efficiencies (CORNETTA et al., 1993; JAFFEE et al., 1993; LEAVITT et al., 1994;

BUNNELL et al., 1995; KAVANAUGH et al., 1995).

The second major defect in the Mo-MLV vector system is its inability to infect non-dividing cells (MILLER and ROSMAN, 1989; MILLER, 1992). Thus T cells fractionated from peripheral blood must be activated with OKT3 and Il-2 prior to transduction. Consequently, the use of Mo-MLV-based vectors for transduction of human CD4$^+$ T cells from HIV infected individuals necessitates the concurrent use of antiretroviral drugs such as nevirapine and AZT during the transduction process (LEAVITT et al., 1994). The corollary of the failure of Mo-MLV-based vectors to transduce non-dividing cells is, that they also fail to reach targets for HIV infection including monocytes, dendritic cells or glial cells (SHNITTMAN et al., 1989; LANGHOFF et al., 1991; WEINBERG et al., 1991). Finally, the generation of replication competent retrovirus and the potential for insertional mutagenesis induced by Mo-MLV is a stochastic event, although this has not been realized in the vast majority of the *in vitro* and *in vivo* experiments performed to date with Mo-MLV vectors (McGARRITY, 1994).

2.2.2 Novel Retroviral and Adeno-Associated Virus (AAV) Based Vectors

Investigators are attempting to overcome the drawbacks of Mo-MLV-based vectors by developing novel viral vector systems which include the development of other retroviral- and AAV-based vectors which might ultimately be suitable for the delivery of anti-viral constructs to primary human CD4$^+$ T cells and monocyte/macrophages (POZNANSKY et al., 1991; SHIMADA et al., 1991; ALI et al., 1994).

An HIV-based vector has the conceptual advantage of selectively targeting and ultimately protecting precisely those cells which are susceptible to infection by wild-type virus (POZNANSKY et al., 1991). Furthermore, an HIV-based vector would be able to transduce non-dividing cells, such as dendritic cells and monocytes (WEINBERG et al., 1991; NALDINI et al., 1996). However, progress towards an efficient packaging defective HIV-based vector has been dogged by the complex packaging system that is utilized by HIV (LEVER et al., 1989; RICHARDSON et al., 1993; ALDOVINI and YOUNG, 1990). In addition, attempts are still under way to create a stable packaging cell line for an HIV vector (NALDINI et al., 1996). The current generation of HIV vectors has only been shown to be capable of delivering selectable marker genes at low titer (POZNANSKY et al., 1991; RICHARDSON et al., 1993; HASELHORST et al., 1994; BUCHSCHACHER and PANGANIBAN, 1994). The fear of any generation of replication competent virus in an HIV vector system would seriously limit its potential usage. Therefore, an HIV-based vector would pose considerable ethical problems before its use in humans is approved.

Other retroviruses such as the human foamy retrovirus (HFV) are currently being considered for the use as vectors. These amphotropic retroviruses are capable of infecting a wide variety of hematopoietic and neural tissues and are, therefore, being considered for the use in protecting cells in these lineages that are susceptible to HIV infection (BIENIASZ et al., 1997). Work is currently underway to define the packaging sequence of the HFV and to design vectors based on this virus.

Adeno-associated virus, AAV-2, is a defective human parvovirus which requires a helper adenovirus to package its genome. AAV-based vectors may prove useful in the delivery of anti-viral constructs to primary human lymphoid cells susceptible to infection by HIV (YU et al., 1994; ALI et al., 1994; BERNS and HAUSWIRTH, 1983; CARTER and LAUGHLIN, 1983). Packaging systems for AAV-based vectors have been designed which allow the generation of AAV-based vectors to titers up to 10^{12} L^{-1} (CARTER and LAUGHLIN, 1983). Moreover, AAV like MoMLV causes no known disease in humans, despite being capable of integrating its genome into a wide variety of dividing and non-dividing human primary lymphoid and hematopoietic cells in a site-specific manner and at high transduction efficiencies. Helper virus is simply and completely inactivated *in vitro* and transduction of human cells by the AAV vec-

tor has been shown to result in sustained expression of the transgene (CARTER and LAUGHLIN, 1983; SAMULSKI et al., 1982; HERMONAT and MUZYCAK, 1984; TRATSCHIN et al., 1985; WALSH et al., 1992; PHILLIP et al., 1994). The limitations of the vector are that only 4.7 kb of new genetic material can be inserted into its genome (CARTER and LAUGHLIN, 1983). In addition, a recent study demonstrated that AAV-based vectors achieved only low transduction efficiencies in primary human cells (RUSSELL et al., 1994; ALEXANDER et al., 1994).

CHATERGEE et al., (1992) were the first to use the AAV-based vector system to deliver anti HIV constructs to T cells *in vitro*. They demonstrated that the transduction of human CD4$^+$ T cell lines by an AAV-based vector encoding an antisense RNA species homologous to the TAR sequence of HIV caused a 1000-fold reduction in HIV replication in the transduced cells following challenge by wild-type HIV. Although, this preliminary use of an AAV-based vector encoding an anti-viral construct appears promising, further work is necessary to demonstrate the reproducibility and efficacy of this transduction strategy in inhibiting HIV replication in primary T cells.

3 Strategies Augmenting the Immune Response to HIV

Extensive *in vitro* and animal studies suggest that CTL directed against HIV infected cells and neutralizing antibodies directed against the CD4-binding domain and the principal neutralizing domain (V3 loop) of gp120 are protective against HIV infection (ZOLLA-PAZNER and GORNY, 1992; WALKER et al., 1987, 1990; TSUBOTA et al., 1989; BYRNE and OLDSTONE, 1984; KUOP and HO, 1994). Two main gene therapy-based approaches to developing immunotherapies for HIV are explored: the transfer of genes for expression of immunogenic HIV proteins *in vivo* and the use of genes encoding selectable markers in order to expand HIV reactive CTL populations *ex vivo* before administration to patients (YU et al., 1994; WANG et al., 1993; WARNER et al., 1991; RIDDELL et al., 1992; GALPIN et al., 1994).

The first of the two approaches involves host cell expression of immunogenic HIV proteins, such as gp120 and gp160. This is achieved either by direct injection of expression vector DNA encoding the proteins into muscle or skin or the implantation of autologous fibroblasts which have been transduced with genes encoding immunogenic HIV proteins *ex vivo* (WANG et al., 1993; WARNER et al., 1991). In both instances expression of HIV gp120 in transfected or transduced cells in mice resulted in the generation of neutralizing antibody responses and HIV specific CTL responses. In the second approach it has been demonstrated that HIV specific CTL may be selected from PBMC derived from HIV infected individuals (RIDDELL et al., 1992). These CTLs are then transduced with a Mo-MLV-based vector encoding a selectable marker gene and the thymidine kinase (TK) gene derived from herpes simplex virus. Expression of the TK gene would permit selective destruction of the HIV specific CTL by treating the patients with gancyclovir, in the unlikely event of excessive proliferation or deleterious effects of those cells (HEYMAN et al., 1989; MOOTTEN and WELL, 1990).

These marked CTLs are then selectively expanded in the presence of OKT3, the CD4-binding monoclonal antibody, and IL-2 *in vitro*. The potential remains to return these expanded HIV specific CTLs to the autologous donor in order to augment the CTLs response to HIV *in vivo* (RIDDELL et al., 1992). Adoptive CTL transfer, as this approach has become known, has been utilized with considerable promise in the prevention of CMV infection during clinially induced immunosuppression following bone marrow transplantation (BMT) (REUSSER et al., 1994). In this case CMV specific CTLs have been expanded *ex vivo* pre transplantation and then reinfused post transplantation. Use of adoptive CTL transfer in this way has significantly reduced the occurrence of CMV infection post bone marrow transplantation (REUSSER et al., 1994; GREENBERG, 1995). A variant of this

approach is being developed which involves engineering CTLs *ex vivo* for improved recognition of HIV infected cells (YANG et al., 1997).

4 Ongoing Phase I Clincal Trials of Gene Therapies for HIV Infection

The work described above highlights two principles:

1. Human CD4$^+$ T cells transduced with genes encoding anti-viral constructs are protected from HIV infection and/or HIV replication *in vitro*.
2. Gene therapy may be used to augment CTL or to neutralize antibody responses to HIV in animal models *in vivo*.

As a result of this proof of principle, clinical protocols for testing the safety and efficacy of gene therapies for HIV have been initiated. These include phase I clinical trials to assess the dominant-negative *rev* m10 mutant, the implantation of autologous transduced fibroblasts expressing HIV gp160 and the use of adoptive HIV specific CTL transfer (RIDDELL et al., 1992; GALPIN et al., 1994; NABEL et al., 1994; YANG et al., 1997). Several other protocols currently considered by regulatory bodies include the use of hairpin ribozyme to the leader sequence of HIV and the use of intracellular antibodies to gp120 as anti-viral constructs.

NABEL et al. (1994) are exploiting the *rev* M10 transdominant mutant which inhibits replication of both laboratory and primary isolates of HIV-1 when stably expressed in human T cell lines and primary CD4$^+$ cells *in vitro* (MALIM et al., 1992; NABEL et al., 1994). The protocol has involved obtaining $2 \cdot 10^8$ primary CD4$^+$ T cells from the peripheral blood from HIV infected individuals by leukopheresis. The cells were then activated with CD3 and IL-2 in the presence of nevirapine and AZT and transduced with a murine retroviral vector encoding the Rev m10 mutant protein driven by the Mo-MLV LTR and the G418 resistance gene driven by the SV40 promoter (KORMAN et al., 1987). Transduced cells expanded in the presence of IL-2 and neomycin were reinfused into the autologous donor. NABEL et al. (1994) estimate that the percentage of circulating cells containing the revM10 gene in these circumstances will range from 0.1–0.0015% of circulating PBMC. Various virological, immunological and clinical parameters were followed in the treated patients up to 20 weeks post infusion. The protocol seeks to assess whether revM10 expression in PBMC improves cell survival in comparison to control cells transduced with ΔrevM10, an inactive frameshift mutant of revM10 with no *in vitro* anti-viral activity. In addition, the investigators plan to assess the effects of the treatment on viral replication and immune function *in vivo*.

The second protocol involved the use of genetically modified CD8$^+$ HIV-specific T cells for HIV$^+$ patients undergoing allogeneic BMT for malignant lymphoma (RIDDELL et al., 1992). PBL and HIV *gag* specific CD8$^+$ CTL selected and transduced with a Mo-MLV-based vector encoding the hygromycin resistance gene for cell expansion and the HSV TK suicide gene are obtained from these patients prior to bone marrow ablation. The transduced HIV specific CD8$^+$ T cell clones are selectively expanded *in vitro* and reinfused into the autologous patient.

A third immunotherapy-based protocol involves augmenting the immune response to HIV-1 in asymptomatic HIV infected individuals by HIV-1 IIIB *env*-transduced autologous fibroblasts (HIV-IT TAF) from skin biopsies (GALPIN et al., 1994). HIV specific CTL can be induced by autologous cells expressing HIV immunogenic antigens in mice and macaques (POPOVIC et al., 1984). HIV-IT TAF are selected and reimplanted into the autologous human donor. An assessment of safety and tolerability of administering three successive doses of HIV-IT TAF will be made in this protocol. At the time of writing, no clear information concerning the efficacy of these approaches was available.

4.1 Biosafety and Ethical Issues Involved in the First Generation of Phase I Trials

Inevitably, the first generation of phase I clinical protocols raise important biosafety and ethical questions. The three protocols described in Sect. 4 utilize Mo-MLV-based vectors to transduce human cells. The use of these retroviral vectors is associated with three main safety concerns: the generation of replication competent retrovirus (RCR) through the recombination of packaging and vector sequences in the packaging cell line; the induction of malignant transformation through insertional mutagenesis induced by retroviral vectors and the theoretical possibility of transduction and mutation of the germ line by retroviral vectors (MILLER, 1990; RIDDELL et al., 1992; GALPIN et al., 1994; NABEL et al., 1994). All clinical-grade supernatants containing packaged retroviral vector and derived from packaging cell lines must be screened for the presence of RCR prior to their use in human trials (RIDDELL et al., 1992; GALPIN et al., 1994; NABEL et al., 1994). Clinical grade supernatants containing Mo-MLV-based retroviral vector are routinely screened by several methods for recombinant RCR (MILLER, 1990). To date RCR have been found in three batches of clinical-grade supernatant generated from the PA 317 packaging cell line transfected with a G1Na-based vector (OTTO et al., 1994). These RCR containing batches of supernatant were obviously not used for human trials. No RCR have been detected to date in clinical grade supernatants from the AM12 cell line (MCGARRITY, 1994). The stringent conditions used to check supernatants for RCR and the molecular engineering applied to reduce the chances of recombination between vector and packaging sequences is justified by the finding that monkeys and mice who received large quantities of RCR show an increased occurrence of T cell lymphomas (DONAHUE et al., 1992; ANDERSON et al., 1993). Reassuringly, patients involved in the human gene therapy trials for the treatment of diseases such as ADA deficiency, LDLR deficiency and the protocols utilizing genetically manipulated tumor infiltrating lymphocytes (TIL) have shown no evidence of the emergence of RCR or an increased incidence of cancer, although follow-up time at this stage is short (BLAESE, 1993; VILE and RUSSELL, 1994; STRAUSS, 1994).

The possibility that those protocols utilizing retroviral vectors may result in cancer due to insertional mutagenesis remains entirely theoretically, since the use of these vectors in nude mice and immunosuppressed suckling mice has failed to show evidence of tumor formation *in vivo*. FEARON and VO-GELSTEIN (1990) demonstrated that upwards of five insertional events are required to transform a cell. This is an extremely unlikely event, given the level of transduction efficiencies of the current generation of retroviral vectors. Furthermore, there is no evidence to suggest that the Mo-MLV-based vectors used in the protocols above result in the transformation of transduced primary human T cells *in vitro* (RIDDELL et al., 1992; GALPIN et al., 1994; NABEL et al., 1994). A retroviral vector delivered systemically may also be capable of transducing germ line cells. Although this remains a powerful argument against *in vivo* gene therapy with retroviral vectors extensive studies of germ line tissues in monkeys and primates after treatment with Mo-MLV-based replication defective vectors suggest this risk is very low (RIDDELL et al., 1992; GALPIN et al., 1994; NABEL et al., 1994).

5 Refining Gene Therapy Protocols

5.1 Optimized Vector Design and Tissue-Specific Expression of the Transduced Gene

The LTR and packaging sequences of the Mo-MLV form the basis for the most commonly used vectors and for the transduction of human lymphoid and hematopoietic cells (MILLER and ROSMAN, 1989; MILLER, 1990; MILLER, 1992). Modifications to the LTR of Mo-MLV-based vectors have been shown to alter the expression of the gene cassette encoded by the vector (MILLER, 1992). Tissue specific promoters are of particular use in driving expression of transduced genes *in vitro* and *in vivo* in a safe manner. COUTURE et al. (1994) if the U3 region of the Mo-MLV vector is replaced by the U3 region of the SL3-3 murine leukemia retrovirus, expression of the gene cassette is augmented in human lymphoid cells (COUTURE et al., 1994; CELANDER and HASELTINE, 1984). Attempts have also been made to improve the promoter activity of the Mo-MLV LTR by inserting human lymphoid specific promoter/enhancer sequences (MOORE et al., 1994). MOORE et al. (1994) demonstrated that the insertion of the enhancer from the α chain of the human T cell receptor (TCR) failed to increase the expression of the transduced ADA gene above that achieved in the native vector when the expression of the transduced gene was driven by the Mo-MLV LTR. More recently VILE and HART (1994) demonstrated that although powerful viral promoters such as the CMV or SV40 promoters were active in cells transduced *in vitro*, tissue specific enhancers were more active in transduced cells *in vivo*.

Locus control regions (LCR) derived from human and murine genes have also been tested in retroviral vectors. LCR are DNase I hypersensitive regions, up to 30 kb in length, which have been found located either 5′ or 3′ to many genes including the human β globin, human α globin and human a1(I) collagen gene clusters (GROSVELD et al., 1987; RYAN et al. 1989; HIGGS et al., 1989; HIGGS et al., 1990; BARSCH et al., 1984; ELSENBERG and ELGIN, 1991; ELGIN, 1991; PRUZINA et al., 1994). The LCR of the human α globin and β globin gene loci appear to play an important role in maintaining a chromatin structure which facilitates the constitutive and high level expression of the globin genes within the loci (HIGGS et al., 1989; HIGGS et al., 1990; BARSCH et al., 1984; ELSENBERG and ELGIN, 1991). Consequently it has been proposed that sequences within LCR of constitutively expressed genes may be of use in vector design for gene therapy to maintain stable and high level expression of a transduced gene (MCCUNE and TOWNES, 1994; SIPPEL et al.,

1994). McCune and Townes (1994) demonstrated that a 1.1 kb HS2 sequence derived from 5'LCR of the human β globin gene conveyed enhancer activity and position independent expression on a human β globin gene in transgenic mice. However, when the HS2 β globin construct is flanked by Mo-MLV vector sequences, expression of the β globin gene is repressed in transgenic mice. Consequently, further engineering of retrovical vector sequences or development of other vectors for stable chromosome insertion are required before sequences derived from LCR can be used to improve the stability of transduced gene expression. In the longer term, however, this is a very promising approach to activating sustained expression of inhibitory molecules, and LCR giving expression specific for T cells or macrophages have been identified (Bonifer et al., 1990).

5.2 Improved Transduction Protocols for Primary Human T Cells

Progress towards optimal vector production and transduction efficiencies has been driven by the need for high titer and highly efficient vectors for clinical trials. Scientists from both laboratories and biotechnology companies have designed methods for generating high titer vectors and the transduction of primary human cells at high efficiencies (Aboud et al., 1982; Cepvko, 1992; Hammer et al., 1989; Paul et al., 1993; Kotani et al., 1994). These methods involve virus concentration by centrifugation or precipitation of viral particles, but have given variable results and an associated reduction in the transduction efficiency of concentrated vector (Paul et al., 1993; Kotani et al., 1994; Karlsson, 1991). However, McGarrity (1994) has consistently generated high titer preparations using a combination of a bioreactor, a tangential flow filtration system and lyophilization. In this way Mo-MLV-based vectors were generated from up to 200 L of supenatant at titers up to 10^{10} L^{-1} on primary human C cells. Paul et al. (1993) have demonstrated that hollow-fiber technology can

be utilized to concentrate viable viral particles.

5.3 Gene Therapy for Progenitor Cells and Stem Cells from HIV Infected Individuals

A gene therapy for HIV infection which involves gene transfer into the hematopoietic stem cell or progenitor cell which in turn is capable of repopulating the bone marrow and peripheral circulation with mature CD4$^+$ T cells protected from HIV infection, is an attractive concept for the future. Gene transfer into the stem cell or progenitor cell compartment would overcome the principal limitation of current approaches which involve the protection of a population of mature peripheral blood T cells with a limited lifespan of between 6 months and 1 year and, therefore, necessitating the use of repeated gene transfer procedures during the life of the patient.

Gene transfer into human hematopoietic stem cells and progenitor cells derived from peripheral and umbilical cord blood has been performed as part of the development of gene therapies for the correction of genetic defects associated with diseases such as β thalassemia, sickle cell anemia, Lesch-Nyhan syndrome, ADA deficiency and most recently for the protection of cells from infection by HIV (Karlsson, 1991; Nienhuis et al., 1991; Moritz et al., 1993; Bahnson et al., 1994). Wong-Staal (1994) proposed a human protocol for gene transfer into CD34$^+$ progenitor cells derived from umbilical cord blood of neonates delivered to known HIV$^+$ mothers. CD34$^+$ cells would be stored until it was clear that the baby was infected by HIV and at this point the CD34$^+$ cells would be retrieved and transduced with anti-viral constructs. These transduced pluripotent cells would then be used to reconstitute the progenitor cell population of the autologous HIV infected infant, thereby generating a population of mature CD4$^+$ T cells protected from HIV infection (Wong-Staal, 1994). In order to allow the implementation of protocols such as this methods have been developed which optimize the procurement of stem and progenitor cells from peripheral blood and bone marrow and the transduction and expansion of these hematopoietic stem cells (MORITZ et al., 1993; LU et al., 1995; BERENSON et al., 1991; LAPIDOT et al., 1992).

5.4 The Next Generation of Gene Therapy Trials for HIV Infection

The aim of any therapy for HIV infection is that it permanently inhibits HIV replication and consequently halts the progression from asymptomatic HIV infection to AIDS. The current generation of anti-retroviral drugs including reverse transcriptase inhibitors and protease inhibitors appears to modify the progression of HIV infection to AIDS and reduce virus levels in peripheral blood, but with notable drug induced side effects (Delta Coordinating Committee, 1995; CHOO, 1995). These therapies also have the disadvantages of high cast and poor patient compliance.

Gene therapy offers a way forward to inhibit HIV replication at various stages in its life cycle in a way that may ultimately be longer lasting and be free of side effects. A diverse range of anti-viral constructs including dominant-negative mutant HIV proteins, RNA decoys, ribozymes and intracellular antibodies inhibits HIV replication without detrimental effects on the target cell *in vitro*. The safety and efficacy of these anti-viral constructs are currently under assessment in humans in phase I trials as indicated in Sects. 4 and 4.1. In addition, the possibility of stem cell gene therapy offers the hope of a one-step treatment for HIV infection in which transfected hematopoietic stem cells regenerate a population of CD4$^+$ T cells and monocyte/macrophages protected from HIV infection by the fact that they express anti-HIV constructs and thereby extend the lifetime of the HIV infected individual. However, the science of gene therapy is still in the beginning and there is still much to be done to optimize anti-viral constructs, vectors and clinical protocols.

The diversity of anti-viral constructs has given rise to the concept of combination gene therapy for HIV, involving the delivery of a number of different constructs to the same cell. For example, a combination of a *gag* specific ribozyme and a TAR decoy RNA was shown to enhance the inhibition of HIV replication in transduced human CD4$^+$ T cells above that seen with the TAR decoy RNA alone (LISKIEWICZ et al., 1993). Subsequently, CHANG et al. (1994) utilized expression vectors encoding combinations of antisense RNA to *tat* and TAR decoy RNA (CHANG et al., 1994). *Tat* antisense RNA or TAR decoy RNA alone could only partially inhibit *tat* driven HIV gene expression in their system whereas a combination of the two anti-viral constructs completely blocked *tat* function (CHANG et al., 1994). Similar results have been achieved when cells have been engineered to express combinations of TAR decoy RNA, antisense *gag* RNA and a dominant negative mutant of *gag* (LORI et al., 1994). MARASCO and coworkers have further advanced the concept of combination gene therapy by incorporating an internal ribosomal entry site (IRE) into their LN-based Mo-MLV retroviral vectors (MARASCO, 1995, personal communication). In this way a retroviral vector has been developed which generates a bicistronic mRNA encoding two anti-viral constructs, in this case intracellular antibodies directed against *tat* and gp120.

In addition to the promise of anti-viral approaches, the use of gene therapy in immunotherapeutic approaches to the treatment of HIV infection has also generated encouraging results. There is clear proof of principle from both laboratory animals and man that gene therapy can be used to initiate or augment an immune response to HIV. The use of direct delivery of DNA encoding immunogenic HIV protein is of particular interest because it may one day represent a simple clinically applicable mode of eliciting protective immunity to HIV. The development of a gene therapy for HIV infection goes hand in hand with our growing understanding of the pathogenesis and the immune response to HIV in humans. Work on viral dynamics, long-term survivors with HIV infection and HIV negative seroconvertors from high-risk areas in the world may yet reveal which aspect of the HIV replicative cycle to target and which part of the human immune response to augment by gene therapy (Ho et al., 1995; ROWLAND-JONES et al., 1995).

These avenues of research must continue to expand in order that the full clinical potential of gene therapy for HIV infection is realized. A clinically viable gene therapy for the treatment of HIV infection may well be at hand within the next decade.

6 References

ABOUD, M., WOLFSON, M., HASSAN, Y., HULEI-HEL, M. (1982), Rapid purification of extracellular and intracellular Moloney leukemia virus, *Arch. Virol.* **71**, 185–195.

ALDOVINI, A., YOUNG, R. A. (1990), Mutations and protein sequences involved in human immunodeficiency virus type 1 packaging result in production of noninfectious virus, *J. Virol.* **64**, 1920–1926.

ALEXANDER, I. E., RUSSELL, D. W., MILLER, A. D. (1994), DNA-damaging agents greatly increase the transduction of non-dividing cells by adeno-associated virus vectors, *J. Virol.* **68**, 8282–8287.

ALI, M., LEMOINE, N. R., RING, C. I. A. (1994), The use of DNA viruses as vectors for gene therapy, *Gene Ther.* **1**, 367–384.

ANDERSON, W. F. (1984), Prospects for human gene therapy, *Science* **226**, 401–409.

ANDERSON, W. F. (1994), Gene therapy for AIDS, *Hum. Gene Ther.* **5**, 149–150.

ANDERSON, W. F., McGARRITY, G. I., MOEN, R. C. (1993), Report to the NIH recombinant DNA Advisory Committee on murine replication-competent retrovirus assays, *Hum. Gene Ther.* **4**, 311–321.

BAHNSON, A. B., NIMGAONKAOR, M.., FEI, Y., BOGGS, S. S., ROBBINS, P. D. et al. (1994), Transduction of CD34+ enriched cord blood and Gaucher bone marrow cells by a retroviral vector carrying the glucocerebrosidase gene, *Gene Ther.* **1**, 176–184.

BALTIMORE, D. (1988), Intracellular immunisation, *Nature* **335**, 395–396.

BARSCH, G. S., ROUSH, C. L., GELINAS, R. E. (1984), DNA and chromatin structure of the human α1-(I) collagen gene, *J. Biol. Chem.* **259**, 14906–14913.

BERENSON, R. J., BENSINGER, W. I., HILL, R. S. et al. (1991), Engraftment after infusion of CD34+ marrow cells in patients with breast cancer and neuroblastoma, *Blood* **77**, 1717–1722.

BEVEC, D., VOLC-PLATZER, B., ZIMMERMANN, K., DOBROVNIK, M., HAUBER, J. (1994), Constitutive expression of chimeric Neo-rev response element transcripts suppresses HIV-1 replication in human CD4+ T lymphocytes, *Hum. Gene Ther.* **5**, 193–201.

BERNS, K. I., HAUSWIRTH, W. W. (1983), Adeno-associated virus DNA structure and replication, in: *The Parvoviruses* (BERNS, K. I., Ed.), pp. 1–31. New York: Plenum Publishing Corp.

BIENAISZ, P., ERLWEIN, O., AGUZZI, A., RETHWILM, A., McCLURE, M. O. (1997), Gene transfer using replication-defective human foamy virus vectors, *Virology* **235**, 65–72.

BLAESE, R. M. (1993), Development of gene therapy for immunodeficiency: adenosine deaminase deficiency, *Paediatr. Res.* **33** (Suppl.), s49–s55.

BONIFER, C., VIDAL, M., GROSVELD, F., SIPPEL, A. E. (1990), Tissue-specific and position-independent expression of the complete gene domain for chicken lysozyme in transgenic mice, *EMBO J.* **9**, 2843–2848.

BORDIER, B., HELENE, C., BARR, P. J., LITVAK, S., SARIH-COTTIN, L. (1992), *In vitro* effect of antisense oligonucleotides on HIV reverse transcription, *Nucleic Acids Res.* **20**, 5999–6006.

BORDIGNON, C., MAVILLO, F., FERRARI, G., SERVIDA, P., UGAZIO, A. G. et al. (1993), Clinical protocol: transfer of the ADA gene into bone marrow cells and peripheral blood lymphocytes for the treatment of patients affected by ADA-deficient SCID, *Hum. Gene Ther.* **4**, 513–520.

BUCHSCHACHER, G. L., PANGANIBAN, A. T. (1994), Human immunodeficiency virus vectors for inducible expression of foreign genes, *J. Virol.* **66**, 2731–3739.

BUCHSCHACHER, G. L., FREED, E. O., PANGANIBAN, A. T. (1992), Cells induced to express human immunodeficiency virus type 1 envelope gene mutant inhibit spread of wild type virus, *Hum. Gene Ther.* **3**, 391–397.

BUNNELL, B. A., MUUL, L. M., DONAHUE, R. E., BLAESE, R. M., MORGAN, R. A. (1995), High efficiency retroviral-mediated gene transfer into human and non-human primate peripheral blood lymphocytes, *Proc. Natl. Acad. Sci. USA* **82**, 7739–7743.

BYRNE, J. A., OLDSTONE, M. B. A. (1984), Biology of cloned cytrotoxic T lymphocytes specific for lymphocyte choriomeningitis virus: clearance of virus *in vivo*, *J. Virol.* **51**, 682.

CARTER, B. J., LAUGHLIN, C. A. (1983), Adeno-associated virus defectiveness and the nature of helper function, in: *The Parvoviruses* (BERNS, K. I., Ed.), pp. 67–127. New York: Plenum Publishing Corp.

CELANDER, D., HASELTINE, W. A. (1984), Tissue-specific transcription preference as a determinant of cell tropism and leukaemogenic potential of murine retroviruses, *Nature* **312**, 159–162.

CEPKO, C. (1992), Large-scale preparation and concentration of retrovirus stocks, in: *Current Protocols in Molecular Biology* (AUSUBEL, F. M., BRENT, R., KINGSTON, R. E., MORE, D. P., SEIDMAN, J. G. et al., Eds.), Suppl. 17, pp. 9.12.1–9.12.5. New York: John Wiley & Sons.

CHANG, H.-K., GENDELMAN, R., LISKIEWICZ, GALLO, R. C., ENSOLI, B. (1994), Block of HIV-1 infection by a combination of antisense *tat* RNA and TAR decoys: a strategy for control of HIV-1, *Gene Ther.* **1**, 208–216.

CHATERGEE, S., JOHNSON, P. R., WONG, K. K. (1992), Dual target inhibition of HIV-1 *in vitro* by means of an adeno-associated virus antisense vector, *Science* **258**, 1485–1488.

CHEN, R. J., BANERJEA, A. C., HARRISON, G. G., HAGLUND, K., SCHUBERT, M. (1992), Multi-target ribozyme directed to cleave up to nine highly conserved HIV-1 *env* RNA regions inhibit HIV-1 replication: potential effectiveness against most presenting sequential HIV-1 isolates, *Nucleic Acids Res.* **20**, 4581–4589.

CHEN, S.-Y., BAGLEY, J., MARASCO, W. (1994), Intracellular antibodies as a new class of therapeutic molecules for gene therapy, *Hum. Gene Ther.* **5**, 595–601.

CHOO, V. (1995), Combination chemotherapy superior to zidovudine in delta trial, *Lancet* **346**, 895.

COHEN, E. A., SUBRAMANIAN, R. A. (1994), The accesory proteins of HIV-1, *J. Virol.* **68**, 6831–6835.

CORNETTA, K., NGUYEN, N., MORGAN, R. A., MUENCHAU, D. D., HARTLEY, J. W. et al. (1993), Infection of human cells with murine amphotropic replication-competent retroviruses, *Hum. Gene Ther.* **4**, 579–588.

COUTURE, L. A., MULLEN, C. A., MORGAN, R. (1994), Retroviral vectors containing chimeric promoter/enhancer elements exhibit cell-type-specific gene expression, *Hum. Gene Ther.* **5**, 667–677.

DALGLEISH, A. G., BEVERLEY, P. C., CLAPHAM, P. R., CRAWFORD, D. H., GREAVES, M. F., WEISS, R. A. (1984), The CD4 (T4) antigen is an essential component of the receptor for the AIDS retrovirus, *Nature* **312**, 763–767.

Delta Coordinating Committee (1995), *Preliminary Analysis of Delta Trial of Combination Chemotherapy for the Treatment of HIV Infection,* Medical Research Council, UK.

DONAHUE, R. E., KESSLER, S. W., BODINE, D. et al. (1992), Helper virus induced T cell lymphoma in non-human primates after retroviral mediated gene transfer, *J. Exp. Med.* **176**, 1125–1135.

DROPULIC, B., JEANG, K.-T. (1994), Gene therapy for human immunodeficiency virus infection: genetic antiviral strategies and targets for intervention, *Hum. Gene Ther.* **5**, 927–939.

DROPULIC, B., LIN, N. H., MARTIN, M. A., JEANG, K.-T. (1992), Functional characterisation of a U5 ribozyme: intracellular supression of human immunodeficiency virus type 1 expression, *J. Virol.* **66**, 1432–1441.

DRUMM, M. L. (1990), Correction of the cystic fibrosis defect *in vitro* by retrovirus-mediated gene transfer, *Cell* **62**, 1227–1233.

DUAN, L., BAGASRA, O., LAUGHLIN, M. A., OAKES, J. W., POMERANTZ, R. J. (1995), Potent inhibition of HIV-1 replication by an intracellular anti-*rev* single chain antibody, *Proc. Natl. Acad. Sci. USA* **91**, 5075–5079.

ELGIN, S. C. R. (1991), Chromatin structure and gene expression, *Curr. Opin. Cell Biol.* **2**, 437–445.

ELSENBERG, J. C., ELGIN, S. C. R. (1991), Boundary functions in the control of gene expression, *Trends Genet.* **7**, 335–340.

FAUCI, A. S. (1993), The multifactorial nature of HIV disease: implications for therapy, *Science* **262**, 1011–1018.

FIELDS, B. (1994), AIDS: time to turn to basic science, *Nature* **369**, 95–96.

FRIEDMANN, A. D., TREIZENBERG, S. J., McKNIGHT, S. L. (1988), Expression of a truncated viral transactivator selectively impedes lytic infection by its cognate virus, *Nature* **355**, 452–454.

GALPIN, J. E., CASCIATO, D. A., RICHARDS, S. B. et al. (1994), Clinical protocol: a phase 1 clinical trial to evaluate the safety and biological activity of HIV-IT (TAF) (HIV-1IIIB *env*-transduced, autologous fibroblasts) in asymptomatic HIV-1 infected subjects, *Hum. Gene Ther.* **5**, 997–1017.

GREEN, P. J., PINES, O., INOUYE, M. (1986), The role of antisense RNA in gene regulation, *Ann. Rev. Biochem.* **55**, 569–597.

GREENBERG, P. (1995), Adoptive T cell transfer and the treatment of HIV infection, in: *Advancing a New Agenda for AIDS Research.* London: Ciba Foundation.

GROSSMAN, M., WILSON, J. M. (1993), Retroviruses: delivery vehicle to the liver, *Curr. Opin. Genet. Dev.* **3**, 110–114.

GROSVELD, F., ANTONIOU, M., VAN ASSENDELFT, G. B., DE BOER, E., HURST J. et al. (1994), The regulation of expression of human beta-globin genes, *Prog. Clin. Biol. Res.* **251**, 133–144.

HAMMER, L., ERIKSSON, S., MALIM, K., MOREIN, B. (1989), Concentration and purification of feline leukaemia virus (FeLV) and its outer envelope protein gp70 by aqueous two-phase systems, *J. Virol. Methods* **24**, 91–102.

HASELHORST, D., KAYE, J. F., LEVER, A. M. L. (1994), Stable packaging cell lines and HIV-1 based retroviral vector systems, *Gene Ther.* **1**, S14 (Suppl. 2).

HASELTINE, W. A. (1990), *The Molecular Biology of the AIDS Virus.* New York: Raven Press Ltd.

HASELTINE, W. A. (1991), Molecular biology of human immunodeficiency virus, type 1, *FASEB J.* **5**, 2349–2360.

HATTA, T., KIM, S. G., NAKASHIMA, H., YAMAMOTO, N., SAKAMOTO, K., YOKOYAMA, S., TAKAKU, H. (1993), Mechanism of the inhibition of reverse transcription by unmodified and modified antisense oligonucleotides, *FEBS Lett.* **330**, 161–164.

HERMONAT, P. L., MUZYCAK, N. (1984), Use of adeno-associated virus as a mammalian cloning vector: transduction of neomycin resistance into mammalina tissue culture cells, *Proc. Natl. Acad. Sci. USA* **81**, 6466–6470.

HEYMANN, R. A., BORELLI, E., LESLEY, J. et al. (1989), Thymidine kinase obliteration: creation of transgenic mice with controlled immune deficiency, *Proc. Natl. Acad. Sci. USA* **86**, 2698–2792.

HIGGS, D. R., VICKERS, M. A., WILKIE, A. O. M., PRETORIUS, I. M., JARMAN, A. P., WEATHERALL, D. J. (1989), A review of the molecular genetics of the human α-globin gene cluster, *Blood* **73**, 1081–1104.

HIGGS, D. R., WOOD, W. G., JARMAN, A. P., SHARPE, J., LIDA, J. et al. (1990), A major positive regulatory region located far upstream of the human α-globin gene locus, *Genes Dev.* **4**, 1588–1601.

HO, D. D., NEUMANN, A. U., PERELSON, A. S., CHEN, W., LEONARD, J. M., MARKOWITZ, M. (1995), Rapid turnover of plasma virions and CD4 lymphocytes in HIV-1 infection, *Nature* **373**, 123–126.

JAFFEE, E. M., DRANOFF, G., COHEN, L. K., HAUDA, K. M., CLIFT, S. et al. (1993), High-efficiency gene transfer into primary human tumor explants without selection, *Cancer Res.* **53**, 2221–2226.

JEANG, K. T., CHANG, Y. N., BERKHOUT, B., HAMMARSKOLD, M. L., REKOSH, D. (1991), Regulation of HIV expression. Mechanism of *tat* and *rev*, *AIDS* **5**, s3–s14.

JOHNSON, M. I., HOTH, D. F. (1993), Present status and future prospects for HIV therapies, *Science* **260**, 1286–1293.

KANTOFF, P. W., KOHN, D. B., MITSUYA, H. et al. (1986), Correction of ADA deficiency in cultured human T and B cells by retrovirus mediated gene transfer, *Proc. Natl. Acad. Sci. USA* **73**, 6563–6567.

KARLSSON, S. (1991), Treatment of genetic defects in haematopoietic cell function by gene transfer, *Blood* **78**, 2481–2492.

KAVANAUGH, M. P., MILLER, D. G., ZHANG, W., LAW, W., KOZAK, S. L. et al. (1995), Cell-surface receptors for gibbon ape leukaemia virus and amphotropic murine retrovirus are inducible sodium-dependent phosphate symporters, *Proc. Natl. Acad. Sci. USA* **91**, 7071–7075.

KINCHINGTON, D., GALPIN, S., JAROSZEWSKI, J. W., GHOSH, K., SUBASINGHE, C., COHEN, J. S. (1992), A comparison of *gag, pol* and *rev* antisense oligodeoxynucleotides as inhibitors of HIV-1, *Antivir. Res.* **17**, 53–62.

KORMAN, A. J., FRANTZ, J. D., STROMINGER, J. L., MULLIGAN, R. C. (1987), MuLV-vector, *Proc. Natl. Acad. Sci. USA* **84**, 2150.

KOTANI, H., NEWTON, P. B., ZHANG, S., CHIANG, Y. L., OTTO, E. et al. (1994), Improved methods of retroviral vector transduction and production for gene therapy, *Hum. Gene Ther.* **5**, 19–28.

KOUP, R. A., HO, D. D. (1994), Shutting down HIV, *Nature* **370**, 416.

KRUGER, K., GRABOWSKI, P. J., ZUAG, A. J., SANDS, J., GOTTSHUNG, D. E., CECH, T. R. (1982), Self-splicing RNA: autoexcision and autocyclisation of the ribosomal RNA intervening sequence of *Tetrahyema, Cell* **31**, 147–157.

LANGHOFF, E., TERWILLIGER, E., BOS, H., KALLAND, K., POZNANSKY, M. C. et al. (1991), Replication of HIV-1 in primary dendritic cell cultures, *Proc. Natl. Acad. Sci. USA* **88**, 7988–8002.

LAPIDOT, T., PFLUMIO, F., DOEDENS, M., MURDOCH, B., WILLIAMS, D. E., DICK, J. E. (1992), Cytokine stimulation of multilineage hematopoiesis from immature human cells engrafted in SCID mice, *Science* **255**, 1137–1141.

LEAVITT, M. C., YU, M., YAMADA, O., KRAUS, G., LOONEY, D. et al. (1994), Transfer of an anti-HIV-1 ribozyme gene into primary human lymphocytes, *Hum. Gene Ther.* **5**, 1115–1120.

LEE, T. C., SULLENGER, B. A., GALLARDO, H. F., UNGERS, G. E., GILBOA, E. (1992), Overexpression of RRE derived sequences inhibits HIV-1 replication in CEM cells, *New Biol.* **4**, 66–74.

LEVER, A. M., GOTTLINGER, H., HASELTINE, W., SODROSKI, J. (1989), Identification of a sequence required for efficient packaging of human immunodeficiency virus type 1 RNA into virions, *J. Virol.* **63**, 4085–4087.

LISKIEWICZ, J., SUN, D., SMYTHE, J., LUSSO, P., LORI, F. et al. (1993), Inhibition of HIV-1 replication by regulated expression of a polymeric *tat* activation response RNA decoy as a strategy for AIDS gene therapy, *Proc. Natl. Acad. Sci. USA* **90**, 8000–80004.

LORI, F., LISKIEWICZ, J., SMYTHE, J., CARA, A., BUNNAG, T. et al. (1994), Rapid protection against human immunodeficiency virus type 1 (HIV-1) replication mediated by high efficiency non-retroviral delivery of genes interfering with HIV-1 *tat* and *gag, Gene Ther.* **1**, 27–31.

LOZIER, J. N., BRINKHOUS, K. M. (1994), Gene therapy and the haemophilias, *JAMA* **271**, 47–51.

LU, M., MARUYAMA, M., ZHANG, N., LEVINE, F., FRIEDMANN, T., HO, A. (1995), High efficiency retroviral mediated gene transduction into CD34$^+$ cells purified from peripheral blood of breast cancer patients primed with chemotherapy and GM-CSF, *Hum. Gene Ther.* **8**, 375-381.

MALIM, M. H., BOHNLEIN, S., HAUBER, J., CULLEN, B. R. (1989), Functional dissection of the HIV-1 Rev transactivator. Derivation of a trans-dominant repressor of Rev function, *Cell* **58**, 205–214.

MALIM, M. H., FREIMUTH, J., LIU, J., BOYLE, T. J., LYERLY, H. K., CULLEN, B. R. (1992), Stable expression of transdominant *rev* protein in human T cells inhibits HIV replication, *J. Exp. Med.* **176**, 1197–1201.

MANN, D. L., GARTNER, S., LESANE, F., BUCHOW, H., POPOVIC, M. (1990), HIV-1 transmission and function of virus-infected monocytes and macrophages, *J. Immunol.* **144**, 2152–2158.

MARASCO, W. A., HASELTINE, W. A., CHEN, S.-Y. (1993), Design, intracellular expression and activity of a human anti-human immunodeficiency virus type 1 gp120 single chain antibody, *Proc. Natl. Acad. Sci. USA* **90**, 7889–7893.

McCUNE, S. L., TOWNES, T. M. (1994), Retroviral sequences inhibit human β-globin gene expression in transgenic mice, *Nucleic Acids Res.* **22**, 4477–4481.

McGARRITY, G. J. (1994), Production and downstream processing of retroviral vectors (Abstract EWGT), *Gene Ther.* Suppl. **2**, A25.

MERRILL, J. F., CHEN, I. S. Y. (1991), HIV-1, macrophages, glial cells and cytokines in AIDS nervous system disease, *FASEB J.* **5**, 2391–2397.

MHASHILKAR, A. M., BAGLEY, J., CHEN, S. Y., SZILVAY, A. M., HELLAND, D. G., MARASCO, W. A. (1995), Inhibition of HIV-1 *tat*-mediated LTR transactivation and HIV-1 infection by anti-*tat* single-chain intrabodies, *EMBO J.* **14**, 1542–1551.

MILLER, A. D. (1990), Retrovirus packaging cells, *Hum. Gene Ther.* **1**, 5–14.

MILLER, A. D. (1992), Retroviral vectors, *Curr. Top. Microbiol. Immunol.* **158**, 1–24.

MILLER, A. D., ROSMAN, G. J. (1989), Improved retroviral vectors for gene transfer and expression, *Biotechniques* **7**, 980–990.

MOOLTEN, F. L., WELL, J. M. (1990), Curability of tumors bearing herpes thymidine kinase genes transferred by retroviral vectors, *JNCI* **82**, 297.

MOORE, K. A., SCARPA, M., KOOYER, S., UTTER, A., CASKEY, C. T., BELMONT, J. W. (1994), Evaluation of lymphoid-specific enhancer addition or substitution in a basic retrovirus vector, *Hum. Gene Ther.* **2**, 307–315.

MORITZ, T., KELLER, D. C., WILLIAMS, D. A. (1993), Human cord blood cells as targets for gene transfer: potential use in genetic therapies of severe combined immunodeficiency disease, *J. Exp. Med.* **178**, 529–536.

MULLIGAN, R. (1993), The basic science of gene therapy, *Science* **260**, 926–932.

NABEL, G. J., FOX, B. A., POST, L., THOMPSON, C. B., WOFFENDIN, C. (1994), Clinical protocol: a molecular genetic intervention of AIDS – effect of a transdominant negative form of *rev, Hum. Gene Ther.* **5**, 79–92.

NALDINI, L., BLOMER, U., GALLAY, P., ORY, D., MULLIGAN, R. et al. (1996), *In vivo* gene delivery and stable transduction of nondividing cells by a lentiviral vector, *Science* **272**, 263–267.

NIENHUIS, A. W., McDONAGH, K. T., BODINE, D. M. (1991), Gene transfer into hematopoietic stem cells, *Cancer* (Suppl.) 2700–2704.

OTTO, E., JONES-TROWER, A., VANIN, E. F., STAMBAUGH, K., MUELLER, S. N. et al. (1994), Characterisation of replication-competent retrovirus resulting form recombination of packaging and vector sequences, *Hum. Gene Ther.* **5**, 567–575.

PATTERSON, S., KNIGHT, S. C. (1987), Susceptibility of human peripheral blood dendritic cells to infection by human immunodeficiency virus, *J. Gen. Virol.* **68**, 1177–1181.

PAUL, R., MORRIS, D., HESS, B. W., DUNN, J., OVERELL, R. W. (1993), Increased viral titer through concentration of viral harvests from retroviral packaging lines, *Hum. Gene Ther.* **4**, 609–615.

PHILIP, R., BRUNETT, E., KILINSKI, L. (1994), Efficient and sustained gene expression in primary T lymphocytes and primary and cultured tumor cells mediated by adeno-associated virus plasmid DNA complexed to cationic liposomes, *Mol. Cell. Biol.* **14**, 2411–2418.

POPOVIC, M., SARNGADHARAN, M. G., READ, E., GALLO, R. C. et al. (1984), Detection, isolation, and continuous production of cytopathic retroviruses (HTLV-III) from patients with AIDS and pre-AIDS, *Science* **224**, 497–500.

POSNER, M., HIDESHIMA, T., CANNON, T., MUK-HERJEE, M., MAYER, K. H., BYRN, R. A. (1991), An IgG monoclonal antibody that reacts with HIV-1/gp120, inhibits virus binding to cells, and neutralizes infection, *J. Immunol.* **146**, 4325–4332.

POZNANSKY, M. C., LEVER, A. M., BERGERON, L., HASELTINE, W., SODROSKI, J. (1991), Gene transfer into human lymphocytes by a defective human immunodeficiency type 1 vector, *J. Virol.* **65**, 532–536.

POZANSKY, M. C., FOXALL, R., MHASILKAR, A., COKER, R., JONES, S. et al. (1988), Inhibition of Human immunodeficiency virus replication and growth advantage of CD4$^+$ T cells from HIV-infected individuals that express intracellular antibodies against HIV-1 gp 120 or Tat, *Hum. Gene Ther.* **9**, 487–496.

PRUZINA, S., ANTONIOU, M., HURST, J., GROS-VELD, F., PHILIPSEN, S. (1994), Transcriptional activation by hypersensitive site three of the human beta-globin locus control region in murine erythroleukemia cells, *Biochim. Biophys. Acta* **1219**, 351–360.

RATNER, L., HASELTINE, W., PATARCA, R. et al. (1985), Complete nucleotide sequence of the AIDS virus, HTLV III, *Nature* (London) **313**, 277–283.

REUSSER, P., RIDDELL, S. R., MYERS, J. D. et al. (1994), Cytotoxic T-cell response to cytomegalovirus following human allogeneic bone marrow transplantation: pattern of recovery and correlation with cytomegalovirus infection decrease, *Blood*.

RHODES, A., JAMES, W. (1991), Inhibition of heterologous strains of HIV by antisense RNA, *AIDS* **5**, 145–151.

RICHARDSON, J. H., CHILD, L. A., LEVER, A. M. L. (1993), Packaging of human immunodeficiency virus type 1 RNA requires *cis*-acting sequences outside the 5′ leader region, *J. Virol.* **67**, 3997–4005.

RIDDELL, S. R., GREENBERG, P. D., OVERELL, R. W. et al. (1992), Protocol: phase 1 study of cellular adoptive immunotherapy using genetically modified CD8$^+$ HIV-specific T cells for HIV seropositive patients undergoing allogeneic bone marrow transplant, *Hum. Gene Ther.* **3**, 319–338.

RITNER, K., SZACKIEL, G. (1991), Identification of antisense RNA target regions in human immunodeficiency virus type 1, *Nucleic Acids Res.* **19**, 1421–1426.

ROSENFELD, M. A., YOSHIMURA, K., TRAPNELL, B. C., YONEYAMA, K., ROSENTHAL, E. R. et al. (1992), *In vivo* transfer of the human cystic fi-brosis transmembrane conductance regulator gene to the airway epithelium, *Cell* **68**, 143–155.

ROWLAND-JONES, S., SUTTON, J., ARIYOSHI, K., DONG, T., GOTCH, F. et al. (1995), HIV-specific cytotoxic T-cells in HIV-exposed but uninfected Gambian women, *Nature Medicine* **1**, 59–64.

RUSSELL, D. W., MILLER, A. D., ALEXANDER, I. E. (1994), Adeno-associated virus vectors preferentially transduce cells in S phase, *Proc. Natl. Acad. Sci. USA* **91**, 8915–8919.

RYAN, T. M., BEHRINGER, R. R., MARTIN, N. C., TOWNES, T. M., PALMITERI, R. D., BRINSTER, R. L. (1989), A single erythroid specific super-hypersensitive site activates high levels of human β-globin gene expression in transgenic mice, *Gene Dev.* **3**, 314–323.

SAMULSKI, R. J., BERNS, K. I., TAM, M., MUZYCZ-KA, N. (1982), Cloning of adeno associated virus into pBR322: rescue of intact virus from the recombinant virus in human cells, *Proc. Natl. Acad. Sci. USA* **79**, 2077–2081.

SARVER, N., CANTIN, E. M., CHANG, P. S., ZAIA, J. A., LADNE, P. A. et al. (1990), Ribozymes as potential anti-HIV therapeutic agents, *Science* **247**, 1222–1225.

SHIMADA, T. H., FUJII, H., MITSUYA, H., NIEN-HUIS, A. W. (1991), Targeted and highly efficient gene transfer into CD4$^+$ cells by a recombinat human immunodeficiency virus retroviral vector, *J. Clin. Invest.* **88**, 1043–1047.

SHNITTMAN, S. M., PSALLADOPILOS, M. C., LANE, H. C., THOMPSON, L., BASELER, M. et al. (1989), The reservoir for HIV-1 in human peripheral blood is a T cell that maintains expression of CD4, *Science* **245**, 305–308.

SIPPEL, A. E., SAUERASSIG, FAUST, N., SHAFER, G., HUBER, M., BONIFER, C. (1994), The mechanism of macrophage specific transgene expression, *Gene Ther.* Suppl. **2**, A5.

SMITH, C., LEE, S. W., SULLENGER, B., GALLAER-DO, H., UNGERS, G., GILBOA, E. (1993), Intracellular immunisation against HIV using RNA decoys, in: *Third International Symposium on Catalytic RNAs and targetted Gene Therapy for the Treatment of HIV-Infection.* London, UK.

STRAUSS, M. (1994), Liver directed gene therapy: prospects and problems, *Gene Ther.* **1**, 156–164.

SULLENGER, B. A., CECH, T. R. (1993), Tethering ribozymes to a retroviral packaging for the destruction of viral RNA, *Science* **262**, 1566–1569.

SULLENGER, B. A., GALLARDO, H. F., UNGERS, G. E., GILBOA, E. (1993), Overexpression of TAR sequences renders cells resistant to HIV replication, *Cell* **63**, 601–608.

TAYLOR, N. R., ROSSI, J. J. (1991), Ribozyme mediated cleavage of an HIV-1 *gag* RNA: the effects of non-targeted sequences and secondary structure on ribozyme cleavage activity, *Antisense Res. Dev.* **1**, 174–186.

TRATSCHIN, J. D., MILLER, I. L., SMITH, M. G., CARTER, B. J. (1985), An adeno-associated virus vector for high frequency integration and rescue of genes in mammalian cells, *Mol. Cell Biol.* **5**, 3251–3260.

TSUBOTA, H., LOND, C. I., WATKINS, D. I., MORIMOTO, C., LETVIN, N. (1989), A cytotoxic lymphocyte inhibits acquired immune deficiency syndrome virus replication in peripheral blood lymphocytes, *J. Exp. Med.* **169**, 1421–1434.

VICKERS, T., BAKER, B. F., COOK, P. D., ZOUNES, M., BUCKHEIT, R. W. et al. (1991), Inhibition of HIV-LTR gene expression by oligonucleotides targeted to the TAR element, *Nucleic Acids Res.* **19**, 3359–3368.

VILE, R., HART, I. R. (1994), Targeting of cytokine gene expression to malignant melanoma cells using tissue specific promoter sequences, *Ann. Oncol.* **5** (Suppl. 4), 59–65.

VILE, R., RUSSELL, S. J. (1994), Gene transfer technologies for the gene therapy of cancer, *Gene Ther.* **1**, 88–98.

WALKER, B. D., CHAKRABARTI, S., et al. (1987), HIV-specific cytotoxic T-lymphocytes in seropositive individuals, *Nature* **328**, 328–345.

WALKER, B. D., PLATA, F. (1990), HIV-1/gp120, inhibits virus binding to cells, and neutralizes infection, *J. Immunol.* **146**, 4325–4332.

WALSH, C. E., JOHNSON, M. L., XIAO, X., YOUNG, N. S., NIENHUIS, A. W., SAMULSKI, R. (1992), Regulated high-level expression of a human γ-globin gene introduced into erythroid cells by adeno-associated virus vector, *Proc. Natl. Acad. Sci. USA* **89**, 7257–7261.

WANG, S., DOLNICK, B. J. (1993), Quantitative evaluation of intracellular sense: antisense RNA hybrid duplexes, *Nucleic Acids Res.* **21**, 4383–4391.

WANG, B., UGEN, D., SRIKANTAN, V. et al. (1993), Gene inoculation generates immune responses against HIV-1, *Proc. Natl. Acad. Sci. USA* **90**, 4156–4160.

WARNER, J. F., ANDERSON, C. G., LAUBE, L. et al. (1991), Induction of HIV specific CTL and antibody responses in mice using retroviral vector transduced cells, *AIDS Res. Hum. Retrov.* **7**, 645–655.

WEATHERALL, D. (1992), A brief history of gene therapy, *Nature Genetics* **2**, 93–98.

WEATHERALL, D. (1994), Heroic gene surgery, *Nature Genetics* **6**, 325.

WEERASINGHE, M., LIEM, S. E., ASAD, S., READ, S. E., JOSHI, S. (1991), Resistance to human immunodeficiency virus type 1 (HIV-1) infection in human CD4⁺ lymphocyte derived cell lines conferred by using retroviral vectors expressing an HIV-1 RNA specific ribozyme, *J. Virol.* **65**, 5531–5534.

WEINBERG, J. B., MATTHEWS, T. J., CULLEN, B. R., MALIM, M. H. (1991), Productive HIV-1 infection of non-proliferating human monocytes, *J. Exp. Med.* **174**, 1477–1482.

WINTER, G., MILSTEIN, C. (1991), Man-made antibodies, *Nature* **349**, 293–299.

WOFFENDIN, C., RANGA, U., YANG, Z., VERMA, S. R., XU, L., NABEL, G. J. (1995), Molecular genetic interventions for AIDS, in: *Advancing a New Agenda for AIDS Research*. London: Ciba Foundation.

WOLFF, J. A., LEDERBERG, J. A. (1994), An early history of gene transfer and therapy, *Hum. Gene Ther.* **5**, 469–480.

WONG-STAAL, F. (1991), Human immunodeficiency virus and its replication, in: *Fundamental Virology*. (FIELD, B. N., KNIPE, D. M., Eds.), pp. 709–723. New York: Raven Press.

WONG-STAAL, F., GALLO, R. C. (1985), Human T-lymphotropic retroviruses, *Nature* **317**, 395–403.

WONG-STAAL, F., HASELTINE, W. A. (1992), Regulatory genes of human immunodeficiency viruses, *Mol. Genet. Med.* **2**, 189–219.

YAMADA, O., YU, M., YEE, J.-K., KRAUS, G., LOONEY, D., WONG-STAAL, F. (1994), Intracellular immunization of human T cells with a hairpin ribozyme against human immunodeficiency virus type 1, *Gene Ther.* **1**, 38–45.

YANG, O. D., TRAN, A. C., KALAMS, S. A., JOHNSON, R. P., ROBERTS, M. R., WALKER, B. D. (1997), Lysis of HIV-I infected cells and inhibition of viral replication by universal receptor T cells, *Proc. Natl. Acad. Sci. USA* **94**, 11478–11483.

YU, M., OJWANG, J., YAMADA, O., HAMPEL, A., RAPAPORT, J. et al. (1993), A hairpin ribozyme inhibits expression of diverse strains of human immunodeficiency virus type 1, *Proc. Natl. Acad. Sci. USA* **90**, 6340–6344.

YU, M., PEOSCHLA, E., WONG-STAAL, F. (1994), Progress towards gene therapy for HIV infection, *Gene Ther.* **1**, 13–26.

ZOLLA-PAZNER, S., GORNY, M. K. (1992), Passive immnunization for the prevention and treatment of HIV infection, *AIDS* **6**, 1235–1247.

Regulatory and Economic Aspects

22 Regulation of Antibodies and Recombinant Proteins

JULIE FOULKES

Burnham, UK

GILLIAN TRAYNOR

Cambridge, UK

1 Introduction   497
2 Registration of Biotechnology Products in the USA   498
  2.1 Introduction and History   498
  2.2 The Review Process   501
    2.2.1 Priority Review   501
    2.2.2 Advisory Committees   501
  2.3 The Well-Characterized Product   502
  2.4 *Food and Drug Administration Modernization Act 1997*   503
  2.5 Investigational New Drugs (INDs)   504
    2.5.1 Treatment INDs   504
    2.5.2 Compassionate Use/Investigator INDs   505
  2.6 Orphan Drug Legislation   505
3 Registration of Biotechnology Products in Europe   506
  3.1 History of Regulatory Control of High Technology, Biotechnology Products   506
  3.2 The European Regulatory Environment and the Role of the EMEA   506
  3.3 The Centralized Procedure for the Registration of Biotechnology-Derived Products   509
    3.3.1 Phase I: Presubmission   509
    3.3.2 Phase II: Validation of the Application   510
    3.3.3 Phase III: Prereview Period   510
    3.3.4 Phase IV: Review of Application   510
    3.3.5 Phase V: Post-Opinion Phase   510
    3.3.6 Phase VI: Decision-Making Phase   510
    3.3.7 Phase VII: Post-Authorization Phase   510
    3.3.8 Responsibilities of the MA Holder   512

3.4 Experience with the Centralized Procedure   512
    3.4.1 Referrals and Public Health   512
    3.4.2 Crisis Management   513
    3.4.3 Trademarks   513
    3.4.4 Comarketing or Copromotion   513
3.5 Orphan Drug Legislation   514
4 Conclusions and the Way Forward for the Regulation of Biotechnology-Derived Products   514
5 References   515

List of Abbreviations

AR	assessment report
BLA	biologics licence application
BRMAC	Biological Response Modifier Advisory Committee
CBER	Center for Biologics Evaluation and Research
CDER	Center for Drug Evaluation and Research
CDHR	Center for Devices and Radiological Health
CDRH	Center of Devices and Radiological Control
CPMP	Committee for Propietary Medicinal Products
CTD	Common Technical Document
CVM	Center for Veterinary Medicine
CVMP	Committee for Veterinary Medicinal Products
ELA	establishment licence
EMEA	European Medicines Evaluation Agency
EPAR	European Public Assessment Report
EU	European Union
FDA	Food and Drug Administration
FDAMA	Food and Drug Administration Modernization Act
GCP	good clinical practice
GLP	good laboratory practice
GMP	good manufacturing practice
HibTITER	*Haemophilus influenzae* B conjugate and diphteria vaccine used for the immunization of infants and children
ICH	International Conference on Harmonization
IND	investigational new drug
MA	marketing authorization
MAA	marketing authorization applications
MHW	Ministry of Health and Welfare (Japan)
NCE	new chemical entity
NDA	new drugs application
NIH	National Institutes of Health
PDUFA	Prescription Drug User Fee Act
PLA	product licence
PTC	points to consider
QP	qualified person
SmPC	summary of product characteristics
SRM	specified risk materials
TSA	Therapeutic Substances Act
WCP	well-characterized product

1 Introduction

In the last 10–20 years, the pharmaceutical industry has developed and in some cases now registered, a significant number of biotechnology-derived products. These products are usually large proteins (>100 amino acids) manufactured by fermentation and chromatography-based purification techniques to provide relatively small quantities of highly potent finished products and contrast with the more traditional lower molecular weight synthetically produced chemical products which are produced in much larger quantities.

The early development of biologicals in the early 1900s consisted of crude preparations of vaccines and antibiotics such as the smallpox vaccine, which were derived by immunizing animals with an immunogenic challenge to elicit an immune response. These crude animal-derived sera were administered to humans prophylactically early in the development of the disease.

Advances in the understanding of the human immune system led to the development of inactivated forms of pathogenic organisms or toxins, known as bacterial vaccines. By the 1940s, viral vaccines were being produced using animal tissue or egg cultures rather than whole animals. Killed viral vaccines and attenuated live viral vaccines were developed and manufactured using these techniques. In comparison to the development process for conventional synthetic products, using either synthetic chemical reactions or extraction/purification from plants or other sources, where control procedures concentrated on end product testing rather than in-process control, the products derived from these new manufacturing processes required the development of appropriate characterization methods and control procedures.

One obvious difference between the two classes of products was that the purity of synthetic products could be determined but that of the biological products could not. The activity of the biological product resided in the generation of an immune response and the potent active substance could not be isolated and, therefore, its purity could not be determined. The manufacturers and regulators soon realized that the control of these products had to be process-based, as did the methods to assess the safety and potency of the products. Thus unlike the synthetic products where end product testing was the primary control, the process-based control of biological products became key to their regulation.

It was obvious from the development of biological products and later the biotechnology-derived products, that the regulatory controls had to be different to those used for synthetic products. Equally, the development of each biological product had to be considered individually, dependent on the particular characteristics of the product and its disease target. Consequently, guidelines, rather than rules have evolved to provide a framework for development. Within this framework the appropriate data package for each product is created from a good scientific rationale based on knowledge of the product. This case by case approach to development is in contrast to the more formalized requirements for the traditional new chemical entities (NCE). The case by case approach is applied in both Europe and the USA with the advantage that data requirements can be similarly acceptable in both jurisdictions. The disadvantage of this approach is that, without clear rules, the risk of satisfying neither EU nor US requirements is also present. However, a comparative review of EU guidelines and US points to consider indicates a level of agreement between the regulatory authorities.

A discussion of specific data issues may be found in earlier chapter of this volume. This chapter outlines the development of the regulatory environment in the USA and Europe and explains the regulatory processes currently in use together with the latest initiatives from these regulatory authorities. The concluding section highlights the current global harmonization for the regulation of all phar-

maceuticals by the International Conference on Harmonization, where biotechnology-derived products still have their own specific guidelines.

2 Registration of Biotechnology Products in the USA

2.1 Introduction and History

In St. Louis in 1902 children were vaccinated against diphtheria using a horse serum antitoxin contaminated with *C. tetani*. The children died and as a result the US Congress passed the *Federal Virus, Serum and Toxin Act* of 1902, to ensure that any biological drug was safe and effective. The *Pure Food and Drugs Act* of 1906 and the *Federal Food and Cosmetic Act* of 1936 which provided legislation for classical drugs, did not refer to this law but imposed separate legislation. In 1944, however, the 1902 Act was incorporated into the *Public Health Services Act*. This act required that a manufacturer of a biological product obtained both an establishment licence (ELA) and a product licence (PLA) for any product. This procedure remained until 1996.

According to US legislation the definition of a biological product is, "… any virus, therapeutic serum, toxin, antitoxin, vaccine, blood, blood component, or derivative, allergenic product or analogous product …". This contrasts with the definition in Europe of, "… any product which cannot be assayed by chemical or physical means …". Thus in Europe the term is broadly applied to any product which requires a biological assay and in the US it is restricted to certain categories of product. In both territories the definition has been applied to all monoclonal antibodies and gene therapy products but interestingly in the US does not apply to hormones which are historically defined as drugs.

In 1937, the US Government established the Division of Biologics Control at the National Institutes of Health (NIH) to ensure the safety of vaccines and the nations blood supply and in 1972 this became the Bureau of Biologics within the Food and Drug Administration, although it remained on the NIH campus. In 1982 it merged formally within the FDA's drug review bureau to form the Center for Drugs and Biologics which, in 1987, was divided into two centers specializing in drugs and biologics. The centers were named the Center for Drugs Evaluation and Research (CDER) and the Center for Biologics Evaluation and Research (CBER).

The FDA has introduced a guideline to explain which center has jurisdiction for the assessment of a product (Guideline for Center Jurisdiction). The factors that determine which center will assess a biological include nature of the end product, the manufacturing process, the intended use of the product, the characteristics of the product and the most appropriate resources for the assessment of the product. In order to apply for center jurisdiction a company is advised to write to FDA and provide details of the product, its use and characterization and to give the company's preference for the center for assessment. In some cases the assessment is not given to the

center of the company's choice. In practice, particularly with new and complex biotechnology products such as antibody toxins, antibody radiolabeled products or device/biotechnology products any center may invite experts from the other center or from any other part of the FDA, [e.g., the Center of Devices and Radiological Control (CDRH)], for advice on an application.

The FDA has not set up a separate review process in the same way that the Europeans have set up the centralized system for biotechnology and high technology products. In 1984, a working group in biotechnology was set up to coordinate decision making on products derived from biotechnology. The group concluded that unless specifically necessary they would not produce separate regulations for biotechnology but that they would issue Points To Consider (PTC) documents where appropriate.

In 1992 Dr. KATHRYN ZOON was appointed as head of CBER and at this time the center was reorganized to anticipate the expected increase in submissions for biotechnology products. The structures of the FDA and CBER are provided in Figs. 1 and 2. Shortly

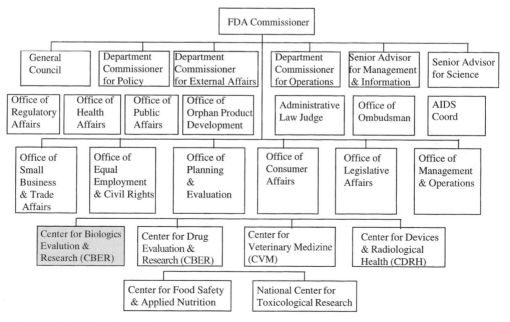

Fig. 1. Structure of the FDA.

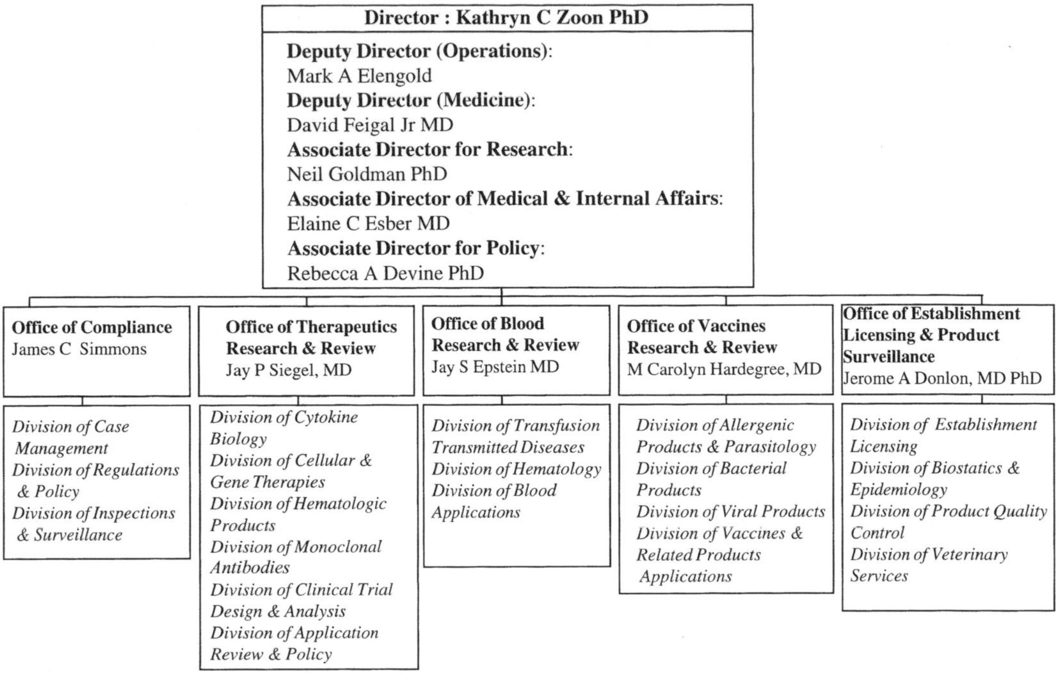

Director : Kathryn C Zoon PhD
Deputy Director (Operations): Mark A Elengold **Deputy Director (Medicine)**: David Feigal Jr MD **Associate Director for Research**: Neil Goldman PhD **Associate Director of Medical & Internal Affairs**: Elaine C Esber MD **Associate Director for Policy**: Rebecca A Devine PhD

Office of Compliance James C Simmons	Office of Therapeutics Research & Review Jay P Siegel, MD	Office of Blood Research & Review Jay S Epstein MD	Office of Vaccines Research & Review M Carolyn Hardegree, MD	Office of Establishment Licensing & Product Surveillance Jerome A Donlon, MD PhD
Division of Case Management *Division of Regulations & Policy* *Division of Inspections & Surveillance*	*Division of Cytokine Biology* *Division of Cellular & Gene Therapies* *Division of Hematologic Products* *Division of Monoclonal Antibodies* *Division of Clinical Trial Design & Analysis* *Division of Application Review & Policy*	*Division of Transfusion Transmitted Diseases* *Division of Hematology* *Division of Blood Applications*	*Division of Allergenic Products & Parasitology* *Division of Bacterial Products* *Division of Viral Products* *Division of Vaccines & Related Products Applications*	*Division of Establishment Licensing* *Division of Biostatics & Epidemiology* *Division of Product Quality Control* *Division of Veterinary Services*

Fig. 2. Center for Biologics Evaluation and Research.

after, in 1993, the *Prescription Drug User Fee Act* (PDUFA) was passed which resulted in a period of some reform in all centers in the FDA. The system was initiated to prevent companies from "queue jumping," by filing an application with interim clinical data and subsequently submitting a supplement to the application once final data were available. This had historically caused problems at the FDA since as soon as the review was complete another set of data would arrive which would mean that the review would need to be restarted. Typically, the time to an approval or approvable letter was 18 months to 2 years. The act investigated a new process for filing and payment for review at FDA, initially for a fixed period of time (October 1992–September 1997) and has now been extended by the Food and Drug Administration Modernization Act of 1997 (FDAMA) passed in November 1997. This is discussed in detail later.

The process for filing is as follows: the application is submitted to FDA with 50% of the fee as determined by FDA. The FDA car-

ries out a preliminary check of the file to ensure that it is complete. If the data are incomplete, e.g., there are missing final study reports, then the FDA returns the application, without the fee, with a "refusal to file" letter, i.e., the FDA would refuse the dossier and not allow the filing to proceed.

In order to justify this FDA has been obliged to shorten approval times and by charging a fee at submission it was able to commit to a standard assessment time of 12 months. The user fees have allowed FDA to employ more staff in all centers, to increase salaries to attract good staff and to upgrade the computer systems. Although the user fees contribute a significant income to FDA they do not allow FDA to be entirely self funding and FDA receives a large budget from the Federal Government.

Further, the FDA and White House announced in November 1995 a new venture called the "ReGo Initiative". "ReGo" is an abbreviation for reinventing government and was instigated after the success of the PDUFA and as a result of lobbying by industry for more reforms at FDA. Out of this has come the most recent and important development for the evaluation of biotechnology products in the USA, the introduction in 1995 of a new regulatory concept of the Well Characterized Product. This has led to a complete revolution in CBERs approach to biotechnology drugs, which can be defined as well characterized, so that certain regulatory requirements laid down by CBER were relaxed. The details on the well characterized product (WCP) and the consequences of this reform are described in Sect. 2.3.

2.2 The Review Process

The overall process for assessment of a biologic at CBER is essentially the same as that for a drug at CDER and begins when a product is still in development and the subject of an IND (investigational new drug) application (see Sect. 2.4).

In 1994 CBER established a procedure for a managed review process in response to the user fees (PDUFA) legislation. CBER specifically appointed teams in the organization responsible for tracking and coordination of particular products while ensuring sufficient team discussion was generated to remove any bias.

Under the managed review process, sponsors are asked to contact CBER prior to a pivotal trial to arrange a meeting. At this point the IND team will discuss in some detail with the sponsor the design of the study, the endpoints and objectives, statistical analysis, proposed indication, the product's use in medical practice in comparison to established practice (i.e., is there a clinical benefit for this product) and will consider any ongoing studies. They will also discuss any manufacturing and toxicology issues and plans for electronic submissions.

Following this meeting the sponsors have a good indication that they have a good scientific rationale for their program, and if they follow the FDAs comments and meet their objectives, then it is likely that this product will be "approvable". This can provide a level of confidence for the development of biotechnology products which is important since each

product is reviewed on a case-by-case basis and precedents to guide development may not be available.

The IND team follow the product and work with the sponsor until the time of a pre-biologics licence application meeting or pre-BLA meeting. At this point all communication is transferred to the BLA team. At this meeting the sponsor would be expected to provide a detailed account of the key pivotal studies and the safety and efficacy data. The FDA would discuss the scheduling of any inspections necessary and identify the filing date with the sponsor. Further, the FDA would arrange tentative review schedules and decide if an advisory committee meeting was necessary and obtain potential target dates. Subsequent to filing of the BLA, the team would follow the application to ensure that review times were in accordance with PDUFA, ensure compliance with PDUFA and arrange a "refusal to file" meeting (if appropriate) within 45 d of the submission, issue action letters, review responses and follow the application to approval or refusal.

This process allows the application to be streamlined. It ensures a broad team review inside FDA and PDUFA compliance.

2.2.1 Priority Review

The FDA has a general policy for so-called priority review where products providing sufficient therapeutic gain are given faster review than the 12 months scheduled for a standard product. This may be as short as 3 months. Products defined as having therapeutic gain are those with greater safety or efficacy, those which offer therapy for previously inadequately controlled diseases or drugs for AIDS treatment or other serious life threatening diseases.

2.2.2 Advisory Committees

During the assessment process the FDA uses various committees called advisory committees. Advisory committee opinions are not binding on the part of the FDA and are strictly "advisory" in nature. These committees are often open to the public, at least in part, or can be closed to include only the representatives of the applicant/company. An open advisory committee meeting is a valuable source of information for competitors and patients, on the process that has been followed to reach the FDA decision to approve or not approve a new product. Strong lobbying from patient groups at open advisory committees often plays a significant role in the drug approval decision. Advisory committees exist for all relevant therapeutic areas and in addition a number of advisory committees has been set up with a more scientific remit, such as the Biological Response Modifier Advisory Committee (BRMAC), the Vaccines and Related Biological Products Advisory Committee and the Allergenic Products Advisory Committee. These committees consist of external experts who offer advice and consider the benefits and risks of a new therapeutic prior to approval. They are often involved in review of novel biotechnology products including monoclonal antibodies or gene therapy products and in many instances points to consider documents are released after an advisory committee has met and considered a novel or unusual biotechnology product. They are also used to resolve disputes that arise within FDA, to generate increased acceptance of FDA decisions, to give input into clinical trial design and safety issues during clinical trials; and to accelerate the drug review process. In this context they can be involved in the IND process to identify drugs that would qualify for expedited review such as those with an important therapeutic advance, novel delivery, or those with a potential safety hazard.

2.3 The Well-Characterized Product

In 1995 the FDA as part of the "ReGo" initiative described in Sect. 2.1, introduced an additional reforming initiative, the concept of a well characterized product for biologicals and the need for less rigid control of such defined products.

The document published in the federal regulations (60 FR 63048, 1995) (FDA, 1995b) defined as well characterized product an entity whose identity, purity, impurities potency and quantity can be determined and controlled. The FDA has listed a number of recombinant proteins as well-characterized products and has put the burden of proof on industry to provide the data to enable other products to be defined in the same way. The granting of the WCP status is conferred by the FDA as a result of data from manufacturers and is based on a detailed knowledge and understanding of the manufacturing process and validation procedures for these types of products; the understanding of the characterization techniques which allow recombinant proteins to be fully characterized to the level of conventional molecules (e.g., NCEs) and the safety of the process and resultant product.

For biological products given WCP status the need for certain testing and data is reduced or removed. Specifically, the need for an establishment licence (ELA as discussed in Sect. 2.1) for the manufacture of the product is removed and batch release by CBER is no longer required. It is also easier to make changes to the manufacturing process and scale of production since the requirements for testing on scale-up and for process changes are reduced.

The following criteria have been agreed by industry and FDA as appropriate to confer WCP status. There must be a robust, reproducible and validated manufacturing process; sufficient process validation to demonstrate a safe product manufactured to a consistent standard (e.g., show removal of impurities, adventitious agents with validated analytical methods); the product must be safe from a toxicological perspective and there must be appropriate testing and characterization work to confirm identity, purity and potency of the product.

This has brought the regulation of WCPs more in line with convential NCEs and removed considerable regulatory burdens from the manufacturers.

2.4 *Food and Drug Administration Modernization Act 1997*

The most significant reform of the FDA since 1984 Waxmann Hatch was passed in November 1997. This incorporated a number of substantial FDA reforms and included the awaited extension of the *Prescription Drug User Fees Act* which had expired on the 30th September 1997 and been temporarily extended as previously described.

The user fees had raised up to US $100 million a year from industry and had substantially augmented the FDA budget enabling better approval times and more staff. Figures published by FDA show that the review times for NDAs and BLAs had fallen over the period of 1993–1996 (Tab. 1).

The major provisions of the Act were as follows:

- continuing PDUFA for an additional 5 years,
- as sought Dept of Health and Human Services the marketing exclusivity for new pediatric indications was extended for 6 months,
- formalization of the mechanism for fast tracking approvals of drugs for serious and life threatening diseases is already in place and previously described,
- allowing manufacturers to distribute peer reviewed journal articles to support off-label used of drugs, provided they have pre-

viously agreed with FDA to file supplemental NDAs for the unapproved uses,
- restricting FDA authority in regulating economic claims submitted to formulary committees or managed care organizations,
- authorizing the FDA to approve an NDA on the basis of one well controlled clinical investigation (this was previously unofficially possible but the Act has formalized this),
- granting antibiotics market exclusivity previously limited to new drugs under the Waxmann Hatch legislation,
- requiring that manufacturers of life saving drugs notify FDA at least 6 months prior to discontinuation,
- allowing FDA to contract out for expert review of biological applications.

In addition the Act has given a legal formalization to a number of regulatory initiatives already taken by FDA such as the well-characterized product philosophy previously described and legislation formalizing labeling of OTC and prescription drugs.

The consequence of the Act will be that FDA will have to pass a number of regulations, guidance documents, notices some of which are now emerging such as guidance notes on labeling.

Although at the time of this review a new Commissioner for the FDA had not been appointed to replace Dr. KESSLER and the FDA have a challenging future.

Tab. 1. Summary Statistics of Original NDAs and BLAs 1993–1996[a]

	Average total times [months]			
	1993	1994	1995	1996
Priority NDAs	13.9	12.9	11.6	6.7[b]
Standard NDAs	23.8	22.2	16.9	13.6[b]
Priority PLAs	15.4	12.2[c]	15.1	9.2
Standard PLAs	30.7	20.1	18.6	[d]

[a] Anonymous (1997), Parexel International Corporation (1997).
[b] At time of collation of figures 34 NDAs were still under review.
[c] Only one application.
[d] No applications approved.

2.5 Investigational New Drugs (INDs)

In order to use any drug prior to marketing approval, whether it be a conventional pharmaceutical, biotechnology product or gene–therapy product, the applicant often referred to as the sponsor must apply to the relevant center at FDA for permission to conduct a clinical study by submitting an investigational new drug application or IND. CBER controls IND approvals for most biotechnology products and in general adopts a flexible case-by-case approach to consideration of the data and requirements at each stage.

The actual data requirements are outlined in the *Code of Federal Regulations 21 CFR 312* and associated *Guidelines and Points to Consider* and can be discussed with center staff, but the procedures described are standard.

Prior to submitting the IND the sponsor is advised to meet with the FDA, particularly if the product is derived from biotechnology, a monoclonal antibody or a gene therapy product where a flexible approach is required. At this time a pre-IND meeting is held and the sponsor is expected to present a summary of the key manufacturing and preclinical data and any issues he, as sponsor wishes to discuss with FDA. In addition the sponsor is expected to present any clinical data already generated (e.g., from a country outside the USA) and the future clinical plans. At this point the FDA offers its advice on any additional data that the sponsor should consider for the IND application and its suggestions for the design of the clinical study at phase I, II and perhaps for the ultimate submission of the new drugs application (NDA) or biologics licence application if appropriate. At this meeting, and subsequently, if the company wishes via telephone, the FDA will play an active role in the development of the clinical program giving its input and advice as how to conduct the studies. This approach has been both applauded and criticized from the industry. The advantage is that if a sponsor wishes, he can have complete FDA input into the design of clinical studies. The disadvantage is that the reviewers may be different at the

BLA stage and have a different opinion to the IND reviewers.

The approach to FDA advice represents an important strategic consideration for any company starting studies in the USA. On the one hand a sponsor should listen to the FDA and ensure that his development program meets with FDA requirements; on the other hand it should be a program that the sponsor is entirely satisfied with and the sponsor should be prepared to present his own philosophy for the product and to negotiate with the FDA to achieve the best development strategy. This allows a sponsor to clearly identify exactly what the FDA considers as important, and if a meeting with FDA is sought early, can allow the preclinical development to be focused and the clinical program to be initiated as early as possible.

The FDA philosophy for pre-IND meetings and involvement in the design of studies is in contrast to that in Europe where the regulatory authorities of individual member states deal with clinical trial applications and in the main prefer applicants to file applications for clinical trials without a prior meeting to discuss the data and plans. In these circumstances the company is expected to rely on their own interpretation and consideration of published guidelines.

The EMEA in Europe will, however, consider requests for scientific advice for issues not covered by guidelines. This facility is useful for products derived from biotechnology, mAbs and gene therapy products and is discussed in Sect. 3.3.

2.5.1 Treatment INDs

The purpose of a treatment IND is to make a particular drug available to desperately ill patients during the development phase and prior to NDA approval, by laying down criteria defining the selection of the treatment with the new drug. The drug must be for a life threatening condition such as AIDS, Alzheimer's disease or multiple sclerosis and must be under active clinical investigation under a normal IND in adequately controlled studies or where studies are complete and the approval is pending. FDA will not approve a treatment IND unless the sponsor is actively pursuing development of the drug and the drug shows some promise or therapeutic benefit. The FDA reviews both the original IND and the treatment IND application to ensure that adequate enrollment is ongoing in the clinical studies and there is sufficient data to evaluate the product as promising. Usually the FDA would expect an early consultation with the sponsor, a risk benefit analysis (i.e., a 'mini' integrated summary of benefits and risks) and agreement on timescales for future studies and BLA submission.

Prior to making the decision to move ahead with a treatment IND, companies must consider a number of ethical and scientific issues and the advantages and disadvantages of following this course. For example the availability of a drug under treatment IND can jeopardize the enrollment of ongoing trials, particularly placebo controlled studies. However, it does allow sponsors to recover their costs for such products by charging for the drug and this may be an important consideration for small companies developing biotechnology products. However, the sponsor must show that the sale of the investigational drug is necessary to continue clinical research and the cost must not be at full commercial price.

2.5.2 Compassionate Use/Investigator INDs

There are no regulations which officially permit compassionate use IND on the US statute. It is however possible for physicians to request their own IND. This is only available if an IND is not in place and the individual physician takes responsibility for the use of the drug. The physician must contact FDA directly and request an IND. The product is prohibited from commercialization.

2.6 Orphan Drug Legislation

In the USA it is possible to recoup a portion of a company's US development cost if so-called orphan drug status can be granted. In Europe the orphan drug legislation is not currently in place (although a council regulation is in draft).

Orphan drug status is limited to rare diseases such as Huntington's chorea, amyotrophic lateral sclerosis, Tourette's syndrome etc. The initial act was signed in 1983 but the final rule incorporating the legislation into the *Code of Federal Regulations* was not passed until 1992 (CFR 21 part 316). A product can be defined as an orphan drug if it affects less than 200,000 patients in the USA or if it affects more than 200,000 but there is no reasonable expectation of recouping development costs. It applies to drugs, biologicals and antibiotics but not foods, devices or veterinary products.

Orphan drug status provides to sponsors: assistance from FDA in protocol development; 7 years protection from a second competitive BLA or NDA approval in the USA, i.e., market exclusivity; up to 50% of US clinical development costs credited against tax and grants (usually to academic institutions) for development of a product and a 20% tax credit for increased research expenditures.

An application for orphan drug "designation" can be made at any time prior to the NDA or BLA filing but the official status is only granted on approval of the BLA/PLA. Orphan drug status is only given to one company for any particular product containing the same active moiety for a particular use. The definition of an active moiety is detailed and has been written to try to define proteins and products of biotechnology. The interpretation of the *Orphan Drug Act* by the FDA results in an apparent difference in the way NCEs and biotechnology products are considered. A minor change in structure of a chemical entity is enough to consider it as a new, different product and eligible for orphan drug status under the Act. However, even large differences in macromolecules may not be deemed sufficient to distinguish them as separate products and thus different protein molecules with the same therapeutic target may be considered the same under the Act and not equally eligible for orphan drug status. Definition of difference between two protein molecules may currently depend on demonstrating clinical superiority either in terms of efficacy or safety.

Thus while this could clearly be valuable legislation for small biotechnology companies developing high cost high technology products for rare diseases, the problem of definition of biotechnology products under the Act may be a barrier.

3 Registration of Biotechnology Products in Europe

3.1 History of Regulatory Control of High Technology, Biotechnology Products

Each of the individual Member States which comprise the European Union (EU) have their own national regulatory requirements, with their own history of development. In the UK, e.g., biologicals were controlled by the *Therapeutic Substances Acts* (TSA) of 1925 and 1956 in which they were defined as "substances whose purity or potency cannot be tested by chemical or physical means". Until the introduction of the *Medicines Act* 1968 into which these acts were incorporated, biologicals were the only controlled medicines in the UK. Post 1973, the UK legislation covering medicinal products was supplemented by EC directives which sought to regulate products of biotechnology differently to conventional chemical entities.

The concertation procedure (forerunner of the centralized procedure for biotechnology and high technology products) (Council Directive 87/22/EEC) was introduced in the EU in July 1987 to promote the development of innovative products (EEC, 1986). The concertation procedure was introduced (in addition to the multi-state and national submission routes), to minimize the delays which novel products may have suffered due to insufficient expertise within a single national authority, thereby maximizing the time to recoup sales to cover the high development costs. Opinions on products were made by the Committee for Proprietary Medicinal Products, the CPMP, which consisted of representatives of all Member States. At that time it was also possible to submit a biotechnology application nationally to one country only, but subsequent submissions to other Member States were precluded for a minimum period of 5 years effectively making the concertation procedure mandatory for this group of products.

In addition to Council Directive 87/22/EEC (1986), a number of other European directives were written to promote the rapid development of innovative products. Over the years the cooperation between the member states progressed and this was duly reflected in Council Directive 93/39/EEC of June 14, 1993, which amended directives 65/65/EEC, 75/318/EEC and 75/319/EEC to adopt the future European Community registration system for medicinal products (future systems) (ECC, 1965; 1975a, b; 1993b; 1997e). These directives introduced the mutual recognition of licences by Member States with a binding central arbitration process. By this time even greater changes were on the horizon, intended to bring about a supranational system within Europe. Plans included the establishment of the European Medicines Evaluation Agency, the EMEA, and the expansion of the CPMP to enable it to draw on expertise from across all Member States.

In line with this, Council Directive 93/41/EEC (EEC, 1993a) incorporated the repeal of the concertation procedure and provided for the transfer of applications to a new centralized procedure for biotechnology and high technology products. This new procedure came into force when the EMEA was established on January 1, 1995 and all outstanding applications were transferred to the new procedure.

The remit, relevant legislation and the procedures of the new centralized procedure are discussed in detail in the following sections.

3.2 The European Regulatory Environment and the Role of the EMEA

The pharmaceutical legislation in the EU has two key objectives: the protection of public health and the free movement of medicinal products. With this in mind, the European Commission announced in a white paper on the completion of the internal market, that it would establish proposals for the *"Future System for the Authorisation of Medicinal Products"* (whether human pharmaceutical or veterinary). These proposals were issued in November 1989 *"Future System for the Authorisation of Medical Products within the European Community"* (EEC, 1989).

Following the publication of this document, the European Commission drafted a number of regulations and directives which were to provide the basis for a new system for the registration of pharmaceuticals in Europe. These Council Directives and Regulations were formally adopted on 14th June 1993 (the council directives) and on 22 July 1993 (the regulation) and the procedure came into effect on January 1, 1995. The legislation can be summarized as follows:

- Council Directive 93/39/EEC amending directives 65/65/EEC, 75/318/EEC and 75/319/EEC in respect of medicinal products to (ECC, 1965; 1975a, b; 1993b, 1997e):
  - set up a system for CPMP arbitration on the area of disagreement in cases where applications in more than one Member State are not mutually recognized,
  - require Member States to systematically prepare assessment reports for exchange and allow Member States to suspend evaluation where the application is under active consideration in another Member State,
  - replace the previous requirement for import testing of all products from third countries (i.e., those outside the EC) by a waiver exempting the need for such testing where there are appropriate arrangements made with the exporting country in respect of GMP,
  - set up a pharmacovigilance system to coordinate the monitoring of adverse reactions in the Member States.
- Council Directive 93/41/EEC (ECC, 1993a) repealing directive 87/22/EEC (ECC, 1986) on the approximation of national measures relating to the placing on the market of high technology medicinal products, particularly those derived from biotechnology.

- Council Regulation No. 2309/93 (ECC, 1993c) laying down community procedures for the authorization and supervision of medicinal products for human and veterinary use and establishing a European Agency for the Evaluation of Medicinal Products (EMEA). The other principle objectives of this regulation were:
  - to set up a centralized procedure for the scientific evaluation of medicinal products,
  - to ensure a working relationship and cooperation between the EMEA and scientists working within the Member States,
  - to strengthen the scientific role and independence of the Committee for Proprietary Medicinal Products (CPMP) and the Committee for Veterinary Medicinal Products (CVMP),
  - to install a permanent technical and administrative secretariat at the location of the EMEA to support the CPMP and CVMP,
  - to set up a pharmacovigilance system to monitor adverse drug reactions,
  - to coordinate supervisory responsibilities of the Member States in respect of GMP, GLP and GCP, and
  - to establish a system for monitoring the environmental risk of products.

Council Regulation No. 2309/93 (ECC, 1993c) also allowed for any application referred to the CPMP before 1995 and not given an opinion to be considered under the centralized procedure.

On January 1, 1995 the EMEA came into being with its offices located at Canary Wharf, London, UK. FERNAND SAUER was appointed as the first executive director and the first staff was appointed in early 1995. The agency has continued to recruit, mainly from the national agencies and the European pharmaceutical industry.

The EMEA comprises three main components: the management board, the secretariat headed by the executive director and two scientific committees, one for human (CPMP) and one for veterinary (CVMP) products. The director reports to the management board comprising:

- 2 representatives from each Member State (one for human medicines and one for veterinary medicines)
- 2 commission representatives, and
- 2 representatives appointed by the European Parliament.

The management board has overall responsibility for the program of work, the annual accounts and reporting for the previous year and the budgets for the coming year.

The scientific committees comprise:

- the CPMP and CVMP scientific members from the Member States,
- the panels of experts (over 1,500, including the internal and external assessors from the national agencies) drawn from the Member States.

The EMEA is responsible for:

- the coordination of the evaluation of marketing authorization applications (MAA)
- the production of assessment reports, summary of product characteristics (SmPC) and labeling/package inserts
- the continued supervison of medicinal products authorized for use within the European Community
- coordination of manufacturing and testing requirements – GMP, GLP, GCP, batch controls etc.
- advice on all questions relating to medicinal products
- promoting cooperation between the public control laboratories and the European pharmacopoeia
- acting as a coordinator for the national pharmacovigilance centers.

Figs. 3 and 4 provide an outline structure of the EMEA secretariat and the more detailed organogram of the unit for the evaluation of human medicines, where the assessment of MAAs for biotechnology-derived products is undertaken.

The function of the EMEA can be compared to that of the FDA in that they both receive applications to market pharmaceutical products. They differ in a fundamental respect: the FDA is a licensing authority, the EMEA is not. The EMEA, through the CPMP, forms an opinion on a MAA but it is the responsibility of the Eropean Commission to convert this opinion into a decision and thereby authorize the marketing of a product.

They are also different in that the FDA is involved in a review of drugs through the de-

Fig. 3. Outline structure of the EMEA secretariat.

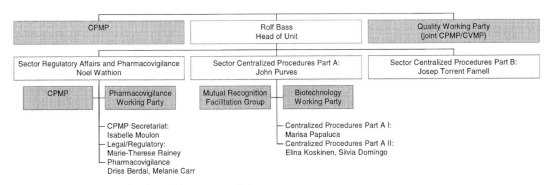

Fig. 4. Unit for the evaluation of human medicines.

velopment process and not just at registration. The EMEA has currently no remit to review the clinical trials procedures for products and these remains with the individual Member States of the EU. Consequently biotechnology products, in common with the drugs, are subject to the different procedures of each Member State at the clinical trial stage and thus no unified advice can be obtained during development. However, while regulation of clinical trials is a national responsibility it is possible to obtain technical or scientific advice from the CPMP, when such guidance is not available in formal guidance documents. To obtain such advice, applicants need to submit a written request to the EMEA secretariat, including an explanation of why the existing guidelines are inadequate. Generally, CPMP appoints a member to coordinate the process and inform CPMP of the issues. It is CPMP which provides the guidance. The advice is offered in good faith (scientific developments can alter the value of this advice) and it is not binding towards eventual MAAs. A draft amendment to Council Regulation No. (EEC) 297/95 (CEE, 1995b) relating to fees, introduces a fee of 60,000 European Currency Units to cover this service (expected in 1998) which was, until now, free.

The procedure for the submission and approval of a MAA for a biotechnology-derived product is referred to as the centralized procedure and is described in the following section.

3.3 The Centralized Procedure for the Registration of Biotechnology-Derived Products

Within the European regulatory framework, biotechnology-derived products are regulated under Council Regulation No. (ECC) 2309/93 of July 22, 1993 (ECC, 1993c) previously described, and defined in part A of the annex of this regulation, as medicinal products developed by means of one of the following biotechnological processes:

- recombinant DNA technology;
- controlled expression of genes coding biologically active proteins in prokaryotes and eukaryotes, including transformed mammalian cells;
- hybridoma and monoclonal antibody methods.

On the basis of this regulation it is mandatory to use the centralized procedure for biotechnology products.

Under this procedure, applications are submitted to the EMEA and the whole registration process can be divided into 7 phases:

- phase I: presubmission
- phase II: validation of the application
- phase III: prereview period
- phase IV: review of application
- phase V: post-opinion phase
- phase VI: decision making phase
- phase VII: post-authorization phase

Briefly, each of the phases involves the following activities.

3.3.1 Phase I: Presubmission

Prior to the submission of the MAA, the sponsor/company is advised to liaise as early as possible with EMEA staff regarding procedural and scientific questions. A presubmission meeting is arranged to discuss additional issues such as the suitability of the trademark and inspections of the bulk and finished product manufacturing sites.

The presubmission application is made about 4–6 months prior to the actual application. At this stage the sponsor nominates his rapporteurs. The sponsor is expected to propose up to 4 potential rapporteurs. Currently, the CPMP appoints the rapporteurs, at every second CPMP meeting, and in doing so will take into account the wishes of the company but balance these with the distribution of rapporteurship within its members. Once the rapporteurs are assigned, EMEA will nominate a project manager from the staff of the medicines evaluation unit. A key role of the project manager will be to liaise with the company on the progress of the application, advise on the procedure, validate the documentation submitted in the MAA and monitor compliance with the timeframe scheduled for the application process.

3.3.2 Phase II: Validation of the Application

Within 10 working days of the submission of the application, the project manager will validate the application against the checking procedure which includes verification that the application includes translations of the SmPC, the label and the user package leaflet in the 11 official languages of the European Union. The project manager may also need to liaise with the rapporteur and corapporteur regarding the need for action relating to matters other than those directly related to the evaluation of the dossier, such as GMP inspections, samples for analysis, GCP inspections, or legal problems such as trademarks.

3.3.3 Phase III: Prereview Period

At the end of the validation process and provided the rapporteur and corapporteur have confirmed that they have received the dossier, the secretariat starts the clock on the procedure. The start of the procedure is delayed until this confirmation is received from the rapporteurs and from other members of the committee that they also have received all necessary parts of the application. The project manager along with the rapporteurs will forward the proposed timetable for evaluation to the committee. The clock start for the application is coordinated with a monthly CPMP meeting.

3.3.4 Phase IV: Review of Application

The procedure lays down set times for the review of an MAA which must be met. The timetable is shown in Tab. 2.

The result of the evaluation will be either a negative opinion, a positive opinion or a conditional opinion from CPMP. A negative opinion means that the application does not meet the criteria set out in Council Regulation No. (ECC) 2309/93 (EEC, 1993c). A conditional opinion means that the committee considers that the summary of product characteristics needs to be amended and/or the labeling does not comply with Council Directive 92/27/EEC (EEC, 1992) (directive covering labeling and package inserts).

If the opinion is negative, the applicant has 15 d to lodge an appeal with the EMEA and forward details within 60 d. CPMP has 60 d to consider the appeal.

3.3.5 Phase V: Post-Opinion Phase

Before day 240 the final CPMP assessment report is sent to the applicant together with the opinion which is also sent to the European Commission and all member states in all languages. Positive opinions are accompanied by the approved SmPC, labeling, user leaflet and any conditions on the use or supply of the product.

3.3.6 Phase VI: Decision-Making Phase

Within 30 d of the receipt of the valid opinion, the European Commission prepares a draft decision (yes or no) on the MAA and submits this to the Member States and applicant. The Member States then have 30 d to forward written observations to the commission on its draft decision. If this raises important new questions it may refer the matter back to CPMP.

The final decision, if positive, is published in the Official Journal of the European Community.

3.3.7 Phase VII: Post-Authorization Phase

Before day 300 the European Public Assessment Report (EPAR) is finalized in con-

Tab. 2. Timetable for the Review of Application

Time	Action
Day 1	Committee receives a valid application. Experts/working parties for the evaluation are nominated. The CPMP secretariat advises the applicant of the timetable
Day 70	Rapporteur and corapporteur circulate their draft assessment reports (AR) to the CPMP members, secretariat and the applicant. The applicant receives the AR without recommendations
Day 100	Rapporteur an corapporteur receive comments from other CPMP members on the draft AR
Day 115	Rapporteur and corapporteur send a draft list of questions (including overall conclusions and overview of the scientific data) to other CPMP members and a list of questions which have been left aside
Day 120	CPMP validates the list of questions as well as the overall conclusions and review of the scientific. EMEA sends this to the applicant at which time the assessment stops (clock stop)
Day 121	Applicant submits written response to the CPMP (clock on). There may be a need for a working party *ad hoc* meeting and the CPMP draft assessment report is prepared
Day 150	Rapporteur/corapporteur submits to CPMP a further assessment report with comments and clarification from the applicant
Day 170	Deadline for comments back from CPMP members to the rapporteurs
Day 180	CPMP discusses the assessment report and decides whether oral explanation by the applicant to the CPMP is necessary (clock stop)
Day 181	Oral explanation (clock on)
Day 185	Final draft of English text of SmPC, label and leaflets from the applicant to the rapporteurs, EMEA and other CPMP members
Before day 210	CPMP concludes its evaluation and adopts an opinion with agreed text of the SmPC, labels and leaflet(s) in all 11 European Community languages
Day 5 at the latest after opinion	Translations into the 11 European Community languages of the SmPC, label and leaflet and sent to CPMP members and EMEA for comment before day 210
Day 15 after opinion	Final revised translation of SmPC, labeling and package leaflets in response to comments from EMEA and CPMP
Day 20 at the latest after opinion	Applicant provides EMEA with final translations of SmPC, labeling and package leaflets in all official languages of the EU. Final full color mock-ups for all Member States submitted

sultation with the rapporteurs, CPMP and the company.

In accordance with article 12 of Council Regulation No. (EEC) 2309/93 (EEC, 1993c), the EPARs are made publicly available. They are easily accessible from the EMEA homepage on the Internet and are a valuable source of competitor information.

European marketing authorizations granted through the centralized procedure are:

• valid for 5 years;
• renewable on application at least 3 months in advance, after consideration by the EMEA of an up-to-date file on pharmacovigilance; and

• can only be amended by the European Commission or Council after the CPMP has been consulted.

The licence holder also benefits from a 10-year period of exclusivity during which an applicant for a licence for a similar product shall not be eligible.

At present it is obligatory to submit applications for products derived from biotechnology via the centralized procedure but as experience develops with biotechnology products, the European Commission and EMEA may well decide that the mutual recognition route can be used. Recently industry groups have sent a proposal to the European Commission that requests recombinant DNA technology used in the manufacture of well characterized products should not lead to the mandatory use of the centralized procedure.

3.3.8 Responsibilities of the MA Holder

In order to hold an EU centralized licence, a company must be resident in the EU and fulfil two key criteria. The first relates to product safety or pharmacovigilance. The licence holder must have a person (often nominated as drug safety executive) who is suitably qualified and permanently and continuously available to report all relevant information about suspected adverse reactions to the EMEA. The licence holders are responsible for setting up and running a data capture system which ensures that data are evaluated and accessible at a single point within the (ECC, 1995a).

The second relates to product release and requires that a qualified person (QP), is permanently and continuously available to release and distribute the product. Once received in the EU a product can be freely distributed to all other Member States.

In additon the MA holder may need to update the MA by way of variations to allow for additional sites of manufacture, changes to the method of manufacture, etc.

Finally, the MAA holder is responsible for the provision of scientific advice under council Directive 92/27/EEC (EEC, 1992).

3.4 Experience with the Centralized Procedure

How many products have been approved by the European Commission? Since it began work in January 1995, up to November 1997, the EMEA received a total of 129 centralized applications, 18 of which were transferred from the previous concertation procedure. 47 of these applications referred to part A of which 19 applications were still undergoing review, 26 had received positive opinions

from the CPMP, 24 marketing authorizations had been granted by the commission and 2 products have been withdrawn.

A number of issues has arisen in regard to the implementation of the marketing authorizations and the interpretation of the rules of centralized procedure for product licensing some of which may have commercial implications for product marketing. Some of the key issues are discussed in the following sections but it is only with more experience of the regulation of products through the centralized procedure that many of these issues will be clarified.

3.4.1 Referrals and Public Health

Authorization of a product through the centralized procedure implies agreement to the marketing of that product by all Member States and objections by any Member States should have been dealt with during the assessment procedure prior to the CPMP opinion and decision. However, instances where disagreement could arise post-authorization are set out in Council Directive 75/319/EEC (CEE, 1975b), as amended and include reference to articles 10 (risk to health), 11 (divergent decisions), 12 (community interest) or 15a (protection of public health). Under article 15a of Council Directive 75/319/EEC (EEC, 1975b), a Member State is allowed to suspend a product in exceptional cases where urgent action is essential to protect public health and a referral of the issue is made to the CPMP. An example of this occurred in January 1997 in Italy. The Italian health authorities suspended the marketing of HibTIT-ER (a *Haemophilus influenzae* B conjugate and diphtheria vaccine used for the immunization of infants and children) which had first been given a positive opinion under the old concertation procedure in 1993. Marketing was suspended by the Italian authorities because of concerns over the possibility of BSE transmission, since bovine materials are used in its manufacture. In response, the CPMP issued a position statement saying that the bovine material is used only to cultivate bacteria and is not an ingredient of the finished product. The issue was referred to the April 1997

CPMP meeting, at which a majority opinion in favor of continued marketing of the product was adopted. As a result of the referral an amended harmonized SmPC was agreed. The press release following the April 1997 CPMP meet noted that the CPMP found the product to be safe and that there was no reason with regard to quality, safety or efficacy to withdraw the product from the European market (EEC, 1997d).

The European Commission refused to accept CPMPs opinion that HibTITER was safe. This affair was right at the heart of the public health concerns which dominated the European pharmaceutical market for much of 1997. Fears of the spread of transmissible spongiform encephalopathies led to a ban on the use of all high-risk ruminant materials in products such as gelatin and tallow. Continued fear of mad cow disease put the European Commission under pressure from the European Parliament to take action, which it did. In Decision 97/534/EC the European Commission announced a blanket ban on all specified risk materials (SRM), even if they were not from BSE or scrapie affected herds (EEC, 1997a). The European Commission finally issued an amendment to the ban in December 1997, to delay the implementation date of these measures until April 1998. Industry has expressed the view that the delay will permit an assessment of the effect the ban will have on pharmaceuticals containing animal-derived products. At present 85% of pharmaceuticals are affected by the ban. It is anticipated that the measures will be introduced over an 18–24 month period to allow industry time to establish sources of SRM-free materials.

3.4.2 Crisis Management

The EMEA has recently developed a crisis management plan to deal with emergencies relating to products approved under the centralized procedure (EEC, 1997b). EMEA said that it needs to be able to handle public relations sensitively and quickly if any safety or other problems with centrally-approved products arise.

3.4.3 Trademarks

The centralized procedure only allows for the submission of a single trademark for a product, which is in line with the EU objective of a single market throughout the member states. The limitation of one trademark for one product does not apply to the authorization of drugs via the decentralized procedure where different trademarks can be granted in each Member State. Thus, biotechnology products which must be processed through the centralized procedure are being subject to constraints not applied to other drugs. This has been presented as a major difficulty by parts of the pharmaceutical industry because of issues related to registration of trademarks with EU. Approval of trademarks by the EU Trademark Authority does not preclude objections being raised by an individual Member State at a later stage. Such objections could theoretically lead to a requirement to change the trademark throughout the EU and amend the MAA even after the product has been launched, with obvious commercial consequences. The European Commission, contrary to its previous strong opposition to multiple applications, claiming that the industry would use them to partition the single market, has allowed affillate companies to submit the same application for the same product with different trademarks. The contents of the MAAs are identical except for the trademarks. Some companies now routinely identify and register at least two trademarks well in advance of submitting the MAA. Industry claims that the second licence is necessary so that in the event of a Member State contesting the first trademark then the second product can be marketed immediately.

3.4.4 Comarketing or Copromotion

The trademark issue as it relates to the single EU market also has implications for comarketing or promoting of products and poses particular difficulties for biotechnology products. Many biotechnology-derived products are the result of research in small bio-technology companies with the late phase clinical work, registration and marketing of the products often licensed-out, or subject to joint-licensing arrangements, with larger pharmaceutical companies. In these circumstances, the companies may have contractual agreements in place to market or comarket the product and this would normally require two MAAs with two trademarks for the same product. However, the one trademark for one product rule of the centralized procedure does not appear to allow for this.

A way around this is for a single marketing authorization (MA) to be held by the larger company with the biotechnology company having distribution rights only, in one or more Member States, which could then be named on the labeling as the local representative for those Member States. In this situation the MA holder has total responsibility for the product in all Member States and retains total liability for the product throughout the EU. The local representative has none of the legal responsibilities of the MA holder and has to rely on the MA holder to interact with the regulatory authorities and communicate effectively with them. The local representative would be required to have procedures in place for reporting adverse events in their market to the MA holder and must be informed of any ongoing safety issues by the MAA holder. This requires the implementation and maintenance of compatible pharmacovigilance systems between companies as well as good cooperation on other marketing issues. However, using this route, a small bioetechnology company will not have the opportunity to market under its own name or product mark.

3.5 Orphan Drug Legislation

A number of biotechnology products has been developed for use in severe or life threatening disease where the patient population is small and for many companies producing such products, recognition of an orphan drug status in Europe, as well as the USA will be a benefit. The European Commission has adopted a parliament and council regulation

on orphan medicinal products for comment. Public funded finance and scientific help with the development of a designated orphan product, as well as a reduced MA fee is likely to result from adoption of an orphan drug regulation. In addition, a product if granted orphan status under this regulation will be entitled to an exclusive marketing period of 10 years. The prevalence of the disease condition which defines the product as an orphan under this legislation is less than 5/10,000 of the population. It is also expected that products that have orphan status in the USA and Japan will also have orphan status in Europe. This regulation will come into force in 2000.

4 Conclusions and the Way Forward for the Regulation of Biotechnology-Derived Products

The regulatory processes in the USA began back in 1902 with the *Federal Virus, Serum and Toxin Act* (US Congress, 1902) and it was the development of regulations for biologicals that was the basis for the development of the FDA. In Europe, all Member States have developed their own national processes to regulate pharmaceuticals although the era of biotechnology-derived products has had a significant influence on the development of the supranational regulatory authority in Europe, the EMEA.

The regulatory authorities of the USA, Europe and of course the Ministry of Health and Welfare (MHW) in Japan, have always had their own particular requirements for the registration of pharmaceuticals, some common to all the authorities but many idiosyncratic to each individual one. The formation of the EMEA has led to harmonization within Europe with respect to the content and format of the applications, through issuing guidelines and directives and a streamlining of the regulatory approval procedure. On a worldwide basis, the International Conference on Harmonization (ICH), established in 1990, was intended to provide a forum to discuss ways in which all three major regulatory bodies (FDA, EMEA and the MHW) and their respective industry bodies could harmonize the approach to drug development and make these products available to the public with the minimum of delay. The ICH process has achieved success because it is based on scientific consensus developed between industry and regulatory experts and because of the commitment of all the regulatory partners to implement the ICH harmonized guidelines and recommendations.

The 4th ICH meeting, held in Brussels in July 1997, marked the conclusion of the first phase of ICH. Significant progress has been made in reducing duplication of data in the process of developing new medicinal products and submitting technical data for registration. The six founder members of ICH have agreed to a second phase of harmonization activities to continue after ICH4.

The ICH process has resulted in the issuing of many harmonized guidelines specifically targeting the data requirements for biotechnology-derived products, which help to progress the move towards a "single data package" for all pharmaceutical MAAs. At ICH4 the steering committee agreed to a new topic, M4, the Common Technical Document (CTD) which is a harmonization of the format and content of the document requirements for MAAs in all three regions. The first step of the CTD process will concentrate on the order in which the data are presented and the level of detail in the reports. It will not include expert reports or administrative data. A maximum period of 18–24 months has been set to achieve this agreement on the content of the CTD so the "single" application is not far off.

5 References

21CFR 136: *Orphan Drugs.*

Anonymous (1997), US *Regulatory Reporter* **14** (6).

EEC (1965), *Council Directive 65/65/EEC of 26 January 1965 on the Approximation of Provisions Laid down by Law, Regulation of Administrative Action Relating to Medicinal Products.* (OJ No 22 of 9. 2. 1965).

EEC (1975a), *Council Directive 75/318/EEC of 20 May 1975 on the Approximation of the Laws of Member States Relating to Analytical, Pharmacotoxicological and Clinical Standards and Protocols in Respect to the Testing of Medicinal Products* (OJ No L 147 of 9. 6. 1975).

ECC (1975b), *Council Directive 75/319 of 20 May 1975 on the Approximation of Provisions Laid down by Law, Regulation of Administrative Action Relating to Medicinal Products* (OJ No L 147 of 9. 6. 1975).

EEC (1986), *Council Directive 87/22/EEC of 22nd December 1986 on the Approximation of National Measures Relating to the Placing on the Market of High-Technology Medicinal Products, Particularly Those Derived from Biotechnology* (OJ L 15 of. 17. 1. 1987).

EEC (1989), *European EC Commission (1989), Future Systems for the Authorisation of Medicinal Products within the European Community* (III/ 8276/89 rev 1).

ECC (1992), *Council Directive 92/27/EEC of 31 March 1992 on the Labelling of Medicinal Products for Human Use and on Package Leaflets* (OJ No L 113 of 30. 4. 1992).

EEC (1993a), *Council Directive 93/41/EEC of 14 June 1993 Repealing Directive 87/22/EEC on the Approximation of National Measures Relating to the Placing on the Market of High Technology Medicinal Products, Particularly Those Derived from Biotechnology* (OJ No L 214 of 24. 8. 1993).

EEC (1993b), *Council Directive 93/39/EEC of 14 June 1993, Amending Directives 65/65/EEC, 75/ 318/EEC and 75/319/EEC in Respect of Medicinal Products* (OJ No L 214 of 24. 3. 1993).

EEC (1993c), *Council Regulation No (EEC) 2309/ 93 of 22 July 1993 Laying down Community Procedures for the Authorisation and Supervision of Medicinal Products for Human and Verterinary Use and Establishing a European Agency for the Evaluation of Medicinal Products* (OJ No L 214 of 24. 8. 1993).

EEC (1995a), *Commission Regulation (EC) No 540/95 of 10 March 1995 Laying down the Arrangements for Reporting Suspected Unexpected*

Adverse Reactions which are not Serious, whether Arising in the Community or in a Third Country, to Medicinal Products for Human or Veterinary Use Authorized in Accordance with the Provisions of Council Regulation (EEC) No 2309/93 (OJ L 55/5, 11. 3. 95).

EEC (1995b), *Council Regulation No (EEC) of 10 February 1995 on Fees Payable to the European Agency for the Evaluation of Medicinal Products* (OJ L 35/1, 15. 2. 1995).

EEC (1996), *Proposal for a European Parliament and Council Regulation on Orphan Medicinal Products,* August 1996.

EEC (1997a), *Commission Decision 97/534/EC of 30 July 1997 on the Prohibition of the Use of Material Presenting Risks as Regards Transmissible Spongiform Encephalopathies* (OJ, L 261/95, 8. 8. 1995).

EEC (1997b), *Draft CPMP Guideline on a Crisis Management Plan Regarding Centrally Authorised Products for Human Use* (CPMP/388/97), 24. 9. 97.

EEC (1997c), *Commission Decision 97/866/EC of 16 December 1997 Amending Decision 97/534/EC on the Prohibition of the Use of Material Presenting Risks as Regards Transmissible Spongiform Encephalopathies* (OJ L 351/69, 23. 12. 97).

EEC (1997d), *EMEA Press Release Following the 25th Plenary Meeting on 18–20 March 1997.* CPMP/234/97, 21. 3. 97).

EEC (1997e), *Council Directive 75/318/EEC of 20 May 1997 on the Approximation of the Laws of Member States Relating to Analytical, Pharmacotoxicological and Clinical Standards and Protocols in Respect of the Testing of Medicinal Products* (OJ No L 147 of 9. 6. 1975).

FDA (1995a), *National Performance Review: Reinventing the Regulations of Drugs Made by Biotechnology,* White House/FDA Joint Press Statement, 9 November 1995.

FDA (1995b), *60 FR 63048 Interim Definition and Elimination of Lot-by-Lot Release for Well Characterized Therapeutic Recombinant DNA Derived and Monoclonal Antibody Products,* Federal Register, 8 December 1995.

Parexel International Corporation (1997), *Biologics and Biotech Regulatory Report.*

US Congress (1902), *Federal Virus, Serum and Toxin Act,* 32 Stat. 728 (1902).

23 Regulation of Human Gene Therapy

JAMES PARKER

Stow, MA, USA

1 Gene Therapy and Biotechnology 518
2 Biotechnology Evolution 519
3 Regulation of Human Gene Therapy 520
   3.1 Historical Developments and Early Involvement 520
   3.2 Current Regulations and Guidelines 523
   3.3 Regulation of Human Gene Therapy in the European Union 526
4 Conclusions 528
5 References 528

List of Abbreviations

CRADA	Cooperative Research and Development Agreement
EMEA	European Medicines Evaluation Agency
EU	European Union
FDA	Food and Drug Administration
FOIA	Freedom of Information Act
GOCO	government owned contractor operated
GTAC	Gene Therapy Advisory Committee
GTPC	gene therapy policy conferences
HGP	Human Genome Project
HUGO	Human Genome Organization
IBC	Institutional Biosafety Committee
IRB	Institutional Review Board
NIH	National Institutes of Health
PDUFA	Prescription Drug User Fee Act
RAC	Recombinant DNA Advisory Committee
r-DNA	recombinant DNA
UNESCO	United Nations Educational, Social, and Cultural Organization

1 Gene Therapy and Biotechnology

The greatest potential contribution of modern biotechnology is in the promise of gene therapy. Many researchers, scientists, and organizations continue to have unprecedented insight into etiology and potential therapies for diseases and conditions previously untreated in conventional medical practice. Clinical research is expanding rapidly in this area. As of this writing, the National Institutes of Health (NIH) of the United States lists 231 clinical trials protocols underway with over 1,900 patients, which accounts for over 70% of all protocols and 75% of all patients worldwide (Countries conducting gene therapy trials, < http://www.wiley.co.uk/gene-therapy/countries.htlm >, visited May 24, 1998; Human gene therapy protocols, < http://www.nih.gov/od/orda/protocol/htm > visited May 27, 1998). The preclinical and clinical development research may lead to dramatic improvement in life, longevity, and quality of life, but to fulfill the promise of gene therapy the domestic and international regulatory control authorities must play an important role. They must not only remain vigilant in helping to safeguard the public health, but also be innovative in the processes and procedures of drug review and approval. Safe and effective therapies from innovative technologies must be supported and facilitated by registration authorities for the new treatments to reach patients. The manner and types of regulatory controls applied by the United States government to biotechnology and products of biotechnology, e.g., appear minimally intrusive and often conducive to drug and therapy development in this rapidly growing area.

In general, biotechnology may be considered to be the application of various biological processes to make, alter, or improve new or pre-existing products, organisms, and services. A very broad and wide ranging field, biotechnology encompasses old technologies, such as plant and animal breeding, alongside new technologies, such as recombinant DNA (r-DNA) and human gene therapy. These approaches have application in many different fields, ranging from drug and medical device manufacturing and marketing practices to environmental protection, agriculture, and human and animal health and safety.

Due to the enormous scope of this field, biotechnology offers numerous benefits and risks to society. Many of the benefits exist in the agricultural industry. For example, genetic engineering and r-DNA processes enable scientists to modify certain climate-sensitive plants to withstand changing weather conditions by introducing desired traits in them, such as frost or drought resistance (ALLEN, 1990). Other agricultural advances include the development of insect, disease, and herbicide resistant crops and the genetic manipulation of plants and animals to increase size and growth rate. Biotechnology has produced similar benefits in environmental protection and conservation. For example, microbiologists have developed bacteria which attach to and break down oil molecules thereby facilitating oil spill cleanups and preventing prolonged damage to coastlines and wildlife (ALLEN, 1990). Biotechnology has led to outstanding and unprecedented advances in genetic engineering and human gene therapy. Through gene identification and gene mapping scientists have developed, and continue to refine, the ability to treat, correct, and prevent certain diseases and medical conditions by isolating and altering or replacing a defective gene with a desired one.

Although biotechnology has many positive applications, risks or potential dangers attending advances and processes in this area must be considered as well. Releasing a new species of plant, animal, or pesticide into the environment always carries some potential of causing unforeseen, unintended, and harmful results. Similarly, certain new drugs, disease treatment techniques, and medical devices, while often very beneficial, may pose real and obvious dangers to human health.

2 Biotechnology Evolution

Although biotechnology has progressed rapidly over the last three decades, the United States Federal Government generally has not placed many stringent industry-wide regulations on either the experimentation or the application of biotechnology. Rather, because the scope of the field encompasses so many different and unique technologies and practices, the government, through the various agencies and departments, regulates biotechnology products on a case-by-case basis (Food and Drug Administration Notice, 1986). Federal regulation of the biotechnology industry originated within the scientific community in the early 1970s once the potential health and safety risks of conducting certain biotechnology experiments involving viruses became evident. In 1974, in response to these common, indeed global, concerns, the

whole scientific community accepted a voluntary moratorium on all experiments involving r-DNA until proper standards and regulations for r-DNA experimentation had been established.

The first evidence of significant federal interaction with the newly emerging and virtually unknown field of biotechnology in the United States was the creation of the Recombinant DNA Advisory Committee (RAC) by the NIH. One of the first organizations involved in biotechnology, the NIH may be referred to properly as "the father of the biotechnology industry in the United States" (McGARITY, 1995). The NIH is a scientific administrative agency, and, unlike other agencies such as the Federal Food and Drug Administration (FDA), it possesses limited regulatory authority. The NIH conducts extensive research in the industry, particularly in the area of human gene therapy, and endows many public and private institutions, such as universities and laboratories, with billions of dollars annually for biotechnology research and development in health and environmental sciences. The NIH established the RAC to serve as a scientific advisory board to industry concerning the ethics, safety, and efficacy of new r-DNA techniques and procedures which began to appear more frequently at this time.

Toward this end, in 1976 the RAC set forth the *NIH Guidelines for Research Involving Recombinant DNA Molecules* (NIH Guidelines) prescribing standards and practices for the research, manipulation, and disposition of r-DNA molecules and organisms throughout the biotechnology industry (Food and Drug Administration Notice, 1976; Domestic Policy Council Working Group of Biotechnology, 1984). Although devoid of mandatory compliance authority, the NIH mandates and achieves compliance with the NIH Guidelines among all institutions that receive NIH funding (Actions under the NIH Guidelines, <*http://www.nih.gov/od/orda/10-97 act.htm*>, pp. 1–40, visited May 25, 1998; Food and Drug Administration Notice, 1997). It remains one of the largest financial supporters of academic and private organizations engaged in r-DNA research and the risk of loss of funding dollars in the event of noncompliance is persuasive. Other non-NIH funded entities engaged in biotechnology research endeavors followed suit voluntarily (KIN, 1996).

In 1984, in light of the rapid progression of the biotechnology industry, particularly in the area of genetic engineering, and in view of the diversity of pre-existing federal authorities, members from 17 major federal agencies and departments formed the Domestic Policy Council Working Group on Biotechnology (Working Group) (1984), (Food and Drug Administration Notice, 1984). The Working Group was established to organize and promulgate a comprehensive federal policy for ensuring the safety of biotechnology research and products. In 1986, the White House Office of Science and Technology Policy issued the *Coordinated Framework for the Regulation of Biotechnology* (Coordinated Framework), which organized and detailed the largely confusing and unorganized body of federal biotechnology law (Food and Drug Administration Notice, 1986). The Coordinated Framework provided a matrix coordinating the activities and jurisdictions of all federal agencies involved in regulation of biotechnology, clarified the policies and missions of these agencies, and designed an advisory committee to deal with various biotechnology issues, such as safety and ethics (Food and Drug Administration Notice, 1986). The Coordinated Framework represented a significant and unprecedented initiative by the Federal Government in its involvement in both the regulation and promotion of biotechnology in the United States, however, much of the old regulatory scheme predating the Coordinated Framework still remained intact.

3 Regulation of Human Gene Therapy

3.1 Historical Developments and Early Involvement

The Federal Government generally does not interfere with the practice of medicine in the United States (FURROW, 1996). The courts, rather than the government, through the development of product warranty law and negligence theories are responsible for the vast majority of restraints on drug prescription and treatment between a physician and an individual patient. Comment k of the Restatement (Second) of Torts gives much deference to drugs, treatments, and vaccines that, while sometimes "unavoidably unsafe" in terms of their inevitable side effects, may offer substantial benefits to the public [Restatement (Second) of Torts, 1965].

Relative to the rapid growth of the biotechnology industry, federal regulation has been minimally intrusive. For example, physician treatment of an individual patient with human gene therapy techniques is generally restricted only by proper medical practice and the general requirement of informed patient consent. Certain clinical protocol requirements recently proposed by the RAC concerning the ethics and safety of human gene therapy, however, place heightened scrutiny even in the private physician–patient relationship (Actions under the NIH Guidelines, <*http://www.nih.gov/od/orda(10-97act.htm*>, pp. 1–40, visited May 25, 1998). Federal regulations apply and increase, however, in the commercial context once a physician or company seeks to treat or market biotechnology products to a group of patients (Actions under the NIH Guidelines, <*http://www.nih.gov/od/orda(10-97act.htm*>, pp. 1–40, visited May 25, 1998). A commercial enterprise is often found particularly when the activities reach across state and federal borders and appear not to be part of the doctor–patient relationship.

Human gene therapy is arguably the most active component within biotechnology today. The United States, by demonstrating strong leadership and minimal regulatory intrusion, have established approaches within existing registration systems which most advanced nations now employ (MALINOWSKY and O'ROURKE, 1996). International interest in human gene therapy accelerated with the Human Genome Project (HGP), an international effort commenced by the United States to identify every gene in the human body in order to treat currently incurable genetically related diseases and conditions (KESSLER et al., 1996).

Representing the latest in a long line of biological developments dating back to the early 1970s, human gene therapy and other genetic research involve the study of inherited genetic compositions in humans, locating certain mutations in the DNA structures, and manipulating or altering those structures through various biological procedures designed to correct adverse conditions. Human gene therapy involves two "therapies" or procedures, *in vivo* and *ex vivo* gene therapies. *In vivo* gene therapy, the more widely used method, involves repairing the defective gene inside the patient's body by way of gene delivering carriers, or vectors, which contain "healthy" or reparative genes. These vectors are often inactive viruses capable of penetrating cells, but may also be nonviral in origin, and are administered through inhalation, injection, oral ingestion, or topical application. *Ex vivo* gene therapy, the less common method, entails isolating and altering cells with dysfunctional genes outside of the patient's body and then reintroducing the "corrected" cells into the body. Furthermore, gene therapy protocols are geared toward two processes of repairing gene defects in cells: somatic cell gene therapy, which involves modification of nonhereditary traits to alleviate a disease or condition in an individual, and germ line gene therapy, or germ cell therapy, which involves modification of hereditary traits to prevent the transmission of defective genes to an individual's progeny. Somatic cell therapy involves the "prevention, treatment, cure, diagnosis, or mitigation of disease or injuries in humans by the administration of autologous, allogeneic, or xenogeneic cells that have been manipulated or altered *ex vivo*" (Food and Drug Administration Notice, 1993). Germ line gene therapy may be defined as "a specific attempt to introduce genetic changes into the germ (reproductive) cells of an individual, with the aim of changing the set of genes passed on to the individual's offspring." (Food and Drug Administration Notice, 1996).

Within just the last few years scientists and researchers have made enormous advances in identifying the DNA sequences of genes in the human body and in matching these sequences to genetic functional defects associated with many diseases, such as cystic fibrosis, Alzheimer's disease, and Huntington's disease, as well as medical conditions, such as obesity and diabetes. From a patient's perspective human gene therapy represents the potential for new cures and vaccines for common, rare, and untreatable afflictions, such as tuberculosis, HIV, and Ebola. For scientists, human gene therapy represents unlimited possibilities in areas such as genetic engineering and cloning.

The human gene therapy industry in the United States comprises a conglomeration of alliances among federal agencies, academic institutions, private funding sources, multinational and small pharmaceutical companies, and other profit and nonprofit private and public institutions (MALINOWSKI and O'ROURKE, 1996). As was previously noted, the Federal Government generally does not interfere with or regulate the area of physician treatment of an individual patient, even when involving human gene therapy techniques. Rather, Congress, through legislation, and by developing close bonds with industry and academia, has effectively promoted an environment conducive to rapid and unprecedented, yet relatively safe, advances in human gene therapy and r-DNA technologies.

Early government support of human gene therapy began with federal funding of academic institutions engaged in genetic research and development and r-DNA studies (MALINOWSKI and O'ROURKE, 1996). As institutions began making rapid advances in gene therapy, the commercial and economic aspects of the science became more and more evident. This marketing potential quickly attracted venture capital from the private sector geared toward product patents and marketability, thus marking the beginning of the genotechnology industry. With this rapid shift from government participation to private sector involvement heavy pressure came from the private sector. As the companies which were engaged in human gene therapy were pressured to develop products more quickly, the FDA felt the urgency to approve them more rapidly. As the commercialization of gene therapy began to gain momentum, many researchers and scientists began to leave their government and academic positions in favor of more lucrative, venture capital-fueled, private industry opportunities. Although this huge influx of private sector funding has caused the government to play a smaller role in promoting the development of practical applications of human gene therapy, the early federal support and funding of research and development of human gene therapy, and biotechnology in general, remains responsible for the rapid advancement of the human gene therapy industry and the unprecedented advances and technologies being used.

In addition to its initial funding of gene therapy research and development at the academic level, the Federal Government created intense national and international interest and activity in human gene therapy development through the initiation and promotion of the Human Genome Project in 1988 (LAWTON, 1997). The HGP is a government-initiated effort to identify and map all 23 pairs of human chromosomes to develop cures and treatments for presently incurable medical and genetic diseases, conditions, and disorders. Moreover, it is an intentional effort, in recognition of the overwhelming interest in the commercialization and marketability of human gene therapy, to cooperate and work with, rather than compete with, the private sector. The Federal Government has contributed more than $100 million each year to the HGP and will continue to do so until the year 2005 (MALINOWSKI and O'ROURKE, 1996). With many countries, including the European Union, financing parallel projects, this congressional effort has sparked a global genetic "arms race."

In addition to directly funding gene therapy research and development, the Federal Government has provided substantial incentives for private sector funding through the strong policy of technology transfer. In general, technology transfer involves the movement of any knowledge, technology, or information from government laboratories to private industry. Toward this end, the *Stevenson–Wydler Technology Innovation Act* of 1980 and the *Bayh-Dole Act* of 1980 allow small businesses and non-profit organizations to claim ownership to any innovations, devices, or techniques developed with federal funds, but these groups may only do so if they intend to

patent and commercialize the specific technology acquired. The *Federal Technology Transfer Act* of 1986 amended the *Stevenson–Wydler Act* by requiring federal laboratories to actively seek opportunities to transfer technology to industry, universities, and state and local governments. It also gives employees of government agencies the right to patent technologies if the agency does not intend to do so. The *National Competitiveness Technology Transfer Act* of 1989 further amended the *Stevenson–Wydler Act* and made technology transfer the official policy and mission of government owned, contractor operated (GOCO) laboratories and their employees. As applied to biotechnology and human gene therapy, these statutory enactments have allowed for significant improvements and advances in the field, and they exhibit the unfailing willingness and desire of the Federal Government to foster a productive and energetic science geared toward improving the quality of human life and health.

There are many mechanisms for achieving federal technology transfer policy. One of the most widely used methods is through cooperative research and development agreements (CRADAs). CRADAs are contractual agreements between a government agency and a private company, university, non-profit organization, or state or local government to work together on researching and developing a matter of mutual interest. Under these agreements, the federal laboratory will often provide personnel, equipment, and facilities to the other party in furtherance of project research and development. CRADAs are prevalent in the area of human gene therapy and allow both parties to maximize their budgets, share and reduce costs, and exchange knowledge, expertise, and information.

A second common method for enabling technology transfer is through technical assistance. Under this method, companies, universities, or state or local governments may request assistance on a technically challenging problem from a federal agency or federal laboratory. Once the request is made, the scientist or other member who is most qualified to assist contacts the party making the request. In this manner, the federal agency shares its expertise and knowledge with the nonfederal entity, and the problem is most often solved quickly and cost-effectively.

A third common approach to facilitating federal technology transfer policy is through educational partnerships. An educational partnership is an agreement entered into between a government agency or laboratory and a non-profit university or other educational agency. Similar to CRADAs, the government agency under this system may loan or donate laboratory equipment to the educational group and make personnel available to teach or assist in laboratory research and development projects. The Federal Government may also achieve the free-flow exchange of technology and expertise to these institutions through federal grants. Grants are contractual agreements between a federal agency or laboratory under which the government provides money and/or property to an educational or non-profit institution to aid in research and development in human gene therapy.

In addition to promulgating federal technology transfer policy, the Federal Government has been instrumental in fostering innovations in human gene therapy through manipulation and alteration of the patent system. Unlike many other nations, the United States recognize intellectual property rights to gene therapies (United States Constitution, 1787; Patent Act, 1952). The United States Constitution provides that "Congress shall have power to ... promote the progress of ... useful arts, by securing for limited times to ... inventors the exclusive rights to their ... discoveries" (United States Constitution, 1787). Although a controversial practice, recognition and protection of patents attract a large percentage of the much needed venture capital by offering a significant incentive to private industries and companies to develop, patent, and market certain technologies and products.

The *Orphan Drug Act* of 1983 further illustrates the Federal Government's initiative in encouraging the development of the human gene therapy industry (*Orphan Drug Act,* 1993). The act was intended to facilitate the development of drugs for rare diseases or conditions, defined as those diseases or conditions which affect less than 200,000 persons in the United States. By granting certain compa-

nies 7 year monopolies and tax credits up to 50% of their total expenses incurred in clinical testing on animals, the act encourages companies to develop drug products (of all types) for rare medical conditions (United States Code, 1996; Code of Federal Regulations, 1996).

3.2 Current Regulations and Guidelines

To date, both the NIH and the FDA employ their own separate methods for reviewing somatic cell human gene therapy protocols in the United States. Current guidelines for human gene therapy protocols under the NIH are rooted primarily in the NIH Guidelines (Food and Drug Administration Notice, 1976; Domestic Policy Council Working Group on Biotechnology, 1984) and the RAC. The NIH Guidelines and RAC review, while not regulatory in nature and binding only on those investigators and institutions receiving NIH funding, embody a series of directives and "points to consider" for individual human gene therapy protocols (Actions under the NIH Guidelines, <*http://www.nih.gov/od/orda/10-97act.htm*>, pp. 1–40, visited May 25, 1998). The NIH Guidelines, provide, in pertinent part, that "[a]ny recombinant DNA experiment, which according to the NIH Guidelines requires approval by NIH, must be submitted to NIH or to another Federal agency that has jurisdiction for review and approval" (Actions under the NIH Guidelines, <*http://www.nih.gov/od/orda/10-97act.htm*>, pp. 1–40, visited May 25, 1998). Furthermore, all submissions of somatic cell gene therapy protocols to the NIH must conform to a specific set of criteria. For those individuals and institutions receiving NIH funding and for whom RAC review is required, all proposed protocols must include certain elements. These elements include the objectives of the proposed research, a risk benefit analysis, a description of the overall research format, the proposed process for patient selection, and a provision for informed consent. Once these preliminary elements have been satisfied and the protocol has received initial approval from the Institutional

Review Board (IRB) and the Institutional Biosafety Committee (IBC), the RAC reviews the proposed protocol and makes a recommendation to the NIH Director whether to approve the protocol.

The RAC oversees, proposes, and amends gene therapy protocols in the United States and identifies which gene transfer experiments are sufficiently novel to merit full RAC "public discussion" and analysis. Public discussion of human gene therapy experiments serves to inform the public about the technical aspects of the experiments, the meaning and significance of the research, relevant safety issues, and the ethical implications of the research. Such public disclosure helps ensure safe and ethical conduct of gene therapy experiments, and it facilitates public understanding of this novel area of biomedical research. Similarly, the *Freedom of Information Act* of 1994 (FOIA), which requires all federal agencies to disclose certain information to the public, further reinforces and supports this objective (*Freedom of Information Act*, 1994, § 522).

In determining the novelty of a proposed somatic cell gene transfer experiment, RAC reviewers assess the scientific rationale and content relative to other proposals reviewed by the RAC, the sufficiency of preliminary data, and the resolution of questions relating to the safety, effectiveness, and ethics of the proposed experiment (*Freedom of Information Act*, 1994, § I(B)(1)(b)). While only somatic cell gene therapy research has been widely accepted, the RAC presently does not entertain proposals for germ line experiments (*Freedom of Information Act*, 1994, appendix M). The RAC particularly scrutinizes the type of vector system, or provirus, proposed for use in delivering the gene *in vivo* into the target cell area. Most protocols traditionally have involved retroviral vectors, inactive viruses capable of penetrating cells and delivering genetically desirable genes into target cells containing dysfunctional genes. Many protocols studying cystic fibrosis, however, involve adenovirus vectors and adeno-associated viruses. The RAC requires that all protocols involving a retrovirus be accompanied with data demonstrating the resistance or susceptibility of both the vector and the provirus

to alteration, variation, recombination, or mutation. After this initial assessment, the RAC, NIH, and FDA jointly review the proposed experiment to determine whether it offers adequate assurance that their consequences will not go beyond their intended purpose. Prior to this review, investigators and researchers must submit certain materials to the NIH as part of the protocol approval process. These include:

(1) a scientific abstract,
(2) a non-technical abstract,
(3) IBC and IRB approvals and their deliberations pertaining to the individual protocol,
(4) a description of the proposal, informed consent, and privacy and confidentiality issues,
(5) the clinical protocol (as approved by the IBC and IRB),
(6) an informed consent document (approved by the IRB),
(7) appendices (including tables, figures, and manuscripts), and
(8) biographical information for each key professional person involved in the proposed protocol (*Freedom of Information Act*, 1994, § 552, Appendix M-1).

Like other federal review procedures of biotechnology products, this review is geared primarily toward protecting the health and safety of prospective human subjects and patients.

Current guidelines for human gene therapy protocols under the FDA jurisdiction are derived from Federal Regulations and a "Points to Consider" document, or "Guidance for Industry" (Guidance for human somatic cell therapy and gene therapy, *<http://www.fda.gov/cber/guidelines.htm>*, pp. 1–27, visited May 27, 1998). Like the NIH equivalent, the FDA's guidelines are voluntary and ethically oriented, not regulatory. They provide researchers and manufacturers with current information, policies, and ethical considerations regarding regulatory concerns for production, quality control testing, administration of recombinant vectors for gene therapy, and preclinical testing of both cellular therapies and vectors. One of the guidelines addresses the area of collection of cells. It provides that the origin of all cell populations proposed for administration to human patients should be classified as autologous, allogeneic, or xenogeneic, and all relevant identifying information, such as tissue source, should be included as well. This guideline further suggests that all relevant donor characteristics, such as age and sex, be included. Also, "exclusion criteria," such as the presence of or likelihood of infection by HIV, hepatitis, or other viruses should be included. Furthermore, the FDA suggests that manufacturers submit a description of the procedures for cell collection, including the location of the facility and any devices or materials used.

A second industry guideline considers cell culture procedures, which include quality control procedures, culture media, adventitious agents in cell cultures, monitoring of cell identity and heterogeneity, characterization of therapeutic entity, and culture longevity. First, the FDA stresses that cell culture operations should be carefully conducted in terms of quality of materials, manufacturing controls, and equipment validation and monitoring. Second, the Agency urges the establishment of minimum acceptance criteria for all media and components, including validation of serum additives and growth factors, as well as verification of the absence of adventitious agents. Third, the guideline states that manufacturers should verify that cells are handled, propagated, and subjected to laboratory procedures under conditions which minimize contamination with adventitious agents, such as bacteria, yeast, mold, mycoplasma, and adventitious viruses. Moreover, cells subjected to long-term culturing should be inspected periodically for contamination by such agents. Fourth, the agency suggests the implementation of manufacturing and testing procedures to ensure quality control of cell cultures with regard to identity and heterogeneity. Fifth, if the intended therapeutic effect is based on a particular molecular species synthesized by the cells, manufacturers should submit enough structural and biological information to show the presence of an appropriate and biologically active form. Finally, the guideline urges the submission of a profile of the essential characteristics of the cultured cell population.

A third industry guideline treats cell banking system procedures. Given that cell banking systems are appropriate for use with some somatic cell therapy products that are made repeatedly from the same cell source, the FDA stresses that these cell banks should be managed by a uniform, cell banking system. Such a model system should:

(1) describe the origin and history of the cells,
(2) characterize the identity of the cells by appropriate genotypic and/or phenotypic markers,
(3) test for contaminating organisms and other biological agents,
(4) ensure cell expiration dating, and
(5) conduct tests on thawed cells.

A fourth consideration addresses the need to identify materials used during manufacturing and cell manipulation procedures. The FDA requests that particular scrutiny should be given to materials of animal origin in which adventitious agents may persist. Furthermore, manufacturers should establish limits for the concentrations of all production components that may remain in the final product and detail the methods used to remove any such components.

The two primary concerns of the RAC, NIH, and FDA focus on the potential adverse genetic and health consequences of the proposed human gene therapy experiment: the vertical transmission of genetic changes from a human subject to children, and the horizontal transmission of viral infection to other persons with whom the individual comes in contact, such as medical personnel or relatives. If adopted by the NIH, the RAC proposals will form the basis of the NIH Guidelines, and apply to all r-DNA research that is either conducted at, or sponsored by, an institution that receives support for r-DNA research from the NIH, or which involves the testing in humans of materials containing r-DNA developed with NIH funds. Toward this end of helping to ensure the safety of human subjects, and in order to foster public discussion relevant to the scientific, safety, social, and ethical implications of gene therapy research, the NIH director may regularly convene Gene Therapy Policy Conferences (GTPCs). GTPCs, which consist of members and representatives from other agencies and industry organizations, focus on broad, overreaching policy and scientific issues related to gene therapy research.

In response to a recent request made by the National Task Force on AIDS Drug Development to streamline these separate but parallel review tracks, the NIH and the FDA have proposed to consolidate their review processes in the interest of eliminating overlap and expediting protocol review and approval times. Under the proposed review process, all human gene therapy protocols will be simultaneously submitted to the NIH and the FDA, and the RAC will serve, as needed, as the advisory body to both groups concerning human gene therapy protocols. In determining whether full review by the RAC is warranted, staff members of the NIH, the FDA, and the RAC will address whether the protocol proposes the use of new vectors, targets a new disease for experimentation, involves a unique application of gene therapy, and raises any new ethical concerns requiring public review. Beyond joint review of individual protocols by the NIH and the FDA, the new proposal also calls for joint review and assessment of larger, global concerns, such as the ethics of a gene therapy patient registry.

3.3 Regulation of Human Gene Therapy in the European Union

Although the United States are internationally recognized as the pre-eminent leader in human gene therapy with more than 230 clinical trials, 22 other nations are currently engaged in just under 100 protocols collectively (Countries conducting gene therapy trials, < *http://www.wiley.co.uk/genetherapy/ countries.htlm* >, visited May 24, 1998). Among nations outside of the United States, the European Union (EU) leads the way with 70 active protocols involving over 450 patients. Many of these studies focus on cystic fibrosis and cancer. Historically, however, the

EU as a whole has not kept pace with the American regulatory efforts concerning human gene therapy. As a result, the individual countries have established their own guidelines for human gene therapy regulation (DEJAGER, 1995).

The relative absence of Union wide standards and the prevalence of individual, nation wide regulations have hindered the overall development of a comprehensive system of regulation in Europe (DEJAGER, 1995). In response to this situation, and in the interest of spurring global competition, the Commission of the European Communities (1993) (the Commission) issued the *White Paper* in which it proposed to consolidate and coordinate the individual efforts of the Union countries in the area of biotechnology and human gene therapy. Recognizing the potential challenges and benefits of human gene therapy, the *White Paper* stressed the need for uniformity and harmony in regulation of this new field. In the same year, in response to a request from the Commission for an evaluation of the current status of gene therapy in Europe, the *Archer Report* confirmed much of what the *White Paper* proposed. Concluding that somatic gene therapy remained a new and still much unknown science, the report stressed the importance of subjecting all protocols and research studies to established ethical standards (Group of Advisors on Ethical Implications of Biotechnology of the European Commission, 1994). Like the *White Paper,* the *Archer Report* further suggested the utmost need for broad, uniform regulations applicable to all countries in the EU. To date, with guidelines and regulations still only existing in the individual countries, few of these much needed Union wide regulations have been enacted since 1993.

While germ line gene therapy research in the EU is currently limited to nonhuman studies, countries in the EU have expressed great enthusiasm for and approval of somatic gene therapy (Actions under the NIH guidelines, < *http://www.nih.gov/od/orda/10- 97act.htm* >, pp. 1–40, visited May 25, 1998). Countries such as Germany, Italy, and France have embodied the initial global interest in human gene therapy with their respective creations of the Bundestag Commission of Inquiry on Genetic Engineering in Germany, the Commission of Inquiry on Genetic Engineering in Italy, and the Comité Consultative National D'Ethique in France, all of which have given their approval of somatic gene therapy research (DeJAGER, 1995). Together with other nations such as the United Kingdom and the Netherlands, these countries constitute the forerunners of European nations in human gene therapy regulation.

Two international organizations, the United Nations Educational, Social, and Cultural Organization (UNESCO) and the Human Genome Organization (HUGO) are primarily responsible for the international coordination of gene therapy research and development. Both organizations, comprised primarily of scientists, coordinate the activities of scientists and researchers involved in the HGP and encourage ongoing discussion and dissemination of knowledge among the nations concerning issues such as the ethics, legality, and patentability of human gene therapy. Several smaller, national efforts and organizations have contributed positively to the field as well. These include the Human Genome Mapping Project in the United Kingdom, the Science and Technology Agency of Japan, the International Human Genome Diversity Project, and three French efforts: the French Muscular Dystrophy Association, Genethon, and Le Centre d'Etude du Polymorphisme Humaine. Several nations, including France, Spain, and Sweden, do not have any official guidelines for gene therapy. In addition to these efforts, the European Medicines Evaluation Agency (EMEA), consisting of members from individual countries, was established in 1995 to promulgate uniform, Union wide marketing authorization procedures for gene therapy products. These proposals are then translated into directives or law by the Commission.

Mirroring to a large extent the American system, the United Kingdom has the most comprehensive guidelines among European nations and has made the most significant advances in the area (KESSLER et al., 1996). The Gene Therapy Advisory Committee (GTAC), the NIH equivalent, serves as the main overseer of gene therapy research and promulgator of human gene therapy guide-

lines in the United Kingdom. Similar to the FDA and the NIH, the GTAC reviews individual protocols on a case-by-case basis and addresses larger issues and concerns relating to the clinical practice of human gene therapy, such as ethics, morality, and legality (LLOYD, 1994). Predating the GTAC, the Clothier Committee on the Ethics of Gene Therapy (the Committee), like the RAC, serves as the national ethical watchdog to all human gene therapy studies and experiments in the United Kingdom. The Committee consists of individuals with backgrounds in numerous scientific and medical disciplines, including genetics, molecular biology, virology, oncology, toxicology, clinical psychology, and immunology, as well as other non-medical fields, such as the law, the media, and industry. In a 1992 report, the Committee decreed that specific gene therapy techniques and procedures must meet certain established ethical thresholds before being studied in clinical settings on human subjects. Thus, similar to the RAC–NIH protocol review process, no protocol or experiment proposal reaches the GTAC without first passing ethical muster under the Committee.

Encompassing broad, common issues encountered in gene therapy research, the GTAC guidelines are similar to those of the NIH and place similar requirements on protocol submissions. The elements to be addressed in a protocol include:

(1) the objective and rationale of the proposed study,
(2) a risk–benefit analysis, including safety and efficacy concerns,
(3) the mode and criteria for human subject selection,
(4) a detailed informed consent provision which takes into consideration different levels of understanding and comprehension among human subjects, and
(5) the well-being of the human subjects and their potential offspring (LLOYD, 1994).

These considerations exist not only to safeguard the human participants in clinical trials, but also to ease the ethical and moral concerns of greater public. While there is considerable pressure from the private sector to expedite review and approval of protocols, the review and approval time period of the GTAC depends largely on the completeness of the submitted proposal in compliance with GTAC guidelines. The GTAC gives an initial review to the protocol once it is complete and then forwards it to the Committee for a full review.

The GTAC promotes a policy of information exchange similar to the American federal technology transfer policy. Mindful of the strong public fears surrounding ethics as well as the results which may be achieved through national and international discussion, the GTAC strongly believes in the dissemination of information, knowledge, and expertise among researchers and the greater public (LLOYD, 1994). The GTAC encourages such an exchange of ideas and knowledge through press releases which present accurate, realistic views of human gene therapy advances and benefits. Toward this end, the GTAC encourages researchers to publish their knowledge, results, and important discoveries in peer review journals.

4 Conclusions

The evolution of gene therapy regulation has fostered rather than restricted advances in this important area of research and development. This result has been possible, no doubt, due to the prior existence of domestic and international drug and device control authorities with considerable experience which could be applied quickly to the new area. The results might have been different if registration and regulatory systems had to be formulated and then impressed on an already existing and rapidly expanding biotechnology field.

The development of standards and practices has been marked by cooperative analyses, promotion of ethical reviews, and encouragement of technology transfers and intellectual property. Full safety and ethical discussions with broad disclosure have helped to alleviate public concern about gene therapy research and application.

In the United States, government reviews of potential products proceed quickly. The various initiatives making timely reviews possible include the *FDA Export Reform Act* of 1996, the *Prescription Drug User Fee Act* (PDUFA), the *Orphan Drug Act,* the *FDA Modernization Act,* executive regulatory reforms, and expedited and fast track designations. It is reasonable to expect to see the first human gene therapy product approved within a year.

Acknowledgement

The legal research and writing assistance of JEREMY S. DAVID, Suffolk University Law School student and Legal Intern, is gratefully acknowledged.

5 References

ALLEN, W. (1990), The current federal regulatory framework for release of genetically altered organisms into the environment, *Florida Law Rev.* **42**, 531–532.

Bayh-Dole Act (1980), *Public Law Number 96–517.*

Code of Federal Regulations (1996), *26, § 1.28-1,* Washington, D.C.: U.S. Government Printing Office.

Commission of the European Communities, Growth, Competitiveness, Employment (1993), *The Challenges and Ways Forward into the 21st Century* (1993), *White Paper,* COM(93) 700 Final, Publication Number 1993/704, Issued by Secretariate-General for the Commission, Published by the Office of Official Publications of the European Communities.

DEJAGER, C. F. (1995), The development of regulatory standards for gene therapy in European Union, *Fordham Int. Law J.* **18**, 1303–1305, 1316–1321, 1337–1338.

Domestic Policy Council Working Group on Biotechnology (1984), NIH Guidelines for research involving recombinant DNA molecules (NIH Guidelines), *Federal Register* **41**, 27920–921.

Federal Technology Transfer Act (1986), *Public Law Number 99-502.*

Food and Drug Administration Notice (1976), NIH Guidelines for Research Involving Recombinant DNA Molecules, *Federal Register* **41**, 27920–921.

Food and Drug Administration Notice (1984), Coordinated Framework for the Regulation of Biotechnology, *Federal Register* **49**, 50856.

Food and Drug Administration Notice (1986), Statement of Policy for Regulating Biotechnology, *Federal Register* **51**, 23302, 23310, § 2181.

Food and Drug Administration Notice (1993), Application of Current Statutory Authorities to Human Somatic Cell Therapy Products and Gene Therapy Products, *Federal Register* **58**, 53250.

Food and Drug Administration Notice (1996), Recombinant DNA Research: Proposed Actions Under the Guidelines, *Federal Register* **61**, 53251.

Food and Drug Administration Notice (1997), Recombinant DNA Research: Proposed Actions under the Guidelines, *Federal Register* **62**, 44387.

Freedom of Information Act (1994), *United States Code 5, §§ 552, I(B)(1)(b), Appendixes M, M-1,* Washington, D.C.: U.S. Government Printing Office.

FURROW, B. R. (1996), Enterprise liability for bad outcomes from drug therapy: the doctor, the hospital, the pharmacy, and the drug firm, *Drake Law Rev.* **44**, 377–437.

Group of Advisors on Ethical Implications of Biotechnology of the European Commission (1994), Report on ethical aspects of gene therapy 1.

KESSLER, D. A., HASS, A. E., FEIDEN, K. L., LUMPKIN, M., TEMPLE, R. (1996), Approval of new drugs in the United States, *J. Am. Med. Assoc.* **276**, No. 22.

KIN, C. A. (1996), Coming soon to the genetic supermarket near you, *Stanford Law Rev.* **48**, 1573–1604.

LAWTON, A. (1997), Regulating genetic destiny: a comparative study of legal constraints in Europe and the United States, *Emory Int. Law Rev.* **11**, 365–418.

LLOYD, D. J. (1994), *Gene Therapy 1*, pp. 341–342, New York: Macmillan.

MALINOWSKI, M. J., O'ROURKE, M. A. (1996), A false start? The impact of federal policy on the gene technology industry, *Yale J. Regul.* **13**, 163–249.

McGARITY, T. O. (1995), Peer review in awarding federal grants in the arts and sciences, *9 High Technol. Law J.* **1**, 7.

National Competitiveness Technology Transfer Act (1989), *Public Law Number 101–189.*

Orphan Drug Act (1993), *Public Law Number 97-414.*

Patent Act (1952), United States Code Annotated (1984), *35, §§ 1–293.*

Restatement (Second) of Torts (1965), *§ 402A, comment k.*

Stevenson-Wydler Technology Innovation Act (1980), *Public Law Number 96-480.*

United States Constitution (1787), *Art. I, § 8, cl. 8.*

United States Code (1996), *26, § 45C,* Washington, D.C.: U.S. Government Printing Office.

24 Economic Considerations

Ian J. Nicholson

Abingdon, UK

1 Introduction 532
2 Industry Background 532
3 Economic Significance of Biotechnology-Derived Medicines 534
   3.1 Drive towards Single Sourced Innovative Medicines 536
   3.2 Industry Consolidation to Achieve Economic and Structural Advantage 537
   3.3 Increased Use of Alliances Outsourcing to Speed the Drug Discovery and Development Process 537
   3.4 Growth in the CRO Industry 538
4 Biopharmaceutical Manufacturing Costs – A Key Economic Issue in Product Development 539
5 Manufacturing Technology 540
6 Biomanufacturing Trends 540
7 Long-Term Perspectives on the Nature of Biopharmaceutical Development 541
8 Conclusions 541
9 References 542

List of Abbreviations

CMO	contract manufacturing organization
CRO	contract research organization, clinical research organization
DHFR	dihydrofolate reductase
FDA	Food and Drug Administration
GMP	good manufacturing practice
IND	investigational new drug
NCE	new chemical entity
OTC	over the counter
PPL	Pharmaceutical Proteins Ltd.

1 Introduction

The biotechnology revolution now into its third decade represents a decisive technological landmark for the 20th century. Over the past 20 years the techniques of modern molecular biology and immunology have opened up therapeutic and diagnostic solutions previously undreamed of. The use of genetic engineering and hybridoma technology supported by modern manufacturing technology has lead to the discovery and development of many important medicines, and through new developments in gene therapy, genomics and bioinformatics provides the promise of significant advances well into the next millennium. The industry in 1996 had grown to a market capitalization of over $80 billion and had product sales of over $10 billion. This chapter will cover the building of the biotechnology industry and the economic factors which need to be addressed in the development, manufacture and marketing of these products both now and in the future. This chapter deals with the commercial issues involved in the development of biopharmaceuticals be they monoclonal antibodies, recombinant proteins or gene therapy-based approaches; it is not intended to deal with a product or company focused review. Although biotechnology has found applications in many fields outside the pharmaceutical arena, this chapter will focus on biopharmaceuticals.

The economic environment for companies involved in biopharmaceutical research and development has changed over the early part of this decade in both a fundamental and rapid way. The costs involved in the discovery and development of a new product are increasing rapidly. The clinical development process for a new protein-based drug now takes 7–12 years and can cost from $150–$250 million. Not only costs are accelerating but increased competition has led to an overall decrease in product life cycles as competing products are developed. The period following expiration of patent coverage is now a battleground for market share as the ability of generic competitors becomes ever stronger. The need for single source innovative new products has never been greater as a means to sustain growth and diversify risk from portfolios overdependent on a limited number of blockbuster drugs. It is not surprising then that the influence of biotechnology-derived medicines is key to the future of many pharmaceutical companies. The drive to get these products through the clinic and into the market is fierce.

2 Industry Background

The story of the biotechnology industry to this point is a volatile one – product news is good or bad, financial markets on which so much depends open and close and the regulatory authorities decide upon product approval. The biotech industry itself has now undoubtedly overcome the critics to emerge as one of the key industries of the next century bringing with it a clearly established utility and the prospect of transforming the field of human healthcare with novel therapeutics and means of diagnosis.

In the mid to late 1970s the emergence of recombinant DNA technology did not immediately realize this potential, and its impact on pharmaceutical research at that time was limited. The use of the technology was restricted essentially to the production of useful proteins that were available only in limited quantities – examples of this were human growth hormone and insulin. These proteins, although well known, had the disadvantage of being available in extracted form from animal sources at very low yields. The emergence of a technology which enabled production of relatively high yields of pure proteins from a highly controlled process, lead to the launch of recombinant human insulin in 1982 and recombinant human growth hormone in 1986 (Tab. 1). The sales of recombinant human growth hormone amounted to approximately $500 million worldwide in 1995 (Tab. 2). These "replacement therapy" recombinant products represented the first wave of biotechnology-derived medicines to reach the market.

The potential of recombinant technology really emerged in the 1980s as it became pos-sible to identify genes and produce proteins of therapeutic interest that had previously been difficult or impossible to isolate. The cytokines, erythropoietin, granulocyte colony stimulating factor, interleukin-2, are all examples of this category of recombinant proteins. These products have so far proved to be some of the most successful biotechnology products on the market with sales of the red blood cell stimulator erythropoietin currently in excess of $1 billion worldwide. Many analysts believe that erythropoietin will be the largest selling pharmaceutical by the year 2000. By the mid-1990s several recombinant protein-based drugs had achieved blockbuster status with sales in excess of $500 million per annum. A summary of these is provided in Tab. 2.

Tab. 1. Biotechnology Product Approvals (BURRILL and LEE, 1994)

1982	1983	1984	1985	1986	1987	1988	1989
Human insulin			Human growth hormone	OKT3	t-PA	EPO	IL-2
First recDNA product			Second genetically engineered drug	First mono-clonal antibody approved			Approved in parts of Europe

Tab. 2. Leading Biotechnology Drugs on the Market

Product	Product type	Developer	Marketer	1995 Worldwide sales ($m)	% Change from 1994
Neupogen	biological response modifier	Amgen	Amgen	936.0	13.0
Epogen	erythropoiesis enhancer	Amgen	Amgen	882.6	22.0
Procrit	erythropoiesis enhancer	Amgen	Ortho Biotech	795.0	19.5
Humulin	antidiabetic agent	Genentech	Eli Lilly	794.1	19.4
Engerix-B	hepatitis B vaccine	Genentech	SmithKline Beecham	624.1	3.7
Intron A	biological response modifier	Biogen	Schering-Plough	433.0	1.6
Activase	clot dissolver	Genentech	Genentech	301.0	7.0
Betaferon	multiple sclerosis therapy	Schering AG	Schering AG	279.7	n/a
Humatrope	human growth hormone	Eli Lilly	Eli Lilly	269.1	19.2
Protropin	human growth hormone	Genentech	Genentech	219.4	(2.9)

Note: Dollars are in millions (Pharma Business, 1996)

Although beset by early problems and some notable product failures, the development of therapeutic monoclonal antibodies (in particular those based on the exploitation of new technologies in protein engineering) is now showing encouraging signs of success. Most notable is the anti-platelet monoclonal ReoPro developed by the US biotechnology company Centocor used in the prevention of blood clots following angioplasty. Sales of the product in 1996 were approximately $150 million. Monoclonal antibodies have also already found utility in the efficient purification of certain marketed recombinant proteins notably α-interferon and recombinant factor VIII for the treatment of hemophilia.

But perhaps the greatest impact that biotechnology will have on the development of pharmaceutical products is yet to come – the identification and sequencing of the full complement of human genes (currently thought to be between 80,000 and 100,000). This new ability to learn about the gene's structure, function, and regulatory changes, inherited or acquired, that are involved in disease onset and progression, provides a fertile and previously untapped source of new drug targets and diagnostic tools. The development of gene therapy protocols has to date been hampered by problems associated with vector design and delivery. It is widely expected that the development of gene therapy-based approaches to therapy will have a broad applicability in the management of human disease. It now appears that the technology will not be restricted to single gene disorders such as cystic fibrosis but may have wide ranging utility in major disease conditions such as cancer, cardiovascular disease and diabetes. The methodology used in the clinical studies to date, however, is far from optimal and the effective *in vivo* delivery of genetic material is still an obstacle which needs to be overcome. The prospect, though, that gene therapy will become a reality having significant advantages over conventional approaches looks promising as these problems are overcome.

The understanding of the correlation between gene function and disease provides the basis for the discovery of novel therapeutic approaches. So powerful and fundamentally important is the information emerging from the so-called genomics toolbox that most major pharmaceutical companies are investing heavily in the field, either through external partnerships and alliances or in-house programs. The importance of this new paradigm in the drug discovery world will be discussed in Sect. 7.

The biotechnology revolution then has demonstrated huge potential to fundamentally alter human therapy through the provision of tools, targets and molecules which impact pharmaceutical development.

3 Economic Significance of Biotechnology-Derived Medicines

The number of approved biopharmaceuticals continues to grow at an encouraging rate with many new products gaining regulatory approval and already marketed products exceeding prior years sales. The commercialization of biotechnology-derived medicines is now very much a business reality. Total sales of biotech products have reached a total of $10.8 billion (LEE and BURRILL, 1997). The industry has a full pipeline with over 700 products from 167 companies in U.S. clinical trials (RYAN and CROWE, 1994).

In recent years there has been a significant number of product approvals – key products are given in Tab. 3. The high number of recent product approvals is a result of two main factors, firstly the large number of products in clinical trials and secondly reform of approval policies most notably by the U.S. FDA which has reduced the renewal times (Fig. 1). The pipeline of the industry is fuelled primarily from two sources, the biotech companies themselves and the traditional research-based pharmaceutical industry. It is interesting to note that the Pharmaceutical Manufacturers Association survey of biotechnology medicines revealed that 33% of research programs in major pharmaceutical companies were based on biotechnology compared with 2% in 1980. Many studies have evaluated the impact of these new medicines on the global pharma-

Tab. 3. Significant Recent Product Approvals

Product	Developer	Indication	Date
ReoPro	Centocor	high-risk coronary intervention	December 1994
Epivir	Biochem Pharma	AIDS	November 1995
Neupogen	Amgen	PBPC mobilization	December 1995
Invirase	Hoffmann La-Roche	AIDS	December 1995
Respigam	MedImmune	RSV	January 1996
Avonex	Biogen	multiple sclerosis	May 1996
Vistide	Gilead Sciences	CMV retinitis	June 1996
Myoscint	Centocor	cardiac imaging	June 1996
Activase	Genentech	stroke	June 1996
Seprafilm	Genzyme	postoperative scarring	August 1996

ceutical market and estimates of future growth of biotech drugs based on filed INDs and patent filings indicate strong growth for the sector. Clearly then the economic significance of the biotechnology industry is enormous.

It is useful to evaluate the drivers for the huge investment in the biotech industry, an industry which now has a market capitalization of greater than $80 billion employs over 118,000 personnel and uses up over $7.9 billion in research and development costs (LEE and BURRILL, 1997).

The economic environment of the research-based pharmaceutical industry in the 1990s is vastly different and more hostile than that of the recent past. Coupled to the increasing investment outlays for innovative research and development are the rapid and fundamental changes in the healthcare marketplace. The costs involved in the discovery and development have escalated rapidly in real terms. The intense competition in both the generic market and the impact of following drug candidates on product life cycles has put ever increasing pressure on pharmaceutical companies to innovate and get to market with all haste. Recent estimates have suggested that towards the year 2000 the share of the world market taken by prescription drugs would drop from about 75–55%. The generics share would rise from 13–25% and the over-the-counter (OTC) share would rise from 12–20% (DOWER, 1996). Areas of unmet clinical need are the battleground for novel drug development as companies compete to develop therapies for diseases such as cancer, arthritis and AIDS.

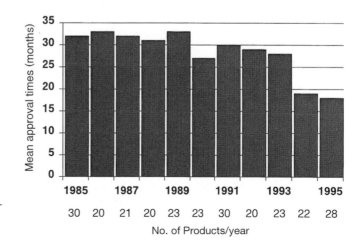

Fig. 1. Expedited approval times (mean) of the FDA in months, numbers below years are numbers of products per year.

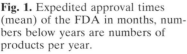

There can be no mistaking the fact that health care has become too expensive resulting in major cost reduction programs on a global basis. The science of pharmacoeconomics has been born out of a need to assess the economic outcomes of the therapeutic armory that modern medicine provides. It is now imperative that incremental economic benefit can be demonstrated from a novel treatment when compared to existing therapies. Health care providers, regulatory agencies and governments themselves are demonstrating a sensitivity to health care cost reduction which is having enormous impact on the way in which pharmaceutical research and development is carried out. The industry is under severe price pressure and is adopting a new *modus operandi* for the next millennium. The key features of this cost-driven paradigm shift are summarized below:

- drive towards single sourced innovative medicines
- industry consolidation to achieve economic and structural advantage
- increased use of alliances and outsourcing to speed the drug discovery and development process

In the context of economic considerations in the development of biotech medicines it is valuable to delve further into each of these strategic outputs.

3.1 Drive towards Single Sourced Innovative Medicines

Staying ahead in the pharmaceutical industry today requires intensive effort in the development of novel and economically viable medicines that meet a clinical need which is currently poorly served. The whole research-based industry (the generics industry has been ignored for our purposes) is in a race to beat competitors to market with such products. The stakes are high and the resources needed to compete are vast. Companies cannot afford to keep running on the spot but must generate innovative pipelines to survive. The major pharmaceutical companies are driving to launch at least one new significant product per year. This is not an easily achievable goal as the elegant study of Prof. JUERGEN DREWS (DREWS, 1995) demonstrates. Using projected product success rates and assumptions of revenues, growth and life cycle, the outlook for research and development success was evaluated in the top 50 pharmaceutical companies. On the basis of new chemical entities (NCEs) alone, the outlook for achieving the desired goal of one significant product launch per annum looks poor. In fact the analysis suggested that projected launches of novel NCEs would result in reduction in sales over time of 10%. This problem is compounded by the shortening of product life cycles by fast followers and the reality of reduced patent exclusivity periods.

The study of DREWS (1995) clearly indicates both the need for new sources of innovative therapies and a conclusion that the industry will consolidate further to achieve a sustainable and supportable size. The economic expectation then that in-house sources of NCEs will not provide sufficient growth has led to a massive strategic rethinking and subsequent changes within the industry.

The answer adopted to date lies with the extensive restructuring of the pharmaceutical industry and the growing pipeline of novel therapies emerging from the biopharmaceutical industry. If we assume that the 700 biotech products in clinical development carry risks not unlike that of traditional NCEs in successfully making it to market, then it is quite feasible that between 20 and 30 new biotechnology-derived therapies will reach the market on an annual basis from 2000 onwards. This forecast does not seem unrealistic in view of the number of significant approvals gained in the last 2 years (Tab. 3). The array of technologies used in drug discovery continues to advance at a very rapid pace – a new toolbox has emerged, a toolbox which includes genomics, bioinformatics, and combinatorial chemistry. Pharmaceutical companies are aware of the benefits of collaborating with academia and of striking alliances with small entrepreneurial companies. It seems likely that the level of biotech–pharma alliances is likely to increase as big companies strive to fill research and development pipelines with single sourced (i.e., non-generic) innovative therapies.

3.2 Industry Consolidation to Achieve Economic and Structural Advantage

The major research-based pharmaceutical companies then are in a global race to acquire and develop both products and technology to compete in the next millennium. This wealth of product opportunities, however, may not be enough to halt the massive restructuring and consolidation which has indelibly altered the competitive landscape of the industry in the 1990s. The scale of consolidation is dramatic as the number of recent mergers and acquisitions demonstrates. The whole hierarchy of the industry has now shifted with many industry observers believing that the emergence of 10–20 truly global companies will be the hallmark of the industry in the early part of the next century. The necessity for global presence, competitive product offerings and novel research and development pipelines to effectively compete in a cost-effective manner is driving the creation of a new elite band of leaders. One industry executive recently forecast that by the year 2000 there will be just 15 truly global companies with a turnover of more than $7 billion; 35 international companies with a turnover of $3–$7 billion; about 50 regional companies with revenues of $1–$3 billion; and about 400 national companies generating revenues of less than $1 billion.

3.3 Increased Use of Alliances Outsourcing to Speed the Drug Discovery and Development Process

The same pressures that have forced the massive consolidation and restructuring upon the pharmaceutical industry have led to a major trend towards the outsourcing of numerous operations functions. These include the increasing use of contract research organizations (CROs) and contract manufacturing organizations (CMOs). This trend has also led to the creation of so-called virtual pharmaceutical companies. In view of the acceleration in the mid-1990s of the number of drugs losing patent protection and the increasing influence of the managed care companies in North America, the emphasis of filling pharmaceutical pipelines with innovative medicines has never been greater. The drivers behind this trend to outsourcing is the need to achieve greater efficiencies by enhancing revenues, reducing costs and improving the speed at which innovative therapies can be brought to market. In particular, pharmaceutical companies are increasingly adopting ways to minimize fixed costs and better manage variable costs. Additional factors towards this outsourcing trend are listed below:

- the increasing complexity and global nature of clinical trials
- the increasing number of biopharmaceutical products in clinical development
- the range of novel drug discovery technologies now available (rational drug design, genomics, etc.)
- the strategic focus by pharmaceutical companies on the core competencies of clinical development and marketing

By considering that it takes on the average $250 million to get a biopharmaceutical product to market, the need to ensure rapid market launch cannot be underestimated. A 12-month reduction in development time can mean incremental sales of many hundreds of millions of dollars. It is thus not surprising that a new generation of companies has sprung up to cater for this new demand. The range of contract services opportunities in the pharmaceutical industry can be summarized as follows:

- contract market research
- contract research (e.g., preclinical/toxicology testing)
- contract development: regulatory/advisory phase I–IV clinical trials, pharmacoeconomics
- contract manufacturing

Tab. 4. Clinical Research Organization Activities

Study/protocol design	monitoring/data collection
Case/report form development	project management
Site/investigator recruitment	data entry and verification
Site initiation	biostatistical analysis
Patient enrollment	regulatory affairs
Quality assurance	pharmacoeconomics

- contract distribution
- contract sales and marketing

At the time of writing it would appear that the two main beneficiaries of the outsourcing trend will be the contract research organizations and the contract manufacturing organizations. It is not yet clear whether the use of external sales and marketing functions will provide a third major area of outsourcing.

3.4 Growth in the CRO Industry

Clinical research organizations, although somewhat of a misnomer, help pharmaceutical companies better manage their costs by enabling them to minimize fixed costs (including fixed headcount) and allowing variable costs to be managed on an as "as-needed" basis. They act as coordinators and managers of clinical trials and regulatory filings helping to coordinate and speed the global registration process. Typical activities undertaken by CROs are listed in Tab. 4. In addition to providing pharmaceutical companies with a flexible clinical development resource (and thus enabling them to reduce internal resource to a minimum level), the major CROs are able to offer expertise and excellence in selected therapeutic areas and in the case of the large full-service companies manage large-scale multinational trials.

The CRO industry, which was largely established in the last decade, has experienced tremendous growth in the last few years as more and more pharmaceutical companies have built up trust in the use of their services. Large-scale clinical trials involving thousands of patients across several continents may generate several million dollars of revenues for the CRO, although in some cases only a small element of a clinical development program will be undertaken. In 1993, revenue estimates for the combined industry were approximately $1 billion, by 1995 some industry observers estimated revenues as $2.5 billion. This established and rapidly growing field is still, however, a fragmented industry with hundreds of competitors present although there is now a noticeable trend towards consolidation in the industry with several large companies emerging with a dominant market share.

It should also be noted that there is an increasing trend on the part of both large and small life science companies to outsource both process development and manufacturing activities. A number of studies has been undertaken in this area (Bio/Pharmaceutical Outsourcing Report, 1997; NICHOLSON and LATHAM, 1994) and provide the reader with useful background. Again, the underlying rationale for this trend can be found in the need for companies to focus on core competencies, improve efficiency and minimize internal headcount. It is envisaged that recent structural changes in the industry, with leading outsourcing firms from the traditional pharmaceutical outsourcing industry (e.g., Lonza and Gist-Brocades) taking a stake in biopharmaceutical manufacture, will continue.

4 Biopharmaceutical Manufacturing Costs – A Key Economic Issue in Product Development

In considering the economic issues facing the development of biotechnology-derived medicines the subject of manufacturing must be treated with the priority it deserves. All too often the considerations of production costs are left to a late stage in product development when really these should have been considered a lot earlier in the product's development history. Understanding the fundamentals of biotechnology manufacturing is especially important when it is generally recognized that the cost of making protein-derived medicines is significantly higher than chemical synthesis. The characteristics of biotechnology processes that account for this difference are as follows: costs of process development are high, highly skilled labor is essential; both expensive and extensive product and process testing are needed; recovery of product is costly; plant carries a relatively high capital cost. Evaluation of cost-effective manufacturing strategies can thus be critical for the success or otherwise of promising clinical development candidates.

In a very simplistic model, biopharmaceutical manufacturing can be divided into 3 key steps: upstream processing, downstream processing and analytical/quality functions. Whether recombinant protein or monoclonal antibody the fundamental choice of host cell of either eukaryotic or microbial origin needs to be considered. Generally eukaryotic cells are used to ensure accurately folded and glycosylated bioactive proteins, but yields from these cells are usually about 5–20% of that achieved in microbial hosts. Bacterial cells, such as *E. coli*, on the other hand, tend to accumulate the protein in an improperly folded, non-native form in inclusion bodies. Given the generally accepted lower manufacturing costs for bacterial-derived products, many companies have chosen to use bacterial production over mammalian cell culture but only where glycosylation is not important for bioactivity. It should, however, be noted that many of the largest selling biopharmaceutical products are derived from mammalian cell culture.

To address these cost and yield issues efforts are under way to coax bacterial cells to express correctly folded and glycosylated proteins. There is still a considerable amount of effort needed to achieve this routinely in a production setting. The key issue for mammalian cell culture lies not in achieving appropriate product quality, but rather in increasing the expression levels of about 5–15% of their *in vivo* secretory capacity in order to compete on a cost-effective basis with bacterial fermentation. Efforts to improve the widely used DHFR expression system have been made with some success by several companies, most notably the glutamine synthetase system™, developed by Lonza Biologics plc and the IDEC expression technology developed at IDEC Pharmaceuticals Inc. Both of these systems have reported product yields in excess of 1.5 g L^{-1} which, while encouraging, still leave room for significant further optimization work.

A recent development in the manufacture of biopharmaceuticals has been the development of transgenic animals for the production of proteins of interest. Several companies have been formed to commercialize the technology which promises to deliver high yields and correctly processed product. Early data suggest that very high yields of certain recombinant proteins and monoclonal antibodies may be possible with this novel approach although some significant issues regarding the length of time to reach a production herd and the regulations surrounding transgenic products will need to be addressed. Transgenic companies, notably Pharmaceutical Proteins Ltd. (PPL), are developing technologies to reduce the time taken to generate a production herd. In the case of PPL the often publicized enabling technology used to produce the first cloned sheep (named Dolly) is being harnessed for this purpose. The acceptance of this technology particularly for the expression of complex proteins is increasingly presenting a realistic alternative to traditional mammalian cell culture.

5 Manufacturing Technology

The key differentiation in choice of biopharmaceutical manufacturing technology rests with the decision on host cell (Tab. 5) and consequently the selection of a microbial process or a mammalian cell culture-based process. Microbial technology largely developed for antibiotic manufacture is well proven and a great deal of manufacturing experience exists in the industry. For mammalian cell culture-based processes, a large number of bioreactor designs has been developed. However, most large companies have chosen to install and utilize large-scale suspension fermentation equipment (predominantly of airlift or stirred tank design). Large-scale microbial fermentation which has already been scaled to capacities of hundreds of cubic meters presents less challenge than the scale-up of mammalian cell culture. At the time of writing the largest mammalian cell culture vessels for biopharmaceutical manufacture are believed to be at the 10,000 L scale.

On the downstream side, most biopharmaceutical products utilize cumbersome multistep chromatography steps. Much effort is being devoted to streamlining these complex processes – a notable example of this is the use of perfusion-based chromatography to provide simpler high-throughput methods.

One of the key reasons for underestimating manufacturing costs is the lack of understanding of the quality costs involved in producing parenteral biopharmaceutical products. Necessitated by FDA and other regulatory agencies, the requirements for GMP plant operation, validation and ongoing product, process and environmental testing place a significant cost burden on overall manufacturing costs. In some cases this can be as high as 25–40% of total cost and thus needs to be factored in to cost estimates at an early stage.

6 Biomanufacturing Trends

Despite the significant advances of recent years in both arenas of mammalian and microbial production there remains a significant differential between the costs involved in the synthesis of small molecule chemical drugs and biopharmaceuticals. It is inevitably difficult to foresee with any degree of accuracy the degree of progress that both commercial and academic groups will make towards closing this gap. Neither is it clear how newer technologies such as transgenics will penetrate the biopharmaceutical field. For the foreseeable future though it is unlikely that chemical synthesis costs will be matched by biopharmaceutical processes and that the existing technologies will continue to dominate the manufacturing scene for the short term at least.

Tab. 5. Proteins Produced by Genetic Engineering (DAVIES, 1988)

Protein	Production system	Application
Insulin	*E. coli, S. cerevisiae*	diabetes
Interferon-α	*E. coli*	cancer
Human growth hormone	*E. coli*, mammalian cells	dwarfism
Interleukin 2	*E. coli*	cancer
Colony stimulating factors	*E. coli*	cancer
Erythropoietin	mammalian cells	anemia
Plasminogen activators	mammalian cells	cardiovasular disease
Hepatitis B surface proteins	*S. cerevisiae,* mammalian cells	vaccine
Monoclonal antibodies	mammalian cells	various

7 Long-Term Perspectives on the Nature of Biopharmaceutical Development

With the economic drivers of massive well-funded demand for novel single sourced pharmaceutical products from the pharmaceutical industry well established, it seems clear that there will be major investment in technologies that achieve this goal. If we set our sights into the future then a combination of technologies only recently established, looks set to yield results through a potent mix of biotechnology and novel chemistry. The mix is that of genomics, high-throughput screening and combinatorial chemistry.

Genomics or genomic science is currently spuming an entire industry in its own right, ranging from the mechanistic aspects of sequencing the human genome to functional genomics, the science of identifying the function of novel genes. Perhaps surprisingly there is only a relatively small number of genes coding for secreted proteins (those proteins most likely to have pharmaceutical importance via signaling pathways etc.) perhaps 500–1,000. The identity of these genes having a known function provides a potent starting point for drug development. This is truly where the skills of parallel developments in chemistry really provide a synergy for explosive discovery of new small molecule drugs. The ability to identify novel targets via genomic science increases pressure on the traditional discovery and development role of the pharmaceutical medicinal chemistry department. However, advances in chemical technology have led to a new science of combinatorial or high-speed chemistry. This technology enables large numbers of structural analogs and derivatives to be rapidly generated around a template chosen for its drug-like characteristics via rational drug design or medicinal chemistry. These compounds are then screened in a rapid fashion in so-called high-throughput screening assays to generate lead compounds for preclinical and clinical study.

This potent combination of developments in biotechnology coupled to significant advances in traditional chemistry-based drug development promises to deliver a host of innovative therapies to satiate the hunger of the pharmaceutical giants of the next millennium.

8 Conclusions

In reviewing the economic aspects of biopharmaceutical product development it becomes clear that the challenges facing the pharmaceutical industry and the biotechnology industry are vast. The pressures to generate successful single sourced medicines have increased significantly with the likelihood of more consolidation and contraction in the pharmaceutical industry ever more likely. With this paradigm shift in the industry the creation of pharmaceutical global giants seems inevitable as we approach the next millennium. These global companies, despite large research and development budgets, will increasingly rely on biotechnology as a source of global competitiveness. It is clear that, while the protein-based clinical development pipeline is laden with promising biopharmaceutical candidates, the explosive growth needed to fulfil pharmaceutical company goals of 1–3 new products per year is likely to be derived from a potent combination of biotechnology and chemistry, both traditional and modern. Not only does it seem likely that there will be an ongoing stream of recombinant proteins and monoclonal antibody-based therapeutics, but our increased understanding of the human genome and functional genomics coupled to rapid advances in our ability to generate lead discovery libraries of novel chemical molecules (combinatorial chemistry) and high-throughput screening should rapidly increase the number and quality of novel single sourced therapeutics as we enter the new millenium.

9 References

Bio/Pharmaceutical (1997), *Bio/Pharmaceutical Outsourcing Report* **2**, 2–3.

BURRILL, G. S., LEE, K. B. (1994), *Biotech '94 Long Term Value, Short Term Hurdles, Ernst & Young's 8th Annual Report on the Biotechnology Industry.*

DAVIES, J. C. (1988), *TIBTECH.* **6**, 67–511.

DOWER, M. (1996), *Script Magazine.*

DREWS, J. (1995), The impact of cost containment on pharmaceutical research and development, *Tenth CMR Annual Lecture,* London.

LEE, K. B., BURRILL, G. S. (1997), *Biotech '97: Alignment, Ernst & Young's 11th Annual Report on the Biotechnology Industry.*

NICHOLSON, I. J., LATHAM, P. (1994), When "make or buy", means "make or break", *Biotechnology* **12**, 473–477.

Pharma Business (1996), September/October 1996.

RYAN, D. R., CROWE, K. R. (1994), *Contract Pharmaceutical Services (Nov. 9th 1994)*, pp. 1–11, San Francisco, CA: Robertson Stevens & Co.

Index

A

Abciximab *see* ReoPro
abzymes (catalytic antibodies) 221, 233
actin filaments 31
activin 165
acute myelogenous leukemia (AML), phase I
  study, preliminary results – 376
– treatment of 358ff
ADA *see* adenosine deaminase deficiency
ADAPT (antibody-dependent abzyme prodrug
  therapy) 221
adeno-associated virus (AAV), biology 413f
– genome 414f
– integration site 414
– life cycle 413f
– serotypes 414
– structure 414f
– type 2 414
adeno-associated virus vectors 397f, 413ff
– cell entry 415
– clinical studies 421
– design 416
– gene cassettes, design rules 417
– – limitations 417
– genome 416
– host responses 420
– host tropism 415
– in HIV gene therapy 480f
– manufacturing 417ff
– – complementation systems 417
– – recombination events 417
– of cystic fibrosis 421
– packaging limit 417
– packaging systems 417f
– particle structure effects 417
– persistence and integration 415f
– preclinical studies 419ff
– – *in vitro* 419ff
– – *in vivo* 420
– properties 417
– purification 419
– regulatory elements, design rules 417
– – limitations 417
– replication 415f
– replication competent – 418
– target cells 420
– toxicity 420
– wild-type generation 418
adenosine deaminase deficiency (ADA) 472
– gene therapy of 444
adenovirus, biology 406
– genome 406f
– latency 408
– life cycle 407f
– persistence 408
– replication competent – 410f
– structure 406f
– type 5 genome 407
adenovirus vectors 397ff, 406ff
– animal models 411
– cell targeting 411f
– clinical studies 413
– design 408f
– expression *in vivo* 412
– Food and Drug Administration requirements
  411
– genome 408
– immune responses 410, 412
– inflammation 412
– manufacturing 409ff
– – complementation systems 409
– packaging cells 409f
– persistence *in vivo* 412
– preclinical studies 411ff
– – *in vitro* – 411
– – *in vivo* – 411f
– production systems 409
– properties 408f
– purification 411
– RCA contamination 410f
– recombination 410
– scale-up of production 411
– therapy of cystic fibrosis 413
– therapy of tumors 413

– toxicity 412
– transgene expression cassettes 409
ADEPT (antibody-directed enzyme prodrug therapy) 221, 321
ADEPT/ADAPT 233
adult respiratory distress syndrome (ARDS), use of liposomes in gene therapy 433
AIDS *see also* HIV
– gene therapy with retroviruses 405
– monoclonal antibody therapy 317
– treatment with fusion proteins 169
– use of interleukin-4 antagonist 152
– use of recombinant hormones 164
– vaccine 138
Aldesleukin 151
allergy, mAb products 222
– medical applications of recombinant proteins 140, 145
– monoclonal antibody therapy 316
allotransplantation 145f
– medical applications of recombinant proteins 145
Alteplase 129
AML *see* acute myelogenous leukemia
amphotropic murine leukemia virus, use of vector systems, in HIV gene therapy 479f
α-amylases 192ff
– biology 192
– commercial products 193
– detergent applications 196ff
– – automatic dishwashing 196f, 199
– development 197f
– – for the detergent industry 197
– – high-fructose corn sirup 197
– in paper production 195f
– in starch liquefaction 193f
– in textile desizing 194f, 198
– industrial applications 192ff
– properties of technical – 193
– reaction catalyzed by – 192
– stability 192ff
– – against oxidants 197f
– thermostability 198
analytical ultracentrifugation, of protein complexes 91
angina, unstable –, treatment with ReoPro 376
angiogenesis 133f
– medical applications of recombinant proteins 133
angiostatin 134
antibodies *see also* monoclonal antibodies, therapeutic antibodies
– activity of 295ff
– – preclinical determination 295f
– binding specificity 312
– chimeric – 248
– choice of starting antibody 292f
– for sepsis 329ff

– immunogenicity 289ff, 299f
– intracellular –, in HIV gene therapy 477f
– kinetics 289ff
– pharmacokinetics 297f
– pharmacology 289ff
– preclinical testing of 289ff
– – affinity 292
– – animal models 296ff
– – carcinogenicity studies 309
– – cell-based assays 291
– – conditions of administration 308
– – cross-reactivity 295, 306f
– – determination of activity 295
– – fragments 294f
– – genotoxicity studies 309
– – guidance documents 305
– – immunoassays 291
– – limit of detection 292
– – local tolerance 309
– – oncogenicity studies 309
– – pharmacokinetics 297f
– – pharmacology 295
– – regulatory requirements 305
– – reproduction toxicity studies 308f
– – safety aspects 303ff
– – safety testing program 305
– – selection of isotypes 293f
– – species used 308
– – timing in relation to clinical trials 306
– – tissue specificity 306f
– – use of transgenic models 297
– – use of xenografts 297
– – validation of assays 291f
– – whole antibodies 294f
– product selection 292f
– recombinant – 248f
– regulation 495ff
– therapeutic –, manufacture of 245ff
– use for immunopurification 275ff
antibody–antigen complex 35, 99f
antibody-based products, manipulation of half-life 234f
antibody binding sites 221ff
– binding strength 230
– bispecific antibodies 231
– complementary determining regions 223
– conformation predictions 223
– formats 229f
– generation of 221ff
– human anti-murine antibody response 227
– humanized – 226ff
– low molecular weight – 229
– murine – 226f
– number per product 230
– primatized – 228
– sources 226ff
– structures 221f
antibody cross-reactivity 306f

antibody-dependent abzyme prodrug therapy *see* ADAPT
antibody-directed enzyme prodrug therapy *see* ADEPT
antibody–effector conjugates, ADAPT 233
– ADEPT 233
– cytokines 232
– DNA (antibody targeted gene) therapy 234
– drugs 233
– enzymes 232
– immunotoxins 232
– radioisotopes 234
antibody–effector fusions 232
antibody engineering 219ff
– effector mechanisms 230f
– – ADAPT 233
– – ADEPT 233
– – antibody–effector conjugates 232
– – antibody–effector fusions 232
– – antibody Fc functions 231f
– – bispecific antibodies 231
– – cytokines 232
– – drugs 233
– – enzymes 232f
– – immunotoxins 232f
– half-life of products 234f
– production of recombinant antibodies 235ff
– – *E. coli* fermentation 236
– – glycosylation 236
– – GRAS organisms 237
– – mammalian expression 236
– – use of insect cells 237
– – use of transgenic animals 237
– – use of transgenic plants 237
antibody expression 219ff
– effector mechanisms 230f
– – ADAPT 233
– – ADEPT 233
– – antibody binding sites 234
– – antibody–effector conjugates 232
– – antibody–effector fusions 232
– – antibody Fc functions 231f
– – bispecific antibodies 232
– – cytokines 232
– – DNA (antibody targeted gene) therapy 234
– – drugs 233
– – enzymes 232f
– – immunotoxins 232f
– – radioisotopes 234
– half-life of products 234f
– production of recombinant antibodies 235ff
antibody Fc functions 321f
antibody fragments 283f, 293ff
– human/mouse chimeric monoclonal – 365ff
– therapeutic applications 313
antibody humanization 226ff
antibody light-chain loci 224ff
antibody primatization 228

antibody product development 292
antibody purification 263ff
– DNA contamination 265
– economic aspects 266
– methods 267f
– – affinity chromatography 267
– – ion exchange chromatography 267
– – SDS-PAGE 267
– – size exclusion chromatography 267f
– protein A matrices 267
– protein contaminants 264f
– purity specification 264
– requirements 264f
– virus contamination 265
antibody selection 292
– of isotypes 293f
– surface plasmon resonance analysis 293
antibody targeted chemotherapy 355ff
antibody tissue specificity 306f
anticytokine therapy, in sepsis 333f
antigens, human –, expressed in transgenic mice 297
antisense RNA, in HIV gene therapy 474f
anti-TNF antibody 332f, 343ff
antiviral constructs, for HIV gene therapy 473ff
– – antisense RNA 474f
– – decoy RNA 476
– – delivery of – 479ff
– – protein inhibitors 476f
– – ribozymes 474ff
– – vector design 479ff
apolipoproteins 15f
– in human plasma, function of 16
archaebacteria, genome 51
Archer Report 526
ARDS *see* adult respiratory distress syndrome
aspartyl proteases 207
Aspergillus phytase 210f
aspirin 374
– in bleeding events 354, 369
asthma, mAb products 222
– medical applications of recombinant proteins 145
– monoclonal antibody therapy 316
atherectomy 375f
– therapy with ReoPro 375
ATPases, cation transport 30
– functions 30
autoimmune diseases 38
– mAb products 222
– medical applications of recombinant proteins 140
– monoclonal antibody therapy 315, 318

B

baby hamster kidney cell lines (BHK), in manufacturing of antibodies 253

– – serum-free media 257
bacteriotoxins 34
baloon angioplasty *see* PTCA
Bayh-Dole Act 522
binding specificity, of antibodies 312
biodistribution studies 361f
– hCTM01 361
– tumors 361f
biologics license applications, statistics 503
biomanufacturing trends 540
biopharmaceutical development, long-term perspectives 541
biopharmaceutical manufacturing costs, transgenic animals 539
bioreactor types, in manufacture of therapeutic antibodies 261f
biotechnology-derived medicines 535ff
– approval times 535
– clinical research organizations 538
– commercialization 534
– contract manufacturing organizations 537
– contract research organizations 537
– costs 535f
– economic environment 535
– economic significance 534f, 537
– industry consolidation 537
– market 535
– single sourced – 536f
– virtual pharmaceutical companies 537
– woldwide sales 533
biotechnology evolution 519f
biotechnology products 531ff
– approvals 533ff
– approval times 535
– cell culture 539f
– characterization methods 497
– clinical research organizations 538
– commercialization 534
– contract manufacturing organizations 537, 537
– contract research organizations 537
– control procedures 497
– costs 535
– crisis management in Europe 513
– development, long-term perspectives 541
– economic considerations 531ff
– economic environment 532, 535
– economic significance 534f, 537
– European Council regulation 509
– genetic engineering 540
– industry background 532f
– industry consolidation 537
– manufacturing costs 539f
– manufacturing steps 539
– manufacturing technology 540
– market 535
– – capitalization 532
– marketing authorization holder, responsibilities 512

– marketing in Europe 510ff
– – organization 510f
– registration of 509
– – advisory committees 499, 501ff
– – European Medicines Evaluation Agency 506f
– – filing process 500
– – Food and Drug Administration 499ff
– – Food and Drug Administration Modernization Act 503
– – in Europe 506ff, 511
– – in the UK 506
– – orphan drug legislation 505f, 514
– – review process 501ff
– – well-characterized product 502f
– sales 534
– single sourced – 536f
– specific guidelines 498
– timetable for review of application, in Europe 511
– trademarks in Europe 513
– transgenic animals 539
– use of 534
– virtual pharmaceutical companies 537
– worldwide regulation 514
bispecific antibodies 321
blood clotting cascade 34
– intrinsic pathway 32
blood group antigens 10
bone marrow transplantation, monoclonal antibody therapy 318
bone morphogenetic proteins 162f
– function 163
– in therapy 162f
Borrelia burgdorferi 140
brain-derived neurotrophic factor 159
brain natriuretic peptide 165
breast cancer, monoclonal antibody therapy 321
Brookhaven Protein Database 66, 92

C
c7E3 Fab *see* ReoPro
Caenorhabditis elegans, genome 51
calcitonin 165
calicheamicin 356ff
– binding to DNA 356
– bioreduction 357
– conjugation with monoclonal antibodies 358
– disulfide reactions 357f
– functionalization 357
– structure 356
– triggering 357
Calin 130
calorimetry of protein complexes 90f
Cambridge Structural Database 92
cancer *see also* tumor
– chemotherapy 321

– mAb products   222
– monoclonal antibody therapy   321
– treatment with fusion proteins   169
– use of recombinant enzymes   168
cancer vaccines   135, 138
CAPTURE trial   376ff
cardiopulmonary bypass, monoclonal antibody therapy   316
cardiovascular disorders   129f, 315f
– mAb products   222
– therapeutic applications of recombinant proteins 129f
– use of ReoPro   366ff
casein   209
catalase   209
catalytic antibodies *see* abzymes
catalytic mechanisms of enzymes   25f
catenates   446
CC-1065   321
cDNA sequences   50f
CDP571, cross-reacting antibodies   351
– elimination profiles   348, 351
– generation of   345
– immunogenicity   345, 348
– in chronic therapy   343ff
– in rheumatoid arthritis   347ff
– – immunogenicity   348ff
– – pharmacokinetics   348ff
– – repeated-dose study   347f
– – results of treatment   349f
– preclinical studies, in non-human primates   345f
– – in patients   346ff
CDR *see* complementary determining regions
CD4 receptor   472ff
cell banking, in manufacturing of antibodies   257f
– – documentation   258
– – function of   258
– – good manufacturing practice   257
– – origin of   257
– – preparation of   257f
– – storage   258
cell banking systems, regulations   525
cell-based assays, testing of antibodies   291
cell culture   539f
cell lines, in therapeutic antibody manufacture 248ff, 258f
– – stability parameters   259
– – use of recombinant DNA technology   259
cellobiase *see* β-glucosidase
cellobiohydrolases   203ff
cellular assays, testing of antibodies, relative potencies   293
cellulases, activity   203f
– cloning   204ff
– effects   204
– industrial applications   202ff
– in textile industry   205f
– microbial –   202ff

– properties   202f
– recombinant   204
– – in detergent industry   204
– – in textile industry   204f
cellulosomes   203
Center for Biologics Evaluation and Research (CBER)   499ff
– clinical studies   504f
– investigational new drugs   504f
– plasmid DNA, regulations   459
– plasmid DNA therapeutics, guidelines   445
– structure of   500
Center for Drugs Evaluation and Research (CDER)   499ff
– review process   501
cerebral ischemia, medical applications of recombinant proteins   130f
cerebrovascular thromboembolic disorders   129f
– therapeutic uses of recombinant proteins   129
cervical cancer, vaccines   138
cheese production, use of cloned calf chymosin 209
chemokines   165f
– classes   166
– function   166
– in therapy   165f
– receptors   166
chemotherapy, antibody-targeted –   355ff
chimeric antibodies *see* antibodies
Chinese hamster ovary cell line (CHO)   248f, 251f
– in manufacturing of antibodies, serum-free media   257
chloroquine, transfection level enhancement   436f
cholesteroloxidase, reaction catalyzed by –   99
chorionic gonadotropin   164
chronic inflammatory disorders, monoclonal antibody therapy   315
chronic therapy, with CDP571   343ff
chylomicrons   15
chymosin, cloning of   209
– industrial applications   209
– manufacturing of cheese   209
cilia, mobility system   32
ciliary neurotrophic factor   159
clinical trials
– adeno-associated virus vectors   421
– adenovirus vectors   413
– CMA-676   360f
– CMB-401   362
– macrophage stimulating factor   157
– ReoPro   365ff
– sepsis   335ff
Clothier Committee on the Ethics of Gene Therapy   527
clotting factors   34
– in therapy   168f
CMA-676 treatment   356ff
– adverse events   359f

– development   358
– efficacy data   361
– immune reaction to   361
– of acute myeloid leukemia   358ff
– patient enrollment   359
– phase I study   360ff
– – preliminary results   360
– safety data   359
CMB-401   356ff, 361f
– clinical studies   362
– development   358
– imaging studies   361f
– in tumor treatment   361
Code of Federal Regulations   462
coenzymes   24
coiled-coils   67f
colitis, ulcerative –   319
collagens   4, 32f
– functions   33f
colony stimulating factors   155ff
– genetic engineering of   540
– in therapy   155
colorectal carcinoma, monoclonal antibody thera-
    py   320
complement, in therapy   166
complement receptors   166
complement system   36
– regulation of   38
complementary determining regions (CDR)   223
concatemers   446
congestive heart failure, medical applications of
    recombinant proteins   132
connective tissue growth factor   132
contractile proteins   31ff
Coordinated Framework for the Regulation of
    Biotechnology   520
corn syrup, high-fructose –   194, 197
coronary diseases   366ff
– treatment with ReoPro   366
coronary interventions, use of ReoPro   367ff, 371
coronary stenting, therapy with ReoPro   375
coronary syndromes, treatment with ReoPro   371
corticosteroids, in sepsis therapy   334
COS *see* monkey kidney cell lines
Crohn's disease, monoclonal antibody therapy
    315, 319
– role of tumor necrosis factor   346
CTM01, biodistribution studies   361f
cysteine proteases   207
cystic fibrosis   472
– gene therapy with adenovirus vectors   413, 421
– use of liposomes in gene therapy   433
– use of recombinant enzymes   167
cytochrome c   20
cytokine antagonists, in therapy   140
cytokines   135ff, 146ff
– – *see also* interferons, interleukins
– definition   146

– in IBD therapy   144
– medical applications of recombinant proteins
    146
cytomegalovirus   472
– monoclonal antibody therapy   317
cytotoxic lymphocytes, HIV reactive –   481f

D
database searches of protein sequences   53ff
Debye–Hückel parameter   94
defense proteins   34f
– – *see also* antibodies
– blood clotting   34
– clotting factors   34
– of the immune system   35
– preventing penetration of disease causing fac-
    tors   34
denim processing, use of industrial cellulases   204
detergents, use of, α-amylases   196
– – cellulases   204
– – lipases   200f
– – recombinant proteases   208
DHFR (dihydrofolate reductase) expression sys-
    tems   250ff
diabetes mellitus   134f, 161, 164
– antibody therapy   314, 319
– insulin dependent –   144
– medical applications of recombinant proteins
    134, 144
– non-insulin dependent –   134f
– treatment with CDP517   346f
DNA *see also* plasmid DNA
– binding with polylysine   435
– calcium phosphate precipitation   430f
– compaction of –   435
– content of various cells   4
– direct transfer   429f
– for transgene expression   437
– immunization using particle bombardment   432
– *in vivo* targeted delivery   435
– microinjection   430
– particle bombardment for transfection   432
– plasmid, gene delivery   435
– receptor-mediated endocytosis   434f
– targeting ligands   436
– use of, anionic liposomes for transfection   434
– – cationic liposomes for transfection   432
– – cationic polymers for transfection   431
– – electroporation for transfection   431
– – neutral liposomes for transfection   434
DNA manufacturing   443ff
DNA vaccination   430ff
Domestic Policy Council Working Group on Bio-
    technology   520
drug master file, plasmid DNA manufacturing
    457f
dynein   4

E
EBV *see* Epstein Barr virus
E. coli, expression of, human insulin   112
– – recombinant antibodies   236
– – somatostatin   112
– genome   3
– in plasmid DNA manufacturing   457f
– in plasmid production   453
– plasmid vectors, guidelines   452
– removal of endotoxins   455, 465
E. coli K12, in plasmid DNA manufacturing
   447ff
EGF *see* endothelial growth factor
elastin fibers   33
electron transfer proteins   20
electroporation   431
ELISA *see* enzyme-linked immunosorbant assay
EMEA *see* European Medicines Evaluation Agency
enceophalomyelitis, monoclonal antibody therapy
   319
endoglucanase   203ff
endopeptidases   206
endosome release, transfection level enhancement   436f
endostatin   134
endothelial growth factor (EGF), in therapy   141
endotoxin, bacterial, composition of   331
enzyme cofactors, metal ions   19
– trace elements   19
enzyme–inhibitor complexes   96f
enzyme inhibitors, medical applications   167
enzyme kinetics   22f
enzyme-linked immunosorbant assay (ELISA)
   290
enzymes   22ff
– α-amylases   192ff
– catalytic effect   23
– catalytic mechanisms   25f
– coenzymes   24
– hydrolases   25
– induced fit hypothesis   22
– industrial applications   189ff
– isomerases   25
– ligases   25
– lock and key hypothesis   22
– lyases   25
– market   191
– medical applications   167
– Michaelis-Menten equation   23f
– nomenclature   24ff
– prosthetic group   24
– reactions catalyzed by –   22
– recombinant technology   191
– synthetases   25
– therapeutic recombinant –   167f
– transferases   25
enzyme–substrate complex   22

EPIC trial   367ff
epidermal growth factor, in therapy   162
– receptor   162
EPILOG trials   374ff
EPO *see* erythropoietin
Epstein-Barr virus (EBV), vaccines   139
erythroid differentiation factor   156
erythropoietin (EPO)   155f
– genetic engineering of   540
– in therapy   155
– market   533
– sales   533
eubacteria, genome   51
eukaryotes, genome   51
eukaryotic cells, DNA content   4
– organelles   6ff
– protein compartmentalization   6
– protein content   4
– secreted proteins   9
European Agency of Medicinal Products, good
   manufacturing practice   460ff
– – definition   460
European Commission, number of approvals of
   biotechnology products   512
European Community, regulations of gene therapy   461
European Medicines Evaluation Agency
   (EMEA)   506ff, 513, 527
– clinical studies   504f
– function of   507f
– investigational new drugs   504f
– organization of   507
– structure of   508
European Union   506ff
– pharmaceutical legislation   506ff
evolution, of antibodies   3
– of DNA   2f
– of enzymes   3
– of metabolic pathways   3
– of multicellular organisms   3
– of proteins   2f
– of RNA   2f
– of self-replicating systems   2f
– of the genetic code   3
exopeptidases   206

F
Fab fragments   224
factor IX   283
factor VIII   283
FAD *see* flavin adenine dinucleotide
FDA *see* Food and Drug Administration
Federal Food and Cosmetic Act   498
Federal Technology Transfer Act   522
Federal Virus, Serum and Toxin Act   498
feed industry, use of phytase   210ff
ferritin   19, 34

fibrin   34
fibrinogen   34
fibroblast growth factors   133
– acidic –   162
– androgen-induced –   162
– basic –   162
– glia-activating –   162
– in therapy   162
– keratinocyte growth factor   162
– receptors   162
– use in therapy   131
fibronectin   33
fibrosis   132f
– medical applications of recombinant proteins
   132
flagella, mobility system   32
flavin adenine dinucleotide (FAD)   20f
flavoproteins   20ff
– cofactor   20
– structure   20
fluid mosaic model   7
folding catalysts, use in inclusion body folding
   120f
folding types, of proteins   64
– – prediction of   66
follicle stimulating hormone   164
food, removal of bacteria   285
Food and Drug Administration (FDA)   445f,
   499ff, 534, 540
– adenovirus vectors, testing of   411
– advisory committees   501
– approval times   535
– clinical studies   504f
– drug master file   457f, 461
– gene therapy regulation   523f
– good manufacturing practice   460ff
– – definition   460
– guidelines   523f
– investigational new drugs   504f
– plasmid DNA, quality issues   460
– plasmid DNA therapeutics, guidelines   445
– plasmid vectors   452
– priority review   501f
– process for filing   500
– regulation of biotechnology products   519f
– regulation of gene therapy   521
– retrovirus vectors, testing of   404
– structure of   499f
Food and Drug Administration Export Reform
   Act   528
Food and Drug Administration Modernization
   Act   500, 503ff
force field methods, energy calculations   93f
Fourier spectral analysis, automated evaluation
   58
Freedom of Information Act   524
free energy of binding   95
fusion proteins, medical applications   169f

G
gancyclovir   482
Gaucher's disease, use of recombinant enzymes
   168
Gelsolin   167
gene addition therapy   384, 392
gene amplification, in manufacturing of antibodies   251f
– – cell lines   251ff
– – DHFR   251
– – glutamine synthetase   251f
– – markers   251ff
– – methotrexate   251f
– – selection procedures   251
gene delivery   385ff, 438
– approaches   385f
– barriers to   429
– vehicles   386f
– – non-viral –   386, 429
– – physical –   386
– – viral –   386
gene-directed enzyme prodrug therapy   387
gene expression, barriers to   429
– in *Escherichia coli,* human insulin   112
– – somatostatin   112
– non-viral –   429
– stability of, cell line dependence   254
– – influencing factors   254
– transient nature of   437
gene expression systems   250f
– in manufacturing of antibodies   250
gene gun   386, 388, 392, 466
gene replacement, recombinase driven –   253
gene replacement therapy   384, 392
gene therapy   383ff
– access to target cells   387f
– anionic liposomes   434
– antibody-targeted –   234
– approved clinial protocols   385f
– Archer Report   526
– calcium phosphate precipitation   430f
– candidate diseases   385f
– cationic liposomes   432ff
– – adult respiratory distress syndrome   433
– – cystic fibrosis   433
– – limitations   433
– – stability   433
– – toxicity   433
– clinical status   386f
– cost issues   391f
– definitions   384
– delivered DNA components   437
– development   387ff
– direct gene transfer   429ff
– – expression   430
– electroporation   431
– *ex vivo*   384ff, 522
– for HIV infection   471ff

– – future concepts 485ff
– – optimization of – 484ff
– – phase I clinical trials 482ff
– future of 392f
– gene delivery 385ff
– – approaches 385f
– – costs 392
– – vehicles 386f
– gene expression 437
– guidance for industry 524f
– guidelines 461, 523f
– history 386
– immune responses 391
– in animals 170f
– *in vivo* 384ff, 521
– inflammatory reactions 387, 391
– liposome–HVJ complexes 434
– locus control regions 389f
– long-term transgene persistence 390
– microinjection of DNA 430
– neutral liposomes 434
– non-targeted cationic polymers 431
– non-viral vectors 427ff
– of AIDS 405
– of cystic fibrosis 413, 421
– of germ cells 521
– of inoperable gliomas 405
– of somatic cells 521
– of tumors 405f, 413
– overview 383ff
– particle bombardment, DNA immunization
432
– plasmid vector design 452
– promoters used 437
– receptor-mediated endocytosis, DNA-binding
435
– – DNA-compaction 435
– – efficiencies 437
– – endosome release 436
– – targeting ligands 436
– – targeting moiety 435
– receptor-mediated endocytosis 434f
– regulating acts 522f
– regulations 460f, 517ff
– – history 520f
– – in the European Union 526ff
– regulatory committees 520f
– replicating vectors 389
– safety issues 391f
– target cells 385, 390
– technology transfer approaches 522
– trafficking of vectors 388
– transfection efficiency 388
– transfection systems 429ff
– transgene expression 389f
– – level 437
– transgene size limitation 389
– vector systems 386

– – for *in vivo* delivery 438
– – non-viral – 427ff
– – viral – 395ff
– White Paper 526
Gene Therapy Advisory Committee, guidelines
527
gene therapy products, registration 498ff
– – authorities 498ff
– – in Europe 498
– – in the USA 498ff
– – laws 498ff
gene therapy research, germ line experiments 524
– somatic cells 524
genetic code 49f
– evolution of 3, 50
gene transfection efficiency, non-viral vectors 388
– proliferating cells 388
gene transfer, direct – 429f
generic drugs, market share 535
genomes, annotated sequences 49
– *Escherichia coli* 3
– human 52
– large-scale sequencing projects 51
– of archaebacteria 51
– of *Caenorhabditis elegans* 51
– of eubacteria 51
– of eukaryotes 51
– of *Haemophilus influenzae* 51
– phylogenetic tree 51
– sizes of 51
– yeast 3
germ cell therapy 521
Gibbs free energy 89f
glial cell growth factor 2 160
glial-derived neurotrophic factor 159
glial maturation factor 160
gliomas, gene therapy with retroviruses 405
GLP *see* good laboratory practice
glucoseoxidase 209f
β-glucosidase (cellobiase) 203
glutamine synthetase 250ff
– expression systems 250
– vector system 249
glycoconjugates 9
glycoproteins 9ff
– function of 9
– glycosylphosphatidyl inositol membrane-
anchored proteins 13
– N-linked – 12
– – primary structure 13
– O-glycosidic attachment 10
– therapeutic – 14
GMP *see* good manufacturing practice
good laboratory practice (GLP) 307
good manufacturing practice (GMP) 510f
– cell banking 257f
– in manufacture of therapeutic antibodies 268f
– – quality assurance 268

– – quality control   268
– in manufacturing of antibodies   261
– manufacturing facilities   269
– – inspection of required interactions   269
– plasmid DNA   445, 452
– regulatory authorities   269
granulocyte colony stimulating factor   156
– clinical trials   157
– production by *in vitro* folding   113
granulocyte macrophage colony stimulating factor,
   in therapy   157
GRAS (generally regarded as safe) organisms
   237
growth and differentiation factor-1   160
growth factors, interaction with cell surface receptors   39
– therapeutic use   155ff
growth hormone releasing factor   165
growth hormones   163f, 283
Guidance for Industry, somatic cell therapy   524f

H

Haemophilus influenzae genome   51
β-hairpins   62
HAMA *see* human anti-mouse antibody
HDL   15
Helicobacter pylori vaccine   139
α-helix   59ff
helix types, in protein structure   60ff
hematopoietic growth factors, in tumor therapy
   137
hemoglobin   20
hemoproteins   20ff
heparin   371, 374
– in bleeding events   369, 376
hepatitis A virus, monoclonal antibody therapy
   317
hepatitis B surface proteins, genetic engineering
   of   540
hepatitis B vaccination   137f
hepatitis B virus, monoclonal antibody therapy
   317
hepatitis virus vaccines   139
hepatocyte growth factor   163
herpes simplex vaccines   138f
herpes simplex virus (HSV), monoclonal antibody
   therapy   317
HFV *see* human foamy retrovirus
HIV (human immunodeficiency virus) *see also*
   AIDS
– infection   471ff
– life cycle   473f
– protein   473f
HIV gene therapy   471ff
– antiviral constructs, antisense RNA   475
– – design of   473ff
– – protein inhibitors   476ff

– – RNA decoys   476
– – RNA inhibitors   474ff
– future concepts   485ff
– – next generation of clinical trials   486
– immunotherapics   481f
– – expression of immunogenic HIV proteins
   482
– – selectable marker genes   482
– optimization of –   484ff
– – tissue-specific gene expression   484
– – transduction protocols   485
– – vectors   484
– phase I clinical trials   482ff
– – biosafety issues   483
– – ethical issues   483f
– vector design   479ff
HIV protease   96ff
HIV protease–inhibitor complex   97f
homology searches, in protein databases   57ff
hormones   163ff
– interaction with cell surface receptors   39
– therapeutic recombinant-, under development
   164
HSV *see* herpes simplex virus
human anti-mouse antibody (HAMA)   290
– development of   248
human embryonic kidney cell line   249
human foamy retrovirus (HFV), use as vector
   481
Human Genome Organization   528
Human Genome Project   520ff
human growth hormone, genetic engineering of
   540
– production by *in vitro* folding   113
human immunodeficiency virus *see* HIV
human insulin like growth factor I   134
human papilloma virus vaccines   139
hybridomas   128
– in manufacturing of antibodies, serum-free media   257
hydrolases   25f

I

IBD *see* inflammatory bowel disease
IgGκ   225ff
– domain   225
immobilization of antibodies   278ff
immune deviation, medical applications of recombinant proteins   142
immune responses   35f
– cellular –   35
– humoral –   35
– in gene therapy   391
immunoaffinity chromatography   276, 280f
immunoassays, for preclinical testing of antibodies   291

immunoconjugates *see also* calicheamicin, CMA-676, CMB-401
- in therapy 356ff
immunogenicity, of antibodies, preclinical testing 289ff
- of CDP571 348ff
immunoglobulin G, structure 12
immunoglobulins 36
- classes 37
immunopurification, choice of antibodies 276f
- clinical applications 285
- detection, aflatoxin contamination 284
- - cannabis metabolites 284
- for substance measurements 284
- immobilization of antibodies 277f
- - chemical coupling techniques 278
- - properties of desirable matrices 278
- immunoaffinity chromatography procedures 280ff
- monoclonal antibodies 276
- of growth hormones 283
- of interferons 283
- of tumor necrosis factor 283
- of whole cells 284
- polyclonal reagents 276
- removal of bacteria 285
- ricin removal 284
- site specific immobilization of antibodies 279f
- specific removal of substances 284
- use of antibodies 275ff, 283
- - applications 282f
- - clincal applications 285
- - single domain antibodies 283f
- use of antibody fragments 283f
immunosuppression 317f
- monoclonal antibody therapy 317
- polyclonal antibody therapy 317
immunotherapies for HIV, gene therapy-based – 481f
inclusion body proteins 111ff
- formation, role of overexpression 113
- - suppression of 113
- in cytosol 113
- *in vitro* aggregation 115
- *in vitro* disulfide bond formation 115f
- *in vitro* folding 111ff, 115ff
- - effect of denaturant concentrations 118
- - effect of L-arginine 117
- - effect of Na$_2$SO$_4$ 118
- - effect of Tris buffer 117
- - immobilized catalysts 120
- - improvement of 116f
- - low-molecular weight additives 116
- - pH dependence 116
- - transfer optimization 117f
- - transfer procedures 118
- - use of folding catalysts 120f
- - use of molecular chaperones 120
- - yield 118f
- isolation 113f
- - by cell lysis 113
- solubilization of 114f
- - by chaotrophs 114
inclusion body renaturation 121
induced fit hypothesis 22
industrial enzymes 189ff
- α-amylases 192ff
- cellulases 202ff
- chymosin 209
- lipases 199ff
- market 191, 214
- oxidoreductases 209ff
- phytase 210f
- proteases 205ff
infectious disease, mAb products 222
- monoclonal antibody therapy 316
inflammation, use of recombinant enzymes 167
inflammatory bowel disease (IBD) 144f
- medical applications of recombinant proteins 144
- role of tumor necrosis factor 344, 346
influenza hemagglutinin proteins, use as vaccines 138
insulin 163
- expression in *E. coli* 112
- genetic engineering of 540
- production by *in vitro* folding 113
insulin-dependent diabetes mellitus, medical applications of recombinant proteins 144
insulin like growth factors 133, 163f
insulin restistance 134f
- medical applications of recombinant proteins 134
integrin antagonists 134
integrin receptor 130
intercellular adhesion molecule-1 131
interferon α, genetic engineering of 540
interferon β, in therapy of multiple sclerosis 142
interferon γ, in tumor therapy 136
interferons 283
- in therapy 147ff
- type I 148f
- - indications for – 147
- type II 148
interleukins 135f, 149ff
- antagonists 150, 152
- antibodies to 135
- genetic engineering of 540
- in therapy 135, 144, 148ff
- in tumor therapy 136
- receptors 149
- recombinant –, indications for – 149
- - under development 149
- therapeutic potential 149
- use in therapy 130
- - of osteoporosis 133

investigational new drugs 504f
– clinical studies 504f
– regulatory aspects 504f
ischemia, use of ReoPro 377f
isomerases 25ff

K
Kaplan-Meier event rates, in myocardial infarction 378
α-keratin 33
kinetics, of antibodies 289ff, 297f
– – preclinical testing 298
Kozak sequence 402

L
β-lactamase, crystal structure 78
lactic dehydrogenase, *in vitro* folding 118
laundry detergents, use of lipases 200f
LDL 15
Leuferon 151
leukemia, AML 358ff
– monoclonal antibody therapy 320
leukemia inhibitory factor 158
Leukine 157
Leukoctropin 157
ligases 25ff
– acid–base catalysis 25
lipases 199ff
– biology 199
– characteristics 200
– developments 202
– function 199f
– industrial applications 199ff
– in laundry detergents 200f
– reaction catalyzed by – 200
– recombinant – 200
– sources 199f
– stability 202
– use of household detergents, effect of wash cycles 201
lipid-linked proteins 16ff
lipocalins 29
lipopolysaccharide-binding protein 135
lipoproteins 14ff
– chylomicrons 15
– function 14
– high density – (HDL) 15
– low density – (VDL) 15
– non-covalent aggregates 14
– of human plasma, major classes of 15
– very low density – (VLDL) 15
liposomes, as transfection systems 432ff
lock and key hypothesis 22
London energy 89
lupus erythematodes, antibody therapy 314, 319
luteinizing hormone 164

lyases 25f, 26
Lyme disease, vaccine 138, 140
lymphomas, monoclonal antibody therapy 321

M
mAb *see* monoclonal antibodies
macrophage colony stimulating factor, clinical trials 157
– in therapy 157
mad cow disease 513
major histocompatibility complex (MHC) 141
major non-immune-based diseases 129f
– therapeutic uses of recombinant proteins 129
malaria vaccine 139
mammalian cell culture, antibody manufacture 249
master cell bank 452f
– quality control 453
melanoma vaccines 138
melatonin 165
membrane protein anchoring 7f
membrane structure, fluid mosaic model 7
memprane proteins 6ff
metalloproteases 207f
metalloproteins 19ff
methotrexate 250
MHC *see* major histocompatibility complex
MI *see* myocardial infarction
Michaelis–Menten equation 23f
microinjection of DNA 430
modeling *see* protein modeling
Modernization Act 528
molecular chaperones, use in inclusion body folding 120f
Moloney murine leukemia virus (Mo-MLV), genome structure 399
Moloney murine sarcoma virus (Mo-MSV) 397
monkey kidney cell lines (COS) 248f
monoclonal antibodies (MAb) *see also* antibodies, therapeutic antibodies
– against tumor antigens 294
– anti-endotoxin – 331ff
– anti-TNF 134, 332f
– CDP571 343ff
– conjugation with calicheamicin 358
– cross-reactivity 295
– engineering of 343ff, 540
– half-life, animal models 298
– – in humans 298
– – stability 298
– human – 228ff
– – *in vitro* production 229
– – production in transgenic animals 228
– human anti-mouse antibody response 299
– human/mouse chimeric fragment 365ff
– immunogenicity 299f
– in therapy 21, 290ff, 299

– – animal models 296ff
– in xenotransplantation 146
– medical applications of recombinant proteins 134
– pharmaco-toxicological testing 307ff
– preclinical testing, use of transgenic models 297
– – use of xenografts 297
– reduction of pyrexia 294
– registration of 498ff
– – authorities 498ff
– – in Europe 498
– – in the USA 498ff
– – laws 498ff
– serotype-specific – 331
– therapeutic applications 311ff
– – targets for intervention 314
– therapeutic grade –, purity 264
– therapy, clinical trials 335
– treatment of chronic diseases 290
– use in HIV gene therapy 477f
– use in sepsis therapy 330ff
– use in tumor therapy 135
monoclonal antibody anti-intercellular adhesion molecule-1 131
monoclonal antibody-based products 222f
mucin 12
multiple sclerosis 142f
– medical applications of recombinant proteins 140, 142
– monoclonal antibody therapy 319
– role of tumor necrosis factor 344
– vaccines 138
muscle contraction, filaments 28
– schematic diagram of 28
muscle proteins 31
myasthenia gravis, monoclonal antibody therapy 319
myeloma cells (NSO) 248f
– in manufacturing of antibodies 252f
– – serum-free media 257
myocardial infarction (MI), Kaplan-Meier event rates 378
– use of ReoPro 367ff, 374, 377f
myoglobin 20
myosin filaments 31

N
Na$^+$-K$^+$pump 29
National Competitiveness Technology Transfer Act 522
National Institutes of Health (NIH), Division of Biologics Control 498
– gene therapy regulations 520, 523f
– Guidelines 519f, 523f
– regulation of biotechnology products 519f
Natuphos 212

nerve growth factor 158ff
Neupogen 158
neurological diseases, monoclonal antibody therapy 319
neurotrophic factor, brain-derived – 159
– ciliary – 159
– glial-derived – 159
neurotrophin-3 160
new drugs applications, statistics 503
NIDDM *see* non-insulin-dependent diabetes mellitus
NIH *see* National Institutes of Health
nitric oxide 135
NMR spectroscopy 96
– of protein complexes 92f
NMR structures, refinement of 94
noggin 160
non-insulin-dependent diabetes mellitus (NIDDM) 134f
– medical applications of recombinant proteins 134
NSO *see* myeloma cells
nuclear Overhauser effect (NOE) 93
nutrient proteins, functions 33
– of animals 33
– of plants 33
Nutropin 164

O
OKT3 317
Oncostatin M 159
organ transplantation, medical applications of recombinant proteins 141
orphan drug, definition 505
Orphan Drug Act 523, 528
orphan drug legislation 505ff, 514
osteoblastic cells, synthesis of phosphoproteins 17
osteoporosis, medical applications of recombinant proteins 133
ovalbumin 4f
over-the-counter drugs, market share 535
oxidoreductases 24ff, 209ff
– industrial applications 209

P
pancreatic ribonuclease, unfolding in urea 112
paper production, use of α-amylases 195f
papilloma virus, monoclonal antibody therapy 317
parathyroid hormone 165
particle bombardment 432
peptidases, classification 206
percutaneous transluminal coronary angioplasty *see* PTCA

pertussis vaccines   138
pharmacokinetics, of antibodies, preclinical testing   298
– of CDP571   348ff
pharmacology, of antibodies   289ff, 297f
– – preclinical testing   298
pharmaco-toxicological testing, of antibodies   307ff
– – tissues used   308
phase I clinical trials, CMA-676   360f
– of HIV gene therapy   482ff
phosphoproteins   17ff
– secreted into milk   17
phytase   210f
– characteristics   211
– industrial applications   210f
– of *Aspergillus*   210f
– – properties   210
– reaction catalyzed by –   210
– recombinant –   212
– substrate specificities   211
– use in feed industry   210ff
phytic acid, in feeds   210
plasma membrane, lipid-linked proteins   17
plasmid DNA manufacturing, fermentation of cells   448f
plasmid-containing cells, fermentation of   448f
– – high-performance technology   448
plasmid DNA *see also* DNA
– cell lysis   450
– clinical-grade –   465
– clinical trials   460
– drug master file   461
– fermentation   446ff
– good manufacturing practice   445, 452ff
– in vaccination   444, 465f
– manufacturing conditions, quality assurance   445
– – quality control   445
– pharmaceutical requirements   460f
– pilot production   453
– purification   446ff, 450f
– quality control   458ff
– – DNA quantity   459
– – restriction analysis   459
– – sequencing   459
– – sterility   459
– therapeutic –, future of   465f
– therapeutic use, guidelines   461
plasmid DNA manufacturing, design   449ff
– documentation   465
– downstream processing   454f
– drug master file   457f
– *E. coli* host cells   447
– facility requirements   461f
– fermentation   453f
– flow diagram   454
– guidelines   461

– harvesting   453f
– host cells   449
– in-process control   463
– large scale –   443ff
– master cell bank   452f
– plasmid isolation   447f
– – laboratory scale –   447f
– plasmid stability   451f
– process types   458
– process validation   462f
– purification   447f, 453, 455f, 458
– – lysis procedures   447f
– quality assurance   463f
– quality control   463f
– safety aspects   449
– strain selection   452
– upscaling   456f
– use of antibiotic resistance genes   446
– vector selection   452
– yield   449, 458
plasmid DNA therapeutics, good manufacturing practice   445
– guidelines   445
– manufacturing conditions, quality assurance   445
– – quality control   445
– regulatory aspects   445f
– safe manufacturing process   445f
– system development process   445
plasmid preparation, research-grade –   448
plasmid stability   451ff
– effect on plasmid yield in manufacturing processes   451
– in manufacturing   451
– testing of   451
plasmid vectors, use in gene therapy   452
– use in vaccination   452
plasminogen activators, genetic engineering of   540
– in therapy   168f
platelet-derived growth factor   161f
– in therapy   161
platelet endothelial cell adhesion molecule   145
platelet receptor blockage, with ReoPro   367
Poisson–Boltzmann equation   71, 94
polyclonal antibodies   317
polypeptide backbone   60
polypeptides, conformational properties   60
porins   3
prebiotic reactions   2
preclinical testing, of antibodies   289ff, 303ff
– – assays   291ff
– of CDP571   345f
prediction, of antibody conformations   223
– of protein interactions   93ff
– of protein structure and function   55ff, 65ff, 88
– – antigenic sites   67f
– – coiled-coil segments   67f

– – quaternary structure 77f
– – secondary structure 65ff
– – solvent accessibility 76f
– – tertiary structure 73ff
– – transmenbrane regions 67f
prescription drugs, market share 535
Prescription Drug User Fee Act 500ff, 528
primary structure *see* protein structure
product consistency, of therapeutic antibodies 254f
product recovery, in manufacture of therapeutic antibodies 262f
prokaryotic cells, DNA content 4
– lack of Golgi apparatus 14
– protein compartmentalization 6
– protein content 4
Proleukin 151
proprietary medicinal products, committee for – 509
proteases 205ff
– actions 205
– classification 206ff
– – aspartyl proteases 207
– – cysteine proteases 207
– – detergent proteases 208
– – metalloproteases 207
– – serine proteases 206f
– in detergents 208
– industrial applications 205ff
– production by genetically modified microorganisms 208
– substrate specificity 206
– types 205
protein–carbohydrate interaction 98
protein classification, chemical constituents 6
– compartilization 6ff
– – in eukaryotic cells 6
– – in prokaryotic cells 6
– functional – 4f
– structure 6
protein complexes 96ff
– enzyme–inhibitor complexes 96
– recognition of targets 96
– with HIV protease inhibitors 97f
protein crystallography 96
– Fourier maps 92
– of protein complexes 92
protein database searches 49, 53, 57ff, 66
protein–DNA docking 96
protein–DNA interactions 100
protein docking, binding sites for small ligands 95
– *de novo* design of enzyme inhibitors 95
– docking algorithms 94
– simulation 94
– structure predictions 94f
– with macromolecules 95
protein folding 111ff

– *in vitro* 111, 115ff
– of inclusion bodies 111ff
protein functionality 47ff
protein functions 1f, 5, 22ff, 52ff
– definition 54
– enzymes 22ff
– knowledge-based prediction 55ff
– multiplicity 4
protein glycosylation 9
protein inhibitors, in HIV gene therapy 476ff
– – dominant-negative proteins 476f
– – intracellular antibodies 477b
protein interactions 87ff
– analytical ultracentrifugation 91
– association constant 88
– calorimetric methods 90f
– continuum models for – 94
– crystallography 92
– electrostatic – 89
– hydrogen bonds 89
– hydrophobic – 89
– NMR spectroscopy 92f
– prediction, force field methods 93f
– quantitative description 88f
– strengths 88
– surface–plasmon resonance 92
– theoretical methods 93ff
protein–ligand complexes 94, 96f
– enzyme–inhibitor complexes 96
protein–ligand docking, ligand design 95
– ligand optimization 95
protein localization 5ff
protein–macromolecule docking 95f
protein modeling 73ff
– homology modeling 75ff
protein modification, posttranscriptional – 49
– posttranslational – 49ff
protein–nucleic acid complexes 100f
protein–protein complexes, antibody–antigen complexes 99f
– antigen–antibody complexes 99
– protease inhibitors 99
protein–protein docking 77, 94ff
protein–RNA interactions 101
protein sequence 43ff
– analysis 50ff, 55ff
– cDNA 50f
– database-aided homology search 57ff
– databases 49
– – errors 54ff
– – identification algorithm 54f
– – quality 53ff
– motif searches 58
– profile searches 58
– significance of weak homologies 58
– similarity searches 56
– WWW pointers 56
protein sequence–structure relationships 59

protein size 49
protein splicing, posttranslational – 50f
protein–steroid complexes 98f
protein structure 43ff
– Brookhaven Protein Database 66, 92
– chemical – 50
– conformational searches 74
– crystallographic 62
– determination, force field methods 93f
– helix types 60ff
– hierarchial description 47f
– investigation of primary – 50ff
– knowledge-based prediction 55ff
– polypeptide backbone 60
– prediction of 88
– quaternary– 77ff
– – phenomenology 77
– – prediction 77f
– Ramachandran plot 59
– secondary – 59ff
– – computer algorithm 63
– – extended – 61f
– – folding types 64
– – loop segments 62f
– – polypeptide segments 62
– – predictions 65ff
– – repetitive – 60ff
– – structural classes 63ff
– – types 59ff
– solvation energy 71
– structural comparisons, techniques 72
– tertiary – 68ff
– – construction principles 69ff
– – modeling 73
– – phenomenology 68ff
– – predictions 73ff
– – solvent accessibility 76f
– – structural families 71ff
– three-dimensional – 72
– torsional angle 59
protein surfaces 70f
protein topologies 74
protein types 1ff, 5ff
protein variability 5
proteins, conjugated – 9ff
– contractile – 31ff
– content of various cells 4
– evolution of 2f
– fatty acylated – 17
– *in vitro* folding 112
– isoprenylated – 18
– recombinant –, medical applications 125ff
– – large-scale production 112
– – market 533
– – regulations 495
– – sales 533
– regulatory – 38ff
prourokinase, use in therapy 131

psoriasis, medical applications of recombinant
    proteins 143
– monoclonal antibody therapy 319
– vaccination 143
PTCA (balloon angioplasty) 315, 375ff
– monoclonal antibody therapy 316
– therapy with ReoPro 375
– use of ReoPro 376
Public Health Services Act 498
pulmonary inflammatory disorders, monoclonal
    antibody therapy 315
Pulmozyme 167
Pure Food and Drugs Act 498

Q
quality control, of plasmid DNA 458ff
– – DNA quantity 459
– – restriction analysis 459
– – sequencing 459
– – sterility 459
quaternary structure *see* protein structure

R
RA *see* rheumatoid arthritis
radiotherapy, monoclonal antibody therapy 320
Ramachandran plot 59, 62
receptor-mediated endocytosis 30f
recombinant antibodies *see* antibodies
Recombinant DNA Advisory Committee (RAC)
    519f, 525f
– gene therapy regulation 520, 523f
– guidelines 523f
– regulation of biotechnology products 519f
recombinant proteins *see* proteins
recombinant technology, potential 533
registration, of biotechnology products 498ff
regulation, of biotechnology products 519
– of human gene therapy 517
regulatory proteins 38ff
– cell surface receptors 38
– signal transduction pathways 40
– transcription factors 40
relaxin 133
renal failure 164
ReoPro (Abciximab, c7E3 Fab) 130, 230, 315
– approval process 366, 370
– blinding 369
– CAPTURE trial 376ff
– clinical development 365ff
– clinical trials 367
– coronary interaction 367
– coronary interventions 371
– cost-effectiveness 370f, 373
– effect on blood clotting time 374
– EPIC trial 367ff
– EPILOG trial 374ff

– high-risk patients 368ff, 375
– in bleeding events 369, 374, 376
– indication 366
– intention-to-treat analysis 368f, 377
– ischemia 367ff, 377
– low-risk patients 374f, 375
– myocardial infarction 367ff, 377f
– phase II trials 366
– platelet receptor blockage 367
– PTCA 376
– reduction in hospitalizations 370ff
– sales 534
– stent implantation 375
– treatment of, acute coronary syndromes 371
– – coronary diseases 366ff
– – intracoronary thrombus 371, 377
– use in, atherectomy 375f
– – balloon angioplasty 375
– – coronary stenting 375
– – saphenous vein grafts 371f
– – unstable angina 376
respiratory chain 20
restenosis 131f
– medical applications of recombinant proteins 131
– monoclonal antibody therapy 316
Reteplase 130, 168
retrovirus, replication competent – 403f
– – generation of 404
retrovirus vectors 397ff
– cell entry 400
– clinical studies 405f
– design of 400
– gene expression 399
– genome 400f
– host range 401
– host responses 405
– host tropism 400
– in HIV gene therapy 405, 480f
– integration 400
– in therapy of inoperable gliomas 405
– in tumor therapy 405f
– Kozak sequence 402
– life cycle 399
– manufacturing 402ff
– – complementation systems 402f
– packaging cells 403
– packaging limit 401
– particle structure 401
– persistence 400
– preclinical studies, *in vitro* 404
– – *in vivo* 404f
– producer cells 402
– properties 401
– purification 404
– RCR contamination 403f
– resistance to human complement – 401
– specificity 401

– stability in therapy 401
– structure 399
– subfamilies 399
– transgene expression cassettes, design 402
– – promoters 402
– with two genes 402
Revasc 130, 169
rheumatoid arthritis (RA) 140, 296, 318f
– animal models 344f
– antibody therapy 344
– CDP571 347ff
– medical applications of recombinant proteins 143
– monoclonal antibody therapy 315, 318
– role of tumor necrosis factor 344, 346
– treatment with CDP571 346ff
– – repeated dose studies 347ff
– – results 349f
– vaccines 138
ribozymes, in HIV gene therapy 474ff
ricin 34
RNA decoys, for HIV gene therapy 476
RNA editing 49
RNA inhibitors, in HIV gene therapy, antisense RNA 474f
– – decoy RNA 476
– – ribozymes 474f

S

safety aspects, preclinical testing of antibodies 303ff
saphenous vein grafts 371f
– treatment with ReoPro 371
sarcomere 31
Sargrastim 157
Saruplase 168
secondary structure *see* protein structure
sepsis 134f, 330f
– animal models 335f
– anticytokine therapy 333
– clinical trial design 335ff
– corticosteroid therapy 334
– definitions 336f
– medical applications of recombinant proteins 134
– monoclonal antibody therapy 316f, 329ff
– – anti-endotoxin monoclonal antibodies 331f
– – anti-tumor necrosis factor 332f
– – clinical experience 330
– – clinical trials 333
– – rationale for 330f
– preclinical studies 335f
– therapy 296
– use of polyclonal antibodies 330
septic shock 344
– treatment with CDP571 346f
sequence analysis *see* protein sequence

sequence databases 53ff
serine proteases 206f
serotonin 34
serum-free media 256f
β-sheet 59ff
shock, clinical definition 337
similarity searches, in protein databases 56
SIRS *see* systemic inflammatory response syndrome
solvation energy, of proteins 71
species cross-reativity 295ff
spongiform encephalopathies, as risk factors in
  biological products 513
staphylokinase, use in therapy 130
starch liquefaction 193f
– increase in dextrose equivalents 195
– use of α-amylases 193
stem cell factor 158
stent implantation, therapy with ReoPro 375
Stevenson–Wydler Technology Innovation Act
  522
stone-washing, use of industrial cellulases 204
storage proteins 33
streptokinase, use in therapy 130
striated muscle 31
stroke 130ff
– medical applications of recombinant proteins
  130f
structural families, of proteins 71ff
structural proteins 32ff
subtilisin 206, 208
surface–plasmon resonance, of protein complexes
  92
suspension adaptation 256f
Svedberg relation 91
systemic inflammatory response syndrome
  (SIRS) 316f
– monoclonal antibody therapy 316
systemic vasculitis 318f
– monoclonal antibody therapy 318

T
T lymphocytes 35
– CD4$^+$ 472f
TAR decoys, antiviral action 476
Teceleukin 151
tertiary structure *see* protein structure
textile desizing, protective layer 194
– stages 194f
– use of α-amylases 194f
textile industry, use of cellulases 204f
therapeutic antibodies *see also* antibodies, monoclonal antibodies
– glycosylation 255
– manufacture of, antibody purification 263ff
– – bioreactor types 261f
– – cell banking 257

– – cell culture media 256
– – cell line 248f, 257
– – cell line stability 258f
– – cell removal 263
– – choice of host cell 255
– – cloning techniques 256
– – copy numbers 252f
– – culture conditions 255, 258, 262f
– – debris removal 263
– – DHFR expression systems 250
– – economic aspects 266
– – gene amplification 251f
– – gene expression systems 250f
– – glutamine synthetase expression systems 250
– – good manufacturing practice 268f
– – high expression 251ff
– – integration into genome 252f
– – key parameters 260
– – Kozak sequence 253
– – limits 260
– – methods 252f
– – methotrexate 253f
– – pilot scale fermentations 260
– – process consistency 256
– – process definition 259f
– – process monitoring 261
– – product consistency 254f
– – product recovery 262f
– – production systems 261f
– – reproducibility of bioprocessing 255
– – selectable marker 253
– – selectable marker genes 250
– – selection 256f
– – selection of suitable cell lines 251ff
– – serum-free media 256f
– – stability of expression 254
– purification of 263ff
– – DNA contamination 265
– – protein contaminants 264f
– – purification methods 267f
– – requirements 264f
– – virus contamination 265
– robustness 265f
therapeutic applications, of antibody fragments
  313
– of monoclonal antibodies 311ff
– – autoimmune disorders 318f
– – clinical applications 315f
– – immunosuppression 317f
– – infection 316
– – rheumatoid arthritis 318f
– – sepsis syndrome 316
– – targets for intervention 314
– – transplantation 317f
therapeutic proteins *see also* proteins
– antibodies
– – clinical applications 311ff
– – engineering and expression 219ff

– – for sepsis 329ff
– – in chemotherapy 371ff
– – in chronic therapy 359ff
– – manufacture of 245ff
– – preclinical testing 289ff
– – ReoPro 343ff
– blood products 21
– classes 21
– economics 531
– functions 21ff
– growth factors under development 156
– in humans and animals 125ff
– in tumor therapy 135ff
– regulations 495ff
– regulatory factors 21
– under development 171
– use as vaccines 21, 137ff
– use in immunosuppression 141
thermolysin 207
thrombopoietin 157f
– in therapy 157
thrombosis, treatment with ReoPro 366, 371
– urokinase therapy 373
thromboxane 34
thyroid stimulating hormone 165
tissue plasminogen activator (tPA), use in therapy 129f
tissue-type plasminogen activator, *in vitro* disulfide bond formation 115f
– *in vitro* folding 115ff
TNF *see* tumor necrosis factor
tPa *see* tissue plasminogen activator
transfection systems 429ff
– anionic liposomes 434
– calcium phosphate precipitation 430f
– cationic liposomes 432ff
– – transfection levels 432
– cationic lipsomes, efficiencies 432
– direct gene transfer 429ff
– electroporation 431
– liposome–HVJ complexes 434
– microinjection of DNA 430
– neutral liposomes 434
– non-targeted cationic polymers 431
– particle bombardment, DNA immunization 432
– receptor-mediated endocytosis, DNA-binding 435
– – DNA-compaction 435
– – efficiencies 437
– – endosome release 436
– – targeting ligands 436
– – targeting moiety 435
– receptor-mediated endocytosis 434f
transferases 25f
transferrin 19, 30
transforming growth factor β 132, 160f
– animal experiments 161
– binding proteins 161

– in therapy 140, 161
transgenes, expression of 389f
– persistence of 390
– size 389
transgenic animals 539
transmembrane proteins 29
transplantation 314, 317f
– mAb products 222
– monoclonal antibody therapy 317f
– role of tumor necrosis factor 344
transport proteins 26ff
– as transmembrane proteins 29
– ATPases 29
– erythrocyte glucose transporter 28
– facilitated transport 27
– functions 26ff
– receptor-mediated endocytosis 30
tubulin 4
– contractile properties 32
tumor *see also* cancer
– biodistribution studies 361f
– gene therapy with adenovirus vectors 413, 421
– gene therapy with retroviruses 405f
– monoclonal antibody therapy 315, 320ff
– radiotherapy 320
– treatment of 361
tumor necrosis factor (TNF) 283, 291f, 295f, 332f
– antibodies 293ff
– – in therapy 296
– antibodies against –, CDP571 345ff
– function 344
– half-life in blood 334
– in Crohn's disease 346
– in inflammatory bowel disease 346
– in tumor therapy 136
– in ulcerative colitis 346f
– monoclonal antibody therapy, anti-tumor necrosis factor 332f
– response to inflammatory stimuli 334
– testing of antibodies 291
– TNFα, antibodies to – 135
– – receptor 135
– – use in therapy 130, 134, 144
tumor specific antigens, in tumor therapy 136
tumor suppressor protein, in tumor therapy 136
tumor therapy 135ff, 294f
– antibody fragments 294f
– hematopoietic growth factors 137
– medical applications of recombinant proteins 135ff
– use of monoclonal antibodies 135f, 294f
β-turn 62

U

ulcerative colitis, monoclonal antibody therapy 319
– role of tumor necrosis factor 346

– treatment with CDP571   346f
urokinase, in therapy of thrombus   373

V
vaccination, plasmid vector design   452
– use of plasmid DNA   444
vaccines   21, 137ff
– AIDS   138
– bacterial   139ff
– cancer   135, 138
– hepatitis virus   139
– herpes simplex   138f
– malaria   139
– melanoma   138
– plasmid DNA   465
– recombinant proteins   137ff
– viral –   137f
van der Waals energy   88f
van't Hoff equation   90
varicella-zoster virus, monoclonal antibody therapy   317
vascular endothelial cell growth factor   133f
vector design, in HIV gene therapy   479ff
– – adeno-associated virus based –   480f
– – amphotropic murine leukemia virus based –
   479f
– – novel retroviral and based –   480f
– – optimization of –   484
vectors, design for gene therapy   387ff
– – non-viral –   427ff
– viral –   395ff
viral-based vector systems, in HIV gene therapy
   479ff

viral vectors, adeno-associated –   397f
– adenovirus   397ff, 406ff
– for gene therapy   395ff
– retrovirus   397ff
virus vectors, adeno-associated –   413ff
VLDL   15

W
Waxmann Hatch legislation   503
White Paper   526
WHO   445f
– plasmid DNA, quality issues   460
– plasmid DNA therapeutics, guidelines   445
– plasmid vectors, guidelines   452
working cell bank   452
– quality control   453

X
xenotransplantation, medical applications of recombinant proteins   146
– monoclonal antibodies   146

Y
yeast genome   3

Z
Zenapax   318
zinc finger motifs   19